Handbook of Fourier Analysis & Its Applications

Handbook of Fourier Analysis & Its Applications

Robert J. Marks II

2009

OXFORD
UNIVERSITY PRESS

Oxford University Press, Inc., publishes works that further
Oxford University's objective of excellence
in research, scholarship, and education.

Oxford   New York
Auckland   Cape Town   Dar es Salaam   Hong Kong   Karachi
Kuala Lumpur   Madrid   Melbourne   Mexico City   Nairobi
New Delhi   Shanghai   Taipei   Toronto

With offices in
Argentina   Austria   Brazil   Chile   Czech Republic   France   Greece
Guatemala   Hungary   Italy   Japan   Poland   Portugal   Singapore
South Korea   Switzerland   Thailand   Turkey   Ukraine   Vietnam

Copyright © 2009 by Oxford University Press, Inc.

Published by Oxford University Press, Inc.
198 Madison Avenue, New York, New York 10016

www.oup.com

Oxford is a registered trademark of Oxford University Press

All rights reserved. No part of this publication may be reproduced, stored in a
retrieval system, or transmitted, in any form or by any means, electronic,
mechanical, photocopying, recording, or otherwise, without the prior
permission of Oxford University Press.

Library of Congress Cataloging-in-Publication Data

Marks, Robert J.
Handbook of fourier analysis & its applications / Robert J. Marks II.
p. cm.
Includes bibliographical references and index.
ISBN 978-0-19-533592-7
1. Fourier analysis. I. Title.
QA403.5.M47 2008
515'.2433—dc22
2008019802

1 3 5 7 9 8 6 4 2

Printed in the United States of America
on acid-free paper

There is one God, the Father, ever-living, omnipresent, omniscient, almighty, the maker of heaven and earth, and one Mediator between God and man, the man Jesus Christ...
Isaac Newton [400].

There is a God-shaped vacuum in the heart of every man, which only God can fill through His Son Jesus Christ.
Blaise Pascal [1098, 1415].

Christ [is] the power of God and the wisdom of God. The Christian religion is a revelation and that revelation is the Word of God.
Michael Faraday [1031].

For the invisible things of Him from the creation of the world are clearly seen, being understood by the things that are made, even His eternal power and Godhead; so that they are without excuse.
Romans 1:20.

Dedicated to the Christ Jesus who makes things clear.

Preface

> The last thing one knows when writing a book is what to put first.
> Blaise Pascal

> Someone told me that each equation I included in the book [**A Brief History of Time**] would halve its sales.
> Stephen Hawking. [585]

Audience

The emphasis of this text is the broad field of Fourier analysis and some of its immediate outgrowths including convolution, systems theory, Shannon sampling theory and the modelling of random variables and stochastic processes. The book is written at the level of the advanced senior or first year graduate student who has an introductory foundation in Fourier analysis and stochastic processes. Rudimentary familiarity with probability and image processing is helpful but not necessary.

Exposition Philosophy

There are a number of features that distinguish this book from others [247, 398, 652, 1268, 1402] on the topics of Fourier series [233, 328, 362, 363, 580, 802, 1122, 1368, 1401, 1553, 1563, 1590], Fourier transforms [127, 196, 330, 579, 584, 717, 827, 1200, 1243, 1244, 1295, 1269], or Fourier analysis [153, 386, 454, 627, 631, 760, 767, 1002, 1156, 1157, 1267, 1059, 1382, 1410, 1411, 1457, 1488]. Consistent with goal of a handbook, the topical content is broad. Yet the presentation format is crafted to allow use as a text. Cross referencing is extensive, a comprehensive indexed bibliography with over 1500 references is provided, and over 400 figures illustrate the material.

This book is aimed at practitioners rather than mathematicians. Two classic texts aimed at practitioners were published in 1965: Athanasios Papoulis' **Probability, Random Variables, and Stochastic Processes** [1075], and Ronald N. Bracewell's **The Fourier Transform and its Applications** [148]. These books popularized a pedagogy that balanced rigor and intuition. The lucid style pioneered by Papoulis and Bracewell is used in this book. Doing so allows presentation of concepts without the obfuscation of unnecessary detail. For example, little attention is given to the status of the Dirac delta as a distribution rather than a function, or the need for measure theory in the formulation of Hilbert spaces. Likewise, absolute convergence in infinite sums (integrals) is a requirement for interchanging the order of summation (integration) but is not invoked in every instance. The question of "When can we interchange the order of integration?" has herein an answer of "always". Rigorously, this is of course not true. However, in the rare cases where such a switch is inappropriately made, the math, by becoming ludicrous or inconsistent, will often communicate the error. Although we point to the availability of mathematical rigor in most cases, meticulousness is not painstakingly applied. Doing so would mask recognition of the beautiful forest with the distraction of small pesky trees. In cases where a derivation is of monumental importance, steps can always be retraced with mathematical detail.

Like Papoulis' books [1074, 1075, 1076, 1077, 1078, 1082, 1087, 1088], this book cannot be read casually. Illumination, rather, comes through familiarization and rumination. To assist in this process, solutions to selected exercises are included at the end of each chapter. The exercises are often not repetitive illustration of chapter contents, but contain insightful generalizations, alternate viewpoints and even new material. Ancillary problems on tangential albeit related material with a somewhat recreational bent are occasionally included. They are marked with a "§".

Curricular Considerations

This book can effectively serve as either a rich source of material for self study or as the primary text for at least four courses.

- **Introduction to Fourier Analysis** [124, 386, 454, 545, 627, 631, 717, 767, 1002, 1157, 1267, 1268, 1269, 1457, 1472]. Cover Chapter 2 on *Fundamentals of Fourier Analysis* followed by Chapter 3 on *Fourier Analysis in Systems Theory* and Chapter 5 on *The Sampling Theorem*. The first half of the material in Chapter 10 on *Signal Recovery* is also appropriate.
- **Multidimensional Signal Processing** [387, 1239, 1423, 1509]. After review of Chapter 2, cover Chapter 8 on *Multidimensional Signal Analysis*[1] and Chapter 11 on *Signal and Image Synthesis: Alternating Projections Onto Convex Sets* [1333]. Section 13.2 provides an application of multidimensional signals and systems to *Fourier Transforms in Optics and Wave Propagation*. Chapters 9 on *Time-Frequency Representations* [312] is also appropriate material.
- **Introduction to Shannon Sampling and Interpolation Theory** [92, 605, 606, 915, 1577]. First, review Chapter 2. Then cover Chapter 5 on *The Sampling Theorem*, Chapter 6 on *Generalizations of the Sampling Theorem* [957, 916], and Chapter 7 on *Noise and Error Effects*.[2] Chapter 10 on *Signal Recovery* covers continuous sampling and the multidimensional sampling theorem is covered in Section 8.9.
- **Advanced Topics in Fourier Analysis.** After review of Chapter 2, cover Chapter 9 on *Time-Frequency Representations* [312], Chapter 4 on *Fourier Transforms in Random Variables and Stochastic Processes*, Chapter 10 on *Signal Recovery*, and Chapter 12 on *Mathematical Morphology and Fourier Analysis on Time Scales*. The topics in Section 13.1 on *The Wave Equation, its Fourier Solution and Harmony in Western Music*, Section 13.3 on *Heisenberg's Uncertainty Principle*, and Section 13.4 on *Elementary Deterministic Finance* provide specific applications.

---

1. Chapter 4 on *Fourier Transforms in Probability, Random Variables and Stochastic Processes* contains the required background on stochastic processes.
2. Prerequisite material on stochastic processes in Chapter 4 on *Fourier Transforms in Probability, Random Variables and Stochastic Processes* is required here.

Web Site and Errata

The web site www.HandbookOfFourierAnalysis.com contains supplementary material to this text, including power point presentations, simulations and animations.

Also, the author has taken great pains to keep the book error free, but is aware the goal has not been achieved. All errors herein are his responsibility. If you find errors in the text, please e-mail details to HandbookOfFourierAnalysis@gmail.com. Errata will be posted on www.HandbookOfFourierAnalysis.com.

On the Shoulders of Giants

From Pythagoras to Shannon, accomplishments of the great minds who have contributed to the foundational material in this book should be held in jaw dropping awe. "The dwarf sees farther than the giant, when he has the giant's shoulder to mount on." Samuel Taylor Colerage (1772—1834) [311].

Much of the work presented in this volume was performed with friends and valued colleagues whose contributions, directly and indirectly, are gratefully acknowledged. They include Payman Arabshahi, Les E. Atlas, Walter L. Bradley, Kwan Fai Cheung, Paul S. Cho, Jai J. Choi, John M. Davis, William A. Dembski, Mohamed A. El-Sharkawi, Warren L.J. Fox, Marc H. Goldburg, Ian A. Gravagne, Marion O. Hagler, Douglas G. Haldeman, Michael W. Hall, Dmitry Kaplan, E. Lee Kral, Thomas F. Krile, H.G. Kuterdem, B. Randall Jean, John N. Larson, Loren Laybourn, Shinhak Lee, Michael J. Meyer, Robert T. Miyamoto, Alan C. Nelson, Seho Oh, Dong Chul Park, Jiho Park, Hal Philipp, James W. Pitton, Dmitry Radbel, Brian Ricci, Ceon Ramon, Dennis Sarr, David K. Smith, Michael J. Smith, Benjamin B. Thompson, Shiao-Min Tseng, John F. Walkup, John L. Whited, Gary L. Wise, Wen–Chung Stewart Wu, Donald C. Wunsch, and Yunxin Zhao. Thanks also to image models Lenore Marks, Robert (Jack) Marks, Connie Lynn J. Marks, Ray A. Marks, Jeremiah J. Marks, Joshua J. Marks, and Marilee M. Marks.

Acronym List

Acronym	
AITS	additively idempotent time scales
AM	amplitude modulation
a.k.a.	also known as
BIBO	bounded input–bounded output stability
BE	bounded energy signals
BL	bandlimited signals
BPF	bandpass filter
BS	bounded signals
CA	constant area
CAT	computed axial tomography
CD	compact disk
CP	constant phase
CRT	cathode ray tube
CTFT	continuous time Fourier transform
dB	decibel
DCT	discrete cosine transform
DFT	discrete Fourier transform
DSP	digital signal processing
DTFT	discrete time Fourier transform
DTS	discrete time signal
DVD	digital versatile disc or digital video disc
FIR	finite impulse response
FFT	fast Fourier transform
FM	frequency modulation
GTFR	generalized time-frequency representations
IIR	infinite impulse response
IM	signals with identical middles
JPEG	Joint Photographic Experts Group [702]
JPG	variation of JPEG
LED	light emitting diode
LPF	low pass filter
LPK	low passed kernel

ACRONYM LIST

LTI system	linear time invariant system
LV	linear variety
PGA	Papoulis-Gerchberg algorithm
PI	pseudo-inverse
PIA	piecewise invariant approximation
POCS	(alternating) projections onto convex sets
PSNR	peak signal-to-noise ratio
PSWF	prolate spheroidal wave functions
SSSC	single side band suppressed carrier
sup	supremum. For example, $\sup(\mathrm{sinc}(t)) = 1$ and $\sup\left(1 - e^{-t}\right)\mu(t) = 1$.
TFR	time-frequency representations

Notation

The following is a list of notation and variable used in the book. The most common uses are listed, but may also be used in different contexts.

Notation

$\forall$	"for all"
$\ni$	"such that"
$\exists$	"there exists"
$:=$	"equal to by definition"
$\equiv$	"is equivalent to"
$\times$	Cartesian product. See (6.17).
$\times$	multiplication. See (8.25).
$\times$	vector curl. See (13.11).
$*$	convolution. See Section 2.3.4.5.
$\circ$	opening. See (12.3).
$\square$	marks the conclusion of a proof.
$\|\cdot\|$	vector norm or signal norm. See (8.2) and (11.3).
$\bullet$	closing. See (12.4).
$\star$	deterministic autocorrelation. See Table 2.3.
$\oplus$	dilation or Minkowski addition.
$\ominus$	Minkowski subtraction.
$\lfloor a \rfloor$	the greatest integer not exceeding a, for example $\lfloor 6.98 \rfloor = 6$.
$\angle$	a signal's phase. In polar coordinates, $x(t) = \|x(t)\|e^{j\angle x(t)}$.
§	ancillary problems on tangential albeit related material.
$\perp$	perpendicular, used to describe null spaces. See Section 11.3.1.3.
$\sharp$	musical notation (sharp). See Section 13.1.
$\flat$	musical notation (flat). See Section 13.1.
$(\cdot)^*$	complex conjugation. See the solution to Exercise 3.12.
†	Sections that can be skipped. See Section 2.1.
‡	Sections that can be skipped. See Section 2.1.
¶	Sections that can be skipped. See Section 2.1.
$\leftrightarrow$	Fourier transform pair. See (2.11).
$\underset{z}{\leftrightarrow}$	z-transform pair. See (2.130).
$\overset{\mathbb{L}}{*}$	convolution on the the time scale $\mathbb{L}$.
$\overset{\xi}{*}$	convolution with respect to the variable ξ.

Symbol	Description
$\aleph_N$	transfinite number of order N. See Exercise 3.32.
$\mathbb{A}^c$	set complement - $\mathbb{A}^c = \{a \mid a \notin A\}$.
A./**B**	divide each element in the matrix **A**. by the corresponding element in **B**.
A.∗**B**	multiply each element in the matrix **A** by the corresponding element in **B**.
$a := b$	a is equal to b by definition.
$A = \{x \mid x \in B\}$	"A is equal to the set of all x such that x is in the set B." See (11.1).
array($\cdot$)	the array function. Section 2.3.5.9.
$B(b, c)$	the beta function. See (2.68).
B	the bandwidth of a bandlimited signal. See Section 5.1.1.
cas(ζ)	cas(ζ) = cos(ζ) − sin(ζ). See (2.126).
comb(t)	the comb function. See (2.53).
$d_p(\cdot)$	derivative kernel. See (6.67).
$\delta(t)$	Dirac delta function. See Section 2.3.5.1.
$\delta[n]$	Kronecker delta function. See (2.43).
$d_p(t)$	p derivatives of the sinc. See entry in Table 2.4.
$d\vec{t}$	multidimensional differential. (8.7).
$E[g(X)]$	the expected value of $g(X)$. See (4.4).
$e(t)$	an even function. See (2.6).
${}_1F_1$	confluent hypergeometric function of the first kind. See Exercise (4.9).
$f_X(x)$	probability density function for a random variable X. See Section 4.2.1.
$F_X(x)$	cumulative distribution function for a random variable X. See (4.1).
$\mathcal{F}$	Fourier transform operator. See (3.41).
$\mathcal{F}^a$	fractional Fourier transform operator. See (3.47).
f	frequency variable for discrete time signals. See (2.5).
f	focal length of a lens. See Section 13.2.4.
$f(\cdot)$	in some instances such as Exercise (6.12), the "f" denotes a signal we wish to *find*.
$g(\cdot)$	in some instances such as Exercise (6.12), the "g" denotes a signal that is *given*.
$\overline{g(X)}$	alternate notation for $E[g(X)]$
γ	the Euler-Mascheroni constant. See (14.21).
$\Gamma(z)$	the gamma function. See (2.64).
$\mathcal{H}$	Hilbert or Hankel transform operator. See (2.98).
$\mathcal{H}_k$	Hankel transform operator. See (8.41).
$H_n(t)$	Hermite polynomial of order n. See (2.88).
$h(\cdot)$	continuous time impulse response. See (3.13).
$h[\cdot]$	discrete time impulse response. See (3.13).
$\hbar$	Planck's constant. See Section 13.3.
$I_n(t)$	modified Bessel function of the first kind. See (2.75).
$\Im$	the imaginary part of
iff	contraction of "if and only if"
i	$-j$
j	$\sqrt{-1}$
jinc(t)	the jinc function. See (2.72).
$J_\nu(\cdot)$	Bessel function of order ν. See (2.69).
$j_n(\cdot)$	spherical Bessel function of order n. See (2.74).
k	the magnitude of the propagation vector. See (13.13).
k	Boltzmann's constant. See Section 4.4.3.1.
$K_n(t)$	Hermite-Gaussian function of order n. See (2.92).

L_1	the set of all Lebesgue measurable finite area continuous time signals. See Footnote 1 in Section 2.2.
L_2	the set of all Lebesgue measurable finite energy continuous time signals. See Footnote 1 in Section 2.2.
L_∞	the set of all Lebesgue measurable bounded continuous time signals. See Footnote 1 in Section 2.2.
ℓ_1	the set of all Lebesgue measurable finite area discrete time signals. See Footnote 1 in Section 2.2.
ℓ_2	the set of all Lebesgue measurable finite energy discrete time signals. See Footnote 1 in Section 2.2.
ℓ_∞	the set of all Lebesgue measurable bounded discrete time signals. See Footnote 1 in Section 2.2.
λ	wavelength. See (13.13).
λ_n	eigenvalues. See Section 10.5.
$\Lambda(t)$	the triangle function. See (2.57).
$\mathcal{M}$	a set of integers. See Section (6.2.2.3).
$\mu(t)$	continuous time unit step function. See (2.49).
$\mu[n]$	discrete time unit step function. See (2.51).
$\mathbb{N}$	natural numbers, $\{1, 2, 3, \cdots\}$
$n!$	n factorial $= n(n-1)(n-2) \times \cdots \times 3 \times 2 \times 1$.
$n!!$	n double factorial $= n(n-2)(n-4) \cdots$.
$((n))_m$	n mod m. See (12.20).
$\binom{n}{k}$	binomial coefficient, a.k.a. "n choose k" $= n!/\{k!(n-k)!\}$. See Section 14.4.1.
$o(t)$	an odd function. See (2.7).
$P_n(t)$	Legendre polynomial of order n. See (2.81).
$\Phi_X(u)$	characteristic function for a random variable X. See (4.6).
$\Pi(t)$	continuous time rectangle function. See (2.45).
$\Pi[n]$	discrete time rectangle function. See (2.46).
$\Psi_X(u)$	second characteristic function for a random variable X. See (4.11).
r	sampling rate parameter. See (6.3).
$R_X(t, \tau)$	the autocorrelation of the continuous time stochastic process $X(t)$. See (4.60).
$R_\xi[n]$	the autocorrelation of the discrete time stochastic process $\xi[n]$. See (4.70).
$\mathbb{R}$	the set of all real numbers.
$\Re$	the real part of.
$\text{sgn}(t)$	the signum function. See (2.47).
$S\{\cdot\}$	a system operator. See (3.1).
$S^{-1}\{\cdot\}$	inverse system operator. See Section 3.2.1.9.
$\text{Si}(\cdot)$	the sine integral. See (2.77).
$S_\xi(\cdot)$	the power spectral density of the stochastic process ξ. See Section 4.4.2.
$\text{sinc}(t)$	the sinc function.
$\text{sinc}_k(t)$	generalized sinc function. See (2.44).
σ_X^2	the variance of the random variable X. See (4.12).
T	the period of a periodic function. See Section 2.3.2.
t	time. See Section 2.3.1.
$\vec{t}$	a vector. See (6.16) and (8.1).
$\mathbb{T}$	time scale or set.
$\hat{\mathbb{T}}$	the negative of a set.

$T_n(\cdot)$	Chebyshev polynomial of order n. See (2.84).
u	temporal frequency. See Section 2.3.1.
W	a frequency interval $\geq$ a bandlimited signal's bandwidth, B. See Section 6.2.1.
$x(t)$	a continuous time signal.
$\hat{x}(t)$	the analytic signal corresponding to $x(t)$. See Exercise 2.14.
$\widetilde{x(t)}$	convolution of $x(t)$ with a jitter probability density function. See (7.71).
$x_T(t)$	a single period of a periodic function. See Section 2.3.2.
$x[n]$	a discrete time signal.
$X_{\text{hart}}(u)$	Hartley transform of $x(t)$. See (2.125).
$X_{\cos}(u)$	cosine transform of $x(t)$. See (2.120).
$X_L(s)$	Laplace transform of $x(t)$. See (2.22).
$X_M(s)$	Mellin transform of $x(t)$. See (3.60).
$X_{\sin}(u)$	sine transform of $x(t)$. See (2.124).
$X_z(z)$	z transform of $x[n]$. See (2.129).
$\overline{X^k}$	the kth moment of the random variable X. See (4.5).
$\mathbb{Z}$	the set of integers, $\{\cdots, -2, -1, 0, 1, 2, \cdots\}$.

Contents

Preface   vi
Acronym List   ix
Notation   xi

1 Introduction   3
　1.1 Ubiquitous Fourier Analysis   3
　1.2 Jean Baptiste Joseph Fourier   4
　1.3 This Book   6
　　1.3.1 Surveying the Contents   6

2 Fundamentals of Fourier Analysis   10
　2.1 Introduction   10
　2.2 Signal Classes   11
　2.3 The Fourier Transform   13
　　2.3.1 The Continuous Time Fourier Transform   14
　　2.3.2 The Fourier Series   14
　　　2.3.2.1 Parseval's Theorem for the Fourier Series   16
　　　2.3.2.2 Convergence   16
　　2.3.3 Relationship Between Fourier and Laplace Transforms   18
　　2.3.4 Some Continuous Time Fourier Transform Theorems   18
　　　2.3.4.1 The Derivative Theorem   18
　　　2.3.4.2 The Convolution Theorem   19
　　　2.3.4.3 The Inversion Theorem   19
　　　2.3.4.4 The Duality Theorem   19
　　　2.3.4.5 The Scaling Theorem   20
　　2.3.5 Some Continuous Time Fourier Transform Pairs   20
　　　2.3.5.1 The Dirac Delta Function   20
　　　2.3.5.2 Trigonometric Functions   21
　　　2.3.5.3 The Sinc and Related Functions   27
　　　2.3.5.4 The Rectangle Function   28
　　　2.3.5.5 The Signum Function   29
　　　2.3.5.6 The Gaussian   30
　　　2.3.5.7 The Comb Function   31
　　　2.3.5.8 The Triangle Function   32
　　　2.3.5.9 The Array Function   32
　　　2.3.5.10 The Gamma Function[†‡]   35
　　　2.3.5.11 The Beta Function[‡]   36

    2.3.5.12 Bessel Functions†‡ 37
    2.3.5.13 The Sine Integral 39
    2.3.5.14 Orthogonal Polynomials¶ 40
  2.3.6 Other Properties of the Continuous Time Fourier Transform 45
    2.3.6.1 The Signal Integral Property 45
    2.3.6.2 Conjugate Symmetry of a Real Signal's Spectrum 45
    2.3.6.3 The Hilbert Transform 46
    2.3.6.4 Hilbert Transform Relationships Within a Causal Signal's Spectrum 46
    2.3.6.5 The Power Theorem 47
    2.3.6.6 The Poisson Sum Formula 47
2.4 Orthogonal Basis Functions¶ 48
  2.4.1 Parseval's Theorem for an Orthonormal Basis¶ 48
  2.4.2 Examples¶ 49
    2.4.2.1 The Cardinal Series¶ 49
    2.4.2.2 The Fourier Series¶ 49
    2.4.2.3 Prolate Spheroidal Wave Functions¶ 49
    2.4.2.4 Complex Walsh Functions§ 50
    2.4.2.5 Orthogonal Polynomials¶ 50
    2.4.2.6 Fourier Transforms of Orthonormal Functions are Orthonormal¶ 50
2.5 The Discrete Time Fourier Transform 50
  2.5.1 Relation of the DFT to the Continuous Time Fourier Transform 50
2.6 The Discrete Fourier Transform 52
  2.6.1 Circular Convolution 53
  2.6.2 Relation to the Continuous Time Fourier Transform 54
  2.6.3 DFT Leakage 55
2.7 Related Transforms 56
  2.7.1 The Cosine Transform† 56
  2.7.2 The Sine Transform† 57
  2.7.3 The Hartley Transform 57
  2.7.4 The z Transforms 58
2.8 Exercises 60
2.9 Solutions for Selected Chapter 2 Exercises 72

3 Fourier Analysis in Systems Theory 104
3.1 Introduction 104
3.2 System Classes 104
  3.2.1 System Types 104
    3.2.1.1 Homogeneous Systems 104
    3.2.1.2 Additive Systems 105
    3.2.1.3 Linear Systems 105
    3.2.1.4 Time-Invariant Systems 106
    3.2.1.5 LTI Systems 106
    3.2.1.6 Causal Systems 107
    3.2.1.7 Memoryless Systems 107
    3.2.1.8 Stable Systems 107
    3.2.1.9 Invertible Systems 107
  3.2.2 Example Systems 107
    3.2.2.1 The Magnifier 108
    3.2.2.2 The Fourier Transformer 109
    3.2.2.3 Convolution 110

- 3.3 System Characterization   112
  - 3.3.1 Linear System Characterization   112
    - 3.3.1.1 Continuous Time Systems   112
    - 3.3.1.2 Discrete Time Systems   113
  - 3.3.2 Causal Linear Systems   113
  - 3.3.3 Linear Time Invariant (LTI) Systems   114
    - 3.3.3.1 Convolution Algebra   115
    - 3.3.3.2 Convolution Mechanics   117
    - 3.3.3.3 Circular Convolution Mechanics   118
    - 3.3.3.4 Step and Ramp Response Characterization   118
    - 3.3.3.5 Characterizing an LTI System   120
    - 3.3.3.6 Filters   120
- 3.4 Amplitude Modulation   122
  - 3.4.1 Coherent Demodulation   122
    - 3.4.1.1 Loss of Coherence and Fading   124
  - 3.4.2 Envelope Demodulation   124
- 3.5 Goertzel's Algorithm for Computing the DFT   125
- 3.6 Fractional Fourier Transforms   126
  - 3.6.1 Periodicity of the Fourier Transform Operator   126
  - 3.6.2 Fractional Fourier Transform Criteria   127
  - 3.6.3 The Weighted Fractional Fourier Transform   128
- 3.7 Approximating a Linear System with LTI Systems   129
  - 3.7.1 Examples of the Piecewise Invariant Approximation   130
    - 3.7.1.1 The Magnifier   130
    - 3.7.1.2 The Fourier Transformer   131
- 3.8 Exercises   132
- 3.9 Solutions for Selected Chapter 3 Exercises   139

4 Fourier Transforms in Probability, Random Variables and Stochastic Processes   151

- 4.1 Introduction   151
- 4.2 Random Variables   151
  - 4.2.1 Probability Density Functions, Expectation, and Characteristic Functions   152
    - 4.2.1.1 Discrete Random Variables   152
    - 4.2.1.2 Moments from the Characteristic Functions   154
    - 4.2.1.3 Chebyshev's Inequality   157
  - 4.2.2 Example Probability Functions, Characteristic Functions, Means and Variances   158
    - 4.2.2.1 The Uniform Random Variable   159
    - 4.2.2.2 The Triangle Random Variable   159
    - 4.2.2.3 The Gaussian Random Variable   159
    - 4.2.2.4 The Exponential Random Variable   160
    - 4.2.2.5 The Laplace Random Variable   161
    - 4.2.2.6 The Hyperbolic Secant Random Variable   161
    - 4.2.2.7 The Gamma Random Variable   162
    - 4.2.2.8 The Beta Random Variable   164
    - 4.2.2.9 The Cauchy Random Variable   166
    - 4.2.2.10 The Deterministic Random Variable   168
    - 4.2.2.11 The Bernoulli Random Variable   169
    - 4.2.2.12 The Discrete Uniform Random Variable   169

- 4.2.2.13 The Binomial Random Variable   170
- 4.2.2.14 The Poisson Random Variable   171
- 4.2.3 Distributions of Sums of Random Variables   173
  - 4.2.3.1 The Sum of Independent Random Variables   174
  - 4.2.3.2 Distributions Closed Under Independent Summation   174
  - 4.2.3.3 The Sum of i.i.d. Random Variables   176
  - 4.2.3.4 The Average of i.i.d. Random Variables   176
- 4.2.4 The Law of Large Numbers   177
  - 4.2.4.1 Stochastic Resonance   178
- 4.2.5 The Central Limit Theorem   179
  - 4.2.5.1 The Central Limit Theorem Applied to Randomly Generated Probability Density Functions   187
  - 4.2.5.2 Random Variables That Are Asymptotically Gaussian   189
- 4.3 Uncertainty in Wave Equations   190
- 4.4 Stochastic Processes   193
  - 4.4.1 First and Second Order Statistics   194
    - 4.4.1.1 Stationary Processes   195
  - 4.4.2 Power Spectral Density   196
  - 4.4.3 Some Stationary Noise Models   197
    - 4.4.3.1 Stationary Continuous White Noise   197
    - 4.4.3.2 Stationary Discrete White Noise   198
    - 4.4.3.3 Laplace Autocorrelation   198
    - 4.4.3.4 Ringing Laplace Autocorrelation   198
  - 4.4.4 Linear Systems with Stationary Stochastic Inputs   198
  - 4.4.5 Properties of the Power Spectral Density   200
    - 4.4.5.1 Second Moment   200
    - 4.4.5.2 Realness   200
    - 4.4.5.3 Nonnegativity   200
- 4.5 Exercises   201
- 4.6 Solutions for Selected Chapter 4 Exercises   204

5 The Sampling Theorem   217

- 5.1 Introduction   217
  - 5.1.1 The Cardinal Series   217
  - 5.1.2 History   218
- 5.2 Interpretation   220
- 5.3 Proofs   220
  - 5.3.1 Using Comb Functions   221
    - 5.3.1.1 Aliasing   222
    - 5.3.1.2 The Resulting Cardinal Series   222
  - 5.3.2 Fourier Series Proof   222
  - 5.3.3 Papoulis' Proof   223
- 5.4 Properties   223
  - 5.4.1 Convergence   223
    - 5.4.1.1 For Finite Energy Signals   224
    - 5.4.1.2 For Bandlimited Functions with Finite Area Spectra   224
  - 5.4.2 Trapezoidal Integration   225
    - 5.4.2.1 Of Bandlimited Functions   225
    - 5.4.2.2 Of Linear Integral Transforms   226

    5.4.2.3 Derivation of the Low Passed Kernel 227
    5.4.2.4 Parseval's Theorem for the Cardinal Series 231
  5.4.3 The Time-Bandwidth Product 231
5.5 Application to Spectra Containing Distributions 232
5.6 Application to Bandlimited Stochastic Processes 233
5.7 Exercises 234
5.8 Solutions for Selected Chapter 5 Exercises 236

6 Generalizations of the Sampling Theorem 242

6.1 Introduction 242
6.2 Generalized Interpolation Functions 242
  6.2.1 Oversampling 243
  6.2.2 Restoration of Lost Samples 243
    6.2.2.1 Sample Dependency 243
    6.2.2.2 Restoring a Single Lost Sample 243
    6.2.2.3 Restoring M Lost Samples 245
    6.2.2.4 Direct Interpolation from M Lost Samples 246
    6.2.2.5 Relaxed Interpolation Formulae 247
  6.2.3 Criteria for Generalized Interpolation Functions 248
    6.2.3.1 Interpolation Functions 248
  6.2.4 Reconstruction from a Filtered Signal's Samples 250
    6.2.4.1 Restoration of Samples from an Integrating Detector 250
6.3 Papoulis' Generalization 252
  6.3.1 Derivation 252
  6.3.2 Interpolation Function Computation 256
  6.3.3 Example Applications 257
    6.3.3.1 Recurrent Nonuniform Sampling 257
    6.3.3.2 Interlaced Signal–Derivative Sampling 258
    6.3.3.3 Higher Order Derivative Sampling 259
    6.3.3.4 Effects of Oversampling in Papoulis' Generalization 261
6.4 Derivative Interpolation 261
  6.4.1 Properties of the Derivative Kernel 262
6.5 A Relation Between the Taylor and Cardinal Series 265
6.6 Sampling Trigonometric Polynomials 266
6.7 Sampling Theory for Bandpass Functions 267
  6.7.1 Heterodyned Sampling 268
  6.7.2 Direct Bandpass Sampling 270
6.8 A Summary of Sampling Theorems for Directly Sampled Signals 271
6.9 Lagrangian Interpolation 272
6.10 Kramer's Generalization 273
6.11 Exercises 274
6.12 Solutions for Selected Chapter 6 Exercises 278

7 Noise and Error Effects 288

7.1 Introduction 288
7.2 Effects of Additive Data Noise 288
  7.2.1 On Cardinal Series Interpolation 288
    7.2.1.1 Interpolation Noise Level 289
    7.2.1.2 Effects of Oversampling and Filtering 290

        7.2.2   Interpolation Noise Variance for Directly Sampled Signals   293
                 7.2.2.1   Interpolation with Lost Samples   294
                 7.2.2.2   Bandpass Functions   301
        7.2.3   On Papoulis' Generalization   302
                 7.2.3.1   Examples   304
                 7.2.3.2   Notes   304
        7.2.4   On Derivative Interpolation   306
        7.2.5   A Lower Bound on the NINV   308
                 7.2.5.1   Examples   309
  7.3   Jitter   313
  7.4   Filtered Cardinal Series Interpolation   314
        7.4.1   Unbiased Interpolation from Jittered Samples   314
        7.4.2   Effects of Jitter In Stochastic Bandlimited Signal Interpolation   316
                 7.4.2.1   NINV of Unbiased Restoration   318
                 7.4.2.2   Examples   318
  7.5   Truncation Error   320
        7.5.1   An Error Bound   320
  7.6   Exercises   322
  7.7   Solutions for Selected Chapter 7 Exercises   324

8 Multidimensional Signal Analysis   326
  8.1   Introduction   326
  8.2   Notation   327
  8.3   Visualizing Higher Dimensions   328
        8.3.1 N Dimensional Tic Tac Toe   328
        8.3.2   Vectorization   329
  8.4   Continuous Time Multidimensional Fourier Analysis   330
        8.4.1   Linearity   332
        8.4.2   The Shift Theorem   332
        8.4.3   Multidimensional Convolution   332
                 8.4.3.1   The Mechanics of Two Dimensional Convolution   333
        8.4.4   Separability   334
        8.4.5   Rotation, Scale and Transposition   336
                 8.4.5.1   Transposition   337
                 8.4.5.2   Scale   337
                 8.4.5.3   Rotation   337
                 8.4.5.4   Sequential Combination of Operations   338
                 8.4.5.5   Rotation in Higher Dimensions   339
                 8.4.5.6   Effects of Rotation, Scale and Transposition on
                            Fourier Transformation   340
        8.4.6   Fourier Transformation of Circularly Symmetric Functions   341
                 8.4.6.1   The Two Dimensional Fourier Transform of a Circle   343
                 8.4.6.2   Generalization to Higher Dimensions   343
                 8.4.6.3   Polar Representation of Two Dimensional Functions that are
                            Not Circularly Symmetric   344
  8.5   Characterization of Signals from their Tomographic Projections   345
        8.5.1   The Abel Transform and Its Inverse   345
                 8.5.1.1   The Abel Transform in Higher Dimensions   347
        8.5.2   The Central Slice Theorem   348
        8.5.3   The Radon Transform and Its Inverse   349

8.6 Fourier Series 352
    8.6.1 Multidimensional Periodicity 352
    8.6.2 The Multidimensional Fourier Series Expansion 356
    8.6.3 Multidimensional Discrete Fourier Transforms 359
        8.6.3.1 Evaluation 360
        8.6.3.2 The Role of Phase in Image Characterization 360
8.7 Discrete Cosine Transform-Based Image Coding 360
    8.7.1 DCT Basis Functions 362
    8.7.2 The DCT in Image Compression 362
8.8 McClellan Transformation for Filter Design 366
    8.8.1 Modular Implementation of the McClellan Transform 372
    8.8.2 Implementation Issues 373
8.9 The Multidimensional Sampling Theorem 373
    8.9.1 The Nyquist Density 376
        8.9.1.1 Circular and Rectangular Spectral Support 377
        8.9.1.2 Sampling Density Comparisons 379
    8.9.2 Generalized Interpolation Functions 380
        8.9.2.1 Tightening the Integration Region 380
        8.9.2.2 Allowing Slower Roll Off 381
8.10 Restoring Lost Samples 382
    8.10.1 Restoration Formulae 382
        8.10.1.1 Lost Sample Restoration Theorem 382
        8.10.1.2 Restoration of a Single Lost Sample 383
    8.10.2 Noise Sensitivity 383
        8.10.2.1 Filtering 385
        8.10.2.2 Deleting Samples from Optical Images 385
8.11 Periodic Sample Decimation and Restoration 387
    8.11.1 Preliminaries 387
    8.11.2 First Order Decimated Sample Restoration 390
    8.11.3 Sampling Below the Nyquist Density 392
        8.11.3.1 The Square Doughnut 392
        8.11.3.2 Sub-Nyquist Sampling of Optical Images 393
    8.11.4 Higher Order Decimation 393
8.12 Raster Sampling 395
    8.12.1 Bandwidth Equivalence of Line Samples 397
8.13 Exercises 398
8.14 Solutions for Selected Chapter 8 Exercises 404

9 Time-Frequency Representations 411
9.1 Introduction 411
9.2 Short Time Fourier Transforms and Spectrograms 413
9.3 Filter Banks 413
    9.3.1 Commonly Used Windows 415
    9.3.2 Spectrograms 415
    9.3.3 The Mechanics of Short Time Fourier Transformation 416
        9.3.3.1 The Time Resolution Versus Frequency Resolution Trade Off 416
    9.3.4 Computational Architectures 417
        9.3.4.1 Modulated Inputs 418
        9.3.4.2 Window Design Using Truncated IIR Filters 419
        9.3.4.3 Modulated Windows 423

9.4 Generalized Time-Frequency Representations 424
    9.4.1 GTFR Mechanics 425
    9.4.2 Kernel Properties 426
    9.4.3 Marginals 427
        9.4.3.1 Kernel Constraints 428
    9.4.4 Example GTFR's 431
        9.4.4.1 The Spectrogram 431
        9.4.4.2 The Wigner Distribution 434
        9.4.4.3 Kernel Synthesis Using Spectrogram Superposition 434
        9.4.4.4 The Cone Shaped Kernel 434
        9.4.4.5 Cone Kernel GTFR Implementation Using Short Time Fourier Transforms 435
        9.4.4.6 Cone Kernel GTFR Implementation for Real Signals 438
        9.4.4.7 Kernel Synthesis Using POCS 438
9.5 Exercises 439
9.6 Solutions for Selected Chapter 9 Exercises 442

10 Signal Recovery 447
10.1 Introduction 447
10.2 Continuous Sampling 447
10.3 Interpolation From Periodic Continuous Samples 450
    10.3.1 The Restoration Algorithm 451
        10.3.1.1 Trigonometric Polynomials 453
        10.3.1.2 Noise Sensitivity 456
    10.3.2 Observations 461
        10.3.2.1 Comparison with the NINV of the Cardinal Series 461
        10.3.2.2 In the Limit as an Extrapolation Algorithm 462
        10.3.2.3 Application to Interval Interpolation 463
10.4 Interpolation of Discrete Periodic Nonuniform Decimation 463
    10.4.1 Problem Description 466
        10.4.1.1 Degree of Aliasing 467
        10.4.1.2 Interpolation 467
        10.4.1.3 The $\mathbf{A}[N, P]$ Matrix 470
    10.4.2 The Periodic Functions, $\Psi_M(\nu)$ 470
    10.4.3 Quadrature Version 472
10.5 Prolate Spheroidal Wave Functions 473
    10.5.1 Properties 473
    10.5.2 Application to Extrapolation 475
    10.5.3 Application to Interval Interpolation 476
10.6 The Papoulis-Gerchberg Algorithm 477
    10.6.1 The Basic Algorithm 477
        10.6.1.1 An Alternate Derivation of the PGA Using Operators 479
        10.6.1.2 Application to Interpolation 480
    10.6.2 Proof of the PGA using PSWF's 480
    10.6.3 Remarks 482
10.7 Exercises 482
10.8 Solutions for Selected Chapter 10 Exercises 486

11 Signal and Image Synthesis: Alternating Projections Onto Convex Sets 495
11.1 Introduction 495

- 11.2 Geometical POCS   496
  - 11.2.1 Geometrical Convex Sets   496
  - 11.2.2 Projecting onto a Convex Set   497
  - 11.2.3 POCS   498
- 11.3 Convex Sets of Signals   501
  - 11.3.1 The Hilbert Space   502
    - 11.3.1.1 Convex Sets   503
    - 11.3.1.2 Subspaces   503
    - 11.3.1.3 Null Spaces   503
    - 11.3.1.4 Linear Varieties   504
    - 11.3.1.5 Cones   504
    - 11.3.1.6 Convex Hulls and Convex Cone Hulls   504
  - 11.3.2 Some Commonly Used Convex Sets of Signals   504
    - 11.3.2.1 Matrix equations   505
    - 11.3.2.2 Bandlimited Signal   507
    - 11.3.2.3 Duration Limited Signals   507
    - 11.3.2.4 Real Transform Positivity   508
    - 11.3.2.5 Constant Area Signals   508
    - 11.3.2.6 Bounded Energy Signals   511
    - 11.3.2.7 Constant Phase Signals   511
    - 11.3.2.8 Bounded Signals   512
    - 11.3.2.9 Signals With Identical Middles   513
- 11.4 Example Applications of POCS   514
  - 11.4.1 Von Neumann's Alternating Projection Theorem   514
  - 11.4.2 Solution of Simultaneous Equations   514
  - 11.4.3 The Papoulis-Gerchberg Algorithm   514
  - 11.4.4 Howard's Minimum-Negativity-Constraint Algorithm   516
  - 11.4.5 Associative Memory   516
    - 11.4.5.1 Template Matching   517
    - 11.4.5.2 POCS Based Associative Memories   517
    - 11.4.5.3 POCS Associative Memory Examples   517
    - 11.4.5.4 POCS Associative Memory Convergence   521
    - 11.4.5.5 POCS Associative Memory Capacity   526
    - 11.4.5.6 Heteroassociative Memory   527
  - 11.4.6 Recovery of Lost Image Blocks   527
  - 11.4.7 Subpixel Resolution   532
    - 11.4.7.1 Formulation as a Set of Linear Equations   532
    - 11.4.7.2 Subpixel Resolution using POCS   533
  - 11.4.8 Reconstruction of Images from Tomographic Projections   533
  - 11.4.9 Correcting Quantization Error for Oversampled Bandlimited Signals   535
  - 11.4.10 Kernel Synthesis for GTFR's   537
    - 11.4.10.1 GTFR Constraints as Convex Sets   537
    - 11.4.10.2 GTFR Kernel Synthesis Using POCS   543
  - 11.4.11 Application to Conformal Radiotherapy   545
    - 11.4.11.1 Convex Constraint Sets   547
    - 11.4.11.2 Example Applications   557
- 11.5 Generalizations   558
  - 11.5.1 Iteration Relaxation   558
  - 11.5.2 Contractive and Nonexpansive Operators   558
    - 11.5.2.1 Contractive and Nonexpansive Functions   559

11.6 Exercises   563
11.7 Solutions for Selected Chapter 11 Exercises   567

12 Mathematical Morphology and Fourier Analysis on Time Scales   570
  12.1 Introduction   570
  12.2 Mathematical Morphology Fundamentals   570
   12.2.1 Minkowski Arithmetic   570
   12.2.2 Relation of Convolution Support to the Operation of Dilation   572
   12.2.3 Other Morphological Operations   573
   12.2.4 Minkowski Algebra   576
  12.3 Fourier and Signal Analysis on Time Scales   583
   12.3.1 Background   586
    12.3.1.1 The Hilger Derivative   586
    12.3.1.2 Hilger Integration   587
   12.3.2 Fourier Transforms on a Time Scale   587
   12.3.3 The Minkowski Sum of Time Scales   590
   12.3.4 Convolution on a Time Scale   592
    12.3.4.1 Convolution on Discrete Time Scales   592
    12.3.4.2 Time Scales in Deconvolution   594
   12.3.5 Additively Idempotent Time Scales   595
    12.3.5.1 AITS Hulls   596
    12.3.5.2 AITS Examples   596
    12.3.5.3 AITS Conservation   597
    12.3.5.4 Asymptotic Graininess of $\mathbb{A}_{\xi\eta}$ 597
   12.3.6 Discrete Convolution of AITS'   601
    12.3.6.1 Applications   601
   12.3.7 Multidimensional AITS Time Scales   604
  12.4 Exercises   608
  12.5 Solutions for Selected Chapter 12 Exercises   609

13 Applications   610
  13.1 The Wave Equation, Its Fourier Solution and Harmony in Western Music   610
   13.1.1 The Wave Equation   610
   13.1.2 The Fourier Series Solution   612
   13.1.3 The Fourier Series and Western Harmony   613
   13.1.4 Pythagorean Harmony   614
    13.1.4.1 Melodies of Harmonics: Bugle Tunes   616
    13.1.4.2 Pythagorean and Tempered String Vibrations   617
   13.1.5 Harmonics Expansions Produce Major Chords and the Major Scale   618
    13.1.5.1 Subharmonics Produce Minor Keys   620
    13.1.5.2 Combining the Harmonic and Subharmonic Expansions Approximates the Tempered Chromatic Scale   621
   13.1.6 Fret Calibration   621
    13.1.6.1 Comments   623
  13.2 Fourier Transforms in Optics and Wave Propagation   623
   13.2.1 Scalar Model for Wave Propagation   624
    13.2.1.1 The Wave Equation   624
    13.2.1.2 Solutions to the Wave Equation   626
   13.2.2 The Angular Spectrum   627

                13.2.2.1 Plane Waves as Frequencies   628
                13.2.2.2 Propagation of the Angular Spectrum   629
                13.2.2.3 Evanescent Waves   630
        13.2.3 Rayleigh-Sommerfield Diffraction   630
                13.2.3.1 The Diffraction Integral   633
                13.2.3.2 The Fresnel Diffraction Integral   633
                13.2.3.3 The Fraunhofer Approximation   634
        13.2.4 A One Lens System   638
                13.2.4.1 Analysis   638
                13.2.4.2 Imaging System   640
                13.2.4.3 Fourier Transformation   641
                13.2.4.4 Implementation of the PGA Using Optics   642
        13.2.5 Beamforming   643
                13.2.5.1 Apodization   644
                13.2.5.2 Beam Steering   644
    13.3 Heisenberg's Uncertainty Principle   645
    13.4 Elementary Deterministic Finance   646
        13.4.1 Some Preliminary Math   647
        13.4.2 Compound Interest on a One Time Deposit   647
                13.4.2.1 Yield Increases with Frequency of Compounding   648
                13.4.2.2 Continuous Compounding   649
                13.4.2.3 Different Rates and Compounding Periods – Same Yield   649
                13.4.2.4 The Extrema of Yield   650
                13.4.2.5 Effect of Annual Taxes   650
                13.4.2.6 Effect of Inflation   651
        13.4.3 Compound Interest With Constant Periodic Deposits   652
                13.4.3.1 Continuous Time Solution   653
                13.4.3.2 Be a Millionaire   653
                13.4.3.3 Starting With a Nest Egg   654
        13.4.4 Loan and Mortgage Payments   655
                13.4.4.1 Monthly Payment   655
                13.4.4.2 Amortization   656
                13.4.4.3 Monthly Payment Over Long Periods of Time   656
    13.5 Exercises   656
    13.6 Solutions for Selected Chapter 13 Exercises   658

14 Appendices   660
    14.1 Schwarz's Inequality   660
    14.2 Leibniz's Rule   661
    14.3 A Useful Limit   661
    14.4 Series   662
        14.4.1 Binomial Series   662
        14.4.2 Geometric Series   662
                14.4.2.1 Derivation   623
                14.4.2.2 Trigonometric Geometric Series   663
    14.5 Ill-Conditioned Matrices   663
    14.6 Other Commonly Used Random Variables   664
        14.6.1 The Pareto Random Variable   665
        14.6.2 The Weibull Random Variable   666
        14.6.3 The Chi Random Variable   667

        14.6.4   The Noncentral Chi-Squared Random Variable   667
        14.6.5   The Half Normal Random Variable   668
        14.6.6   The Rayleigh Random Variable   668
        14.6.7   The Maxwell Random Variable   669
        14.6.8   The Log Random Variable   669
        14.6.9   The Von Mises Variable   670
        14.6.10   The Uniform Product Variable   671
        14.6.11   The Uniform Ratio Variable   671
        14.6.12   The Logistic Random Variable   672
        14.6.13   The Gibrat Random Variable   673
        14.6.14   The F Random Variable   673
        14.6.15   The Noncentral F Random Variable   673
        14.6.16   The Fisher-Tippett Random Variable   675
        14.6.17   The Gumbel Random Variable   675
        14.6.18   The Student's t Random Variable   676
        14.6.19   The Noncentral Student's t Random Variable   676
        14.6.20   The Rice Random Variable   677
        14.6.21   The Planck's Radiation Random Variable   677
        14.6.22   The Generalized Gaussian Random Variable   677
        14.6.23   The Generalized Cauchy Random Variable   678

15 References   680

    Index   745

Handbook of Fourier Analysis & Its Applications

1

Introduction

> The profound study of nature is the most fertile source of mathematical discoveries.
> Jean Baptiste Joseph Fourier [743]

> About Fourier: It was, no doubt, partially because of his very disregard for rigor that he was able to take conceptual steps which were inherently impossible to men of more critical genius.
> Rudoph E. Langer [354]

> Fourier is a mathematical poem.
> William Thomson (Lord Kelvin), (1824–1907) [1392]

1.1 Ubiquitous Fourier Analysis

Jean Baptiste Joseph Fourier's powerful idea of decomposition of a signal into sinusoidal components has found application in almost every engineering and science field. An incomplete list includes acoustics [1497], array imaging [1304], audio [1290], biology [826], biomedical engineering [1109], chemistry [438, 925], chromatography [1481], communications engineering [968], control theory [764], crystallography [316, 498, 499, 716], electromagnetics [250], imaging [151], image processing [1239] including segmentation [1448], nuclear magnetic resonance (NMR) [436, 1009], optics [492, 514, 517, 1344], polymer characterization [647], physics [262], radar [154, 1510], remote sensing [84], signal processing [41, 154], structural analysis [384], spectroscopy [84, 267, 724, 1220, 1293, 1481, 1496], time series [124], velocity measurement [1448], tomography [93, 1241, 1242, 1327, 1330, 1325, 1331], weather analysis [456], and X-ray diffraction [1378], Jean Baptiste Joseph Fourier's last name has become an adjective in the terms like Fourier series [395], Fourier transform [41, 51, 149, 154, 160, 437, 447, 926, 968, 1009, 1496], Fourier analysis [151, 379, 606, 796, 1472, 1591], Fourier theory [1485], the Fourier integral [395, 187, 1399], Fourier inversion [1325], Fourier descriptors [826], Fourier coefficients [134], Fourier spectra [624, 625] Fourier reconstruction [1330], Fourier spectrometry [84, 355], Fourier spectroscopy [1220, 1293, 1438], Fourier array imaging [1304], Fourier transform nuclear magnetic resonance (NMR) [429, 1004], Fourier vision [1448], Fourier optics [419, 517, 1343], and Fourier acoustics [1496].

Applied Fourier analysis is ubiquitous simply because of the utility of its descriptive power. It is second only to the differential equation in the modelling of physical phenomena. In contrast with other linear transforms, the Fourier transform has a number of physical manifestations. Here is a short list of everyday occurrences as seen through the lens of the Fourier paradigm.

- Diffracting coherent waves in sonar and optics in the far field[1] are given by the two dimensional Fourier transform of the diffracting aperture.[2] Remarkably, in free space, the physics of spreading light naturally forms a two dimensional Fourier transform.
- The sampling theorem,[3] born of Fourier analysis, tells us how fast to sample an audio waveform to make a discrete time CD or an image to make a DVD.
- Some audio equipment contains a number of adjacent vertical bars of LED's that, like the level in a mercury thermometer, bounce up and down in illumination as the audio is playing. The vertical bars on the left correspond to low frequencies and the bars on the right to high. The bars are capturing a spectrogram-like time-frequency representations.[4]
- Likewise, musical scores can be construed as notation for spectrogram-like time-frequency representations. The musical score tells the musician what note, i.e., frequency, to play at what time.
- The eyes decompose a portion of electromagnetic frequencies into a Fourier spectrum of color.
- When the neighbor has the stereo volume too loud, the throbbing bass notes come through the wall better than the high notes. We are experiencing a mechanical low pass filter as modelled by Fourier analysis.
- Our ear's cochlea is designed to decompose incident acoustic waves into its Fourier spectrum. We can therefore differentiate frequencies and enjoy music.
- Backwards rotating wagon wheels on old western movies are an example of temporal aliasing described through spectral overlap in the temporal Fourier transform.
- In low resolution television, the large floating patterns on a shirt with a fine structured pattern is an example of spatial aliasing characterized by the spectral overlap in the spatial Fourier transform.
- JPEG image encoding, used to reduce the file size of an image, is based of the discrete cosine transform (DCT)—a close relative of the Fourier transform.

The Fourier transform in everyday occurrences is ubiquitous.

1.2 Jean Baptiste Joseph Fourier

Fourier analysis has its foundation in the paper *On the Propagation of Heat in Solid Bodies* [455]. Jean Baptiste Joseph Fourier (1768–1830) read it to the Paris Institute on 21 December 1807. Fourier's work is still in print [455].

Here are highlights of Fourier's life.[5]

- Fourier was born on March 21, 1768 in Auxurre, France. He was the ninth child of Joseph Fourier, a tailer. Fourier's mother, Edmie Fourier, was the second wife of Joseph Senior.

---

1. Also known as the Fraunhofer diffraction region.
2. See Section 13.2 on *Fourier Transforms in Optics and Wave Propagation*.
3. See Chapter 5 on *The Sampling Theorem*.
4. See Chapter 9 on *Time-Frequency Representations*.
5. For the interested reader, numerous books [24, 547, 595] and articles [327, 546, 594, 1274, 1374, 1420, 1548] have been written on the life of Fourier.

FIGURE 1.1. Jean Baptiste Joseph Fourier. ("Portraits et Histoire des Hommes Utiles, Collection de Cinquante Portraits," Societe Montyon et Franklin, 1839–1840.)

- Fourier, named after Saint Joseph and John the Baptist, initially trained for the priesthood at the Benedictine abbey of Saint Benoît-sur-Loire. His love of mathematics, however, led him to abandon his plans to take religious vows. Fourier graduated with honors from the military school in Auxerre. He joined the staff of the École Normale and then the École Polytechnique in Paris. Fourier was recommended to the Bishop of Auxerre and, as a result of this introduction, was educated by the Benvenistes.
- In 1793, Fourier was attracted to politics and joined a local revolutionary committee. His political passion resulted in his arrest, imprisonment and release in 1794.
- In 1797, Fourier succeeded J.L. Lagrange in appointment to the École Polytechnique.
- In 1798, Fourier travelled to Egypt with Napoleon, who made him Governor of Lower Egypt. Fourier returned to France in 1801 and returned to his post at the École Polytechnique. In 1807, Fourier completed his landmark work *On the Propagation of Heat in Solid Bodies*. The committee for his reading on December 21, 1807 consisted of esteemed mathematicians J.L. Lagrange, P.S. Laplace, Gaspard Monge and Sylvestre Lacroix. Fourier's presentation was received with skepticism. Both Laplace and Lagrange objected to Fourier's expansions of functions as trigonometrical series, what we now call Fourier series. They were skeptical the smooth sinusiod could be used to represent function discontinuities. They were wrong.
- Fourier was made a baron by Napoleon in 1808.
- Fourier was a major contributor to the 21 volumes of *Description de l'Egypte* written from 1808 to 1825. The work contained the cultural and scientific results

of Napoleon's Egypt invasion. *Description de l'Egypte* established Egyptology as a discipline.
- Fourier died in Paris in 1830.

Fourier has another scientific claim to fame. He was the first to identify the greenhouse effect. Fourier wondered how the Earth stays warm. He hypothesized energy from the sun is trapped between the Earth's surface and the blanket of atmospheric gases.

In engineering, science and mathematics, Fourier will always be remembered as the first to represent temporal signals as superposition of sinusoids in the frequency domain.

1.3 This Book

No single volume can present all of the applications for Fourier's theory. The material, like a collection of the theory and applications of differential equations, is simply too voluminous. This book is written largely from the perspective of the engineer and the scientist with and an eye towards application.

1.3.1 Surveying the Contents

Chapter 1 contains a brief biographical sketch of Jean Baptiste Joseph Fourier, the transform that bears his name, and examples of the enormous impact of Fourier analysis in the fields of physics, medicine, engineering, finance, and music.

The foundations of Fourier analysis are recounted in Chapter 2 at a level appropriate for a review at an advanced level. Emphasis is placed on Fourier analysis, including to higher transcendental functions. Relationships between transforms of discrete and continuous time signals are firmly established. Some of the material in this chapter, as discussed in Section 2.1, can be skipped on a first reading.

Chapter 3 deals with application of Fourier analysis to system modelling. Systems are characterized as the response of a black box to a stimuli. For mathematical traction, constraints must be placed on the black box system. Examples include causality, stability, linearity and time invariance. Fractional Fourier transforms, Goertzel's algorithm, and amplitude modulation are presented in Chapter 3 as examples of linear systems.

Use of the Fourier transform in probability, random variables and stochastic processes is the topic of Chapter 4. Emphasis on the utility of characteristic functions to define random variables is made as is use of the Fourier transform in establishing the central limit theorem. Foundational material on Fourier treatment of stochastic processes is also presented. A type of stochastic resonance is used to illustrate the weak law of large numbers. Foundational concepts of the modelling of stochastic processes are also introduced.

Shannon sampling and interpolation theory is among the most well developed topics in this book. Some of this material was contained in the author's previous book [915] on sampling theory. Included in Chapter 5 are foundational proofs and interpretation of the sampling theorem. Generalizations of the sampling theorem, including those of Papoulis, Kramer, and Lagrange, are the topic of Chapter 6. The consequences of noise on restoration from samples is analyzed in Chapter 7.

Chapter 8 contains an exposition on Fourier analysis of multidimensional signals including images. Emphasis is placed on multidimensional Fourier analysis that is not a straightforward extension of the one-dimensional counterpart. This includes the operations of separability, rotation, and multidimensional symmetry. Applications of multidimensional Fourier analysis include use of the discrete cosine transform for image compression, characterizations of signals from their tomographic projections, McClellan transformation for multidimensional filter design, restoring lost samples, and raster sampling. The multidimensional extension of the Shannon sampling theorem and its generalizations are also presented.

The most familiar example of the time-frequency representations, the topic of Chapter 9, is the musical score. Time flows from measure to measure. At each point in time, the musician is asked to play a certain note or, for our purposes, frequency. The musical score therefore contains a temporal sequence of frequency. The challenge of the time-frequency representation is the transformation of a temporal signal into a time-frequency representation, or musical score. The classic approach is generation of a spectrogram. This approach, though, suffers from a time versus frequency resolution tradeoff. This problem is addressed, in part, by Cohen's *generalized time frequency representation* (GTFR). Properties and implementations of the GTFR are developed in Chapter 9.

Fourier analysis has a strong role in signal recovery. This is the topic of Chapter 10. Attention is restricted to restoration of continuously sampled signals and the Papoulis-Gerchberg algorithm. Likewise, Chapter 11 introduces the *alternating projection onto convex sets*, or POCS paradigm. Many important synthesis and recovery problems can be placed in the POCS rubric including solution of simultaneous equations, associative memory formation, recovery of lost blocks in JPG images, subpixel resolution, tomography, correcting quantization error, synthesizing time-frequency representations, and conformal radiotherapy.

Times scales is a theory which bridges continuous and discrete time signals and systems models. Continuous and discrete time are both special cases of time scales. After establishing foundational properties of mathematical morphology, an introduction to time scales is presented in Chapter 12. The relationship between time scale Fourier analysis, multidimensional convolution and certain operations in mathematical morphology is also established.

Chapter 13 contains discourses on applications of Fourier analysis to a plurality of selected disciplines. These include

- The role of the Fourier series in solution of boundary value problems and the establishment of the tempered scale and western music (Section 13.1).
- The natural occurrence of the two-dimensional Fourier transform in diffraction phenomena (Section 13.2).
- The role of the Fourier transform in establishment of Heisenberg's uncertainty principle (Section 13.3).
- Convolution applications in elementary deterministic finance (Section 13.4.)

Other applications are sprinkled throughout the text. Here is a partial list.

- *Stochastic resonance* wherein either too much or too little noise renders a process impotent (Section 4.2.4.1).
- Lost sample restoration (See Sections 6.2.2 for the one dimensional and 8.10 for the multi-dimensional case).
- Use of DCT's in image compression (Section 8.7).
- Foundations of reconstruction of multidimensional signals from their projections, i.e., tomography (Sections 8.5, 8.5, and 11.4.8).
- Subpixel resolution (Section 11.4.7).
- McClellan transform methods for the design of multidimensional digital filters (Section 8.8).

- Restoration of linear motion blur (Exercise 3.25).
- Correcting quantization error (Section 11.4.9).
- Beam synthesis in conformal radiotherapy (Section 11.4.11).
- Modulation (Section 3.4).
- Beamforming (Section 13.2.5).

For the book browser, here is a list of some of the more fascinating material covered in this volume.

- The phase of a signal's Fourier transform is generally more important than its magnitude. See Section 8.6.3.2.
- High resolution pictures can be synthesized from images from a low resolution camera. See Section 11.4.7.
- The central limit theorem says convolution of a number of functions, under loose conditions, approaches a Gaussian. See Section 4.2.5.
- The Fourier transform is not only a useful tool in signal analysis, it occurs naturally in nature. Coherent waves, for example, take on the form of the Fourier transform of the source in the far field. See Section 13.2.
- Western music's chromatic scale is a compromise between maintaining the integer ratios of Fourier harmonics and the ability to change keys. See Section 13.
- Fractional derivatives, such as $\left(\frac{d}{dt}\right)^{1/2} x(t)$, are well defined and arise in the modelling of the restoration of images from their projections. See Exercise 8.19.
- There are also fractional Fourier transforms. See Section 3.6.
- There is a unification theory, dubbed *time scales*, that include both the discrete and continuous time Fourier transforms as special cases. See Section 12.3.
- A bandlimited signal's samples can be connected by lines and placed through a filter to restore the original bandlimited signal. See Section 6.2.3.1.
- Heisenberg's uncertainty principle is expressed using Fourier analysis. See Section 13.3
- Simple personal finance equations, including mortgage payments, savings accrual and amortization schedules, can be derived using the solution to a simple first order difference equation. See Section 13.4.
- For example: (a) At 12% interest, a monthly penny deposit will accumulate to a million dollars in 116 years. See Exercise 13.10. (b) The monthly mortgage payment of a 5% interest mortgage over 100 years is reduced by less that one percent when paid over a period of 1000 years. The monthly payment over 10,000 years is, to the penny, the same as for 1000 years. See Section 13.4.4.3.
- In the absence of noise, knowledge of an arbitrarily short section of a bandlimited signal is sufficient to specify the signal everywhere. Presence of noise or other uncertainty, however, renders this extrapolation ill-posed. See Section 10.2.
- In the absence of noise, if a signal is sampled in excess of the Nyquist rate, an arbitrarily large but finite number of lost samples can be regained from those remaining. See Section 6.2.2.
- If a bandlimited signal is oversampled, the sinc is a viable interpolation function. So is the difference between two $sinc^2$ functions. See, *e.g.*, Exercise 6.2. So are a plethora of other functions. See Section 6.2.2.5.
- The real and imaginary parts of the Fourier transform of a causal signal are Hilbert transform pairs. See (2.99).

- If the integral over all time of a signal is finite, then the integral over all time of its derivative is zero. See (2.94).
- For continuously sampled signals, aliased data can be restored without ambiguity. See Section 10.2.
- The Fourier transform of Legendre polynomials are spherical Bessel functions. See (2.82).
- Weighed Hermite polynomials are their own Fourier transforms. See (2.90).
- The Fourier series and the sampling theorem are Fourier duals of each other. See Section 5.3.2.
- The coefficients of a Fourier series can be obtained by sampling the Fourier transform of a single period of the periodic function. See (2.17).
- If functions vary slowly enough, samples of position and velocity taken simultaneously at a fast enough rate can be used to restore both the continuous position and velocity as a function of time. If, though, the samples are taken alternately, the continuous functions can't be reconstructed. See Section 6.3.3.
- Shifted sinc functions form an orthogonal basis function for bandlimited signals. See (2.110).
- Sampling densities can lie below the Nyquist density in higher dimensions without imposing irreversible aliasing. The original multidimensional function can be recovered from the aliased samples. See Section 8.11.
- Sampling continuous white noise does not result in discrete white noise. See 4.4.3.
- $n!$ ends in about $\frac{5^n-1}{4}$ zeros. See Exercise 2.7 in Chapter 2.
- No one has yet defined a real system that is additive but not homogeneous. No one has identified a homogeneous system that is not additive. See Sections 3.2.1.1 and 3.2.1.2 and Exercise 3.12.
- For Gibb's phenomenon in the Fourier series, the overshoot becomes smaller and smaller but never disappears. See Exercise 2.51 in Section 2.8.
- The geometric series gives rise to interesting *repunit* and *Mersenne numbers*. See Appendix 14.4.2.
- Dividing a uniform random variable by a second independent uniform random variable results in a random variable for whom all moments, including the mean, are infinite. See Appendix 14.6.11.

Two topics not covered in this book are

(1) fast Fourier transforms (FFT's), [160, 161, 322, 323, 625, 1111], and
(2) wavelets [93, 296, 318, 347, 520, 1249, 1362, 1459, 1468, 1517, 1591], except for filter banks [451, 805, 977, 1348, 1429, 1430] in Section 9.3.

2

Fundamentals of Fourier Analysis

> I have tried, with little success, to get some of my friends to understand my amazement that the abstraction of integers for counting is both possible and useful. Is it not remarkable that 6 sheep plus 7 sheep makes 13 sheep; that 6 stones plus 7 stones make 13 stones? Is it not a miracle that the universe is so constructed that such a simple abstraction as a number is possible? To me this is one of the strongest examples of the unreasonable effectiveness of mathematics. Indeed, I find it both strange and unexplainable.
>
> Richard W. Hamming (1915–1998) [573]

> To the extent that there is some correspondence between [the natural world and its mathematical modelling] ... we have the miracle of modem science - the deepening understanding of our universe, and the bounty and ease of the technological society in which we live. A second-order miracle, little recognized or appreciated, is that this first miracle could arise from such a really ragged fit between the [natural world and its mathematical modeling].
>
> David Slepian [1283]

> Finally, two days ago, I succeeded - not on account of my hard efforts, but by the grace of the Lord. Like a sudden flash of lightning, the riddle was solved. I am unable to say what was the conducting thread that connected what I previously knew with what made my success possible.
>
> Karl Friedrich Gauss (1777–1855) [422]

> One cannot escape the feeling that these mathematical formulas have an independent existence and an intelligence of their own, that they are wiser than we are, wiser even than their discoverers, that we get more out of them than was originally put into them.
>
> Heinrich Hertz (1857–1894) [89]

2.1 Introduction

This chapter contains foundational material for modelling of signals and systems. Section 2.2 introduces classes of functions useful in signal processing and analysis. The Fourier transform, in Section 2.3, begins with the Fourier integral and develops the Fourier series, the discrete time Fourier transform and the discrete Fourier transform as special cases.

The following material in this chapter can be skipped on a first reading.

† denotes material relevant to multidimensional signals in Chapters 8 and 11.
‡ denotes material relevant to probability and stochastic processes in Chapter 4.
¶ denotes material used in continuous sampling in Chapter 10.

2.2 Signal Classes

There are a number of signal classes to which we will make common reference. Continuous time signals are denoted with their arguments in parentheses, e.g., $x(t)$. Discrete time signals will be bracketed, e.g., $x[n]$.

(a) **Periodic Signals**

A continuous time signal, $x(t)$, is periodic if there exists a T such that

$$x(t) = x(t - T)$$

for all t. The function $x(t) = constant$ is periodic.

A discrete time signal, $x[n]$, is periodic if there exists a positive integer N such that

$$x[n] = x[n - N]$$

for all n. The function $x[n] = constant$ is periodic.

(b) **Finite Energy Signals**[1]

If

$$E = \int_{-\infty}^{\infty} |x(t)|^2 \, dt < \infty, \tag{2.1}$$

then the continuous time signal, $x(t)$, is said to have finite energy,[2] E. A discrete time signal, $x[n]$, has finite energy if

$$E = \sum_{n=-\infty}^{\infty} |x[n]|^2 < \infty. \tag{2.2}$$

(c) **Finite Area Signals**[3]

If

$$A = \int_{-\infty}^{\infty} |x(t)| \, dt < \infty, \tag{2.3}$$

then the continuous time signal $x(t)$ is said to have finite area A. For discrete time signals, $x[n]$ has finite area if

$$A = \sum_{-\infty}^{\infty} |x[n]| < \infty.$$

---

1. The classes of continuous time, finite energy, finite area, and bounded signals, when Lebesgue measurable, are recognized respectively as L_2, L_1 and L_∞ signals.
   Likewise, the respective discrete time signals are ℓ_2, ℓ_1 and ℓ_∞ signals [854, 1020].
2. Why is this quantity referred to as an *energy*? For a one ohm resister, the instantaneous dissipated power is $|x(t)|^2$ if $x(t)$ is either a voltage or a current. The total energy used by the resistor is the integral of the power and is given by (2.1). There is no analogous interpretation for the discrete time finite energy signal. Dubbing (2.2) *energy* is a simple carryover from the continuous case.
3. See Footnote 1.

(d) **Bounded Signals**[4]
If, for C a constant,

$$|x(t)| \leq C < \infty, \tag{2.4}$$

then the continuous time signal, $x(t)$, is said to be bounded. A discrete time signal, $x[n]$, is bounded if

$$|x[n]| \leq C < \infty.$$

(e) **Bandlimited Signals**
If there exists a finite *bandwidth*, B, such that

$$\int_{-\infty}^{\infty} x(t)e^{-j2\pi ut}\, dt = 0\,;\ |u| > B,$$

then the continuous time signal, $x(t)$, is said to be bandlimited in the low pass sense.[5] A discrete time signal is bandlimited if there is a $B < \frac{1}{2}$ such that

$$\sum_{n=-\infty}^{\infty} x[n]e^{-j2\pi fn} = 0\,;\ B < |f| \leq \frac{1}{2}. \tag{2.5}$$

For continuous time signals, where t has units of time, the bandwidth, B, has units of Hertz or, equivalently, cycles per second. For discrete time signals, the bandwidth B is unitless.

(f) **Analytic Signals**
If, for every finite complex number φ, we have equality in the Taylor series

$$x(z) = \sum_{n=0}^{\infty} \frac{x^{(n)}(\varphi)\,(z-\varphi)^n}{n!}$$

where the signal's nth derivative is

$$x^{(n)}(z) = \left(\frac{d}{dz}\right)^n x(z)$$

then $x(z)$ is said to be *analytic everywhere* in the closed z plane. Such signals are also called *entire*.

(g) **Causal Signals**
A signal is *causal* if it is identically zero for negative argument. Thus, $x(t)$ is causal if $x(t) = 0$ for all $t < 0$.

(h) **Symmetric Signals**
By *symmetric signals*, we mean signals that are either even or odd. A signal, $e(t)$, is even if

$$e(t) = e(-t) \tag{2.6}$$

---

4. See Footnote 1.
5. In disciplines such as circuit design, the bandwidth is alternately defined as the frequency interval over which the magnitude of the signal's spectrum exceeds $1/\sqrt{2}$, or 3 dB, of its maximum value. Our definition here is different.

TABLE 2.1. A summary of the properties of continuous and discrete time even and odd functions. The asterisk, ∗, denotes either continuous or discrete time convolution. (See entries in Tables 2.3 and 2.6 for convolution definitions.) The double arrow, ↔ denotes either continuous or discrete time Fourier transformation. (See (2.11) and (2.112).) For proofs of these properties, see Exercise 2.21 and its solution.

(a)	$e + e$	=	e	(k)	$\int_{-T}^{T} o \, dt$	=	0
(b)	$o + o$	=	o	(l)	$\sum_{0}^{n} o$	=	o
(c)	$e \times e$	=	e	(m)	$\sum_{0}^{n} e$	=	e
(d)	$o \times o$	=	e	(n)	$\sum_{-N}^{N} e$	=	$e[0] + 2\sum_{n=1}^{N} e[n]$
(e)	$o \times e$	=	o	(o)	$\sum_{-N}^{N} o$	=	0
(f)	$\frac{d}{dt} e$	=	o	(p)	$e \ast e$	=	e
(g)	$\frac{d}{dt} o$	=	e	(q)	$o \ast o$	=	e
(h)	$\int_{0}^{t} o \, d\tau$	=	e	(r)	$e \ast o$	=	o
(i)	$\int_{0}^{t} e \, d\tau$	=	o	(s)	e	↔	e
(j)	$\int_{-T}^{T} e \, dt$	=	$2\int_{t=0}^{T} e(t) \, dt$	(t)	o	↔	jo

and $o(t)$ is odd if

$$o(t) = -o(-t). \tag{2.7}$$

Every signal, $c(t)$, can be expressed as the sum of an even and odd function.

$$c(t) = e(t) + o(t).$$

The even component is

$$e(t) = \frac{1}{2}(c(t) + c(-t)) \tag{2.8}$$

and the odd component is

$$o(t) = \frac{1}{2}(c(t) - c(-t)). \tag{2.9}$$

A summary of properties of even and odd functions is in Table 2.1.

2.3 The Fourier Transform

In this section, as listed in Table 2.2, we review the basic properties of continuous time Fourier transforms (CTFT), the Fourier series, the discrete time Fourier transform (DTFT), and the discrete Fourier transform (DFT).

TABLE 2.2. Fourier transform types and their mappings. The entries are for (a) CTFT = continuous time Fourier transform, (b) the Fourier series, (c) the DTFT = discrete time Fourier transform, and (d) DFT = the discrete Fourier transform.

Transform	Time	→	Frequency
CTFT	continuous	→	continuous
Fourier Series	continuous	→	discrete
DTFT	discrete	→	continuous
DFT	discrete	→	discrete

2.3.1 The Continuous Time Fourier Transform

The CTFT of a continuous time signal, $x(t)$, is[6]

$$X(u) := \int_{-\infty}^{\infty} x(t) \, e^{-j2\pi ut} \, dt. \tag{2.10}$$

where the notation ":=" is read "is equal to by definition." CTFT pairs will be denoted by[7]

$$x(t) \longleftrightarrow X(u). \tag{2.11}$$

The inverse Fourier transform is[8]

$$x(t) = \int_{-\infty}^{\infty} X(u) \, e^{j2\pi ut} \, du. \tag{2.12}$$

2.3.2 The Fourier Series

Periodic functions with period T can be expressed as a Fourier series,

$$x(t) = \sum_{n=-\infty}^{\infty} c_n \, e^{j2\pi nt/T}, \tag{2.13}$$

where the Fourier series coefficients are

$$c_n = \frac{1}{T} \int_T x(t) \, e^{-j2\pi nt/T} \, dt, \tag{2.14}$$

and integration is over any arbitrary single period. That is, for an arbitrary τ, we define

$$\int_T := \int_{t=\tau}^{\tau+T}. \tag{2.15}$$

---

6. When t has units of seconds, the frequency variable, u, has units of Hertz. The frequency variable u is used in this book over the more conventional $\omega = 2\pi u$ for the following reasons. (1) ω has units of radians-per-second and, relative to u, is opaque to interpretation, e.g., what is the physical meaning of $\omega = 377$ radians-per-second? One typically divides by 2π to obtain $u = 60$ Hertz which is a more straightforward and meaningful measure. (2) When using ω, dual Fourier transform theorems typically require correct placement of a factor of 2π. For example, the Fourier transform, $X(\omega) = \int_t x(t) \exp(-j\omega t) dt$ has as its (dual) inversion $x(t) = \frac{1}{2\pi} \int_\omega X(\omega) \exp(j\omega t) d\omega$. Using u, on the other hand, results in the transform in (2.10) and its inversion in (2.12). There is no troublesome placing of 2π. The same 2π trouble free expressions are manifest in numerous other Fourier transform theorem duals thereby reducing needless bookkeeping overhead in the manipulation of equations. The primary reasons for the continuing use of ω are (a) historical inertia, and (b) the less effort and space required to write ω than $2\pi u$.

7. Fourier transform pairs are unique at all points of continuity. A finite or countable infinite number of displaced points does not effect a transform. For example, if $x(t) = \Pi(t)$ for $t \neq 3$ and $x(3) = 100$, then the Fourier transform of $f(t)$ is still $\text{sinc}(u)$. Inverting $\text{sinc}(u)$ gives $\Pi(t)$ and the isolated displaced point is removed.

8. Points of discontinuity are inverse transformed to the arithmetic mean. For example, the function $y(t) = \Pi(t)$ for $t \neq 1/2$ and $y\left(\frac{1}{2}\right) = 100$, when transformed and inverse transformed, will result in $\Pi(t)$ where the originally displaced point reemerges at the arithmetic mean. $\Pi\left(\frac{1}{2}\right) = \frac{1}{2}\Pi\left(\frac{1}{2}+\right) + \Pi\left(\frac{1}{2}-\right) = \frac{1}{2}[1+0] = \frac{1}{2}$. For this reason, we choose to define functions with discontinuities, such as the unit step in (2.49), the sgn function in (2.47) and the rectangle function in (2.45) using the arithmetic mean at points of discontinuity. The same property applies for the Fourier series. (See Problem 2.40.)

TABLE 2.3. Continuous time Fourier transform (CTFT) theorems. The Fourier transform, typically complex, can be expressed in rectangular (Cartesian) coordinates as $X(u) = R(u) + jI(u)$, or in polar coordinates, $X(u) = |X(u)|e^{j\angle X(u)}$. Convolution is defined by $x(t) * h(t) = \int_{-\infty}^{\infty} x(\tau) h(t-\tau) d\tau$. The deterministic correlation integral is $x(t) \star h(t) = \int_{-\infty}^{\infty} x(\tau) h^*(\tau - t) d\tau = x(t) * h^*(-t)$. When $x = h$, this operation is deterministic autocorrelation.

Transform	$x(t)$	$\leftrightarrow X(u)$
scaling	$x(at)$	$\leftrightarrow \frac{1}{|a|} X\left(\frac{u}{a}\right)$
shift	$x(t-\tau)$	$\leftrightarrow X(u) e^{-j2\pi u \tau}$
modulation	$x(t) e^{j2\pi vt}$	$\leftrightarrow X(u-v)$
scale then shift	$x\left(\frac{t-\tau}{a}\right)$	$\leftrightarrow \frac{1}{|a|} X\left(\frac{u}{a}\right) e^{-j2\pi u \tau}$
shift then scale	$x\left(\frac{t}{a} - b\right)$	$\leftrightarrow \frac{1}{|a|} X\left(\frac{u}{a}\right) e^{-j2\pi uab}$
derivative	$\left(\frac{d}{dt}\right)^n x(t)$	$\leftrightarrow (j2\pi u)^n X(u)$
integral	$\int_{-\infty}^{t} x(\tau) d\tau$	$\leftrightarrow \frac{X(u)}{j2\pi u} + \frac{1}{2} X(0) \delta(u)$
conjugate	$x^*(t)$	$\leftrightarrow X^*(-u)$
transpose	$x(-t)$	$\leftrightarrow X(-u)$
inversion	$\int_{-\infty}^{\infty} X(u) e^{j2\pi ut} du$	$\leftrightarrow X(u)$
duality	$X(t)$	$\leftrightarrow x(-u)$
linearity	$ax_1(t) + bx_2(t)$	$\leftrightarrow aX_1(u) + bX_2(u)$
convolution	$x(t) * h(t)$	$\leftrightarrow X(u) H(u)$
correlation	$x(t) \star h(t)$	$\leftrightarrow X(u) H^*(u)$
real signals	if $x(t)$ is real	$\leftrightarrow X(u) = X^*(-u)$,
	$\Rightarrow R(u) = R(-u)$,	$I(u) = -I(-u)$
	$\|X(u)\| = \|X(-u)\|$,	$\angle\{X(u)\} = -\angle\{X(-u)\}$.
causal signals	$x(t) = x(t)\mu(t)$	$\leftrightarrow X(u) = \frac{-j}{\pi u} * X(u)$
	$\Rightarrow I(u) = \frac{-1}{\pi u} * R(u)$,	$R(u) = \frac{1}{\pi u} * I(u)$
Fourier series	$\sum_n c_n e^{j2\pi nt/T}$	$\leftrightarrow \sum_n c_n \delta\left(u - \frac{n}{T}\right)$
sampling theorem	$\sum_n x_n \operatorname{sinc}(2Bt - n)$	$\leftrightarrow \sum_n x_n e^{-j\pi u/B} \Pi\left(\frac{u}{2B}\right)$

We can use the arbitrary integration over a period in (2.14) for the following reason. For any periodic function, $y(t)$, with period T, the integral $\int_{t=\tau}^{\tau+T} y(t) dt$ is independent of τ. For any τ, the result is simply the area of a single period. The product of two periodic functions with period T results in a periodic function with period T. Since $x(t)$ has period T and $e^{-j2\pi nt/T}$ has period T, the product, $y(t) = x(t) e^{-j2\pi nt/T}$, has a period of T. Thus, integration over any period, independent of where the integration begins, gives the same value. The integration notation in (2.14) follows.

If we define $x_T(t)$ to be any single period of $x(t)$:

$$x_T(t) = \begin{cases} x(t) \,; \, \tau < t \leq \tau + T \\ 0 \quad ; \text{ otherwise} \end{cases} \quad (2.16)$$

(τ is arbitrary), then (2.14) can be written

$$c_n = \frac{1}{T} \int_{-\infty}^{\infty} x_T(t)\, e^{-j2\pi nt/T}\, dt$$

$$= \frac{1}{T} X_T\left(\frac{n}{T}\right). \tag{2.17}$$

where

$$x_T(t) \longleftrightarrow X_T(u). \tag{2.18}$$

The Fourier coefficients can thus be determined by sampling the CTFT of any period of the periodic function.

2.3.2.1 Parseval's Theorem for the Fourier Series

The energy in a single period of a periodic function is related to its Fourier coefficients via Parseval's theorem for the Fourier series.

$$\int_T |x(t)|^2 dt = T \sum_{n=-\infty}^{\infty} |c_n|^2. \tag{2.19}$$

Proof.

$$\int_T |x(t)|^2 dt = \int_T x(t) x^*(t) dt$$

$$= \int_T \left[\sum_{n=-\infty}^{\infty} c_n e^{j2\pi nt/T} \sum_{m=-\infty}^{\infty} c_m^* e^{-j2\pi mt/T} \right] dt$$

$$= \sum_{n=-\infty}^{\infty} \sum_{m=-\infty}^{\infty} c_n c_m^* \left[\int_T e^{j2\pi (n-m)t/T} dt \right]$$

$$= \sum_{n=-\infty}^{\infty} \sum_{m=-\infty}^{\infty} c_n c_m^* \left[T\, \delta[n-m] \right]$$

$$= T \sum_{n=-\infty}^{\infty} |c_n|^2$$

2.3.2.2 Convergence

The partial Fourier series sum is

$$x_N(t) = \sum_{n=-N}^{N} c_n\, e^{j2\pi nt/T}. \tag{2.20}$$

A single period of a truncated Fourier series with unit period is shown in Figure 2.1. On the interval $[-\frac{1}{2}, \frac{1}{2}]$, the signal is one for $|t| \leq \frac{1}{3}$ and zero otherwise. As the number of terms increases with N, the overshoot at discontinuity gets smaller and smaller but,

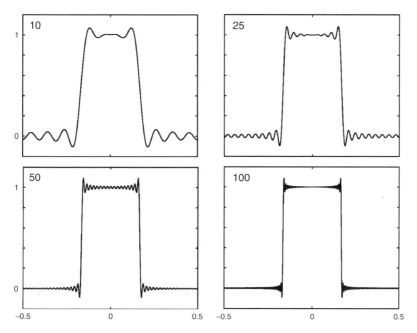

FIGURE 2.1. A truncated Fourier series in (2.20) for a square wave equal to one for $t \leq 0.2$. The coefficients are $c_n = \alpha \operatorname{sinc}(\alpha n)$ for $\alpha = 0.4$. Plots are shown for $N = 10, 25, 50$ and 100.

in the limit, does not disappear.[9] (See Exercise 2.51.) This property, dubbed the *Gibb's phenomenon*, reveals that the Fourier series does not converge uniformly at every point. It, rather, converges in the mean square sense. Equation (2.13) is thus better written as

$$\lim_{N \to \infty} \int_T |x(t) - x_N(t)|^2 \, dt = 0 \qquad (2.21)$$

Dirichlet Conditions. The Dirichlet conditions[10] specify three sufficient conditions for a periodic function to have a Fourier series representation. They are

(a) $x(t)$ must have a finite number of extrema on the interval $0 \leq t < T$. An example of a violating function is $x(t) = \sin(1/t)$.
(b) $x(t)$ must have a finite number of discontinuities. Violating functions include (1) $x(t) = 1$ when t is rational and zero otherwise and (2) $x(t) = \operatorname{sgn}(\sin(1/t))$.
(c) $x(t)$ must have a finite area on the interval $0 \leq t < T$. The function $x(t) = 1/t$ violates this.

The functions that fail the Dirchlet conditions are largely pathological.

---

9. Albert Michelson is most often associated with the 1881 Michelson-Morley experiment proving light travels at a constant speed in all inertial systems of reference. Michelson also designed a mechanical apparatus in 1898 to compute Fourier series. It performed well except when there were discontinuities. J. Willard Gibb's first explained the *Gibb's phenomenon* in 1899. As the Michelson-Morley experiment proved foundational to Einstein's development of relativity, Michelson's Fourier series machine motivated the theoretical establishment of the Gibb's phenomenon.
10. Named for Johann Peter Gustav Lejeune Dirichlet (1805–1859), a contemporary of Fourier's.

2.3.3 Relationship Between Fourier and Laplace Transforms

A wealth of Fourier transform pairs can be obtained from *Laplace transforms* [5, 544]. The (unilateral) Laplace transform is defined as

$$X_L(s) := \int_0^\infty x(t) e^{-st} dt \; ; \; s = \sigma + j2\pi u. \qquad (2.22)$$

If

(1) $x(t)$ is *causal* (i.e., $x(t) = x(t)\mu(t)$), and
(2) $X_L(s)$ converges for $\sigma = 0$,

then the Laplace transform evaluated at $\sigma = 0$ is the Fourier transform of $x(t)$. That is,

$$X_L(j2\pi u) = X(u). \qquad (2.23)$$

A sufficient condition for the second criterion is for $x(t)$ to have finite area.

$$A = \int_{-\infty}^\infty |x(t)| \, dt < \infty.$$

This follows from the inequality

$$\left| \int_{-\infty}^\infty y(t) dt \right| \leq \int_{-\infty}^\infty |y(t)| dt \qquad (2.24)$$

which, when applied to the Laplace transform in (2.22) yields

$$|X_L(j2\pi u)| = \left| \int_{-\infty}^\infty x(t) e^{-j2\pi ut} dt \right| \leq \int_{-\infty}^\infty |x(t)| dt < \infty.$$

2.3.4 Some Continuous Time Fourier Transform Theorems

Here are some useful Fourier transform theorems.

2.3.4.1 The Derivative Theorem

Performing a derivative in time corresponds to multiplying the Fourier transform by $j2\pi u$. More generally

$$\left(\frac{d}{dt} \right)^n x(t) \longleftrightarrow (j2\pi u)^n X(u) e^{j2\pi ut}.$$

Proof.

$$\left(\frac{d}{dt} \right)^n x(t) = \int_{-\infty}^\infty X(u) \left(\frac{d}{dt} \right)^n e^{j2\pi ut} du$$

$$= \int_{-\infty}^\infty (j2\pi u)^n X(u) e^{j2\pi ut} du.$$

Thus

$$\left(\frac{d}{dt}\right)^n x(t) \longleftrightarrow (j2\pi u)^n X(u).$$

2.3.4.2 The Convolution Theorem

Convolution in time corresponds to multiplication in the frequency domain.

$$x(t) * h(t) \longleftrightarrow X(u)H(u).$$

Proof. Convolution of the continuous time functions $x(t)$ and $h(t)$ is defined as

$$x(t) * h(t) := \int_{-\infty}^{\infty} x(\tau) h(t-\tau) \, d\tau. \tag{2.25}$$

Thus

$$x(t) * h(t) \longleftrightarrow \int_{-\infty}^{\infty} \left[\int_{-\infty}^{\infty} x(\tau) h(t-\tau) d\tau\right] e^{-j2\pi ut} dt.$$

Reverse integration orders and applying the shift theorem gives

$$x(t) * h(t) \longleftrightarrow H(u) \int_{-\infty}^{\infty} x(t) e^{-j2\pi ut} dt$$

or, finally

$$x(t) * h(t) \longleftrightarrow X(u)H(u). \tag{2.26}$$

2.3.4.3 The Inversion Theorem

To invert from the Fourier transform domain to the time domain, the inverse Fourier transform in (2.12) is used.

Proof.

$$\int_{-\infty}^{\infty} X(v) e^{j2\pi vt} dv \longleftrightarrow \int_{-\infty}^{\infty} \left[\int_{-\infty}^{\infty} X(v) e^{j2\pi vt} dv\right] e^{-j2\pi ut} du.$$

Reverse integration order and note

$$\int_{-\infty}^{\infty} e^{j2\pi \xi t} dt = \delta(\xi).$$

The proof concludes after application of the sifting property.

2.3.4.4 The Duality Theorem

The forward and inverse Fourier transforms in (2.10) and (2.12) differ only in the sign of the exponential. This similarity leads to the duality theorem.

$$\{x(t) \leftrightarrow X(u)\} \rightarrow \{X(t) \leftrightarrow x(-u)\}.$$

Proof.

$$X(t) \longleftrightarrow \int_{-\infty}^{\infty} X(t) e^{j2\pi(-u)t} dt$$

$$= x(-u).$$

2.3.4.5 The Scaling Theorem

Stretching a signal in the time domain corresponds to a reciprocal shrinking in the frequency domain.

$$x(at) \longleftrightarrow \frac{1}{|a|} X\left(\frac{u}{a}\right).$$

Proof.

$$\begin{aligned}
x(at) &\longleftrightarrow \int_{-\infty}^{\infty} x(a\hat{t}) e^{-j2\pi u \hat{t}} \, d\hat{t} \\
&= \begin{cases} \dfrac{1}{a} \displaystyle\int_{-\infty}^{\infty} x(t) e^{-j2\pi ut/a} dt; & a > 0 \\ \dfrac{1}{a} \displaystyle\int_{\infty}^{-\infty} x(t) e^{-j2\pi ut/a} dt; & a < 0 \end{cases} \\
&= \frac{1}{|a|} \int_{-\infty}^{\infty} x(t) e^{-j2\pi \frac{u}{a} t} dt. \\
&= \frac{1}{|a|} X\left(\frac{u}{a}\right).
\end{aligned}$$

2.3.5 Some Continuous Time Fourier Transform Pairs

A list of CTFT pairs is in Table 2.4.

2.3.5.1 The Dirac Delta Function

For continuous time, the Dirac delta function,[11] also called an impulse, denotes an infinite energy, unit area event of infinitesimal duration. It can be defined by its sifting property. If $x(t)$ is continuous at $\tau = t$, then[12]

$$x(t) = x(t) * \delta(t) = \int_{-\infty}^{\infty} x(\tau) \delta(t - \tau) \, d\tau. \tag{2.27}$$

(a) **Fourier transform**

From (2.27), the Fourier transform of a Dirac delta is

$$\begin{aligned}
\delta(t) &\longleftrightarrow \int_{-\infty}^{\infty} \delta(t) e^{-j2\pi tu} dt \\
&= \left. e^{-j2\pi tu} \right|_{t=0} \\
&= 1.
\end{aligned}$$

---

11. Rigorously, the Dirac delta, $\delta(t)$, is a distribution [158, 240, 620, 1159, 1582] and not a function.
12. Use of Dirac deltas in engineering math is typically trouble free. An exception is when the Dirac delta sits on the end point of an interval. For example, the integral $\int_0^1 \delta(t) dt$ simply has no answer. Is $\delta(t)$ included in the interval, excluded, or partially included? There is no answer. The question must be answered either in context to the problem, or implied notation. The $0-$ in the notation $\int_{0-}^{1} \delta(t) dt$ indicates, for example, inclusion of the delta in the interval of integration.

(b) Dirac delta area

The area of a Dirac delta can be found from the sifting property by setting $x(\tau) = 1$ and $t = 0$. Recognizing further that $\delta(t) = 0$ except at the origin leads us to conclude that for any $\varepsilon > 0$

$$\int_{-\varepsilon}^{\varepsilon} \delta(\tau)\, d\tau = 1. \tag{2.28}$$

The Dirac delta's area is thus one.

(c) As a limit

The Dirac delta can be viewed as the limit of any one of a number of unit area functions which approach zero width and infinite height at the origin. For example, as illustrated in Figure 2.2,

$$\delta(t) = \lim_{A \to \infty} A\, \Pi(At). \tag{2.29}$$

where the rectangle function is defined in (2.45). Alternately, using the definition of the sinc in (2.41)

$$\delta(t) = \lim_{A \to \infty} A\, \text{sinc}(At).$$

(d) Dirac delta scaling

From the Dirac delta characterization in (2.29), the area of $\delta(at) = \lim_{A \to \infty} A\, \Pi(aAt)$ is $1/|a|$. This is also the area of $\delta(t)/|a|$. We therefore infer the scaling property of the Dirac delta function.

$$\delta(t) = |a|\, \delta(at).$$

(e) Duality

From duality and the modulation theorem,

$$\exp(j2\pi \xi t) \longleftrightarrow \delta(u - \xi). \tag{2.30}$$

From (2.13), we conclude that the spectrum of a periodic signal can be written as a string of weighted Dirac deltas.

$$X(u) = \sum_{n=-\infty}^{\infty} c_n\, \delta\left(u - \frac{n}{T}\right).$$

2.3.5.2 Trigonometric Functions

Euler's formula is

$$e^{j\theta} = \cos(\theta) + j\sin(\theta). \tag{2.31}$$

The even and odd components, from (2.8) and (2.9), are

$$\cos(\theta) = \frac{1}{2}\left(e^{j\theta} + e^{-j\theta}\right), \tag{2.32}$$

and

$$j\sin(\theta) = \frac{1}{2}\left(e^{j\theta} - e^{-j\theta}\right). \tag{2.33}$$

TABLE 2.4. Some Fourier transform pairs. Additional Fourier transform pairs are in Tables 2.5, 4.1, 4.2 and 4.3. (Continued on the next page.)

$x(t)$	$\longleftrightarrow$	$X(u)$
$\delta(t)$	$\longleftrightarrow$	1
$\exp(j2\pi t)$	$\longleftrightarrow$	$\delta(u-1)$
$\cos(2\pi t)$	$\longleftrightarrow$	$\frac{1}{2}[\delta(u+1) + \delta(u-1)]$
$\sin(2\pi t)$	$\longleftrightarrow$	$\frac{j}{2}[\delta(u+1) - \delta(u-1)]$
$\Pi(t)$	$\longleftrightarrow$	$\text{sinc}(t)$
$\exp(-t)\mu(t)$	$\longleftrightarrow$	$(1+j2\pi u)^{-1}$
$\cos(\pi t)\Pi(t)$	$\longleftrightarrow$	$\frac{1}{2}\left(\text{sinc}\left(u+\frac{1}{2}\right) + \text{sinc}\left(u-\frac{1}{2}\right)\right)$
$\text{sinc}_k(t)$	$\longleftrightarrow$	$\Pi(u)/\text{sinc}^k(u)$
$\text{sinc}(t)\,\text{sgn}(t)$	$\longleftrightarrow$	$-\frac{j}{\pi}\log\left|\frac{u+\frac{1}{2}}{u-\frac{1}{2}}\right|$
$d_p(t) := \left(\frac{d}{dt}\right)^p \text{sinc}(t)$	$\longleftrightarrow$	$(j2\pi u)^p\,\Pi(u)$
$\text{sinc}(t)\mu(t)$	$\longleftrightarrow$	$\frac{1}{2}\Pi(u) - \frac{j}{2\pi}\log\left|\frac{u+\frac{1}{2}}{u-\frac{1}{2}}\right|$
$\Lambda(t)$	$\longleftrightarrow$	$\text{sinc}^2(u)$
$\Lambda(t)\,\text{sgn}(t)$	$\longleftrightarrow$	$\frac{-j}{2\pi u}(1 - \text{sinc}(2u))$
$\text{sgn}(t)$	$\longleftrightarrow$	$-j/(\pi u)$
$\frac{1}{t}$	$\longleftrightarrow$	$-j\pi\,\text{sgn}(u)$
$\frac{1}{t}\Pi\left(\frac{t}{2}\right)$	$\longleftrightarrow$	$-j2\,\text{Si}(2\pi u)$
$|t|^{-1/2}$	$\longleftrightarrow$	$-|u|^{-1/2}$
$|t|^{-1/2}\,\text{sgn}(t)$	$\longleftrightarrow$	$-j|u|^{-1/2}\text{sgn}(u)$
$t^{-1/2}\,\mu(t)$	$\longleftrightarrow$	$(-j2u)^{1/2}$
$\mu(t)$	$\longleftrightarrow$	$\frac{1}{2}\left(\delta(u) - \frac{j}{\pi u}\right)$
$e^{-|t|}$	$\longleftrightarrow$	$2\left(1 + (2\pi u)^2\right)^{-1}$
$e^{-|t|}\text{sgn}(t)$	$\longleftrightarrow$	$-j4\pi u\left(1 + (2\pi u)^2\right)^{-1}$
$e^{-\pi|t|}\text{sinc}(t)$	$\longleftrightarrow$	$\frac{1}{\pi}\arctan\left(\frac{1}{2u^2}\right)$
$|t|e^{-a|t|}$	$\longleftrightarrow$	$\sqrt{\frac{2}{\pi}}\,\frac{a^2 - (2\pi u)^2}{(a^2 + (2\pi u)^2)^2}$

Equations (2.32) and (2.33) also follow immediately from adding and subtracting the conjugate of $e^{j\theta}$ from $e^{j\theta}$ in (2.31) respectively.

(a) **Taylor series characterization**

To show Euler's formula in (2.31), consider the Taylor series

$$e^z = \sum_{n=0}^{\infty} \frac{z^n}{n!}. \tag{2.34}$$

TABLE 2.5. Some Fourier transform pairs. (Continued from Table 2.4.)

$x(t)$	$\longleftrightarrow$	$X(u)$
$\text{Si}(\pi t)$	$\longleftrightarrow$	$\dfrac{\Pi(u)}{j2u}$
$\cos(\pi t)/(\pi t)$	$\longleftrightarrow$	$j\mu\left(-u-\tfrac{1}{2}\right) - j\mu\left(u-\tfrac{1}{2}\right)$
$J_0(2\pi t)$	$\longleftrightarrow$	$\Pi\left(\tfrac{u}{2}\right)/\sqrt{1-u^2}$
$J_0(2\pi t)\text{sgn}(t)$	$\longleftrightarrow$	$\dfrac{j}{\pi\sqrt{u^2-1}}\left(\Pi\left(\tfrac{u}{2}\right)-1\right)\text{sgn}(u)$
$\text{jinc}(t)$	$\longleftrightarrow$	$\sqrt{1-u^2}\,\Pi\left(\tfrac{u}{2}\right)$
$\dfrac{J_\nu(2\pi t)}{(2t)^\nu};\ \nu > -\dfrac{1}{2}$	$\longleftrightarrow$	$\dfrac{(\pi/2)^\nu}{\sqrt{\pi}\,\Gamma(\nu+\tfrac{1}{2})}(1-u^2)^{\nu-\tfrac{1}{2}}\Pi\left(\tfrac{u}{2}\right)$
$\text{comb}(t)$	$\longleftrightarrow$	$\text{comb}(u)$
$\sum_{n=0}^{\infty}\delta(t-n)$	$\longleftrightarrow$	$\tfrac{1}{2}(1-j\cotan(\pi u))$
$\sum_{n=-N}^{N}\delta(t-n)$	$\longleftrightarrow$	$(2N+1)\,\text{array}_{2N+1}(u)$
$\exp(-\pi t^2)$	$\longleftrightarrow$	$\exp(-\pi u^2)$
$\exp(jat^2)$	$\longleftrightarrow$	$\sqrt{j\pi/a}\,\exp(-j(\pi u)^2/a)$
$\text{sech}(\pi t)$	$\longleftrightarrow$	$\text{sech}(\pi u)$
$\text{sech}(\pi t)\tanh(\pi t)$	$\longleftrightarrow$	$-j\pi u\,\text{sech}(\pi u)$
$\text{sech}^2(\pi t)$	$\longleftrightarrow$	$2u\,\text{cosech}(\pi u)$
$\tanh(\pi t)$	$\longleftrightarrow$	$-j\,\text{cosech}(\pi u)$
$t^{k-1}e^{-at}\mu(t);\ a>0$	$\longleftrightarrow$	$\dfrac{\Gamma(k)}{(a+j2\pi u)^k}$
$J_n(2\pi t)$	$\longleftrightarrow$	$\dfrac{1}{\pi}\dfrac{j^{-n}}{\sqrt{1-u^2}}T_n(u)\Pi\left(\tfrac{u}{2}\right)$
$j_n(2\pi t)$	$\longleftrightarrow$	$\tfrac{1}{2}j^{-n}P_n(u)\Pi(u/2)$
$e^{-\pi t^2}H_n(\sqrt{2\pi}\,t)$	$\longleftrightarrow$	$j^{-n}e^{-\pi u^2}H_n(\sqrt{2\pi}\,u)$

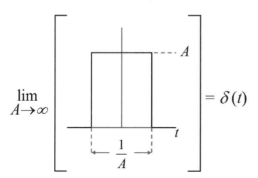

FIGURE 2.2. The limit of this rectangle function as $A \to \infty$ is a Dirac delta function with zero width, infinite height and unit area.

Expand $e^{j\theta}$ in (2.31). All of the even terms ($n = 2k$) in this series are real.[13] The sum of the real components of the expansion is recognized as the Taylor series

$$\cos(\theta) = \sum_{k=0}^{\infty} \frac{(-1)^k \theta^{2k}}{(2k)!} \qquad (2.35)$$

while the imaginary odd components ($n = 2k + 1$) sum to the Taylor series

$$\sin(\theta) = \sum_{k=0}^{\infty} \frac{(-1)^k \theta^{2k+1}}{(2k + 1)!}. \qquad (2.36)$$

Equation (2.31) follows as a consequence.

(b) **Fourier transforms of trigonometric functions**

For the Fourier transform of $\cos(2\pi t)$, we apply Euler's formula in (2.32) and apply the Fourier transform in (2.30) to obtain

$$\cos(t) \longleftrightarrow \frac{1}{2}(\delta(t-1) + \delta(t+1)).$$

Similarly

$$\sin(t) \longleftrightarrow \frac{1}{j2}(\delta(t-1) - \delta(t+1)).$$

(c) **Hyperbolic trig functions**

The hyperbolic cosine, sine, and tangent are

$$\cosh(z) := \cos(jz)$$
$$= \frac{1}{2}\left(e^z + e^{-z}\right),$$
$$\sinh(z) := j \sin(-jz)$$
$$= \frac{1}{2}\left(e^z - e^{-z}\right),$$

and

$$\tanh(z) := \frac{\sinh(z)}{\cosh(z)},$$
$$= \frac{e^{2z} - 1}{e^{2z} + 1}.$$

The hyperbolic secant and cosecant are

$$\text{sech}(x) := \frac{1}{\cosh(x)},$$

---

13. See Exercise 2.5 for a condition required for this discussion: absolute convergence.

and
$$\operatorname{cosech}(x) := \frac{1}{\sinh(x)}.$$

A useful series is

$$\operatorname{sech}(x) = \pi \sum_{k=0}^{\infty} \frac{(-1)^k (2k+1)}{\pi^2 \left(k + \frac{1}{2}\right)^2 + x^2}. \tag{2.37}$$

Linear and logarithmic plots of the hyperbolic functions are shown in Figures 2.3 and 2.4.

(d) **Fourier transforms of some hyperbolic trig functions**

The hyperbolic secant is even and has a Fourier transform of

$$\begin{aligned}
\operatorname{sech}(\pi t) \longleftrightarrow \int_{-\infty}^{\infty} \operatorname{sech}(\pi t) e^{-j2\pi ut} dt \\
= \int_{-\infty}^{\infty} \operatorname{sech}(\pi t) \cos(2\pi ut) dt \\
= 2 \int_{0}^{\infty} \operatorname{sech}(\pi t) \cos(2\pi ut) dt \\
= \Re \left(2 \int_{0}^{\infty} \operatorname{sech}(\pi t) e^{j2\pi ut} dt \right)
\end{aligned} \tag{2.38}$$

where $\Re$ denotes "the real part of". Note that, for $t > 0$,

$$\begin{aligned}
\operatorname{sech}(\pi t) &= \frac{2}{e^{\pi t} + e^{-\pi t}} \\
&= \frac{2e^{-\pi t}}{1 + e^{-2\pi t}}
\end{aligned}$$

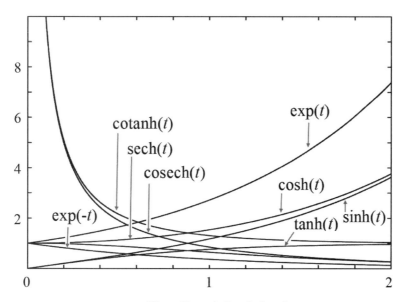

FIGURE 2.3. Plots of hyperbolic trig functions.

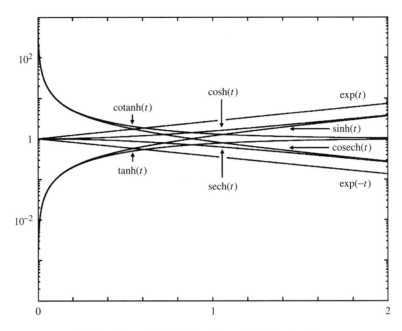

FIGURE 2.4. Logarithmic plots of hyperbolic trig functions.

$$= 2e^{-\pi t} \lim_{\alpha \to 1-} \left(\frac{1}{1 + \alpha e^{-2\pi t}} \right)$$

$$= 2e^{-\pi t} \lim_{\alpha \to 1-} \left(\sum_{k=0}^{\infty} (-1)^k \alpha^n e^{-2k\pi t} \right)$$

where, in the last step, we have used a convergence factor, α, and a geometric series.[14] The notation "$\lim_{\alpha \to 1-}$" means α approaches 1 from the the left, e.g., as $0 \to 1$. Substituting into (2.38) gives

$$\text{sech}(\pi t) \longleftrightarrow 4 \lim_{\alpha \to 1-} \left[\sum_{k=0}^{\infty} (-1)^k \alpha^n \Re \left(\int_0^\infty e^{-\pi(2k+1)t} \right) e^{j2\pi ut} dt \right]$$

$$= 4 \sum_{n=0}^{\infty} \Re \left[\lim_{\alpha \to 1-} \left(\frac{\alpha^n (2n+1)}{\pi \left[(2n+1)^2 + (2u)^2 \right]} \right) \right]$$

$$= \pi \sum_{k=0}^{\infty} \frac{(-1)^k (2k+1)}{\pi^2 \left(k + \frac{1}{2} \right)^2 + (\pi u)^2}.$$

Comparing this with the series for the hyperbolic secant in (2.37) gives the result listed in Table 2.5.

$$\text{sech}(\pi t) \longleftrightarrow \text{sech}(\pi u). \qquad (2.39)$$

---

14. See Appendix 14.4.2.

Since

$$\frac{d}{dt}\text{sech}(t) = -\text{sech}(t)\tanh(t), \qquad (2.40)$$

the Fourier transform entry for $\text{sech}(\pi t)\tanh(\pi t)$ in Table 2.5 immediately follows.

2.3.5.3 The Sinc and Related Functions

We define the sinc function by

$$\text{sinc}(t) := \frac{\sin(\pi t)}{\pi t}. \qquad (2.41)$$

One advantage of this notation is that, for n an integer,

$$\text{sinc}(n) = \delta[n], \qquad (2.42)$$

where the *Kronecker delta* is

$$\delta[n] := \begin{cases} 1 \; ; \; n = 0 \\ 0 \; ; \; n \neq 0. \end{cases} \qquad (2.43)$$

The sinc function is the $k = 0$ plot in Figure 2.5. For inverse filtering applications, we will find useful the following generalization of the sinc.

$$\text{sinc}_k(t) := 2\int_0^{\frac{1}{2}} \frac{\cos(2\pi u t)}{\text{sinc}^k(t)} dt. \qquad (2.44)$$

Note that

$$\text{sinc}_k(t) \longleftrightarrow \frac{1}{\text{sinc}^k(u)} \Pi(u).$$

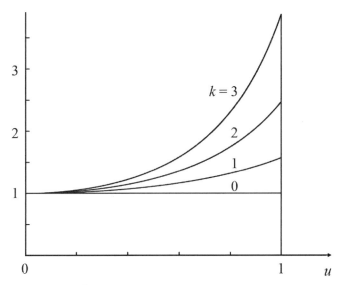

FIGURE 2.5. Plots of $\Pi(u)/\text{sinc}^k(u)$ for $k = 0, 1, 2, 3$. The Fourier transforms of these functions are $\text{sinc}_k(t)$ shown in Figure 2.6.

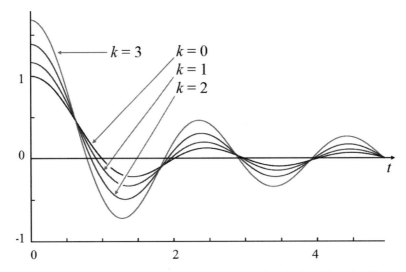

FIGURE 2.6. Plots of $\text{sinc}_k(t)$ for $k = 0, 1, 2, 3$. For $k = 0$, $\text{sinc}_k(t) = \text{sinc}(t)$.

Plots are in Figures 2.5 and 2.6. The sinc is a special case since

$$\text{sinc}_0(t) = \text{sinc}(t).$$

2.3.5.4 The Rectangle Function

For continuous time,

$$\Pi(t) := \begin{cases} 1 & ; \; |t| < \frac{1}{2} \\ 0 & ; \; |t| > \frac{1}{2} \\ \frac{1}{2} & ; \; |t| = \frac{1}{2}. \end{cases} \quad (2.45)$$

The Fourier transform of the rectangle function is

$$\Pi(t) \longleftrightarrow \int_{t=-\frac{1}{2}}^{\frac{1}{2}} e^{-j2\pi ut} \, dt$$

$$= \frac{1}{-j2\pi u} e^{-j2\pi ut} \Big|_{-\frac{1}{2}}^{\frac{1}{2}}.$$

$$= \text{sinc}(u)$$

For discrete time, we adopt the definition

$$\Pi\left[\frac{n}{2L}\right] := \begin{cases} 1 & ; \; |n| \leq L \\ 0 & ; \; \text{otherwise.} \end{cases} \quad (2.46)$$

L need not be an integer.

2.3.5.5 The Signum Function

The function

$$\text{sgn}(t) := \begin{cases} -1 \, ; \, t < 0 \\ 0 \, ; \, t = 0 \\ 1 \, ; \, t > 0 \end{cases} \tag{2.47}$$

is denoted by a contraction of the word *sign* and is pronounced *signum* to avoid confusion with the trigonometric *sin*.

(a) **Fourier transform of sgn(*t*)**

As shown in Figure 2.7, the function sgn(*t*) can be written as

$$\text{sgn}(t) = \lim_{\alpha \to 0} \text{sgn}(t) e^{-\alpha |t|}.$$

Discarding the odd component of the Fourier transform integral of $\text{sgn}(t)e^{-\alpha|t|}$ gives

$$\text{sgn}(t)e^{-\alpha|t|} \leftrightarrow -j \int_{-\infty}^{\infty} \text{sgn}(t) e^{-\alpha|t|} \sin(2\pi u t) \, dt$$

where we have expanded the exponential via Euler's formula and have recognized the odd component of the integrand integrates to zero. The evenness of the remaining integrand can be exploited to write

$$\text{sgn}(t)e^{-\alpha|t|} \leftrightarrow -j2 \int_0^{\infty} \sin(2\pi u t) \, e^{-\alpha t} \, dt$$

$$= -j2\Im \left[\int_0^{\infty} e^{j2\pi u t} \, e^{-\alpha t} \, dt \right]$$

$$= -j2\Im \left[\frac{1}{\alpha - j2\pi u} \right]$$

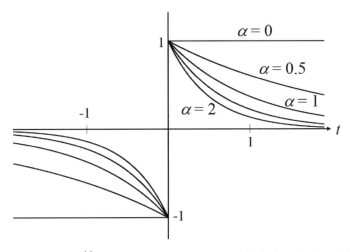

FIGURE 2.7. Plots of $\text{sgn}(t)e^{-\alpha|t|}$ for various values of α. For $\alpha = 0$, the function is sgn(*t*). To evaluate the Fourier transform of sgn(*t*), we evaluate the Fourier transform of $\text{sgn}(t)e^{-\alpha|t|}$ and take the limit as $\alpha \to 0$.

where $\Im$ denotes the imaginary component operation. Evaluating in the limit as $a \to 0$ gives the desired result.

$$\text{sgn}(t) \leftrightarrow \frac{1}{j\pi u}. \tag{2.48}$$

(b) **The Unit Step Function**

Note that the *unit step function* can be written for continuous time as

$$\mu(t) := \frac{1}{2}(\text{sgn}(t) + 1). \tag{2.49}$$

Its Fourier transform is thus

$$\mu(t) = \frac{1}{2}(\text{sgn}(t) + 1) \leftrightarrow \frac{1}{2}\left(\delta(t) - \frac{j}{\pi u}\right). \tag{2.50}$$

For discrete time, the unit step is

$$\mu[n] := \begin{cases} 1 \,; \, n \geq 0 \\ 0 \,; \, \text{otherwise.} \end{cases} \tag{2.51}$$

2.3.5.6 The Gaussian

The Gaussian function, shown in Figure 2.8, is its own Fourier transform.

Proof. Differentiating the expression for the Fourier transform of the Gaussian, $x(t) = \exp(-\pi t^2)$, gives

$$\frac{dX(u)}{du} = \int_{-\infty}^{\infty} (-j2\pi t) \exp[-\pi(t^2 + j2ut)] \, dt$$

$$= -j \int_{-\infty}^{\infty} \pi(2t + j2u) \exp\left(-\pi(t^2 + j2ut)\right) dt - 2\pi u X(u).$$

The resulting integral can be evaluated in closed form.

$$\frac{dX(u)}{du} = j \exp\left(-\pi(t^2 + j2ut)\right)\Big|_{-\infty}^{\infty} - 2\pi u X(u) = -2\pi u X(u)$$

or

$$\frac{dX(u)}{X(u)} = -2\pi u \, du.$$

Integrating both sides gives

$$\ln(X(u)) = -\pi u^2.$$

After exponentiation, we obtain the Fourier transform of the Gaussian.[15]

$$e^{-\pi t^2} \longleftrightarrow e^{-\pi u^2}. \tag{2.52}$$

---

15. An alternate derivation is in Exercise 4.5.

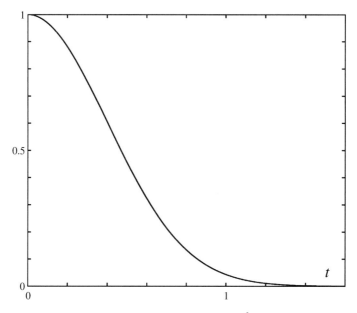

FIGURE 2.8. The Gaussian function, $e^{-\pi t^2}$, for $t \geq 0$.

2.3.5.7 The Comb Function

We define

$$\text{comb}(t) = \sum_{n=-\infty}^{\infty} \delta(t-n). \tag{2.53}$$

(a) **Use in characterization of periodic functions**

Any periodic function can be written as

$$x(t) = x_T(t) * \frac{1}{T} \text{comb}\left(\frac{t}{T}\right) \tag{2.54}$$

where $x_T(t)$ is any period of $x(t)$.

(b) **The comb function's Fourier series and Fourier transform**

Since $\text{comb}(t)$ is periodic, it can be expressed in terms of a Fourier series in (2.13). From the expression for Fourier series coefficients in (2.14) and the sifting property of the Dirac delta, the coefficients are $c_n = 1$ for all n. Thus

$$\text{comb}(t) = \sum_{n=-\infty}^{\infty} e^{j2\pi nt}. \tag{2.55}$$

Since, from (2.30), the Fourier transform of a complex sinusoid is a shifted Dirac delta, we obtain the Fourier transform pair

$$\text{comb}(t) \longleftrightarrow \sum_{n=-\infty}^{\infty} \delta(u-n) = \text{comb}(u).$$

(c) **The causal comb**

The *causal comb* is a string of integer spaced Dirac deltas for nonnegative time and is written $\sum_{n=0}^{\infty} \delta(t-n)$. The causal comb's Fourier transform can be derived using the following steps.

$$\sum_{n=0}^{\infty} \delta(t-n)$$

$$\leftrightarrow \sum_{n=0}^{\infty} e^{-j2\pi nu} \qquad ; \text{ Fourier transform}$$

$$= \lim_{\alpha \to 0} \left[\sum_{n=0}^{\infty} e^{-\alpha n} e^{-j2\pi nu} \right] \quad ; \text{ convergence factor}$$

$$= \lim_{\alpha \to 0} \left[1 - e^{-\alpha} e^{-j2\pi u} \right]^{-1} \quad ; \text{ geometric series}$$

$$= \frac{1}{1 - e^{-j2\pi u}} \qquad ; \text{ set } \alpha = 0 \qquad (2.56)$$

$$= \frac{e^{j\pi u}}{e^{j\pi u} - e^{-j\pi u}} \qquad ; \text{ factor}$$

$$= \frac{1}{j2} \frac{\cos(\pi u) + j\sin(\pi u)}{\sin(\pi u)} \quad ; \text{ Euler's formula}$$

$$= \frac{1}{2} (1 - j \cot(\pi u))$$

2.3.5.8 The Triangle Function

Define the triangle function as

$$\Lambda(t) := (1 - |t|) \, \Pi\left(\frac{t}{2}\right). \qquad (2.57)$$

Note that, using the convolution theorem,

$$\Lambda(t) = \Pi(t) * \Pi(t) \longleftrightarrow \operatorname{sinc}^2(u). \qquad (2.58)$$

2.3.5.9 The Array Function

As illustrated in Figure 2.9, the *array function* of order M is defined by

$$\operatorname{array}_M(t) := \frac{\sin(\pi M t)}{M \sin(\pi t)}. \qquad (2.59)$$

(a) **Periodicity**

The function $\operatorname{array}_M(t)$ is periodic. When M is odd, its period is one. If M is even, its period is two.

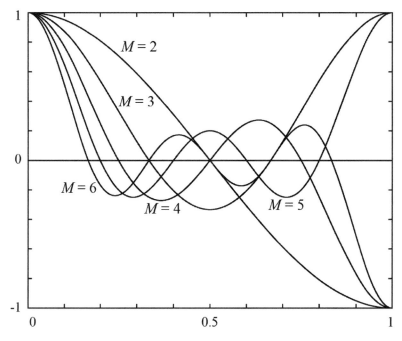

FIGURE 2.9. Plots of $\text{array}_M(t)$.

Proof. From the definition,

$$\text{array}_M(t+1) = \frac{\sin(\pi M(t+1))}{M \sin(\pi(t+1))}$$

$$= (-1)^{M-1} \text{array}_M(t).$$

(b) **Maximum values**

The maximum of $|\text{array}_M(t)|$ is 1 and occur at integer arguments.

$$\text{array}_M(n) = (-1)^{(M+1)n}. \tag{2.60}$$

Proof. The proof follows from application of L'Hopital's rule to (2.59).

Note that, for odd indices, we have $\text{array}_{2N+1}(n) = 1$.

(c) **Zero crossings**

The array function in (2.59) is zero for all $t = n/M$ when n/M is not an integer. This occurs when the numerator in (2.59) is zero and the denominator isn't.

(d) **Relation to the sinc function**

In the limit, the array function approaches a sinc function.

$$\lim_{M \to \infty} \text{array}_M \left(\frac{t}{M}\right) = \text{sinc}(t). \tag{2.61}$$

Proof. We note, from the definition in (2.59), that

$$\text{array}_M \left(\frac{t}{M}\right) = \frac{\sin(\pi t)}{M \sin\left(\frac{\pi t}{M}\right)}.$$

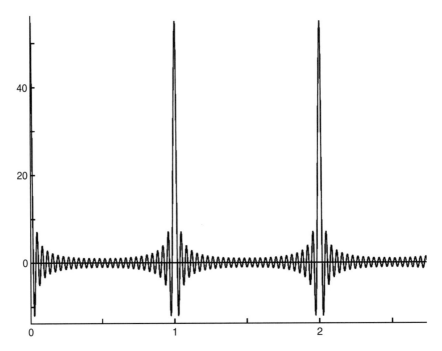

FIGURE 2.10. A plot of array$_{55}$(55u), as predicted by (2.63), approaches comb(u).

Equation (2.61) immediately follows after recognizing

$$\lim_{\theta \to 0} \sin(\theta) = \theta.$$

Equation (2.61) is illustrated in Figure 2.10.

(e) **The Fourier transform of the array function**

Using a geometric series,[16] we can generate the following Fourier transform pair.

$$\sum_{n=-N}^{N} \delta(t-n) \longleftrightarrow \sum_{n=-N}^{N} \left(\int_{-\infty}^{\infty} \delta(t-n) e^{-j2\pi ut} dt \right)$$

$$= \sum_{n=-N}^{N} e^{-j2\pi nt} \qquad (2.62)$$

$$= \sum_{n=-N}^{N} \left(e^{-j2\pi t} \right)^{n}$$

$$= (2N+1)\text{array}_{2N+1}(u)$$

(f) **Relation to the comb function**

In the limit,

$$\lim_{N \to \infty} (2N+1) \times \text{array}_{2N+1}(u) = \text{comb}(u). \qquad (2.63)$$

---

16. See (14.8) in Appendix 14.4.2.

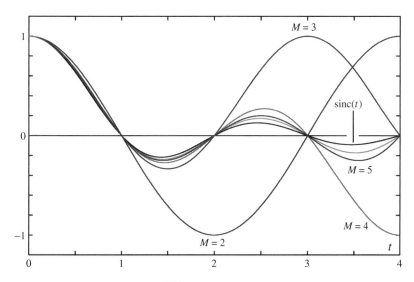

FIGURE 2.11. Plots of $\text{array}_M\left(\frac{t}{M}\right)$ that, according to (2.61), approaches sinc(t).

Proof. The proof follows directly from (2.62) and the Fourier transform pair comb(t) ↔ comb(u).

An illustration is in Figure 2.11.

2.3.5.10 The Gamma Function[†‡]

The *gamma function* for real positive argument can be defined as

$$\Gamma(\xi) := \int_0^\infty \tau^{\xi-1} e^{-\tau} d\tau \; ; \quad \Re\xi > 0. \tag{2.64}$$

The gamma function can be extended to include negative and complex arguments [544].
When $\xi = n$ is a non-negative integer, $\Gamma(n+1) = n!$. When $\xi = n + \frac{1}{2}$,

$$\Gamma\left(n + \frac{1}{2}\right) = \frac{(2n-1)!! \sqrt{\pi}}{2^n} \tag{2.65}$$

where the double factorial is defined as[17]

$$m!! := \prod_{\{k|k \geq 0, m-2k > 0\}} (m - 2k). \tag{2.66}$$

For example 7 double factorial = $7!! = 7 \cdot 5 \cdot 3 \cdot 1$. Every other integer is skipped.

Equation (2.65) can be derived from the property

$$\Gamma(\xi + 1) = \xi \, \Gamma(\xi) \tag{2.67}$$

and $\Gamma(\frac{1}{2}) = \sqrt{\pi}$. Linear and semi-log plots of the gamma function are shown in Figures 2.12 and 2.13.

---

17. See Exercise 2.8 for an alternate definition.

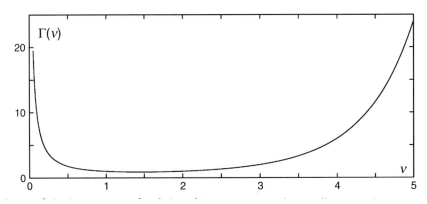

FIGURE 2.12. The gamma function, $\Gamma(\nu)$, as defined by (2.65). The gamma function is also defined for negative and complex arguments [5].

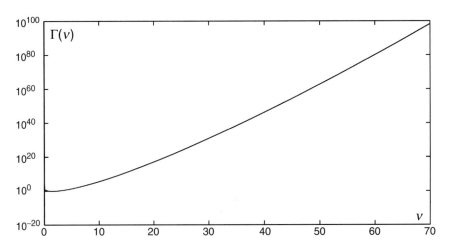

FIGURE 2.13. A semi-log plot of the gamma function, $\Gamma(\nu)$ as defined by (2.65).

2.3.5.11 The Beta Function[‡]

The *beta function* illustrated in Figure 2.14, is

$$B(b, c) := \frac{\Gamma(b)\Gamma(c)}{\Gamma(b+c)}. \tag{2.68}$$

The beta function can be viewed as a continuous version of the *binomial coefficient*

$$\binom{n}{k} := \frac{n!}{k!(n-k)!}.$$

Indeed,

$$\frac{1}{B(k+1, n-k+1)} = \frac{\Gamma(n+2)}{\Gamma(k+1)\Gamma(n-k+1)}$$

$$= \frac{(n+1)!}{k!(n-k)!} = (n+1)\binom{n}{k}.$$

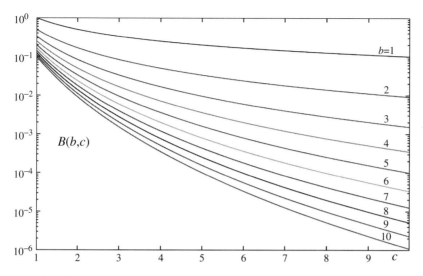

FIGURE 2.14. A semi-log plot of the beta function, $B(b, c)$. Since $\Gamma(0+) = \infty$, all of these curves peak to ∞ as $c \to 0$.

2.3.5.12 Bessel Functions†‡

For $\nu > -\frac{1}{2}$, Bessel functions of the first kind can be defined by the integral [5]

$$J_\nu(z) := \frac{2(\frac{z}{2})^\nu}{\sqrt{\pi}\,\Gamma(\nu + \frac{1}{2})} \int_0^1 (1 - u^2)^{\nu - \frac{1}{2}} \cos(zu)\, du. \qquad (2.69)$$

Example plots are shown in Figure 2.15.

(a) **Fourier transforms of Bessel functions**
From Exercise 2.41, we have the transform pair

$$\frac{J_\nu(2\pi t)}{(2t)^\nu} \longleftrightarrow \frac{(\pi/2)^\nu}{\sqrt{\pi}\,\Gamma(\nu + \frac{1}{2})}(1 - u^2)^{\nu - \frac{1}{2}}\Pi(u/2). \qquad (2.70)$$

The functions $J_0(2\pi t)$ for $\nu = 0$ and $\mathrm{jinc}(t)$ for $\nu = 1$ are special cases. A plot is in Figure 2.16. For $\nu = 0$, (2.69) becomes

$$J_0(2\pi t) = \frac{2}{\pi} \int_0^1 \frac{\cos(2\pi ut)}{\sqrt{1 - u^2}}\, du. \qquad (2.71)$$

A plot is shown in Figure 2.16.

(b) **The jinc function**
We define [149]

$$\mathrm{jinc}(t) := \frac{J_1(2\pi t)}{2t}. \qquad (2.72)$$

Using (2.69) with $\nu = 1$ gives

$$\mathrm{jinc}(t) = 2 \int_0^1 \sqrt{1 - u^2}\, \cos(2\pi ut)\, du.$$

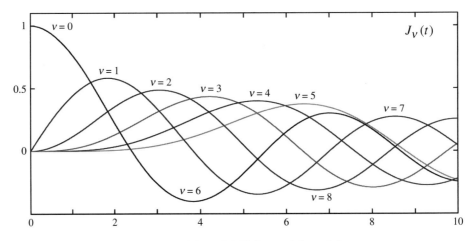

FIGURE 2.15. Plots of $J_\nu(t)$ for $\nu = 1$ through 8.

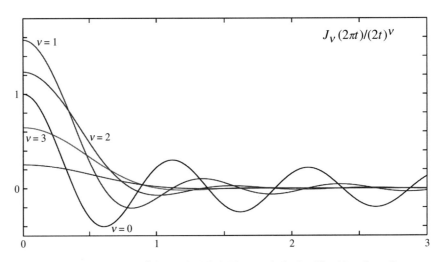

FIGURE 2.16. Plots of $J_\nu(2\pi t)/(2t)^\nu$ for $\nu = 0, 1, 2, 3$. The $\nu = 1$ plot is of jinc(t) and $\nu = 0$ corresponds to $J_0(2\pi t)$.

Since

$$2 \int \sqrt{1 - u^2} \, du = u\sqrt{1 - u^2} + \arcsin(u), \qquad (2.73)$$

it follows that

$$\text{jinc}(0) = \frac{\pi}{2}.$$

The zero crossings of the jinc are identical to the well tabulated $t \neq 0$ zero crossings of $J_1(2\pi t)$.

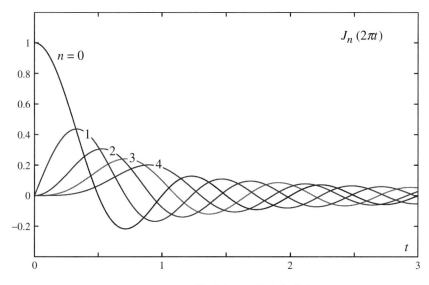

FIGURE 2.17. $J_n(2\pi t)$ for $n = 0, 1, 2, 3, 4$.

(c) **Spherical Bessel functions**
Spherical Bessel functions of the first kind are Bessel functions with half integer argument.[18]

$$j_n(z) := \sqrt{\frac{\pi}{2z}} J_{n+\frac{1}{2}}(z). \tag{2.74}$$

Plots are shown in Figure 2.17. Spherical Bessel functions are finite energy bandlimited functions.

(d) **Modified Bessel functions**
An nth order modified Bessel function of the first kind, as illustrated in Figure 2.18, can be defined as [5, 544]

$$I_n(z) = \frac{1}{\pi} \int_0^\pi e^{z\cos(\theta)} \cos(n\theta) d\theta. \tag{2.75}$$

Modified Bessel functions obey

$$I_n(t) := j^{-n} J_n(jt). \tag{2.76}$$

2.3.5.13 The Sine Integral

The *sine integral function* is

$$\text{Si}(t) := \int_0^t \frac{\sin(z)}{z} dz. \tag{2.77}$$

Equivalently

$$\text{Si}(\pi t) = \pi \int_0^t \text{sinc}(\tau) d\tau. \tag{2.78}$$

---

18. The Fourier transform of the spherical Bessel function is in (2.82).

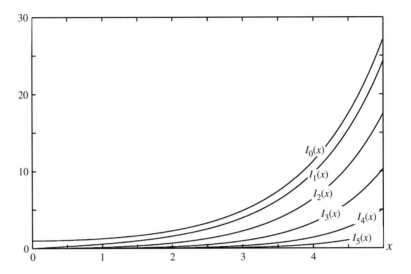

FIGURE 2.18. Modified Bessel functions of the first kind as defined in (2.75).

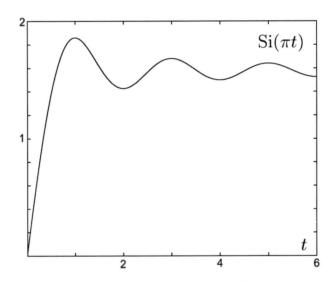

FIGURE 2.19. Plot of Si(πt). As t becomes large, Si(πt) $\leftrightarrow \frac{\pi}{2}$. Also, Si($\pi$) = 1.851937.

A plot of Si(πt) is shown in Figure 2.19.

2.3.5.14 Orthogonal Polynomials[¶]

There exist numerous classes of orthogonal polynomials [5, 651, 544]. Here are three of them.

1. **Legendre polynomials**
   Legendre polynomials on the interval $[-1,1]$ can be defined by

$$P_n(t) := \frac{1}{2^n n!} \left(\frac{d}{dt}\right)^n (t^2 - 1)^n. \qquad (2.79)$$

(a) **Properties**
With initializations $P_0(t) = \Pi\left(\frac{t}{2}\right)$ and $P_1(t) = t\Pi\left(\frac{t}{2}\right)$, Legendre polynomials obey the recursion relationship

$$(n+1)P_{n+1}(t) = (2n+1)tP_n(t) - nP_{n-1}(t). \tag{2.80}$$

The Lengendre polynomials are orthogonal.

$$\int_{t=-1}^{1} P_n(t)P_m(t) = \frac{2}{2n+1}\delta[n-m]. \tag{2.81}$$

A plot of the first few Legendre polynomials are in Figure 2.20. The functions become more oscillatory with increasing index.

(b) **The Fourier transform of the Legendre polynomial**
Legendre polynomials and spherical Bessel functions are Fourier transform pairs.

Proof. Using the definition of the spherical Bessel function in (2.74), the definition of the Bessel function in (2.69) and the definition of the Legendre polynomial in (2.79), we have

$$j_n(2\pi t) = \frac{1}{2\sqrt{t}} J_{n+\frac{1}{2}}(2\pi t)$$

$$= \frac{1}{2\sqrt{t}} \left[\left(\frac{j}{\pi}\right)^n (2t)^{\frac{1}{2}} (-j2\pi t)^n \frac{J_{n+\frac{1}{2}}(2\pi t)}{(2t)^{n+\frac{1}{2}}} \right]$$

$$= \frac{1}{\sqrt{2}} \left(\frac{j}{\pi}\right)^n \left[(-j2\pi t)^n \frac{J_{n+\frac{1}{2}}(2\pi t)}{(2t)^{n+\frac{1}{2}}} \right]$$

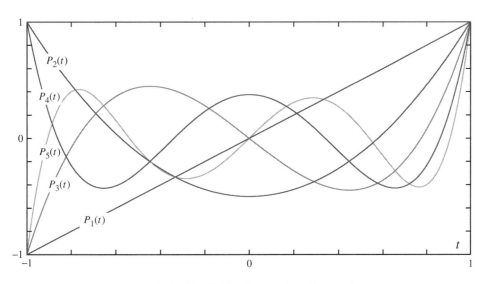

FIGURE 2.20. The first few Lengendre polynomials.

$$\longleftrightarrow \frac{1}{\sqrt{2}} \left(\frac{j}{\pi}\right)^n \left(\frac{d}{du}\right)^n \left[\frac{(\pi/2)^{n+\frac{1}{2}}}{\sqrt{\pi}n!} \left(1-u^2\right)^n\right] \Pi(u/2)$$

$$= \frac{1}{2n!} \left(\frac{-j}{2}\right)^n \left(\frac{d}{du}\right)^n \left[\left(u^2-1\right)^n\right] \Pi(u/2) \qquad (2.82)$$

$$= \frac{1}{2} j^{-n} P_n(u) \Pi(u/2).$$

Thus,

$$j_k(2\pi t) \leftrightarrow \frac{1}{2} j^{-k} P_k(u) \Pi\left(\frac{u}{2}\right). \qquad (2.83)$$

2. **Chebyshev polynomials**

   Chebyshev polynomials of the first kind on the interval $[-1,1]$ are defined by

   $$T_n(\cos\theta) := \cos n\theta. \qquad (2.84)$$

   The first few are shown in Figure 2.21.

   (a) *Recursion relationship.* Chebyshev polynomials obey the recurrence relationship

   $$T_{n+1}(t) = 2tT_n(t) - T_{n-1}(t). \qquad (2.85)$$

   The polynomials can then be generated by this recursion initiated, from (2.84), $T_0(t) = \Pi\left(\frac{t}{2}\right)$ and $T_1(t) = t\,\Pi\left(\frac{t}{2}\right)$.
   To show (2.85), we will assume it true and show it reduces to an identity. Applying (2.84) to (2.85) asks the question

   $$\text{is } [\cos(n+1)\theta = 2\cos\theta\,\cos n\theta - \cos(n-1)\theta] \text{ ?}$$

   The affirmative answer follows immediately from recognizing the equation is the real part of

   $$e^{j(n+1)\theta} = 2\cos\theta\,e^{jn\theta} - e^{j(n-1)\theta}$$

   which, when simplified, reduces to Euler's formula.

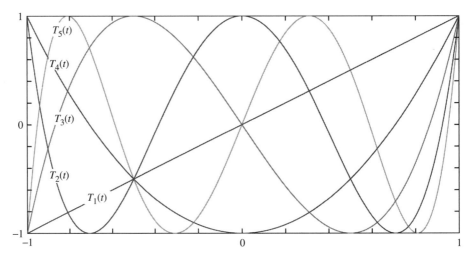

FIGURE 2.21. The first few Chebyshev polynomials, $T_n(t)$.

(b) Orthogonality. Chebyshev polynomials are orthogonal with respect to the *weighting function*, $(1-t^2)^{-1/2}$.

$$\int_{t=-1}^{1} \frac{T_n(t)T_m(t)}{\sqrt{1-t^2}} \, dt = \begin{cases} \frac{1}{2} m \, \delta[n-m] \; ; & n \neq 0 \text{ and } m \neq 0 \\ m & ; \; n = m = 0 \end{cases} \quad (2.86)$$

(c) Fourier transform. The Fourier transform of a Chebyshev polynomials on $[-1,1]$ is an nth order Bessel function

$$J_n(2\pi t) \longleftrightarrow \frac{1}{\pi} \frac{(-j)^n}{\sqrt{1-u^2}} T_n(u) \Pi\left(\frac{u}{2}\right). \quad (2.87)$$

3. **Hermite polynomials**
Hermite polynomials can be defined by

$$H_n(t) := (-1)^n e^{t^2} \left(\frac{d}{dt}\right)^n e^{-t^2} \quad (2.88)$$

The first few Hermite polynomials are shown in Figure 2.22.
(a) Recursion relationship. Hermite polynomials obey the recursion relationship

$$H_{n+1}(t) = 2t H_n(t) - 2n H_{n-1}(t) \quad (2.89)$$

with initialization $H_0(t) = 1$ and $H_1(t) = 2t$.

(b) Orthogonality. Hermite polynomials are orthogonal on the real line with respect to the weighting function, e^{-t^2}.

$$\int_{-\infty}^{\infty} H_n(t) H_m(t) e^{-t^2} \, dt = 2^n \, n! \, \sqrt{\pi} \, \delta[n-m].$$

(c) Fourier transform. Weighted Hermite polynomials, called Hermite-Gaussian functions, are their own Fourier transforms.

$$e^{-\pi t^2} H_n(\sqrt{2\pi} \, t) \longleftrightarrow (-j)^n e^{-\pi u^2} H_n(\sqrt{2\pi} \, u). \quad (2.90)$$

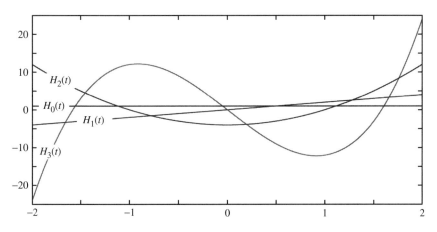

FIGURE 2.22. The first few Hermite polynomials, $H_n(t)$.

Equivalently,

$$K_n(t) \longleftrightarrow (-j)^n K_n(u). \qquad (2.91)$$

where the *Hermite-Gaussian function* is

$$K_n(t) := e^{-\pi t^2} H_n(\sqrt{2\pi}\, u). \qquad (2.92)$$

The Fourier transform of the Gaussian in (2.52) is a special case corresponding to $K_0(t)$. Plots of $K_n(t)$ are in Figures 2.23 and 2.24.

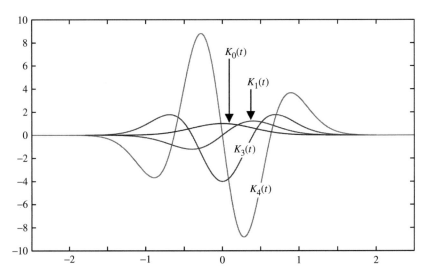

FIGURE 2.23. Hermite-Gaussian functions, $K_n(t)$, are their own Fourier transforms. See (2.91). Additional plots are in Figure 2.24.

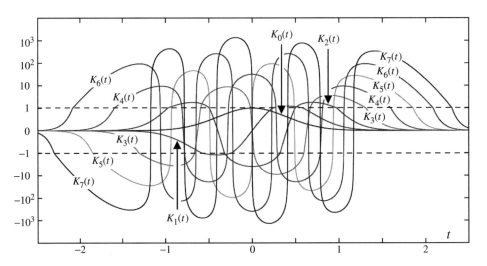

FIGURE 2.24. Hermite-Gaussian functions, $K_n(t)$, are their own Fourier transforms. See (2.91). These plots are linear for the range $(-1,1)$ and are, otherwise, logarithmic.

2.3.6 Other Properties of the Continuous Time Fourier Transform

2.3.6.1 The Signal Integral Property

A useful property emerges from evaluating the inverse Fourier transform in (2.10) at $u = 0$.

$$X(0) = \int_{t=-\infty}^{\infty} x(t)\,dt. \tag{2.93}$$

The Fourier transform evaluated at the origin is equal to the definite integral over all time of the transformed signal.

Definite integration of derivatives property. The signal integral property can be used to show that, if the integral in (2.93) is finite, then

$$\int_{t=-\infty}^{\infty} \left(\frac{d}{dt}x(t)\right) dt = 0. \tag{2.94}$$

Proof. Let $y(t) = \frac{d}{dt} x(t)$. Then, according to the derivative theorem

$$y(t) \longleftrightarrow Y(u) = (j2\pi u)X(u). \tag{2.95}$$

According to the signal integral property in (2.93),

$$\int_{-\infty}^{\infty} y(t)\,dt = Y(0) = (j2\pi u)X(u)|_{u=0}.$$

Thus, assuming $X(0) < 0$, (2.94) follows.

2.3.6.2 Conjugate Symmetry of a Real Signal's Spectrum

The continuous time Fourier transform of a real function is *conjugately symmetric*, (or Hermetian). That is, if $x(t)$ is real, then

$$X(u) = X^*(-u). \tag{2.96}$$

Proof. From the Fourier transform in (2.10), we write

$$X(-u) = \int_{-\infty}^{\infty} x(t)\,e^{j2\pi ut}\,dt.$$

Conjugating both sides and, recognizing a real $x(t)$ dictates $x(t) = x^*(t)$, reduces this equation to the Fourier transform in (2.10).

The spectrum can be separated into its real and imaginary components.

$$X(u) = R(u) + jI(u). \tag{2.97}$$

Equating the real and imaginary components of the Fourier transform of a real signal in (2.96) reveals the real part of the transform is an even function

$$R(u) = R(-u),$$

and the imaginary part is an odd function.

$$I(u) = -I(-u).$$

Similarly, the spectrum can be expressed in polar form as

$$X(u) = |X(u)|e^{j\angle X(u)}.$$

Using (2.96) and equating the magnitude reveals the magnitude of the Fourier transform is an even function.

$$|X(u)| = |X(-u)|,$$

whereas equating the argument gives an odd phase.

$$\angle X(u) = -\angle X(-u).$$

2.3.6.3 The Hilbert Transform

The Hilbert transform of a signal $z(t)$ is the convolution $\frac{-1}{\pi t} * z(t)$. Since[19] $\frac{-1}{\pi t} \longleftrightarrow j\,\mathrm{sgn}(u)$, the Fourier transform of the Hilbert transform is

$$\mathcal{H}z(t) = \frac{-1}{\pi t} * z(t) \longleftrightarrow j\,\mathrm{sgn}(u)\,Z(u). \tag{2.98}$$

where $\mathcal{H}$ is the Hilbert transform operator.

2.3.6.4 Hilbert Transform Relationships Within a Causal Signal's Spectrum

If a signal is causal, the real and imaginary components of its spectrum are Hilbert transform pairs. Specifically

$$I(u) = \mathcal{H}R(u) = \frac{-1}{\pi u} * R(u). \tag{2.99}$$

Proof. We start with the property of a causal signal

$$x(t) = x(t)\mu(t).$$

Since $\mu(t) \longleftrightarrow \frac{1}{2}\left(\delta(u) - \frac{j}{\pi u}\right)$, Fourier transforming both sides gives

$$X(u) = \frac{1}{2}X(u) - \frac{j}{2\pi u} * X(u),$$

or

$$X(u) = -\frac{j}{\pi u} * X(u).$$

Substituting the real and imaginary component equation for the spectrum in (2.97) and equating the imaginary portions of both sides gives the promised result in (2.99).

Likewise, equating the real parts gives the inverse Hilbert transform relation

$$R(u) = \frac{1}{\pi u} * I(u). \tag{2.100}$$

---

19. See, e.g., Table 2.4.

2.3.6.5 The Power Theorem

The *power theorem* states that the inner product of two functions in time is equal to the inner product of their Fourier transforms in frequency.

$$\int_{-\infty}^{\infty} x(t)y^*(t)dt = \int_{-\infty}^{\infty} X(u)Y^*(u)du. \qquad (2.101)$$

An important special case is *Parseval's theorem*[20] that states the energy of a signal in time is equal to the energy of its Fourier transform in frequency.

$$\int_{-\infty}^{\infty} |x(t)|^2 dt = \int_{-\infty}^{\infty} |X(u)|^2 du.$$

Proof. Using the definition of the inverse Fourier transform, we write

$$\int_{-\infty}^{\infty} x(t)y^*(t)dt = \int_{-\infty}^{\infty} \left[\int_{-\infty}^{\infty} X(u)e^{j2\pi ut} du \right] y^*(t)dt.$$

Reversing integration order and recognizing the resulting transform completes the proof.

2.3.6.6 The Poisson Sum Formula

The *Poisson sum formula* is

$$T \sum_{n=-\infty}^{\infty} x(t - nT) = \sum_{n=-\infty}^{\infty} X\left(\frac{n}{T}\right) e^{j2\pi nt/T}. \qquad (2.102)$$

The left side of (2.102) is periodic with period T, although $x(t)$ need not be a period. The right hand side is the Fourier series of this periodic function.

Proof.

$$\sum_{n=-\infty}^{\infty} x(t - nT) = x(t) * \frac{1}{T} \text{comb}\left(\frac{t}{T}\right)$$

$$= x(t) * \frac{1}{T} \sum_{n=-\infty}^{\infty} e^{j2\pi nt/T}$$

where we have used (2.55). Recognizing

$$x(t) * e^{j2\pi vt} = X(v)e^{j2\pi vt} \qquad (2.103)$$

completes the proof.

Note that, for $t = 0$, the Poisson sum formula directly relates the sum of the signal and spectral samples.

$$T \sum_{n=-\infty}^{\infty} x(nT) = \sum_{n=-\infty}^{\infty} X\left(\frac{n}{T}\right). \qquad (2.104)$$

---

20. When applied to the Fourier transform, Parseval's theorem is also referred to as *Rayleigh's theorem* [149, 1162].

2.4 Orthogonal Basis Functions¶

The Fourier series is a special case of an orthogonal basis function expansion [854, 1020]. For a given interval, I, functions $x(t)$ and $y(t)$ are said to be *orthogonal* if

$$\int_I x(t)\, y^*(t)\, dt = 0. \tag{2.105}$$

The functions are said to be *orthonormal* if both $x(t)$ and $y(t)$ have unit energy on the interval I.

Each element in an orthonormal basis set, $\{\varphi_n(t) | -\infty < n < \infty;\ t \in I\}$, [21] has unit energy on the interval I and is orthogonal to every other element in the set.

$$\int_I \varphi_n(t)\, \varphi_m^*(t)\, dt = \delta[n - m]. \tag{2.106}$$

Corresponding to a given signal class, C, of finite energy functions, a basis set is said to be *complete* if, for every $x(t) \in C$, we can write

$$x(t) = \sum_{n=-\infty}^{\infty} \alpha_n\, \varphi_n(t)\ ;\ t \in I \tag{2.107}$$

where equality is at least in the mean square sense.

$$\lim_{N \to \infty} \int_I \left| x(t) - \sum_{n=-N}^{N} \alpha_n \varphi_n(t) \right|^2 dt = 0. \tag{2.108}$$

The expansion coefficients, α_n, can be found by multiplying both sides of (2.107) by $\varphi_m^*(t)$ and integrating over I. Using (2.106), we find that

$$\alpha_n = \int_I x(t)\varphi_n^*(t)\, dt. \tag{2.109}$$

2.4.1 Parseval's Theorem for an Orthonormal Basis¶

Parseval's theorem for a complete set of orthonormal basis functions relates the signal's energy to its expansion coefficients.

$$\int_I |x(t)|^2 dt = \sum_n |\alpha_n|^2.$$

Proof.

$$\int_I |x(t)|^2 dt = \int_I x(t) x^*(t)\, dt$$

$$= \int_I \left[\sum_n \alpha_n \varphi_n(t) \sum_m \alpha_m^* \varphi_n^*(t) \right] dt$$

---

[21]. For certain orthogonal basis sets such as the prolate spheroidal wave functions the index n runs from 0 to ∞. See Section 10.5.

FUNDAMENTALS OF FOURIER ANALYSIS

$$= \sum_n \alpha_n \sum_m \alpha_m^* \int_I \left[\varphi_n(t) \varphi_n^*(t) \right] dt$$

$$= \sum_n \alpha_n \sum_m \alpha_m^* \, \delta[n-m]$$

$$= \sum_n |\alpha_n|^2$$

Parseval's theorem for the Fourier series in (2.19) and the sampling theorem in (5.34) are special cases.

2.4.2 Examples¶

In each of the following examples, C is subsumed in the class of finite energy (L_2) functions.

2.4.2.1 The Cardinal Series¶

The cardinal series, the topic of Chapter 5, is an orthonormal expansion for all signals whose Fourier transforms are bandlimited with bandwidth B. For

$$I = \{\, t \mid -\infty < t < \infty \,\},$$

the orthonormal basis functions are

$$\varphi_n(t) = \sqrt{2B} \, \text{sinc}(2Bt - n). \tag{2.110}$$

Note, then, that[22]

$$\alpha_n = \frac{1}{\sqrt{2B}} \, x\!\left(\frac{n}{2B}\right). \tag{2.111}$$

As we will see in Chapter 5, the cardinal series displays uniform convergence which is stronger than that the mean square convergence in (2.108).

2.4.2.2 The Fourier Series¶

The Fourier series is an orthonormal expansion for signals over the interval

$$I = \left\{\, t \mid -\frac{T}{2} < t < \frac{T}{2} \,\right\}.$$

The orthonormal basis functions here are:

$$\varphi_n(t) = \frac{1}{\sqrt{T}} \exp\!\left(\frac{j 2\pi n t}{T}\right); \; t \in I.$$

It follows that $\alpha_n = \frac{1}{\sqrt{T}} X_T(\frac{n}{T})$.

2.4.2.3 Prolate Spheroidal Wave Functions¶

The *prolate spheroidal wave functions* in Section 10.5 can be used to represent bandlimited functions.

---

22. See Exercise 2.50 for details.

2.4.2.4 Complex Walsh Functions§

Complex Walsh functions in Exercise 6.27 are also examples of orthogonal basis sets on a finite interval. Such sets will later prove useful in the understanding of interpolation from continuous samples and Kramer's generalization of the sampling theorem.

2.4.2.5 Orthogonal Polynomials¶

Orthogonal polynomials, including the Legendre in (2.79), the Chebyshev in (2.86) and the Hermite in (2.88), are, as indicated by the name, orthogonal.

2.4.2.6 Fourier Transforms of Orthonormal Functions are Orthonormal¶

If the set of functions, $\{x_n(t) \mid 0 \leq n < \infty\}$, are orthonormal, then the Fourier transforms of these functions, $\{X_n(u) \mid 0 \leq n < \infty\}$, are also orthonormal. The orthonormality of the functions can be expressed as

$$\int_{-\infty}^{\infty} x_n(t) x_m^*(t) dt = \delta[n-m].$$

From the power theorem in (2.101),

$$\int_{-\infty}^{\infty} x_n(t) x_m^*(t) dt = \int_{-\infty}^{\infty} X_n(u) X_m^*(u) du.$$

Therefore, orthonormality in t dictates orthonormality in u.

2.5 The Discrete Time Fourier Transform

The discrete time Fourier transform (DTFT) of a discrete time sequence, $x[n]$, is

$$X_D(f) := \sum_{n=-\infty}^{\infty} x[n] e^{-j2\pi nf}. \tag{2.112}$$

The frequency variable, f, is unitless. The DTFT is recognized as a Fourier series in f with unit period and Fourier coefficients $c_n = x[n]$. The inversion is therefore

$$x[n] = \int_1 X_D(f) e^{j2\pi nf} df \tag{2.113}$$

where $\int_1$ denotes integration over any unit interval.
Properties of the DTFT are listed in Table 2.6.

2.5.1 Relation of the DFT to the Continuous Time Fourier Transform

To relate discrete time Fourier transform to the continuous time Fourier transform, let the sequence $x[n]$ be the samples of a continuous time signal, $x(t)$ at intervals of T.

$$x[n] = x(nT).$$

TABLE 2.6. Discrete time Fourier transform (DTFT) theorems. Discrete time convolution is defined by $x[n] * h[n] = \sum_{m=-\infty}^{\infty} x[m] h[n-m]$. Correlation is $x[n] \star h[n] = \sum_{m=-\infty}^{\infty} x[m] h^*[n-m] = x[n] * h^*[-n]$. The notation "$\int_1$" means integration over any unit interval.

Transform periodicity	$x[n]$	$\leftrightarrow$	$X_D(f) = \sum_{n=-\infty}^{\infty} x[n] e^{-j2\pi nf}$ $= X_D(f-p); p = 0, \pm1, \pm2, \ldots$
shift	$x[n-k]$	$\leftrightarrow$	$X_D(f) e^{-j2\pi kf}$
modulation	$x[n] e^{j2\pi nv}$	$\leftrightarrow$	$X_D(f-v)$
cumulative sum	$\sum_0^n x[k]$; $x[n]$ is causal	$\leftrightarrow$	$\frac{X_D(u)}{e^{j2\pi f} - 1}$
conjugate	$x^*[n]$	$\leftrightarrow$	$X_D^*(-f)$
transpose	$x[-n]$	$\leftrightarrow$	$X_D(-f)$
inversion	$x[n] = \int_1 X(f) e^{j2\pi nf} df$	$\leftrightarrow$	$X_D(f)$
linearity	$ax[n] + by[n]$	$\leftrightarrow$	$aX_D(f) + bY_D(f)$
convolution	$x[n] * h[n]$	$\leftrightarrow$	$X_D(f) H_D(f)$
circular convolution	$x[n] h[n]$	$\leftrightarrow$	$\int_{-\infty}^{\infty} X_D(v) \Pi(v) H_D(f-v) dv$ $= \int_1 X_D(v) H_D(f-v) dv$
correlation	$x(t) \star h(t)$	$\leftrightarrow$	$X_D(u) H_D^*(u)$

Define the signal of samples as

$$s(t) = \sum_{n=-\infty}^{\infty} x[n] \delta(t - nT).$$

Note that

$$s(t) = x(t) \sum_{n=-\infty}^{\infty} \delta(t - nT)$$

$$= x(t) \frac{1}{T} \text{comb}\left(\frac{t}{T}\right). \quad (2.114)$$

Using the Fourier transform for the comb in Table 2.4 and the convolution property[23] reveals that sampling replicates the spectrum.

$$s(t) \longleftrightarrow S(u) = X(u) * \text{comb}(Tu)$$

$$= X(u) * \frac{1}{T} \sum_{n=-\infty}^{\infty} \delta\left(u - \frac{n}{T}\right)$$

$$= \frac{1}{T} \sum_{n=-\infty}^{\infty} X\left(u - \frac{n}{T}\right).$$

This replication, pictured in Figure 2.25, is periodic in the frequency domain with a period of $\frac{1}{T}$. The Fourier series of this periodic function is obtained from the Fourier transform

---

23. See Table 2.3.

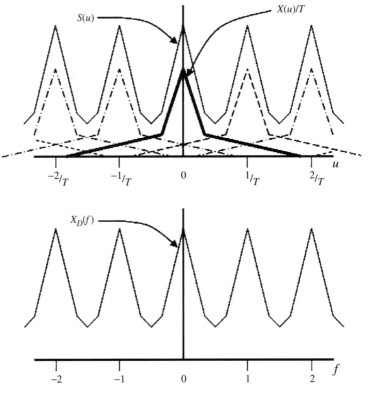

FIGURE 2.25. Replication of a spectrum when a signal is sampled.

of (2.114). Equating gives

$$S(u) = \sum_{n=-\infty}^{\infty} x(nT)e^{-j2\pi nTu} = \frac{1}{T} \sum_{n=-\infty}^{\infty} X\left(u - \frac{n}{T}\right).$$

This is the Fourier dual of the Poisson sum formula in (2.102). Comparing to the DTFT in (2.112) gives the relation between the spectrum of the signal, $x(t)$, and the DTFT of its samples.

$$X_D(f) = S\left(\frac{f}{T}\right) = \sum_{n=-\infty}^{\infty} X\left(\frac{f-n}{T}\right).$$

If $X(u)$ is zero for $u > \frac{1}{2T}$, the spectra do not overlap and $x(t)$ can uniquely be reconstructed from $S(u)$ and therefore from $X_D(f)$.

2.6 The Discrete Fourier Transform

The discrete Fourier transform (DFT) [159, 1367] is ubiquitous in digital signal processing. The fast Fourier transform (FFT) is a computationally efficient algorithm to compute the DFT [160, 322, 323, 1595]. Goertzel's algorithm[24] can also be used.

---

24. See Section 3.5.

The DFT is a special case of the sampled DTFT. The samples in the frequency domain need not be uniform [41]. Let $x[n]$ be defined only on the interval $0 \leq n < N$ and sample the frequency at locations f_k where k is the index of the sample. Then the sampled DTFT in (2.112) becomes

$$X_D(f_k) = \sum_{n=0}^{N-1} x[n] e^{-j2\pi n f_k}. \tag{2.115}$$

If, for example, a logarithmic plot of the spectrum is desired with 100 points per decade[25] we would use, for an appropriate range of k's,

$$f_k = f_0 10^{\frac{k}{100}}.$$

where f_0 is a reference frequency. If the range of k is finite, the sampled DTFT in (2.115) can be evaluated using a matrix vector multiplication.

The DFT with uniform spacing is a special case of the sampled DTFT where $f_k = \frac{k}{N}$. Using the notation

$$X[k] = X_D\left(\frac{k}{N}\right),$$

the DFT of a sequence of N values, $\{ x[n] \mid n = 0, 1, 2, \ldots, N-1 \}$, is

$$X[k] = \sum_{n=0}^{N-1} x[n] e^{-j2\pi nk/N}. \tag{2.116}$$

Since $X[k] = X[k+N]$, the DFT is periodic with period N. Thus, only the first N samples require computing. The principle values for $\{k = 0, 1, 2, \ldots, N-1\}$ are normally used. We will likewise assume the sequence $x[n]$ is periodic with period N. Unless otherwise specified, the DFT is assumed to be uniformly sampled in the frequency domain.

The inverse DFT is

$$x[n] = \frac{1}{N} \sum_{k=0}^{N-1} X[k] e^{j2\pi nk/N}. \tag{2.117}$$

2.6.1 Circular Convolution

Let

$$Y[k] = X[k] H[k] \tag{2.118}$$

where $X[k]$ and $H[k]$ are DFT's of signals $x[n]$ and $h[n]$ both of which are periodic with period N. Then the inverse DFT of $Y[k]$ is given by the *circular convolution*

$$y[n] = \sum_{m=0}^{N-1} x[m] h[n-m]. \tag{2.119}$$

---

25. The frequency one decade above f_0 is $10 f_0$.

Proof. Using the inverse DFT in (2.117) for y[n] and (2.118) gives.

$$y[n] = \frac{1}{N} \sum_{k=0}^{N-1} X[k]H[k]e^{j2\pi nk/N}.$$

Using the DFT expression in (2.116),

$$y[n] = \frac{1}{N} \sum_{k=0}^{N-1} \left[\sum_{m=0}^{N-1} x[m]e^{-j2\pi mk/N} \right] H[k]e^{j2\pi nk/N}$$

or, rearranging summation order

$$y[n] = \sum_{m=0}^{N-1} x[m] \left[\frac{1}{N} \sum_{k=0}^{N-1} H[k]e^{j2\pi(n-m)k/N} \right].$$

Applying the inverse DFT formula in (2.117) gives the promised circular convolution in (2.119).

The mechanics of circular convolution are treated in Section 3.3.3.3.

2.6.2 Relation to the Continuous Time Fourier Transform

We explain the DFT's relationship to the continuous time Fourier transform using Figure 2.26. On the left are functions of time and, on the right, their Fourier transforms. On the top row, we are reminded that the Fourier transform of a periodic signal is a string of Dirac delta functions in the frequency domain. The second row shows the Fourier dual. The Fourier transform of a string of equally spaced Dirac deltas has, as a Fourier transform, a periodic function. Indeed, when couched in discrete time, $(T = 1)$, this is the discrete time Fourier transform. The discrete time Fourier transform has a spectrum that is periodic with a period of one.

The last row illustrates the DFT and makes use of the properties of the first two rows. The Fourier transform of a periodic string of Dirac deltas is a periodic string of Dirac deltas. Furthermore, the number of Dirac deltas in a period in both the time and frequency domains is N.

In the time domain, if the first N values are $x[n]$, then the first N samples in the frequency domain are $X[k]/(NT)$ where $X[k]$ is given by the DFT in (2.116).

To show this from the perspective of continuous time in Figure 2.26, recall, using (2.17), that the Fourier transform of a periodic function with period NT is

$$x(t) \leftrightarrow \frac{1}{NT} \sum_{k=-\infty}^{\infty} X_{NT}\left(\frac{k}{NT}\right) \delta\left(u - \frac{k}{NT}\right)$$

where $X_{NT}(u)$ is the Fourier transform of $x(t)$ over any single period. In our case, the duration of a temporal period is NT. For the periodic sequence of Dirac deltas,

$$X_{NT}(u) = \int_{-\infty}^{\infty} \left[\sum_{n=0}^{N-1} x[n]\delta(t - nT) \right] e^{-j2\pi ut} dt$$

$$= \sum_{n=0}^{N-1} x[n]e^{-j2\pi unT}$$

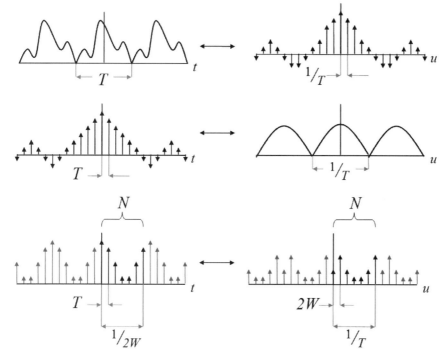

FIGURE 2.26. Top row: The Fourier transform of a periodic function is a sequence of Dirac deltas. Middle row: The Fourier transform of a string of Dirac deltas is a periodic function. This is the Fourier dual of the property of the property shown in the top row. Bottom row: If a function is both (a) periodic and (b) an equally spaced sequence of Dirac deltas, then its Fourier transform will be (a) an equally spaced sequence of Dirac deltas that is (b) periodic. The period in each domain consists of N Dirac deltas. The relation between the samples in each of the two domains is the DFT.

and

$$X_{NT}\left(\frac{k}{NT}\right) - \sum_{n=0}^{N-1} x[n]e^{-j2\pi kn/N} := X[k].$$

Thus

$$x(t) \leftrightarrow \frac{1}{NT} \sum_{k=-\infty}^{\infty} X[k]\delta\left(u - \frac{k}{NT}\right).$$

Therefore, in continuous time, a periodic replication of N samples riding on a string of Dirac delta functions has a Fourier transform of a periodic replication of N values also riding on a string of Dirac deltas. The relationship between the N samples in the time domain and the N samples in the frequency domain is the DFT.

2.6.3 DFT Leakage

The DFT reproduces the DTFT at the frequencies $f = \frac{k}{2N}$. At intermediate frequencies, however, there is leakage due to aliasing.[26] This is illustrated in Figure 2.27 where the

---

26. See Section 5.3.1.1.

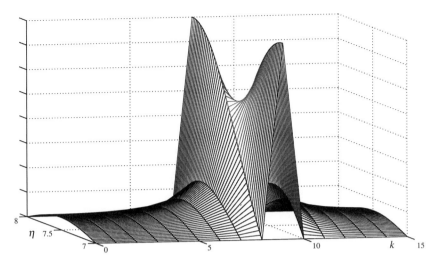

FIGURE 2.27. Illustration of leakage in a DFT.

magnitude of an $N = 8$ DFT is shown for $x[n] = exp(-j2\pi \eta/N)$ is shown for $7 \leq \eta \leq 8$. Points are linearly connected. At $\eta = 7$, the DFT gives a value at $k = 7$ and zero otherwise. As η is increased towards 8, the values of the DFT spread to adjacent bins. This is leakage. At $\eta = 8$, the DFT again returns to a single value for $k = 8$ and zero otherwise. A common approach to address leakage is the use of windows. The DFT for[27] $x[n] = \sin^2(\pi k/N) \exp(-j2\pi \eta/N)$ for the same range of η is shown in Figure 2.28. Use of the window diminishes the nonuniformity of the leakage as the cost of greater spread at the DFT points.

2.7 Related Transforms

There are many transforms closely related to the Fourier transform. In the introductory material to follow, transforms will be subscripted according to type. In later applications, the subscripts will be dropped and the transform type will be clear in the context of its use.

2.7.1 The Cosine Transform[†]

The unilateral *cosine transform*, in continuous time, is[28]

$$X_{\cos}(u) := 2 \int_0^\infty x(t) \cos(2\pi u t) dt. \qquad (2.120)$$

The *discrete cosine transform* (DCT) [116, 929, 1158] is

$$X_{\cos}[k] := \frac{C[k]}{2} \sum_{n=0}^{N-1} x[n] \cos\left(\frac{\pi(2n+1)k}{2N}\right) \qquad (2.121)$$

---

27. This is a Hanning window. Other commonly used windows are in Table 9.1.
28. See Exercise 2.22 for the inverse.

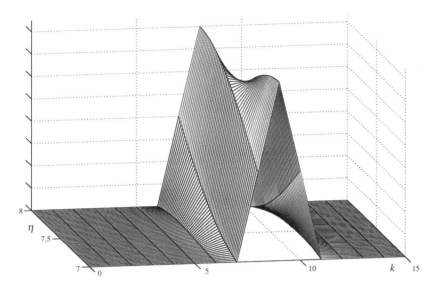

FIGURE 2.28. Illustration of leakage in a windowed DFT.

where

$$C[k] = \begin{cases} 2^{-\frac{1}{2}} & ; k = 0 \\ 1 & ; \text{otherwise} \end{cases} \tag{2.122}$$

The DCT is used in JPEG image compression.[29]

The inverse DCT is

$$x[n] = \sum_{k=0}^{N-1} \frac{C[k]}{2} X_{\cos}[k] \cos\left(\frac{\pi(2n+1)k}{2N}\right). \tag{2.123}$$

2.7.2 The Sine Transform[†]

The unilateral *sine transform*, in continuous time, is[30]

$$X_{\sin}(u) := \int_0^\infty x(t) \sin(2\pi u t) dt. \tag{2.124}$$

2.7.3 The Hartley Transform

The *Hartley transform* [150] of a signal is

$$X_{\text{hart}}(u) := \int_{-\infty}^\infty x(t) \text{cas}(2\pi u t) dt \tag{2.125}$$

where

$$\text{cas}(\zeta) = \cos(\zeta) - \sin(\zeta). \tag{2.126}$$

---

29. See Section 8.7.
30. See Exercise 2.22 for the inverse.

Note that

$$e^{-j\zeta} = \cos(\zeta) - j\sin(\zeta)$$

so the Hartley transform is closely related to the Fourier transform. Indeed, for the Fourier transform pair $x(t) \leftrightarrow X(u) = \Re X(u) + j\Im X(u)$, we have, when $x(t)$ is real,

$$X_{\text{hart}}(u) = \Re X(u) + \Im X(u). \tag{2.127}$$

The Hartley transform has the property that $X(u)$ is real when $x(t)$ is real. Its inverse is

$$x(t) = \int_{-\infty}^{\infty} X_{\text{hart}}(u)\text{cas}(2\pi ut)du. \tag{2.128}$$

2.7.4 The z Transforms

The *unilateral* z-transform of a sequence $x[n]$ is[31]

$$X_z(z) := \sum_{n=0}^{\infty} x[n]\, z^{-n}. \tag{2.129}$$

The transform pair can be written in short hand as

$$x[n] \underset{z}{\longleftrightarrow} X_z(z). \tag{2.130}$$

For example[32]

$$a^n \mu[n] \underset{z}{\longleftrightarrow} \frac{1}{1 - az^{-1}} \tag{2.131}$$

where $\mu[n] = 1$ for $n \geq 0$ and zero otherwise is the unit step.

The shift theorem for the unilateral z transform is

$$x[n+1] \underset{z}{\longleftrightarrow} zX_z(z) - z\, x[0] \tag{2.132}$$

To prove this, we write

$$\sum_{k=0}^{\infty} x[k+1]\, z^{-k} = z\left\{-x[0] + \sum_{n=0}^{\infty} x[n]\, z^{-n}\right\}$$

where we have made the variable substitution $n = k - 1$. Equation 2.132 follows.

When the Laplace transform in (2.22) converges on the imaginary axis of the s plane, the slice of the Laplace transform along the imaginary axis is the Fourier transform (see (2.23)). Similarly, if the z transform converges on the unit circle, $z = e^{j2\pi f}$, then the slice

---

31. When the n summation is over the interval $(-\infty, \infty)$, the z transform is said to be *bilateral*.
32. In some applications, the region of convergence of the z transform is of significance. In (2.131), the region of convergence is for $|z| > |a|$.

of the z transform along the circle is the DTFT of the discrete time signal. Specifically, from (2.129),

$$X_z\left(e^{j2\pi f}\right) = \sum_{n=0}^{\infty} x[n]\, e^{j2\pi nf} \qquad (2.133)$$

From (2.112), we clearly have $X_z\left(e^{j2\pi f}\right) = X_D(f)$.

The z transform is useful in the characterization of *finite impulse response* (FIR) filters and *infinite impulse response* (IIR) filters.[33] FIR filters contain no feedback, are always stable and have an impulse response equal to zero after some finite time interval. IIR filters have feedback and generally have impulse responses that last forever. The FIR and IIR filters can be represented via the difference equation

$$\sum_{p=0}^{N} b_p y[n-p] = \sum_{q=0}^{N} a_q x[n-q] \qquad (2.134)$$

where x is the input and y the output. By convention, we set $b_0 = 1$. Then, equivalently, we can write

$$y[n] = \sum_{q=0}^{N} a_q x[n-q] - \sum_{p=1}^{N} b_p y[n-p]. \qquad (2.135)$$

The order of the filter is N. The b_p's are the coefficients of feedback for previous outputs contributing to the current output. If $\{b_p = 0 \mid 1 \leq p \leq N\}$, there is no feedback and the filter is FIR. Otherwise, the filter is IIR. A standard implementation of (2.135) is shown in Figure 2.29 [511, 1053, 1054].

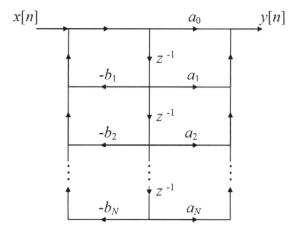

FIGURE 2.29. An implementation of the IIR filter of the difference equation in (2.135). If all of the b_n's are zero, the filter is FIR. The graph is read as follows. A branch labelled as function of n, such as the input, $x[n]$, or output, $y[n]$, denotes the signal flowing on that branch. Branches labelled z^{-1} denote a unit delay. Other labelled branches, such as a_1 and $-b_N$, correspond to signal multiplies. An unlabelled branch corresponds to a unit multiply. When two branches merge, the signals are added. When a branch splits, duplicate versions of the signal are sent down each branch.

---

33. See, for example, Section 9.3.4.2.

With zero initial conditions, the z transform of (2.134) is

$$Y_z(z) \sum_{p=0}^{N} b_p z^{-p} = X_z(z) \sum_{q=0}^{N} a_q z^{-q}.$$

The corresponding transfer function of the filter, $H_z(z) = Y_z(z)/X_z(z)$, is

$$H_z(z) = \frac{\sum_{q=0}^{N} a_q z^{-q}}{\sum_{p=0}^{N} b_p z^{-p}}. \tag{2.136}$$

The frequency response of the filter follows as

$$H_z\left(e^{j2\pi f}\right) = \frac{\sum_{q=0}^{N} a_q e^{j2\pi fq}}{\sum_{p=0}^{N} b_p e^{j2\pi fq}}.$$

Application of the z transform to a simple difference equation in some elementary deterministic problems in finance is in Section 13.4.

2.8 Exercises

2.1. Let the units of $x(t)$ be volts. Let $x[n] = x(nT)$. What are the units of
   (a) the CTFT, $X(u)$, in (2.10)?
   (b) the Fourier series coefficients, c_n, in (2.13)?
   (c) the DTFT transform, $X_D(f)$, in (2.112)?
   (d) the DFT, $X[k]$, in (2.116)?
   (e) the Hilbert transform of $x(t)$ using (2.98)?

2.2. For the signals in Section 2.2, specify any signal classes that are subsumed in any other signal class.

2.3. § Consider the following "proof" that $n = 2n$. Clearly

$$\frac{d}{dn} n^2 = 2n.$$

Also

$$\frac{d}{dn} n^2 = \frac{d}{dn} n \times n$$

$$= \frac{d}{dn}(n + n + \cdots n)$$

$$= \frac{d}{dn}(1 + 1 + \cdots 1)$$

$$= n.$$

Equating gives $n = 2n$.

Specifically identify what is incorrect in *this* derivation of a nonsensical result.

2.4. Absolutely convergent sums I. A series

$$\sum_{m=0}^{\infty} a_m$$

is said to be *absolutely convergent* if the sum

$$S = \sum_{m=0}^{\infty} |a_m|$$

is finite. The terms in an absolutely convergent series can be arbitrarily rearranged without affecting the sum. The Taylor series expansion of $x(t)$ about the (real) number τ is

$$x(t) = \sum_{m=0}^{\infty} \frac{(t-\tau)^m}{m!} x^{(m)}(\tau).$$

Show that if $x(t)$ has finite energy and is bandlimited, then this series is absolutely convergent for all $|t| < \infty$.

Hint: Show that

$$S(t) < \sqrt{2B E} e^{2\pi B |t-\tau|}.$$

2.5. § Absolutely convergent sums II.

For $z = e^{j\theta}$, Euler's formula uses the even and odd terms in (2.34) to generate the Taylor series for the cosine in (2.35) and the sine in (2.36). Consider doing a similar decomposition for the Taylor series expansion

$$\ln(2) = \sum_{n=1}^{\infty} \frac{(-1)^n}{n}. \tag{2.137}$$

Expanding and rearranging gives

$$\ln(2) = 1 - \frac{1}{2} + \frac{1}{3} - \frac{1}{4} + \frac{1}{5} - \frac{1}{6} + \frac{1}{7} - \cdots$$

$$= \left(1 + \frac{1}{3} + \frac{1}{5} + \frac{1}{7} + \cdots\right) - \left(\frac{1}{2} + \frac{1}{4} + \frac{1}{6} + \frac{1}{8} \cdots\right)$$

$$= \left(1 + \frac{1}{3} + \frac{1}{5} + \frac{1}{7} + \cdots\right) + \left(\frac{1}{2} + \frac{1}{4} + \frac{1}{6} + \frac{1}{8} \cdots\right)$$

$$- 2\left(\frac{1}{2} + \frac{1}{4} + \frac{1}{6} + \frac{1}{8} \cdots\right).$$

Adding the first two series gives

$$\ln(2) = \left(1 + \frac{1}{2} + \frac{1}{3} + \frac{1}{4} + \frac{1}{5} + \frac{1}{6} + \cdots\right)$$
$$- 2\left(\frac{1}{2} + \frac{1}{4} + \frac{1}{6} + \frac{1}{8} + \cdots\right)$$
$$= \left(1 + \frac{1}{2} + \frac{1}{3} + \frac{1}{4} + \frac{1}{5} + \frac{1}{6} + \cdots\right)$$
$$- \left(1 + \frac{1}{2} + \frac{1}{3} + \frac{1}{4} + \frac{1}{5} + \frac{1}{6} + \cdots\right)$$
$$= 0.$$

We therefore have the obviously incorrect result that ln(2)=0. The reason is that, in order to rearrange the elements of an infinite summation without changing the sum, such as $\sum_{n=0}^{\infty} a_n$, the sum must be *absolutely convergent*.

$$\sum_{n=0}^{\infty} |a_n| < \infty.$$

(a) Show that the Taylor series in (2.34) is absolutely convergent, therefore allowing the decomposition into the cosine and sine components in (2.35) and (2.36).

(b) Show that the Taylor series in (2.137) is not absolutely convergent.

2.6. § **The Pythagorean theorem**. The magnitude of a complex function is determined from its real and imaginary parts using the Pythagorean theorem. Prove the Pythagorean theorem.

2.7. § A review of the geometric series is given in Appendix 14.4.2.

(a) Show that $(5^n)!$ ends in exactly $(5^n - 1)/4$ zeros. For example

$(5^3)! = 125! = 18$

82677176888926099743767702491600857595403648714924

25887598231508353156331613598866882932889495923133646

40544593005774063016191934138059781888345755854705

55243263755650071317708800000000000000000000000000

0000

ends in 31 zeros.

(b) Show that the number of zeros in $N!$ for arbitrary N is

$$Z[N] = \sum_{k=1}^{\lfloor \log_5 N \rfloor} \left\lfloor \frac{N}{5^k} \right\rfloor \qquad (2.138)$$

where $\lfloor a \rfloor$ denotes "the largest integer not exceeding a". Thus, for example, $\lfloor \pi \rfloor = 3$.

(c) How many zeros are at the end of 2000! ?
(d) Derive an expression for the number of zeros at the end of $N!! = N$ double factorial.
(e) Stirling's formula approximating the factorial is

$$n! \approx \sqrt{2\pi} n^{n+\frac{1}{2}} e^{-n}.$$

or, equivalently,

$$\log n! = \frac{1}{2}\log(2\pi) + \left(n + \frac{1}{2}\right)\log n - n.$$

Is this a good approximation for the 50! ? For 2000! ?

2.8. Show that $m!!$ in (2.66) can also be written as

$$m!! = \begin{cases} 2^{m/2}\left(\frac{m}{2}\right)! & ; \text{ when } m \text{ is even} \\ \dfrac{m!}{2^{(m-1)/2}\left(\frac{m-1}{2}\right)!} & ; \text{ when } m \text{ is odd} . \end{cases}$$

2.9. Evaluate $\int_{-\infty}^{\infty} x(t)dt$ for the following functions.
   (a) $\text{sinc}(t)$
   (b) $\text{jinc}(t)$
   (c) $J_0(2\pi t)$
   (d) $\frac{J_\nu(2\pi t)}{(2t)^\nu}$.
   (e) $\text{sinc}_k(t)$.
   (f) $\text{sinc}(t)\text{jinc}(t)$.
   (g) $\text{sinc}(2t)\text{sinc}(t-1)$.
   (h) $\cos(2\pi \nu t)$.

2.10. Simplify
   (a) $x(t) = \sin\left(\frac{\pi}{2}\Pi(t)\right)$.
   (b) $x(t) = \Pi\left(\sin^2(2\pi t)\right)$.
   (c) $x(t) = \Pi(\Pi(t))$.
   (d) $x(t) = \text{sgn}(\sin(2\pi t))$.
   (e) $x(t) = \sin\left(\frac{\pi}{2}\text{sgn}(t)\right)$.
   (f) $x(t) = \Pi(\Lambda(t))$.
   (g) $x[n] = \delta[\text{sinc}(n)]$.
   (h) $x[n] = \delta[\delta[n]]$.
   (i) $x[n] = \text{sinc}(\delta[n])$.
   (j) $x[n] = \delta[\text{jinc}(n)]$.

2.11. Consider the definite integration of derivatives property in (2.94) which applies when $X(0) < \infty$. Does this property apply when
   (a) $x(t) = e^{-t}\mu(t)$?
   (b) $x(t)$ is an odd function?
   (c) $x(t) = \delta(t)$?

2.12. Show that

$$\int_{-\infty}^{\infty} e^{-t^2} dt = 2\int_{-\infty}^{\infty} t^2 e^{-t^2} dt.$$

Hint: Apply the definite integration of derivatives property in (2.94).

2.13. Show that

$$\frac{f(0)}{\int_{-\infty}^{\infty} f(t)dt} = \frac{\int_{-\infty}^{\infty} F(u)du}{F(0)}. \qquad (2.139)$$

2.14. **Analytic signals.** If $x(t) \longleftrightarrow X(u)$ then the inverse Fourier transform of $X(u)\mu(t)$ is called the *analytic signal*, $\hat{x}(t)$. That is

$$\hat{x}(t) \leftrightarrow X(u)\mu(u). \tag{2.140}$$

(a) What is the analytic signal when $x(t) = \sin(2\pi f_0 t)$ where f_0 is a given frequency?
(b) Express the analytic signal in terms of $x(t)$ and its Hilbert transform.
(c) What is the connection between the analytic signal and the Hilbert transform relationship between the real and imaginary components of the Fourier transform of a causal signal as discussed in Section 2.3.6.4?

2.15. (a) Evaluate the Fourier transform of

$$x(t) = |\text{sinc}(t)|\mu(t)$$

Hint: $|\text{sinc}(t)|\mu(t) = \text{sinc}(t) \times$ a square wave $\times \mu(t)$. The following Hilbert transform may prove useful.

$$\mathcal{H}\Pi(t) = \frac{-1}{\pi t} * \Pi(t) = \ln \left| \frac{t - \frac{1}{2}}{t + \frac{1}{2}} \right|. \tag{2.141}$$

(b) Are the real and imaginary parts of this causal signal Hilbert transform pairs? If not, why?

2.16. Define the Dirac delta, $\delta(t)$, as the limit of a
 (a) triangle function.   (c) sinc squared.
 (b) jinc.   (d) Gaussian

2.17. The function $X(u) = u\delta(u)$, when evaluated at $u = 0$, gives $0 \times \infty$. What is $X(u) = u\,\delta(u)|_{u=0}$?

2.18. **Bilateral Laplace transform.** The *bilateral Laplace transform* of a function $x(t)$ is defined as

$$X_L(s) := \int_{-\infty}^{\infty} x(t)e^{-st}\,dt.$$

(a) Show that a sufficient condition for $X(u) = X_L(j2\pi u)$ is that $x(t)$ has finite area.
(b) For $\Re\left(v + \frac{1}{2}\right) > 0$, the Laplace transform of

$$(1 - t^2)^{n - \frac{1}{2}} \Pi\left(\frac{t}{2}\right)$$

is [544]

$$X_L(s) = \left(\frac{2}{s}\right)^v \Gamma\left(v + \frac{1}{2}\right) \Gamma\left(\frac{1}{2}\right) I_n(s) \tag{2.142}$$

where $I_v(s)$ is a modified Bessel function of the first kind defined in (2.75). Can we use the Laplace transform in (2.142) to evaluate the Fourier transform of $x(t)$ by replacing s with $j2\pi u$?

2.19. Let $x(t)$ be causal.
  (a) What are the even and odd function components of $x(t)$?
  (b) Specify a function that is both causal and even. Causal and odd.
2.20. Is an odd function evaluated at the origin *always* zero?
2.21. Prove the properties of even and odd functions listed in Table 2.1.
2.22. (a) When $x(t)$ is causal, show that the inverse of the unilateral cosine transform in (2.120) is

$$x(t) = \left[2 \int_0^\infty X(u) \cos(2\pi ut) du \right] \mu(t).$$

  (b) The bilateral cosine transform is

$$\hat{X}_{\cos}(u) = 2 \int_{-\infty}^\infty x(t) \cos(2\pi ut) dt.$$

   Show this transform cannot be uniquely inverted.
  (c) Repeat (a) and (b) for the unilateral sin transform in (2.124) and its inverse

$$x(t) = 2 \int_0^\infty X_{\sin}(u) \sin(2\pi ut) du, \qquad (2.143)$$

   and the bilateral sin transform.
  (d) Show that the unilateral cosine and sin transforms are Hilbert transform pairs.
2.23. Derive the entries in Table 2.4 for
  (a) $\delta(t)$. (d) $\mu(t)$.
  (b) $\Pi(t)$. (e) $\text{jinc}(t)$.
  (c) $\Lambda(t)$.
2.24. The derivative of a discontinuity is a Dirac delta.
  (a) Using the derivative theorem and the Fourier transform of $\text{sgn}(t)$, derive the Fourier transform of $\frac{d}{dt}\text{sgn}(t)$.
  (b) Do the same for the unit step function.
2.25. Does the Dirac delta have (a) finite area? (b) finite energy?
2.26. **The array function**
  (a) Derive the Fourier transform pair in Equation (2.62).
  (b) Show that, when $N = QM$,

$$\text{array}_N(t) = \text{array}_Q(Mt) \times \text{array}_M(t)$$
$$= \text{array}_Q(t) \times \text{array}_M(Qt)$$

  (c) Show that the extrema of $\text{array}_{2N+1}(t)$ occur when t is a root of the equation

$$\frac{\text{array}_{2N+1}(t)}{\text{array}_{2N+1}\left(t + \frac{1}{2}\right)} = (-1)^N (2N+1).$$

  (d) **Pointwise versus mean square convergence.** Convergence in (2.61) is said to be *pointwise*. For any specified value of t, a value of M can be found to make the array function arbitrarily close to the sinc. A function $x_M(t)$ is said to converge to $y(t)$.

- pointwise if
$$\lim_{M\to\infty} x_M(t) = y(t),$$
- in the mean square sense if
$$\lim_{M\to\infty} \int |y(t) - x_M(t)|^2 dt = 0.$$

(i) Does $\operatorname{array}_M\left(\frac{t}{M}\right)$ converge to $\operatorname{sinc}(t)$ in the mean square sense?
(ii) Does mean square convergence imply pointwise convergence or visa versa?
(iii) Does $\lim_{M\to\infty} e^{-\pi(t-M)^2}$ converge to zero pointwise? In the mean square sense?

2.27. Since
$$e^{-\alpha t}\mu(t) \longleftrightarrow \frac{1}{\alpha + j2\pi u},$$
we can interpret α as a convergence factor and
$$\lim_{\alpha\to 0} e^{-\alpha t}\mu(t) = \mu(t) \longleftrightarrow \lim_{\alpha\to 0} \frac{1}{\alpha + j2\pi u} = \frac{1}{j2\pi u}. \quad (2.144)$$

But, according to (2.48), this is the Fourier transform of $\operatorname{sgn}(t)$. Rather, from (2.50),
$$\mu(t) \longleftrightarrow \frac{1}{2}\left[\delta(t) - \frac{j}{2\pi u}\right]. \quad (2.145)$$

There is a contradiction between (2.144) and (2.145). What is wrong here?

2.28. **Dirac deltas with functional arguments.** Assume $g(t)$ has a zero crossing at $t = t_0$, i.e., $g(t_0) = 0$. Let $\frac{dg(t_0)}{dt} = a$.
 (a) Simplify the expression for $\delta(g(t))$.
 (b) Evaluate $\delta(\ln(t/b))$ for both t and b positive.
 (c) Evaluate $\pi\delta(\sin(\pi t))$.
 (d) Evaluate $\delta(at^2 + bt + c)$ where a, b and c are real coefficients.

2.29. Evaluate the following transforms listed in Table 2.5.
 (a) $\operatorname{sinc}(t)$.
 (b) $\tanh(\pi t)$.
 (c) $\operatorname{sech}^2(\pi t)$.
 (d) $\operatorname{Si}(\pi t)$.
 (e) $\sum_{n=0}^{\infty} \delta(t-n)$

2.30. Find the area of
 (a) $\operatorname{sinc}^4(t)$.
 (b) $\operatorname{jinc}^2(t)$.
 (c) $J_0^2(2\pi t)$.

2.31. Evaluate
 (a) $\int_{-\infty}^{\infty} j_n(2\pi t)dt$.
 (b) $\int_{-\infty}^{\infty} j_n(2\pi t)j_m(2\pi t)dt$.
 (c) $\int_{-1}^{1} \frac{T_n(t)}{\sqrt{1-t^2}} dt$.
 (d) $\sum_{n=-\infty}^{\infty} j_n(2\pi(t-nT))$ for $T < 1$.

2.32. Evaluate the series $\sum_{n=-\infty}^{\infty} a_n$ when $a_n =$

(a) $\text{sinc}(t-n)$.
(b) $\text{sinc}^2(t-n)$.
(c) $\text{jinc}(t-n)$.
(d) $\text{sinc}(a(t-n))$.
(e) $\text{jinc}(2n)$.
(f) $\text{jinc}^2(\frac{n}{4})$.
(g) $n \exp[-\pi(nT)^2]$.

2.33. (a) Find the first few extrema of $\text{sinc}(t)$.
(b) For large t, find a good approximation for the extrema of $\text{sinc}(t)$.

2.34. For a continuous time periodic function, $x(t)$, with period T, the function $x_T(t)$ in (2.16) is a function of τ. Thus, so is its Fourier transform, $X_T(u)$. Show, however, that the sampled transform, $X_T\left(\frac{n}{T}\right)$, is not a function of τ.

2.35. As illustrated in Figure 2.30, each period of a periodic function, $x(t)$, is divided into three sections. We extract three intervals, T_1, T_2 and T_3 from different periods but do so in such a manner that, together, covers an entire period. Show that the Fourier series coefficients can be written as

$$c_n = \frac{1}{T}\left(\int_{T_1} + \int_{T_2} + \int_{T_3}\right) x(t) e^{-j2\pi nu/T}\, dt$$

where T is the signal's period.

2.36. Specify a periodic function that violates all three Dirichlet conditions.

2.37. For a period of $T = 1$, does $x(t) = \log(t)$ satisfy the Dirichlet conditions? If so, graph the convergence of the Fourier series for the first few terms.

2.38. (a) Evaluate the Fourier series of the periodic function

$$x(t) = e^{z\cos(t)} \qquad (2.146)$$

Hint: Use (2.75).

(b) Evaluate the Fourier series of

$$x(t) = e^{-jz\cos(t)}. \qquad (2.147)$$

Hint: Use (2.76).

2.39. **Frequency Modulation (FM).** A bounded signal,

$$|x(t)| \leq 1, \qquad (2.148)$$

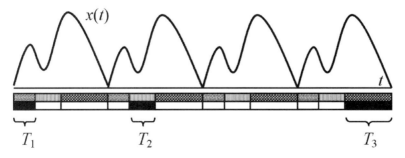

FIGURE 2.30. Illustration of a periodic function for Exercise 2.35.

is frequency modulated[34] on a carrier of frequency of f_0 Hertz as

$$y(t) = A\cos\left(2\pi \int_0^t [f_0 + f_\Delta x(\tau)d\tau]\right)$$

where $f_\Delta \ll f_0$ is the *frequency deviation*. The function

$$f(t) = f_0 + f_\Delta x(t)$$

is the *instantaneous frequency*. From (2.148), one might then expect $y(t)$ to be constrained to the frequency band $f_0 - f_\Delta \leq u \leq f_0 - f_\Delta$. This is not true.

(a) Compute the Fourier transform of $y(t)$ when the signal is a pure tone.[35]

$$x(t) = \cos(2\pi \nu t)$$

where ν is a baseband (audio) frequency.

(b) Show that $y(t)$ in part (a) is not a bandlimited signal.

(c) What is the bandwidth of $y(t)$ as defined as the 3 dB down points of the spectrum as measured from its maximum?

2.40. At a finite discontinuity, a Fourier series converges to the arithmetic midpoint. To illustrate, let

$$y(t) = \sum_{n=-\infty}^{\infty} \Pi\left(\frac{t - nT}{\tau}\right).$$

Express $y(t)$ as a Fourier series and evaluate $y(\frac{\tau}{2})$ using that series. Let $\alpha = \tau/T < 1$.

2.41. Evaluate the Fourier transform of

$$\frac{J_\nu(2\pi t)}{(2t)^\nu} \; ; \; \nu > -\frac{1}{2}.$$

Simplify your result for $\nu = 0$ and $\nu = 1$ thus deriving two of the entries in Table 2.4.

2.42. (a) Derive the recurrence relation in (2.89) for Hermite polynomials as defined in (2.88).

(b) Show the following recurrence relationship for Hermite polynomials

$$H_n'(t) = 2nH_{n-1}(t).$$

(c) Derive the Fourier transform in (2.90).

2.43. A function, $x(t)$, is zero for $|t| > 1$ and can be expressed as a Legendre polynomial expansion.

$$x(t) = \sum_{k=0}^{\infty} \alpha_k P_k(t) \Pi\left(\frac{t}{2}\right). \tag{2.149}$$

---

34. Amplitude modulation (AM) is the topic of Section 3.4.
35. Hint: See Exercise 2.38.

(a) Find the coefficients, α_k, for the Legendre polynomial expansion

(b) The Fourier series of this function, using (2.13), is

$$x(t) = \sum_{n=-\infty}^{\infty} c_n e^{j\pi n t} \Pi\left(\frac{t}{2}\right).$$

Express the Fourier coefficients, c_n, in terms of the Legendre polynomial coefficients, α_k. Your answers in (b) and (c) should be in terms of spherical Bessel functions.

(c) Express the Legendre polynomial coefficients, α_k, in terms of the Fourier coefficients, c_n.

(d) Substitute your answer in (b) into (c) to find an orthogonality expression for spherical Bessel function samples, $j_k(\pi n)$, with respect to k.

(e) Substitute your answer in (c) into (b) to find an orthogonality expression for spherical Bessel function samples, $j_k(\pi n)$, with respect to n.

2.44. Repeat Exercise 2.43 for Chebyshev polynomial expansions on $[-1,1]$.

$$x(t) = \sum_{k=0}^{\infty} \beta_k T_k(t) \Pi\left(\frac{t}{2}\right).$$

Instead of spherical Bessel functions, your answer will be in terms of nth order Bessel functions of the first kind.

2.45. For the Chebyshev polynomials, establish
 (a) the orthogonality condition in (2.86).
 (b) the Fourier transform pair in (2.87).

2.46. (a) Show that Legendre polynomials are evenly odd.

$$P_n(t) = (-1)^n P_n(-t).$$

 (b) Are Hermite polynomials evenly odd?
 (c) Chebyshev polynomials?
 (d) Spherical Bessel functions?
 (e) Bessel functions $J_n(t)$?

2.47. **Chebyshev polynomials of the second kind**, for $t = \cos\theta$, are defined as [5]

$$U_n(\cos\theta) := (n+1) \text{array}_{n+1}\left(\frac{\theta}{\pi}\right). \qquad (2.150)$$

(a) Evaluate $U_n(x)$ for $n = 0$ and $n = 1$.
(b) Establish the trig identity

$$\sin(n+2)\theta = 2\cos\theta \sin(n+1)\theta - \sin n\theta. \qquad (2.151)$$

(c) Use the identity in (2.151) to establish the recurrence relationship

$$U_{n+1}(t) = 2t U_n(t) - U_{n-1}(t).$$

(d) Using the recurrence relationship, plot the first few Chebyshev polynomials of the second kind on $[-1,1]$.

2.48. Show that

$$J_m(t) = \sum_{n=0}^{\infty} \frac{(-1)^n \left(\frac{t}{2}\right)^{2n+m}}{n!\,(m+n)!}. \qquad (2.152)$$

2.49. Show that, if $x(t)$ is a real periodic signal, then the corresponding Fourier coefficients are conjugately symmetric.

$$c_n = c_{-n}^*.$$

2.50. Show that the sincs in (2.110) are, indeed, orthogonal and that the expansion coefficients are given by the samples in (2.111).

2.51. **Gibb's phenomenon.** From the truncated Fourier series example in Figure 2.1, we see that the overshoot at discontinuities becomes smaller in duration, but does not disappear. This property is known as *Gibb's phenomenon*. To show this, from Exercise 2.40, let $\alpha = \frac{1}{2}$ and

$$z(t) = 2y\left(t - \frac{\tau}{2}\right) - 1.$$

The truncated Fourier series for $z(t)$ is $z_N(t)$ which will display overshoot, Δ_N, as shown in Figure 2.31. Compute this overshoot and its exact value as $N \to \infty$. Hint: $\mathrm{Si}(\pi) = 1.851937$ where the sine integral is defined in (2.77).

2.52. Show that

$$\mathrm{comb}\left(t - \frac{1}{2}\right) \leftrightarrow \frac{1}{2}\left[\mathrm{comb}\left(\frac{u}{2}\right) - \mathrm{comb}\left(\frac{u-1}{2}\right)\right].$$

2.53. **The generalized comb function.** [1022]. Let $h(t)$ be any function that integrates to A, so

$$\int_{-\infty}^{\infty} h(t)\,dt = A.$$

Define the generalized comb function

$$\mathrm{comb}_h(t) := \sum_{n=-\infty}^{\infty} \frac{1}{|n|} h\left(\frac{t}{n}\right)$$

where we assume the $n = 0$ term in the sum is $A\delta(t)$.

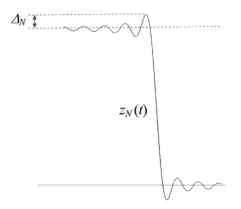

FIGURE 2.31. Illustration of Gibb's phenomenon. See Exercise 2.51.

(a) Show that

$$\text{comb}_h(t) \longleftrightarrow \text{comb}_K(u) \quad (2.153)$$

where

$$K(u) = \frac{1}{|u|} h\left(\frac{1}{u}\right). \quad (2.154)$$

Hint: First, show that

$$\text{comb}_h(t) = \int_{\xi=-\infty}^{\infty} h(\xi) \left[\frac{1}{|\xi|} \text{comb}\left(\frac{t}{\xi}\right)\right] d\xi. \quad (2.155)$$

(b) Identify a function for which the generalized comb function becomes a regular comb function.

2.54. (a) Let $x(t) = t^{-\frac{3}{4}}$. Compute both the area and energy of $x(t)$
  (i) over the interval $(0, 1)$ and,
  (ii) over the interval $(1, \infty)$.
(b) Use your result in (a) to show that finite area does not imply finite energy and vice versa.
(c) If a continuous time signal's area is finite, can we conclude its energy is also finite? What about the converse?

2.55. (a) If $x(t)$ has a spectrum with finite area, show that $x(t)$ is bounded. Let

$$|x(t)| \leq C < \infty.$$

Express C in terms of

$$A = \int_{-\infty}^{\infty} |X(u)| du < \infty.$$

(b) Show that the converse of (a) is not true. That is, there exist bounded signals whose spectra do not have finite area.

2.56. Show that a finite energy bandlimited signal, $x(t)$, must be bounded. Specifically,

$$|x(t)|^2 \leq 2BE.$$

Hint: Apply Schwarz's inequality[36] to the inversion formula.

2.57. (a) A finite energy signal $x(t)$ is bandlimited. Is its pth derivative also a finite energy bandlimited function?
(b) Repeat part (a) substituting the word "area" for "energy".
(c) Substitute "bounded".

2.58. (a) Apply Schwarz's inequality[37] to the derivative theorem to show that, if $x(t)$ is bandlimited, its Mth derivative is bounded as

$$|x^{(M)}(t)|^2 \leq \frac{(2\pi B)^{2M+1} E}{2M+1}. \quad (2.156)$$

---

36. See Appendix 14.1.
37. See Appendix 14.1.

(b) Show that a bandlimited function is smooth in the sense that

$$|x(t+\tau) - x(t)|^2 \leq \frac{(2\pi B)^3 |\tau|^2 E}{3}. \tag{2.157}$$

2.59. Prove (2.103).

2.60. Consider the general linear integral transform

$$g(t) = \int_{-\infty}^{\infty} f(\tau) h(t, \tau) d\tau.$$

For a given t, assume that the kernel, $h(t, \tau)$, as a function of τ, has finite energy,

$$E_h(t) = \int_{-\infty}^{\infty} |h(t, \tau)|^2 d\tau.$$

If $f(t)$ has finite energy, show that $g(t)$ is bounded.

2.61. Prove the inverse DFT in (2.113) inverts.

2.62. Prove the inverse DCT in (2.117) correctly inverts the DCT in (2.121).

2.63. Prove that (2.128) inverts the Hartley transform.

2.64. Prove that relation between the Hartley and Fourier transforms in (2.127).

2.9 Solutions for Selected Chapter 2 Exercises

2.3. Although there are many parallels between continuous and discrete time signals, there are also important differences. One of these differences is illustrated here. Differentiation is not defined for discrete time systems. Although derivatives can be approximated by, say, finite differences, the derivative,

$$\frac{dx(t)}{dt} = \lim_{\Delta t \to \infty} \frac{x(t) - x(t - \Delta t)}{\Delta t}.$$

is simply meaningless when applied to discrete time signals. The foundational premise for this problem, differentiation of a discrete time signal, is therefore faulty.

2.4. **Absolutely convergent sums I.** The series converges absolutely if

$$S(t) \equiv \sum_{m=0}^{\infty} \frac{|t - \tau|^m}{m!} |x^{(m)}(\tau)| < \infty.$$

From the derivative theorem for Fourier transforms,

$$S(t) = \sum_{m=0}^{\infty} \frac{|t-\tau|^m}{m!} \left| \int_{-B}^{B} (j 2\pi u)^m X(u) e^{j 2\pi ut} du \right|$$

$$\leq \sum_{m=0}^{\infty} \frac{|t-\tau|^m}{m!} \left[\int_{-B}^{B} (2\pi u)^{2m} du \right]^{1/2}$$

$$\times \left[\int_{-B}^{B} |X(u)|^2 du \right]^{1/2}$$

where, in the second step, we have used Schwarz's inequality.[38] Since $x(t)$ has finite energy,

$$E = \int_{-B}^{B} |X(u)|^2 \, du$$

is finite. Thus

$$S(t) \leq \sqrt{2B} \, E \sum_{m=0}^{\infty} \frac{(2\pi B \, |t-\tau|^2)^m}{m! \, \sqrt{2m+1}}$$

$$< \sqrt{2B} \, E \sum_{m=0}^{\infty} \frac{(2\pi B \, |t-\tau|)^m}{m!}$$

$$= \sqrt{2B} \, E \, e^{2\pi B |t-\tau|}.$$

where we have used the Taylor series expansion for e^z. This bound is finite for all finite t and τ.

2.5. **Absolutely convergent sums II.** Terms in a sum can be shuffled if the sum is an *absolutely convergent sum*.
   (a) For $Z = \exp(j\theta)$ in (2.34),

$$e^{j\theta} = \sum_{n=0}^{\infty} \frac{(j\theta)^n}{n!}.$$

It follows that the sum of the corresponding absolute values is

$$\sum_{n=0}^{\infty} \frac{|(j\theta)^n|}{n!} = \sum_{n=0}^{\infty} \frac{|\theta^n|}{n!}$$

$$= e^{\theta}.$$

   (b) Since

$$\sum_{n=1}^{\infty} \frac{1}{n} = \infty,$$

the series is not absolutely convergent.

2.6. **The Pythagorean theorem.** One straightforward geometrical proof of the Pythagorean theorem is shown in Figure 2.32. A $c \times c$ square is placed inside an $(a+b) \times (a+b)$ square. The big square's area can be computed by adding the area of the small square plus the area of the four small triangles. The result is

$$(a+b)^2 = c^2 + 4\left(\frac{1}{2}ab\right).$$

Solving gives the Pythagorean theorem: $c^2 = a^2 + b^2$.

2.7. (a) A number ending in Z zeros is divisible by 10^Z. In computing a factorial, a factor of 10 occurs every time factors of 5 and 2 occur. The number of positive

---

38. See Appendix 14.1.

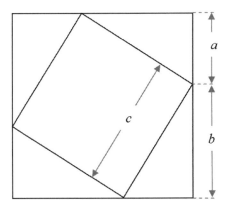

FIGURE 2.32. Geometry to illustrate the Pythagorean theorem.

integers divisible by 5 that are less than N is about $\frac{N}{5}$ when N is large. Since the density of numbers divisible by two is much larger (50%), we conclude that every time we pass a number divisible by 5, we multiply the emerging factorial by 10 and add another zero to the end of the factorial. Every time we hit numbers like $25 = 5^2$ and 50 that are divisible by 5 *twice*, we add two more zeros to the factorial. When $125 = 5^3$, 250 etc. are passed, three zeros are added, and so forth. For $N = 5^n$, the count of numbers divisible by 5 is 5^{n-1}, by 25 is 5^{n-2}, by 125 = 5^3 is 5^{n-3}, etc. The total number of zeros is thus given by the geometric series[39]

$$\sum_{k=0}^{n-1} 5^k = \frac{5^n - 1}{4}.$$

(b) The number of integers below N divisible by 5 is $\lfloor \frac{N}{5} \rfloor$. The number of integers below N divisible by 5^2 is $\lfloor \frac{N}{5^2} \rfloor$, etc. The biggest value of n for which $5^n \leq N$ is $\lfloor \log_5 N \rfloor$. The number of 5's in the factorizations of $N!$ is therefore given by (2.138).

(c) As an example, let $N = 2000$. Then

$$Z[2000] = \left\lfloor \frac{2000}{5} \right\rfloor + \left\lfloor \frac{2000}{25} \right\rfloor + \left\lfloor \frac{2000}{125} \right\rfloor + \left\lfloor \frac{2000}{625} \right\rfloor$$
$$= 400 + 80 + 16 + 3 \qquad (2.158)$$
$$= 499$$

This is verified in Table 2.7.

(d) $N!!$ never ends in a zero.

(e) Stirling's formula for $n = 2000$ is

$$\log_{10} n! = \left(\tfrac{1}{2} \log(2\pi) + \left(n + \tfrac{1}{2}\right) \log n - n \right) / \log 10$$
$$= 5{,}735.5206.$$

---

39. See Appendix 14.4.2.

TABLE 2.7. Tabulation of 2000! As predicted in (2.158), 2000! ends in 499 zeros.

2000! = 33
16275092450633241175393380576324038281117208105780394571935437060380779056008224002732308597 3259225540
23529412258341092580848174152937961313866335263436889056340585561639406051172525718706478563 9354404540
52439574670376741087229704346841583437524315808775336451274879954368592474080324089465615072 3325065279
76557571796715367186893590561128158716017172326571561100042140124204338425737127001758835477 9689992128
35289966658534055798549036573663501333865504011720121526354880382681521522469209952060315644 1856548067
59464970515522882052348999957264508140655366789695321014676226713320268315522051944944616182 3927520402
65297226315025747520482960647509273941658562835317795744828763145964503739913273341772636088 5249009350
66216101444597094127078213137325638315723020199499149583164709427744738703279855496742986088 3937632682
41524788343874695958292577405745398375015858154681362942179499723998135994810165565638760342 2731291225
03847098729096266224619710766059315502018951355831653578714922909167790497022470946119376077 8516511068
44322559056487362665303773846503907880495246007125494026145660722541363027549136715834060978 3107494528
22174907813477096932415561113398280513586006905946199652573107411770815199225645167785714580 5660218565
47609523774630166794224884444857983498015480326208298909658573817518886193766928282798884535 8463989659
42139529844652910920091037100461494499158285880507618679294463851808798745128914080193400746 2592005709
87295785996436506558956124102310186905560603087836291105056012459089988383410799367902052076 85866918347
79065585447001486926569246319333376124280974200671728463619392496986284687199934503938893672 7048712717
27345617003548674775091029555239535479411074219133013568195410919414627664175421615876252628 5808980122
24438902486771820549594157519917012717675717874958616196659318788551418357820926014820717733 173539603
43049690820705899587013819808130355901607629083885745612882176981361824835767392183031184147 1913398689
28423440007792466912097667316514334944374732356365720488444478331854941693030124531676232745 36787932284
74738244850922831399525097325059791270310476836014811911022292533726976938236700575656124002 9057604385
28529029376064795334581796661238396052625491071866638693576766108455046981020840506358276765 2658949239
32495196859541716724193295306836734955440045863598381610430594498266275306054235807558941082 7888042782
59510898806354105679179509740177806887828698102190109001483520616888837202503106659220686014 8364983053
27820882635365580436056867812841692171330471411763121758957771226375847531235172309905498292 1013468730
42058980144180638753826641698977042377594062808772537022654265305808623793014226758211871435 0291863763
63403001732518182620760397473695952026426323641454468511134272021504583838510101369413130348 5622191663
16238926327658153550112763078250599691588245334573454378636831737306732965893551996944582368 7350883027
86577008797498899923435555662406828347637846518384497364887395247510322422211056120129582965 7191136810
86938254757641188868793467251912461921511447388362695916436724900716534282281526612478004639 2254494517
03637236279407577845420910483054616561906221742869816029733240465202019928138548826819510072 82869 70107
07375009276664875021747753727423515087482467202741700315811228058961781221607474379475109506 2093855667
45812525183766821577128078614992558761323529504223463878795485088576446613629039412766597804 420209228
13379871159008962648789424132104549250035666706329094415793729867434214705072135889320195807 230 6478149
84295225955890127548239717733257229103257609297907332995450536836264047465024508080946911607 26 3208749
41439730007041114185955302788273576548191820024496977611113463181952827615909641897909581173 8627 20608
89104329452449785351470141124421430554860896395783783473252359576329143892528839398625627324 286277556
31404638303891684216331134456363095719659784663385514923161963356753551384034258041629198378 2226690952
17701531753387302846108418865541383291719513321178957285416620848236828179325129312375215419 2697026970
32994776438233864830088715303734056663838682940884877307217622688490308493466119426018027261 380210800
50782157410060548482013478595781027707077806555127725405016743323960662532164150048087724030 4761192903
22101543853531368555384864245570790795341176519571188683739880683895792743749683498142923292 196309 77709
01439368436553335930782018131299345502420604456334057860696247196150560339489952332180043435 9967 25662
39271964354028720554750120798543319706747973131268135236537440856622632067688375851327828962 5233328434
18129776246970795434360034923431592396747636389121152854066577836462139112474470512552263427 0123952701
81270454916480459322481088586746009523067931759677555810116799400052498063037631413444122690 3703498735
57999160092592480750524855415682662817608154463083054066774126301244418642041083731190931300 0115447056
02777737243780671888997708510567272767812471988328576958442175888951604678682048100100478164 6235822083
85324881342708340798684866321627202088233087278190853788454691315560217288731219073939652092 6022910147
75270809308653649798585540105774502792898146036884318215086372462169678722821693473705992862 77112 44769
09209029883201668301702734202597656717098633112163495021712644268271196502640542282317596308 7447530184
71940955242634114984695080733900800000000000000000000000000000000000000000000000000000000000 00000000000
0000000000000000000000000000000000000000000000000000000000000000000000000000000000000000000000000000
0000000000000000000000000000000000000000000000000000000000000000000000000000000000000000000000000000
0000000000000000000000000000000000000000000000000000000000000000000000000000000000000000000000000000
0000000000000000000000000000000000000000000000000000000000000000000000000000000000000000000000000000
0000000000000000000000000000000000000000000000000000000000000000000000000000000000000000000000000000
00000000000000000000

The mantissa of the log is 0.5206 and the characteristic is 5,735. Since $10^{0.5206} = 0.3316$, we conclude that $n! \approx 0.3316 \times 10^{5,735}$. Comparing with the correct number in Table 2.7, the approximation is thus very good.

2.9. (a) Since

$$X(0) = \int_{-\infty}^{\infty} x(t)dt,$$

the transforms in Table 2.4 can be used to solve this and the next few problems. For this problem,

$$\int_{-\infty}^{\infty} \text{sinc}(t)dt = \Pi(u)\Big|_{u=0} = 1.$$

(b) Similarly,

$$\int_{-\infty}^{\infty} \text{jinc}(t)dt = \sqrt{1-u^2}\,\Pi(u)\Big|_{u=0} = 1.$$

(c)

$$\int_{-\infty}^{\infty} J_0(2\pi t)dt = \frac{1}{\sqrt{1-u^2}}\Pi(u)\Big|_{u=0} = 1.$$

(d) We conclude from (2.164), that

$$\int_{-\infty}^{\infty} \frac{J_\nu(2\pi t)}{(2t)^\nu}dt = \frac{(\pi/2)^\nu}{\sqrt{\pi}\,\Gamma(\nu + \frac{1}{2})}.$$

(e) From (2.44),

$$\int_{-\infty}^{\infty} \text{sinc}_k(t)dt = \frac{\Pi(u)}{\text{sinc}^k(u)}\Big|_{u=0} = 1.$$

(f) These last problems are easily solved using the power theorem in (2.101).

(g)

$$\int_{-\infty}^{\infty} \text{sinc}(2t)\text{sinc}(t-1)dt = \int_{-\infty}^{\infty} \left[\frac{1}{2}\Pi\left(\frac{u}{2}\right)\right]\left[\Pi(u)e^{-j2\pi u}\right]^* du$$

$$= \frac{1}{2}\int_{-\frac{1}{2}}^{\frac{1}{2}} e^{j2\pi u}du$$

$$= 0.$$

(h) We have the Fourier transform pair

$$1 \leftrightarrow \delta(u).$$

In integral form

$$\int_{-\infty}^{\infty} e^{-j2\pi tv}dt = \delta(v).$$

FUNDAMENTALS OF FOURIER ANALYSIS 77

Taking the real part of both sides gives the desired result.

$$\int_{-\infty}^{\infty} \cos(2\pi tv)dt = \delta(v).$$

2.10. (a)

$$\sin\left(\frac{\pi}{2}\Pi(t)\right) = \begin{cases} \sin\left(\frac{\pi}{2}\right) ; |t| < \frac{1}{2} \\ \sin\left(\frac{\pi}{4}\right) ; |t| = \frac{1}{2} \\ \sin(0) ; |t| > \frac{1}{2} \end{cases}$$

$$= \begin{cases} 1 ; |t| < \frac{1}{2} \\ \frac{1}{\sqrt{2}} ; |t| = \frac{1}{2} \\ 0 ; |t| > \frac{1}{2} \end{cases}$$

$$= \sqrt{\Pi(t)}$$

(c) $\Pi(\Pi(t)) = 1 - \Pi(t)$.
(d) $\text{sgn}(\sin(2\pi t))$ is a square wave.
(e)

$$\sin\left(\frac{\pi}{2}\text{sgn}(t)\right) = \begin{cases} \sin(-\pi/2) = -1 ; t < 0 \\ \sin(0) = 0 ; t = 0 \\ \sin(\pi/2) = 1 ; t > 0 \end{cases}$$

$$= \text{sgn}(t)$$

(f) $\Pi(\Lambda(t)) = 1 - \Pi(t)$.
(g) $\delta[\text{sinc}(n)] = 1 - \delta[n]$.
(h) Since $\text{sinc}(n) = \delta[n]$, this is the same as (g).
(i) $\text{sinc}(\delta[n]) = 1 - \delta[n]$.
(j) $\delta[\text{jinc}(n)] = 0$ because $\text{jinc}(n) \neq 0$ for all n.

2.13 Using the signal integral property in (2.93) and its Fourier dual, the numerators of both sides of (2.139) are equal as are the denominators.

2.14. **Analytic signals.**
(a) Since

$$x(t) = \sin(2\pi f_0 t) \longleftrightarrow X(u) = \frac{1}{j2}(\delta(u - f_0) - \delta(u + f_0)),$$

it follows that $X(u)\mu(u) = \frac{1}{j2}\delta(u - f_0)$. The corresponding analytic signal is thus

$$\hat{x}(t) = \frac{1}{j2} e^{j2\pi f_0 t}.$$

(b) Since $\mu(u) = (\text{sgn}(u) + 1)/2$ it follows that $X(u)\mu(u) = \frac{1}{2}[X(u) + X(u)\text{sgn}(u)]$. Using the Fourier transform of the Hilbert transform in (2.98), we inverse Fourier transform to obtain a general expression for the analytic signal

$$\hat{x}(t) = \frac{1}{2}(x(t) - j\mathcal{H}x(t)).$$

(c) They are Fourier transform duals, i.e., one follows from the other from the duality theorem.[40]

2.15.
$$|\text{sinc}(t)|\mu(t) = \text{sinc}(t)\left[-1 + 2\sum_{k=0}^{\infty} \Pi\left(t - k - \frac{1}{2}\right)\right]\mu(t).$$

Thus
$$|\text{sinc}(t)|\mu(t) \leftrightarrow \Pi(u) * \left[-\delta(u) + 2\,\text{sinc}(u)e^{-j\pi u}\sum_{k=0}^{\infty} e^{-j2\pi ku}\right].$$

The sum over k is recognized as the Fourier transform of the causal comb in (2.56). Thus
$$|\text{sinc}(t)|\mu(t) \leftrightarrow \Pi(u) * \left[-\delta(u) + \text{sinc}(u)\,e^{-j\pi u}(1 - j\cot(\pi u))\right]$$

or, equivalently,
$$|\text{sinc}(t)|\mu(t) \leftrightarrow -\Pi(u) + \Pi(u) * \left[\frac{\sin(\pi u)}{\pi u}\,e^{-j\pi u}\left(1 - j\frac{\cos(\pi u)}{\sin(\pi u)}\right)\right]$$
$$= -\Pi(u) - j\Pi(u) * \left[\frac{1}{\pi u}\,e^{-j\pi u}(\cos(\pi u) + j\sin(\pi u))\right]$$
$$= -\Pi(u) + j\left(\frac{-1}{\pi u}\right) * \Pi(u)$$
$$= -\Pi(u) + j\ln\left(\left|\frac{t - \frac{1}{2}}{t + \frac{1}{2}}\right|\right)$$

where, in the last step, we have used the Hilbert transform in (2.141).

2.16. (a) $\delta(t) = \lim_{A \to \infty} A\Lambda(At)$.
(b) $\delta(t) = \lim_{A \to \infty} A\text{jinc}(At)$.
(c) $\delta(t) = \lim_{A \to \infty} A\text{sinc}^2(At)$.

In each case, the limit of the transform approaches one. For example
$$A\,\text{jinc}(At) \longleftrightarrow \sqrt{1 - (u/A)^2}\,\Pi\left(\frac{u}{2A}\right)$$
$$\longrightarrow 1 \text{ as } A \to \infty.$$

2.20. If $o(t) = -o(-t)$, then $o(0) = -o(0)$ always requires that $o(0) = 0$. So the answer is yes. An odd function evaluated at the origin must always be zero.

2.21. Here we prove the entries in Table 2.1 concerning even and odd functions.
(a) Let $e_3(t) = e_1(t) + e_2(t)$. Then
$$e_2(-t) = e_1(-t) + e_2(-t) = e_1(t) + e_2(t) = e_3(t).$$

---

40. See Section 2.3.4.4.

(b) Let $o_3(t) = o_1(t) + o_2(t)$. Then
$$-o_3(-t) = -[o_1(-t) + o_2(-t)] = -o_1(-t) - o_2(-t)$$
$$= o_1(t) + o_2(t) = o_3(t).$$

(c) Let $e_3(t) = e_1(t) e_2(t)$. Then
$$e_3(-t) = e_1(-t) e_2(-t) = e_1(t) e_2(t) = e_3(t).$$

(d) Let $e_3(t) = o_1(t) o_2(t)$. Then
$$e_3(-t) = o_1(-t) o_2(-t) = [-o_1(-t) (-o_2(-t))]$$
$$= -[o_1(t) o_2(t)] = e_3(t).$$

(e) Let $o_3(t) = o_1(t) e_2(t)$. Then
$$-o_3(-t) = -o_1(-t) e_2(-t) = o_1(t) e_2(t) = o_3(t).$$

(f) Let $o(t) = \frac{d}{dt} e(t)$. Then
$$-o(-t) = -\frac{d}{d(-t)} e(-t) = \frac{d}{dt} e(t) = o(t).$$

(g) Let $e(t) = \frac{d}{dt} o(t)$. Then
$$e(-t) = \frac{d}{d(-t)} o(-t) = \frac{d}{dt} o(t) = e(t).$$

(h) Let $e(t) = \int_0^t o(\tau) \, d\tau$. Then
$$e(-t) = \int_0^{-t} o(\tau) \, d\tau = -\int_0^t o(-\tau) \, d\tau = \int_0^t o(\tau) \, d\tau = e(t).$$

(i) Let $o(t) = \int_0^t e(\tau) \, d\tau$. Then
$$o(-t) = \int_0^{-t} e(\tau) \, d\tau = -\int_0^t e(-\tau) \, d\tau = \int_0^t e(\tau) \, d\tau = o(t).$$

(j) Let $z = \int_{-T}^T e(t) \, dt$. Then
$$z = \left[\int_{-T}^0 + \int_0^T\right] e(t) \, dt = -\int_0^T e(-t)(-dt) + \int_0^T e(t) dt$$
$$= \int_0^T e(t) dt + \int_0^T e(t) dt = 2 \int_0^T e(t) dt.$$

(k) Let $z = \int_{-T}^T o(t) \, dt$. Then
$$z = \left[\int_{-T}^0 + \int_0^T\right] o(t) \, dt = -\int_0^T o(-t)(-dt) + \int_0^T o(t) dt$$
$$= -\int_0^T o(t) dt + \int_0^T o(t) dt = 0.$$

(l) Let $o_1[n] = \sum_{m=0}^{n} o[m]$. Then

$$-o_1[-n] = -\sum_{m=0}^{-n} o[m] = -\sum_{p=0}^{n} -o[-p] = \sum_{p=0}^{n} o[p] = o_1[n].$$

(m) Let $e_1[n] = \sum_{m=0}^{n} e[m]$. Then

$$e_1[-n] = \sum_{m=0}^{-n} e[m] = \sum_{p=0}^{n} e[-p] = \sum_{p=0}^{n} e[p] = e_1[n].$$

(n) Let $z = \sum_{n=-N}^{N} e[n]$. Then

$$z = \sum_{n=-N}^{-1} e[n] + e[0] + \sum_{n=1}^{N} e[n] = \sum_{n=1}^{N} e[-n] + e[0] + \sum_{n=1}^{N} e[n]$$

$$= \sum_{n=1}^{N} e[n] + e[0] + \sum_{n=1}^{N} e[n] = e[0] + 2\sum_{n=1}^{N} e[n].$$

(o) Let $z = \sum_{n=-N}^{N} o[n]$. Then

$$z = \sum_{n=-N}^{-1} o[n] + o[0] + \sum_{n=1}^{N} o[n] = \sum_{n=1}^{N} o[-n] + e[0] + \sum_{n=1}^{N} e[n]$$

$$= \sum_{n=1}^{N} -o[n] + e[0] + \sum_{n=1}^{N} e[n] = 0.$$

(p) Let $e_3 = e_2 * e_1$. We consider the continuous case where

$$e_3(t) = \int_{-\infty}^{\infty} e_2(\tau) e_1(t-\tau) d\tau.$$

Thus

$$e_3(-t) = \int_{\tau=-\infty}^{\infty} e_2(\tau) e_1(-t-\tau) d\tau.$$

Set $\xi = \tau$ and

$$e_3(-t) = \int_{\xi=-\infty}^{\infty} e_2(-\xi) e_1(-t+\xi) d\xi$$

$$= \int_{\xi=-\infty}^{\infty} e_2(\xi) e_1(t-\xi) d\xi = e_3(t).$$

(q) Let $o_3 = o_1 * o_2$. We consider the continuous case where

$$o_3(t) = \int_{-\infty}^{\infty} o_1(\tau) o_2(t-\tau) d\tau.$$

Thus

$$o_3(-t) = \int_{\tau=-\infty}^{\infty} o_1(\tau)o_2(-t-\tau)d\tau.$$

Set $\xi = \tau$ and

$$o_3(-t) = \int_{\xi=-\infty}^{\infty} o_1(-\xi)o_2(-t+\xi)d\xi$$

$$= \int_{\xi=-\infty}^{\infty} (-o_1(\xi))(-o_2(t-\xi))\,d\xi$$

$$= -o_3(t)$$

(r) Let $e_1 = e * o$. For discrete time convolution.

$$e_1[n] = \sum_{k=-\infty}^{\infty} e[k]\,o[n-k].$$

Thus

$$e_1[-n] = \sum_{k=-\infty}^{\infty} e[k]\,o[-n-k].$$

Set $p = -k$ and

$$e_1[-n] = \sum_{p=-\infty}^{\infty} e[-p]\,o[-n+p]$$

$$= \sum_{p=-\infty}^{\infty} e[p]\,(-o[n-p])$$

$$= -e_1[n].$$

(s) For the continuous case, we use the continuous time Fourier transform in (2.11) and write

$$e(t) \longleftrightarrow E(u)$$

$$= \int_{-\infty}^{\infty} e(t)e^{-j2\pi ut}\,dt$$

$$= \int_{-\infty}^{\infty} e(t)\,(\cos(2\pi ut) + j\,\sin(2\pi ut))\,dt.$$

Since $\sin(2\pi ut)$ is an odd function of t and $e(t)$ is even, from (e), $e(t)\,\sin(2\pi ut)$ is odd and, using (k),

$$\int_{-\infty}^{\infty} e(t)\,\sin(2\pi ut)dt = 0$$

and

$$E(u) = \int_{-\infty}^{\infty} e(t)\,\cos(2\pi ut)dt.$$

Thus, setting $\xi = -t$,

$$E(-u) = \int_{t=-\infty}^{\infty} e(t) \cos(2\pi ut) dt$$

$$= \int_{\xi=-\infty}^{\infty} e(-\xi) \cos(2\pi u\xi) dt$$

$$= \int_{\xi=-\infty}^{\infty} e(\xi) \cos(2\pi u\xi) dt$$

$$= E(-u).$$

Thus, the Fourier transform of a possibly complex even function is an even function.

(t) For the discrete case, using the DTFT in (2.112), let

$$o[n] \longleftrightarrow \Theta(f)$$

be DTFT pairs. Then

$$\Theta(f) = \sum_{k=-\infty}^{\infty} o[k] e^{-j2\pi f k} dt$$

$$= \sum_{k=-\infty}^{\infty} o[k] \left(\cos(2\pi f k) - j \sin(2\pi f k) \right).$$

Since $\cos(2\pi f k)$ is an even function of t and $o[k]$ is odd, from (e), $o[k] \cos(2\pi f k)$ is odd and, using (o),

$$\sum_{k=-\infty}^{\infty} o[n] \cos(2\pi f k) = 0$$

and

$$\Theta(f) = -j \sum_{k=-\infty}^{\infty} o[n] \sin(2\pi ut) dt.$$

Thus, setting $p = -k$,

$$\Theta(-f) = \sum_{k=-\infty}^{\infty} -j\, o[k] \sin(2\pi f k)$$

$$= \sum_{p=-\infty}^{\infty} -j\, o[-p] \sin(2\pi f(-p))$$

$$= \sum_{p=-\infty}^{\infty} -j\, (-o[p])(-\sin(2\pi f k))$$

$$= -j\, \Theta(f).$$

Thus, the Fourier transform of a possibly complex odd function is an odd function.

2.22. (a) Using the inverse Fourier transform of a Dirac delta, we can write

$$\delta(u) = \int_{-\infty}^{\infty} e^{j2\pi ut} dt$$

$$= \lim_{\alpha \to 0} \int_{-\infty}^{\infty} e^{-\alpha|t|} e^{j2\pi ut} dt \quad ; \quad \text{convergence factor}$$

$$= \lim_{\alpha \to 0} \int_{-\infty}^{\infty} e^{-\alpha|t|} \cos(2\pi ut) dt \quad ; \quad \text{eliminate odd integrand}$$

$$= \lim_{\alpha \to 0} 2 \int_{0}^{\infty} e^{-\alpha|t|} \cos(2\pi ut) dt \quad ; \quad \text{even integrand}$$

$$= 2 \int_{0}^{\infty} \cos(2\pi ut) dt.$$

Substituting the proposed inversion into the forward cosine transform gives

$$x(t) \stackrel{?}{=} \left[2 \int_{0}^{\infty} X(u) \cos(2\pi ut) du \right] \mu(t)$$

$$= \left[2 \int_{u=0}^{\infty} \left[2 \int_{\tau=0}^{\infty} x(\tau) \cos(2\pi u\tau) d\tau \right] \cos(2\pi ut) du \right] \mu(t)$$

$$= \left[4 \int_{\tau=0}^{\infty} x(\tau) \left[\int_{u=0}^{\infty} \cos(2\pi u\tau) \cos(2\pi ut) du \right] dt \right] \mu(t)$$

$$= \left[2 \int_{\tau=0}^{\infty} x(\tau) \left[\int_{u=0}^{\infty} (\cos(2\pi u(t-\tau)) + \cos(2\pi u(t+\tau))) du \right] dt \right] \mu(t)$$

$$= \left[\int_{\tau=0}^{\infty} x(\tau) [\delta(t-\tau) + \delta(t+\tau)] dt \right] \mu(t)$$

$$= [x(t) + x(-t)] \mu(t)$$

$$= x(t).$$

(b) Let $z_o(t)$ be any odd function of t. Then, since the bilateral cosine transform of $z_o(t)$ is zero, we see that the bilateral cosine transform of any $x(t)$ is the same as that of $x(t) + z_o(t)$. Many functions therefore map to $X_{\cos}(u)$ which cannot be uniquely inverted.

2.24. (a)

$$\frac{d}{dt} \text{sgn}(t) = 2\delta(t) \leftrightarrow (j2\pi u) \frac{1}{j\pi u} = 2.$$

(b) Since

$$u\delta(u) = 0,$$

we conclude that

$$\frac{d}{dt} \mu(t) = \delta(t) \leftrightarrow (j2\pi u) \times \frac{1}{2} \left[\delta(t) + \frac{1}{j\pi u} \right] = 1.$$

2.25. (a) The Dirac delta has finite area since

$$\int_{-\infty}^{\infty} \delta(t) = 1 < \infty.$$

(b) The Dirac delta has infinite energy. From the sifting property in (2.27),

$$\int_{-\infty}^{\infty} \delta(\tau)\delta(t-\tau)\,d\tau = \delta(t).$$

Thus,[41] for $t = 0$,

$$\int_{-\infty}^{\infty} \delta^2(\tau)\,d\tau = \delta(0) = \infty.$$

2.28. (a) $g(t) \approx a(t - t_0)$ around the neighborhood $t = t_0$ where $a = \frac{dg(t_0)}{dt}$. Since $\delta(t)$ is nonzero only when $\xi = 0$, we conclude

$$\delta(g(t)) = \delta(a(t - t_0))$$
$$= \frac{1}{|a|}\delta(t - t_0).$$

(b)

$$g(t) = \ln(t/b); \quad t_0 = b$$
$$g'(t) = 1/t \longrightarrow a = 1/b$$

and

$$\delta(\ln(t/b)) = b\delta(t - b).$$

(c) The argument has an infinite number of zero crossings:

$$\sin(\pi t_n) = 0; \quad t_n = 0, \pm 1, \pm 2, \ldots.$$

Since

$$g'(t_n) = \pi \cos(\pi t_n)$$
$$= (-1)^n \pi$$
$$= a_n,$$

we have

$$\pi \delta(\sin(\pi t)) = \pi \sum_{n=-\infty}^{\infty} \frac{1}{|a_n|} \delta(t - t_n)$$
$$= \text{comb}(t).$$

---

41. Generating a *true* Dirac delta current or voltage waveform requires more power than is available from all of the world's power generation plants. Physically, the Dirac delta is a good approximation for a short high value waveform.

(d) From the quadratic formula, the quadratic equation

$$q(t) = at^2 + bt + c$$

has zeros at

$$t_\pm = \frac{1}{2a}\left(-b \pm \sqrt{b^2 - 4ac}\right).$$

There are three cases of interest, depending on the discriminant,

$$d = b^2 - 4ac.$$

(i) If $d < 0$, there are no real solutions of the quadratic and

$$\delta(at^2 + bt + c) = 0.$$

(ii) If $d > 0$, there are two zeros at

$$t_\pm = \frac{1}{2a}\left(-b \pm \sqrt{b^2 - 4ac}\right).$$

Then

$$q(t) = (t - t_-)(t - t_+).$$

The derivatives at these two points are

$$\frac{dq(t_-)}{dt} = t_- - t_+$$

and

$$\frac{dq(t_+)}{dt} = t_+ - t_-.$$

Thus

$$\delta(at^2 + bt + c) = \frac{\delta(t - t_+) + \delta(t - t_-)}{t_+ - t_-}. \qquad (2.159)$$

(iii) If $d = 0$, there is a double zero at $t_+ = t_- = \frac{-b}{2a}$ and

$$q(t) = \left(t + \frac{b}{2a}\right)^2.$$

Thus

$$q\left(\frac{-b}{2a}\right) = 0$$

and

$$\frac{dq\left(\frac{-b}{2a}\right)}{dt} = 0.$$

The delta function with argument $q(t)$ is then not defined in the sense it has infinite area.

$$\delta(at^2 + bt + c) = \infty \times \delta\left(t + \frac{b}{2a}\right).$$

This is the same result we would obtain in (2.159) by letting $t_+ \to t_-$.

2.29. (a) From duality

$$\text{sinc}(t) \longleftrightarrow \Pi(u).$$

(b) We have established (see (2.39))

$$\text{sech}(\pi t) \longleftrightarrow \text{sech}(\pi u) = \int_{-\infty}^{\infty} \text{sech}(\pi t) \, e^{-j2\pi ut} dt.$$

Thus

$$\psi(t) \equiv \text{sech}(\pi t) \, e^{\pi t} \longleftrightarrow \Psi(u) = \int_{-\infty}^{\infty} \text{sech}(\pi t) e^{-j2\pi\left(u + \frac{j}{2}\right)t} dt.$$

$$= \text{sech}\left(\pi\left(u + \frac{j}{2}\right)\right).$$

$$= \frac{2}{e^{\pi\left(u+\frac{j}{2}\right)} + e^{-\pi\left(u+\frac{j}{2}\right)}}$$

$$= \frac{2}{j\,e^{\pi u} - j\,e^{-\pi u}}$$

$$= -j\,\text{cosech}(\pi u).$$

Since $\psi(-t) \longleftrightarrow \Psi(-u)$, sech is even, and cosech is odd, and

$$\psi(-t) = \text{sech}(\pi t) \, e^{-\pi t} \longleftrightarrow \Psi(-u) = j\,\text{cosech}(\pi u).$$

It follows that

$$\frac{1}{2}[\psi(t) - \psi(-t)] = \frac{1}{2}\left(e^{\pi t} - e^{-\pi t}\right)\text{sech}(\pi t)$$

$$= \sinh(\pi t)\,\text{sech}(\pi t) = \frac{\sinh(\pi t)}{\cosh(\pi t)}$$

$$= \tanh(\pi t)$$

$$\longleftrightarrow \frac{1}{2}[\Psi(u) - \Psi(-u)]$$

$$= -j\,\text{cosech}(\pi u).$$

(c) Recall that

$$\frac{d}{dt}\tanh(t) = \text{sech}^2(t).$$

Using the derivative theorem applied to the tanh transform gives the desired result.

(d) For the sine integral, we first note, using the signal integral property, that

$$\int_{t=-\infty}^{\infty} \operatorname{sinc}(t) = \Pi(0) = 1.$$

Using the integral theorem in Table 2.3 applied to the pair $\operatorname{sinc}(t) \leftrightarrow \Pi(u)$ gives

$$\int_{-\infty}^{t} \operatorname{sinc}(\tau) d\tau \longleftrightarrow \frac{\Pi(u)}{j2\pi u} + \frac{1}{2}\delta(u).$$

Thus, using (2.78),

$$\operatorname{Si}(\pi t) = \pi \int_{0}^{t} \operatorname{sinc}(\tau) d\tau$$

$$= \pi \left[-\frac{1}{2} + \int_{\tau=-\infty}^{t} \operatorname{sinc}(\tau) d\tau \right]$$

$$\longleftrightarrow \pi \left[-\frac{1}{2}\delta(u) + \left\{ \frac{\Pi(u)}{j2\pi u} + \frac{1}{2}\delta(u) \right\} \right]$$

$$= \frac{\Pi(u)}{j2u}.$$

(e) Using a convergence factor

$$\sum_{n=0}^{\infty} \delta(t-n) = \lim_{\alpha \to 1-} \sum_{n=0}^{\infty} \left[\alpha^n \delta(t-n) \right].$$

Then

$$\lim_{\alpha \to 1-} \left[\sum_{n=0}^{\infty} \alpha^n \delta(t-n) \right] \longleftrightarrow \lim_{\alpha \to 1-} \left[\sum_{n=0}^{\infty} \alpha^n e^{-j2\pi nu} \right]$$

$$= \lim_{\alpha \to 1-} \left[\frac{1}{1 - \alpha e^{-j2\pi u}} \right]$$

$$= \frac{1}{1 - e^{-j2\pi u}}$$

$$= \frac{e^{j\pi u}}{e^{j\pi u} - e^{-j\pi u}}$$

$$= \frac{1}{j2} \frac{e^{j\pi u}}{\sin(\pi u)}$$

$$= \frac{-j}{2} \frac{\cos(\pi u) + j\sin(\pi u)}{\sin(\pi u)}$$

$$= \frac{1}{2} [1 - j\cot(\pi u)]$$

where

$$\cot(z) = \frac{1}{\tan(z)} = \frac{\cos(z)}{\sin(z)}.$$

2.30. (a) Use Parseval's theorem with $x(t) = \text{sinc}^2(t)$,

$$\int_{-\infty}^{\infty} \text{sinc}^4(t)\,dt = \int_{-1}^{1} \Lambda^2(u)\,du$$

$$= 2\int_0^1 (1-u)^2\,du$$

$$= 2/3.$$

(b) Same approach, but $x(t) = \text{jinc}(t)$:

$$\int_{-\infty}^{\infty} \text{jinc}^2(t)\,dt = \int_{-1}^{1} (1-u^2)\,du$$

$$= 4/3.$$

(c) Parseval's theorem gives

$$\int_{-\infty}^{\infty} J_0^2(2\pi t)\,dt = \frac{1}{\pi^2}\int_{-1}^{1} \frac{du}{1-u^2}$$

$$= \infty.$$

2.31. (a) Using (2.83), we have

$$\int_{-\infty}^{\infty} j_n(2\pi t)\,dt = \frac{1}{2} j^{-n} P_n(0).$$

From (2.46), the Legendre polynomials are evenly odd. Thus $P_n(0) = 0$ when n is odd and

$$\int_{-\infty}^{\infty} j_n(2\pi t)\,dt = 0 \text{ when } n \text{ is odd}.$$

For even n, we use the recursion equation for the Legendre polynomial in (2.80). Assuming n is odd, we have

$$(n+1)P_{n+1}(0) = -nP_{n-1}(0)$$

or, for $m = n+1$ even,

$$P_m(0) = -\frac{m-1}{m} P_{m-2}(0). \tag{2.160}$$

Since $P_0(0) = 1$, the first few values are

$$P_2(0) = -\frac{1}{2} P_0(0) = -\frac{1}{2},$$

$$P_4(0) = -\frac{3}{4} P_2(0) = \frac{3 \cdot 1}{4 \cdot 2},$$

$$P_6(0) = -\frac{5}{6} P_4(0) = -\frac{5 \cdot 3 \cdot 1}{6 \cdot 4 \cdot 2}.$$

From this pattern, we deduce

$$P_{2m}(0) = (-1)^m \frac{(m-1)!!}{m!!}.$$

Thus

$$P_n(0) = \begin{cases} 0 & ; \; n \text{ odd} \\ \dfrac{((n/2)-1)!!}{(n/2)!!} & ; \; n \text{ even} \end{cases} \qquad (2.161)$$

and

$$\int_{-\infty}^{\infty} j_n(2\pi t)dt = \begin{cases} 0 & ; \; n \text{ odd} \\ \dfrac{1}{2} \dfrac{((n/2)-1)!!}{(n/2)!!} & ; \; n \text{ even}. \end{cases}$$

(b) Using the spherical Bessel function Fourier transform in (2.83) and the power theorem

$$\int_{-\infty}^{\infty} j_n(2\pi t) j_m(2\pi t) dt = j^{n-m} \int_{-1}^{1} P_n(u) P_m(u) du.$$

Using the orthogonality property of the Legendre polynomial in (2.81) gives the final answer

$$\int_{-\infty}^{\infty} j_n(2\pi t) j_m(2\pi t) dt = \frac{2}{2n+1} \delta[n-m].$$

(c) Using the Chebyshev polynomial Fourier transform in (2.87)

$$\int_{-1}^{1} \frac{T_n(t)}{\sqrt{1-t^2}} dt = j^n J_n(0).$$

The series for the Bessel function in (2.152) is

$$J_m(t) = \frac{1}{m!} \left(\frac{t}{2}\right)^m - \frac{1}{(m+1)!} \left(\frac{t}{2}\right)^{m+1} + \cdots$$

Clearly, $J_m(0) = 0$ for $m \neq 0$. For $m = 0$, we have $J_m(0) = 1$. Thus

$$J_m(0) = \delta[m].$$

This is graphically evident in Figure 2.15. In conclusion, we have

$$\int_{-1}^{1} \frac{T_n(t)}{\sqrt{1-t^2}} dt = \delta[n].$$

(d) Use the Fourier transform of the spherical Bessel function in (2.83) as applied to the Poisson sum formula in (2.102). The result is

$$T \sum_{n=-\infty}^{\infty} j_k(2\pi(t-nT)) = \sum_{n=-\infty}^{\infty} j^{-k} P_k\left(\frac{n}{T}\right) \Pi\left(\frac{n}{2T}\right) e^{j2\pi nt/T}. \qquad (2.162)$$

If $T < 1$,
$$\Pi\left(\frac{n}{2T}\right) = \delta[n]$$

and (2.162) becomes

$$\sum_{n=-\infty}^{\infty} j_k(2\pi(t-nT)) = \frac{1}{T} P_k(0).$$

Using the equations for $P_k(0)$ in (2.161)

$$\sum_{n=-\infty}^{\infty} j_k(2\pi(t-nT)) = \begin{cases} 0 & ; \; k \text{ odd} \\ \dfrac{(n/2-1)!!}{T(n/2)!!} & ; \; k \text{ even}. \end{cases}$$

2.32. (a) Use the Poisson sum formula.

$$\sum_{n=-\infty}^{\infty} \operatorname{sinc}(t-n) = \sum_{n=-\infty}^{\infty} \Pi(n) e^{j2\pi nt}$$

$$= \sum_{n=-\infty}^{\infty} \delta[n] e^{j2\pi nt}$$

$$= 1$$

since $\Pi(n) = \delta[n]$.
(b) Same result since $\Lambda(n) = \delta[n]$.
(c) Same result since $\sqrt{1-n^2}\,\Pi(n/2) = \delta[n]$.
(d) $\operatorname{sinc}(at) \longleftrightarrow \Pi(u/a)/a$. Thus

$$\sum_{n=-\infty}^{\infty} \operatorname{sinc}(a(t-n)) = \frac{1}{a} \sum_{n=-\infty}^{\infty} \Pi\left(\frac{n}{a}\right) e^{j2\pi nt}$$

$$= \frac{1}{a} \sum_{n=-N}^{N} e^{j2\pi nt}$$

where N is the greatest integer not exceeding $a/2$. Using a geometric series

$$\sum_{n=-\infty}^{\infty} \operatorname{sinc}(a(t-n)) = \frac{\sin[(2N+1)\pi t]}{a \sin(\pi t)}$$

$$= \frac{2N+1}{a} \operatorname{array}_{2N+1}(t).$$

Part (a) is a special case for $a = 1$ ($N = 0$).

$$\sum_{n=-\infty}^{\infty} \operatorname{sinc}(a(t-n)rrr) = \frac{1}{a}\left[1 + 2\sum_{n=1}^{N} \cos(2\pi nt)\right].$$

Part (a) is special case for $a = 1 (N = 0)$.

(e) Use (2.104) with $T = 2$

$$\sum_{n=-\infty}^{\infty} \text{jinc}(2n) = \frac{1}{2} \sum_{n=-\infty}^{\infty} \sqrt{1 - (n/2)^2}\ \Pi(n/4)$$
$$= (1 + \sqrt{3})/2.$$

(f) Use (2.104) with $T = 1/4$.

$$\sum_{n=-\infty}^{\infty} \text{jinc}^2(n/4) = 4 \sum_{n=-\infty}^{\infty} C(4n)$$

where

$$C(u) = \sqrt{1 - u^2}\ \Pi(u/2) * \sqrt{1 - u^2}\ \Pi(u/2)$$
$$= C(u)\Pi(u/4).$$

Thus

$$\sum_{n=-\infty}^{\infty} \text{jinc}^2(n/4) = 4C(0)$$

where

$$C(0) = \int_{-1}^{1} (1 - u^2)\,du$$
$$= 4/3.$$

(g) The $n > 0$ and $n < 0$ terms cancel. The $n = 0$ term is zero.

2.33. (a) $\frac{d}{dt} \text{sinc}(t) = 0 \implies \theta = \tan(\theta)$; $\theta = \pi t$. The nth extrema location, θ_n, can be founded iteratively by $\theta_n[m] \longrightarrow \theta_n$ as $m \longrightarrow \infty$ with

$$\theta_n[m + 1] = n\pi + \arctan(\theta_n[m])\ ;\ n = 0, \pm1, \pm2, \ldots.$$

The first few locations and the corresponding extrema are listed in the Table 2.8.

(b) $\theta_n \longrightarrow n + 1/2$ as $n \longrightarrow \infty$ and

$$\text{sinc}(t_n) \longrightarrow \frac{(-1)^n}{\pi(n + \frac{1}{2})}.$$

To justify, simply compare plots of θ and $\tan(\theta)$ vs. θ.

2.34. Using (2.18), the Fourier transform of the periodic function, $x_T(t)$, is

$$X_T(u) = \int_{t=\tau}^{\tau+T} x(t)e^{-j2\pi ut}\,dt.$$

Thus

$$X_T\left(\frac{n}{T}\right) = \int_{t=\tau}^{\tau+T} x(t)e^{-j2\pi nt/T}\,dt.$$

TABLE 2.8. The first few extrema of sinc(t).

n	θ_n	sinc(θ_n)
1	4.4934	−0.21723
2	7.7253	0.12837
3	10.9041	−0.09133
4	14.0662	0.07091
5	17.2208	−0.05797
6	20.3713	0.04903
7	23.5194	−0.04248
8	26.6661	0.03747
9	29.8116	−0.03353
10	32.9564	0.03033
11	36.1006	−0.02769
12	39.2444	0.02547

Using Leibniz's rule (see Appendix 14.2)

$$\frac{d}{d\tau}X_T\left(\frac{n}{T}\right) = x(\tau + T)e^{-j2\pi n(\tau+T)/T} - x(T)e^{-j2\pi nT/T}$$
$$= 0.$$

Therefore, $X_T\left(\frac{n}{T}\right)$ is not a function of τ.

2.38. (a) The function in (2.146) is periodic with period 2π. Using (2.14), its Fourier coefficients are

$$c_n = \frac{1}{2\pi}\int_{-\pi}^{\pi} e^{z\cos(t)} e^{-jnt} dt.$$

Since $e^{z\cos(t)}$ is an even function of t,

$$c_n = \frac{1}{2\pi}\int_{-\pi}^{\pi} e^{z\cos(t)} \cos(nt) dt.$$

The integrand is even, so

$$c_n = \frac{1}{\pi}\int_{0}^{\pi} e^{z\cos(t)} \cos(nt) dt.$$

Using (2.75), the Fourier coefficients are seen to be modified Bessel functions.

$$c_n = I_n(z).$$

The Fourier series is thus

$$e^{z\cos(t)} = \sum_{n=-\infty}^{\infty} I_n(z) e^{jnt}.$$

(b) From the Bessel function identity in (2.76), we find that

$$J_n(z) = j^n I_n(-jz).$$

The integral in defining the modified Bessel function in (2.75) becomes

$$J_n(z) = \frac{j^n}{\pi} \int_0^\pi e^{-jz\cos(\theta)} \cos(n\theta) d\theta. \tag{2.163}$$

The periodic function, $x(t) = e^{-jz\cos(t)}$ has a period of 2π and, from (2.14), has Fourier series coefficients

$$c_n = \frac{1}{2\pi} \int_{-\pi}^{\pi} e^{-jz\cos(\theta)} e^{-jn\theta} d\theta.$$

Since $e^{-jz\cos(\theta)}$ is an even function of θ, this is equivalent to

$$c_n = \frac{1}{2\pi} \int_{-\pi}^{\pi} e^{-jz\cos(\theta)} \cos(n\theta) d\theta.$$

The integrand is even. We can therefore write

$$c_n = \frac{1}{\pi} \int_0^\pi e^{-jz\cos(\theta)} \cos(n\theta) d\theta.$$

Using (2.163), we therefore conclude the Fourier series coefficients of the periodic function in (2.147) are

$$c_n = j^{-n} J_n(z).$$

2.40. The Fourier series is

$$y(t) = \alpha \sum_{n=-\infty}^{\infty} \mathrm{sinc}(\alpha n) e^{j2\pi nt/T}.$$

Thus

$$y(\tau/2) = \alpha \sum_{n=-\infty}^{\infty} \mathrm{sinc}(\alpha n) \cos(n\pi\alpha).$$

Write sinc as $\sin(\pi x)/(\pi x)$ and use a trigonometric identity

$$y(\tau/2) = \alpha \sum_{n=-\infty}^{\infty} \mathrm{sinc}(2\alpha n).$$

We evaluate using (2.104) which, for $x(t) = \mathrm{sinc}(t)$, can be written

$$\alpha \sum_{n=-\infty}^{\infty} \mathrm{sinc}(2\alpha n) = \frac{1}{2} \sum_{n=-\infty}^{\infty} \Pi\left(\frac{n}{2\alpha}\right)$$

$$= 1/2.$$

2.41. Since the integrand in (2.104) is even

$$\frac{J_\nu(2\pi t)}{(2t)^\nu} = \frac{(\pi/2)^\nu}{\sqrt{\pi}\Gamma(\nu+\frac{1}{2})} \int_{-1}^{1} (1-u^2)^{\nu-\frac{1}{2}} \cos(2\pi ut)\, du.$$

Thus we have the transform pair

$$\frac{J_\nu(2\pi t)}{(2t)^\nu} \longleftrightarrow \frac{(\pi/2)^\nu}{\sqrt{\pi}\Gamma(\nu+\frac{1}{2})} (1-u^2)^{\nu-\frac{1}{2}} \Pi\left(\frac{u}{2}\right). \qquad (2.164)$$

The entries in Table 2.4 for $\nu = 0$, 1 follow immediately using (2.65).

2.43. (a) We first find the coefficients, α_k, for the Legendre polynomial expansion. From (2.149),

$$P_i(t)x(t) = \sum_{k=0}^{\infty} \alpha_k P_i(t) P_k(t) \Pi\left(\frac{t}{2}\right).$$

Thus

$$\int_{-1}^{1} P_i(t)x(t)\,dt = \sum_{k=0}^{\infty} \alpha_k \int_{-1}^{1} P_i(t)P_k(t)\, dt$$

$$= \sum_{k=0}^{\infty} \alpha_k \frac{2}{2k+1} \delta[i-k]$$

$$= \frac{2\alpha_i}{2i+1},$$

or

$$\alpha_k = \left(k+\frac{1}{2}\right) \int_{-1}^{1} x(t) P_k(t)\, dt. \qquad (2.165)$$

(b) For a period of $T = 2$, the Fourier series coefficients are

$$c_n = \frac{1}{2} \int_{-1}^{1} x(t) e^{j\pi nt}\, dt$$

$$= \frac{1}{2} \int_{-1}^{1} \left[\sum_{k=0}^{\infty} \alpha_k P_k(t)\right] e^{j\pi nt}\, dt \qquad (2.166)$$

$$= \frac{1}{2} \sum_{k=0}^{\infty} \alpha_k \left[\int_{-1}^{1} P_k(t) e^{j\pi nt}\, dt\right]$$

From (2.83),

$$j_k(2\pi t) \leftrightarrow \frac{1}{2} j^{-k} P_k(u) \Pi\left(\frac{u}{2}\right) \qquad (2.167)$$

so that the inverse Fourier transform is

$$\int_{-1}^{1} P_k(u) e^{j2\pi ut}\, du = 2j^k j_k(2\pi t).$$

or, substituting $t \leftarrow u$ and $u \leftarrow \frac{n}{2}$ gives

$$\int_{-1}^{1} P_k(t) e^{j\pi nt} dt = 2j^k j_k(\pi n). \tag{2.168}$$

Therefore, (2.166) becomes

$$c_n = \sum_{k=0}^{\infty} j^k \alpha_k j_k(\pi n). \tag{2.169}$$

(c) Here, we express the Legendre polynomial coefficients, α_k, in terms of the Fourier coefficients, c_n. From (2.165),

$$\begin{aligned} \alpha_k &= \left(k + \frac{1}{2}\right) \int_{-1}^{1} x(t) P_k(t) dt. \\ &= \left(k + \frac{1}{2}\right) \int_{-1}^{1} \left[\sum_{n=-\infty}^{\infty} c_n e^{j\pi nt}\right] P_k(t) dt \\ &= \left(k + \frac{1}{2}\right) \sum_{n=-\infty}^{\infty} c_n \left[\int_{-1}^{1} P_k(t) e^{j\pi nt} dt\right] \\ &= (2k+1) j^k \sum_{n=-\infty}^{\infty} c_n j_k(\pi n) \end{aligned} \tag{2.170}$$

where, in the last step, we have used (2.168).

(d) Substitute (2.169) into (2.170) gives

$$\alpha_k = (2k+1) j^k \sum_{n=-\infty}^{\infty} \left(\sum_{p=0}^{\infty} j^p \alpha_p j_p(\pi n)\right) j_k(\pi n)$$

$$= (2k+1) j^k \sum_{p=0}^{\infty} j^p \alpha_p \left(\sum_{n=-\infty}^{\infty} j_p(\pi n) j_k(\pi n)\right).$$

For this to be an identity, we must have the following orthogonality condition for the spherical Bessel function with respect to the function's index.

$$(2k+1) \sum_{n=-\infty}^{\infty} j^{p+k} j_p(\pi n) j_k(\pi n) = \delta[k-p].$$

(e) Substitute (2.170) into (2.169) gives

$$\begin{aligned} c_n &= \sum_{k=0}^{\infty} j^k \left[(2k+1) j^k \sum_{m=-\infty}^{\infty} c_m j_k(\pi m)\right] j_k(\pi n) \\ &= \sum_{m=-\infty}^{\infty} c_m \left[\sum_{k=0}^{\infty} (-1)^k (2k+1) j_k(\pi m) j_k(\pi n)\right]. \end{aligned}$$

We therefore have the following orthogonality condition for samples of the spherical Bessel function.

$$\sum_{k=0}^{\infty}(-1)^k(2k+1)j_k(\pi m)j_k(\pi n) = \delta[n-m].$$

2.46. (a) Using the definition of the Legendre polynomial in (2.81), we have

$$P_n(-t) = \frac{1}{2^n n!}\left(\frac{d}{d(-t)}\right)^n (t^2-1)^n$$

$$= (-1)^n \frac{1}{2^n n!}\left(\frac{d}{dt}\right)^n (t^2-1)^n$$

$$= (-1)^n P_n(t).$$

(b) Using the definition of the Hermite polynomial in (2.88) gives

$$H_n(-t) = (-1)^n e^{t^2}\left(\frac{d}{d(-t)}\right)^n e^{-t^2}$$

$$= (-1)^n \left[(-1)^n e^{t^2}\left(\frac{d}{dt}\right)^n e^{-t^2}\right]$$

$$= (-1)^n H_n(t).$$

Yes, Hermite polynomials are evenly odd.

(c) The Chebyshev polynomials recursion in (2.85) is repeated here

$$T_{n+1}(t) = 2tT_n(t) - T_{n-1}(t). \tag{2.171}$$

We know $T_0(t) = 1$ is even and $T_1(t) = t$ is odd. When $T_n(t)$ is even, $2tT_n(t)$ in (2.171) is odd, and, for $T_{n-1}(t)$ odd, we see $T_{n+1}(t)$ in (2.171) is odd. A similar argument can be made for even recurrence, and we have established that Chebyshev polynomials are evenly odd.

2.47. Chebyshev polynomials of the second kind can be defined by

$$U_n(\cos\theta) = \frac{\sin(n+1)\theta}{\sin\theta}.$$

(a)
$$U_0(\cos\theta) = 1 \quad \to \quad U_0(t) = 1.$$

$$U_1(\cos\theta) = \frac{\sin 2\theta}{\sin\theta} \tag{2.172}$$

but

$$\sin 2\theta = \Im e^{j2\theta}$$
$$= \Im e^{j\theta} e^{j\theta}$$
$$= \Im(\cos\theta + j\sin\theta)(\cos\theta + j\sin\theta)$$
$$= 2\cos\theta\,\sin\theta$$

so that (2.172) becomes[42]

$$U_1(\cos\theta) = \frac{2\cos\theta \sin\theta}{\sin\theta} = 2\cos\theta.$$

Thus

$$U_1(t) = 2t.$$

(b) Equation 2.151 is the imaginary part of the equation

$$e^{j(n+2)\theta} = 2\cos\theta e^{j(n+1)\theta} - e^{jn\theta}.$$

Multiply through by $e^{-j(n+1)\theta}$ and this reduces to Euler's formula.

(c) Divide (2.151) through by $\sin\theta$.

$$\frac{\sin(n+2)\theta}{\sin\theta} = 2\frac{\cos\theta \sin(n+1)\theta}{\sin\theta} - \frac{\sin n\theta}{\sin\theta},$$

or, equivalently,

$$U_{n+1}(\cos\theta) = \cos\theta \, U_n(\cos\theta) - U_{n-1}(\cos\theta)$$

from which the recurrence relationship for Chebyshev polynomials of the second kind in (2.150) follows.

2.49. From (2.14), since $x(t)$ is real,

$$c_n^* = \frac{1}{T}\int_T x(t) \, e^{j2\pi nt/T} \, dt.$$

Thus

$$c_{-n}^* = \frac{1}{T}\int_T x(t) \, e^{-j2\pi nt/T} \, dt = c_n.$$

2.51. With attention to Figure 2.33:

$$z_T(t) = \Pi\left(\frac{t-\frac{T}{4}}{T/2}\right) - \Pi\left(\frac{t+\frac{T}{4}}{T/2}\right).$$

Thus

$$Z_T(u) = -j \, \text{sinc}\left(\frac{Tu}{2}\right) \sin\left(\frac{\pi ut}{2}\right).$$

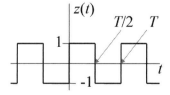

FIGURE 2.33. See Exercise 2.51.

---

42. Alternately, we could have looked up a trig identity.

The Fourier coefficients follow as

$$c_n = \frac{1}{T} Z_T(n/T)$$
$$= -j\,\text{sinc}(n/2)\sin(\pi n/2).$$

Thus, since $z(t)$ is real,

$$z(t) = \sum_{n=-\infty}^{\infty} -j\,\text{sinc}(n/2)\,\sin(\pi n/2)\,e^{j2\pi nt/T}$$
$$= \sum_{n=-\infty}^{\infty} \text{sinc}(n/2)\,\sin(\pi n/2)\,\sin\left(\frac{2\pi nt}{T}\right).$$

The truncated series is

$$z_N(t) = \sum_{n=-N}^{N} \text{sinc}(n/2)\,\sin(\pi n/2)\,\sin\left(\frac{2\pi nt}{T}\right).$$

Extrema come from

$$0 = \frac{d}{dt} z_N(t)$$
$$= \frac{2\pi}{T}\sum_{n=-N}^{N} \text{sinc}(n/2)\,\sin(\pi n/2)\,n\,\cos\left(\frac{2\pi nt}{T}\right)$$
$$= \frac{4}{T}\sum_{n=-N}^{N} \sin^2(\pi n/2)\,\cos\left(\frac{2\pi nt}{T}\right).$$

But

$$\sin^2(\pi n/2) = \frac{1-(-1)^n}{2}$$

and

$$0 = \frac{2}{T}\sum_{n=-N}^{N} [1-(-1)^n]\,\cos\left(\frac{2\pi nt}{T}\right)$$
$$= \frac{2}{T}\Re\sum_{n=-N}^{N} [1-(-1)^n]\,e^{j2\pi nt/T}.$$

Applying a geometric series[43] gives

$$0 = \frac{\sin(N+\tfrac{1}{2})\theta}{\sin(\theta/2)} - \frac{(-1)^N\,\cos(N+\tfrac{1}{2})\theta}{\cos(\theta/2)}$$

---

43. See Appendix 14.4.2.

or, for N even,

$$\tan\left(N+\frac{1}{2}\right)\theta = \tan(\theta/2).$$

Therefore

$$\left(N+\frac{1}{2}\right)\theta = \frac{\theta}{2} + p\pi; \quad p = 0, \pm 1, \pm 2, \ldots$$

or

$$\theta = \frac{2\pi t}{T} = \frac{p\pi}{N} \implies t = \frac{pT}{2N}.$$

The first extrema is at $p = 1$ corresponding to $t = T/2N$. Substituting gives

$$z_N\left(\frac{T}{2N}\right) = \sum_{n=-N}^{N} \frac{\sin^2(\pi n/2)}{\pi n/2} \sin(\pi n/N)$$

$$= 2 \sum_{n=1}^{N} \frac{\sin^2(\pi n/2)}{\pi n/2} \sin(\pi n/N)$$

$$= \frac{2}{\pi} \sum_{n=1}^{N} \frac{1-(-1)^n}{n} \sin(\pi n/N).$$

Let $2m + 1 = n$

$$z_N\left(\frac{T}{2N}\right) = \frac{4}{\pi} \sum_{m=0}^{N/2} \frac{\sin\frac{\pi(2m+1)}{N}}{2m+1}.$$

Define

$$h(t) = \sum_{m=0}^{M} \frac{\sin(2m+1)t}{2m+1}; \quad M = N/2$$

so that

$$z_N\left(\frac{T}{2N}\right) = \frac{4}{\pi} h(\pi/N).$$

Note that $h(0) = 0$. Now

$$\frac{dh(t)}{dt} = \sum_{m=0}^{N} \cos(2m+1)t$$

$$= \frac{\sin 2(M+1)t}{2\sin(t)}$$

$$= 2(M+1)\,\text{array}_{2(M+1)}(t/\pi).$$

Thus
$$h(t) = 2(M+1)\int_0^t \text{array}_{2(M+1)}\left(\frac{\tau}{\pi}\right) d\tau$$

and
$$z_N\left(\frac{T}{2N}\right) = \frac{2(N+1)}{\pi}\int_0^{\pi/N} \text{array}_{N+1}\left(\frac{\tau}{\pi}\right) d\tau.$$

Since the interval of integration gets smaller and smaller, $\sin\tau \longrightarrow \tau$ and as $N \to \infty$

$$z_N\left(\frac{T}{2N}\right) \longrightarrow \frac{4}{\pi}\int_0^{\pi/N} \frac{\sin(N+2)\tau}{2\tau} d\tau$$

$$= \frac{2}{\pi}\int_0^{\pi+\frac{2\pi}{N}} \frac{\sin(\xi)}{\xi} d\xi.$$

Since $\text{Si}(\pi) = 1.8519370$, we conclude that, as $N \to \infty$,

$$\Delta \longrightarrow 1 - \frac{2}{\pi}\text{Si}(\pi) = 0.1789797.$$

This is the Gibb's phenomenon overshoot of the Fourier series expansion of the square wave.

2.52. Using the shift theorem

$$\text{comb}(t - 1/2) \longleftrightarrow \text{comb}(u)\, e^{-j\pi u}$$

$$= \sum_{n=-\infty}^{\infty} \delta(u-n)\, e^{-j\pi n}$$

$$= \sum_{n=-\infty}^{\infty} (-1)^n\, \delta(u-n)$$

$$= \sum_{\text{even } n} \delta(u-n) - \sum_{\text{odd } n} \delta(u-n)$$

$$= \sum_{m=-\infty}^{\infty} \delta(u-2m) - \sum_{m=-\infty}^{\infty} \delta(u-(2m+1))$$

$$= \frac{1}{2}\left[\sum_{m=-\infty}^{\infty} \delta\left(\frac{u}{2}-m\right) - \sum_{m=-\infty}^{\infty} \delta\left(\frac{u-1}{2}-m\right)\right]$$

$$= \frac{1}{2}\left[\text{comb}\left(\frac{u}{2}\right) - \text{comb}\left(\frac{u-1}{2}\right)\right].$$

2.54. This problem illustrates that finite area does not dictate finite energy. Neither does finite energy dictate finite area.
(a) See Table 2.10.
(b) On the interval $(0, 1)$, $x(t)$ has finite area and infinite energy. On the interval $(0, \infty)$, the area is infinite and the energy is finite.

TABLE 2.10. See the solution to Exercise 2.54.

	(0, 1)	(1, ∞)
A	4	∞
E	∞	2

(c) $A < \infty \Longrightarrow F < \infty$. This follows from the proof that the space of ℓ_1 sequences is subsumed in l_2 [854, 1020]. The converse is not true. Consider

$$x(t) = \text{sinc}(Bt).$$

Clearly, $E = 1/B$. However, since

$$\left| x\left(\frac{n}{2B}\right) \right| = \begin{cases} 1 & ; \quad n = 0 \\ \frac{2}{\pi |n|} & ; \quad \text{odd } n \\ 0 & ; \quad \text{otherwise} \end{cases}$$

we have the divergent series

$$A = 1 + \frac{2}{\pi} \sum_{k=0}^{\infty} \frac{1}{2k+1} = \infty.$$

2.55. (a) A spectrum with finite area has a bounded inverse Fourier transform, but the converse is not true.

$$|x(t)| = \left| \int_{-\infty}^{\infty} X(u) \, e^{j2\pi ut} \, du \right|$$

$$\leq \int_{-\infty}^{\infty} |X(u)| \, du$$

$$= A.$$

Thus, $C = A$. Note, then, that $x(t)$ is bounded if $X(u)$ has finite area.

(b) A counter example is $x(t) = \text{sgn}(t)$. Since $|x(t)| \leq 1$, the function $x(t)$ is bounded. The Fourier transform of $x(t)$, though, does not have finite area.

2.56. To show all finite energy bandlimited functions are bounded, we write

$$|x(t)|^2 = \left| \int_{-B}^{B} X(u) \, e^{j2\pi ut} \, du \right|^2$$

$$\leq 2B \int_{-B}^{B} |X(u)|^2 \, du$$

$$= 2BE.$$

2.57. (a) Yes, the pth derivative of a finite energy bandlimited function has finite energy and is bandlimited. Applying Parseval's theorem to the derivative theorem gives

$$E_p = \int_{-\infty}^{\infty} |x^{(p)}(t)|^2 \, dt$$

$$= \int_{-B}^{B} |(j2\pi u)^p X(u)|^2 \, du.$$

Since $(u/B)^{2p} < 1$ over the interval $|u| < B$,

$$E_p = (2\pi B)^{2p} \int_{-B}^{B} (u/B)^{2p} |X(u)|^2 \, du$$

$$\leq (2\pi B)^{2p} \int_{-B}^{B} |X(u)|^2 \, du$$

$$= (2\pi B)^{2p} E$$

where E is the energy of $x(t)$. Also, $x^{(p)}(t)$ is clearly bandlimited.

(c) Yes, the pth derivative of a bounded bandlimited function is bounded and bandlimited.

$$\left| x^{(p)}(t) \right| = \left| \int_{-B}^{B} (j2\pi u)^p X(u) \, e^{j2\pi ut} \, du \right|$$

$$\leq \int_{-B}^{B} |2\pi u|^p \, |X(u)| \, du$$

$$= (2\pi B)^p \int_{-B}^{B} |u/B|^p \, |X(u)| \, du$$

$$\leq (2\pi B)^p A$$

where, in the last step, we have recognized $|u/B| \leq 1$ over the interval if integration, and A is the area of $X(u)$.

2.58. (a) From the derivative theorem

$$\left| x^{(M)}(t) \right|^2 = \left| \int_{-B}^{B} (j2\pi u)^M X(u) \, du \right|^2$$

$$\leq 2(2\pi)^{2M} E \int_{-B}^{B} u^{2M} \, du$$

which, when evaluated, gives (2.156).

(b) Clearly

$$|x(t+\tau) - x(t)| = \left| \int_{t}^{t+\tau} x'(\xi) \, d\xi \right|$$

$$\leq \int_{t}^{t+\tau} |x'(\xi)| \, d\xi$$

$$\leq \sqrt{\frac{(2\pi B)^3 E}{3}} \int_{t}^{t+\tau} d\xi$$

which, when evaluated, gives (2.157).

2.60. Define
$$E_f = \int_{-\infty}^{\infty} |f(t)|^2 \, dt.$$

Then, applying Schwarz's inequality to the integral gives
$$|g(t)|^2 \leq E_f \, E_h(t).$$

2.61 To show the inversion of the DFT, substitute the DFT in (2.116) into the inverse DFT in (2.117).

$$\frac{1}{N} \sum_{k=0}^{N-1} X[k] e^{j2\pi nk/N} = \frac{1}{N} \sum_{k=0}^{N-1} \left[\sum_{m=0}^{N-1} x[m] e^{j2\pi mk/N} \right] e^{j2\pi nk/N}$$

$$= \frac{1}{N} \sum_{m=0}^{N-1} x[m] \left[\sum_{k=0}^{N-1} e^{j2\pi (n-m)k/N} \right]$$

$$= \frac{1}{N} \sum_{m=0}^{N-1} x[m] \left[N \, e^{j\pi (N-1)(n-m)/N} \right. \quad (2.173)$$

$$\left. \times \operatorname{array}_N \left(\frac{n-m}{N} \right) \right]$$

where, in the last step, we have used (14.9). Since

$$\operatorname{array}_N \left(\frac{n-m}{N} \right) = \delta[n-m] \text{ for } 0 \leq m \leq N-1,$$

(2.173) becomes $x[n]$.

3

Fourier Analysis in Systems Theory

> Nature laughs at the difficulties of integration
> Pierre-Simon de Laplace (1749–1827) [779]
>
> God does not care about our mathematical difficulties. He integrates empirically.
> Albert Einstein (1879–1955) [645]
>
> Does anyone believe that the difference between the Lebesgue and Riemann integrals can have physical significance, and that whether say, an airplane would or would not fly could depend on this difference? If such were claimed, I should not care to fly in that plane.
> Richard W. Hamming (1915–1998) [1189]

3.1 Introduction

In the most general sense, any process wherein a stimulus generates a corresponding response can be dubbed a *system*. For a temporal system with single input, $f(t)$, and single output, $g(t)$, the relation can be written as

$$g(t) = S\{f(t)\} \tag{3.1}$$

where $S\{\cdot\}$ is the *system operator*. This is illustrated in Figure 3.1.

3.2 System Classes

There exist numerous system types. We define them here in terms of continuous signals. The equivalents in discrete time are given as an exercise.[1]

3.2.1 System Types

3.2.1.1 Homogeneous Systems

For homogeneous systems, amplifying or attenuating the input likewise amplifying or attenuating the output. For any constant, a,

$$S\{a f(t)\} = a S\{f(t)\} \tag{3.2}$$

This is illustrated in Figure 3.2.

---

1. See Exercise 3.11.

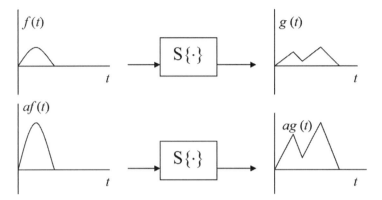

FIGURE 3.1. A temporal system with a single input and output.

FIGURE 3.2. For homogeneous systems, amplifying the input corresponds to an equal amplification of the output.

3.2.1.2 Additive Systems

If the response of the sum is the sum of the responses, the system is said to be *additive*. Specifically,

$$S\{f_1(t) + f_2(t)\} = S\{f_1(t)\} + S\{f_2(t)\} \tag{3.3}$$

3.2.1.3 Linear Systems

Systems that are both homogeneous and additive are said to be *linear*. The criteria in (3.2) and (3.3) can be combined into a single necessary and sufficient condition for linearity.

$$S\{a f_1(t) + b f_2(t)\} = aS\{f_1(t)\} + bS\{f_2(t)\} \tag{3.4}$$

where a and b are constants.

All linear systems produce a zero output when the input is zero.

$$S\{0\} = 0. \tag{3.5}$$

To show this, we use (3.4) with $a = -b$ and $f_1(t) = f_2(t)$.

Note that, because of (3.5), the system defined by

$$g(t) = b f(t) + c$$

where b and $c \neq 0$ are constants, is not linear. It is not homogeneous since

$$S\{a f\} = b f + c$$
$$\neq aS\{f\} = a(b f + c).$$

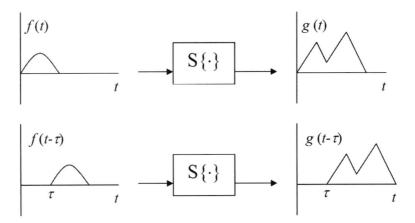

FIGURE 3.3. For time-invariant systems, shifting the input corresponds to an equal shift in the output. Thus, the response today will be the same as the response tomorrow. The system has not changed. It is *invariant* over time.

Neither is it additive since

$$S\{f_1 + f_2\} = b(f_1 + f_2) + c$$
$$\neq S\{f_1\} + S\{f_2\} = (bf_1 + c) + (bf_2 + c).$$

3.2.1.4 Time-Invariant Systems

If, as illustrated in Figure 3.3, shifting any input by, say, τ results in the output being shifted by the same amount, the system is said to be *time–invariant*. Let D_τ be a delay operator so that

$$D_\tau\{f(t)\} := f(t - \tau).$$

The system is time-invariant if D_τ commutes with S. That is

$$D_\tau\{S\{f(t)\}\} = S\{D_\tau\{f(t)\}\}.$$

Time-invariance models systems that do not change with respect to time, i.e., all system parameters are fixed and do not change in time. A circuit with a defined input and output that uses fixed parameter diodes, resistors, transistors, and capacitors is a time invariant system. Systems changing with respect to time are *time–variant* or *dynamic*. A circuit that contains a thermal resistor whose resistance changes with ambient temperature is time variant. For spatial systems, like imaging systems, the time-invariance property is dubbed *shift invariant* or *isoplanatic*.

3.2.1.5 LTI Systems

Linear, time-invariant (LTI) systems, as their name indicates, are systems that are both time-invariant and linear. Thus, input signal superposition, amplification and shift results in equivalent superposition, amplification and shift of the output. Specifically, for shifts τ_1 and τ_2,

$$S\{a f_1(t - \tau_1) + b f_2(t - \tau_2)\} = a g_1(t - \tau_1) + b g_2(t - \tau_2) \qquad (3.6)$$

where $g_k(t) = S\{f_k(t)\}$, $k = 1, 2$.

3.2.1.6 Causal Systems

As is illustrated in Figure 3.4, the output of a causal system must be "caused" by the input. The output $g(\xi)$ at time ξ is due only from the input $f(t)$ for times $t \leq \xi$. Equivalently, the system output cannot be a function of an input which has not yet been fed into the system. Causal systems are also referred to as *nonanticipatory* systems since the output cannot anticipate a future input.

Causality applies only to temporal systems and is nearly nonsensical when applied to spatial systems. In optical imaging, for example, requiring an image output at a point to a function of only points to the left and below the point is a property rarely if ever observed. A temporal signal, when stored in memory such as a CD or a WAV file on a hard drive, is transformed into a spatial signal.

Real time filters, both digital [574, 650, 991, 1053] and analog [193, 1443], feed signals, as they are received, into the filter. Causality plays a monumental role in the synthesis of such systems. A low pass filter, for example, has an impulse response of a sinc function. The sinc is not a causal signal and thus can't serve as the impulse response of a causal filter. We can, however, shift the sinc down the positive time axis and accept a linear delay in the filter output. To insure causality, however, the portion of the shifted sinc for $t < 0$ must be set to zero thereby altering the desired rectangular shape of its Fourier transform. Therefore, a compromise must be made between the causality required by real time filters and the approximation to the shape of the ideal low pass filter's sinc function impulse response. Hamming, Chebyshev and other filters explore the tradeoffs of this compromise [1235, 1353, 1383, 1501]. When the signal is spatial, such as when stored in memory, a low pass filter is readily accomplished by truncation of a Fourier transform followed by inversion.

3.2.1.7 Memoryless Systems

A system is memoryless if the output at time t is determined only by the input at time t or, more generally, a diminishingly small neighborhood around time t such as required by a derivative. See Figure 3.4.

3.2.1.8 Stable Systems

There are numerous ways to define a stable system. A commonly used definition is *bounded input - bounded output* (BIBO) stability. An input is bounded if it is a bounded signal.[2] If, for all such signals, there is a finite $D < \infty$, such that $|y(t)| \leq D$, then the system is said to be BIBO stable. In other words, all bounded inputs to the system produce bounded outputs.

Like causality, stability is a temporal property. An unstable spatial system, with no reference to time, is conceptually nonsensical.

3.2.1.9 Invertible Systems

If a system in invertible, its input can be determined by its output. If $S\{f(t)\} = g(t)$ then there exists an operator S^{-1} such that $S^{-1}\{g(t)\} = f(t)$.

3.2.2 Example Systems

We consider three example systems: the magnifier, the Fourier transformer, and convolution. Each is categorized according to its properties.

---

2. See (2.4).

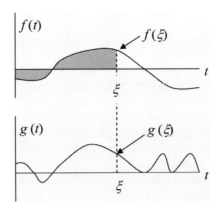

FIGURE 3.4. For a system to be causal, the output cannot anticipate the input. Thus, $g(\xi)$, the output at time $t = \xi$, is determined only by the input, $f(t)$, for $t \leq \xi$. This portion of the input is shaded. For a memoryless systems, the output at time $t = \xi$, denoted $g(\xi)$, is determined only by $f(\xi)$, the input at $t = \xi$. A more general definition, allowing derivatives to be memoryless operations, allows contribution of input around a diminishingly small neighborhood around $t = \xi$.

3.2.2.1 The Magnifier

The magnifier with magnification $M \neq 0$ is defined by the input-output relationship

$$g(t) = S\{f(t)\} = \frac{1}{M} f\left(\frac{t}{M}\right). \tag{3.7}$$

- The magnifier is homogeneous since

$$S\{af(t)\} = \frac{1}{M}\left[af\left(\frac{t}{M}\right)\right]$$

$$= a\left[\frac{1}{M}f\left(\frac{t}{M}\right)\right]$$

$$= a\, S(f(t))$$

- The magnifier is additive since

$$S\{f_1(t) + f_2(t)\} = \frac{1}{M}\left[f_1\left(\frac{t}{M}\right) + f_2\left(\frac{t}{M}\right)\right]$$

$$= \frac{1}{M} f_1\left(\frac{t}{M}\right) + \frac{1}{M} f_2\left(\frac{t}{M}\right)$$

$$= S\{f_1(t)\} + S\{f_2(t)\}.$$

- The magnifier is linear since it is both homogeneous and additive.
- The magnifier is not time-invariant since shifting a magnification is not the same as magnifying a shift. Shifting $f(t)$ to $f(t - \tau)$ followed by magnification gives $f\left(\frac{t}{M} - \tau\right)$. On the other hand, if we magnify to $f\left(\frac{t}{M}\right)$ followed by shifting, we

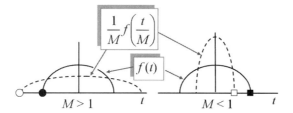

FIGURE 3.5. Examples of magnification for (left) $M > 1$ and (right) $0 < M < 1$.

obtain the different answer $f\left(\frac{t-\tau}{M}\right)$. The two results are equal only in the degenerate case of $M = 1$.
- Since the magnifier is not time-invariant, it is not an LTI system.
- The magnifier is not, in general, causal. If $M > 1$, then, as shown on the left in Figure 3.5, the function spreads. The output at time $t = \circ$ occurs before the corresponding input at $t = \bullet$. The relationship is not causal. When $M > 1$, as illustrated on the right in Figure 3.5, the operation is still not causal. The output at $t = \square$ occurs before the corresponding point on the input.

If a system is limited to causal inputs and $M > 1$, the input-output relationship of the magnifier is causal. The input at any time $t > 0$ dictates the output only at time $Mt > t$.
- The magnifier is not memoryless for all $M \neq 1$.
- The magnifier is stable since, if $|x(t)| \leq C$, then $|y(t)| \leq C/M = D$.
- Solving for f in (3.7) gives

$$f(t) = M\, g(Mt).$$

Since the input can be uniquely determined from the output, the magnifier is invertible.

3.2.2.2 The Fourier Transformer

The Fourier transformer is defined be the input-output pair

$$g(t) = S\{f(t)\} = \int_{-\infty}^{\infty} f(\tau) e^{-j2\pi t\tau}\, d\tau. \tag{3.8}$$

The operation is identical to the Fourier transform, except t is used in lieu of u in order to remain consistent with the systems notation we have established.

- The Fourier transformer is homogeneous since

$$S\{af(t)\} = \int_{-\infty}^{\infty} a f(\tau) e^{-j2\pi t\tau}\, d\tau$$

$$= a \int_{-\infty}^{\infty} f(\tau) e^{-j2\pi t\tau}\, d\tau$$

$$= a\, S\{f(t)\}.$$

- The Fourier transformer is additive since

$$S\{f_1(t)+f_2(t)\} = \int_{-\infty}^{\infty} [f_1(\tau)+f_2(\tau)] e^{-j2\pi t\tau} d\tau$$

$$= \int_{-\infty}^{\infty} f_1(\tau)e^{-j2\pi t\tau} d\tau + \int_{-\infty}^{\infty} f_2(\tau)e^{-j2\pi t\tau} d\tau$$

$$= S\{f_1(t)\} + S\{f_2(t)\}.$$

- The Fourier transformer is linear since it is both homogeneous and additive.
- The Fourier transformer is not time-invariant since shifting a Fourier transform is not the same as Fourier transforming a shift. Let $f(t) \leftrightarrow F(u)$. Shifting $f(t)$ to $f(t-\xi)$ followed by a Fourier transform gives $F(t)\exp(-j2\pi\xi t)$. On the other hand, if we Fourier transform to $F(t)$ followed by shifting, we obtain the different answer $F(t-\xi)$. The two results are equal only in the degenerate case of a zero shift corresponding to $\xi = 0$.
- Since the Fourier transformer is not time-invariant, it is not an LTI system.
- The Fourier transformer is not causal. Indeed, the value of the input at time t contributes to the output at all times.
- For the same reason, the Fourier transformer is not memoryless.
- The Fourier transformer is not stable in the BIBO sense. Consider the bounded input, $x(t) = C \cos(2\pi t)$. Hence, $|x(t)| \leq C$. No matter how small we make C, the Fourier transform contains unbounded Dirac delta functions.
- Since the Fourier transform has an inverse, the Fourier transformer is invertible.

3.2.2.3 Convolution

For a given $h(t)$, we define the system operation of convolution, from (2.25), as

$$g(t) = f(t) * h(t) = \int_{-\infty}^{\infty} f(\tau) h(t-\tau) d\tau. \tag{3.9}$$

- Convolution is homogeneous since

$$S\{a f(t)\} = \int_{-\infty}^{\infty} a f(\tau) h(t-\tau) d\tau$$

$$= a \int_{-\infty}^{\infty} f(\tau) h(t-\tau) d\tau$$

$$= a S\{f(t)\}.$$

- Convolution is additive since

$$S\{f_1(t)+f_2(t)\} = \int_{-\infty}^{\infty} [f_1(\tau)+f_2(\tau)] h(t-\tau) d\tau$$

$$= \int_{-\infty}^{\infty} f_1(\tau)h(t-\tau) d\tau + \int_{-\infty}^{\infty} f_2(\tau)h(t-\tau) d\tau$$

$$= S\{f_1(t)\} + S\{f_2(t)\}.$$

- Convolution is linear since it is both homogeneous and additive.
- Convolution is time-invariant since shifting a convolution is the same as convolving a shift.
  - *Shifting the convolution* by ξ gives

$$g(t - \xi) = \int_{-\infty}^{\infty} f(\hat{\tau}) \, h(t - \xi - \hat{\tau}) \, d\hat{\tau}. \tag{3.10}$$

  - *Convolving the shift* by ξ gives

$$S\{f(t - \xi)\} = \int_{-\infty}^{\infty} f(\tau - \xi) \, h(t - \tau) \, d\tau.$$

  Making the variable substitution $\hat{\tau} = \tau - \xi$ results in an expression identical to (3.10).
- Since the convolution operation is linear and time-invariant, it is an LTI system. Indeed, in Section 3.3.3 we will show the converse is also true. Any time-invariant system can be expressed as a convolution operation.
- The causality of convolution is determined by the specified value of $h(t)$. If $h(t)$ is a causal signal (i.e., $h(t) = h(t)\mu(t)$), then convolution corresponds to a causal system.[3] To show this, we substitute $h(t)\mu(t)$ for $h(t)$ in the convolution integral in (3.9)

$$\begin{aligned} g(t) &= f(t) * (h(t)\mu(t)) \\ &= \int_{-\infty}^{\infty} f(\tau) \, h(t - \tau)\mu(t - \tau) \, d\tau \\ &= \int_{-\infty}^{t} f(\tau) \, h(t - \tau) \, d\tau \end{aligned} \tag{3.11}$$

  where, in the last step, the unit step has been imposed explicitly in the integration limits. From (3.11) we see that the output $g(t)$ at time t is determined by an integral containing the input, $f(\tau)$, from $-\infty < \tau \leq t$. Thus, the output is determined by no future point of the input and the system is causal.
- Except for some specific choices of $h(t)$, convolution is not generally memoryless.[4]
- Convolution is stable in the BIBO sense when $h(t)$ has a finite area. To show this, we begin with the definition of convolution in (3.9) and use the inequality in (2.24).

$$\begin{aligned} |g(t)| &= \left| \int_{-\infty}^{\infty} f(\tau) \, h(t - \tau) \, d\tau \right| \\ &\leq \int_{-\infty}^{\infty} |f(\tau) \, h(t - \tau)| \, d\tau \\ &= \int_{-\infty}^{\infty} |f(\tau)| \, |h(t - \tau)| \, d\tau \end{aligned}$$

---

3. Because of this property, signals which are zero for negative time are dubbed *causal signals*.
4. A memoryless counterexample is $h(t) = \delta(t)$

To establish BIBO stability, let $|f(t)| \leq C$. Then

$$|g(t)| \leq C \int_{-\infty}^{\infty} |h(t-\tau)|\, d\tau$$

$$= C \int_{-\infty}^{\infty} |h(\tau)|\, d\tau$$

Thus, if the area $A = \int_{-\infty}^{\infty} |h(\tau)|\, d\tau$ is finite, then

$$|g(t)| \leq AC$$

and the convolution system is stable in the BIBO sense.

3.3 System Characterization

A system is characterized if, for any input, the corresponding output can be ascertained. In order to completely characterize the input-output relationship of a system of an unspecified type, every response to all possible inputs must be cataloged. If the system is linear, only the response to all impulsive inputs is required to uniquely specify the input-output relationship. If the system is both linear and time invariant, a single function, dubbed the *impulse response*, is required. The impulse response of linear time variant and linear time invariant system goes by other names. In optics and imaging, it is referred to as the *line spread function* or the *point spread function*. In electromagnetics, the impulse response is called the *Green's function*.

3.3.1 Linear System Characterization

3.3.1.1 Continuous Time Systems

The input output relationship of a system can be written in terms of the sifting property of the Dirac delta in (2.27).

$$g(t) = S\{f(t)\} = S\left\{\int_{-\infty}^{\infty} f(\tau)\delta(t-\tau)d\tau\right\}.$$

Since the integral is a continuous sum, due to the additive property of the linearity, we can write

$$g(t) = S\{f(t)\} = \int_{-\infty}^{\infty} S\{f(\tau)\delta(t-\tau)d\tau\}.$$

The operator, $S\{\ldots\}$, operates only on functions ot t. Thus, $x(\tau)$ and $d\tau$ can be extracted from operator due to the homogeneity property of linearity. The result is

$$g(t) = S\{f(t)\} = \int_{-\infty}^{\infty} f(\tau)S\{\delta(t-\tau)\}d\tau. \tag{3.12}$$

Define the impulse response, $h(t, \tau)$, as the system response to all Dirac delta inputs.

$$h(t, \tau) = S\{\delta(t-\tau)\}. \tag{3.13}$$

This is illustrated in Figure 3.6. When the input impulse is shifted to a different value of τ, the impulse response generally changes its shape.

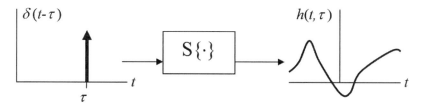

FIGURE 3.6. The linear system is totally characterized by its response to all Dirac delta inputs, $\{\delta(t-\tau)| -\infty < t < \infty, -\infty < \tau < \infty\}$, for all values of (t, τ).

Substituting (3.13) into (3.12) gives the *superposition integral*

$$g(t) = \int_{\tau=-\infty}^{\infty} f(\tau)h(t,\tau)d\tau. \tag{3.14}$$

Knowledge of the impulse response, $h(t, \tau)$, therefore suffices to totally characterize a linear system.

3.3.1.2 Discrete Time Systems

Discrete Time Systems have an analogous characterization. Let S be a linear operator. The steps for derivation of the *superposition sum* characterization of discrete time linear systems follow the same steps as the continuous time case.

$$g[n] = S\{f[n]\}$$

$$= S\left\{\sum_{m=-\infty}^{\infty} f[m]\delta[n-m]\right\} \quad ; \text{Kronecker delta sifting property}$$

$$= \sum_{m=-\infty}^{\infty} S\{f[m]\delta[n-m]\} \quad ; \text{additivity property}$$

$$= \sum_{m=-\infty}^{\infty} f[m]S\{\delta[n-m]\} \quad ; \text{homogeneity property}$$

$$= \sum_{m=-\infty}^{\infty} f[m]h[n,m] \quad ; \text{the } \textit{superposition sum}$$

where the discrete time impulse response is

$$h[n,m] = S\{\delta[n-m]\} \tag{3.15}$$

3.3.2 Causal Linear Systems

The impulse of a causal linear system is zero for $t < \tau$. Hence

$$h(t,\tau) = h(t,\tau)\mu(t-\tau). \tag{3.16}$$

To show this, we make use of the property that a zero input into a linear system produces a zero output. Thus, with reference to Figure 3.7, the input $\delta(t - \tau_1)$, zero from $-\infty < t < \tau$,

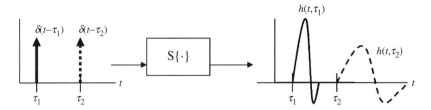

FIGURE 3.7. If a system is both linear and causal, the response to an input Dirac delta at $t = \tau$ must result in a response that is zero for $t < \tau$.

produces an output which is zero on the same interval. No output can occur before a system stimulus is applied. Thus, $h(t, \tau_1)$ in Figure 3.7 is zero for $t < \tau_1$. Similarly, in Figure 3.7, with an input $\delta(t - \tau_2)$ (drawn with a broken line), the impulse response, $h(t, \tau_2)$ (also drawn with a broken line) of a causal linear system must be zero for $t < \tau_2$. These observations substantiate (3.16).

3.3.3 Linear Time Invariant (LTI) Systems

Consider the linear time invariant system in Figure 3.8. Let an impulse, $\delta(t)$, applied at $t = 0$, have a system response of $h(t)$. Since the system is time invariant, shifting the input to $\delta(t - \tau_1)$ similarly shifts the output to $h(t - \tau_1)$. Similarly, $\delta(t - \tau_2)$ produces $h(t - \tau_2)$. Hence, if $S\{\delta(t)\} = h(t)$, then

$$S\{\delta(t - \tau)\} = h(t - \tau). \qquad (3.17)$$

Compating with (3.13), we see that the impulse reponse of an LTI system is only a function of the distance between t and τ. Specifically[5]

$$h(t, \tau) = h(t - \tau) \qquad (3.18)$$

and only a single function, $h(t)$, is required to totally characterize the LTI system.

$$h(t) = S\{\delta(t)\}. \qquad (3.19)$$

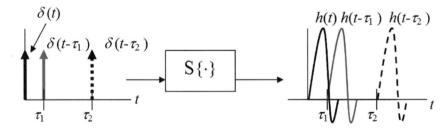

FIGURE 3.8. For a time invariant system, shifting the location of a Dirac delta input results in an equivalent shift in the response of the system.

---

5. The relation in (3.18) is a common abuse of notation. A two-dimensional function, $h(t, \tau)$, cannot be equated to a one-dimensional function of the same name.

When the impulse response in (3.19) is used, the superposition integral in (3.14) becomes the *convolution integral*

$$g(t) = \int_{-\infty}^{\infty} f(\tau)h(t-\tau)d\tau. \quad (3.20)$$

A shorthand notation for convolution is

$$g(t) = f(t) * h(t). \quad (3.21)$$

For discrete time systems, the convolution sum is

$$g[n] = \sum_{m=-\infty}^{\infty} f[m]h[n-m]$$
$$= f[n] * h[n]$$

where the discrete time LTI impulse response is

$$h[n] = S\{\delta[n]\}.$$

3.3.3.1 Convolution Algebra

Convolution algebra properties for continuous time and discrete time operations are in Tables (3.1) and (3.2). Using the definition of convolution and Fourier transformation, these properties are straightforward to show.

The *frequency response* property of convolution indicates a single frequency signal, expressed as a complex exponential, when placed through an LTI system, yields an output of the same frequency albeit with different magnitude and phase. The magnitude and phase are expressed by the frequency response as a complex number. The magnitude of the complex

TABLE 3.1. Continuous time convolution algebra.

Property	Continuous		
definition	$x(t) * h(t)$	$=$	$\int_{\tau=-\infty}^{\infty} x(\tau)h(t-\tau)d\tau$
identity element	$x(t) * \delta(t)$	$=$	$x(t)$
commutative	$x(t) * h(t)$	$=$	$h(t) * x(t)$
associative	$x(t) * \{y(t) * h(t)\}$	$=$	$\{x(t) * y(t)\} * h(t)$
distributive	$x(t) * \{y(t) + h(t)\}$	$=$	$\{x(t) * y(t)\}$
			$+\{x(t) * h(t)\}$
shift	$x(t-\tau) * h(t)$	$=$	$x(t) * h(t-\tau)$
frequency response	$h(t) * e^{j2\pi ut}$	$=$	$H(u)e^{j2\pi ut}$
Fourier transform	$x(t) * h(t)$	$\leftrightarrow$	$X(u)H(u)$
derivative	$\frac{d}{dt}[x(t) * h(t)]$	$=$	$\frac{dx(t)}{dt} * h(t)$
		$=$	$x(t) * \frac{dh(t)}{dt}$

TABLE 3.2. Discrete time convolution algebra.

Property	Discrete		
definition	$x[n] * h[n]$	$=$	$\sum_{-\infty}^{\infty} x[m]h[n-m]$
identity element	$x[n] * \delta[n]$	$=$	$x[n]$
commutative	$x[n] * h[n]$	$=$	$h[n] * x[n]$
associative	$x[n] * \{y[n] * h[n]\}$	$=$	$\{x[n] * y[n]\} * h[n]$
distributive	$x[n] * \{y[n] + h[n]\}$	$=$	$\{x[n] * y[n]\}$
			$+\{x[n] * h[n]\}$
shift	$x[n-m] * h[n]$	$=$	$x[n] * h[n-m]$
frequency response	$h[n] * e^{j2\pi nf}$	$=$	$H_D(f)e^{j2\pi nf}$
Fourier transform	$x[n] * h[n]$	$\leftrightarrow$	$X_D(f)H_D(f)$

number indicated amplification of the input complex exponential and the angle of the complex number indicates the phase shift. Specifically, for an LTI system with impulse response $h(t)$,

$$S\{e^{j2\pi vt}\} = e^{j2\pi vt} * h(t) = H(v)e^{j2\pi vt}. \qquad (3.22)$$

where the frequency response, $H(u)$, is the Fourier transform of the impulse response.

$$h(t) \longleftrightarrow H(u).$$

This property, illustrated in Figure 3.9, reveals the LTI system alters the amplitude and phase of each frequency, but otherwise does not alter the frequency. If a large mechanical body has a single component vibrating at a given frequency, then, if an LTI system, all vibrations in the body are of the same frequency, albeit at different phases and amplitudes. Likewise, in large LTI systems of arbitrarily connected resisters, inductors, capacitors and amplifiers, a single sinusoidal voltage source inserted at any part of the circuit will result in every current and voltage in the circuit oscillating at the same frequency. The phasor used in linear circuit analysis allows the magnitude and phase of these oscillations to be characterized [141, 1184].

The complex frequency response in (3.22) is physically interpreted by its real component. Taking the real part of both sides of (3.22) gives

$$S\{\cos(2\pi vt)\} = \cos(2\pi vt) * h(t) = |H(v)| \cos(2\pi vt + \angle H(v)) \qquad (3.23)$$

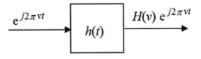

FIGURE 3.9. A sinusoidal input into an LTI system always results in a sinusoidal response of the same frequency. The magnitude change and phase shift are specified by the frequency response, $H(u)$.

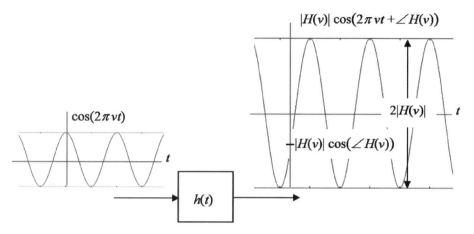

FIGURE 3.10. A cosine into an LTI system will emerge as a cosine of the same frequency with different magnitude and phase.

where

$$H(v) = |H(v)|e^{j\angle H(v)}.$$

keeps tally of the phase shift and amplitude scaling of an input cosine. This is illustrated in Figure 3.10.

3.3.3.2 Convolution Mechanics

The convolution integral in (3.20) can be interpreted using the "flip and shift" mechanics of convolution illustrated in Figure 3.11. The integral is a function of t. To visualize the operation, we begin with the impulse response, $h(\tau)$, shown in the upper left corner in Figure 3.11. We transpose this function to form $k(\tau) = h(-\tau)$ as shown. As a function of τ for a fixed t, the expression $k(\tau - t)$ is recognized as $k(\tau)$ shifted from $\tau = 0$ to $\tau = t$. This is shown in the upper right of Figure 3.11. Clearly, $k(\tau - t) = h(t - \tau)$ and $h(t - \tau)$ is the impulse response contribution to the convolution integral in (3.20). The function $h(t - \tau)$ is recognized as the "flip and shift" of $h(\tau)$.

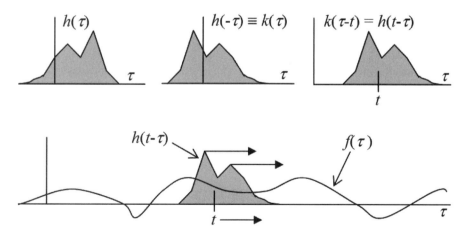

FIGURE 3.11. The "flip and shift" mechanics of convolution.

As t increases, $h(\tau - t)$ moves down the τ axis. The mechanics of the convolution integral are now clearly seen. With $f(\tau)$ fixed, as illustrated on the bottom of Figure 3.11, $h(t - \tau)$ shifts to the right as t increases. For any fixed value of t, the area (integral) of the product $f(\tau)h(t - \tau)$ produces the output, $g(t)$, for the corresponding value of t.

The procedure for discrete time convolution is similar.

3.3.3.3 Circular Convolution Mechanics

Circular convolution is obtained by performing an inverse DFT on the product of two DFT's and is the topic of Section 2.6.1. For convenience, the circular convolution sum is repeated here from (2.119).

$$y[n] = \sum_{m=0}^{N-1} x[m]h[n - m]. \tag{3.24}$$

Relation to Linear Convolution. Circular convolution can be expressed as linear convolution if we define

$$x_N[n] = \begin{cases} x[n] \, ; \, 0 \leq n < N \\ 0 \quad ; \text{ otherwise.} \end{cases}$$

Then the circular convolution in (3.24) can be written in terms of linear convolution

$$y[n] = \sum_{m=-\infty}^{\infty} x_N[m]h[n - m] = x_N[n] * h[n]. \tag{3.25}$$

This is illustrated in Figure 3.12.

Circular Convolution Mechanics. The term "circular" in circular convolutions comes from the ability to interpret (3.24) by performing the "flip and shift" operations of convolution on a circle. The values of $x_N[n]$ are labelled [a,b,c,d] in Figure 3.12 where the period is $N = 4$. Likewise, the values of $h[n]$ for its first period are labelled [A,B,C,D]. Using these same letters, the convolution of $x_N[n]$ with $h[n]$ can be interpreted as shown in Figure 3.13. The letters [a,b,c,d] corresponding to $x_N[n]$ appear on the outer circle and do not move. The [A,B,C,D] letters of $h[n]$, shifted, rotate clockwise on the inner circle. At time $n = 0$, the circular convolution result is

$$y[0] = aA + bD + cC + dB.$$

For time $n = 1$, the inner circle rotates one unit and

$$y[1] = aB + bA + cD + bC.$$

The circular shift continues one unit at a time. Since N shifts bring us back to the beginning, the output $y[n]$ is periodic with period N.

Continuous circular convolution[6] is performed analogously.

3.3.3.4 Step and Ramp Response Characterization

An LTI system's impulse response can be determined from the system's step or ramp response. Define

$$h_n(t) = S\{r_n(t)\} \tag{3.26}$$

---

6. e.g., See Table 2.6.

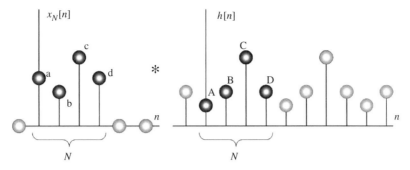

FIGURE 3.12. Illustration of performing circular convolution of two discrete time periodic signals with identical periods, N. One of the functions, $x[n]$, is truncated to a single period and forms $x_N[n]$. The linear convolution of $x_N[n]$ with $h[n]$ is the circular convolution of $x[n]$ and $h[n]$.

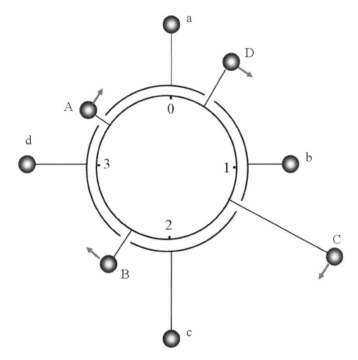

FIGURE 3.13. Illustration of the mechanics of circular convolution. The values [a,b,c,d] and [A,B,C,D] are the same as in Figure 3.12. The two figures illustrate two visualizations of the same circular convolution problem.

where

$$\left(\frac{d}{dt}\right)^n r_n(t) = \delta(t). \tag{3.27}$$

The general solution is, for $n \geq 1$,

$$r_n(t) = \frac{1}{n!} t^{n-1} \mu(t). \tag{3.28}$$

Clearly, $h_0(t) = h(t)$ is the system's impulse response. $h_1(t) = S\{\mu(t)\}$ is the system's step response and $h_2(t) = S\{t\mu(t)\}$ is the system's ramp response.

For LTI systems, differentiation is commutative with the system operator. Differentiating (3.26) and using (3.27), gives

$$\left(\frac{d}{dt}\right)^n h_n(t) = \left(\frac{d}{dt}\right)^n (S\{r_n(t)\})$$

$$= S\left\{\left(\frac{d}{dt}\right)^n r_n(t)\right\} \qquad (3.29)$$

$$= S\{\delta(t)\}$$

$$= h(t)$$

Thus, the impulse response, $h(t)$, can be obtained by differentiating the step response once or the ramp response twice. For continuous time systems, this characterization is particularly important due to the difficulty of realization of an approximation of a Dirac delta.[7] The unit step, on the other hand, can be better realized experimentally without exposing the system under test to high power levels.

3.3.3.5 Characterizing an LTI System

An LTI system is characterized by its impulse response or, equivalently, its frequency response. The frequency response can be experimentally determined by repeated experiments of the type shown in Figure 3.10. The response of the system to a cosinusiod is used to determine the magnitude and phase of the frequency response for a single frequency. Changing the frequency results in determination of a second magnitude and phase. Finding these points for all frequencies constitutes determination of the system's frequency response.[8]

Alternately, the impulse response of an LTI system can be determined from a single experiment either directly by application of a Dirac delta input or indirectly by appropriate differentiation of the system's step or ramp response. The Fourier transform of the impulse response is the system's frequency response.

3.3.3.6 Filters

Commonly used filters are LTI systems characterized by their frequency responses. They are illustrated in Figure 3.14. Ideally, they are zero phase filters in that the phase and, consequently, the imaginary part of the Fourier transform is zero. We will here treat the case for continuous time. The discrete time case is the topic of Exercise 3.16. For the continuous case, the filters are as follows

(a) The low pass filter with bandwidth B has a frequency response of

$$H(u) = \Pi\left(\frac{u}{2B}\right).$$

---

7. A Dirac delta has infinite height and infinite power and "the infinite is nowhere to be found in reality" David Hilbert.
8. For all real systems, negative frequency values follow as the result of the conjugate symmetry of the frequency response.

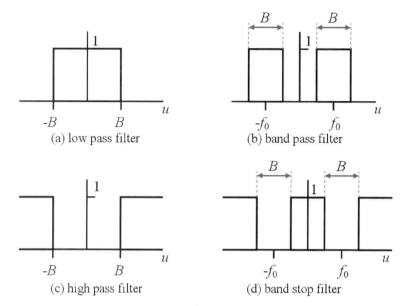

FIGURE 3.14. Commonly used ideal filters include (a) the low pass filter, (b) the bandpass filter, (c) the high pass filter and (d) band stop filter. The band stop filter is also referred to as a *notch* filter.

All frequencies components less than B are set to zero. The corresponding continuous time impulse response follows from the inverse Fourier transform

$$h(t) = 2B \operatorname{sinc}(2Bt).$$

(b) The bandpass filter passes frequencies in a band of width B centered at f_0. The frequency response of the bandpass filter is

$$H(u) = \Pi\left(\frac{u - f_0}{B}\right) + \Pi\left(\frac{u + f_0}{B}\right).$$

It follows that the impulse response of the high pass filter is

$$h(t) = 2B \operatorname{sinc}(2Bt) \cos(2\pi f_0 t).$$

(c) The high pass filter has a frequency response of

$$H(u) = 1 - \Pi\left(\frac{u}{2B}\right).$$

and an impulse response

$$h(t) = \delta(t) - 2B \operatorname{sinc}(2Bt).$$

(d) The band stop (or notch) filter rids the signal of frequency components in a band of length B centered at frequency f_0. The frequency response is

$$H(u) = 1 - \left[\Pi\left(\frac{u - f_0}{B}\right) + \Pi\left(\frac{u + f_0}{B}\right)\right]$$

122 HANDBOOK OF FOURIER ANALYSIS AND ITS APPLICATIONS

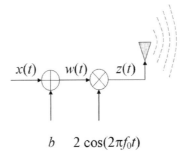

FIGURE 3.15. The process of amplitude modulation. A signal proportional to the modulated signal, $z(t)$, is broadcast by the antenna shown on the right.

and the corresponding impulse response follows as

$$h(t) = \delta(t) - 2B \operatorname{sinc}(2Bt) \cos(2\pi f_0 t).$$

3.4 Amplitude Modulation

AM radio derives from *amplitude modulation* used to multiplex audio frequencies on the AM radio frequency band. The modulation process is shown in Figure 3.15. An audio baseband signal, $x(t)$, has bandwidth B. The signal is added to a bias, b. For coherent demodulation, the value of b is unimportant.

$$w(t) = x(t) + b \geq 0. \tag{3.30}$$

This signal is multiplied by a carrier frequency, f_0, to form

$$z(t) = 2 w(t) \cos(2\pi f_0 t). \tag{3.31}$$

Using the convolution theorem, the spectrum of this signal is

$$Z(u) = W(u) * [\delta(u + f_0) + \delta(u - f_0)]$$

or

$$Z(u) = W(u + f_0) + W(u - f_0) \tag{3.32}$$

where, from (3.30),

$$W(u) = X(u) + b \,\delta(u). \tag{3.33}$$

As is illustrated in Figure 3.16, the baseband signal is shifted or *heterodyned* down the frequency axis. AM radio multiplexes a number of such baseband signals side by side on the AM frequency spectrum by using different carrier frequencies.

3.4.1 Coherent Demodulation

Coherent demodulation of the amplitude modulated signal, $z(t)$, is illustrated in Figure 3.17. The coherent modulator regains the original baseband signal, $x(t)$. For reasons to become obvious, however, it is not used for commercial AM radio.

An antenna receives a broadband signal which is passed through a bandpass filter equal to one for $f_0 - B \leq |u| \leq f_0 + B$ and zero otherwise. From Figure 3.16, this is the band of

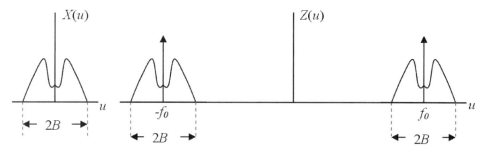

FIGURE 3.16. The spectrum, $X(u)$, of the baseband signal and the spectrum, $Z(u)$, resulting from the amplitude modulation performed in Figure 3.15. See (3.31).

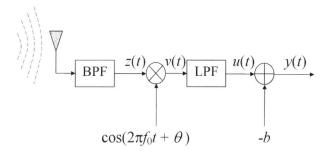

FIGURE 3.17. Coherent demodulation of an AM signal. The bandpass filter (BPF) over the band $f_0 - B \leq |u| \leq f_0 + B$. The low pass filter (LPF) passes the band of frequencies $-B \leq u \leq B$.

frequencies over which then we see the spectrum $Z(u)$. As shown in Figure 3.16, we will assume the output of the bandpass filter is $z(t)$. This signal is then multiplied by a cosinusoid with the same frequency, f_0, as the carrier. If there is coherence (i.e., the phase is $\theta = 0$), then the filter output is

$$v(t) = z(t)\cos(2\pi f_0 t) \tag{3.34}$$

or, in the frequency domain

$$V(u) = Z(u) * \frac{1}{2}[\delta(u+f_0) + \delta(u-f_0)].$$

Using the expression for $Z(u)$ in (3.31) gives

$$V(u) = \frac{1}{2}W(u+2f_0) + W(u) + \frac{1}{2}W(u-2f_0). \tag{3.35}$$

The component of interest is $W(u)$. The other terms are high frequency components and can be removed using the low pass filter (LPF) in Figure 3.17 which is one for $|u| \leq B$ and is otherwise one. The output of the low pass filter is therefore

$$u(t) = w(t)$$

or, from (3.30),

$$u(t) = x(t) + b.$$

Thus, after subtracting the bias in the last step shown in Figure 3.17, the output is the originally transmitted signal

$$y(t) = x(t). \qquad (3.36)$$

The coherent demodulator has therefore regained the original signal from the heterodyned signal, $z(t)$.

3.4.1.1 Loss of Coherence and Fading

Coherent demodulation must maintain the carrier frequency phase $\theta = 0$ in Figure 3.17 and must therefore be synchronized with the phase of the transmitter carrier in Figure 3.15. To illustrate, suppose the phase in the coherent demodulator is $\theta = -\pi/2$. The carrier thus becomes $\cos(2\pi f_0 t - \theta) = \sin(2\pi f_0 t)$ and, instead of (3.34), we have

$$v(t) = z(t) \sin(2\pi f_0 t).$$

Its Fourier transform is

$$V(u) = Z(u) * \frac{1}{j2} [\delta(u - f_0) - \delta(u + f_0)]$$

or, using (3.31)

$$V(u) = \frac{1}{j2} W(u - 2f_0) - \frac{1}{j2} W(u + 2f_0).$$

Note, in comparison to (3.35), the spectrum of the low pass signal, $W(u)$, has been destroyed by destructive interference. The output of the coherent demodulator is thus

$$y(t) = 0. \qquad (3.37)$$

Therefore, as θ changes from zero to $\pi/2$, we see *fading* of the original signal to zero. Greater details about fading are the topic of Exercise 3.6.

3.4.2 Envelope Demodulation

To avoid the synchronization of the transmitter and receiver oscillators required by coherent demodulation, an envelope detector can be used to demodulate $z(t)$. Doing so requires that the bias, b, be chosen such that,[9]

$$w(t) = x(t) + b \geq 0. \qquad (3.38)$$

The modulated signal, $z(t)$, in (3.31) then resembles that shown in Figure 3.18. The signal, $w(t)$, rides the upper *envelope* of $z(t)$. Roughly, the envelope can be found by connecting the relative maxima of the signal.[10]

The necessity of requiring $w(t) \geq 0$ is apparent in Figure 3.19 where $w(t)$ is allowed to go negative. When the envelope hits zero, we are faced with the problem of whether to follow

---

9. Choosing the bias to minimize the average power required by the modulated signal is the topic of Exercise 4.21.
10. For details of recovering $x(t)$ from the envelope, see Exercise 6.15.

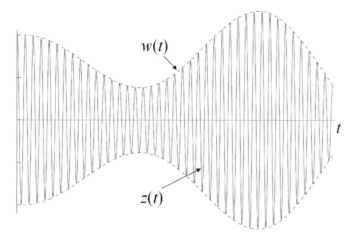

FIGURE 3.18. The modulated signal $z(t)$ in (3.31) when the bias is sufficiently large to assure that $w(t) \geq 0$. The signal, $w(t)$, then rides the upper envelope of the amplitude varying sinusoid.

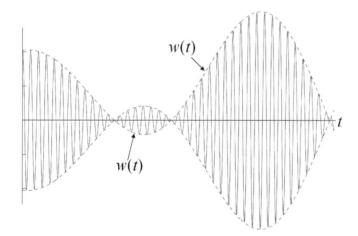

FIGURE 3.19. The modulated signal $z(t)$ in (3.31) when the bias is too small to assure that $w(t) \geq 0$.

the envelope into negative territory, or to keep the envelope positive. The information is there to make this decision. Otherwise, coherent demodulation would not be possible for all values of bias. From the perspective of envelope detection, however, taking advantage of this information is difficult.[11]

3.5 Goertzel's Algorithm for Computing the DFT

Goertzel's algorithm uses an LTI system to compute the discrete Fourier transform (DFT)[12] of signal at a single value of k. The impulse response of the filter for Goertzel's algorithm is

$$h_G[n] = e^{j2\pi(n+1)k/N} \tag{3.39}$$

---

11. See Exercise 3.7.
12. See (2.116).

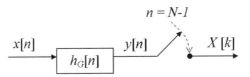

FIGURE 3.20. Illustration of Goertzel's algorithm for computing the DFT. A discrete time signal, $x[n]$, is placed into an LTI filter with an impulse response given by (3.39). The filter's output, $y[n]$, is equal to the DFT value $X[k]$ at time $n = N - 1$.

there N is the length of the DFT. The resulting convolution is

$$y[n] = x[n] * h_G[n]$$
$$= e^{j2\pi(n+1)k/N} \sum_{m=0}^{n} x[m] \, e^{-j2\pi mk/N}. \quad (3.40)$$

The output signal at time $n = N - 1$ is

$$y[N - 1] = \sum_{m=0}^{N-1} x[m] \, e^{-j2\pi mk/N}.$$

Thus, with reference to the DFT in (2.116), the output at this point in time is equal to the DFT of the input at the frequency corresponding to k.

$$y[N - 1] = X[k].$$

This is illustrated in Figure 3.20.

The Goertzel algorithm is more efficient than an FFT when only a few values of Fourier transform are required.

Implementation of Goertzel's algorithm and its use in generating short time Fourier transforms is the topic of Exercise 9.7.

3.6 Fractional Fourier Transforms

The *fractional Fourier transform* [245, 1068, 1101], a generalization of the Fourier transform [1579], was first presented by Namias [1015, 969] and has subsequently found application in sonar [830], radar [661, 1359], beamforming [1546], signal recovery [257, 415], signal synthesis [1147], pattern recognition [1558, 705], time-frequency representations [17, 1066, 1067], ultrasound [95], speech recognition [1206], target detection [1359], quantum mechanics [1015], and optics [775, 1068]. The fractional Fourier transform is also called the *angular Fourier transform* [969].

3.6.1 Periodicity of the Fourier Transform Operator

The Fourier transform operator, $\mathcal{F}$, is defined through the Fourier transformer relation in (3.8),

$$\mathcal{F} x(t) = X(t) = \int_{-\infty}^{\infty} x(\tau) e^{-j2\pi t\tau} d\tau, \quad (3.41)$$

where

$$x(t) \leftrightarrow X(u)$$

are Fourier transform pairs. From the duality theorem, application of the operator a second time gives transposition of the original signal

$$\mathcal{F}^2 x(t) = x(-t). \tag{3.42}$$

A third application gives

$$\mathcal{F}^3 x(t) = X(-t) \tag{3.43}$$

A fourth application returns to us the original signal.

$$\mathcal{F}^4 x(t) = x(t) \tag{3.44}$$

Additional applications of $\mathcal{F}$ repeat the cycle.

The operations $\mathcal{F}^n$ are linear for $n = 1, 2, 3, 4$ and therefore can be written using a superposition integral

$$\mathcal{F}^n x(t) = \int_{-\infty}^{\infty} x(\tau)\phi_n(t, \tau)d\tau \tag{3.45}$$

where the impulse kernel is

$$\phi_n(t, \tau) = \begin{cases} \delta(\tau - t) \, ; & n = 0 \\ e^{-j2\pi t\tau} \, ; & n = 1 \\ \delta(\tau + t) \, ; & n = 2 \\ e^{j2\pi t\tau} \, ; & n = 3 \end{cases} \tag{3.46}$$

The kernel $\phi_n(t, \tau)$ can be viewed as periodic in n with a period of four. Thus, for example, $\phi_9(t, \tau) = \phi_1(t, \tau)$.

3.6.2 Fractional Fourier Transform Criteria

The fractional Fourier transform is motivated by a generalization of (3.45) to operations of the form $\mathcal{F}^a x(t)$ where a is a real number. We require the generalization to be linear. Thus the generalization can be expressed as a superposition integral of the form

$$\mathcal{F}^a x(t) = \int_{-\infty}^{\infty} x(\tau)\phi_a(t, \tau)d\tau \tag{3.47}$$

The generalization then corresponds to identifying the fractional Fourier transform kernel, $\phi_a(t, \tau)$. The solution must

(a) meet the boundary conditions in (3.46) when $a = n$, and
(b) be periodic in a with a period of four.

These criteria do not specify a unique generalization.

3.6.3 The Weighted Fractional Fourier Transform

The *weighted fractional Fourier transform*, one of many possible definitions of the fractional Fourier transform meeting the boundary conditions in (3.46), is defined as [245]

$$\mathcal{F}^a x(t) = \iota_0(a)\, x(t) + \iota_1(a)\, X(t) + \iota_3(a)\, x(-t) + \iota_3(a)\, X(-t) \qquad (3.48)$$

where the weights are

$$\iota_m(a) = \frac{1}{4} \frac{1 - e^{j2\pi a}}{1 - e^{j\pi(a-m)/2}}. \qquad (3.49)$$

The kernel of the weighted fractional Fourier transform follows as

$$\phi_a(t, \tau) = \iota_0(a)\, \delta(\tau - t) + \iota_1(a)\, e^{-j2\pi t\tau} + \iota_3(a)\, \delta(\tau + t) + \iota_3(a)\, e^{j2\pi t\tau} \qquad (3.50)$$

The plots of $\iota_m(a)$ in Figure 3.21 reveal that

$$\iota_m(n) = \delta[n - m],$$

a property we can also demonstrate analytically. At $n = m$, this Kronecker delta sifts the proper term in the kernel in (3.50) and the boundary conditions in (3.46) are satisfied. The weights in (3.49) thus smoothly interpolate the fractional Fourier transform. For all values

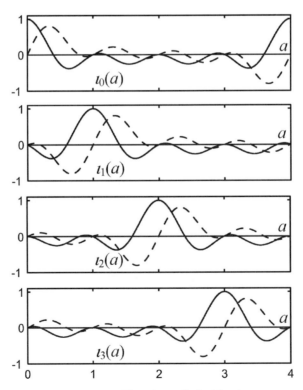

FIGURE 3.21. Plots of the real (solid line) and imaginary (dashed lines) components of $\iota_m(a)$ in (3.49).

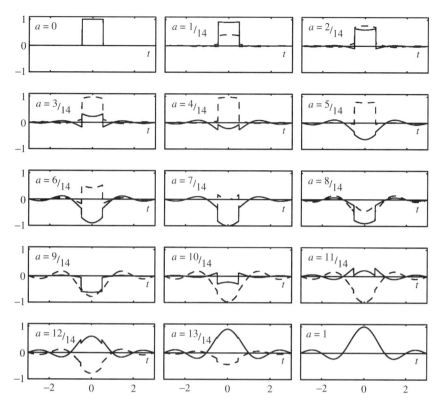

FIGURE 3.22. The weighted fractional Fourier transform of $x(t) = \Pi(t)$ for the parameter a between zero and one in stems of $\frac{1}{14}$. The real (solid line) and imaginary (dashed lines) components in each step show the transition from the rectangle ($a = 0$) function to the sinc ($a = 1$).

of a, the weights also sum to one.[13]

$$\sum_{m=0}^{3} \iota_m(a) = 1 \qquad (3.51)$$

An example of a weighted fractional Fourier transform is shown in Figure 3.22 for the case of $x(t) = \Pi(t)$.

3.7 Approximating a Linear System with LTI Systems

A linear system response can be approximated by a superposition of LTI systems. Here is one way to perform this operation. Sample τ at points $\{\tau_n | -\infty < n < \infty\}$ such that $\tau_n < \tau_{n+1}$. Then the superposition integral in (3.14) can be written as

$$\begin{aligned} g(t) &= \int_{\tau=-\infty}^{\infty} f(\tau)h(t,\tau)d\tau. \\ &= \sum_{n=-\infty}^{\infty} \int_{\tau=\tau_n}^{\tau_{n+1}} f(\tau)h(t,\tau)d\tau. \end{aligned} \qquad (3.52)$$

---

13. See Exercise 3.28.

The input, $f(\tau)$, can be chopped into intervals.

$$f_n(\tau) = \begin{cases} f(\tau) ; & \tau_n \leq \tau < \tau_{n+1} \\ 0 ; & \text{otherwise} \end{cases} \qquad (3.53)$$

and (3.52) becomes

$$g(t) = \sum_{n=-\infty}^{\infty} \int_{\tau=-\infty}^{\infty} f_n(\tau) h(t,\tau) d\tau. \qquad (3.54)$$

Choose a point s_n in each interval such that

$$\tau_n \leq s_n \leq \tau_{n+1}.$$

We would like to use the impulse response, $h(t,\tau)$ at $\tau = s_n$ to approximate an LTI system. If, indeed, the system is LTI, then $h(t,\tau) = k(t-\tau)$ where $k(t)$ is the LTI impulse response. Equivalently, for an LTI system, $h(t+\tau,\tau) = k(t)$. Motivated by this expression, we denote the impulse response assigned to the nth interval in τ by

$$k_n(t) = h(t+s_n, s_n) \qquad (3.55)$$

or

$$k_n(t-\tau) = h(t-\tau+s_n, s_n). \qquad (3.56)$$

The superposition integral in (3.54) is then approximated by

$$\begin{aligned} g(t) &\approx \sum_{n=-\infty}^{\infty} \int_{\tau=-\infty}^{\infty} f_n(\tau) k_n(t-\tau) d\tau \\ &= \sum_{n=-\infty}^{\infty} f_n(t) * k_n(t) \end{aligned} \qquad (3.57)$$

The *piecewise invariant approximation*[14] [884] in (3.57) is a summation of convolutions. As illustrated in Figure 3.23, the time variant linear system is approximated by a superposition of LTI systems. The piecewise invariant approximation meets the boundary condition of becoming exact when the linear system being approximated is time invariant.[15]

3.7.1 Examples of the Piecewise Invariant Approximation

3.7.1.1 The Magnifier

For the linear time variant magnifier in Section 3.2.2.1, the impulse response from the input-output relationship in (3.7) follows as

$$h(t,\tau) = \delta(M\tau - t).$$

---

14. See Exercise 3.19 for the discrete time equivalent.
15. See Exercise 3.18.

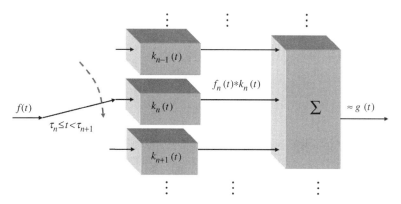

FIGURE 3.23. Block diagram visualization of the piecewise invariant approximation in (3.57). For the time interval, $\tau_n \geq t < \tau_{n+1}$, the input, $f(t)$, is fed into the LTI filter whose impulse response, $k_n(t)$, is determined from a sample of the time varying linear system impulse response, $h(t, \tau)$. The outputs of the LTI filters are summed to obtain an approximation of the linear system output, $g(t)$.

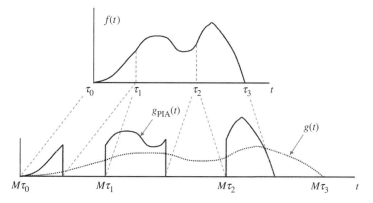

FIGURE 3.24. The piecewise invariant approximation, $g_{\text{PIA}}(t)$, of the linear time variant magnifier with an input-output relationship of $g(t) = \frac{1}{M} f\left(\frac{t}{M}\right)$. In this figure, $M = 2$.

The operation is time variant. Choose $s_n = \tau_n$. Thus, from (3.55),

$$k_n(t) = \delta\left(M\tau_n - (t + \tau_n)\right) = \delta\left((M - 1)\tau_n - t\right)$$

Using the piecewise invariant approximation in (3.57) gives

$$g(t) \approx \sum_n f_n\left(t - (M - 1)\tau_n\right)$$

This is illustrated in Figure 3.24. The piecewise invariant approximation shifts the nth subinterval from the point $t = \tau_n$ to the point $t + (M - 1)\tau_n = M\tau_n$.

3.7.1.2 The Fourier Transformer

The Fourier transformer in (3.8) has an impulse response of

$$h(t, \tau) = e^{-j2\pi t\tau}$$

Thus

$$h(t+\tau,\tau) = e^{-j2\pi(t+\tau)\tau}$$

and

$$k_n(t) = e^{-j2\pi(t+s_n)s_n}.$$

Substituting into the piecewise invariant approximation in (3.57) gives

$$g(t) \approx \sum_{n=-\infty}^{\infty} \int_{-\infty}^{\infty} f_n(\tau) e^{-j2\pi((t-\tau)+s_n)s_n} d\tau$$

$$= \sum_{n=-\infty}^{\infty} e^{-j2\pi(t+s_n)s_n} \int_{-\infty}^{\infty} f_n(t) e^{j2\pi s_n \tau} d\tau \qquad (3.58)$$

$$= \sum_{n=-\infty}^{\infty} e^{-j2\pi(t+s_n)s_n} F_n(-s_n)$$

where we use the Fourier transform pair

$$f_n(t) \longleftrightarrow F_n(u).$$

To illustrate, let $f(t) = \Pi(t)$. Divide the $\left[-\frac{1}{2}, \frac{1}{2}\right]$ interval into $2N+1$ subintervals of length $\Delta = (2N+1)^{-1}$ and center points $\{s_n = n\Delta \mid -N \leq n \leq N\}$. Then

$$f_n(t) = \Pi\left(\frac{t-s_n}{\Delta}\right) \leftrightarrow F_n(u) = \Delta \operatorname{sinc}(u\Delta) e^{-j2\pi s_n u}.$$

Substituting into (3.58) gives

$$g(t) \approx \sum_{n=-N}^{N} e^{-j2\pi(t+s_n)s_n} \Delta \operatorname{sinc}(s_n \Delta) e^{-j2\pi s_n^2}$$

$$= \Delta \sum_{n=-N}^{N} \operatorname{sinc}(s_n \Delta) e^{-j2\pi s_n t} \qquad (3.59)$$

$$= \Delta \left[1 + 2\sum_{n=1}^{N} \operatorname{sinc}(s_n \Delta) \cos(2\pi s_n t)\right].$$

Plots of this piecewise invariant approximation are shown in Figure 3.25 for various values of N. The $N = 50$ plot is very close to the desired result of $\operatorname{sinc}(t)$ which is plotted with a dotted line.

3.8 Exercises

3.1. Evaluate the following convolutions.
   (a) $\operatorname{jinc}(t) * \operatorname{sinc}(at); a \geq 2$.
   (b) $\operatorname{sech}(\pi t) * \operatorname{cosech}(\pi t)$

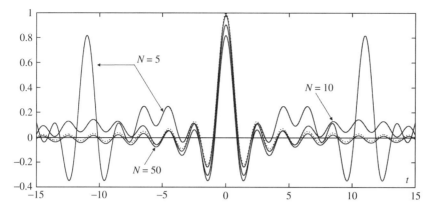

FIGURE 3.25. The piecewise invariant approximation to the Fourier transformer response to $f(t) = \Pi(t)$ given in (3.59) for $N = 5$, 10 and 50. The $N = 50$ curve is close to the desired result of $g(t) = \text{sinc}(t)$ which is plotted with a dotted line.

3.2. (a) Evaluate the autoconvolution of $h(t) = \dfrac{1}{t}$.
   (b) Are the real and imaginary parts of the Fourier transform of the unit step, $\mu(t)$, Hilbert transform pairs?
3.3. Show that, if $x(t)$ is bandlimited in the sense that $X(u) = X(u)\Pi(u/2)$, then [478]
   (a) convolving $x(t)$ with the spherical Bessel function, $2j_0(2\pi t)$, is $x(t)$.
   (b) convolving $x(t)$ with the spherical Bessel function, $j_1(2\pi t)$, gives a result proportional to the derivative of $x(t)$.
3.4. Let $z(t)$ in (3.31) be a system output.
   (a) Do we have a linear system if $x(t)$ is the input? If $w(t)$ is the input?
   (b) If linear, what is the impulse response, $h(t, \tau)$?
   (c) Do we have a time invariant system if $x(t)$ is the input? If $w(t)$ is the input?
   (d) If LTI, what is the impulse response, $h(t)$?
   (e) Do we have a memoryless system if $x(t)$ is the input? If $w(t)$ is the input?
3.5. Repeat Exercise 3.4 except the system output is $v(t)$ in (3.34) rather than $w(t)$.
3.6. **Fading.** Evaluate the strength of the signal $y(t)$ in the coherent amplitude modulation demodulator in Figure 3.17 as a function of θ. From (3.36), the strength is one (the coefficient of $x(t)$) for $\theta = 0$ and, from (3.37), the strength is zero.
3.7. Consider use of an envelope detector on a signal where $w(t)$ dips below zero. An example of such a modulated signal is shown in Figure 3.19. When the envelope hits zero, explain how we can use the phase of $z(t)$ to decide whether to follow the envelope into negative territory, or whether to continue with a positive envelope.
3.8. **Single side band suppressed carrier amplitude modulation (SSSC AM).** Consider amplitude modulation of a real signal, $x(t)$. The spectrum, $Z(u)$, of the heterodyned signal in Figure 3.16 has symmetry in that knowledge of $Z(u)$ for $f_0 < |u| \leq f_0 + B$, called the upper sideband, can be used to construct $Z(u)$ over the lower sideband $f_0 - B \leq |u| < f_0$. If only the upper side band were broadcast, then the number of baseband signals multiplexed on an AM frequency band could

be doubled. The signal $\tilde{z}(t)$ with spectrum[16]

$$\tilde{z}(t) \longleftrightarrow \tilde{Z}(u) = \begin{cases} X(u - f_0) & ; \ f_0 \leq |u| \leq f_0 + B \\ 0 & ; \ \text{otherwise} \end{cases}$$

is referred to as the *single side band suppressed carrier amplitude modulation* encoding of the baseband signal, $x(t)$. Using a bias of $b = 0$, the signal $\tilde{z}(t)$ can be generated as shown in Figure 3.15, except only the frequency band $f_0 < |u| \leq f_0 + B$ of $z(t)$ is broadcast.

(a) Consider the coherent demodulation procedure in Figure 3.17 except that $\tilde{z}(t)$ emerges from the bandpass filter instead of $z(t)$. Do we generate a form of $x(t)$ at the demodulator output when $\theta = 0$?
(b) When $\theta \neq 0$?
(c) Explain in words what happens to the demodulated signal when the oscillator frequency in Figure 3.17 is slightly greater than f_0.
(d) Slightly less than f_0.

3.9. **Mellin convolution.** The Mellin transform of a signal $f(t)$ is

$$F_M(z) = \int_0^\infty f(t) t^{z-1} dt. \tag{3.60}$$

(a) Evaluate the Mellin transform of $f\left(\dfrac{t}{\tau}\right)$ for $\tau > 0$ in terms of $F_M(z)$.
(b) Define the Mellin convolution by

$$g(t) = \int_0^\infty f(\tau) h\left(\frac{t}{\tau}\right) \frac{d\tau}{\tau}. \tag{3.61}$$

Express the Mellin transform of $g(t)$ in terms of the Mellin transforms of $h(t)$ and $f(t)$.

3.10. **Mellin transfer functions.** Similar to the Mellin convolution in (3.61) is the convolution

$$g(t) = \int_0^\infty f(\tau) h(t\tau) d\tau. \tag{3.62}$$

(a) Evaluate the Mellin transform of $f(t\tau)$ for $\tau > 0$ in terms of the Mellin transform in (3.60).
(b) Show that

$$G_M(z) = F_M(1 - z) H_M(z). \tag{3.63}$$

We will call $H_M(z)$ the *Mellin transfer function*.
(c) The Laplace transform is a special case of Mellin convolution. What is its Mellin transfer function?
(d) What is the Mellin transfer function of the Hankel transform[17] given by

$$g(t) = 2\pi \int_0^\infty \tau f(\tau) J_0(2\pi t\tau) d\tau.$$

---

[16]. Note that $\tilde{z}(t)$ is the amplitude modulation of the analytic signal, $\hat{x}(t)$, corresponding to $x(t)$. See Exercise 2.14.

[17]. The Hankel transform is introduced in Chapter 8 in (8.41).

3.11. Let $g[n] = S\{f[n]\}$ denote a discrete time system. Define discrete time systems that are
   (a) homogeneous (see Section 3.2.1.1),
   (b) additive (see Section 3.2.1.2),
   (c) linear (see Section 3.2.1.3),
   (d) time-invariant (see Section 3.2.1.4),
   (e) LTI (see Section 3.2.1.5),
   (f) causal (see Section 3.2.1.6),
   (g) memoryless (see Section 3.2.1.7),
   (h) stable (see Section 3.2.1.8), and
   (1) invertible (see Section 3.2.1.9),
3.12. Specify a system that is
   (a) additive but not homogeneous.
   (b) homogeneous but not additive.
3.13. **IIR Filters.** An IIR (infinite impulse response) filter can be expressed as

$$\sum_{k=0}^{Q} a_k\, y[n-k] = \sum_{i=0}^{P} b_i\, x[n-i] \qquad (3.64)$$

where $x[n]$ is the input and $y[n]$ is the outputs. The coefficients are fixed constants.
   (a) Is this system LTI?
   (b) If so, what is its impulse response and frequency response?
3.14. **Linear differential equations with constant coefficients.** Repeat Exercise 3.13 for the linear differential equation with constant coefficients.

$$\sum_{k=0}^{Q} a_k \left(\frac{d}{dt}\right)^k y(t) = \sum_{i=0}^{P} b_i \left(\frac{d}{dt}\right)^i x(t).$$

3.15. Show that the superposition integral in (3.14) can be written as

$$g(t) = \int_{-\infty}^{\infty} F(u) H(t, u)\, du. \qquad (3.65)$$

where $f(t) \leftrightarrow F(u)$ and

$$H(t, u) = \int_{-\infty}^{\infty} h(t, \tau) e^{j2\pi u \tau}\, d\tau.$$

3.16. The four continuous time filters presented in Section 3.3.3.6 are (a) the low pass filter, (b) the bandpass filter, (c) the high pass filter and (d) band stop filter. Derive the impulse responses, $h[n]$, for these signals for the discrete time equivalent. Assume the frequency responses in Figure 3.14 are functions of f and show a single period of the frequency response on $\left[-\frac{1}{2}, \frac{1}{2}\right]$.

3.17. Unlike space, time only flows in one dimension. As a consequence, all real temporal systems must be causal. The LTI system in Figure 3.26 seems to violate this property. The output is

$$g(t) = \varepsilon(t) + \varepsilon(t - \tau) \qquad (3.66)$$

where

$$\varepsilon(t) = f(t) + g(t).$$

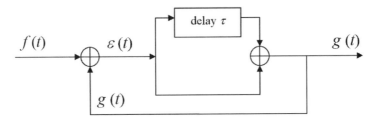

FIGURE 3.26. A system which is apparently noncausal. See Exercise 3.17.

Substituting (3.66) gives

$$\varepsilon(t - \tau) = -f(t).$$

Thus,

$$\varepsilon(t) = -f(t + \tau).$$

Substituting the last two equations into (3.66) gives

$$g(t) = -f(t) - f(t + \tau).$$

Thus, the output at time t depends on the input *at a future time*, $t + \tau$. This appears to not be a causal system. Since all temporal systems must be causal, something here is wrong. What is it?

3.18. Show the piecewise invariant approximation in (3.57) becomes exact for LTI systems.

3.19. Derive the equivalent of the piecewise invariant approximation in (3.57) for discrete time systems.

3.20. Show that, when $s_n = nT$ for a fixed T, the piecewise invariant approximation for the Fourier transformer in (3.58) is periodic with period $1/T$.

3.21. Time variant linear systems a and b have impulse responses $h_a[n, m]$ and $h_b[n, m]$. A signal, $x[n]$, is placed into system a whose output is placed into system b.
   (a) Express the impulse response, $h[n, m]$, of the composite system in terms of $h_a[n, m]$ and $h_b[n, m]$.
   (b) Does the order of the cascade of the systems matter?
   (c) Under what sufficient condition does the cascade order of the systems not matter?
   (d) Can there exist two time variant systems that, when cascaded, results in a time invariant system?

3.22. The magnifier in (3.7) is subjected only to signals that are identically zero for $-\infty < t \leq 0$. If $M = -1$, does the input-output relationship display causality?

3.23. Consider the linear system defined by

$$g(t) = f(t) - \frac{1}{M} f\left(\frac{t}{M}\right) \qquad (3.67)$$

Derive an expression to invert this system, i.e., given $g(t)$, find $f(t)$. A two dimensional illustration of this problem is shown in Figure 3.27.

FIGURE 3.27. Illustration of (3.67) as an image for $M = \frac{1}{2}$. The image on the left, *Monika*, is degraded by superimposing a smaller brighter negative version of the image as shown on the right. (Gray scales are scaled for display purposes.) The restoration problem is to recover the original image from the degradation.

3.24. The input to an LTI system is, for a fixed ξ,

$$f(t) = \begin{cases} 0 \;; & t \leq 0 \\ t \;; & 0 \leq t \leq \xi \\ \xi \;; & t \geq \xi \end{cases}$$

Express the impulse response of the system, $h(t)$, as a series of derivatives of the output. HINT: Consider the solution methodology used in Exercise 3.23.

3.25. **Linear Motion Blur** [297]. A causal signal, $f(t)$ is subjected to uniform motion blur.

$$g(t) = f(t) * \Pi\left(\frac{t-T}{2T}\right) \quad (3.68)$$

where T is a known time duration. An image illustration of this linear motion blur, shown in Figure 3.28. Specify a procedure to invert the system, i.e., find $f(t)$ from knowledge of $g(t)$ and T. Hint: One might perform a Fourier transform of the convolution in (3.68) to obtain

$$G(u) = 2TF(u)\operatorname{sinc}(2Tu)e^{-j2\pi uT}.$$

Then solve for $F(u)$ and inverse transform. The reciprocal of the sinc, however, contains undesirable poles and renders the procedure ill-posed. Here is an alternate approach. Differentiate $g(t)$ and apply the methodology used in Exercise 3.24.

3.26. **Causal convolution**. If both $x(t)$ and $h(t)$ are causal, show that the convolution integral can be written as

$$g(t) = \int_{\tau=0}^{t} x(\tau)h(t-\tau)d\tau.$$

3.27. The following are examples of systems, $g(t) = S\{f(t)\}$. Categorize each as linear and/or time-invariant. Assume that ξ is a given number, p a given integer, and $h(t)$ or $h[n]$ is a specified function. If linear, determine the impulse response.

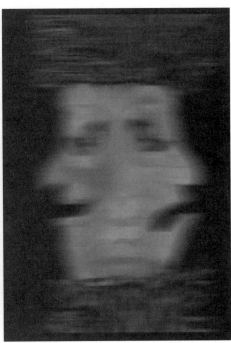

FIGURE 3.28. The image on the left, *Modulo Man*, is degraded by a horizontal linear motion blur from left to right. The restoration problem is to find the original image from the smeared degradation on the right given by (3.68). See Exercise 3.25.

(a) $g[n] = |f[n-p]|^2$.
(b) $g[n] = h[n]f[n]$.
(c) $g[n] = h(f[n])$.
(d) $g[n] = nf[n]$.
(e) $g[n] = f[n] - f[n-1]$.
(f) $g[n] = 3f[n] - 2$.
(g) $g[n] = \lfloor f[n] \rfloor$.
(h) $g[n] = f[p]f[n]$.
(i) $g[n] = f[2n]$.
(j) $g[n] = f[n^2]$.

(k) $g(t) = f(t) + \xi f(t-\xi) - \frac{d}{dt}f(t)$.
(l) $g(t) = \delta(f(t))$.
(m) $g(t) = \int_{-\infty}^{\infty} f(\tau)d\tau$.
(n) $g(t) = f(h(t)))$.
(o) $g(t) = f(f(t))$.
(p) $g(t) = \int_0^{f(t)} h(\tau)d\tau$.
(q) $g(t) = \frac{d}{dt}f(t)$.
(r) $g(t) = 1$.
(s) $g(t) = 0$.

3.28. Show that the sum of the weights in the weighted fractional Fourier transform sum to one as in (3.51).

3.29. Some functions are their own Fourier transforms.
   (a) Evaluate the weighted fractional Fourier transform for $x(t) = e^{-\pi t^2}$.
   (b) Evaluate the weighted fractional Fourier transform for $x(t) = K_n(t)$ where $K_n(t)$ is the Hermite-Gaussian function in (2.92). Hint: From Exercise 2.46, Hermite polynomials are evenly odd in the sense that

$$H_n(t) = (-1)^n H_n(-t).$$

(c) What does the result in (b) say about weighted fractional Fourier transforms that are expressed as a series

$$x(t) = \sum_{n=0}^{\infty} \alpha_n K_n(t).$$

3.30. Let $x(t)$ be bandlimited with bandwidth B. Can the weighted fractional Fourier transform also be a bandlimited function?

3.31. Show that the weighted fractional Fourier transform, $\mathcal{F}^a$ is periodic in a with a period of two when $x(t)$ is real and even.

3.32. § **Transfinite numbers and systems theory.** Georg Cantor (1845-1918), the creator of set theory, also developed the theory of transfinite numbers [7, 244, 482]. Basically, Cantor showed that some infinities are larger than others.[18] An example of a set containing the smallest transfinite number of elements, $\aleph_0$, is the set of nonnegative integers, $\{0, 1, 2, 3, \ldots\}$. An example of the next largest transfinite number, $\aleph_1$, is the number of elements in the set of all real numbers. An example of the next largest, $\aleph_2$, is the set of all functions of a real variable. Interestingly, like comprehending a fourth or higher spatial dimension,[19] comprehension of $\aleph_3$ or higher escapes the intuition [482]. Such sets, however, can be constructed mathematically. The set of all subsets of a set with $\aleph_N$ elements contains $\aleph_{N+1}$ elements.[20]

Consider then, characterizing a continuous time system. If we know nothing about the system, we require the cataloging of every response to every input. Thus we require knowledge of $\aleph_2$ input-output relationships. If the system is linear and time variant, we require the response of the system to a Dirac delta at every instance in the input time domain, i.e., for all real numbers. Thus, we require the cataloging of $\aleph_1$ input-output relationships for complete system characterization. If the system is LTI, a single input-output relationship is required: the response of the system to an input impulse at time $t = 0$. For any given input, the output is given by the convolution with the impulse response thus generated.

(a) What transfinite numbers analogously correspond to characterizing discrete time systems?

(b) Specify a scenario where a continuous time linear system is completely characterized by $\aleph_0$ input-output relationships.

3.9 Solutions for Selected Chapter 3 Exercises

3.1. (a)

$$\text{jinc}(t) * \text{sinc}(at) \longleftrightarrow \sqrt{1-u^2}\ \Pi\left(\frac{u}{2}\right)\left[\frac{1}{a}\Pi\left(\frac{u}{a}\right)\right]$$

$$= \frac{1}{a}\sqrt{1-u^2}\ \Pi\left(\frac{u}{2}\right); a \geq 2.$$

Thus

$$\text{jinc}(t) * \text{sinc}(at) = \frac{1}{a}\text{jinc}(t).$$

---

18. $\aleph_0 \neq \infty$. The symbol ∞, rather, is always used mathematically in terms of a limit.
19. See Section 8.3.
20. Possibilities of sets with $\aleph_{\aleph_0}$ or even $\aleph_{\aleph_{\aleph_0}}$ elements etc., prompted Cantor to correspond with the Vatican on possible theological implications of his theory [331].

(b) From Table 2.5 and the duality theorem

$$\operatorname{cosech}(\pi t) \longleftrightarrow -j \tanh(\pi u).$$

Since $\operatorname{sech}(\pi t) \longleftrightarrow \operatorname{sech}(\pi u)$, we conclude

$$\operatorname{sech}(\pi t) * \operatorname{cosech}(\pi t) \longleftrightarrow -j \operatorname{sech}(\pi u) \tanh(\pi u).$$

From Table 2.5 and duality, we also conclude

$$\pi u \operatorname{sech}(\pi u) \longleftrightarrow -j \operatorname{sech}(\pi u) \tanh(\pi u).$$

Hence

$$\operatorname{sech}(\pi t) * \operatorname{cosech}(\pi t) = \pi u \operatorname{sech}(\pi u).$$

3.2. (a) Since

$$\frac{1}{t} \leftrightarrow j\pi \operatorname{sgn}(t),$$

if follows that

$$\frac{1}{t} * \frac{1}{t} \leftrightarrow -\pi^2.$$

Thus

$$\frac{1}{t} * \frac{1}{t} = -\pi^2 \delta(t).$$

Note: Since Hilbert transformation is equivalent to convolving with $\frac{-1}{\pi t}$, this problem illustrates the Hilbert transforming a Hilbert transformation results in the negation of the original signal. Equivalently, an inverse Hilbert transform is obtained by convolving with $\frac{1}{\pi t}$.

(b) Yes. From (2.99) and (2.100), for a causal signal,

$$I(u) = \frac{-1}{\pi u} * R(u),$$

and

$$R(u) = \frac{1}{\pi u} * I(u).$$

For the unit step (see Table 2.4),

$$\mu(t) \longleftrightarrow X(u) = \frac{1}{2}\left(\delta(u) - \frac{j}{\pi u}\right). \tag{3.69}$$

For $X(u) = R(u) + jI(u)$, we have $I(u) = \frac{-j}{2\pi u}$. Clearly, from (2.99), we have, proceeding as in (a), the Hilbert transform

$$I(u) * \frac{-1}{2\pi u} = \frac{-1}{2\pi u} * \frac{-1}{2\pi u} = \frac{1}{2}\delta(u).$$

From (3.69) this is, indeed, $R(u)$.

3.3. (a) From the Fourier transform of the spherical Bessel function in (2.83),

$$2j_0(2\pi t) \leftrightarrow P_0(u)\Pi(u/2) = \Pi(u/2).$$

Therefore, the zeroth order spherical Bessel function is a sinc.

$$j_0(2\pi t) = 2\,\text{sinc}(2t),$$

Thus

$$x(t) * 2\,j_0(2\pi t) \leftrightarrow \left[X(u)\Pi\left(\frac{u}{2}\right)\right]\Pi\left(\frac{u}{2}\right) = X(u),$$

and, as was to be shown,

$$x(t) * 2\,j_0(2\pi t) = x(t).$$

(b) From (2.83),

$$-4\pi j_1(2\pi t) \leftrightarrow \frac{1}{2} j4\pi P_1(u)\Pi(u/2) = j2\pi u\,\Pi(u/2).$$

Then

$$-4\pi j_1(2\pi t) * x(t) \leftrightarrow [j2\pi u\,\Pi(u/2)]\,[X(u)\Pi(u/2)] = j2\pi u X(u).$$

From the derivative theorem, we therefore conclude

$$-4\pi j_1(2\pi t) * x(t) = \frac{d}{dt}x(t).$$

3.10. **Mellin transfer functions**

(a) Setting $\xi = t\tau$ and using (3.60), the Mellin transform of $f(t\tau)$ is

$$\int_0^\infty f(t\tau)t^{z-1}\,dt = \int_0^\infty f(\xi)\left(\frac{\xi}{\tau}\right)^{z-1}\frac{d\xi}{\tau} = \tau^{-z}F_M(z).$$

(b) Applying this result to (3.62) gives

$$G_M(z) = \int_{\tau=0}^\infty f(\tau)\left[\tau^{-z}H_M(z)\right]d\tau = H_M(z)\int_{\tau=0}^\infty f(\tau)\tau^{(-z+1)-1}\,d\tau$$

from which the desired solution in (3.63) follows.

(c) For the Laplace transform, $h(t) = e^{-t}$. Using the definition of the gamma function in (2.64) gives the following Mellin transfer function for the Laplace transform.

$$H_M(z) = \int_0^\infty t^{z-1}e^{-t}\,dt = \Gamma(z)\,;\;\Re z > 0.$$

3.11. The discrete time system is defined analogous to (3.1) as $g[n] = S\{f[n]\}$. The answers to this problem are straightforward restatements of continuous time counterparts.

(a) From Section 3.2.1.1, a discrete time system is homogeneous if, analogous to (3.2),

$$S\{a\,f[n]\} = aS\{f[n]\} \qquad (3.70)$$

where a is a scalar.

(b) From Section 3.2.1.2, a discrete time system is additive if, analogous to (3.3),

$$S\{f_1[n] + f_2[n]\} = S\{f_1[n]\} + S\{f_2[n]\} \qquad (3.71)$$

(c) From Section 3.2.1.3, a discrete time system is linear if it is both additive and homogeneous.

(d) From Section 3.2.1.4, a discrete time system is time-invariant if the system operator, $S\{\cdot\}$, commutes with all shift operators.

(e) From Section 3.2.1.5, a discrete time system is linear time-invariant (LTI) if is both linear and time-invariant.

(f) From Section 3.2.1.6, a discrete time system is causal if the response at $n = k$, $g[k]$, is determined only by the input $f[n]$ for $n \leq k$.

(g) From Section 3.2.1.7, a discrete time system is memoryless if the response at $n = k$, $g[k]$, is determined only by the input at $n = k, f[k]$.

(h) From Section 3.2.1.8, a discrete time system is stable in the BIBO sense if for all bounded inputs, $|f[n]| \leq N < \infty$, there exists a finite number $M < \infty$ such that $g[n] \leq M$.

(i) From Section 3.2.1.9, a discrete time system is invertible if its input can be determined uniquely by its output. If $S\{f[n]\} = g[n]$ then there exists an operator S^{-1} such that $S^{-1}\{g[n]\} = f[n]$.

3.12. (a) An example of a system that is additive but not homogeneous is the conjugation operation, $S\{f(t)\} = f^*(t)$ is additive since the sum of the conjugate is the conjugate of the sum. It is not, however, homogeneous since $S\{a\,f(t)\} = a^*f^*(t) \neq a\,f^*(t)$. There is no known example when the signals and systems are constrained to be real. (b) A system that is homogeneous but not additive has yet to be identified.

3.13. **IIR Filters**

(a) Yes, the system is LTI.

(b) Perform a DTFT in (2.112) on both sides of (3.64) and use the shift theorem[21] to give the frequency response

$$H_D(f) = \frac{Y_D(f)}{X_D(f)} = \frac{\sum_{k=0}^{Q} a_k e^{-j2\pi k f}}{\sum_{i=0}^{P} i^{-j2\pi i f}}.$$

The corresponding impulse response is obtained by applying the inverse DTFT in (2.113). There is no closed form solution.

3.15. The power theorem in (2.101) can be written

$$\int_{-\infty}^{\infty} f(\tau)k^*(\tau)\,dt = \int_{-\infty}^{\infty} F(u)K^*(u)\,du. \qquad (3.72)$$

---

21. See the entry in Table 2.6.

Set $k^*(\tau) = h(t, \tau)$ for a fixed t. Then

$$K^*(u) = \left[\int_{-\infty}^{\infty} h^*(t, \tau)e^{-j2\pi u\tau} d\tau\right]^*$$

$$= \int_{-\infty}^{\infty} h(t, \tau)e^{j2\pi u\tau} d\tau$$

$$\equiv H(u, t).$$

Hence, (3.72) becomes (3.65).

3.20. When $s_n = nT$ for a fixed T, the piecewise invariant approximation for the Fourier transformer in (3.58) is

$$g(t) \approx \sum_{n=-\infty}^{\infty} e^{-j2\pi(t+s_n)s_n} F_n(-s_n)$$

$$= \sum_{n=-\infty}^{\infty} \left[F_n(-s_n)e^{-j2\pi s_n^2}\right] e^{-j2\pi ntT}.$$

This is a Fourier series with Fourier coefficients

$$c_n = F_n(-s_n)e^{-j2\pi s_n^2}.$$

The period is $1/T$.

3.21. (a)

$$y[n] = \sum_{k=-\infty}^{\infty} h_b[n, k]\hat{x}[k]$$

$$= \sum_{k=-\infty}^{\infty} h_b[n, k] \left[\sum_{m=-\infty}^{\infty} h_a[k, m]x[m]\right]$$

$$= \sum_{m=-\infty}^{\infty} \left[\sum_{k=-\infty}^{\infty} h_b[n, k]h_a[k, m]\right] x[m]$$

$$= \sum_{m=-\infty}^{\infty} \hat{h}[n, m]x[m]$$

where the composite impulse response is

$$\hat{h}[n, m] = \sum_{k=-\infty}^{\infty} h_b[n, k]h_a[k, m].$$

The results are similar to that obtained by multiplying two matrices.

(b) As is the case with multiplying two matrices, order matters.

(c) Order does not matter when the linear systems are time invariant[22].

---

22. There may be preferred ordering in certain cases. A low pass filter, for example, should precede a differentiator so that a smooth signal rather than a potentially spiky signal is differentiated.

(d) Yes. A Fourier transformer followed by an inverse Fourier transformer, for example, gives the degenerate albeit valid counterexample of a time invariant identity operation.

3.22. Yes, the input-output relationship is causal. All (nonzero) response points are for $t \geq 0$ and occur after the stimuli for $t \leq 0$.

3.23. To solve (3.67), define the magnifier operator, $\mathcal{M}$, by

$$\mathcal{M}\{f(t)\} = \frac{1}{M} f\left(\frac{t}{M}\right)$$

so that (3.67) can be written

$$g(t) = (1 - \mathcal{M}) f(t).$$

We will assume the operator is invertible and solve for $f(t)$.

$$f(t) = (1 - \mathcal{M})^{-1} g(t). \tag{3.73}$$

Expand the operator into a geometric series

$$(1 - \mathcal{M})^{-1} = \sum_{k=0}^{\infty} \mathcal{M}^k$$

so that (3.73) becomes

$$f(t) = \sum_{k=0}^{\infty} \mathcal{M}^k g(t) \tag{3.74}$$

or

$$f(t) = \sum_{k=0}^{\infty} \frac{1}{M^k} g\left(\frac{t}{M^k}\right). \tag{3.75}$$

This is our desired answer.

The steps in (3.73) through (3.75) lack a solid mathematical foundation. The problem under consideration, though, is of the type that can be verified by substitution of the result into the problem statement. If the answer works, there is no need to substantiate the steps leading to it.

The mechanics of the solution in (3.75) can be best illustrated by truncating the series to K terms and letting K tend to infinity.

$$f_K(t) = \sum_{k=0}^{K} \frac{1}{M^k} g\left(\frac{t}{M^k}\right). \tag{3.76}$$

The first few terms are

$$f_0(t) = g(t) = f(t) - \frac{1}{M} f\left(\frac{t}{M}\right),$$

$$f_1(t) = \left[f(t) - \frac{1}{M} f\left(\frac{t}{M}\right)\right] + \left[\frac{1}{M} f\left(\frac{t}{M}\right) - \frac{1}{M^2} f\left(\frac{t}{M^2}\right)\right]$$

$$= f(t) - \frac{1}{M^2} f\left(\frac{t}{M^2}\right),$$

$$f_2(t) = \left[f(t) - \frac{1}{M^2} f\left(\frac{t}{M^2}\right)\right] + \left[\frac{1}{M^2} f\left(\frac{t}{M^2}\right) - \frac{1}{M^3} f\left(\frac{t}{M^3}\right)\right]$$

$$= f(t) - \frac{1}{M^3} f\left(\frac{t}{M^3}\right).$$

The pattern is now evident and we can show by induction that

$$f_K(t) = f(t) - \frac{1}{M^K} f\left(\frac{t}{M^K}\right).$$

The limit of the unwanted residual term is

$$\lim_{K \to \infty} \frac{1}{M^K} f\left(\frac{t}{M^K}\right) = \begin{cases} 0 & ; M > 1 \\ A\,\delta(t) & ; M < 1 \end{cases}$$

where[23]

$$A = \int_{-\infty}^{\infty} f(t)dt.$$

The iteration therefore converges for $M > 1$ and converges with a Dirac delta artifact for $M < 1$. Thus, with reference to Figure 3.27, if we attempt to restore the larger signal by first demagnifying and subtracting, there will be an asymptotic delta in the middle of the restored image. If the smaller of the two images on the right side of Figure 3.27 is restored, there will be no artifact.

3.24. The input can be written as

$$f(t) = r(t) - r(t - \xi)$$

where the ramp is $r(t) = t\,\mu(t)$. From (3.29), we have

$$g'(t) = h(t) - h(t - \xi).$$

Define $w(t) := g'(t)$. Then, using the shift theorem

$$w(t) \longleftrightarrow W(u) = \left(1 - e^{j2\pi u\xi}\right) H(u).$$

---

23. See, e.g., Exercise 2.16.

Solving for $H(u)$ and using a geometric series gives

$$H(u) = \frac{W(u)}{1 - e^{j2\pi u \xi}} = \sum_{n=0}^{\infty} W(u) e^{j2\pi n u \xi}. \qquad (3.77)$$

Inverse transforming gives

$$h(t) = \sum_{n=0}^{\infty} w(t - n\xi).$$

The accuracy of this solution becomes more evident when we define

$$h_N(t) := \sum_{n=0}^{N} g'(t - n\xi).$$

so that $h_N(t)$ approaches the impulse response, $h(t)$, as $N \longrightarrow \infty$ in the following pointwise sense.

$$h_0(t) = h(t) \, ; \; 0 \le t < \xi,$$
$$h_1(t) = h(t) \, ; \; 0 \le t < 2\xi,$$
$$h_2(t) = h(t) \, ; \; 0 \le t < 3\xi,$$

so that, in general

$$h_N(t) = h(t) \, ; \; 0 \le t < (N+1)\xi. \qquad (3.78)$$

As is the case in the solution to Exercise 3.23, use of the geometric series in (3.77) should be questioned since

$$|e^{j2\pi u \xi}| \not< 1$$

as required.[24] As is the case in the solution in Exercise 3.23, however, this questionable step becomes moot in light of the success of the solution in (3.78).

3.25. **Linear Motion Blur** [297]. Differentiate (3.68)

$$\frac{d}{dt} g(t) = f(t) * \frac{d}{dt} \Pi \left(\frac{t - T}{2T} \right).$$

or

$$\frac{d}{dt} g(t) = f(t) * [\delta(t) - \delta(t - 2T)] = f(t) - f(t - 2T). \qquad (3.79)$$

The horizontal derivative of the right to left linear motion blur in Figure 3.28 is shown in Figure 3.29. A negative of the original image is shown to the right of the original image. Applying the methodology of the solution to Exercise 3.24 gives

$$f(t) = \sum_{n=0}^{\infty} g'(t - 2nT).$$

If we define

$$f_N(t) = \sum_{n=0}^{N} g'(t - 2nT) \qquad (3.80)$$

---

24. See (14.6).

FIGURE 3.29. Illustration of the horizontal derivative of the linearly blurred image *Modulo Man* shown in Figure 3.28. This is an illustration of (3.79). To restore the original image according to the formula in (3.80), repeatedly shift this image to the right and add.

we have, as in the solution to Exercise 3.24,

$$f_N(t) = f(t) \; ; \; 0 \leq t < 2(N+1)T.$$

Thus, the image in Figure 3.29 is shifted and added to eliminate the negative image on the right. Doing so inserts another negative image farther to the right. The original image is shifted and added to eliminate this image, etc.

3.27. See Table 3.3.

3.28. Substituting (3.49) into (3.51) gives

$$\sum_{m=0}^{3} \iota_m(a) = \frac{1-e^{j2\pi a}}{4} \sum_{m=0}^{3} \frac{1}{1-e^{j\pi(a-m)/2}}$$

$$= \frac{1-e^{j2\pi a}}{4} \left[\frac{1}{1-e^{j\pi a/2}} + \frac{1}{1-e^{j\pi(a-1)/2}} \right.$$

$$\left. + \frac{1}{1-e^{j\pi(a-2)/2}} + \frac{1}{1-e^{j\pi(a-3)/2}} \right]$$

$$= \frac{1-e^{j2\pi a}}{4} \left[\frac{1}{1-e^{j\pi a/2}} + \frac{1}{1+je^{j\pi a/2}} + \frac{1}{1+e^{j\pi a/2}} + \frac{1}{1-je^{j\pi a/2}} \right]$$

(3.81)

TABLE 3.3. Solution to Exercise 3.27.

	Input-output relation	Linear	TI	Impulse response
(a)	$g[n] = |f[n-p]|^2$	no	yes	
(b)	$g[n] = r[n]f[n]$	yes	no	$h[n,k] = r[n]\delta[n-k]$
(c)	$g[n] = r(f[n])$	no	no	
(d)	$g[n] = nf[n]$	no	no	
(e)	$g[n] = f[n] - f[n-1]$	yes	yes	$h[n] = \delta[n] - \delta[n-1]$
(f)	$g[n] = 3f[n] - 2$	no	yes	
(g)	$g[n] = \lfloor f[n] \rfloor$	no	yes	
(h)	$g[n] = f[p]f[n]$	no	no	
(i)	$g[n] = f[2n]$	yes	no	$h[n,k] = \delta[k - 2n]$
(j)	$g[n] = f[n^2]$	yes	no	$h[n,k] = \delta[k - n^2]$
(k)	$g(t) = f(t) + \xi f(t-\xi) - \dfrac{d}{dt}f(t)$	yes	yes	$h(t) = \delta(t) + \xi\delta(t-\xi) + \delta'(t)$
(l)	$g(t) = \delta(f(t))$	no	yes	
(m)	$g(t) = \int_{-\infty}^{\infty} f(\tau)d\tau$	yes	yes	$h(t) = 1$
(n)	$g(t) = f(h(t))$	yes	no'	$h(t,\tau) = \delta(\tau - h(t))$
(o)	$g(t) = f(f(t))$	no	no	
(p)	$g(t) = \int_0^{f(t)} h(\tau)d\tau$	no	yes	
(q)	$g(t) = \dfrac{d}{dt}f(t)$	yes	yes	$h(t) = \delta'(t)$
(r)	$g(t) = 1$	no	yes	
(s)	$g(t) = 0$	yes	yes	$h(t) = 0$

$$= \frac{1-e^{j2\pi a}}{4}\left[\frac{2}{(1-e^{j\pi a/2})(1+e^{j\pi a/2})} + \frac{2}{(1+je^{j\pi a/2})(1-je^{j\pi a/2})}\right]$$

$$= \frac{1-e^{j2\pi a}}{4}\left[\frac{2}{1-e^{j\pi a}} + \frac{2}{1+e^{j\pi a}}\right]$$

$$= \frac{1-e^{j2\pi a}}{4}\left[\frac{4}{(1-e^{j\pi a})(1+e^{j\pi a})}\right]$$

$$= \frac{1-e^{j2\pi a}}{4}\left[\frac{4}{1-e^{j2\pi a}}\right]$$

$$= 1$$

3.29. (a) For $x(t) = e^{-\pi t^2}$, we have $X(t) = x(-t) = X(-t) = e^{-\pi t^2}$. Thus, from (3.48), the weighted fractional Fourier transform is

$$\mathcal{F}^a e^{-\pi t^2} = e^{-\pi t^2} \sum_{m=0}^{3} \iota_m(a)$$

Since, from (3.49), the sum of the weights is one, we see that the weighted fractional Fourier transform of the Gaussian is the same for all a.

$$\mathcal{F}^a \, e^{-\pi t^2} = e^{-\pi t^2}.$$

(b) From (2.91), the Fourier transform of the Hermite-Gaussian function is

$$K_n(t) \longleftrightarrow (-j)^n K_n(u).$$

Thus, taking advantage of the evenly odd property of the Hermite polynomial

$$x(t) = K_n(t),$$

$$X(t) = j^{-n} K_n(t),$$

$$x(-t) = (-1)^n K_n(t),$$

and

$$X(-t) = j^n K_n(t),$$

The weighted fractional Fourier transform follows as

$$\mathcal{F}^a K_n(t) = \varpi_n K_n(t)$$

where

$$\varpi_n = \iota_0(a) + j^{-n} \iota_1(a) + (-1)^n \iota_2(a) + j^n \iota_3(a)$$

$$= \frac{1 - e^{j2\pi}}{4} \left[\frac{1}{1 - e^{j\pi a/2}} + \frac{j^{-n}}{1 + j\, e^{j\pi a/2}} \right.$$

$$\left. + \frac{(-1)^n}{1 + e^{j\pi a/2}} + \frac{j^n}{1 - j\, e^{j\pi a/2}} \right].$$

The coefficients ϖ_n are periodic with period four, i.e., $\varpi_{n+4} = \varpi_n$. Algebraic manipulation similar to (3.81) gives

$$\varpi_n = \frac{1}{4} \left[\left(1 + e^{j\pi a}\right) \left\{ \left(1 + e^{j\pi a/2}\right) + (-1)^n \left(1 - e^{j\pi a/2}\right) \right\} \right.$$

$$\left. + j^n \left(1 - e^{j\pi a}\right) \left\{ \left(1 + j\, e^{j\pi a/2}\right) + (-1)^n \left(1 - j\, e^{j\pi a/2}\right) \right\} \right].$$

Simplifying gives

$$\varpi_n = \begin{cases} 1 & ; n = 0 \\ e^{j3\pi a/2} & ; n = 1 \\ e^{j2\pi a} & ; n = 2 \\ e^{j\pi a/2} & ; n = 3. \end{cases}$$

(c) If

$$x(t) = \sum_{n=0}^{\infty} \alpha_n K_n(t)$$

then

$$\mathcal{F}^a K_n(t) = \sum_{n=0}^{\infty} \alpha_n \varpi_n K_n(t).$$

3.30. When $x(t)$ is bandlimited, $X(t)$ has compact support and, therefore, cannot be bandlimited. Any weighted sum of $x(t)$ and $X(t)$ can therefore not be bandlimited. For $0 \leq a < 4$, the weighted fractional Fourier transform, $\mathcal{F}^a x(t)$, is bandlimited only for $a = 0$ and $a = 2$.

3.32. **Transfinite numbers and systems theory**
 (b) The piecewise invariant approximation in Section 3.7 requires the cataloging of $\aleph_0$ impulse responses.

 For a second example, consider a linear system whose inputs, $x(t)$ are limited to signals of the type [893]

$$x(t) = \sum_{n=0}^{\infty} a_n \phi_n(t) \quad (3.82)$$

where the $\phi_n(t)$'s form a (not necessarily orthogonal) basis of some sort. Examples include (1) periodic signals represented by a Fourier series and (2) bandlimited signals expressed as a superposition of sincs as in the cardinal series.

The output to a linear system when (3.82) is the input is

$$y(t) = S\{x(t)\} = \sum_{n=0}^{\infty} a_n \theta_n(t)$$

where

$$\theta_n(t) = S\{\phi_n(t)\}.$$

Knowledge of the $\theta_n(t)$'s for $n = 0, 1, 2, \ldots$ constitute a catalog of $\aleph_0$ input-output responses required to characterize the linear system.

4

Fourier Transforms in Probability, Random Variables and Stochastic Processes

> I am convinced that He (God) does not play dice.
> Albert Einstein (1879–1955) [1320]
>
> [Einstein]. Quit telling God what to do. (1885–1962).
> Niels Henrik David Bohr [1320]
>
> Not only does God play dice, but he sometimes throws them where they cannot be seen.
> Stephen Hawking [585]
>
> **Pascal's Wager.** Let us weigh the gain and the loss in wagering that God is. Let us consider the two possibilities. If you gain, you gain all; if you lose, you lose nothing. Hesitate not, then, to wager that He is.
> Blaise Pascal (1623–1662) [1098]

4.1 Introduction

In this Chapter, we present application of Fourier analysis to probability, random variables and stochastic processes [1089, 1097, 1387, 1329].

4.2 Random Variables

A random variable, X, is the assignment of a number to the outcome of a random experiment. We can, for example, flip a coin and assign an outcome of a heads as $X = 1$ and a tails $X = 0$. Often the number is equated to the numerical outcome of the experiment, such as the number of dots on the face of a rolled die or the measurement of a voltage in a noisy circuit. The *cumulative distribution function* is defined by

$$F_X(x) = \Pr[X \leq x]. \tag{4.1}$$

The *probability density function* is the derivative $f_X(x) = \frac{d}{dx} F_X(x)$.

Our treatment of random variables focuses on use of Fourier analysis. Due to this viewpoint, the development we use is unconventional and begins immediately in the next section with discussion of properties of the probability density function.

4.2.1 Probability Density Functions, Expectation, and Characteristic Functions

The function, $f_X(x)$ is said to a probability density function of a random variable, X, if it is non-negative[1] and has an area of one. That is,

$$\int_{-\infty}^{\infty} f_X(x)dx = 1, \qquad (4.2)$$

and

$$f_X(x) \geq 0. \qquad (4.3)$$

We define the *expected value* of a function, $g(X)$, by

$$E[g(X)] := \int_{-\infty}^{\infty} g(x) f_X(x)dx. \qquad (4.4)$$

The over bar notation will be used interchangeably for expectation.

$$\overline{g(X)} := E[g(X)].$$

The expectation

$$\overline{X^k} = \int_{-\infty}^{\infty} x^k f_X(x)dx \qquad (4.5)$$

is dubbed the kth moment of the random variable, X.

The function

$$\Phi_X(u) := E\left[e^{-j2\pi uX}\right] \qquad (4.6)$$

is the *characteristic function* of the random variable, X. Clearly, we have the Fourier transform pair[2]

$$f_X(x) \longleftrightarrow \Phi_X(u) = \int_{-\infty}^{\infty} f_X(x)e^{-j2\pi ux}dx.$$

A table of commonly used probability density functions with corresponding characteristic functions, means and variances are in Tables 4.1 and 4.2.

4.2.1.1 Discrete Random Variables

A discrete random variable has a probability density function consisting of point masses

$$f_X(x) = \sum_{k=-\infty}^{\infty} p_k \delta(x - x_k).$$

---

1. Implying, then, that $f_X(x)$ is real.
2. We do not use the capital F as the Fourier transform of f because (a) Φ is the universally used symbol for the characteristic function, and (b) F is reserved for the *cumulative distribution function*, $F_X(x) = \int_{-\infty}^{x} f_X(\xi)d\xi$.

TABLE 4.1. The probability density functions, characteristic functions, means and variance of some common continuous random variables are listed here. The characteristic function, $\Phi_X(u)$, is the Fourier transform of $f_X(x)$. (Continued in Table 4.2.)

Name	Parameters	$f_X(x)$	$\Phi_X(u)$	$\overline{X}$	σ_X^2
Uniform	$\overline{X}, R > 0$	$\frac{1}{R}\Pi\left(\frac{x-\overline{X}}{R}\right)$	$\text{sinc}(Ru)e^{-j2\pi u\overline{X}}$	$\overline{X}$	$\frac{R}{12}$
Triangle	$\overline{X}, R > 0$	$\frac{1}{R}\Lambda\left(\frac{x-\overline{X}}{R}\right)$	$\text{sinc}^2(Ru)e^{-j2\pi u\overline{X}}$	$\overline{X}$	$\frac{R^2}{6}$
Gaussian	$\overline{X}$	$\frac{1}{\sqrt{2\pi}\sigma_X}e^{-\frac{(x-\overline{X})^2}{2\sigma_X^2}}$	$e^{-j2\pi u\overline{X}}$	$\overline{X}$	σ_X^2
	$\sigma_X > 0$		$\times e^{-2(\pi u\sigma_X)^2}$		
Exponential	$\lambda > 0$	$\lambda e^{-\lambda x}\mu(x)$	$\frac{\lambda}{\lambda+j2\pi u}$	$\frac{1}{\lambda}$	$\frac{1}{\lambda^2}$
Laplace	$\overline{X}$	$\frac{\alpha}{2}e^{-\alpha|x-\overline{X}|}$	$\frac{\alpha^2}{(2\pi u)^2+\alpha^2}$	$\overline{X}$	$\frac{2}{\alpha^2}$
	$\alpha > 0$		$\times e^{-j2\pi u\overline{X}}$		
Sech	$\alpha > 0$	$\alpha \,\text{sech}(\pi\alpha x)$	$\text{sech}(\pi u/\alpha)$	0	$\frac{1}{4\alpha^2}$
Pearson III	$\alpha, \beta > 0$	$\frac{1}{\beta\Gamma(p)}\left(\frac{x-\alpha}{\beta}\right)^{p-1}$	$\left(\frac{\beta}{\beta+j2\pi u}\right)^p$	$\alpha + \frac{p}{\beta}$	$p\beta^2$
	$p \in \mathbb{N}$	$\times e^{-\frac{x-\alpha}{\beta}}\mu(x-\alpha)$	$\times e^{-j2\pi u\alpha}$		

TABLE 4.2. Continued from Table 4.1. Notes. (1) The Erlang random variable (See Figure 4.5) is a special case of the gamma random variable with $\alpha = m \in \mathbb{N}$. (2) The chi-square ($\chi^2$) random variable (See Figure 4.4) is also a special case of the gamma with $\alpha = k/2$, $k \in \mathbb{N}$ and $\lambda = 1/2$. (3) $\Upsilon_{a,b}(u) =_1 F_1(b+1, b+c+2, -j2\pi u)$. See Exercise 4.9 for the derivation. (4) The Cauchy random variable has undefined mean and variance.

Name	Parameters	$f_X(x)$	$\Phi_X(u)$	$\overline{X}$	σ_X^2
Gamma	$\lambda > 0, \alpha > 0$	$\frac{\lambda(\lambda x)^{\alpha-1}e^{-\lambda x}\mu(x)}{\Gamma(\alpha)}$	$\frac{1}{(1+j2\pi u/\lambda)^\alpha}$	$\frac{\alpha}{\lambda}$	$\frac{\alpha}{\lambda^2}$
Erlang[1]	$\lambda > 0, m \in \mathbb{N}$	$\frac{\lambda(\lambda x)^{m-1}e^{-\lambda x}\mu(t)}{(m-1)!}$	$\left(\frac{\lambda}{\lambda+j2\pi u}\right)^m$	$\frac{m}{\lambda}$	$\frac{m}{\lambda^2}$
Chi-Square[2]	$k \in \mathbb{N}$	$\frac{x^{-1+k/2}e^{-x/2}\mu(t)}{2^{k/2}\Gamma(k/2)}$	$\left(\frac{1}{1+j4\pi u}\right)^{k/2}$	k	$2k$
Beta[3]	$a, b > 0$	$\frac{x^b(1-x)^c}{B(b+1,c+1)}$	$(b+c+1)$	$\frac{b+1}{b+c+2}$	$\frac{b+1}{(b+c+2)^2}$
		$\times \Pi\left(x - \frac{1}{2}\right)$	$\times \Upsilon_{a,b}(u)$		$\times \frac{b+2}{b+c+3}$
Cauchy[4]	$\alpha > 0$	$\frac{\alpha/\pi}{x^2+\alpha^2}$	$e^{-2\pi\alpha|u|}$	no	no

The p_k's denote the kth *probability mass*. To satisfy the conditions in (4.2) and (4.3), we require

$$\sum_{k=-\infty}^{\infty} p_k = 1$$

and

$$p_k \geq 0 \text{ for all } k.$$

The discrete random variable is said to be of the *lattice type* if the point probability masses are located at only integers. Then

$$f_X(x) = \sum_{k=-\infty}^{\infty} p_k \delta(x-k). \tag{4.7}$$

The characteristic function of such random variables is given by the discrete time Fourier transform

$$\Phi_X(u) = E\left[e^{-j2\pi uX}\right]$$

$$= \sum_{k=-\infty}^{\infty} p_k e^{-j2\pi uk}. \tag{4.8}$$

Some commonly used discrete random variables and their characteristic functions are in Table 4.3.

4.2.1.2 Moments from the Characteristic Functions

Recall the derivative theorem.

$$\left(\frac{d}{du}\right)^k \Phi_X(u) = \int_{-\infty}^{\infty} f_X(x) \frac{d^k}{du^k} e^{-j2\pi ux} dx$$

$$= \int_{-\infty}^{\infty} (-j2\pi x)^k f_X(x) e^{-j2\pi ux} dx.$$

TABLE 4.3. Probability mass functions for common discrete random variables. The probability density function for discrete random variables of the lattice type are in (4.7). The characteristic function is in (4.8). The characteristic function, $\Phi_X(u)$, is the discrete time Fourier transform of p_k. Notes: (1) $0 \leq p \leq 1$ and $q := 1 - p$, (2) $\binom{n}{k} = \frac{n!}{k!(n-k)!}$ (3) See Exercise 4.12. The negative binomial is stated more generally in some sources with probability mass $p_k = \Gamma(\rho + k)p^\rho(1-p)^k/(k!\,\Gamma(\rho))$ for $\rho > 0$. This is also referred to as the *Pólya distribution*. The entry in this table is then relegated to the status of a special case for $\rho = r =$ a positive integer and is called the *Pascal distribution*.

Name	Parameter(s)	p_k	$\Phi_X(u)$	$\overline{X}$	σ_X^2
Deterministic	a	$\delta[x-a]$	$e^{-j2\pi au}$	a	0
Bernoulli[1]	p	$p\,\delta[k] + q\,\delta[k-1]$	$p + qe^{-j2\pi u}$	p	pq
Uniform discrete	$N \in \mathbb{N}$	$\frac{1}{N};\ 0 \leq n < N$	$e^{-j\pi(N-1)u}$ $\times \text{array}_N(u)$	$\frac{N-1}{2}$	$\frac{(N-1)(N+1)}{12}$
Binomial[1,2]	$n > 0, p$	$\binom{n}{k}p^k q^{n-k};$ $0 \leq k \leq n$	$(q + pe^{-j2\pi u})^n$	np	npq
Geometric[1]	p	$pq^k \mu[k]$	$\frac{p}{1-qe^{-j2\pi u}}$	$\frac{1}{p}$	$\frac{q}{p^2}$
Poisson	$\alpha > 0$	$\frac{\alpha^k e^{-\alpha} \mu[k]}{k!}$	$e^{\alpha(e^{-j2\pi u}-1)}$	α	α
Negative binomial[1,2,3]	p, r	$\binom{k-1}{r-1}p^r q^{k-r} \mu[k-r]$	$\left(\frac{p}{e^{j2\pi u}-q}\right)^r$	$\frac{r}{p}$	$\frac{rq}{p^2}$

Thus, we can obtain the kth moment from the kth derivative of the characteristic function

$$\overline{X^k} = \frac{1}{(-j2\pi)^k}\left(\frac{d}{du}\right)^k \Phi_X(0). \tag{4.9}$$

If the characteristic function can be expanded in a Taylor series about the origin, then the kth term of the Taylor series contains the kth moment.[3]

$$\Phi_X(u) = \sum_{k=0}^{\infty} \Phi_X^{(k)}(0) u^k/k!$$

$$= \sum_{k=0}^{\infty} (-j2\pi u)^k \, \overline{X^k}/k! \tag{4.10}$$

The *second characteristic function* is the natural log of the first.

$$\Psi_X(u) := \ln \Phi_X(u). \tag{4.11}$$

Note that $\Phi_X(0) = 1$ and $\Psi_X(0) = 0$. Given $\Psi_X(u)$, $\Phi_X(u)$, or $f_X(x)$, we can find the other two functions.

The *mean* (a.k.a. *first moment*) of the random variable is $\overline{X}$ and, as illustrated in Figure 4.1 is the center of mass of the probability density function, i.e., the point of balance.[4] The width or, more properly, the *dispersion* of the probability density function can be measured by σ_X where

$$\sigma_X^2 := E\left[(X - \overline{X})^2\right] = \overline{X^2} - \overline{X}^2. \tag{4.12}$$

is dubbed the *variance* and σ_X the *standard deviation*.

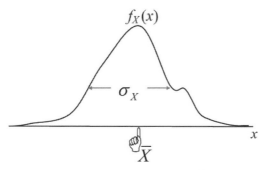

FIGURE 4.1. A probability density function, $f_X(x)$, its mean $\overline{X}$ and standard deviation, $\sqrt{\text{var}(X)}$.

---

3. See, for example, Exercise 4.5.
4. The other measures of *central tendency* of a probability density function are the *median* which is the point the density function is divided into equal areas and the *mode* which is the density function's mass. For the Gaussian, all are the same. For the density function $f_X(x) = e^{-x}\mu(x)$, the mean is one, the median is $\ln(2)$ and the mode is zero.

The mean and variance of a random variable can be determined immediately from the second characteristic function.

$$\frac{d}{du}\Psi_X(u) = \frac{d}{du}\ln \Phi_X(u)$$
$$= \Phi'_X(u)/\Phi_X(u). \qquad (4.13)$$

Thus

$$\frac{d}{du}\Psi_X(0) = \Phi'_X(0)/\Phi_X(0)$$
$$= -j2\pi\overline{X}.$$

and the first moment is determined through the first derivative.

$$\overline{X} = \frac{1}{-j2\pi}\frac{d}{du}\Psi_X(0). \qquad (4.14)$$

For the variance, we use the quotient rule for differentiation of (4.13) and write

$$\left(\frac{d}{du}\right)^2 \Psi_X(u) = \frac{\Phi_X(u)\Phi''_X(u) - (\Phi'_X(u))^2}{\Phi_X^2(u)},$$

and we have the variance directly from the second derivative of the second derivative function.

$$\sigma_X^2 = \frac{-1}{(2\pi)^2}\left(\frac{d}{du}\right)^2 \Psi_X(0). \qquad (4.15)$$

Example 4.2.1. Let[5]

$$f_Y(x) = \frac{1}{S} f_X\left(\frac{x-a}{S}\right) \qquad (4.16)$$

where $S > 0$ and a is real. If f_X has unit area, then so does f_Y. Our goal is to find the mean and standard deviation of Y in terms of the mean and standard deviation of X.

From the scaling theorem

$$\Phi_Y(u) = \Phi_X(Su)e^{-j2\pi au}$$

and

$$\Psi_Y(u) = \Psi_X(Su) - j2\pi au.$$

Thus

$$\frac{d}{du}\Psi_Y(u) = S\Psi'(Su) - j2\pi a,$$

and the mean of Y is related to the mean on X via

$$\overline{Y} = S\overline{X} + a.$$

---

5. This relationship results from the random variable transformation $Y = SX + a$.

For the variance, we have

$$\left(\frac{d}{du}\right)^2 \Psi_Y(u) = S^2 \Psi''(Su)$$

and

$$\sigma_Y^2 = S^2 \sigma_X^2. \tag{4.17}$$

Two random variables, X and Y, are said to be *independent*[6] if, for all functions $g(\cdot)$ and $h(\cdot)$, assuming existence, the expected value of the probability density function product is equal to the product of the expected values.

$$E[g(X)h(Y)] = E[g(X)]E[h(Y)].$$

The random variables X and Y are *identically distributed* if, for all $g(\cdot)$,

$$E[g(X)] = E[g(Y)].$$

If X and Y are both independent and identically distributed, they are said to be *i.i.d.*

4.2.1.3 Chebyshev's Inequality

The variance is revealed as a measure of a probability density function's dispersion by *Chebyshev's inequality*[7] which states

$$\int_{|x-\overline{X}|\leq a} f_X(x)dx \geq 1 - \left(\frac{\sigma_X}{a}\right)^2 \tag{4.18}$$

for any probability density function with defined mean and standard deviation and any[8] $a > 0$. For a fixed a, as the standard deviation, σ_X, increases, the function spreads and the area of the density function in the interval about the mean, $\overline{X} - a \leq x \leq \overline{X} + a$, decreases. This is illustrated in Figure 4.2.

Proof. The definition of the variance is

$$\sigma_X^2 = \int_{-\infty}^{\infty} (x - \overline{X})^2 f_X(x)\, dx.$$

We break up the integration as

$$\sigma_X^2 = \left[\int_{-\infty}^{\overline{X}-a} + \int_{\overline{X}-a}^{\overline{X}+a} + \int_{\overline{X}+a}^{\infty}\right] (x - \overline{X})^2 f_X(x)\, dx.$$

Since the integrand is nonnegative, removal of the middle integral will result in the same or a smaller value.

---

6. This definition of independence is not traditional, but serves our purposes. Traditionally, two random variables X and Y are said to be independent if their joint probability density function is separable, i.e., $f_{XY}(x, y) = f_X(x) f_Y(y)$. Our definition allows establishing independence without introducing the joint probability density function.

7. Also spelled Tchebycheff.

8. For $\sigma_X > a$, the right side of (4.18) is negative. Although the inequality still applies, the result has no use since we already know the probability density function, and thus its integral, is always nonnegative.

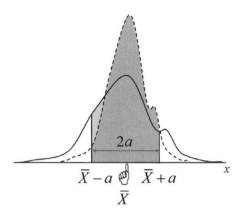

FIGURE 4.2. Illustration of Chebyshev's inequality. The probability density function with the small variance, shown by the dashed line, has a greater percentage of area around the mean, $\overline{X}$, than does the function with larger variance.

Thus

$$\sigma_X^2 \geq \left[\int_{-\infty}^{\overline{X}-a} + \int_{\overline{X}+a}^{\infty} \right] (x - \overline{X})^2 f_X(x)\, dx. \tag{4.19}$$

The integration is over the interval $x < \overline{X} - a$ and $x > \overline{X} + a$, or equivalently over all x for which

$$|x - \overline{X}| \geq a.$$

Therefore, over the integration limit we satisfy

$$(x - \overline{X})^2 \geq a^2. \tag{4.20}$$

Note that the term $(x - \overline{X})^2$ is explicitly in the integrand of (4.19). Substitution of a^2 for this term in (4.20) guarantees the result will, at worst, get larger. Thus, (4.19) becomes

$$\sigma_X^2 \geq a^2 \int_{|x-\overline{X}|>a} f_X(x)\, dx. \tag{4.21}$$

Since the probability density function integrates to one,

$$\int_{|x-\overline{X}|>a} f_X(x)\, dx = 1 - \int_{|x-\overline{X}|\leq a} f_X(x)\, dx.$$

substituting into (4.21) gives Chebyshev's inequality in (4.18).

We will later use Chebyshev's inequality to derive the weak law of large numbers.

4.2.2 Example Probability Functions, Characteristic Functions, Means and Variances

Some commonly used probability density functions are in Tables 4.1, 4.2 and 4.3. Derivation of some of these entries follows. Probability density functions for some other random variables are listed in Appendix 14.6.

4.2.2.1 The Uniform Random Variable

The uniform random variable with mean $\overline{X}$ and variance

$$\sigma_X^2 = \frac{R^2}{12}$$

has a probability density function of

$$f_X(x) = \frac{1}{R}\, \Pi\left(\frac{x - \overline{X}}{R}\right).$$

This uniform random variable is said to be

"uniform on $\left[\overline{X} - \frac{R}{2}, \overline{X} - \frac{R}{2}\right]$."

A randomly chosen angle is oft modelled as a uniform random variable on $[0, 2\pi]$ or $[-\pi, \pi]$.

The characteristic function of the uniform random variable is

$$\Phi_X(x) = \operatorname{sinc}(Ru)\, e^{-j2\pi u \overline{X}}.$$

4.2.2.2 The Triangle Random Variable

The triangle (or uniform difference) random variable with mean $\overline{X}$ and variance

$$\sigma_X^2 = \frac{R^2}{6}$$

has a probability density function of

$$f_X(x) = \frac{1}{R}\, \Lambda\left(\frac{x - \overline{X}}{R}\right).$$

A triangle random variable results from adding two independent random variables uniform on $\left[\frac{\overline{X}-R}{2}, \frac{\overline{X}+R}{2}\right]$ or, equivalently, subtracting a random variable uniform on $\left[-\frac{\overline{X}+R}{2}, \frac{-\overline{X}-R}{2}\right]$ from another uniform on $\left[\frac{\overline{X}-R}{2}, \frac{\overline{X}+R}{2}\right]$. The characteristic function of the triangular random variable is

$$\Phi_X(u) = \operatorname{sinc}^2(Ru)\, e^{-j2\pi u \overline{X}}.$$

4.2.2.3 The Gaussian Random Variable

The probability density function of the Gaussian random variable is commonly referred to as the *bell shaped curve*.[9] The Gaussian random variable is often chosen to be the underlying

---

9. See Figure 4.7.

random variable in experiments-an assumption which can be inappropriate. The central limit theorem[10] has the Gaussian random variable as the limiting model of the sums and averages of many random variables.

The Gaussian random variable is also referred to as the *normal* random variable. The probability density function with mean $\overline{X}$ and variance σ_X^2 is

$$f_X(x) = \frac{1}{\sqrt{2\pi}\,\sigma_X} e^{-\frac{(x-\overline{X})^2}{2\sigma_X^2}}.$$

Some plots are in Figure 4.3.

The characteristic function of the Gaussian follows directly from the Gaussian's Fourier transform.[11] From Table 4.1, the second characteristic function for the Gaussian random variable is

$$\Psi_X(u) = \ln \Phi_X(u) = -j2\pi u \overline{X} - 2(\pi u \sigma_X)^2.$$

The first and second derivatives using (4.14) and (4.15) reveals that the mean of the Gaussian random variable is $\overline{X}$ and its variance is σ_X^2.

4.2.2.4 The Exponential Random Variable

The exponential random variable has many uses, including in the fields of queuing theory and reliability. A time of failure of a light bulb whose operation begins at time zero is often modelled as an exponential random variable. The time from receipt of a given bit packet to the arrival of the next bit packet is often modelled as an exponential random variable.

The exponential random variable with parameter $\lambda > 0$ has a probability density function of

$$f_X(x) = \lambda e^{-\lambda x} \mu(x).$$

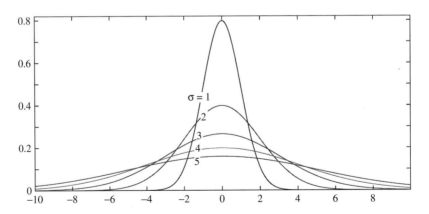

FIGURE 4.3. The Gaussian probability density function for various values of σ.

---

10. See Section 4.2.5.
11. See Section 2.3.5. An alternate derivation is in Exercise 4.5.

Straightforward integration gives the corresponding characteristic function.

$$\Phi_X(u) = \frac{\lambda}{\lambda + j2\pi u}. \tag{4.22}$$

The exponential random variable is a special case of the gamma random variable and the generalized Gaussian random variable.

4.2.2.5 The Laplace Random Variable

With mean $\overline{X}$ and scale parameter α, the Laplace random variable has a probability density function of

$$f_X(x) = \frac{\alpha}{2} e^{-\alpha|x-\overline{X}|}.$$

Laplace noise is used to measure impulsive disturbances such as lightning [890]. From the Fourier transform in Table 2.4 and the scaling theorem, the corresponding characteristic function follows as

$$\Phi_X(u) = \frac{\alpha^2}{(2\pi u)^2 + \alpha^2} e^{-j2\pi u \overline{X}}.$$

The second characteristic function is therefore

$$\Psi_X(u) = 2\ln(\alpha) - \ln\left((2\pi u)^2 + \alpha^2\right) - j2\pi u \overline{X}.$$

The first derivative is then

$$\frac{d}{du}\Psi_X(u) = -\frac{2(2\pi)^2 u}{(2\pi u)^2 + \alpha^2} - j2\pi \overline{X}$$

which dictates, unsurprisingly, that the mean is $\overline{X}$. Differentiating again gives

$$\left(\frac{d}{du}\right)^2 \Psi_X(u) = -2(2\pi)^2 \frac{\left[(2\pi u)^2 + \alpha^2\right] - 2(2\pi u)^2}{\left[(2\pi u)^2 + \alpha^2\right]^2}.$$

Hence

$$\sigma_X^2 = \frac{2}{\alpha^2}.$$

4.2.2.6 The Hyperbolic Secant Random Variable

Hyperbolic secant noise has tails that decay in the same manner as those in Laplace noise and is used to model similar phenomena. With scale parameter, α, its probability density function is

$$f_X(x) = \alpha \operatorname{sech}(\pi \alpha x). \tag{4.23}$$

From the list of Fourier transforms in Table 2.5, we find the following characteristic function.

$$\Phi_X(x) = \operatorname{sech}\left(\frac{\pi u}{\alpha}\right).$$

The second characteristic function is

$$\Psi_X(x) = -\ln\left(\cosh\left(\frac{\pi u}{\alpha}\right)\right).$$

Thus

$$\frac{d}{du}\Psi_X(u) = -\frac{\pi}{\alpha}\frac{\sinh\left(\frac{\pi u}{\alpha}\right)}{\cosh\left(\frac{\pi u}{\alpha}\right)}$$

$$= -\frac{\pi}{\alpha}\tanh\left(\frac{\pi u}{\alpha}\right)$$

Since the probability density function in (4.23) is even, we are not surprised to see the mean is equal to zero.

Since

$$\frac{d}{d\theta}\tanh(\theta) = 1 - \tanh^2(\theta),$$

the second derivative of the second characteristic function is

$$\left(\frac{d}{du}\right)^2 \Psi_X(u) = -\left(\frac{\pi}{\alpha}\right)^2 \left[1 - \tanh^2\left(\frac{\pi u}{\alpha}\right)\right].$$

Using (4.15), the corresponding variance of the hyperbolic secant random variable is

$$\sigma_X^2 = \frac{1}{4\alpha^2}.$$

4.2.2.7 The Gamma Random Variable

The gamma random variable with parameters of λ and α has a probability density function

$$f_X(x) = \frac{\lambda(\lambda x)^{\alpha-1} e^{-\lambda x} \mu(x)}{\Gamma(\alpha)} \qquad (4.24)$$

Special cases of the gamma probability density function are as follows.

1. *The chi-squared random variable*, also written as the χ^2-random variable, is the special case of the gamma random variable for $\lambda = \frac{1}{2}$. The sum of the squares of k i.i.d. zero mean unit variance Gaussian random variables is a chi-squared random variable. Probability density function plots are shown in Figure 4.4.
2. *The Erlang random variable* is a special case of the gamma random variable for $\alpha = k$. The sum of k i.i.d. exponential random variables is an Erlang random variable. Probability density functions, are shown in Figure 4.5.

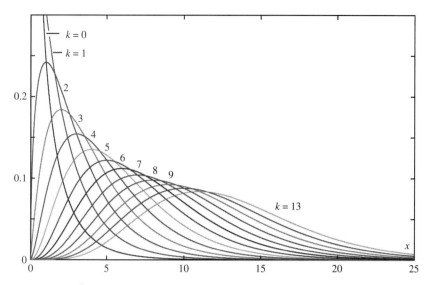

FIGURE 4.4. The χ^2 (chi-square) probability density function with k degrees of freedom. $f_X(x) = (x/2)^{-1+k/2} e^{-x/2} \mu(x) / \Gamma(k/2)$.

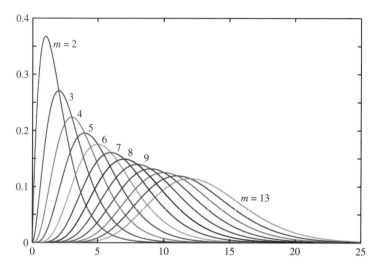

FIGURE 4.5. The normalized probability density function for the Erlang random variable, $f_X(x/\lambda)/\lambda = x^{m-1} e^{-x} \mu(x)/(m-1)!$.

To derive the characteristic function for the gamma random variable in Table 4.1, we explicitly parameterize its characteristic function by α.

$$\Phi_X(u, \alpha) = \int_0^\infty \frac{\lambda(\lambda x)^{\alpha-1} e^{-\lambda x}}{\Gamma(\alpha)} e^{-j2\pi ut} dt.$$

Integrate by parts.

$$\hat{u} = (\lambda x)^{\alpha-1}/\Gamma(\alpha) \quad ; \quad d\hat{v} = e^{-\lambda x} e^{-j2\pi ux} dx$$
$$d\hat{u} = (\alpha-1)\lambda^{\alpha-1} x^{\alpha-2} dx/\Gamma(\alpha) \quad ; \quad \hat{v} = -e^{-\lambda x} e^{-j2\pi ux}/(lambda + j2\pi u).$$
$$= \lambda(\lambda x)^{\alpha-2} dx/\Gamma(\alpha-1)$$

Thus

$$\Phi_X(u, \alpha) = \frac{-\lambda x^{\alpha-1}}{(\lambda + j2\pi u)\Gamma(\alpha)} e^{-\lambda x} e^{-j2\pi ux} \Big|_0^\infty$$

$$+ \int_0^\infty \frac{(\lambda x)^{\alpha-2}}{\Gamma(\alpha-1)} e^{-\lambda x} e^{-j2\pi ux} dx \frac{1}{1 + j2\pi u/\lambda}$$

$$= 0 + \frac{1}{1 + j2\pi u/\lambda} \Phi_X(u, \alpha - 1).$$

or

$$\Phi_X(u, \alpha) = \frac{\Phi_X(u, \alpha - 1)}{1 + j2\pi u/\lambda}.$$

A solution of this recursion is the entry in Table 4.1 which we rewrite here.

$$\Phi_X(u, \alpha) = \left(\frac{1}{1 + j2\pi u/\lambda}\right)^\alpha. \quad (4.25)$$

The gamma random variable for $\alpha = 1$ is the exponential random variable. The characteristic function in (4.22) verifies a boundary condition and therefore the validity of the solution in (4.25) is established.

The second characteristic for the gamma random variable follows as

$$\Psi_X(u) = -\alpha \ln\left(1 + \frac{j2\pi u}{\lambda}\right).$$

The first derivative follows as

$$\frac{d}{du} \Psi_X(u) = \frac{-j2\pi\alpha/\lambda}{1 + j2\pi u/\lambda},$$

and the second derivative is

$$\left(\frac{d}{du}\right)^2 \Psi_X(u) = \frac{-\alpha(2\pi/\lambda)^2}{(1 + j2\pi u/\lambda)^2}.$$

From (4.14) and (4.15), the mean and variance of the gamma random variable follow as

$$\overline{X} = \frac{\alpha}{\lambda} \quad (4.26)$$

and

$$\sigma_X^2 = \frac{\alpha}{\lambda^2}. \quad (4.27)$$

The mean and variance of the exponential, Erlang and chi-squared random variable follow as special cases.

4.2.2.8 The Beta Random Variable

The beta random variable with parameters b and c has a probability density function of

$$f_X(x) = \frac{x^b(1-x)^c}{B(b+1, c+1)} \Pi\left(x - \frac{1}{2}\right) \quad (4.28)$$

where $B(b, c)$ is the beta function defined in (2.68). The uniform random variable on [0, 1] is a special case for $b = c = 0$. Plots of the beta probability density function are in Figure 4.6.

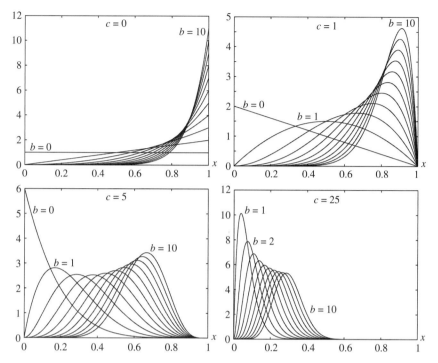

FIGURE 4.6. The probability density function for the beta random variable in (4.28). The beta probability density function has symmetry in the sense that $f_X(x; b, c) = f_X(1 - x; c, b)$.

Since (4.28) is a probability density function that must integrate to one, we have

$$\int_0^1 x^b(1-x)^c\,dx = B(b+1, c+1).$$

Thus

$$\int_0^1 x^{b+n}(1-x)^c\,dx = B(b+n+1, c+1)$$

and the nth moment of the beta random variable follows as

$$\overline{X^n} = \int_0^1 \frac{x^{b+n}(1-x)^c}{B(b+1, c+1)}\,dx$$

$$= \frac{B(b+n+1, c+1)}{B(b+1, c+1)} = \frac{\Gamma(b+n+1)}{\Gamma(b+1)}\frac{\Gamma(b+c+2)}{\Gamma(b+n+c+2)}. \quad (4.29)$$

Using the gamma function recursion in (2.67) gives the mean as

$$\overline{X} = \frac{b+1}{b+c+2}$$

and the second moment is

$$\overline{X^2} = \frac{(b+2)(b+1)}{(b+c+3)(b+c+2)}.$$

The variance follows as

$$\sigma_X^2 = \frac{(b+1)(c+1)}{(b+c+3)(b+c+2)^2}.$$

4.2.2.9 The Cauchy Random Variable

The Cauchy probability density function is

$$f_X(x) = \frac{\alpha/\pi}{x^2 + \alpha^2}. \tag{4.30}$$

The ratio of two independent zero mean Gaussian random variables is a Cauchy random variable. For this reason, the Cauchy random variable is also referred to as the *Gaussian ratio* random variable. Plots are shown in Figure 4.7.

The Cauchy probability density function has much larger tales than the Gaussian probability density function. This is shown in Figure 4.8.

The mean of the Cauchy random variable in Table 4.1 is not defined. We can show this is two ways. The first is through use of its characteristic function in Table 4.1. We take the log to find the second characteristic function.

$$\Psi_X(u) = -2\pi\alpha|u|.$$

The derivative follows as

$$\frac{d}{du}\Psi_X(u) = -2\pi\alpha \, \text{sgn}(u).$$

Using (4.14) to find the mean of the Cauchy random variable necessitates the dangerous practice of evaluating a continuous function at a discontinuity, namely sgn(0). In this sense, the mean of the Cauchy random variable is undefined.

Alternately, we can examine the integral in (4.5) for $k = 1$. First, let's consider some background. Since $1/x$ is an odd function, one might suppose

$$\int_{-1}^{1} \frac{1}{x} \, dx = 0 \quad \leftarrow ? \tag{4.31}$$

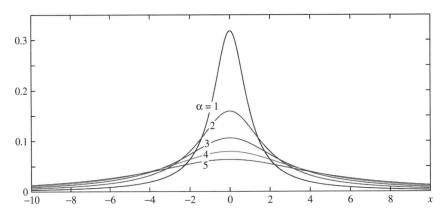

FIGURE 4.7. The probability density function for the Cauchy random variable in (4.30) for various values of α.

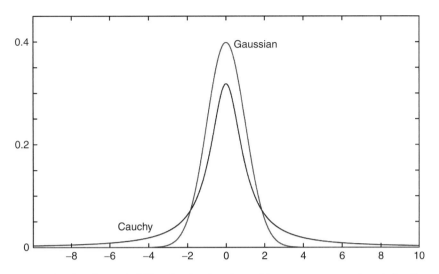

FIGURE 4.8. The Cauchy probability density function with parameter $\alpha = 1$ and the Gaussian probability density function with zero mean and unit variance.

Since, however

$$\int_0^1 \frac{1}{x}\, dx = \infty,$$

we are, in (4.31), making the dangerous assumption that

$$\infty - \infty = 0.$$

We can, though, accurately write

$$\lim_{\epsilon \to 0} \left[\int_{-1}^{-\epsilon} + \int_{\epsilon}^{1} \right] \frac{1}{x}\, dx = 0.$$

This is referred to as the *Cauchy principle value*.

A similar problem occurs in evaluating the first moment of the Cauchy random variable. Using the probability density function in Table 4.1 and (4.5) for $k=1$ gives,

$$\overline{X} = \int_{-\infty}^{\infty} x\, \frac{\alpha/\pi}{x^2 + \alpha^2}\, dx. \tag{4.32}$$

Like (4.31), we might suppose that, since the integrand is odd, this integral goes to zero. Note, though, that (1) since, for large x,

$$x\, \frac{\alpha/\pi}{x^2 + \alpha^2} \to \frac{\alpha/\pi}{x}$$

and (2) the tail of the curve $1/x$ has infinite area, we conclude

$$\int_0^{\infty} x\, \frac{\alpha/\pi}{x^2 + \alpha^2}\, dx = \infty.$$

Thus, like (4.31), we are left with an $\infty - \infty$ situation in evaluating (4.32).

All positive even moments of the Cauchy random variable are infinite. The positive even moments of the Cauchy random variable are, for $k = 1, 2, 3, \ldots$

$$\overline{X^{2k}} = \int_{-\infty}^{\infty} x^{2k} \frac{\alpha/\pi}{x^2 + \alpha^2} dx. \tag{4.33}$$

As $x \to \pm\infty$, the integrand approaches $\alpha x^{2(k-1)}/\pi$. For $k = 1, 2, 3, \ldots$, this corresponds to an integrand with infinite area.

There is an alternative explanation for the infinite second moment of the Cauchy random variable. The second derivative of the Cauchy characteristic function contains a Dirac delta at the origin and the second moment is proportional to $\delta(0) = \infty$.

4.2.2.10 The Deterministic Random Variable

The phrase *deterministic random variable* seems to be an oxymoron. It is not. A model of randomness must contain deterministic events as a special case. The probability density function for a deterministic random variable with parameter a is

$$f_X(x) = \delta(x - a).$$

All of the probability mass of the density function is at $x = a$. The characteristic function of the deterministic random variable is

$$\Phi_X(u) = e^{-j2\pi au}.$$

Thus

$$\Psi_X(u) = -j2\pi au.$$

The mean and variance of the deterministic random variable are

$$\overline{X} = a$$

and

$$\sigma_X^2 = 0.$$

Let $f_Y(x)$ denote the probability density function of a random variable, Y. Then[12]

$$\lim_{r \to \infty} r f_Y(rx) = \delta(x).$$

To see this, Let $Z = Y/r$. The probability density function of Z is

$$f_Z(x) = r f_Y(rx).$$

Independent of the distribution of Y, the random variable Z approaches the deterministic value of zero as $r \to \infty$.

---

12. A similar problem is addressed in Exercise 2.16.

4.2.2.11 The Bernoulli Random Variable

A Bernoulli random variable has success with probability p and failure with probability $q = 1 - p$. A zero is assigned to a failure and one to a success. The probability density function follows as

$$f_X(x) = (1-p)\delta(x) + p\delta(x-1).$$

The corresponding characteristic function is

$$\Phi_X(x) = (1-p) + pe^{-j2\pi ux}.$$

The performance of an experiment to generate a Bernoulli random variable, e.g., the flipping of a fair coin, $p = \frac{1}{2}$, or getting 3 dots on the rolling of a fair die, $p = \frac{1}{6}$, is dubbed a *Bernoulli trial*.

4.2.2.12 The Discrete Uniform Random Variable

The discrete uniform random variable, with parameter N, has probability mass of $\frac{1}{N}$ for $0 \leq k < N$. Its probability density function follows as

$$f_X(x) = \frac{1}{N} \sum_{k=0}^{N-1} \delta(t-n).$$

The Bernoulli random variable is a special case for $N = 2$.

The corresponding characteristic function is obtained using a trigonometric geometric series.[13] The result is

$$\Phi_X(u) = e^{-j\pi(N-1)u} \, \text{array}_N(u).$$

The mean and variance of the discrete uniform random variable is best calculated directly rather than through manipulation of the characteristic function. Since

$$\sum_{n=0}^{N-1} n = \frac{1}{2}(N-1)N,$$

The mean follows as

$$\overline{X} = \frac{N-1}{2}.$$

Similarly, since

$$\sum_{n=0}^{N-1} n^2 = \frac{1}{6}(N-1)N(2N-1),$$

the second moment is

$$\overline{X^2} = \frac{1}{6}(N-1)(2N-1).$$

---

13. See Appendix 14.4.2.2.

The variance of the discrete uniform random variable follows as

$$\sigma_X^2 = \overline{X^2} - \overline{X}^2 = \frac{1}{12}(N-1)(N+1).$$

4.2.2.13 The Binomial Random Variable

The probability density function for the binomial random variable with parameters n and p is

$$f_X(x) = \sum_{k=0}^{n} \binom{n}{k} p^k (1-p)^{n-k} \delta(t-k).$$

The binomial random variable is a discrete random variable of the lattice type. It models the number of successes, k, in n repeated i.i.d. Bernoulli trials with probability of success, p. The probability mass functions are shown in Figures 4.9 and 4.10. (See also Figures 4.23, 4.24 and 4.25.)

The characteristic function of the binomial random variable follows as

$$\Phi_X(u) = \int_{-\infty}^{\infty} f_X(x) e^{-j2\pi ux} dx$$

$$= \sum_{k=0}^{n} \binom{n}{k} p^k q^{n-k} e^{-j2\pi uk}$$

$$= \sum_{k=0}^{n} \binom{n}{k} \left(p e^{-j2\pi u}\right)^k q^{n-k}$$

$$= \left(p e^{-j2\pi u} + q\right)^n$$

where $q = 1 - p$ and, in the last step, we have used the binomial series.[14]

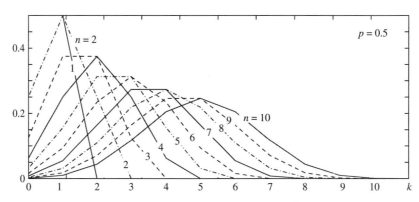

FIGURE 4.9. The binomial probability mass function with for $p = \frac{1}{2}$. Points are connected linearly. See Figure 4.10 for the case of $p = 0.75$. A waterfall plot of the same data is in Figure 4.23.

---

14. See Appendix 14.4.1.

For the mean of the binomial random variable, we evaluate

$$\frac{d}{du}\Phi_X(u) = -j2\pi np\, e^{-j2\pi u}\left(pe^{-j2\pi u} + q\right)^{n-1}.$$

Thus, using (4.9),

$$\overline{X} = np.$$

Differentiating again

$$\left(\frac{d}{du}\right)^2 \Phi_X(u) = (-j2\pi)^2 e^{-j2\pi u}\left(p\, e^{-j2\pi u} + q\right)^{n-2}\left[\left(p\, e^{-j2\pi u} + q\right) + (n-1)p\right].$$

Again using (4.9) gives

$$\overline{X^2} = np + (n-1)np^2.$$

The variance of the binomial random variable is thus

$$\sigma_X^2 = \overline{X^2} - \left(\overline{X}\right)^2 = np(1-p).$$

4.2.2.14 The Poisson Random Variable

The Poisson random variable with parameter $\alpha > 0$ is a discrete random variable of the lattice type. Its probability density function is

$$f_X(x) = e^{-\alpha}\sum_{n=0}^{\infty}\frac{\alpha^k}{k!}\,\delta(x-k).$$

Example plots of the corresponding probability mass functions are shown in Figure 4.11. The corresponding characteristic function is

$$\Phi_X(x) = e^{-\alpha}\sum_{n=0}^{\infty}\frac{\left(\alpha e^{-j2\pi u}\right)^k}{k!}.$$

Using the exponential Taylor series in (2.34) gives the desired result.

$$\Phi_X(x) = e^{-\alpha}\exp\left(\alpha e^{-j2\pi u}\right).$$

The second characteristic function is then

$$\Psi_X(u) = -\alpha + \alpha e^{-j2\pi u}.$$

Since

$$\frac{d}{du}\Psi_X(u) = -j2\pi\alpha e^{-j2\pi u},$$

the mean of the Poisson random variable, using (4.14), is

$$\overline{X} = \alpha. \tag{4.34}$$

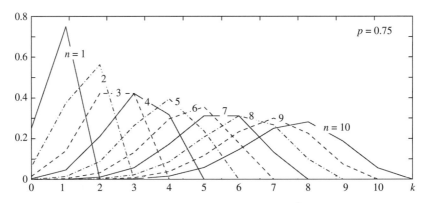

FIGURE 4.10. The binomial probability mass function with for $p = \frac{3}{4}$. Points are connected linearly. See Figure 4.9 for the case of $p = \frac{1}{2}$. A waterfall plot of the same data is in Figure 4.24.

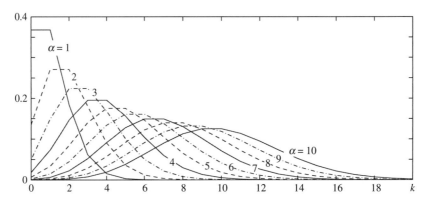

FIGURE 4.11. The Poisson probability mass function for various values of α. Points are connected linearly. A waterfall plot of the same data is in Figure 4.26.

Differentiating again gives

$$\left(\frac{d}{du}\right)^2 \Psi_X(u) = -(2\pi)^2 \alpha e^{-j2\pi u},$$

and the variance follows from (4.15) as

$$\sigma_X^2 = \alpha. \tag{4.35}$$

Interestingly, then, the Poisson random variable's mean is equal to its variance.

The Poisson Approximation. The probability mass of the binomial random variable can, under certain constraints, be approximated by the probability mass of the Poisson random variable. The binomial random variable's probability mass is

$$p_k = \binom{n}{k} p^k (1-p)^{n-k}. \tag{4.36}$$

The approximation criteria are

$$p \ll 1 \tag{4.37}$$

and

$$n \gg 1. \tag{4.38}$$

Therefore, we expect

$$k \ll n. \tag{4.39}$$

As a consequence

$$\binom{n}{k} = \frac{n!}{k!(n-k)!}$$
$$= \frac{1}{k!} \frac{n(n-1)\ldots(n-k+1)(n-k)(n-k-1)\ldots}{(n-k)(n-k-1)\ldots} \tag{4.40}$$
$$\approx \frac{n^k}{k!}$$

Also, since $n \gg k$, we have $(1-p)^{n-k} \approx (1-p)^n$. For $p \ll 1$, we can approximate $(1-p) \approx e^{-p}$. In summary,

$$(1-p)^{n-k} \approx (1-p)^n \approx (e^{-p})^n = e^{-np}.$$

This and the approximation in (4.40) when substituted into (4.36), give

$$p_k = \binom{n}{k} p^k (1-p)^{n-k} \approx e^{-np} \frac{(np)^k}{k!}. \tag{4.41}$$

The binomial probability mass can therefore be approximated by a Poisson probability mass.

4.2.3 Distributions of Sums of Random Variables

We now examine distributions of the sum

$$S_n = \sum_{k=1}^{n} X_k. \tag{4.42}$$

For continuous random variables, the probability density function of the sum is a continuous function. For discrete time, the probability density function is of the form

$$f_{S_n}(x) = \sum_{k=-\infty}^{\infty} p_{S_n}[k]\delta(x - x_k)$$

where $p_{S_n}[k]$ is the probability mass. In either case, the characteristic function of the sum is

$$\Phi_{S_n}(u) = E\left[e^{-j2\pi u S_n}\right]$$

$$= E\left[\exp\left(-j2\pi u \sum_{k=1}^{n} X_k\right)\right]$$

$$= E\left[\prod_{k=1}^{n} \exp(-j2\pi u X_k)\right].$$

4.2.3.1 The Sum of Independent Random Variables

If the random variables are independent, the characteristic function of the sum is

$$\Phi_{S_n}(u) = \prod_{k=1}^{n} E\left[\exp(-j2\pi u X_k)\right]$$

$$= \prod_{k=1}^{n} \Phi_{X_k}(u). \tag{4.43}$$

The corresponding second characteristic function follows.

$$\Psi_{S_n}(u) = \sum_{k=1}^{n} \Psi_{X_k}(u). \tag{4.44}$$

Differentiating both sides and applying (4.14) reveals that the mean of the sum of independent random variables is the sum of the means.

$$\overline{S_n} = \sum_{k=1}^{n} \overline{X_k}. \tag{4.45}$$

Similarly, differentiating (4.44) twice and applying (4.15) shows that the variance of the sum of independent random variables is equal to the sum of the random variables' variances.

$$\sigma_{S_n}^2 = \sum_{k=1}^{n} \sigma_{X_k}^2. \tag{4.46}$$

These results indicate how functions shift and spread when they are convolved. Taking the inverse Fourier transform of (4.43) gives

$$f_{S_n}(x) = f_{X_1}(x) * f_{X_2}(x) * \ldots * f_{X_k}(x) * \ldots * f_{X_n}(x).$$

The center of mass of $f_{S_n}(x)$ is the sum of the means in (4.45). The dispersion of the convolution, σ_{S_n}, becomes larger, as indicated by the sum in (4.46), i.e., the operation of convolution spreads.

4.2.3.2 Distributions Closed Under Independent Summation

We say a random variable is closed under independent summation if the summation of independent random variables in a class results in random variable in that same class.

The sum of independent Gaussian random variables, for example, is a Gaussian random variable. Exponential, uniform, Laplace and geometric random variables are not closed under independent summation. Here are some random variables that are.[15]

- *Gaussian Random Variables.* If n independent Gaussian random variables with parameters $\overline{X_k}$ and $\sigma^2_{X_k}$ are added, the characteristic function of their sum is

$$\Phi_{S_n}(u) = \prod_{k=1}^{n} e^{-j2\pi u \overline{X_k}} e^{-(2\pi u \sigma_{X_k})^2}$$

$$= \exp\left(-j2\pi u \sum_{k=1}^{n} \overline{X_k}\right) \exp\left(-(2\pi u)^2 \sum_{k=1}^{n} \sigma^2_{X_k}\right)$$

$$= e^{-j2\pi u \overline{S_n}} e^{-(2\pi u)^2 \sigma^2_{S_n}}.$$

Therefore, the sum is also distributed as a Gaussian random variable with mean

$$\overline{S_n} = \sum_{k=1}^{n} \overline{X_k}$$

and variance

$$\sigma_{S_n} = \sum_{k=1}^{n} \sigma^2_{X_k}.$$

- *Gamma Random Variables.* If n independent Gamma random variables with parameters λ and α_k are added, the characteristic function of their sum is that of a gamma random variable. Specifically,

$$\Phi_{S_n}(u) = \frac{1}{(1 + j2\pi u/\lambda)^{\alpha_{S_n}}}$$

where

$$\alpha_{S_n} = \sum_{k=1}^{n} \alpha_k.$$

The result extends to Erlang and chi-square random variables which are special cases of the gamma random variable.

- *Cauchy Random Variables.* We add n independent Cauchy random variables. The characteristic function of the sum is

$$\Phi_{S_n}(u) = \prod_{k=1}^{n} e^{-2\pi |u| \alpha_k}$$

$$= \exp\left(-2\pi |u| \sum_{k=1}^{n} \alpha_k\right)$$

$$= e^{-2pi\alpha_{S_n} |u|}$$

---

15. The Pearson III random variable is treated in Exercise 4.10 and the negative binomial random variable is analyzed in Exercise 4.12. Both are closed under independent summation.

The result is a Cauchy random variable with parameter

$$\alpha_{S_n} = \sum_{k=1}^{n} \alpha_k.$$

- *Poisson Random Variables.* The sum of n independent Poisson random variables with parameter α_k is

$$\Phi_{S_n}(u) = \prod_{k=1}^{n} \exp\left(\alpha_k \left(e^{-j2\pi u} - 1\right)\right)$$

$$= \exp\left(\sum_{k=1}^{n} \alpha_k \left(e^{-j2\pi u} - 1\right)\right)$$

$$= \exp\left(\alpha_{S_n} \left(e^{-j2\pi u} - 1\right)\right).$$

The sum is a Poisson random variable with parameter

$$\alpha_{S_n} = \sum_{k=1}^{n} \alpha_k.$$

4.2.3.3 The Sum of i.i.d. Random Variables

We now examine the specific case of the density function of the sum when all components are identically distributed. If, in addition to being independent, the random variables are identically distributed, then all of the characteristic functions of the random variables in the sum are identical and

$$\Phi_{X_k}(u) = \Phi_X(u) \text{ for } 1 \le k \le n.$$

The characteristic function of the sum in (4.43) then becomes

$$\Phi_{S_n}(u) = \Phi_X^n(u).$$

The mean and variance of the sum, from (4.45) and (4.46), are

$$\overline{S_n} = n\,\overline{X}$$

and

$$\sigma_{S_n}^2 = n\,\sigma_X^2.$$

4.2.3.4 The Average of i.i.d. Random Variables

The *average*, or *sample mean*, of n i.i.d. random variables is

$$A_n = \frac{1}{n}\sum_{k=1}^{n} X_k = \frac{1}{n} S_n. \tag{4.47}$$

Following the same procedure as for the sum, we obtain

$$\Phi_{A_n}(u) = \Phi_X^n\left(\frac{u}{n}\right). \tag{4.48}$$

The second characteristic function of the average is

$$\Psi_{A_n}(u) = n\,\Psi_X\left(\frac{u}{n}\right). \tag{4.49}$$

Differentiating once gives the mean of the average as

$$\overline{A_n} = \frac{1}{-j2\pi}\frac{d}{du}\Psi_{A_n}(0)$$

$$= \frac{n}{-j2\pi}\frac{d}{du}\Psi_X\left(\frac{u}{n}\right)\bigg|_{u=0}$$

$$= \overline{X}. \tag{4.50}$$

The average is used to estimate the mean. Therefore, the expected value of the average should be equal to the mean. We have here showed that it, indeed, is.

As the number, n, of samples increases, the accuracy of the average in estimation of the mean should increase. The uncertainty of the average in estimating the mean is the standard deviation, σ_{A_n}. This can be obtained by taking the second derivative of (4.49).

$$\sigma_{A_n}^2 = \frac{-n}{(2\pi)^2}\left(\frac{d}{du}\right)^2 \Psi_X\left(\frac{u}{n}\right)\bigg|_{u=0}$$

$$= \sigma_X^2/n. \tag{4.51}$$

The dispersion, σ_{A_n}, therefore decreases by a factor $1/\sqrt{n}$ and, as $n \to \infty$, goes to zero. Thus, not only does the expected value of the average equal the mean, but the uncertainty of this measure goes to zero as n tends to ∞. Recall the deterministic random variable has zero variance.[16] The result is better stated using a law of large numbers.

4.2.4 The Law of Large Numbers

The law of large numbers states that, as the number of samples used to compute the average in (4.48) increases, the average approaches the mean. Thus, as the number of rolls of a fair die increase without number, the percentage of instances where six dots show will approach $\frac{1}{6}$.

Theorem 4.2.1. *The weak law of large numbers*[17] *states for a random variable with finite mean and variance, for any $\epsilon > 0$, that*

$$\lim_{n\to\infty}\int_{x=\overline{X}-\epsilon}^{\overline{X}+\epsilon} f_{A_n}(x)\,dx = 1. \tag{4.52}$$

Thus, all of the probability mass of the density function for the average squeezes into the interval $\overline{X} \pm \epsilon$. In the limit, the density function resembles that of a deterministic random variable. This is illustrated in Figure 4.12 for the uniform random variable.

---

16. See Section 4.2.2.10.
17. The strong law of large numbers is that, if $\overline{|X|} < \infty$, then $\lim_{n\to\infty} \Pr[A_n = \overline{X}] = 1$, i.e., the average converges *almost surely* to the mean.

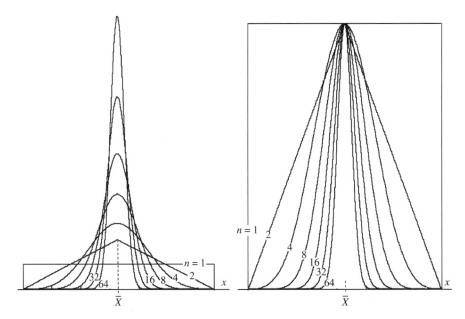

FIGURE 4.12. An illustration of the law of large numbers in (4.52) for a uniform random variable. As the number, n, of samples used in the average increases, the probability density function, as shown on the left, approaches a Dirac delta centered on the mean, $\overline{X}$. An arbitrarily large percentage of the probability mass can thus be contained in any fixed neighborhood around $\overline{X}$. On the left are shown the probability density functions, $f_{A_n}(x)$, for various values of n. The plots on the right are of the same functions, but each has been normalized to have the same maximum value. Note also, as required by the central limit theorem, a Gaussian shape emerges.

Proof. To show the weak law of large numbers, we use the Chebyshev inequality in (4.18).

$$\int_{|x-\overline{A_n}|\leq a} f_{A_n}(x)dx \geq 1 - \left(\frac{\sigma_{A_n}}{a}\right)^2.$$

From (4.50), the expected value of the average is equal to the random variable's expected value. If we also apply the variance relationship in (4.51), Chebyshev's inequality becomes

$$\int_{|x-\overline{X}|\leq a} f_{A_n}(x)dx \geq 1 - \frac{1}{n}\left(\frac{\sigma_X}{a}\right)^2.$$

As $n \longrightarrow \infty$, the right side of the inequality approaches one and the weak law of large numbers in (4.52) follows.

4.2.4.1 Stochastic Resonance

The law of large number establishes that the average converges to the mean. We present an example of *stochastic resonance* that illustrates the law of large numbers [920]. Stochastic resonance is loosely defined as the phenomenon in a detection process wherein the addition of just the right amount and type of noise improves performance. Too little or too much noise result in degraded performance. The phenomena has numerous manifestations [21, 477, 920, 988, 1168, 1597].

Pictures of the type illustrated in Figure 4.13 are used to illustrate a type of stochastic resonance when a simple threshold detector is used [920]. A gray level image is subjected to noise and is then subjected to a threshold. Too little noise renders the picture unrecognizable as does too much noise. When just the right amount of noise is added, a semblance of the original image is evident. The $\alpha = 0.5$ image from the right is at stochastic resonance.

Let a pixel in the image be denoted by x where x is a gray level between zero and one. We choose to corrupt the pixel by noise Ξ with a uniform random variable with probability density function

$$f_\Xi(\xi) = \frac{1}{2\alpha} \Pi\left(\frac{\xi}{2\alpha}\right). \tag{4.53}$$

The pixel now has a value of $Y = x + \Xi$. We threshold this at a level T to obtain either a value of zero (black) or one (white). This value is the random variable

$$W(x) = \mu(Y - T) = \mu(x + \Xi - T).$$

W must either be one or zero. Its expected value is

$$\overline{W}(x) = \frac{1}{2\alpha} \int_{\xi=-\alpha}^{\alpha} \mu(x + \xi - T) d\xi$$

$$= \begin{cases} 1 & ; T - x \leq -\alpha \\ \frac{1}{2\alpha} \int_{\xi=T-x}^{\alpha} d\xi = \frac{\alpha - T + x}{2\alpha} & ; -\alpha \leq T - x \leq \alpha \\ 0 & ; T - x \geq \alpha \end{cases}$$

For

$$\alpha = T = \frac{1}{2}, \tag{4.54}$$

we are at stochastic resonance because, at these values, the expected value of the threshold is equal to the pixel's value.

$$\overline{W}(x) = x.$$

For α and T at any other values, we are not at resonance. Indeed, from simply a visual grading, the $\alpha = 0.5$ image in Figure 4.13 can be argued to be the best representation of the original.

At resonance, the law of large numbers states that the average of a number of W's will approach x in the sense of (4.52). This is illustrated in Figure 4.15 for an image at stochastic resonance. As the number of images used in the average increases, the average looks more and more like the original. The average using 1024 images is indistinguishable from the original.

Figure 4.18 illustrates what happens at noise levels other than resonance. The original of the image *Brother Ray* is shown in the upper left corner. Each of the remaining images is for 1024 averaged images subjected to noise at a level α and then clipped. As in the previous example, resonance occurs at $\alpha = \frac{1}{2}$. $\alpha = 0, \frac{1}{4}$ is too little noise, and $\alpha = \frac{3}{4}, 1$ is too much noise. They do not faithfully reconstruct the original image.

Stochastic resonance, in another context, is the topic of Exercise 4.18.

4.2.5 The Central Limit Theorem

The central limit theorem states that, under loose conditions, the sum of random variables will be a Gaussian random variable. For independent random variables, the central limit

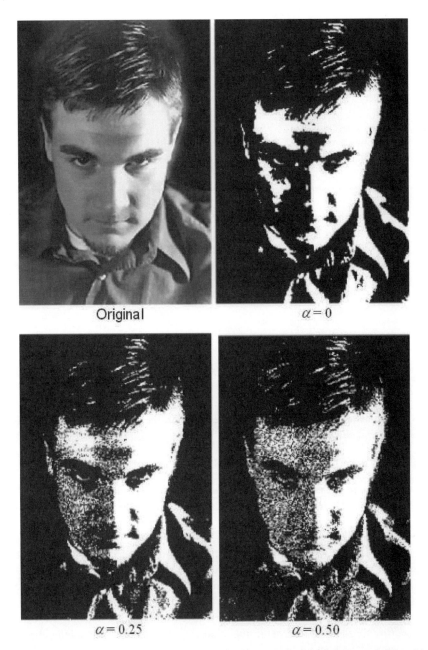

FIGURE 4.13. An example of stochastic resonance. The picture, *Jeremiah*, at the upper left is subjected to additive noise in (4.53) for various values of α and is then thresholded at the mid of the gray scale range. Values greater than 1 are pictured as white and those below zero are set to zero. The $\alpha = 0$ image is simply the original image thresholded. The $\alpha = 0$ corresponds to too little noise and $\alpha = 1.75$ is too much. The optimal amount of noise, corresponding to *stochastic resonance*, lies somewhere in between. As shown in (4.54), $\alpha = 0.5$ gives the best stochastic representation of the original image. (Continued in Figure 4.14.)

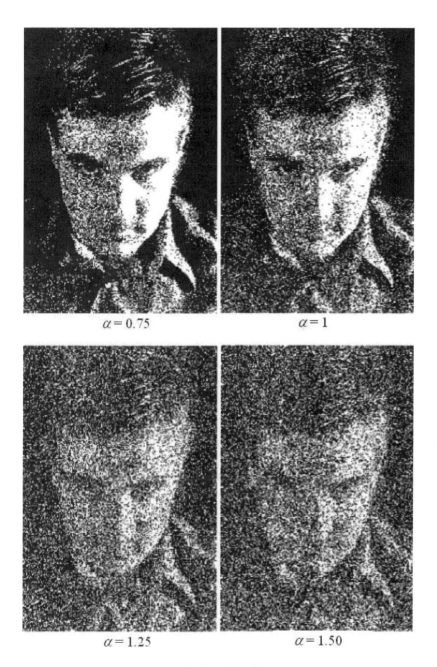

FIGURE 4.14. Continuation of Figure 4.13.

FIGURE 4.15. The average of a number of binary realizations of an image, *Baby Senator*, at stochastic resonance. The original image is shown at the top left. The averages of a number of binary realizations is shown. For the average of 1024 noisy images, the result is nearly indistinguishable from the original image. (Continued in Figures 4.16 and 4.17.)

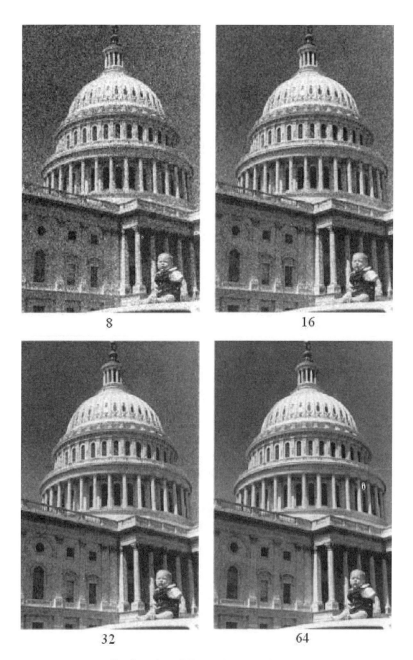

FIGURE 4.16. Continuation of Figure 4.15. Continued in Figure 4.17.

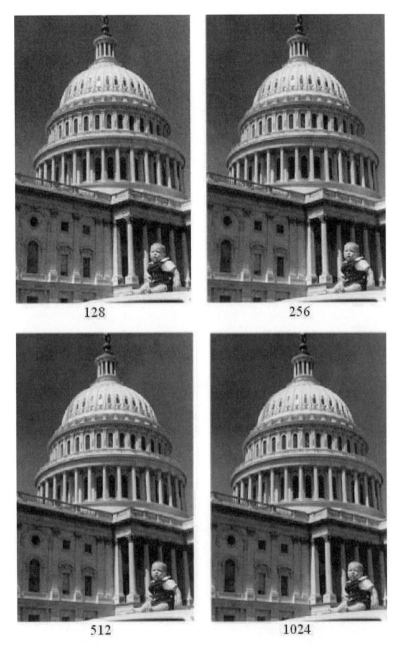

FIGURE 4.17. Continuation of Figures 4.15 and 4.16.

FIGURE 4.18. Image from averaging 1024 clipped noisy images of *Brother Ray* at different noise levels. The $\alpha = 0$ image is simply the original image thresholded at mid gray level. Resonance occurs at $\alpha = \frac{1}{2}$.

theorem results from the property that convolution of a number of functions, under mild conditions, will approach a Gaussian shaped curve. We will illustrate this for the sum of i.i.d. random variables. Define the normalized random variable

$$Z_n = \frac{A_n - \overline{A_n}}{\sigma_{A_n}}. \tag{4.55}$$

It follows that

$$\overline{Z_n} = 0$$

and

$$\sigma_{Z_n} = 1.$$

Our derivation will be made simpler by assuming $\overline{X} = 0$. This is done without loss of generality since shifting a probability density function to the origin to give a zero mean followed by autoconvolution is equivalent to autoconvolving first and then shifting. Since $\overline{X} = 0$, we have $\overline{A_n} = 0$, and the characteristic function of (4.55) is

$$\Phi_{Z_n}(u) = E\left[e^{-j2\pi u Z_n}\right]$$

$$= E\left[\exp\left(-j2\pi u \frac{A_n - \overline{A_n}}{\sigma_{A_n}}\right)\right]$$

$$= E\left[\exp\left(-j2\pi u \frac{A_n}{\sigma_{A_n}}\right)\right]$$

$$= \Phi_{A_n}\left(\frac{u}{\sigma_{A_n}}\right).$$

Using the characteristic function of the average in (4.48) and (4.51) gives

$$\Phi_{Z_n}(u) = \Phi_{A_n}\left(\frac{u}{n\sigma_{A_n}}\right)$$

$$= \Phi_X^n\left(\frac{u}{\sqrt{n}\sigma_X}\right).$$

Use the Taylor series of the characteristic function for the random variable X in (4.10) gives

$$\Phi_{Z_n}(u) = \left(\sum_{k=0}^{\infty} \frac{1}{k!}\left(\frac{-j2\pi u}{\sqrt{n}\sigma_X}\right)^k \overline{X^k}\right)^n$$

$$= \left(1 - \frac{j2\pi u}{\sqrt{n}\sigma_X}\overline{X} - \frac{(2\pi u)^2}{n\sigma_X^2}\frac{\overline{X^2}}{2} + \cdots\right)^n$$

$$= \left(1 - \frac{(2\pi u)^2}{n\sigma_X^2}\frac{\overline{X^2}}{2} + \cdots\right)^n \quad ; \text{ since } \overline{X} = 0$$

$$= \left(1 - \frac{2(\pi u)^2}{n} + \cdots\right)^n \quad ; \text{ since } \overline{X^2} = \sigma_X^2.$$

The higher order terms have powers of n that will tend to zero as n increases. We can now make use of the limit[18] for large n

$$\Phi_{Z_n}(u) \to \exp\left(-2(\pi u)^2\right).$$

From Table 4.1, this is recognized as the characteristic function of a Gaussian random variable with zero mean and unit variance. Note that this derivation of the central limit theorem for the sum of i.i.d. random variables makes the assumption that the mean and variance of the random variable X exist. Such is not the case, for example, with the Cauchy random variable and certain cases of the Pareto random variable.

4.2.5.1 The Central Limit Theorem Applied to Randomly Generated Probability Density Functions

- Adding i.i.d. uniform random variables together results in a probability density function equal to the convolutional concatenation of N rectangle functions. This is illustrated in Figure 4.19. The result is becoming graphically indistinguishable from the Gaussian as N increases.
- The central limit theorem is also applicable when the probability density functions are not identical. This is illustrated in Figure 4.20 where six randomly generated probability density functions are shown. The convolution of these functions is shown in Figure 4.21. The Gaussian with the same mean and variance is shown with dashed lines. The curves are similar. An example of the cumulative convergence is shown in Figure 4.22 where, in each step, a new randomly generated probability density function is added. Graphically, we have close convergence to the Gaussian using six functions.

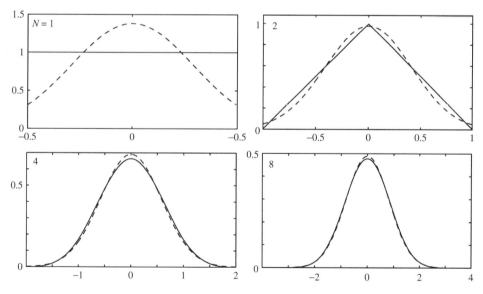

FIGURE 4.19. Illustration of the central limit theorem. Each figure shows the convolution of N rectangles, $\Pi(t)$. Shown with dashed lines is the Gaussian density function with the same mean and variance as the convolutions. In accordance to the central limit theorem, the successive convolutions approach a Gaussian as N increases. The result for $N = 4$ is the Parzen window.

---

18. See (14.3) in Appendix 14.3.

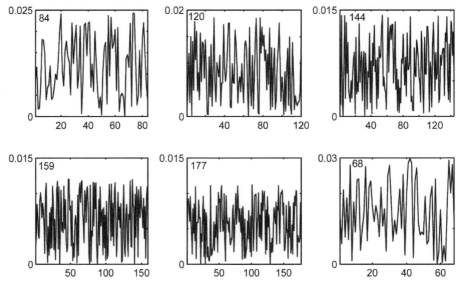

FIGURE 4.20. Six randomly generated probability density functions. The length of each function was chosen randomly on the interval [1, 200]. Over this interval, numbers are chosen randomly on the unit interval for each point. The function is then normalized to unit area. The convolution of these six functions is shown in Figure 4.21.

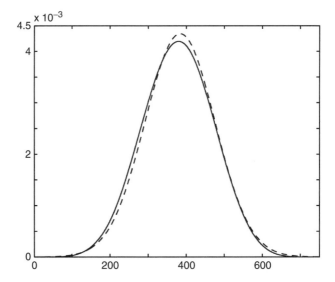

FIGURE 4.21. An illustration of the central limit theorem. Shown is the convolution of the six randomly generated probability density functions in Figure 4.20. The Gaussian with the same mean and variance is shown with the dashed lines.

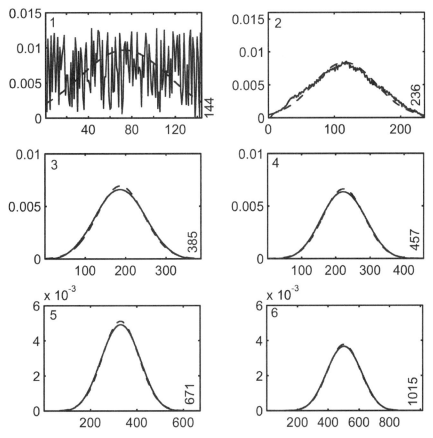

FIGURE 4.22. An illustration of the cumulative convergence of the central limit theorem. In each step, a randomly generated probability density function is added. Each density function is generated in the manner explained in the caption of Figure 4.20. The corresponding Gaussian in each case is shown with dashed lines. Graphically, we have close convergence for six component functions.

4.2.5.2 Random Variables That Are Asymptotically Gaussian

Many of the distributions discussed herein that are closed under addition approach Gaussian random variables. We adopt the following shorthand notation for the Gaussian random variable.

$$N(\overline{X}, \sigma_X^2; x) := \frac{1}{\sqrt{2\pi}\,\sigma_X} \exp\left(-\frac{(x-\overline{X})^2}{2\sigma_X^2}\right).$$

The sum of many i.i.d. random variables asymptotically approaches a Gaussian. We need only identify the mean and variance of the random variable sum. This becomes the mean and variance of the Gaussian. Here is a partial list.

- *Gaussian random variable.*

$$f_{S_n}(x) = \frac{1}{\sqrt{2\pi}\,\sqrt{n}\sigma_X} \exp\left(-\frac{(x-n\overline{X})^2}{2n\sigma_X^2}\right) = N\left(n\overline{X}, n\sigma_X^2; x\right)$$

- *Gamma random variable.*

$$f_{S_n}(x) = \frac{\lambda(\lambda x)^{n\alpha-1}e^{-\lambda x}}{\Gamma(n\alpha)} \to N\left(\frac{n\alpha}{\lambda}, \frac{n\alpha}{\lambda^2}; x\right).$$

- *Erlang random variable.*

$$f_{S_n}(x) = \frac{\lambda(\lambda x)^{nm-1}e^{-\lambda x}}{(nm-1)!} \to N\left(\frac{nm}{\lambda}, \frac{nm}{\lambda^2}; x\right).$$

Indeed, note the $n = 13$ plot of the Erlang probability density function in Figure 4.5 resembles a Gaussian curve.

- *Chi-square random variable.*

$$f_{S_n}(x) = \frac{x^{(nk-2)/2}e^{-x/2}}{2^{nk/2}\Gamma(nk/2)} \to N(k, 2k; x).$$

The $k = 13$ plot for the chi-squared probability density function in Figure 4.4 closely resembles a Gaussian curve.

- *Binomial random variable.* For notational clarity, let the random variable parameters be p and m. As always, $q = 1 - p$.

$$p_{S_n}[k] = \binom{nm}{k} p^k q^{nm} \to N(nmp, nmpq; x).$$

The binomial graphically approaching a Gaussian curve is graphically evident, for $p = \frac{1}{2}$, in Figure 4.23 and, for $p = 075$, in Figure 4.24. The central limit theorem applies to the binomial random variable for $p = 0.95$ but, as evident in Figure 4.24, convergence takes longer.

Application of the central limit theorem to the binomial random variable is oft called the *deMoivre-Laplace theorem*

- *Poisson random variable.*

$$p_{S_n}[k] = \frac{(n\alpha)^k e^{-n\alpha}}{k!} \to N(n\alpha, n\alpha; x).$$

The approach of the Poisson random variable to the Gaussian is graphically evident in Figure 4.26.

- *The negative binomial random variable's* probability density function approaches a Gaussian as $r \to \infty$. See Exercise 4.12.
- *The Pearson III random variable's* probability density function as the parameter $p \to \infty$. See Exercise 4.10.

4.3 Uncertainty in Wave Equations

A function, $\psi_X(x)$, possibly complex, is said to be a *wave function* if $f_X(x) = |\psi_X(x)|^2$ is a probability function. Let $\Psi_V(v)$ be the Fourier transform of $\psi_X(x)$. Then, from Parseval's theorem,

$$\int_{-\infty}^{\infty} |\psi_X(x)|^2 dx = \int_{-\infty}^{\infty} |\Psi_V(v)|^2 dv = 1$$

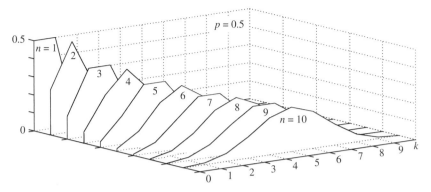

FIGURE 4.23. A waterfall plot of the binomial probability mass function, $\binom{n}{k}p^k q^{n-k}$, for various values of n. In accordance to the central limit theorem, the binomial probability mass function for $p = \frac{1}{2}$ is seen to approach a Gaussian as n increases. For $p = 0.75$, see Figure 4.24 and, for $p = 095$, see Figure 4.25. Points here are connected linearly. A plot of the same data in a different format is shown in Figure 4.25.

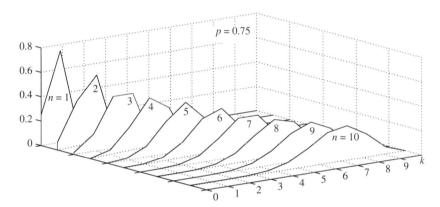

FIGURE 4.24. Continuation of Figure 4.23. Here, $p = 0.75$. The same plot in a different format is shown in Figure 4.10.

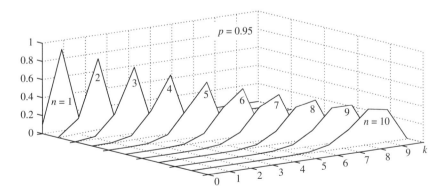

FIGURE 4.25. Continuation of Figure 4.24. Here, $p = 0.95$ will approach a Gaussian as n increases. Compared to the $p = \frac{1}{2}$ plot in Figure 4.23 and the $p = 0.75$ plot in Figure 4.24, a clear Gaussian shape has not yet emerged. The value of n needs to be larger than considered here.

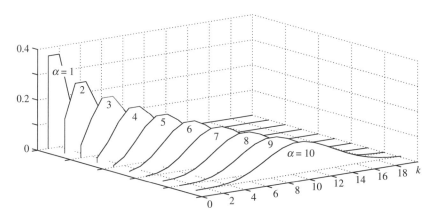

FIGURE 4.26. According to the central limit theorem, the Poisson probability mass function will approach a Gaussian as α increases. This is evident in this waterfall plot. Points are connected linearly. A plot of the same data in a different format is in Figure 4.11.

and $f_V(v) = |\Psi_V(v)|^2$ can also be considered a probability density function. Thus, $\Psi_V(v)$ is also a wave function. The scaling theorem states that, as a function is stretched in one domain, it is squeezed in the other. We thus expect that, as the variance in one domain is increased, the variance in the other will decrease. This property is expressed via the *uncertainty relationship*

$$\sigma_V \sigma_X \geq \frac{1}{4\pi}. \tag{4.56}$$

The uncertainty relationship is used in Section 13.3 to derive Heisenberg's uncertainty principle from Schrödinger's equation.

Proof. To show the uncertainty relationship in (4.56), we will assume, with no loss of generality, that both $\overline{X}$ and $\overline{V}$ are zero.[19] Then $\sigma_X^2 = \overline{X^2}$ and $\sigma_V^2 = \overline{V^2}$.

$$\sigma_X^2 \sigma_V^2 = \int_{-\infty}^{\infty} x^2 f_X(x) dx \int_{-\infty}^{\infty} v^2 f_V(v) dv$$

$$= \int_{-\infty}^{\infty} x^2 |\phi_X(x)|^2 dx \int_{-\infty}^{\infty} v^2 |\Phi_V(v)|^2 dv \tag{4.57}$$

$$= \int_{-\infty}^{\infty} |x\phi_X(x)|^2 dx \frac{1}{(2\pi)^2} \int_{-\infty}^{\infty} \left| \frac{d\psi_X(x)}{dx} \right|^2 dv$$

where, in the last step, we have used the power and derivative theorems. Schwarz's inequality[20] is

$$\left| \Re \int_{-\infty}^{\infty} g(x) h^*(x) dx \right|^2 \leq \int_{-\infty}^{\infty} |g(x)|^2 dx \int_{-\infty}^{\infty} |h(x)|^2 dx$$

---

19. For arbitrary $\overline{V}$, the new wave function $\hat{\psi}_X(x) = \psi_X(x) e^{j2\pi x \overline{V}}$ will give a $\hat{\Psi}_V(v)$ with a zero mean.
20. See Appendix 14.1.

We apply this to (4.57) with
$$g(x) = x\psi_X(x)$$
and
$$h(x) = \frac{d\psi_X(x)}{dx}.$$

The result is
$$\begin{aligned}\sigma_X^2 \sigma_V^2 &\geq \frac{1}{(2\pi)^2} \left| \Re \int_{-\infty}^{\infty} \left(x\psi_X^*(x) \frac{d\psi_X(x)}{dx} \right) \right|^2 \\ &= \frac{1}{(2\pi)^2} \left| \Re \int_{-\infty}^{\infty} \frac{1}{2} x \frac{d}{dx} \left(\psi_X(x) \psi_X^*(x) \right) \right|^2 \quad (4.58) \\ &= \frac{1}{(4\pi)^2} \left| \int_{-\infty}^{\infty} x \frac{d}{dx} f_X(x) dx \right|^2.\end{aligned}$$

If $\overline{X}$ is finite, then[21]
$$x f_X(x)|_{x=\pm\infty} = 0.$$

Integrating by parts
$$\int_{-\infty}^{\infty} x \frac{d}{dx} f_X(x) dx = x f_X(x)|_{x=\pm\infty} + \int_{-\infty}^{\infty} f_X(x) dx$$
$$= 1.$$

Substituting into (4.58) gives
$$\sigma_X^2 \sigma_V^2 \geq \frac{1}{(4\pi)^2}$$
which is the desired result.

4.4 Stochastic Processes

A stochastic process can be either discrete or continuous time signal and can be used to model either noise or a signal. In many applications, the first and second order statistics of a stochastic process suffice for a description.

Example 4.4.1. *Poisson Point Processes.* The Poisson approximation in (4.41) is valuable for modelling random points on a line. Consider the interval $[0, t]$ and a subsumed subinterval of duration T. If a point is placed randomly on $[0, t]$, the probability it lands in the subinterval T is
$$p = \frac{T}{t}.$$

---

21. Since $\overline{X} = \int_{-\infty}^{\infty} x f_X(x) dx$, the integrand must go to zero at $\pm\infty$ if the integral is finite.

Repeat the trial n times, The number of successes is a binomial random variable with parameters n and p. Keep T constant and let t and n be large keeping constant their ratio

$$\lambda = \frac{t}{T}.$$

The criteria in (4.37, 4.38, 4.39) for the Poisson approximation are fulfilled and, using (4.41), we have

$$p_k \approx e^{-\lambda T} \frac{(\lambda T)^k}{k!}. \tag{4.59}$$

As n and t grow without bounds, the approximation becomes exact. The $n = \infty$ points on the interval from zero to ∞ are said to be a *Poisson point process* with parameter λ. This stochastic process is used to measure, for example, random arrival of photons on a photodetector and arrival of messages at a router. It is also a good model for the pops heard when popping popcorn and the audible hits of rain on a metal roof. The cumulative spotting of a sidewalk by light rain is an example of a spatial Poisson process.

4.4.1 First and Second Order Statistics

The mean of a continuous time stochastic process, $\xi(t)$, is

$$\overline{\xi(t)} = E[\xi(t)]$$

where E denotes the expected value operator. The *noise level* is the variance of the process:

$$\text{var } \xi(t) = E[|\xi(t) - \overline{\xi(t)}|^2]$$
$$= \overline{|\xi(t)|^2} - |\overline{\xi(t)}|^2.$$

If a process is *zero mean*, then $\overline{\xi(t)}$ is zero and the noise level is $\overline{|\xi(t)|^2}$. If a process is a signal, this is called the *signal level*.

Both the mean and noise (signal) level are first order statistics since they are only concerned about the process at a single point in time. Second order statistics are concerned about the process at two different points in time. The *autocorrelation*, for example, is

$$R_\xi(t, \tau) = E[\xi(t)\xi^*(\tau)]. \tag{4.60}$$

Note that

$$R_\xi(t, t) = \overline{|\xi(t)|^2}. \tag{4.61}$$

Similarly, the cross correlation between two processes, $\xi(t)$ and $\eta(t)$, is defined as

$$R_{\xi\eta}(t, \tau) = E[\xi(t)\eta^*(\tau)].$$

Example 4.4.2. *Poisson Point Process Mean and Autocorrelation.*

Using the Poisson point process in Example 4.4.1 for a given λ, define the *Poisson counting process* by

$$X(t) = \{\, k \mid k = \text{ the number of point on the interval } (0, t)\}.$$

For a given t, the process $X(t)$ is a Poisson random variable with parameter $\alpha = \lambda t$. Thus

$$E[X(t)] = \lambda t$$

and

$$\sigma^2_{X(t)} = \lambda t. \tag{4.62}$$

It follows that

$$\overline{X^2(t)} = \lambda t\,(1 + \lambda t).$$

For the autocorrelation, we write

$$R_X(t, \tau) = E[X(t)X(\tau)].$$

Choose, for now, $t \geq \tau$ so that the number of Poisson points on $(0, \tau)$ is independent of the number of points on (τ, t). Then

$$E[X(\tau)\,(X(t) - X(\tau))] = E[X(\tau)]\,E[X(t) - X(\tau)]$$
$$= \lambda \tau\, \lambda(t - \tau). \tag{4.63}$$

Since

$$X(t)X(\tau) = X(\tau)\,[X(\tau) + X(t) - X(\tau)]$$

we have

$$R_X(t, \tau) = E[X(\tau)\,(X(\tau) + X(t) - X(\tau))]$$
$$= E\left[X(\tau)\,(X(t) - X(\tau)) + X^2(\tau)\right].$$

Using (4.62) and (4.63), we therefore have

$$R_X(t, \tau) = \lambda \tau\, \lambda(t - \tau) + \lambda \tau\,(1 + \lambda \tau)$$
$$= \lambda^2 t\tau + \lambda \tau.$$

This, however, is valid only for $t \geq \tau$. For the general case, the autocorrelation for the Poisson counting process is

$$R_X(t, \tau) = \lambda^2 t\tau + \lambda \min[t, \tau]. \tag{4.64}$$

4.4.1.1 Stationary Processes

A process that does not change its character with respect to time is said to be *stationary*. A process is said to be *wide sense stationary* (or *stationary in the wide sense*) if it meets two conditions. First, the mean of the process must be a constant for all time.

$$\overline{\xi(t)} = \overline{\xi}.$$

Secondly, the autocorrelation is only a function of the distance between the two points of interest.

$$E[\xi(t)\xi^*(\tau)] = R_\xi(t - \tau). \tag{4.65}$$

For such processes, the autocorrelation can thus be described by a one-dimensional rather than a two-dimensional function.[22]

Example 4.4.3. Let $\xi(t)$ be the current or voltage through a one ohm resistor. The instantaneous power is a stochastic process, $P(t) = \xi^2(t)$. If $\xi(t)$ is stationary, the average power dissipated by the resistor is

$$\overline{P(t)} = \overline{\xi^2}. \tag{4.66}$$

Therefore, the second moment of a stochastic process is associated with power.

Properties of the autocorrelation for a real signal include

(a) *Relation to second moment.* For stationary processes, (4.61) becomes[23]

$$R_\xi(0) = \overline{\xi^2}. \tag{4.67}$$

Similarly, in discrete time

$$R_\xi[0] = \overline{\xi^2}.$$

(b) *Hermetian property.* The autocorrelation obeys the symmetry constraint

$$R_\xi(t) = R_\xi^*(-t). \tag{4.68}$$

Similarly, in discrete time

$$R_\xi[n] = R_\xi^*[-n].$$

(c) *Bounded by the second moment.* The autocorrelation is bounded by the second moment of the process.[24]

$$|R_\xi(\tau)| \leq \overline{\xi^2}. \tag{4.69}$$

The property is also true in discrete time

$$|R_\xi[n]| \leq \overline{\xi^2}.$$

4.4.2 Power Spectral Density

The *power spectral density*, $S_\xi(u)$, of a wide sense stationary process can be defined in continuous time as the Fourier transform of the autocorrelation.

$$R_\xi(t) \longleftrightarrow S_\xi(u).$$

The power spectral density measures the power content of the process at each frequency component.

---

22. For stationary processes, the two-dimensional autocorrelation function becomes one dimensional. The autocorrelation therefore becomes invariant to shifts. This is similar to the linear system wherein, if the impulse response becomes a function of one variable, the system is dubbed time invariant. In a thesaurus, the words *stationary* and *time invariant* could appear as synonyms. Indeed, stationary stochastic processes are processes whose character does not change with respect to time. A time invariant system is a system whose parameters do not change with time.
23. See Exercise 4.20.
24. See Exercise 4.19 for a proof.

For a wide sense stationary discrete process, the autocorrelation is

$$R_\xi[n-m] = E\{\xi[n]\xi^*[m]\}. \tag{4.70}$$

The power spectral density for such processes is given by a Fourier series.

$$S_\xi(f) = \sum_{n=-\infty}^{\infty} R_\xi[n] e^{-j2\pi nf}. \tag{4.71}$$

For discrete time stochastic processes, the power spectral density is therefore periodic with a period of one.

4.4.3 Some Stationary Noise Models

Here we list some commonly used autocorrelation functions. For continuous stochastic processes, the power spectrum's inverse transform is the autocorrelation function with

$$\overline{|\xi(t)|^2} = \int_{-\infty}^{\infty} S_\xi(u)\, du \tag{4.72}$$

or, for a discrete stochastic process,

$$\overline{|\xi[n]|^2} = \int_{-\frac{1}{2}}^{\frac{1}{2}} S_\xi(f)\, df. \tag{4.73}$$

4.4.3.1 Stationary Continuous White Noise

Real stationary white noise has an autocorrelation of

$$R_\xi(\tau) = \overline{\xi^2}\, \delta(\tau). \tag{4.74}$$

Only for white noise does the notation $\overline{|\xi|^2}$ *not* correspond to the second moment of the process. Indeed, the noise level for this process is infinite. White noise is so named because its power spectral density

$$S(u) = \overline{\xi^2} \tag{4.75}$$

has the same energy level at every frequency.

Because continuous white noise requires an infinite amount of power, it does not physically exist. A physical noise which closely approximates continuous white noise is *thermal noise* (or *Johnson-Nyquist noise*). The noise, in volts, associated with resistor R ohms has power spectral density

$$S_\xi(u) = 2RkT\, \frac{\vartheta u}{e^{\vartheta u} - 1}$$

where

$$\vartheta = \frac{h}{kT},$$

k is Boltzmann's constant,[25] h is Planck's constant,[26] and T is temperature in degrees Kelvin. For $\vartheta u \ll 1$, we approximate $e^{\vartheta u} \simeq 1 + \vartheta u$ and

$$S_\xi(u) \simeq 2RkT = \text{constant}; \quad |\vartheta u| \ll 1.$$

Over this interval, thermal noise is therefore approximately white.

4.4.3.2 Stationary Discrete White Noise

This is a discrete process with autocorrelation

$$R_\xi[n] = \overline{\xi^2}\, \delta[n]. \tag{4.76}$$

Note that one does not obtain a discrete white sequence by sampling a continuous white noise process. Here, for example, the noise level is finite.

4.4.3.3 Laplace Autocorrelation

For a given positive parameter λ, the Laplace autocorrelation of a real process is

$$R_\xi(\tau) = \overline{\xi^2}\, e^{-\lambda|\tau|}. \tag{4.77}$$

It follows that

$$S_\xi(u) = \frac{2\lambda\,\overline{\xi^2}}{\lambda^2 + (2\pi u)^2}. \tag{4.78}$$

4.4.3.4 Ringing Laplace Autocorrelation

A special case of the Laplace autocorrelation is the *ringing Laplace autocorrelation* where

$$R_\xi(\tau) = \overline{\xi^2}\, e^{-\lambda|\tau|}\, \cos(2\pi vt)$$

where v is a given frequency. The corresponding power spectral density is

$$S_\xi(u) = \lambda \overline{\xi^2} \left[\frac{1}{\lambda^2 + (2\pi(u-v))^2} + \frac{1}{\lambda^2 + (2\pi(u+v))^2} \right].$$

4.4.4 Linear Systems with Stationary Stochastic Inputs

A stochastic process, $\xi(t)$, is placed through an LTI system with deterministic impulse response $h(t)$. The system output is given by the convolution

$$\eta(t) = \xi(t) * h(t). \tag{4.79}$$

Then the cross correlation between input and output is[27]

$$R_{\eta\xi}(\tau) = R_\xi(\tau) * h(\tau) \tag{4.80}$$

---

25. $k = 1.3807 \times 10^{-16}$ J K^{-1}
26. $h = 6.626068 \times 10^{-34}$ m^{2}kg/sec^2
27. See Exercise 4.17 for proofs of (4.80) and (4.81).

or, in terms of the power spectral densities,

$$S_{\eta\xi}(u) = S_\xi(u)H(u).$$

The output's autocorrelation is

$$R_\eta(t) = h(t) * h^*(-t) * R_\xi(t) = [h(t) \star h(t)] * R_\xi(t) \qquad (4.81)$$

or, in the frequency domain,

$$S_\eta(u) = |H(u)|^2 \, S_\xi(u). \qquad (4.82)$$

The same relationship holds for discrete time.

Example 4.4.4. Poisson point processes and shot noise. Extend the Poisson counting process in Example 4.4.2 to $(-\infty, \infty)$ and differentiate. The example is a sequence of Dirac deltas located at Poisson point locations

$$Y(t) = \sum_{n=-\infty}^{\infty} \delta(t - t_n). \qquad (4.83)$$

The autocorrelation is[28] [1077]

$$R_Y(\tau) = \lambda^2 + \lambda \, \delta(\tau). \qquad (4.84)$$

The mean is

$$\overline{Y(t)} = E\left[\frac{d}{dt}X(t)\right]$$
$$= \frac{d}{dt}E[X(t)]$$
$$= \frac{d}{dt}\lambda t$$
$$= \lambda.$$

Since the mean is constant and the autocorrelation is a function of only τ, the process is therefore wide sense stationary. The power spectral density is the Fourier transform of (4.84).

$$S_Y(u) = \lambda^2 \, \delta(u) + \lambda.$$

Shot noise, for a given $h(t)$, is defined as

$$Z(t) = \sum_{n=-\infty}^{\infty} h(t - t_n).$$

Shot noise can therefore be obtained by placing $Y(t)$ in (4.83) through an LTI filter with impulse response $h(t)$. From (4.82), the power spectral density of shot noise is thus

$$S_Z(u) = |H(u)|^2 S_Y(u)$$
$$= \lambda^2 \, \delta(u) \, |H(0)|^2 + \lambda \, |H(u)|^2.$$

---

28. See Exercise 4.14.

4.4.5 Properties of the Power Spectral Density

The term *power spectral density* contains three words. From (4.85) and Example 4.4.3, integration of $S_\xi(u)$ gives units of power. Thus, $S_\xi(u)$ is a *power density*. Because integration is over frequency, it is a power *spectral* density.

4.4.5.1 Second Moment

We can now substantiate (4.72) and (4.73). From the signal integral property in (2.93) and (4.67), the area of the power spectral density is the second moment of the stochastic process.

$$\overline{\xi^2} = \int_{-\infty}^{\infty} S_\xi(u)du. \tag{4.85}$$

For discrete time stochastic processes, we use Parseval's theorem for the Fourier series in (2.19) and conclude

$$\overline{\xi^2} = \int_{-1/2}^{1/2} S_\xi(f)df.$$

4.4.5.2 Realness

The power spectral density is real. To show this for continuous time, we write the inverse Fourier transform

$$R_\xi(\tau) = \int_{-\infty}^{\infty} S_\xi(u)e^{j2\pi u t}du.$$

It follows that

$$R_\xi^*(-\tau) = \int_{-\infty}^{\infty} S_\xi^*(u)e^{j2\pi u t}du.$$

Because of the Hermetian property in (4.68), we require that $S_\xi(u) = S_\xi^*(u)$ which dictates that $S_\xi(u)$ is real.

The autocorrelation for discrete stochastic processes is also conjugately symmetric.

4.4.5.3 Nonnegativity

The power spectral density is nonnegative. To show this for continuous time, we pass $\xi(t)$ through an LTI system with a frequency response equal to one in the interval $B - \epsilon \leq u \leq B + \epsilon$ and otherwise zero. The output is $\eta(t)$. Using (4.82), the second moment of the filter's output is

$$\overline{\eta^2} = \int_{-\infty}^{\infty} S_\eta(u)du$$

$$= \int_{B-\epsilon}^{B+\epsilon} S_\xi(u)du$$

$$\approx 2\epsilon\, S_\xi(B)$$

The approximation becomes exact as $\epsilon \to 0$. Since $\overline{\eta^2} \geq 0$, we therefore must conclude $S_\xi(B) \geq 0$. This must be true for all B. Thus, $S_\xi(u) \geq 0$ for all u.

4.5 Exercises

4.1. Let X denote a random variable with units of volts.
   (a) What are the units of the mean and variance of X?
   (b) Since the units of the mean and variance are different, how can the mean and variance of the Poisson random variable in (4.34) and (4.35) both be equal to α?

4.2. If $f_X(x)$ is a probability density function, show that

$$\int_{-\infty}^{\infty} \left(\frac{d}{dx} f_X(x) \right) dx = 0.$$

4.3. **Gamma random variable moments**
   (a) Using the characteristic function of the gamma random variable, evaluate the expression for the mth moment of the gamma random variable. Simplify the expression for the special cases of the Erlang and the chi-square random variables.
   (b) Using *only* the definition of the gamma random variable probability density function in (4.24) and the moment definition in (4.5), derive the nth moment of the gamma random variable.
   (c) Verify the mean and variance in (4.26) and (4.27) using the result in (b).

4.4. With reference to (4.55), show that an alternate expression is

$$Z_n = \frac{S_n - \overline{S_n}}{\sigma_{S_n}}. \qquad (4.86)$$

4.5. **Gaussian random variable moments.** Consider the Gaussian probability density function with zero mean and variance σ^2.
   (a) Set

$$\alpha = \frac{1}{2\sigma^2}$$

   so that

$$\sqrt{\frac{\pi}{\alpha}} = \int_{-\infty}^{\infty} e^{-\alpha x^2} dx.$$

   Note, then,

$$\left(\frac{d}{d\alpha} \right)^k \sqrt{\frac{\pi}{\alpha}} = \int_{-\infty}^{\infty} \left(-x^2 \right)^k e^{-\alpha x^2} dx. \qquad (4.87)$$

   Use this to expression to evaluate all the moments of the Gaussian random variable.
   (b) Use the moments in (a) in a Taylor series expansion to derive the characteristic function for the Gaussian random variable. Compare your result to the entry in Table 4.1.

4.6. § **Reliability.** The *reliability* of a random variable, X, is defined as $R_X(x) = \Pr[X > x]$. Note that $R_X(x) = 1 - F_X(x)$ where $F_X(x)$ is the cumulative distribution function of the random variable X. Consider the reliability defined by the transcendental relationship

$$R_X(x) = \begin{cases} e^{-xR_X(x)} & ; x \geq 0 \\ 1 & ; x < 0 \end{cases} \quad (4.88)$$

(a) Provide a plot of $R_X(x)$.
(b) Prove that the corresponding probability density function is bounded.
(c) Show that all of the moments of the random variable X are infinite.
(d) Are there any other random variables with all moments equal to infinity?

4.7. Show that the uncertainty relationship in (4.56) is, for a Gaussian probability density function, an identity.

4.8. Show that the maximum of
(a) the gamma random variable in (4.24) is at

$$x = \frac{\alpha - 1}{\lambda}.$$

(b) the beta probability density function in (4.28) is at

$$x = \frac{b}{b+c}.$$

4.9. **Confluent hypergeometric functions.** The *confluent hypergeometric function of the first kind* can be defined by

$$_1F_1(\alpha, \beta; z) = \frac{\Gamma(\beta)}{\Gamma(\beta - \alpha)\Gamma(\alpha)} \int_0^1 e^{zx} x^{\alpha-1} (1-x)^{\beta-\alpha-1} dx. \quad (4.89)$$

(a) Use this definition to derive the characteristic function of the beta random variable.
(b) Using the Taylor series for the characteristic function in (4.10), derive a Taylor series for the confluent hypergeometric function of the first kind.

4.10. **The Pearson III random variable** is defined in Table 4.1.
(a) Derive the characteristic function shown in Table 4.1 for the Pearson III random variable.
(b) Verify the mean and variance for the Pearson III random variable listed in Table 4.1.
(c) Is the sum of i.i.d. Pearson III random variables a Pearson III random variable?
(d) Evaluate the central limit theorem approximation of the probability density function of sum of i.i.d. Pearson III random variables as the number in the sum becomes large.

4.11. **The Von Mises random variable.** Show that the characteristic function for the Von Mises random variable in (14.16) can be expressed as a sampling theorem expansion using the modified Bessel functions, $I_n(b)$, in the samples. Hint: See Exercise 2.38.

4.12. **Negative binomial random variable.** From Table 4.3, the *negative binomial* or *Pascal* random variable is discrete with probability mass

$$p_k = \binom{k-1}{r-1} p^r q^{k-r} \mu[k-r]$$

where $0 < p < 1$ and $r \in \mathbb{N}$ are given parameters.

(a) Show that the characteristic function is

$$\Phi_X = \left(\frac{p}{e^{j2\pi u} - q}\right)^r.$$

(b) Show that the mean is

$$\overline{X} = \frac{r}{p},$$

and the variance is

$$\sigma_X^2 = \frac{rq}{p^2}.$$

(c) Does the negative binomial random variable result from the addition of n geometric random variables? Assume each has parameter p.
(d) What is the characteristic function of the sum of n independent negative binomial random variables?
(e) What is the characteristic function of the sum of n i.i.d. independent negative binomial random variables?
(f) Using the central limit theorem, approximate the probability mass function of the sum of n negative binomial random variables as a Gaussian.

4.13. **Mixed random variables.** A mixed random variable, with both continuous and discrete components, is

$$f_X(x) = \frac{1}{2}\delta(x+T) + \frac{1}{4}e^{-(x+T)/2}\Pi\left(\frac{x}{2T}\right) + \frac{1}{2}e^{-T}\delta(x-T). \quad (4.90)$$

where $T > 0$ is a given parameter.
(a) Show that the corresponding characteristic function is

$$\Phi_X(u) = e^{-T/2}\left[\cosh(\theta) + \frac{T}{2\theta}\sinh(\theta)\right] \quad (4.91)$$

where

$$\theta = \left(\frac{1}{2} + j2\pi u\right)T.$$

(b) The n-fold autoconvolution of the mixed probability density function has an important application in optimal detection theory. Marks et al. [890] used it to derive the performance of the optimal detector for signals corrupted by Laplace noise [342]. Except for the case of Gaussian noise, it remains the only known closed form optimal detector performance solution.

Evaluate the n-fold autoconvolution of (4.90). Hint: Expand the characteristic function into a binomial expansion. Then expand the sinh and cosh, expressed as exponentials, into binomial series. Use the following Fourier transform pair from Table 2.5.

$$\frac{t^{\ell-1}e^{-at}\mu(t)}{(\ell-1)!} \leftrightarrow \frac{1}{(a+j2\pi u)^\ell}. \quad (4.92)$$

4.14. (a) Show that if a stochastic process, $X(t)$, has autocorrelation $R_X(t, \tau)$, then the auto-correlation of

$$Y(t) = \frac{d}{dt}X(t),$$

if it exists, is

$$R_Y(t, \tau) = \frac{\partial^2}{\partial t\, \partial \tau} R_X(t, \tau).$$

(b) Show that, the autocorrelation of the Poisson counting process, $X(t)$ in (4.64), results in the autocorrelation in (4.84) for $Y(t) = \frac{d}{dt}X(t)$.

4.15. Evaluate the autocorrelation for the wide sense stationary stochastic process

$$\xi(t) = \sqrt{2\,\overline{\xi^2}}\,\cos(2\pi f_0 t - \Theta) \qquad (4.93)$$

where f_0 is given and the random variable Θ is uniform on $[-\pi, \pi]$. If the process is wide sense stationary, determine its power spectral density.

4.16. (a) Derive the bound for the autocorrelation in (4.72).
(b) Derive a similar expression for a discrete stochastic process.

4.17. Prove the cross correlation and autocorrelation relationships in (4.80) and (4.81) for a stochastic process being fed through an LTI system.

4.18. **Stochastic resonance**, in another context than presented in Section 4.2.4.1 [1076], occurs in the filtering of stochastic processes. Consider the high pass filter

$$H(u) = \frac{1}{(j2\pi u + \alpha)(j2\pi u + \beta)}. \qquad (4.94)$$

where $\alpha, \beta > 0$. Let $\xi(t)$ denote the class of wide sense stationary stochastic processes with fixed second moment, $\overline{\xi^2}$.
(a) What stochastic process in this class generates a filter output with maximum power?
(b) What is the maximum power output of the filter?

4.19. Prove the bounding property of the autocorrelation in (4.69).

4.20. Show that the Hermetian symmetry of the autocorrelation in (4.68) is valid.

4.21. Consider amplitude modulation as discussed in Section 3.4. Assume the baseband signal, $x(t)$, is a zero mean bounded wide sense stationary stochastic process with second moment $\overline{x^2}$. Assume $|x(t)| \leq A$. To use envelope detection demodulation,[29] we must keep $w(t) = x(t) + b \geq 0$.
(a) What value of bias minimizes the average power required to transmit $x(t)$ using the amplitude modulation procedure illustrated in Figure 3.15?
(b) What is this average power?

4.6 Solutions for Selected Chapter 4 Exercises

4.1. (a) If X is a random variable with units of volts, then its mean and standard deviation have units of volts. Its variance has units of volts squared. (b) The Poisson random

---

[29]. See Section 3.4.2.

variable is applicable only to unitless quantities, e.g., the *number* of hits on a web site in a given hour.

4.2. See the definite integration of derivatives property in (2.94).

4.3. **Gamma random variable moments.** Our goal is derivation of the moments of the gamma random variable in two ways and verify the mean and variance in (4.26) and (4.27) that were derived using the second characteristic function.

(a) The first few derivatives of the characteristic function for the gamma random variable are

$$\Phi(u) = \frac{1}{(1+j2\pi u/\lambda)^\alpha}$$

$$\frac{d}{du}\Phi(u) = \left(\frac{-j2\pi}{\lambda}\right)\frac{\alpha}{(1+j2\pi u/\lambda)^{\alpha+1}}$$

$$\left(\frac{d}{du}\right)^2\Phi(u) = \left(-\frac{j2\pi}{\lambda}\right)^2\frac{\alpha(\alpha+1)}{(1+j2\pi u/\lambda)^{\alpha+2}}$$

$$\left(\frac{d}{du}\right)^3\Phi(u) = \left(-\frac{j2\pi}{\lambda}\right)^3\frac{\alpha(\alpha+1)(\alpha+2)}{(1+j2\pi u/\lambda)^{\alpha+3}}$$

$$\left(\frac{d}{du}\right)^4\Phi(u) = \left(-\frac{j2\pi}{\lambda}\right)^4\frac{\alpha(\alpha+1)(\alpha+2)(\alpha+3)}{(1+j2\pi u/\lambda)^{\alpha+4}}.$$

We deduce the pattern

$$\left(\frac{d}{du}\right)^m\Phi(u) = \left(\frac{-j2\pi}{\lambda}\right)^m\frac{1}{(1+j2\pi u/\lambda)^{\alpha+m}}\prod_{q=0}^{m-1}(\alpha+q)$$

which can be firmly established by induction. Thus, from (4.9)

$$\overline{X^m} = \lambda^{-m}\prod_{q=0}^{m-1}(\alpha+q) \qquad (4.95)$$

For the Erlang random variable, $\alpha = k$ and

$$\prod_{q=0}^{m-1}(k+q) = k(k+1)(k+2)\ldots(k+m-1) = \frac{k!}{(k+m)!}.$$

Equation (4.95) becomes

$$\overline{X^m} = \frac{(k+m-1)!}{(k-1)!}\lambda^{-m}.$$

The mean and variance of the Erlang random variable follow as

$$\overline{X} = \frac{\alpha}{\lambda}$$

and
$$\sigma_X^2 = \overline{X^2} - \left(\overline{X}\right)^2$$
$$= \frac{\alpha(\alpha+1)}{\lambda^2} - \left(\frac{\alpha}{\lambda}\right)^2$$
$$= \frac{\alpha}{\lambda^2}.$$

These are the same results derived in (4.26) and (4.26) which were derived using the second characteristic function.

For the chi-square random variable, $\alpha = k/2$ and $\lambda = \frac{1}{2}$. Then

$$\prod_{q=0}^{m-1} \left(\frac{k}{2}+q\right) = \frac{k}{2}\left(\frac{k}{2}+1\right)\left(\frac{k}{2}+2\right)\left(\frac{k}{2}+3\right)\ldots\left(\frac{k}{2}+m-1\right)$$
$$= 2^m k(k+2)(k+4)(k+6)\ldots(k+2(m-1))$$
$$= 2^m (k+2(m-1))!!/(k-2)!!$$

where the double factorial is defined in (2.66). Thus, using (4.95),

$$\overline{X^m} = 4^m \frac{(k+2(m-1))!!}{(k-2)!!}.$$

(b) The nth moment of the gamma random variable follows from (4.24) and (4.5) as

$$\overline{X^n} = \int_0^\infty \frac{\lambda x^n (\lambda x)^{\alpha-1} e^{-\lambda x}}{\Gamma(\alpha)} dx$$

which we can also write as

$$\overline{X^n} = \frac{\Gamma(\alpha+n)}{\lambda^n \Gamma(\alpha)} \int_0^\infty \frac{\lambda(\lambda x)^{\alpha+n-1} e^{-\lambda x}}{\Gamma(\alpha+n)} dx.$$

The integration is over a gamma probability density function with parameter $\hat{\alpha} = \alpha + n$ and therefore integrates to one. This leaves

$$\overline{X^n} = \frac{\Gamma(\alpha+n)}{\lambda^n \Gamma(\alpha)}. \tag{4.96}$$

(c) From (4.96),

$$\overline{X} = \frac{\Gamma(\alpha+1)}{\lambda \Gamma(\alpha)} = \frac{\alpha \Gamma(\alpha)}{\lambda \Gamma(\alpha)} = \frac{\alpha}{\lambda}$$

which is the same as the gamma random variable mean calculated in (4.26). Similarly

$$\overline{X^2} = \frac{\Gamma(\alpha+2)}{\lambda^2 \Gamma(\alpha)} = \frac{\alpha(\alpha+2)}{\lambda^2}.$$

Thus

$$\sigma_X^2 = \overline{X^2} - \overline{X}^2 = \frac{\alpha(\alpha+1)}{\lambda^2} - \left(\frac{\alpha}{\lambda}\right)^2 = \frac{\alpha}{\lambda^2}$$

which is identical to the variance computed in (4.27).

4.4. Beginning with (4.86).

$$Z_n = \frac{S_n - \overline{S_n}}{\sigma_{S_n}}$$

$$= \frac{S_n/n - \overline{S_n}/n}{\sigma_{S_n}/n}$$

$$= \frac{A_n - \overline{A_n}}{\sigma_{A_n}}$$

This is (4.55).

4.5. **Gaussian random variable moments**
   (a) The first few derivatives are

$$\frac{d}{d\alpha}\alpha^{-1/2} = -\frac{1}{2}\alpha^{-3/2},$$

$$\left(\frac{d}{d\alpha}\right)^2 \alpha^{-1/2} = \frac{1}{2}\frac{3}{2}\alpha^{-5/2},$$

$$\left(\frac{d}{d\alpha}\right)^3 \alpha^{-1/2} = -\frac{1}{2}\frac{3}{2}\frac{5}{2}\alpha^{-7/2},$$

$$\left(\frac{d}{d\alpha}\right)^4 \alpha^{-1/2} = \frac{1}{2}\frac{3}{2}\frac{5}{2}\frac{7}{2}\alpha^{-9/2}.$$

From this pattern, we deduce

$$\left(\frac{d}{d\alpha}\right)^k \alpha^{-1/2} = (-1)^k \frac{1}{2}\frac{3}{2}\cdots\frac{2k-1}{2}\alpha^{-(2k+1)/2}.$$

or, equivalently,

$$\left(\frac{d}{d\alpha}\right)^k \alpha^{-1/2} = \frac{(2k-1)!!}{(-2)^k}\alpha^{-(2k+1)/2}.$$

Thus, (4.87) becomes

$$\sqrt{\pi}\frac{(2k-1)!!}{2^k}\alpha^{-(2k+1)/2} = \int_{-\infty}^{\infty}\left(-x^2\right)^k e^{-\alpha x^2}dx.$$

Substituting $\alpha = 1/(2\sigma^2)$ and manipulating gives

$$(2k-1)!!\,\sigma^{2k} = \frac{1}{\sqrt{2\pi}\,\sigma}\int_{-\infty}^{\infty} x^{2k} e^{-\frac{x^2}{2\sigma^2}}dx.$$

Thus

$$\overline{X^{2k}} = (2k-1)!!\,\sigma^{2k}.$$

Since the probability density function is even, all of the odd moments are zero. Thus

$$\overline{X^k} = (k-1)!!\,\sigma^k\,\delta[k - \text{even}]. \tag{4.97}$$

(b) Substituting (4.97) in the Taylor series expansion in (4.10) gives the characteristic function

$$\Phi_X(u) = \sum_{k=0}^{\infty} (-j2\pi u)^k \, (k-1)!! \, \sigma^k \, \delta[k-\text{even}]/k!. \qquad (4.98)$$

Since

$$\frac{(k-1)!!}{k!} = \frac{(k-1)(k-3)(k-5)\cdots}{k(k-1)(k-2)(k-3)\cdots} = \frac{1}{k!!},$$

(4.98) becomes

$$\Phi_X(u) = \sum_{k=0}^{\infty} \frac{(-j2\pi u \sigma)^k}{k!!} \, \sigma^k \, \delta[k-\text{even}].$$

Setting $k = 2n$ and noting that

$$(2n)!! = 2n(2n-2)(2n-4)\cdots = 2^n n(n-1)(n-2)\cdots = 2^n n!$$

gives

$$\Phi_X(u) = \sum_{n=0}^{\infty} \frac{\left(-2(\pi u \sigma)^2\right)^n}{n!}.$$

From the Taylor series for the exponential in (2.34), we conclude

$$\Phi_X(u) = e^{-2(\pi u \sigma)^2}.$$

Applying the shift theorem for a nonzero mean gives the Gaussian entry in Table 4.1.

4.6. **Reliability**. Note that, for a range of values of y, the operation $y_{n+1} = e^{-xy_n}$ corresponds to a contractive operator (see Section 11.5.2) and will converge to $y_\infty = e^{-xy_\infty}$. This, however, is not the most straightforward method to generate the data to plot $R_X(x)$.

(a) Rather, we can solve for the inverse of (4.88) to obtain

$$x = \frac{-\ln(R_X(x))}{R_X(x)}.$$

The inverse can be easily plotted and then appropriately rotated and flipped to give $R_X(x)$. The result is shown in Figure 4.27.

(c) Since $f_X(x) = -\frac{d}{dx} R_X(x)$, the kth moment in (4.5) can be written as

$$\overline{X^k} = -\int_0^\infty x^k \frac{dR_X(x)}{dx} dx.$$

Administer the variable substitution $y = R_X(x)$ to obtain

$$\overline{X^k} = \int_0^1 \left(\frac{-\ln(y)}{y}\right)^k dy.$$

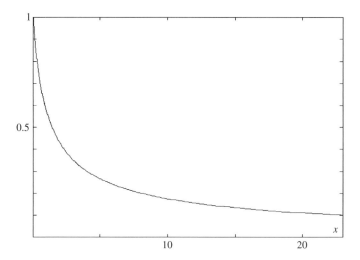

FIGURE 4.27. The plot of the reliability function, $R_X(x)$, defined by (4.88) in Exercise 4.6.

Since the integrand is positive, we have

$$\overline{X^k} \geq \int_0^{e^{-1}} \left(\frac{-\ln(y)}{y}\right)^k dy.$$

Over the interval $0 < y < e^{-1}$ we have $-\ln(y) > 1$ and, for $k \geq 1$,

$$\overline{X^k} \geq \int_0^{e^{-1}} \frac{1}{y^k}\, dy = \infty.$$

All of the moments of X are therefore infinite.

(d) Let $f_W(x)$ denote the probability density function of a Cauchy random variable. Then all of the moments of the random variable $Z = |W|$, with a probability density function of $f_Z(x) = 2f_W(x)\mu(x)$, are infinite. There are other examples.

4.9. **Confluent hypergeometric functions**

(a) To massage the integrand of confluent hypergeometric function definition in (4.89) into the probability density function of the beta random variable in (4.28), we set

$$b = \alpha - 1$$

and

$$c = \beta - \alpha - 1.$$

Thus

$$\alpha = b + 1$$

and

$$\beta = b + c + 2.$$

The integral in (4.89) can then be written as

$$_1F_1(b+1, b+1; z) = \frac{\Gamma(b+c+1)}{\Gamma(c+1)\Gamma(b+1)} \int_0^1 e^{zx} x^b (1-x)^c dx. \qquad (4.99)$$

The characteristic function for the beta random variable is the Fourier transform of the probability density function in (4.28).

$$\Phi_X(u) = \int_0^1 \frac{x^b(1-x)^c}{B(b+1, c+1)} e^{-j2\pi u x} dx.$$

Simplifying using (4.99) gives

$$\Phi_X(u) = (b+c+1)\,_1F_1(b+1, b+c+2, -j2\pi u) \qquad (4.100)$$

(b) Substituting the expression for the moments of the beta random variable in (4.29) into the characteristic function Taylor series expansion in (4.10) gives

$$(b+c+1)\,_1F_1(b+1, b+c+2, -j2\pi u)$$

$$= \sum_{n=0}^\infty (-j2\pi u)^n \frac{\Gamma(b+n+1)}{\Gamma(b+1)} \frac{\Gamma(b+c+2)}{\Gamma(b+n+c+2)}.$$

4.10. The Pearson III random variable

(a) Use the Pearson III random variable defined in Table 4.1 with $\alpha = 0$. We will add in the shift after performing a Fourier transform. From Table 2.5,

$$t^{k-1} e^{-at} \mu(t) \longleftrightarrow \frac{\Gamma(k)}{(a+j2\pi u)^k}$$

Thus, we have

$$\frac{1}{\beta \Gamma(p)} \left(\frac{x}{\beta}\right)^{p-1} e^{-x/\beta} \mu(x) \leftrightarrow \left(\frac{\beta}{\beta + j2\pi u}\right)^p$$

and, after application of the shift theorem, the characteristic function of the Pearson III random variable is that entered in Table 4.1.

(b) The second characteristic function for the Pearson III random variable is

$$\Psi_X(u) = p \ln \beta - p \ln(\beta + j2\pi u) - j2\pi \alpha u.$$

Thus

$$\frac{d}{du} \Psi_X(u) = -j2\pi \left(\frac{p}{\beta + j2\pi u} + \alpha\right)$$

and

$$\overline{X} = \frac{1}{-j2\pi} \frac{d}{du} \Psi_X(0) = \frac{p}{\beta} + \alpha.$$

Continuing with the second derivative

$$\left(\frac{d}{du}\right)^2 \Psi_X(u) = \frac{(-j2\pi)^2 p}{(\beta + j2\pi u)^2}$$

and the variance of the Pearson III random variable is

$$\sigma_X^2 = \frac{1}{-(2\pi)^2} \left(\frac{d}{du}\right)^2 \Psi_X(0) = p\beta^2.$$

(c) The characteristic function of the sum of N i.i.d. Pearson III random variable's with parameters α, β and p is

$$\Psi_S(u) = \Psi_X^N(u) = \left(\frac{\beta}{\beta + j2\pi u}\right)^{Np} p e^{-j2\pi u N\alpha}.$$

The sum is therefore a Pearson III random variable with parameters $N\alpha$, β and Np.

(d) Since the mean of the Pearson random variable is $\overline{X} = \alpha + \frac{p}{\beta}$ and its variance $\sigma_X^2 = p\beta^2$, we expect, from the central limit theorem, the sum to approach a Gaussian probability function for large N with a mean of $N\overline{X}$ and variance $N\sigma_X^2$.

4.13. Mixed random variables

(a) The characteristic function for the mixed probability density in (4.90) is

$$\Phi_X(x) = \frac{1}{2}e^{j2\pi uT} + \frac{1}{4}e^{-T/2}\int_{x=-T}^{T} e^{-x/2}e^{-j2\pi ux}dx$$

$$+ \frac{1}{2}e^{-T}e^{j2\pi uT}.$$

$$= e^{-T/2}\left[\left(\frac{1}{2}e^{T/2}e^{j2\pi uT} + \frac{1}{2}e^{-T/2}e^{-j2\pi uT}\right)\right.$$

$$\left.+ \left(\frac{1}{2}e^{T/2}e^{j2\pi uT} - \frac{1}{2}e^{-T/2}e^{-j2\pi uT}\right)/(1+j4\pi u)\right].$$

Simplifying gives (4.91).

(b) We raise (4.91) to the nth power to obtain

$$\Phi_{S_n}(u) = \Phi_X^n(u)$$

$$= e^{-nT/2}\left[\cosh(\theta) + \frac{T}{2\theta}\sinh(\theta)\right]^n$$

$$= e^{-nT/2}\sum_{k=0}^{n}\binom{n}{k}\left(\frac{T}{2\theta}\right)^k \sinh^k(\theta)\cosh^{n-k}(\theta)$$

$$= e^{-nT/2}\sum_{k=1}^{n}\binom{n}{k}\left(\frac{T}{2\theta}\right)^k \sinh^k(\theta)\cosh^{n-k}(\theta)$$

$$+ e^{-nT/2}\cosh^n(\theta).$$

Express the cosh and sinh as exponentials and expand again using a binomial expansion.

$$\Phi_{S_n}(u) = \frac{1}{2^n} e^{-nT/2} \left(e^\theta + e^{-\theta} \right)^n$$

$$+ e^{-nT/2} \sum_{k=1}^{n} \binom{n}{k} \left(\frac{T}{2\theta} \right)^k$$

$$\times \frac{1}{2^k} \left(e^\theta - e^{-\theta} \right)^k \frac{1}{2^{n-k}} \left(e^\theta + e^{-\theta} \right)^{n-k}$$

$$= \frac{1}{2^n} e^{-nT/2} \sum_{m=0}^{n} \binom{n}{m} e^{m\theta} (e^{-\theta})^{n-m}$$

$$+ e^{-nT/2} \sum_{k=1}^{n} \binom{n}{k} \left(\frac{T}{2\theta} \right)^k \qquad (4.101)$$

$$\times \sum_{p=0}^{k} \binom{k}{p} (-1)^p (e^{-\theta})^p e^{(k-p)\theta}$$

$$\times \frac{1}{2^{n-k}} \sum_{q=0}^{n-k} \binom{n-k}{q} (e^{-\theta})^q e^{(n-k-q)\theta}$$

$$= \frac{1}{2^n} e^{-nT/2} \sum_{m=0}^{n} \binom{n}{m} e^{(2m-n)\theta}$$

$$+ \frac{1}{2^n} e^{-nT/2} \sum_{k=1}^{n} \binom{n}{k} \left(\frac{T}{2\theta} \right)^k$$

$$\times \sum_{p=0}^{k} (-1)^p \binom{k}{p} \sum_{q=0}^{n-k} \binom{n-k}{q} e^{(n-2(p+q))\theta}.$$

Since

$$\theta = \left(\frac{1}{2} + j2\pi u \right) T$$

and, from (4.92),

$$\frac{t^{\ell-1} e^{-at} \mu(t)}{(\ell - 1)!} \leftrightarrow \frac{1}{(a + j2\pi u)^\ell},$$

we then conclude from the shift theorem that

$$\frac{(x + \ell T)^{k-1} e^{-x/2} \mu(x + \ell T)}{T^k (k - 1)!} \leftrightarrow \frac{e^{\ell T \left(\frac{1}{2} + j2\pi u \right)}}{\left[T \left(\frac{1}{2} + j2\pi u \right) \right]^k}.$$

Thus, with $\ell = n - 2(p + q)$, the inverse Fourier transform of (4.101) is our final answer

$$f_{S_n}(x) = \frac{1}{2^n} e^{-nT} \sum_{m=0}^{n} \binom{n}{m} e^{mT} \delta(x + (2m-n)T)$$

$$+ \frac{1}{2^n} e^{-(x+nT)/2} \sum_{k=1}^{n} \binom{n}{k} \frac{1}{2^k (k-1)!}$$

$$\times \sum_{p=0}^{k} (-1)^p \binom{k}{p} \quad (4.102)$$

$$\times \sum_{q=0}^{n-k} (x + (n - 2(p+q))T)^{k-1}$$

$$\times \mu(x + (n - 2(p+q))T).$$

The computational complexity of this probability density function can be simplified dependent on the value of x [342].

4.14. (a) Since, for real $X(t)$,

$$R_X(t, \tau) = E[X(t)X(\tau)],$$

and

$$Y(t) = \frac{d}{dt} X(t),$$

we have

$$R_Y(t, \tau) = E[Y(t)Y(\tau)]$$

$$= E\left[\frac{d}{dt} X(t) \frac{d}{d\tau} X(\tau)\right]$$

$$= E\left[\frac{\partial^2}{\partial t \, \partial \tau} X(t)X(\tau)\right]$$

$$= \frac{\partial^2}{\partial t \, \partial \tau} E[X(t)X(\tau)]$$

$$= \frac{\partial^2}{\partial t \, \partial \tau} R_X(t, \tau).$$

(b) Using (4.64) for $Y(t) = \frac{d}{dt} X(t)$,

$$R_Y(t, \tau) = \frac{\partial^2}{\partial t \, \partial \tau} \left[\lambda^2 t\tau + \lambda \min[t, \tau]\right]$$

$$= \frac{\partial^2}{\partial t \, \partial \tau} \left[\lambda^2 t\tau + \lambda \tau \, \mu(t-\tau) + \lambda t \, \mu(\tau-t)\right]$$

$$= \frac{\partial}{\partial \tau} \left[\lambda^2 \tau + \lambda \tau \, \delta(t-\tau) - \lambda t \, \delta(t-\tau) + \lambda \mu(\tau-t)\right].$$

Since
$$\tau\,\delta(t-\tau) = t\,\delta(t-\tau)$$
we have
$$R_Y(t,\tau) = \frac{\partial}{\partial \tau}\left[\lambda^2 \tau + \lambda \mu(\tau - t)\right].$$
$$= \lambda^2 + \lambda\,\delta(t-\tau).$$

This is equivalent to (4.84)

4.15. First, since Θ is uniform on $[-\pi, \pi]$,
$$\overline{\xi(t)} = \frac{1}{2\pi}\sqrt{2\overline{\xi^2}}\int_{-\pi}^{\pi}\cos(2\pi f_0 t - \theta)\,d\theta = 0.$$

The expected value of the process is thus identically zero and meets the first of two criteria for being stationary in the wide sense.

The autocorrelation of the stochastic process in (4.93) is
$$R_\xi(t,\tau) = 2\overline{\xi^2}\,E\left[\cos(2\pi f_0 t - \Theta)\,\cos(2\pi f_0 \tau - \Theta)\right].$$

Using a trig identity[30]
$$R_\xi(t,\tau) = \overline{\xi^2}\left(\cos(2\pi f_0(t-\tau)) + E\left[\cos(2\pi f_0(t+\tau) - 2\Theta)\right]\right).$$

But, since
$$E\left[\cos(2\pi f_0(t+\tau) - 2\theta)\right] = \frac{1}{2\pi}\int_{-\pi}^{\pi}\cos(2\pi f_0(t+\tau - 2\theta))\,d\theta = 0$$
and
$$R_\xi(t,\tau) = \overline{\xi^2}\,\cos(2\pi f_0(t-\tau)).$$

The stochastic process is therefore stationary with autocorrelation
$$R_\xi(\tau) = \overline{\xi^2}\,\cos(2\pi f_0 \tau).$$

Its power spectral density follows as
$$S_\xi(u) = \frac{\overline{\xi^2}}{2}\left[\delta(u-f_0) + \delta(u+f_0)\right].$$

4.16. (a) $\overline{\xi^2} = R_\xi(0) = \int_{-\infty}^{\infty} S_\xi(u)\,du.$

(b) $\overline{\xi^2} = R_\xi[0]$. Note that (4.71) is a Fourier series with period 1. Thus
$$\overline{\xi^2} = \int_1 S_\xi(u)\,du$$
where integration is over any period.

---

30. Here is a derivation of the trig identity. $\cos(\alpha)\cos(\beta) = \Re\,e^{j\alpha}\cos(\beta) = \frac{1}{2}\Re\,e^{j\alpha}\left(e^{j\beta} + e^{-j\beta}\right)$. Thus $\cos(\alpha)\cos(\beta) = \frac{1}{2}[\cos(\alpha+\beta) + \cos(\alpha-\beta)]$.

4.18. **Stochastic resonance**
(a) The magnitude squared of (4.94) is maximum when

$$\frac{1}{|H(u)|^2} = (v+a)(v+b)$$

and is minimum where $v = (2\pi u)^2$, $a = \alpha^2$, $b = \beta^2$. This occurs when

$$v = \frac{a+b}{2}$$

or

$$u = u_r = \frac{1}{2\pi}\sqrt{\frac{\alpha^2+\beta^2}{2}}.$$

The maximum filter output occurs when all the power spectral density is concentrated at the maximum of $|H(u)|$. This occurs when $X(t)$ has a power spectral density with all of its mass at the resonant frequency, u_r, i.e.,

$$S_X(u) = \frac{\overline{X^2}}{2}\left[\delta(u+u_r) + \delta(u-u_r)\right].$$

A candidate is the stochastic process in Exercise 4.15 with $f_0 = u_r$.

(b) Let $\eta(t)$ be the filter output. Since

$$|H(u_r)|^2 = \frac{4}{\alpha^2+\beta^2},$$

we have, using (4.82), the average power of the filter output as

$$\overline{\eta^2} = \frac{4\overline{\xi^2}}{(\alpha^2+\beta^2)^2}.$$

4.20. (a) From (4.65)

$$R_\xi^*(\tau - t) = E[\xi(\tau)\,\xi^*(t)]^*$$
$$= R_\xi(t-\tau).$$

(b)

$$S_\xi^*(u) = \int_{-\infty}^{\infty} R_\xi^*(t)\, e^{j2\pi ut}\, dt$$

$$= \int_{-\infty}^{\infty} R_\xi^*(-\tau)\, e^{-j2\pi u\tau}\, d\tau$$

$$= \int_{-\infty}^{\infty} R_\xi(\tau)\, e^{-j2\pi u\tau}\, d\tau$$

$$= S_\xi(u).$$

4.21. (a) The bias, b, is the only free parameter. The minimum average power results by minimizing the bias. To keep $w(t) = x(t) + b \geq 0$, we choose $b = A$. Then the modulated signal envelope will touch zero but not go negative. This is illustrated in Figure 4.28

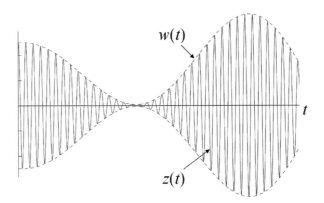

FIGURE 4.28. Full modulation using minimum average power.

5

The Sampling Theorem

> When I am violently beset with temptations, or cannot rid myself of evil thoughts, [I resolve] to do some Arithmetic, or Geometry, or some other study, which necessarily engages all my thoughts, and unavoidably keeps them from wandering.
> Jonathon Edwards (1703–1758) [867].
>
> I tell them that if they will occupy themselves with the study of mathematics they will find in it the best remedy against the lusts of the flesh.
> Thomas Mann (1875–1955) [871].
>
> Life is good for only two things, discovering mathematics and teaching mathematics.
> Siméon Poisson [468].
>
> If I feel unhappy, I do mathematics to become happy. If I am happy, I do mathematics to keep happy.
> Alfréd Rényi (1921–1970) [1419].
>
> I just wondered how things were put together.
> Claude Shannon

5.1 Introduction

Much of that which is ordinal is modelled as analog. Most computational engines, on the other hand, are digital. Transforming from analog to digital is straightforward: we simply sample. Regaining the original signal from these samples or assessing the information lost in the sampling process are the fundamental questions addressed by sampling and interpolation theory.

This chapter deals with understanding, generalizing and extending the cardinal series of Shannon sampling theory. The fundamental form of this series states, remarkably, that a bandlimited signal is uniquely specified by its sufficiently close equally spaced samples.

The cardinal series has many names, including the *Whittaker-Shannon sampling theorem* [514], the *Whittaker-Shannon-Kotelnikov sampling theorem* [679], and the *Whittaker-Shannon-Kotelnikov-Kramer sampling theorem* [679]. For brevity, we will use the terms *sampling theorem* and *cardinal series*.

5.1.1 The Cardinal Series

If a signal has finite energy, the minimum sampling rate is equal to two samples per period of the highest frequency component of the signal. Specifically, if the highest frequency component of the signal is B hertz, then the signal, $x(t)$, can be recovered from the samples by

$$x(t) = \frac{1}{\pi} \sum_{n=-\infty}^{\infty} x\left(\frac{n}{2B}\right) \frac{\sin(\pi(2Bt - n))}{2Bt - n}.$$

The frequency B is also referred to as the signal's bandwidth and, if B is finite, $x(t)$ is said to be bandlimited [1283].

5.1.2 History

The history of Shannon sampling theory and the cardinal series is intriguing. A summary of key events in the development of the cardinal series is listed in Table 5.1.

Luke [856] credits Lagrange for deriving the sampling theorem for the special case of periodic bandlimited signals in 1765.

H.S. Black[119] credits Cauchy for recognition of the mechanics of band-limited signal sampling in 1841 and even offers the following translation from Cauchy's original French text [252].

> IF A SIGNAL IS A MAGNITUDE-TIME FUNCTION, AND IF TIME IS DIVIDED INTO EQUAL INTERVALS SUCH THAT EACH SUBDIVISION COMPRISES AN INTERVAL T SECONDS LONG WHERE T IS LESS THAN HALF THE PERIOD OF THE HIGHEST SIGNIFICANT FREQUENCY COMPONENT OF THE SIGNAL, AND IF ONE INSTANTANEOUS SAMPLE IS TAKEN FROM EACH SUBINTERVAL IN ANY MANNER; THEN A KNOWLEDGE OF THE INSTANTANEOUS MAGNITUDE OF EACH SAMPLE PLUS A KNOWLEDGE OF THE INSTANT WITHIN EACH SUBINTERVAL AT WHICH THE SAMPLE IS TAKEN CONTAINS ALL THE INFORMATION OF THE ORIGINAL SIGNAL.

In a later historical overview, Higgins [603], however, notes that such a quote was not included in the paper by Cauchy that was cited by Black. Higgins, rather, credits

TABLE 5.1. Key events in the development of the cardinal series

1765 - Lagrange derives the sampling theorem for the special case of periodic bandlimited signals [856].
1841 - Cauchy's recognition of the Nyquist rate [252][1].
1897 - Borel's recognition of the feasibility of regaining a bandlimited signal from its samples [136].
1915 - E.T. Whittaker's publishes his highly cited paper on the cardinal series [1482].
1928 - H. Nyquist establishes the time–bandwidth product of a signal [1033].
1929 - J.M. Whittaker coins the term *cardinal series*.
1933 - A. Kotelnikov publishes the sampling theorem in the Soviet literature.
1948 - C.E. Shannon publishes a paper which establishes the field of information theory. The sampling theorem is included [1256].
1959 - H.P. Kramer generalizes the sampling theorem to functions that are bandlimited in other than the Fourier sense [772].
1962 - D.P. Peterson and D. Middleton extend the sampling theorem to higher dimensions [984].
1977 - A. Papoulis publishes his generalization of the sampling theorem [1086]. A number of previously published extension are shown to be special cases.

---
1. Disputed. See text.

Emile Borel [136] in 1897 for the initial recognition of the cardinal series and cites the following passage translated from the original French.

CONSIDER

$$f(z) = \int_{-\pi}^{\pi} \Psi(x)\, e^{jzx} dx$$

AND SUPPOSE THAT THE FUNCTION $\Psi(x)$ SATISFIES THE CONDITIONS OF DIRICHLET. IF ONE KNOWS THE VALUES OF THE FUNCTION, $f(z)$, AT THE POINTS $z = 0$, $\pm 1, \pm 2, \ldots$, THEN THE FUNCTION $\Psi(x)$ IS COMPLETELY DETERMINED AND, CONSEQUENTLY, THE ENTIRE FUNCTION $f(z)$ IS KNOWN WITHOUT AMBIGUITY.

This connection of the Fourier series to the sampling theorem was the same tool of explanation later used by Shannon in his classic paper [1256].

E. T. Whittaker published his highly cited paper on the sampling theorem in 1915 [1482]. In his work, if one function had the same uniformly spaced samples as another, the functions were said to be *cotabular*. The sampling theorem interpolation from these samples resulted in what Whittaker called the *cardinal function*. The interpolation formula was later dubbed the *cardinal series* by Whittaker's son, J. M. Whittaker [1484]. Among other things, the senior Whittaker showed the functions, $x(t)$, to which the cardinal series applied were bandlimited and entire on the finite t plane. He also noted that applicability of the cardinal series to a function was independent of the choice of sampling phase.

The sampling theorem was reported in the Soviet literature in a paper by Kotelnikov in 1933. Shannon [1256] used the sampling theorem to demonstrate that an analog bandlimited signal was equivalent in an information sense to the series of its samples taken at the Nyquist rate. He was aware of the work of Whittaker which he cited. Other noted historical generalizations and extensions of the sampling theorem are listed chronologically in Table 5.1.

Marks published the first volume dedicated entirely to the sampling theorem in 1991 [915]. Other texts have followed [605, 606, 916, 957].

A signal is bandlimited in the low pass sense if there is a $B > 0$ such that

$$X(u) = X(u)\, \Pi\left(\frac{u}{2B}\right). \tag{5.1}$$

That is, the spectrum is identically zero for $|u| > B$. The B parameter is referred to as the signal's *bandwidth*. It then follows that

$$x(t) = \int_{-B}^{B} X(u)\, e^{j2\pi ut}\, du. \tag{5.2}$$

In most cases, the signal can be expressed by the cardinal series

$$x(t) = \sum_{n=-\infty}^{\infty} x\left(\frac{n}{2B}\right) \operatorname{sinc}(2Bt - n). \tag{5.3}$$

The ability to thus express a continuous signal in terms of its samples is the fundamental statement of the sampling theorem.

At other than sample locations, a more computationally efficient form of (5.3) requiring evaluation of only a single trigonometric function is

$$x(t) = \frac{1}{\pi} \sin(2\pi Bt) \sum_{n=-\infty}^{\infty} \frac{(-1)^n x(\frac{n}{2B})}{2Bt - n}. \tag{5.4}$$

5.2 Interpretation

The sampling theorem reduces the normally continuum infinity of ordered pairs required to specify a continuous time function to a countable - although still infinite – set. Remarkably, these elements are obtained directly by sampling.

Bandlimited functions are smooth. Any behavior deviating from "smooth" results in high frequency components which in turn invalidate the required property of being bandlimited. The smoothness of the signal between samples precludes arbitrary variation of the signal there.

Let's examine the cardinal series more closely. Evaluation of (5.3) at $t = m/2B$ and using (2.42) reduces (5.3) to an identity. Thus, only the sample at $t = m/2B$ contributes to the interpolation at that point. This is illustrated in Figure 5.1 where the reconstruction of a signal from its samples using the cardinal series is shown. The value of $x(t)$ at a point other than a sample location (e.g., $t = (m + \frac{1}{2})/2B$) is determined by all of the sample values.

5.3 Proofs

We now present three proofs of the conventional sampling theorem. The first is used most commonly in texts. The second, due to Shannon, exposes the sampling theorem as the

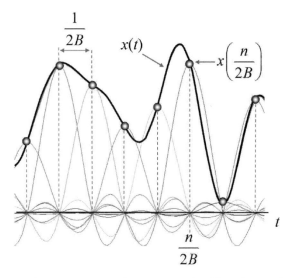

FIGURE 5.1. Illustration of the manner by which $x(t)$ is regained from its samples using the cardinal series. Note that the sinc for a given sample is zero at all other sample locations.

Fourier transform dual of the Fourier series. Finally, an eloquently compact form of proof due to Papoulis [1087] is presented.

5.3.1 Using Comb Functions

In this section we present the standard textbook proof of the sampling theorem. Since it can be nicely illustrated, the proof is quite instructive. It requires only an introductory knowledge of Fourier analysis.

In presenting this proof, we will repeatedly refer to Figure 5.2 where five functions and their Fourier transforms are shown. In (a) is pictured a signal which, as is seen from its transform, is bandlimited with bandwidth B. The sampling is performed by multiplying the signal by the sequence of Dirac deltas shown in (b). The result, shown in (c) is

$$s(t) = x(t)\, 2B \operatorname{comb}(2Bt) \tag{5.5}$$

$$= \sum_{n=-\infty}^{\infty} x\left(\frac{n}{2B}\right) \delta\left(t - \frac{n}{2B}\right). \tag{5.6}$$

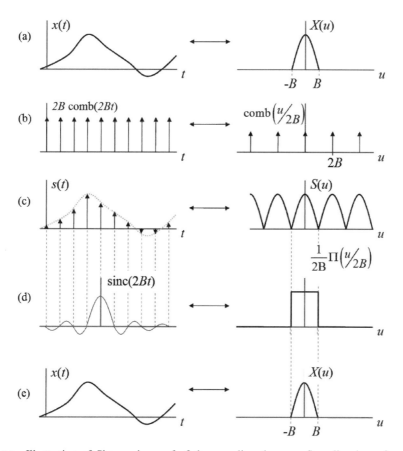

FIGURE 5.2. Illustration of Shannon's proof of the sampling theorem. Sampling is performed by multiplying lines (a) and (b) in time to obtain line (c). The signal is regained by multiplying lines (c) and (d) in frequency to obtain line (e).

Our goal is to recover $x(t)$ from $s(t)$ which is specified only by the signal's samples.

Let's examine what happens in the frequency domain when we sample. Multiplication in the time domain corresponds to convolution in the frequency domain. Since

$$2B \operatorname{comb}(2Bt) \longleftrightarrow \operatorname{comb}\left(\frac{u}{2B}\right),$$

we conclude that the transform of (5.5) is

$$S(u) = X(u) * \operatorname{comb}\left(\frac{u}{2B}\right).$$

Because convolving a function with $\delta(u - a)$ centers that function at a,

$$F(u) * \delta(u - a) = F(u - a),$$

we conclude that

$$S(u) = 2B \sum_{n=-\infty}^{\infty} X(u - 2nB). \tag{5.7}$$

These replications do not overlap.

5.3.1.1 Aliasing

If $X(u)$ was not truly bandlimited or if the sampling rate were below $2B$, then the spectra would overlap. This phenomena is referred to as *aliasing*. Here, the low frequency components of the signal can still be regained although the high frequency components are irretrievably lost. Clearly, sampling can be performed at a rate greater than $2B$ without aliasing. The minimum sampling rate resulting in no aliasing (in this case $2B$) is referred to as the *Nyquist rate*. Sampling above the Nyquist rate can relax interpolation formula requirements and improve noise sensitivity. If there are Dirac deltas in the signal's spectrum, sampling a bit above twice the signal's bandwidth may be required.

5.3.1.2 The Resulting Cardinal Series

Sampling is performed in Figure 5.2 by multiplying lines (a) and (b) in time to obtain line (c). Clearly, if there is no aliasing, multiplication of $S(u)$ on line (c) by the low pass filter on line (d) will result in the original spectrum, $X(u)$. That is

$$X(u) = S(u) \frac{1}{2B} \Pi\left(\frac{u}{2B}\right).$$

The corresponding time domain operation is convolution:

$$x(t) = s(t) * \operatorname{sinc}(2Bt).$$

Substituting (5.6) and evaluating gives the sampling theorem in (5.3).

5.3.2 Fourier Series Proof

Since $X(u)$ is identically zero for $|u| > B$, we can replicate it to form a periodic function in the frequency domain with period $2B$. This periodic function can be expressed as a

Fourier series. The result of the series for $|u| \leq B$ is $X(u)$. The Fourier series

$$X(u) = \sum_{n=-\infty}^{\infty} c_n e^{-j\pi nu/B} \Pi\left(\frac{u}{2B}\right) \qquad (5.8)$$

has Fourier coefficients

$$c_n = \frac{1}{2B} \int_{-B}^{B} X(u) e^{j\pi nu/B} du$$

$$= \frac{1}{2B} x\left(\frac{n}{2B}\right). \qquad (5.9)$$

Substituting into (5.8) and inverse transforming gives the sampling theorem series in (5.3) which, as we see here, is the Fourier transform dual of the Fourier series.

5.3.3 Papoulis' Proof

A eloquent proof [1087] of the sampling theorem begins with the Fourier series expansion of a periodic function with period $2B$ that is equal to $\exp(j2\pi ut)$ for $|u| \leq B$:

$$e^{j2\pi ut} = \sum_{n=-\infty}^{\infty} \text{sinc}(2Bt - n) e^{j2\pi(n/2B)u}. \qquad (5.10)$$

Substituting into the inversion formula in (5.2) gives

$$x(t) = \sum_{n=-\infty}^{\infty} \int_{-B}^{B} X(u) e^{j\pi nu/B} du \ \text{sinc}(2Bt - n)$$

Evaluating the integral by again using (5.2) gives the cardinal series in (5.3).

5.4 Properties

We now present some properties of the cardinal series.

5.4.1 Convergence

Although there are exceptions, the cardinal series generally converges uniformly. That is

$$\lim_{N \to \infty} |x(t) - x_N(t)| = 0 \qquad (5.11)$$

where the truncated cardinal series is

$$x_N(t) = \sum_{n=-N}^{N} x\left(\frac{n}{2B}\right) \text{sinc}(2Bt - n). \qquad (5.12)$$

The validity of (5.11) is obvious at the sample locations ($t = \frac{m}{2B}$) since contributions from adjacent points are zero. Uniform convergence is stronger than, say, the mean square convergence characteristic of Fourier series expansions of functions with discontinuities.

5.4.1.1 For Finite Energy Signals

We first will prove (5.11) for the case where $x(t)$, and therefore $X(u)$, has finite energy [478]. Note that

$$x(t) - x_N(t) = \int_{-B}^{B} [X(u) - X_N(u)] e^{j2\pi ut} \, du \qquad (5.13)$$

where

$$x_N(t) \longleftrightarrow X_N(u)$$

$$= \frac{1}{2B} \sum_{n=-N}^{N} x\left(\frac{n}{2B}\right) e^{-j\pi nu/B} \Pi\left(\frac{u}{2B}\right).$$

Schwarz's inequality[2] can be written

$$\left| \int_{-\infty}^{\infty} A(u) C(u) \, du \right|^2 \leq \int_{-\infty}^{\infty} |A(u)|^2 \, du \int_{-\infty}^{\infty} |C(u)|^2 \, du. \qquad (5.14)$$

Application to (5.13) yields

$$|x(t) - x_N(t)|^2 \leq 2B \int_{-B}^{B} |X(u) - X_N(u)|^2 \, du. \qquad (5.15)$$

The right side approaches zero if $X_N \to X$ in the mean square sense. Since the limit of X_N is the Fourier series of X and the Fourier series displays mean square convergence,[3] the right side of (5.15) tends to zero and our proof is complete.

5.4.1.2 For Bandlimited Functions with Finite Area Spectra

We now present an alternate proof of the cardinal series' uniform convergence for the case where $X(u)$ has finite area:

$$\int_{-B}^{B} |X(u)| \, du < \infty. \qquad (5.16)$$

From Exercise 2.19, this constraint requires that $x(t)$ be bounded. Our proof will, for example, allow cardinal series representation for $J_0(2\pi t)$ and $x(t) = constant$ both of which, due to infinite energy, are excluded in the conditions of the previous proof.

We begin our proof by defining

$$e_N(u; t) = \sum_{n=-N}^{N} \text{sinc}(2Bt - n) e^{j\pi nu/B}. \qquad (5.17)$$

From (5.10), we recognize that $e_N(u; t)$ is a truncated Fourier series of $\exp(j2\pi ut)$ for $|u| < B$. Since, for a fixed t, $\exp(j2\pi ut)$ is continuous for $|u| < B$, the Fourier series

---

2. See Appendix 14.1.
3. See Section 2.3.2.2.

in (5.10) converges pointwise on this interval. Define the error magnitude

$$\varepsilon_N(u; t) = \left| e^{j2\pi ut} - e_N(u; t) \right|.$$

Let Δ_N denote the maximum (or supremum) of $\varepsilon_N(u; t)$ for $|u| < B$. Then we are guaranteed that

$$\lim_{N \to \infty} \Delta_N = 0. \tag{5.18}$$

Next, note that the truncated cardinal series in (5.12) can be written as

$$x_N(t) = \sum_{n=-N}^{N} \int_{-B}^{B} X(u) \, e^{j\pi nu/B} \, du \, \text{sinc}(2Bt - n)$$

$$= \int_{-B}^{B} X(u) \, e_N(u; t) \, du.$$

Using the inequality

$$\left| \int Y(u) \, du \right| \leq \int |Y(u)| \, du \tag{5.19}$$

we find that

$$|x(t) - x_N(t)| = \left| \int_{-B}^{B} X(u) \left[e^{j2\pi ut} - e_N(u; t) \right] du \right|$$

$$\leq \int_{-B}^{B} |X(u)| \, \varepsilon_N(u; t) \, du.$$

Since the integrand is positive,

$$|x(t) - x_N(t)| \leq \Delta_N \int_{-B}^{B} |X(u)| \, du.$$

From (5.16), the integral is finite. Use of (5.18) results in (5.11) and the proof is complete.

The uniform convergence result for the cardinal series should not be surprising. Because of limited frequency constraints, bandlimited functions are inherently smooth. The sinc interpolation function is similarly smooth. There is thus no mechanism by which deviations such as Fourier series' *Gibb's phenomenen* can occur.[4]

5.4.2 Trapezoidal Integration

5.4.2.1 Of Bandlimited Functions

Clearly, if $x(t) \leftrightarrow X(u)$, then

$$X(0) = \int_{-\infty}^{\infty} x(t) \, dt.$$

---

4. See Exercise 2.51.

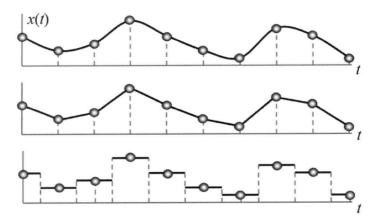

FIGURE 5.3. When sampling is performed at or above the Nyquist rate, the integration of $x(t)$ over all t (top) gives the same result as integrating over a piecewise linear interpolation of the samples, i.e trapezoidal integration (middle) and a piecewise constant representation of the signal (bottom).

Using (5.8) with (5.9), we thus conclude the integral of a bandlimited function can be written directly in terms of its samples:

$$\int_{-\infty}^{\infty} x(t)\, dt = \frac{1}{2B} \sum_{n=-\infty}^{\infty} x\left(\frac{n}{2B}\right).$$

Thus, as illustrated in Figure 5.3, trapezoidal integration of bandlimited signals contains no error due to the piecewise linear approximation of the signal if sampling is at or above the Nyquist rate. Since integration is over all of t, trapezoidal integration gives the same result as piecewise constant (rectangular) integration.

Accurate integration results from a signal's samples can also be obtained when sampling below the Nyquist rate.[5]

5.4.2.2 Of Linear Integral Transforms

Consider numerical evaluation of the linear integral transform

$$g(t) = \int_{-\infty}^{\infty} u(\tau)\, h(t;\tau)\, d\tau \qquad (5.20)$$

where $u(\tau)$ is the input, $g(t)$ is the transform and $h(t;\tau)$ is the transform kernel. Special cases are numerous and include correlation, convolution, and Laplace, Abel, Mellin, Hilbert and Hankel transforms [149].

One popular approach is to evaluate (5.20) by trapezoidal integration:

$$g(t) \approx g_\Delta(t) = \Delta \sum_{n=-\infty}^{\infty} u(n\Delta)\, h(t; n\Delta) \qquad (5.21)$$

where Δ is the input sampling interval. If the output is sampled, (5.21) can be expressed simply as a matrix-vector product.

---

5. See Exercise 5.3.

We will show that by a simple alteration of the transform kernel, the expression in (5.21) can be made exact in the spirit of the sampling theorem [898]. Certain linear operations that cannot be directly evaluated by use of (5.21) because of singularities can be evaluated through this sampling theorem characterization.

5.4.2.3 Derivation of the Low Passed Kernel

Let $u(\tau)$ be bandlimited in the low pass sense with bandwidth B. Let $W > B$. Then $u(\tau)$ is unaffected by low-pass filtering.

$$u(\tau) = 2W \int_{-\infty}^{\infty} u(\eta) \operatorname{sinc}(2W(\tau - \eta)) \, d\eta.$$

Substituting into the expression for the linear integral transform in (5.20) gives

$$g(t) = \int_{-\infty}^{\infty} u(\eta) \, k(t; \eta) \, d\eta \tag{5.22}$$

where the *low-passed kernel* (LPK) is

$$k(t; \eta) = 2W \int_{-\infty}^{\infty} h(t; \tau) \operatorname{sinc}(2W(\tau - \eta)) \, d\tau. \tag{5.23}$$

Even though the kernel in (5.22) is altered, it yields the same result as in (5.20).

Since both the input and the LPK are bandlimited in η, they can be expressed by the cardinal series:

$$u(\eta) = \sum_{n=-\infty}^{\infty} u(n\Delta) \operatorname{sinc}\left(n - \frac{\eta}{\Delta}\right) \tag{5.24}$$

and

$$k(t; \eta) = \sum_{m=-\infty}^{\infty} k(t; m\Delta) \operatorname{sinc}\left(m - \frac{\eta}{\Delta}\right) \tag{5.25}$$

where the input sampling interval must be chosen such that

$$\Delta \leq \frac{1}{2W} \leq \frac{1}{2B}.$$

Substituting (5.24) and (5.25) into (5.22) gives

$$g(t) = \sum_{n=-\infty}^{\infty} \sum_{m=-\infty}^{\infty} u(n\Delta) \, k(t; m\Delta) \tag{5.26}$$

$$\times \int_{-\infty}^{\infty} \operatorname{sinc}\left(n - \frac{\eta}{\Delta}\right) \operatorname{sinc}\left(m - \frac{\eta}{\Delta}\right) d\eta \tag{5.27}$$

$$= \Delta \sum_{n=-\infty}^{\infty} u(n\Delta) \, k(t; n\Delta). \tag{5.28}$$

This is the desired result. Comparing it with (5.21), we conclude that the inaccuracy that is due to trapezoidal integration can be totally eliminated if the LPK is used in lieu of the original linear integral transform kernel.

Example Transforms

To illustrate use of (5.28), we now present example applications for the cases of Laplace and Hilbert transformation.

Laplace Transform. The (unilateral) Laplace transform, from (2.22), can be written as

$$g(t) = \int_0^\infty u(\tau) e^{-t\tau} d\tau.$$

Comparing with (5.20) and (5.22), we have

$$h(t; \tau) = e^{-t\tau} \mu(\tau)$$

and

$$k(t; \eta) = \int_0^\infty e^{-t\tau} \operatorname{sinc}(\tau - \eta) d\tau$$

where $\mu(\cdot)$ denotes the unit step and we have chosen $2W = 1$.

Consider the Laplace transform of $u(t) = \operatorname{sinc}(t)$. We evaluate the resulting Laplace transform integral in three ways.

- DIRECTLY.

  The Laplace integral becomes

  $$g(t) = \int_0^\infty \operatorname{sinc}(\tau) e^{-t\tau} d\tau. \qquad (5.29)$$

  To evaluate this integral, consider

  $$\gamma = \int_0^\infty \frac{\sin(a\tau)}{\tau} e^{-t\tau} d\tau.$$

  Clearly

  $$\frac{\partial \gamma}{\partial a} = \int_0^\infty \cos(a\tau) e^{-t\tau} d\tau$$
  $$= \frac{t}{t^2 + a^2}.$$

  Thus, using the boundary condition that $a = 0 \longrightarrow \gamma = 0$, we conclude that $\gamma = \arctan(\frac{a}{t})$, and the Laplace transform of $\operatorname{sinc}(\tau)$ is

  $$g(t) = \frac{1}{\pi} \arctan\left(\frac{\pi}{t}\right).$$

- USING TRAPEZOIDAL INTEGRATION

  Applying trapezoidal integration to the integral in (5.29) with a step size of Δ gives[6]

  $$g(t) \approx g_\Delta(t) = \frac{\Delta}{2} + \Delta \sum_{n=1}^\infty \operatorname{sinc}(n\Delta) e^{-tn\Delta}$$
  $$= \frac{\Delta}{2} + \frac{1}{\pi} \Im \sigma \qquad (5.30)$$

---

6. The $\frac{\Delta}{2}$ term is replaced by Δ for rectangular integration.

where $\Im$ is the imaginary operator,

$$\sigma = \sum_{n=1}^{\infty} \frac{1}{n} Z^n,$$

and $Z = \exp(-(t - j\pi)\Delta)$. From the geometric series,[7]

$$\frac{\partial \sigma}{\partial Z} = \sum_{m=0}^{\infty} Z^m$$

$$= \frac{1}{1-Z}$$

we conclude, since $Z = 0 \longrightarrow \sigma = 0$, that

$$\sigma = -\ln(1 - Z).$$

For $Z = |Z| e^{j\angle(Z)}$, recall that $\Im \ln Z = \angle(Z)$. Thus, after some substitution, (5.30) becomes

$$g_\Delta(t) = \frac{\Delta}{2} + \frac{1}{\pi} \arctan\left(\frac{\sin(\pi \Delta)}{e^{t\Delta} - \cos(\pi \Delta)}\right). \tag{5.31}$$

Note that, as $\Delta \longrightarrow 0$,

$$\frac{\sin(\pi \Delta)}{e^{t\Delta} - \cos(\pi \Delta)} \longrightarrow \frac{\pi}{t}.$$

Thus, as we would expect,

$$\lim_{\Delta \to 0} g_\Delta(t) = g(t).$$

A plot of $g_\Delta(t)$ for various Δ's is shown in Figure 5.4 along with $g(t)$. Recall that for $\Delta = 1$, the LPK approach, in the spirit of the cardinal series, gives exact results.
- USING THE LPK
  For $u(\tau) = \text{sinc}(\tau)$ and $2W = 2B = 1/\Delta = 1$, the LPK expression in (5.28) becomes

$$g(t) = k(t; 0)$$

where we have used the property that $\text{sinc}(n) = \delta[n]$. Using (5.23) with $\eta = 0$ gives

$$g(t) = \int_{-\infty}^{\infty} h(t; \tau) \text{sinc}(\tau) d\tau$$

which is the integral we wished to evaluate in the first place. The LPK, when applied to $\text{sinc}(t)$, therefore reduces to an identity for Laplace transformation or, for that matter, any other linear transform.

---

7. See Appendix 14.4.2

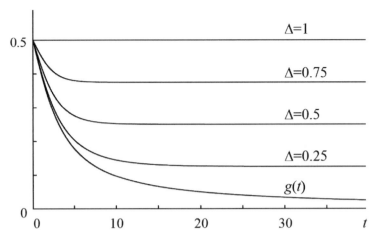

FIGURE 5.4. Evaluation of the Laplace transform of sinc(t) by trapezoidal integration. Shown is $g_\Delta(t)$ for $\Delta = 1, 0.5$ and 0.25. Use of the low passed kernel will converge to the desired $g(t)$ for step sizes, Δ, less than or equal to one.

Hilbert Transform

The Hilbert transform

$$g(t) = -\frac{1}{\pi} \int_{-\infty}^{\infty} \frac{u(\tau)\,d\tau}{t - \tau} \tag{5.32}$$

cannot be accurately evaluated by direct trapezoidal integration because of the singularity at $\tau = t$. Fourier transform analysis, rather, is commonly used. We will now show, however, that through application of the LPK, an accurate matrix-vector characterization of the Hilbert transform is possible. From (5.32)

$$h(t; \tau) = \frac{-1}{\pi(t - \tau)}.$$

If we choose $W = B$, the corresponding LPK is

$$k(t; \eta) = \int_{-B}^{B} \left[-\frac{1}{\pi} \int_{-\infty}^{\infty} \frac{\exp(-j2\pi u\tau)}{t - \tau} d\tau \right] e^{j2\pi u\eta} \, du$$

$$= -j \int_{-B}^{B} \text{sgn}(u)\, e^{-j2\pi u(t-\eta)} \, du$$

$$= -2B \sin \pi B(t - \eta) \, \text{sinc} B(t - \eta). \tag{5.33}$$

A further simplification arises after we note that Hilbert transformation is a shift-invariant operation. Thus, if $u(\tau)$ is bandlimited, so then is the output, $g(t)$. This being true, we need to know g only at the points where $t = m/2B$. Substituting (5.33) into (5.28) with $\Delta = 1/2B$ gives

$$g(m\Delta) = \sum_{n=-\infty}^{\infty} u(n\Delta) \sin\left(\frac{\pi}{2}(m - n)\right) \text{sinc}\left(\frac{1}{2}(m - n)\right).$$

Noting that every other term is zero yields the final desired result.

$$g(m\Delta) = -\frac{2}{\pi} \sum_{m-n \text{ odd}} \frac{u(n\Delta)}{m-n}.$$

This discrete convolution version of the Hilbert transform contains no singularities and is exact for all band-limited inputs.

5.4.2.4 Parseval's Theorem for the Cardinal Series

The energy of a signal is

$$E = \int_{-\infty}^{\infty} |x(t)|^2 \, dt.$$

For a bandlimited signal, we substitute the cardinal series in (5.3) and write

$$E = \sum_{n=-\infty}^{\infty} \sum_{m=-\infty}^{\infty} x\left(\frac{n}{2B}\right) x^*\left(\frac{m}{2B}\right) I_{nm}$$

where

$$I_{nm} = \int_{-\infty}^{\infty} \operatorname{sinc}(2Bt - n) \operatorname{sinc}(2Bt - m) \, dt.$$

Using the power theorem

$$I_{nm} = \frac{1}{(2B)^2} \int_{-B}^{B} e^{-j\pi(n-m)u/B} \, du$$

$$= \frac{1}{2B} \delta[n-m].$$

This gives Parseval's theorem for the sampling theorem:

$$E = \frac{1}{2B} \sum_{n=-\infty}^{\infty} \left|x\left(\frac{n}{2B}\right)\right|^2. \tag{5.34}$$

The signal's energy can thus be determined by summing the square of the magnitude of each sample.

5.4.3 The Time-Bandwidth Product

The cardinal series requires an infinite number of samples. Since all bandlimited functions are analytic, they cannot be identically zero over any finite subinterval except for the degenerate case $x(t) \equiv$ zero. Thus, the number of nonzero samples taken from almost every band-limited function is finite. The only exceptions are signals that can be expressed as the sum of a finite number of uniformly spaced sinc functions.

We can, however, have a "good" representation of the function using a finite number of samples. If a signal has either finite area or finite energy, it must asymptotically approach zero at $t = \pm\infty$. In such cases, there is always an interval of duration T outside of which the samples are negligibly small. If we sample over this interval at the Nyquist rate, $2B$, then a total of

$$S = 2BT \tag{5.35}$$

samples are needed to characterize the signal. This quantity, the *time–bandwidth product*, measures the number of degrees of freedom of the signal. It has also been termed the *Shannon number* [1407].

Choice of T is dictated by the *truncation error* one can tolerate. This topic is treated in Section 7.5.

5.5 Application to Spectra Containing Distributions

The cardinal series is applicable in certain cases where $X(u)$ contains distributions such as the Dirac delta. Indeed $x(t) = 1$ is bandlimited in the sense of (5.1) for all $B > 0$ and falls into the category of bandlimited signals with finite area spectra. The corresponding cardinal series is

$$\sum_{n=-\infty}^{\infty} \text{sinc}(2Bt - n) = 1. \tag{5.36}$$

Similarly, $\cos(2\pi Bt - \phi)$ has a transform containing two Dirac delta functions at $u = \pm B$. To insure both deltas are contained in the replicated sample spectrum, the sampling rate, $\frac{1}{T}$, must *exceed* $2B$ and

$$\cos(2\pi Bt - \phi) = \sum_{n=-\infty}^{\infty} \cos(\pi rn - \phi) \,\text{sinc}\left(\frac{t}{T} - n\right)$$

where

$$r = 2BT < 1. \tag{5.37}$$

If $r = 1$, the bandwidth interval begins and ends at delta function locations. We are confronted with the unanswerable question of what percentage of each delta should be included in the bandwidth interval. Requiring (5.37) to be a strict inequality avoids this problem. Note, also, for $r = 1$, it is possible to have every sample be zero. The resulting interpolation clearly would be identically zero and therefore incorrect.

A distribution whose inverse transform does not have a valid cardinal series is the *unit doublet*–the derivative of the Dirac delta:

$$\delta^{(1)}(u) = \left(\frac{d}{du}\right) \delta(u).$$

From the dual of the derivative theorem,

$$-j2\pi t \longleftrightarrow \delta^{(1)}(u).$$

Thus, $x(t) = t$ is bandlimited in the sense of (5.1). Using the form of the cardinal series in (5.4), we ask the question

$$t \stackrel{?}{=} \frac{1}{\pi} \sin(2\pi Bt) \sum_{n=-\infty}^{\infty} (-1)^n \frac{(n/2B)}{(2Bt - n)}.$$

For a fixed $t \neq m/2B$, the answer is clearly "no" since the nth term in the sum approaches $(-1)^{n+1}$ – an oscillatory and thus divergent series. The truncated cardinal series thus does not asymptotically approach the desired value.

5.6 Application to Bandlimited Stochastic Processes

A real wide sense stationary stochastic process, $f(t)$, is said to be bandlimited if its power spectral density obeys

$$S_f(u) = S_f(u)\, \Pi\left(\frac{u}{2B}\right).$$

As a consequence, the autocorrelation is a bandlimited function. We will use this observation to show mean square convergence of the cardinal series for $f(t)$. Specifically, define

$$\hat{f}(t) = \sum_{n=-\infty}^{\infty} f\left(\frac{n}{2B}\right) \operatorname{sinc}(2Bt - n).$$

Then $\hat{f}(t)$ is equal to $f(t)$ in the mean square sense:

$$E\left[\{f(t) - \hat{f}(t)\}^2\right] = 0. \tag{5.38}$$

Proof. Expand the mean square error expression to give

$$E[\{f(t) - \hat{f}(t)\}^2] = \overline{f^2}$$
$$+ \sum_{n=-\infty}^{\infty} \sum_{m=-\infty}^{\infty} R_f\left(\frac{n-m}{2B}\right) \operatorname{sinc}(2Bt - n) \operatorname{sinc}(2Bt - m)$$
$$- 2 \sum_{n=-\infty}^{\infty} R_f\left(t - \frac{n}{2B}\right) \operatorname{sinc}(2Bt - n)$$
$$= \overline{f^2} + T_2 + T_3. \tag{5.39}$$

In the second term, we make the variable substitution $k = n - m$:

$$T_2 = \sum_{k=-\infty}^{\infty} R_f\left(\frac{k}{2B}\right) \sum_{n=-\infty}^{\infty} \operatorname{sinc}(2Bt - n + k) \operatorname{sinc}(2Bt - n).$$

Application of the cardinal series yields, for arbitrary τ,

$$\sum_{n=-\infty}^{\infty} \operatorname{sinc}(2B\tau - n)\, \operatorname{sinc}(2Bt - n) = \operatorname{sinc} 2B(t - \tau). \tag{5.40}$$

Using the result of substituting $\tau = t + \frac{k}{2B}$ in (5.40) gives

$$T_2 = \sum_{k=-\infty}^{\infty} R_f\left(\frac{k}{2B}\right) \delta[k]$$
$$= \overline{f^2}. \tag{5.41}$$

To evaluate the third term, we recognize that $R_f(\tau - t)$ is bandlimited and can be written as a cardinal series:

$$R_f(\tau - t) = \sum_{n=-\infty}^{\infty} R_f\left(\tau - \frac{n}{2B}\right) \operatorname{sinc}(2Bt - n).$$

Evaluation at $t = \tau$ gives

$$T_3 = -2\overline{f^2}.$$

Substituting this and (5.41) into (5.39) yields (5.38) and our proof is complete.

5.7 Exercises

5.1. A mono audio recording lasts for 5 minutes. Assume the frequency range for human hearing is 20 Hertz to 20,000 Hertz.
   (a) If 20,000 Hertz is the highest frequency in the audio signal, estimate the minimum number of samples required to characterize the signal.
   (b) Assume the audio recording is a duet between a tuba and a flute. Describe what happens if the recording is reproduced from samples taken below the Nyquist rate.

5.2. The derivation of the Poisson sum formula closely parallels Shannon's proof of the sampling theorem. Starting with the Fourier dual of the Poisson sum formula, derive the sampling theorem series.

5.3. The integral of a bandlimited signal

$$I = \int_{-\infty}^{\infty} x(t)\, dt$$

can be determined from signal samples taken *below* the Nyquist rate. Find I from $\{x(nT)|n = 0, \pm 1, \pm 2, \ldots\}$ when $B < 1/T < 2B$ and $X(u) = X(u)\Pi\left(\frac{u}{2B}\right)$.

5.4. Assume $x(t)$ can be expressed by the cardinal series.
   (a) Show that, for any real α,

$$x(t) = \sum_{n=-\infty}^{\infty} x\left(\frac{n}{2B} + \alpha\right) \operatorname{sinc}(2B(t-\alpha) - n) \qquad (5.42)$$

   (b) Since $x(t)$ does not depend on α, is

$$\frac{d}{d\alpha} x(t) = 0\ ?$$

5.5. Investigate application of the cardinal series to a signal whose spectrum is an mth–let:

$$\delta^{(m)}(u) = \left(\frac{d}{du}\right)^m \delta(u).$$

When, if ever, is the cardinal series applicable here? Assume $m > 1$.

5.6. Let $y(t)$ be any well behaved (not necessarily bandlimited) function. We sample $y(t)$ at a rate of $\frac{1}{T}$ and, in the spirit of (5.5), define

$$\hat{s}(t) = y(t) \frac{1}{T} \text{comb}\left(\frac{t}{T}\right).$$

Show that

$$T \int_{-1/T}^{1/T} \hat{S}(u)\, du = \int_{-\infty}^{\infty} Y(u)\, du.$$

5.7. Let $x(t)$ and $y(t)$ denote two finite energy bandlimited functions with bandwidth B. How is the series

$$\sum_{n=-\infty}^{\infty} x\left(\frac{n}{2B}\right) y^*\left(\frac{n}{2B}\right)$$

related to

$$\int_{-\infty}^{\infty} x(t) y^*(t)\, dt?$$

5.8. The spectrum of a real signal is Hermetian or *conjugately symmetric*, i.e., it is equal to the conjugate of its transpose. Thus, if we know the spectrum for positive frequencies, we know it for negative frequencies. Visualize setting the negative frequency components to zero and shifting the remaining portion of the spectrum to be centered about the origin. Clearly, we have reduced the bandwidth by a factor of one half yet have lost no information. Explain, however, why the sampling density of this new signal is the same as that required by the original.

5.9. Squaring a bandlimited function at least doubles its bandwidth and therefore the signal's Nyquist rate.
   (a) Are there finite energy nonbandlimited signals that, when squared, become bandlimited?
   (b) Given a complex bandlimited signal, $x(t)$, show that the same number of samples per time interval is required as for the signal $|x(t)|^2$.
   (c) Is the converse of squaring a bandlimited function doubling its bandwidth true? Does the square root of a positive bandlimited signal result in a signal with half the original signal's bandwidth?[8]

5.10. Apply the low passed kernel technique to Fourier inversion of a bandlimited function. Let $f(t)$ have a bandwidth of B and use the Fourier kernel

$$h(u; t) = \exp(-j2\pi ut).$$

Comment on the usefulness of the result.

---

8. Here is some supporting evidence for the truth of the converse. The square root of any positive number approaches one in the limit. Thus, if $x(t) > 0$, repeated application of positive square roots to any function, in the limit, approaches one. If n denotes the number of square root operations, then

$$\lim_{n \to \infty} (x(t))^{2^{-n}} = 1.$$

In the limit, therefore, the bandwidth of the signal approaches zero.

5.11. **Upsampling**. Let a bandlimited function, $x(t)$, have a bandwidth B. From knowledge of the samples, $x\left(\frac{n}{2B}\right)$, we wish to upsample to M samples per sample interval. Denote these samples by the discrete time signal

$$x[n] = x\left(\frac{n}{2B}\right)$$

and the upsampled signal by

$$y[m] = x\left(\frac{n}{M2B}\right).$$

(a) Show the upsampled signal can be characterized by the discrete time convolution

$$y[m] = x[n] * \operatorname{sinc}\left(\frac{m}{M}\right). \qquad (5.43)$$

(b) Zero pad the original samples and let

$$\hat{x}[m] = \begin{cases} x[n] \; ; \; m = nM \\ 0 \quad ; \; \text{otherwise} \end{cases}$$

Show that $y[m]$ results from placing $\hat{x}[m]$ through a discrete time low pass filter,

$$H(f) = M \ \Pi(Mu).$$

(c) Assume $x[-1] = 2$, $x[0] = 5$, $x[1] = 3$ and all other samples are zero. The global maximum of $x(t)$ apparently lies in the interval $\left(0, \frac{1}{2B}\right)$. Numerically determine the location and maximum by upsampling and identifying the resulting maximum value.

5.8 Solutions for Selected Chapter 5 Exercises

5.1. (a) Use the Shannon number (or time-bandwidth product) in (5.35) with $B = 20,000$ Hertz and $T = 5$ minutes.
(b) The high frequencies will be aliased.

5.2. The Fourier dual of the Poisson sum formula is

$$2B \sum_{n=-\infty}^{\infty} X(u - 2nB) = \sum_{n=-\infty}^{\infty} x\left(\frac{n}{2B}\right) e^{-j\pi nu/B}.$$

Since $X(u) = X(u) \ \Pi(u/2B)$, we multiply both sides by $\Pi(\frac{u}{2B})$ and inverse transform. The sampling theorem series results.

5.3. Use the Fourier dual of the Poisson sum formula again:

$$\sum_{n=-\infty}^{\infty} X\left(u - \frac{n}{T}\right) = T \sum_{n=-\infty}^{\infty} x(nT) \ e^{-j2\pi nuT}.$$

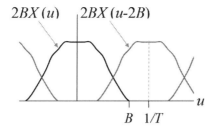

FIGURE 5.5. Samples taken below the Nyquist rate can still be used to compute the area of a signal. See Exercise 5.3.

For $B < 1/T < 2B$, the sum on the left will be the overlapping aliased version in Figure 5.5. Note that no spectra overlap the zeroth order spectrum at $u = 0$. Thus

$$\sum_{n=-\infty}^{\infty} X\left(u - \frac{n}{T}\right)\bigg|_{u=0} = X(0) \ ; B < 1/T < 2B$$

and

$$\int_{-\infty}^{\infty} x(t)\, dt = T \sum_{n=-\infty}^{\infty} x(nT) \ ; T < 1/B.$$

5.4. If $x(t)$ is bandlimited, so is $x(t + \alpha)$. Thus

$$x(t + \alpha) = \sum_{n=-\infty}^{\infty} x\left(\frac{n}{2B} + \alpha\right) \operatorname{sinc}(2Bt - n).$$

(a) Substitute $t - \alpha$ for t and we're done.
(b) We have

$$\frac{dx(t)}{d\alpha} = \sum_{n=-\infty}^{\infty} x'\left(\frac{n}{2B} + \alpha\right) \operatorname{sinc}\left(2B(t - \alpha) - n\right)$$

$$+ \sum_{n=-\infty}^{\infty} x\left(\frac{n}{2B} + \alpha\right)(-2B) \operatorname{sinc}'\left(2B(t - \alpha) - n\right). \qquad (5.44)$$

If $x(t)$ is bandlimited, so is $x'(t + \alpha)$ and

$$x'(t + \alpha) = \sum_{n=-\infty}^{\infty} x'\left(\frac{n}{2B} + \alpha\right) \operatorname{sinc}(2Bt - n).$$

or, setting t to $t - \alpha$,

$$x'(t) = \sum_{n=-\infty}^{\infty} x'\left(\frac{n}{2B} + \alpha\right) \operatorname{sinc}\left(2B(t - \alpha) - n\right). \qquad (5.45)$$

Differentiating (5.42) with respect to t gives

$$x'(t) = \sum_{n=-\infty}^{\infty} x\left(\frac{n}{2B} + \alpha\right) 2B \operatorname{sinc}'(2B(t-\alpha) - n). \qquad (5.46)$$

Substituting (5.45) and (5.46) into (5.44) reveals that

$$\frac{dx(t)}{d\alpha} = x'(t) - x'(t) = 0.$$

5.5. Since

$$(-j2\pi t)^m \longleftrightarrow \delta^{(m)}(t)$$

we use (5.4) and ask the question

$$t^m \stackrel{?}{=} \frac{1}{\pi(2B)^m} \sin(2\pi Bt) \sum_{n=-\infty}^{\infty} \frac{(-1)^n n^m}{2Bt - n}.$$

The result is clearly a divergent series for $m \geq 1$.

5.6.

$$\hat{s}(t) = \sum_{n=-\infty}^{\infty} y(nT)\,\delta(t - nT).$$

Thus

$$\hat{S}(u) = \sum_{n=-\infty}^{\infty} y(nT)\, e^{-j2\pi nuT}$$

$\hat{S}(u)$ is periodic with Fourier coefficients

$$y(nT) = T \int_{-1/T}^{1/T} \hat{S}(u)\, e^{j2\pi nuT}\, du.$$

Thus

$$T \int_{-1/T}^{1/T} \hat{S}(u)\, du = y(0).$$

From the inversion formula

$$y(t) = \int_{-\infty}^{\infty} Y(u)\, e^{j2\pi ut}\, du.$$

Thus

$$y(0) = \int_{-\infty}^{\infty} Y(u)\, du$$

and our exercise is complete.

5.7.
$$\int_{-\infty}^{\infty} x(t)\, y^*(t)\, dt = \frac{1}{2B} \sum_{n=-\infty}^{\infty} x\left(\frac{n}{2B}\right) y^*\left(\frac{n}{2B}\right).$$

5.8. The new signal's spectrum is no longer Hermetian. The signal, therefore, is complex. Although the sampling rate is reduced by a factor of a half, each sample now requires two numbers.

5.9. (a) Yes. An example is $|\text{sinc}(t)|$.

(b) For the complex signal, $x(t)$, there are $4B$ samples per time interval because each sample consists of two numbers: the real and the imaginary portions of the signal. The signal $|x(t)|^2$ has a bandwidth of $2B$ and therefore requires a sampling rate of $4B$. Only one number per sample is needed. Therefore, the number of samples per time interval is the same for $x(t)$ and $|x(t)|^2$.

(c) Squaring a bandlimited function at least doubles its bandwidth. The converse is not true. Taking the square root of a positive bandlimited function does not reduce the bandwidth by a factor of a half. Before showing this, the problem statement needs to be further refined. There is an inherent problem in the original statement. The square root of a positive signal has two values: plus and minus. Therefore, the square root of $\text{sinc}^2(t)$ can be $\text{sinc}(t)$, or, if the positive square root is always taken, the rectified signal, $\text{sinc}(t) \times \text{sgn}[\text{sinc}(t)]$, results. This problem is removed if the bandlimited signal, $x(t)$, is strictly positive. An example of a strictly positive finite energy bandlimited signal is

$$x(t) = \text{sinc}^2\left(t + \frac{1}{3}\right) + \text{sinc}^2\left(t - \frac{1}{3}\right). \tag{5.47}$$

Every place the first sinc^2 goes to zero, the second sinc^2 is positive and visa versa. Let $x_n(t)$ be the result of application of the square root operator n times. Then

$$x_n(t) - [x(t)]^{2^{-n}}. \tag{5.48}$$

Clearly

$$x_0(t) = x(t).$$

A log plot of $x_n(t)$ is shown in Figure 5.6 for $n = 0, 1, 2$. Let

$$x_n(t) \longleftrightarrow X_n(u).$$

Note that

$$X(u) = X_0(u) = 2\Lambda(u) \cos\left(\frac{2\pi u}{3}\right). \tag{5.49}$$

Normalized numerically evaluated linear and log plots of $X_n(u)$ for $n = 0, 1, 2$ are in Figures 5.7 and 5.8. The plots are of $\frac{X_n(u)}{X_n(0)}$. For the $n = 0$ case, the plot corresponding to $X(u) = X_0(u)$, as expected, is identically zero for $u > 1$. For the square root and the fourth root, this is not the case.

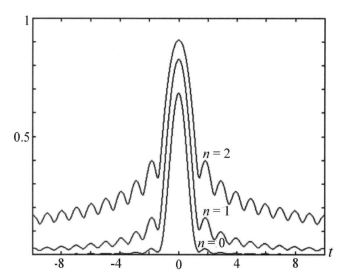

FIGURE 5.6. Plots of $x_n(t)$ in (5.48) for $n = 0, 1, 2$. The $n = 0$ signal is the strictly positive finite energy bandlimited function, $x(t)$, in (5.47).

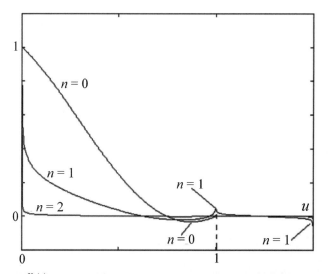

FIGURE 5.7. Plots of $\frac{X_n(u)}{X_n(0)}$ for $n = 0, 1, 2$. $X_n(u)$, the Fourier transforms of $x_n(t)$ in (5.48), are shown. The plot for $n = 0$ is zero for $u > 1$. A log plot is shown in Figure 5.8.

Thus, we have empirically shown that the square root of a strictly positive finite energy bandlimited function not only doesn't produce a bandlimited signal with half of the bandwidth, it does not even produce a bandlimited signal.

Note, interestingly, that as n increased, the spectra in Figure 5.7 becomes more and and more impulsive. This is because

$$\lim_{n \to \infty} x_n(t) = 1 \longleftrightarrow \lim_{n \to \infty} X_n(u) = \delta(u).$$

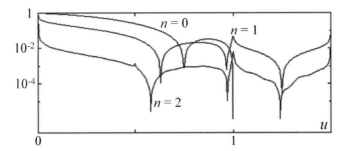

FIGURE 5.8. Log plots of $\frac{X_n(u)}{X_n(0)}$ for $n = 0, 1, 2$. The linear plots are shown in Figure 5.7.

The signal thus clearly approaches a signal with bandwidth zero, but not in a manner wherein intermediate functions are strictly bandlimited.

5.10. The resulting low pass kernel is

$$k(u; t) = \exp(-j2\pi ut) \, \Pi\left(\frac{u}{2B}\right).$$

We are thus assured that $g(u)$, the Fourier transform of $f(t)$, is zero outside of the interval $|u| \leq B$. Since this is the definition of a bandlimited function, the result is of little use. Note that Fourier inversion of (5.23), as we would expect, results in the cardinal series.

5.11. Equation (5.43) results from evaluation of the cardinal series at the points

$$t = \frac{m}{2MB}.$$

6

Generalizations of the Sampling Theorem

> One should always generalize.
> Carl Gustav Jakob Jacobi (1804–1851) [354]
>
> Each problem that I solved became a rule which served afterwards to solve other problems.
> Rene Descartes (1596–1650) [365]
>
> In mathematics you don't understand things. You just get used to them.
> Johann von Neumann (1903–1957) [1598]

6.1 Introduction

There have been numerous interesting and useful generalizations of the sampling theorem. Some are straightforward variations on the fundamental cardinal series. Oversampling, for example, results in dependent samples and allows much greater flexibility in the choice of interpolation functions. In Chapter 7, we will see that it can also result in better performance in the presence of sample data noise.

Bandlimited signal restoration from samples of various filtered versions of the signal is the topic addressed in Papoulis' generalization [1086, 1087] of the sampling theorem. Included as special cases are recurrent nonuniform sampling and simultaneously sampling a signal and one or more of its derivatives.

Kramer [772] generalized the sampling theorem to signals that were bandlimited in other than the Fourier sense. We also demonstrate that the cardinal series is a special case of Lagrangian polynomial interpolation. Sampling in two or more dimensions is the topic of Section 8.9.

6.2 Generalized Interpolation Functions

There are a number of functions other than the sinc which can be used to weight a signal's samples in such a manner as to uniquely characterize the signal. Use of these *generalized interpolation functions* allows greater flexibility in dealing with sampling theorem type characterizations.

6.2.1 Oversampling

If a bandlimited signal has bandwidth B, then it can also be considered to have bandwidth $W \geq B$. Thus,

$$x(t) = \sum_{n=-\infty}^{\infty} x\left(\frac{n}{2W}\right) \operatorname{sinc}(2Wt - n). \tag{6.1}$$

Note, however, since

$$x(t) = x(t) * 2B \operatorname{sinc}(2Bt)$$

we can write

$$x(t) = \sum_{n=-\infty}^{\infty} x\left(\frac{n}{2W}\right) \operatorname{sinc}(2Wt - n) * 2B \operatorname{sinc}(2Bt)$$

$$= r \sum_{n=-\infty}^{\infty} x\left(\frac{n}{2W}\right) \operatorname{sinc}(2Bt - rn) \tag{6.2}$$

where the *sampling rate parameter* is

$$r = B/W \leq 1. \tag{6.3}$$

Equation (6.2) reduces to the conventional cardinal series for $r = 1$. Oversampling can be used to reduce interpolation noise level.

6.2.2 Restoration of Lost Samples

6.2.2.1 Sample Dependency

When a bandlimited signal is oversampled, its samples become dependent. Indeed, in this section we will show that in the absence of noise, any finite number of lost samples can be regained from those remaining. First, using intuitively straightforward arguments, we will illustrate the feasibility of lost sample recovery. Then, an alternate expression with better convergence properties will be derived.

6.2.2.2 Restoring a Single Lost Sample

Consider a finite energy bandlimited signal $x(t)$ and its spectrum as shown at the top of Figure 6.1. From (5.5) and (5.7) the spectrum of the signal of samples

$$s(t) = x(t) \, 2W \operatorname{comb}(2Wt) \tag{6.4}$$

is

$$S(u) = 2W \sum_{n=-\infty}^{\infty} X(u - 2nW). \tag{6.5}$$

As is shown in Figure 6.1, there are intervals identically equal to zero in $S(u)$ when we oversample.

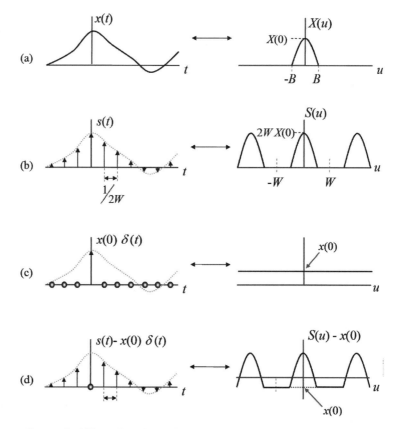

FIGURE 6.1. Geometrical illustration of restoring a single lost sample from an oversampled bandlimited signal.

In Figure 6.1a, the sample of $x(t)$ at the origin has been set to zero. We can view this as the subtraction of $x(0)\,\delta(t)$ in Figure 6.1c from $s(t)$.

$$\hat{s}(t) = s(t) - x(0)\,\delta(t).$$

Given $\hat{s}(t)$, we can regain $x(0)$. Indeed, transforming gives

$$\hat{S}(u) = S(u) - x(0). \tag{6.6}$$

Since $S(u) \equiv 0$ on the interval $B < |u| < 2W - B$,

$$\hat{S}(u) = -x(0) \quad;\quad B < |u| < 2W - B.$$

This is illustrated in Figure 6.1d. An appropriate point to sample in this interval is $u = W$. Thus

$$x(0) = -\hat{S}(W). \tag{6.7}$$

Therefore, the lost sample at the origin can be regained from those samples remaining.

Samples from oversampled signals display an interesting zero sum property. Writing (6.4) in terms of delta functions and transforming gives

$$S(u) = \sum_{n=-\infty}^{\infty} x\left(\frac{n}{2W}\right) e^{-j\pi nu/W}.$$

Since $S(W) = 0$, we conclude that

$$\sum_{n=-\infty}^{\infty} (-1)^n x\left(\frac{n}{2W}\right) = 0\,;\ r < 1. \tag{6.8}$$

Thus, for example,

$$x(0) = -\sum_{n \neq 0} (-1)^n x\left(\frac{n}{2W}\right). \tag{6.9}$$

An alternate formula for restoring a lost sample (with better convergence properties) results directly from inspection of (6.2). Note that, unlike the conventional ($r = 1$) cardinal series, (6.2) does not reduce to an identity when $t = n/2W$. For example, at $t = 0$ we have

$$x(0) = r \sum_{n=-\infty}^{\infty} x\left(\frac{n}{2W}\right) \operatorname{sinc}(rn). \tag{6.10}$$

Isolating the $n = 0$ term in (6.10) and solving for $x(0)$ gives

$$x(0) = \frac{r}{1-r} \sum_{n \neq 0} x\left(\frac{n}{2W}\right) \operatorname{sinc}(rn). \tag{6.11}$$

The sample at the origin is thus completely specified by the remaining samples if $r < 1$. The convergence here is better than in (6.9) due to the $1/n$ decay of the summand from the sinc term. Equation (6.8), on the other hand, does not require knowledge of r.

From the cardinal series, we have

$$x(t) = x(0) \operatorname{sinc}(2Wt) + \sum_{n \neq 0} x\left(\frac{n}{2W}\right) \operatorname{sinc}(2Wt - n).$$

Substituting (6.11) and simplifying gives an interpolation formula not requiring knowledge of the sample at the origin.[1]

$$x(t) = \sum_{n \neq 0} x\left(\frac{n}{2W}\right) \left[\operatorname{sinc}(2Wt - n) + \frac{r}{1-r} \operatorname{sinc}(rn) \operatorname{sinc}(2Wt)\right]. \tag{6.12}$$

Both sides can be low pass filtered to give the corresponding low passed filter expression.[2]

6.2.2.3 Restoring M Lost Samples

The single sample restoration result can be generalized to restoring an arbitrarily large but finite number of lost samples. Let $\mathcal{M}$ denote a set of M integers corresponding to the

---

1. The noise sensitivity and truncation error for this interpolation is explored in Section 7.2.2.1.
2. See Exercise 6.23.

locations of M lost samples. From the data set $\{x\left(\frac{n}{2W}\right) \mid n \notin \mathcal{M}\}$ we wish to find the lost samples, $\{x(\frac{n}{2W}) \mid n \in \mathcal{M}\}$.

To do this, we write (6.2) as

$$x(t) = r \left[\sum_{n \in \mathcal{M}} + \sum_{n \notin \mathcal{M}}\right] x\left(\frac{n}{2W}\right) \operatorname{sinc}(2Bt - rn). \tag{6.13}$$

Evaluating this expression at the M points $\{t = \frac{m}{2W} \mid m \in \mathcal{M}\}$ and rearranging gives

$$\sum_{n \in \mathcal{M}} x\left(\frac{n}{2W}\right) \{\delta[n - m] - r \operatorname{sinc}(r(n - m))\} = g\left(\frac{m}{2W}\right) \; ; \; m \in \mathcal{M} \tag{6.14}$$

where

$$g(t) = r \sum_{n \notin \mathcal{M}} x\left(\frac{n}{2W}\right) \operatorname{sinc}(2Bt - rn) \tag{6.15}$$

can be computed from the known samples.

Equation (6.14) consists of M equations and M unknowns. In matrix form

$$\mathbf{H} \vec{x} = \vec{g} \tag{6.16}$$

where $\{g_m = g\left(\frac{m}{2W}\right) \mid m \in \mathcal{M}\}$, $\{x_n = x\left(\frac{n}{2W}\right) \mid n \in \mathcal{M}\}$ and $\mathbf{H} = \mathbf{I} - \mathbf{S}$ where $\mathbf{S}$ has elements

$$\{s_{nm} = s_{n-m} = r \operatorname{sinc}(r(n - m)) \mid (n, m) \in \mathcal{M} \times \mathcal{M}\}. \tag{6.17}$$

Clearly, we can determine the lost samples if $\mathbf{H}$ is not singular. The matrix is singular when $r = 1$.

6.2.2.4 Direct Interpolation from M Lost Samples

Here we address direct generation of $x(t)$ from the samples $\{x(n/2W) \mid n \notin \mathcal{M}\}$. Using (6.15) the solution of (6.14) can be written as

$$x\left(\frac{q}{2W}\right) = \sum_{p \in \mathcal{M}} a_{pq} \, g\left(\frac{p}{2W}\right)$$

$$= r \sum_{n \notin \mathcal{M}} x\left(\frac{n}{2W}\right) \sum_{p \in \mathcal{M}} a_{pq} \operatorname{sinc}(r(p - n)) \, ;$$

$$q \in \mathcal{M} \tag{6.18}$$

where $\{a_{pq} \mid (p, q) \in \mathcal{M} \times \mathcal{M}\}$ are elements of the inverse of $\mathbf{H}$. The cardinal series can be written:

$$x(t) = \sum_{n \notin \mathcal{M}} x\left(\frac{n}{2W}\right) \operatorname{sinc}(2Wt - n) + \sum_{q \in \mathcal{M}} x\left(\frac{q}{2W}\right) \operatorname{sinc}(2Wt - q).$$

Substituting (6.18) gives

$$x(t) = \sum_{n \notin \mathcal{M}} x\left(\frac{n}{2W}\right) k_n(2Wt) \tag{6.19}$$

where the interpolation function is

$$k_n(t) = \text{sinc}(t-n) + r \sum_{p \in \mathcal{M}} \sum_{q \in \mathcal{M}} a_{pq} \, \text{sinc}\,(r(n-p)) \, \text{sinc}(t-q). \quad (6.20)$$

Alternately, we can pass (6.19) through a filter unity for $|u| < B$ and zero elsewhere. The result is

$$x(t) = \sum_{n \notin \mathcal{M}} x\left(\frac{n}{2W}\right) k_n^{(r)}(2Bt) \quad (6.21)$$

where

$$k_n^{(r)}(2Bt) = k_n(2Wt) * 2B \, \text{sinc}(2Bt).$$

It is straightforward to show that

$$k_n^{(r)}(t) = r\,\text{sinc}(t - rn) + r^2 \sum_{p \in \mathcal{M}} \sum_{q \in \mathcal{M}} a_{pq}\text{sinc}(r(n-p))\text{sinc}(t - rq). \quad (6.22)$$

For $\mathcal{M}$ empty, (6.19) reduces to the cardinal series and (6.22) to the oversampled restoration formula in (6.10).

In the absence of noise we can, in general, restore an arbitrarily large number of lost samples if $r < 1$. In Chapter 7, we demonstrate that restoration becomes more and more unstable as M increases and/or r approaches one. The algorithm is extended to higher dimensions in Chapter 8 where we show restoration of lost samples may be possible even when sampling is performed below the Nyquist density.

6.2.2.5 Relaxed Interpolation Formulae

Here we show oversampled signals can tolerate rather significant perturbations in the interpolation function. Consider (6.5). If $x(t)$ has a bandwidth of W, then the replicated spectra will be separated as is shown in Figure 6.2. Define

$$K_r(u) = \begin{cases} \dfrac{1}{2W} & ; \quad |u| \leq B \\ \text{anything convenient} & ; \quad B < |u| < 2W - B \end{cases}$$

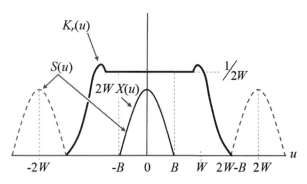

FIGURE 6.2. Replicated spectra for an oversampled signal allows flexibility in interpolation function choice.

where the subscript r is for "relaxed." Clearly, since $S(u) = 0$ for $B < |u| < 2W - B$.

$$X(u) = S(u) K_r(u).$$

Inverse transforming gives

$$x(t) = s(t) * k_r(t)$$

$$= \sum_{n=-\infty}^{\infty} x\left(\frac{n}{2W}\right) k_r\left(t - \frac{n}{2W}\right) \quad (6.23)$$

where we have used (6.4). Both the cardinal series and (6.2) are subsumed in this expression. Also, in practice, we can relax the roll–off in the spectrum of the generalized interpolation function, $k_r(t)$, with no error cost in restoration (in the absence of data noise and truncation error).

6.2.3 Criteria for Generalized Interpolation Functions

Let $k(t)$ now be an arbitrary function and define

$$g(t) = \sum_{n=-\infty}^{\infty} f\left(\frac{n}{2B}\right) k(2Bt - n). \quad (6.24)$$

If $f(t)$ has a bandwidth of B, under what condition can we recover $f(t)$ from $g(t)$?
Transforming (6.24) gives

$$G(u) = \frac{1}{2B} \sum_{n=-\infty}^{\infty} f\left(\frac{n}{2B}\right) e^{-j\pi nu/B} K\left(\frac{u}{2B}\right)$$

$$= \sum_{n=-\infty}^{\infty} F(u - 2nB) K\left(\frac{u}{2B}\right)$$

where we have used the Fourier dual of the Poisson sum formula. The function, $G(u)$, is recognized as the replicated signal spectrum weighted by $K(u/2B)$. Define the transfer function

$$H(u) = \frac{\Pi(u)}{K(u)}. \quad (6.25)$$

Then the signal spectrum can be regained by

$$F(u) = G(u) H\left(\frac{u}{2B}\right).$$

Thus, $f(t)$ can be generated by passing $g(t)$ through a filter with impulse response $h(t)$. The cardinal series is the special case when $H(u) = \Pi(u)$.

Clearly, $H(u)$ does not exist if $K(u)$ is identically zero over any subinterval of $|u| \leq 1/2$. If $K(u)$ passes through zero for $|u| \leq B$, then restoration is still possible but may be ill-posed.

6.2.3.1 Interpolation Functions

The function $k(t)$ in (6.24) is said to be an *interpolation function* if the resulting interpolation passes through the samples. For (6.24), this is equivalent to requiring that

$$k(n) = \delta[n]. \quad (6.26)$$

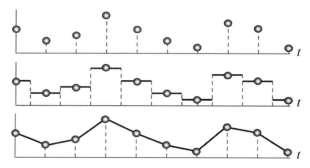

FIGURE 6.3. Sample data (top) with zeroth order sample and hold (middle) and piecewise linear (bottom) interpolation.

This condition assures that the resulting interpolation passes through the sample points. Specifically, for $t = m/2B$, (6.24) reduces to an identity when (6.26) is valid.

Some commonly used interpolation functions follow:

(a) For $k(t) = \text{sinc}(t)$, (6.24) becomes the cardinal series.
(b) **Zeroth order sample and hold**. Consider the interpolation function

$$k(t) = \Pi(t).$$

The resulting interpolation is referred to as a zeroth order *sample and hold*. An example is shown in Figure 6.3. To restore $f(t)$ from this interpolation, we pass the zero order sample and hold data through a filter with frequency response $H(u/2B)$ where

$$H(u) = \frac{\Pi(u)}{\text{sinc}(u)}.$$

Note that $H(u)$ is bounded. It follows that the inverse Fourier transform is[3]

$$h(t) = \text{sinc}_1(t).$$

(c) **Piecewise linear interpolation** uses the interpolation function

$$k(t) = \Lambda(t).$$

The result, as is shown in Figure 6.3, is that the sample points are linearly connected. The signal $x(t)$ can be regained by passing this waveform through a filter with frequency response $H(u/2B)$ where

$$H(u) = \frac{\Pi(u)}{\text{sinc}^2(u)}. \qquad (6.27)$$

Again, $H(u)$ is bounded. The inverse Fourier transform is

$$h(t) = \text{sinc}_2(t).$$

---

3. As defined in (2.44).

6.2.4 Reconstruction from a Filtered Signal's Samples

Here we consider reconstruction of a bandlimited signal $f(t)$ from samples of

$$g(t) = f(t) * h(t)$$

taken at the Nyquist rate. Let

$$2B\,K(u) = \frac{\Pi(\frac{u}{2B})}{H(u)} \qquad (6.28)$$

If $K(u)$ is bounded, we can write

$$F(u) = 2B\,G(u)\,K(u)\,\Pi\left(\frac{u}{2B}\right).$$

Using this and

$$G(u) = \frac{1}{2B} \sum_{n=-\infty}^{\infty} g\left(\frac{n}{2B}\right) e^{-j\pi nu/B} \, \Pi\left(\frac{u}{2B}\right)$$

in the inversion formula gives

$$f(t) = \int_{-B}^{B} F(u) e^{j2\pi ut} du$$

$$= \sum_{n=-\infty}^{\infty} g\left(\frac{n}{2B}\right) k\left(t - \frac{n}{2B}\right) \qquad (6.29)$$

where

$$k(t) = \frac{1}{2B} \int_{-B}^{B} \frac{e^{j2\pi ut}}{H(u)} du \longleftrightarrow K(u). \qquad (6.30)$$

As we will see in Chapter 7, (6.29) is ill posed when $K(u)$ contains a pole.

The signal samples can be regained from (6.29) with the discrete time convolution

$$f\left(\frac{m}{2B}\right) = \sum_{n=-\infty}^{\infty} g\left(\frac{n}{2B}\right) k\left(\frac{m-n}{2B}\right).$$

6.2.4.1 Restoration of Samples from an Integrating Detector

The *integrating detector* is shown in Figure 6.4. Digital cameras capture images using an array integrating detectors. The axis is divided into bins of duration $T = \frac{1}{2B}$. A function, $f(t)$, is integrated in each bin to generate the samples

$$g(nT) = \int_{t=(n-\frac{1}{2})T}^{t=(n+\frac{1}{2})T} f(t) dt. \qquad (6.31)$$

GENERALIZATIONS OF THE SAMPLING THEOREM

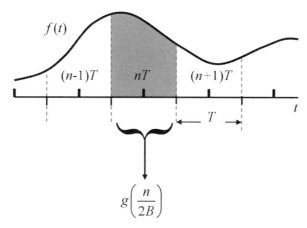

FIGURE 6.4. An integrating detector. The area of $f(t)$ over intervals of duration T are available as the only information from which to regain the bandlimited signal, $f(t)$.

Our problem is to restore the signal $f(t)$ from these samples. To do so, define

$$g(t) = f(t) * \Pi\left(\frac{t}{T}\right)$$
$$= \int_{-\infty}^{\infty} f(\tau) \Pi\left(\frac{t-\tau}{T}\right) d\tau$$
$$= \int_{t-\frac{T}{2}}^{t+\frac{T}{2}} f(\tau) d\tau.$$

Sampling $g(t)$ at $t = \frac{n}{T}$ gives the samples of the integrating detector in (6.31). Using

$$h(t) = \Pi\left(\frac{t}{T}\right)$$

gives the Fourier transform

$$H(u) = T\,\text{sinc}(Tu)$$

and (6.30) gives

$$k(t) = \frac{1}{2TB} \int_{-B}^{B} \frac{e^{j2\pi ut}}{\text{sinc}(Tu)} du.$$

At the Nyquist rate, $T = \frac{1}{2B}$ and

$$k(t) = 2B\,\text{sinc}_1(2Bt).$$

The original samples can therefore be regained through the discrete time convolution

$$f\left(\frac{n}{2B}\right) = g\left(\frac{n}{2B}\right) * 2B\,\text{sinc}_1\left(2B(m-n)\right).$$

6.3 Papoulis' Generalization

There are a number of ways to generalize the manner in which data can be extracted from a signal and still maintain sufficient information to reconstruct the signal. Shannon [1256], for example, noted that one could sample at half the Nyquist rate without information loss if, at each sample location, two sample values were taken: one of the signal and one of the signal's derivative. The details were later worked out by Linden [837] who generalized the result to restoring from a signal sample and samples of its first $N-1$ derivatives taken every N Nyquist intervals.

Alternately, one can choose any N distinct points within N Nyquist intervals. If signal samples are taken at these locations every N Nyquist intervals, we address the question of restoration from *interlaced* or *bunched* samples [1544].

Another problem encountered is restoration of a signal from samples taken at half the Nyquist rate along with samples of the signal's Hilbert transform taken at the same rate.

Remarkably, all of these cases are subsumed in a generalization of the sampling theorem developed by Papoulis [1086, 1087]. The generalization concerns restoration of a signal given data sampled at $1/N^{th}$ the Nyquist rate from the output of N filters into which the signal has been fed. The result is a generalization of the reconstruction from the filtered signal's samples.

In this section, we first present a derivation of *Papoulis' Generalized Sampling Theorem* [1086, 1087]. Specific attention is then given to interpolation function evaluation. Lastly, specific applications of the problems addressed at the beginning of this section are given.

6.3.1 Derivation

Let $\{H_p(u) \mid p=1,2,\ldots,N\}$ be a set of N given filter frequency responses and let $f(t)$ have bandwidth B. As is shown in Figure 6.5, $f(t)$ is fed into each filter. The outputs are

$$g_p(t) = f(t) * h_p(t) \ ; \ 1 \leq p \leq N. \tag{6.32}$$

Each output is sampled at $1/N^{th}$ the Nyquist rate. Define

$$B_N = B/N.$$

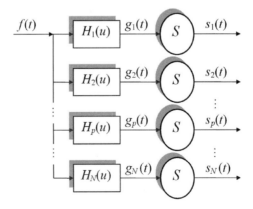

FIGURE 6.5. Generation of sample data from Papoulis' generalized sampling theorem. The encircled S is a sampler.

The signal of samples obtained from the pth filter is

$$s_p(t) = g_p(t)\, 2B_N\, \text{comb}(2B_N t)$$

$$= \sum_{n=-\infty}^{\infty} g_p(nT_N)\,\delta(t - nT_N) \tag{6.33}$$

where $T_N = 1/2B_N$. Our problem is to restore $f(t)$ from this set of functions or, equivalently, the sample set

$$\{g_p(nT_N) \mid 1 \leq p \leq N,\ -\infty < n < \infty\}.$$

We will show that

$$f(t) = \sum_{p=1}^{N} \sum_{n=-\infty}^{\infty} g_p(nT_N)\, k_p(t - nT_N) \tag{6.34}$$

where

$$k_p(t) = \int_{B-2B_N}^{B} K_p(u; t)\, e^{j2\pi u t}\, du \tag{6.35}$$

and the $K_p(u; t)$'s, if they exist, are the solutions of the simultaneous set of equations

$$2B_N \sum_{p=1}^{N} K_p(u; t)\, H_p(u - 2mB_N) = \exp(-j2\pi m t / T_N);$$

$$0 \leq m < N \tag{6.36}$$

over the parameter set $0 \leq m < N$, $B - B_N < u < B$ and $-\infty < t < \infty$. Note that $K_p(u; t)$ and $k_p(t)$ are not Fourier transform pairs.

Proof. From (6.32)

$$G_p(u) = F(u) H_p(u).$$

Fourier transforming (6.33) and using the Fourier dual of the Poisson sum formula gives

$$S_p(u) = 2B_N \sum_{n=-\infty}^{\infty} G_p(u - 2nB_N). \tag{6.37}$$

Clearly, each $S_p(u)$ is periodic with period $2B_N$. Since $G_p(u)$ has finite support, i.e.,

$$G_p(u) = G_p(u)\, \Pi\left(\frac{u}{2B}\right),$$

we conclude (6.37) is simply an aliased replication of $G_p(u)$. Example replications for $N = 2$ and 3 are shown in Figure 6.6.

Note that on the interval $|u| \leq B$, there are $2N - 1$ portions of shifted $G_p(u)$'s. Equivalently, there are $M = N - 1$ spectra overlapping the zeroth order spectrum on both sides of the origin. Accordingly, M is referred to as the *degree of aliasing*. Over the interval

$$B - 2B_N < u < B, \tag{6.38}$$

corresponding to one period of $S_p(u)$, there are a total of N portions of replicated spectra. If we have N varied forms of Nth order aliased data, it makes sense that our signal

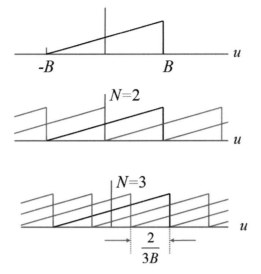

FIGURE 6.6. For the spectrum shown at the top, we have illustration of first order (middle) and second order (bottom) aliasing.

can be recovered. Indeed, on the interval in (6.38),

$$S_p(u) = 2B_N \sum_{n=0}^{N-1} G_p(u - 2nB_N) \tag{6.39}$$

$$= 2B_N \sum_{n=0}^{N-1} H_p(u - 2nB_N)F(u - 2nB_N) \tag{6.40}$$

Here we have N equations and N unknowns. This may be made clearer by viewing (6.40) in matrix form:

$$2B_N \begin{bmatrix} H_1(u) & \ldots & H_1(u - 2nB_N) & \ldots & H_1(u - 2(N-1)B_N) \\ H_2(u) & \ldots & H_2(u - 2nB_N) & \ldots & H_2(u - 2(N-1)B_N) \\ \vdots & & \vdots & & \vdots \\ H_p(u) & \ldots & H_p(u - 2nB_N) & \ldots & H_p(u - 2(N-1)B_N) \\ \vdots & & \vdots & & \vdots \\ H_N(u) & \ldots & H_N(u - 2nB_N) & \ldots & H_N(u - 2(N-1)B_N) \end{bmatrix}$$

$$\times \begin{bmatrix} F(u) \\ F(u - 2B_N) \\ \vdots \\ F(u - 2nB_N) \\ \vdots \\ F(u - 2(N-1)B_N) \end{bmatrix} = \begin{bmatrix} S_1(u) \\ S_2(u) \\ \vdots \\ S_p(u) \\ \vdots \\ S_N(u) \end{bmatrix}$$

or, in short hand notation,

$$2B_N \mathbf{H}\vec{F} = \vec{S}. \tag{6.41}$$

Thus, assuming the **H** matrix is not singular, we can solve for $F(u)$ with knowledge of the set:

$$\{F(u - 2nB_N) \mid B - 2B_N < u < B ; \ 0 \leq n < N\}.$$

Indeed, each $F(u - 2nB_N)$ over the interval $B - 2B_N < u < B$ is a displaced section of $F(u)$. This is illustrated in Figure 6.7 for $N = 3$. The sections $F(u - 2nB_3)$ for $-2B_3 < u < B$ are shown there for $n = 0, 1$ and 2.

Our purpose now is to appropriately put these pieces of $F(u)$ together and inverse transform. Towards this end, let the inverse of the matrix $2B_N \mathbf{H}$ be $\mathbf{Z}$ with elements

$$\{Z_p(u; n) \mid B - 2B_N < u < B ; \ 1 \leq p \leq N, \ 0 \leq n < N\}.$$

That is

$$2B_N \sum_{p=1}^{N} Z_p(u; n) H_p(u - 2mB_N) = \delta[n - m]; \ B - 2B_N < u < B, \ 0 \leq n, m, < N.$$

The solution of (6.40) is thus

$$F(u - 2nB_N) = \sum_{p=1}^{N} S_p(u) Z_p(u; n). \tag{6.42}$$

Consider, then

$$f(t) = \int_{-B}^{B} F(\nu) e^{j2\pi \nu t} d\nu. \tag{6.43}$$

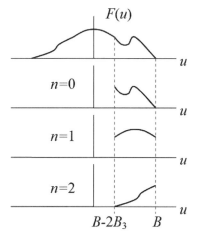

FIGURE 6.7. Illustration of $F(u - 2nB_N)$ on $B - 2B_N < u < B$ for various n.

To facilitate use of (6.42), we divide the integration into N intervals of width $2B_N$:

$$\int_{-B}^{B} = \int_{B_N-2B_N}^{B_N} + \int_{B-4B_N}^{B-2B_N} + \ldots + \int_{-B}^{-B+2B_N}$$

$$= \sum_{n=0}^{N-1} \int_{B-2(n+1)B_N}^{B-2nB_N}.$$

Substitute into (6.43) and make the variable substitution $v = u - 2nB_N$.

$$f(t) = \sum_{n=0}^{N-1} \int_{B-2B_N}^{B} F(u-2nB_N)e^{j2\pi t(u-2nB_N)}du. \tag{6.44}$$

Define

$$K_p(u;t) = \sum_{n=0}^{N-1} Z_p(u;n)\exp(-j4\pi nB_N t). \tag{6.45}$$

Then substitution of (6.42) into (6.44) gives

$$f(t) = \sum_{p=1}^{N} \int_{B-2B_N}^{B} S_p(u)K_p(u;t)e^{j2\pi tu}du. \tag{6.46}$$

Directly transforming (6.33) yields

$$S_p(u) = \sum_{n=-\infty}^{\infty} g_p(nT_N)\exp(-j\pi nu/B_N).$$

Substituting into (6.46) produces Papoulis' generalized sampling theorem.

$$f(t) = \sum_{p=1}^{N} \sum_{n=-\infty}^{\infty} g_p(nT_N)k_p(t-nT_N) \tag{6.47}$$

where

$$k_p(t) = \int_{B-2B_N}^{B} K_p(u;t)e^{j2\pi tu}du. \tag{6.48}$$

Equation (6.47) generates $f(t)$ from the undersampled outputs of each of the N filters.

6.3.2 Interpolation Function Computation

In order to find the k_p's required for interpolation in (6.47), for a given set of frequency responses, we need not invert the $\mathbf{H}$ matrix in (6.41). Rather, we need only to solve N simultaneous equations.

To derive this set of equations, we rewrite (6.45) as

$$\vec{K} = \mathbf{Z}\vec{E} \tag{6.49}$$

where the vector $\vec{K}$ has elements

$$K_p(u; t) \; ; \; B - 2B_N < u < B, \; 1 \le p \le N$$

and $\vec{E}$ has elements

$$\exp(-j4\pi n B_N t) \; ; \; -\infty < t < \infty, \; 0 \le n < N.$$

Multiplying both sides of (6.49) by $2B_N \mathbf{H} = \mathbf{Z}^{-1}$ gives

$$2B_N \mathbf{H} \vec{K} = \vec{E}$$

or, equivalently

$$2B_N \sum_{p=1}^{N} K_p(u; t) H_p(u - 2mB_N) = \exp(-j4\pi m B_N t) \tag{6.50}$$

where $0 \le m < N$, $B - 2B_N < u < B$ and t is arbitrary. The $K_p(u; t)$'s can be determined from this set of equations and the corresponding interpolation functions from (6.48).

6.3.3 Example Applications

6.3.3.1 Recurrent Nonuniform Sampling

As is shown in Figure 6.8, let $\{\alpha_p \mid p = 1, 2, \ldots, N\}$ denote N distinct locations in N Nyquist intervals. A signal is sampled at these points in every N Nyquist intervals. We thus have knowledge of the data

$$\left\{ f\left(\alpha_p + \frac{m}{2B_N}\right) \middle| \; 1 \le p \le N, \; -\infty < m < \infty \right\}.$$

Such sampling is also referred to as *bunched* or *interlaced* sampling.

The generalized sampling theorem is applicable here if we choose for filters

$$H_p(u) = \exp(j2\pi \alpha_p u); \; 1 \le p \le N. \tag{6.51}$$

The corresponding equations in (6.50) can be solved in closed form using Cramer's rule and the Vandermonde determinant[4] [572]. On the interval $(0, T_N)$, the resulting interpolation

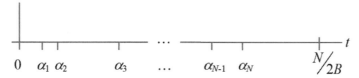

FIGURE 6.8. Illustration of Nth order recurrent nonuniform sampling. In each N Nyquist intervals, samples are taken at these same relative locations.

---

4. See Exercise 6.9.

functions are

$$k_p(t) = \text{sinc}\,(2B_N(t - \alpha_p)) \prod_{\substack{q=1 \\ q \neq p}}^{N} \frac{\sin\,(2\pi B_N(t - \alpha_q))}{\sin\,(2\pi B_N(\alpha_p - \alpha_q))}. \tag{6.52}$$

Note that $k_p(t)$ is a true interpolation function in the sense that

$$k_p(\alpha_n) = \delta[p - n].$$

6.3.3.2 Interlaced Signal–Derivative Sampling

Consider the $N = 2$ case where

$$H_1(u) = e^{j2\pi\alpha u} \tag{6.53}$$

$$H_2(u) = (j2\pi u)^M. \tag{6.54}$$

The output of filter #1 is $f(t + \alpha)$ and that of #2 is the Mth derivative of $f(t)$. The resulting sampling geometry is shown in Figure 6.9 We will derive the spectra of the two corresponding interpolation functions.

The output of the two filters is sampled at a rate of $2B_N = B$. From (6.40), the desired signal spectrum, $F(u)$, satisfies the set of equations

$$\begin{bmatrix} S_1(u) \\ S_2(u) \end{bmatrix} = B \begin{bmatrix} e^{j2\pi\alpha u} & e^{j2\pi\alpha(u-B)} \\ (j2\pi u)^M & [j2\pi(u - B)]^M \end{bmatrix} \begin{bmatrix} F(u) \\ F(u - B) \end{bmatrix} \; ; \; 0 \leq u \leq B.$$

The determinant of the **H** matrix here is

$$\Delta(u) = -(j2\pi)^M \, e^{j2\pi\alpha u} \left[u^M \, e^{-j2\pi\alpha B} - (u - B)^M \right]. \tag{6.55}$$

Solving the two simultaneous equations results gives

$$F(u) = \frac{j2\pi(u - B)^M S_1(u) - e^{j2\pi\alpha(u-B)} S_2(u)}{B\Delta(u)} \; ; \; 0 < u < B$$

and

$$F(u - B) = -\frac{(j2\pi u)^M S_1(u) - e^{j2\pi\alpha u} S_2(u)}{B\Delta(u)} \; ; \; 0 < u < B.$$

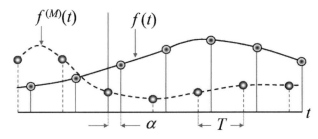

FIGURE 6.9. Interlaced signal-derivative sampling. The signal $f(t)$, and its Mth derivative are both sampled at half the Nyquist rate.

Shifting the second term to the interval $(-B, 0)$ and recognizing that both $S_1(u)$ and $S_2(u)$ are periodic with period B gives:

$$F(u) = K_1(u)S_1(u) + K_2(u)S_2(u),$$

where the spectra of the interpolation functions are

$$K_1(u) = \frac{(j2\pi)^M}{B}\left[-\frac{(u-B)^M}{\Delta(u)}\Pi\left(\frac{u}{B}-\frac{1}{2}\right) + \frac{(u+B)^M}{\Delta(u+B)}\Pi\left(\frac{u}{B}+\frac{1}{2}\right)\right] \quad (6.56)$$

and

$$K_2(u) = \frac{e^{j2\pi\alpha u}}{B}\left[\frac{e^{-j2\pi\alpha B}}{\Delta(u)}\Pi\left(\frac{u}{B}-\frac{1}{2}\right) - \frac{e^{j2\pi\alpha B}}{\Delta(u+B)}\Pi\left(\frac{u}{B}+\frac{1}{2}\right)\right]. \quad (6.57)$$

We will use these results in Chapter 7 to show that interpolation here becomes unstable (ill–posed) when

(a) M is even and $\alpha = 0$, or
(b) M is odd and $\alpha = \frac{1}{2B}$.

Otherwise, interpolation can be tolerant of data noise.

Reconstruction from the $M = 1, \alpha = 0$ data was first addressed by Shannon [1256] and derived by Linden [837]. Inverse transforming (6.56) and (6.57) for this case gives the interpolation functions

$$k_1(t) = \text{sinc}^2(Bt) \quad (6.58)$$

and

$$k_2(t) = t\,\text{sinc}^2(Bt) \quad (6.59)$$

These are pictured in Figure 6.10. It follows that

$$f(t) = \frac{\sin^2(\pi Bt)}{\pi^2}\sum_{n=-\infty}^{\infty}\left[\frac{f(\frac{n}{B})}{(Bt-n)^2} + \frac{f'(\frac{n}{B})}{B(Bt-n)}\right]. \quad (6.60)$$

6.3.3.3 Higher Order Derivative Sampling

Consider sampling a signal and its first $N-1$ derivatives every N Nyquist intervals [837]. We can show that, as $N \to \infty$, the interpolation functions for the restoration approach those used in a Taylor series expansion.

The filters required for our problem are

$$H_p(u) = (j2\pi u)^{p-1}; \quad 1 \le p \le N. \quad (6.61)$$

The solution for the interpolation function for the $N = 1$ case is clearly

$$k_1(t) = \text{sinc}(2Bt).$$

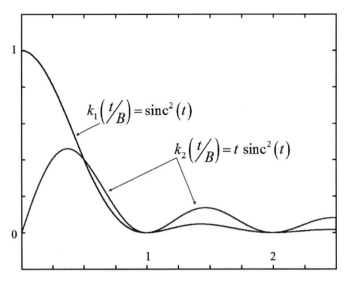

FIGURE 6.10. The functions for interpolating a signal from its samples and samples of its derivatives each taken simultaneously at half the Nyquist rate. The function $k_1(t/B)$ is used for the samples and $k_2(t/B)$ for the derivative samples.

For $N = 2$, the interpolation functions are given by (6.58) and (6.59) which we rewrite here as

$$k_1(t) = \operatorname{sinc}^2\left(\frac{2Bt}{2}\right)$$

$$k_2(t) = t \operatorname{sinc}^2\left(\frac{2Bt}{2}\right)$$

In Exercise 6.26, we show that, for $N = 3$

$$k_1(t) = \operatorname{sinc}^3\left(\frac{2Bt}{3}\right)$$

$$k_2(t) = t \operatorname{sinc}^3\left(\frac{2Bt}{3}\right) \qquad (6.62)$$

$$k_3(t) = \frac{1}{2} t^2 \operatorname{sinc}^3\left(\frac{2Bt}{3}\right).$$

From this pattern, we deduce that, in general,

$$k_p(t) = \frac{t^{p-1} \operatorname{sinc}^N\left(\frac{2Bt}{N}\right)}{(p-1)!}; \quad 1 \le p \le N.$$

Substituting into (6.47) gives the interpolation series

$$f(t) = \sum_{p=1}^{N} \sum_{n=-\infty}^{\infty} \frac{\left(t - \frac{nN}{2B}\right)^{p-1}}{(p-1)!} f^{(p-1)}\left(\frac{nN}{2B}\right) \operatorname{sinc}^N\left(\frac{2Bt}{N} - n\right).$$

Since
$$\lim_{N\to\infty} \mathrm{sinc}^N\left(\frac{2Bt}{N} - n\right) = \delta[n],$$
we conclude that
$$\lim_{N\to\infty} f(t) = \sum_{p=1}^{\infty} \frac{t^{p-1}}{(p-1)!} f^{(p-1)}(0)$$
which is recognized as the Taylor series expansion of $f(t)$ about $t = 0$.

6.3.3.4 Effects of Oversampling in Papoulis' Generalization

Suppose $f(t)$ has bandwidth B and is sampled at a rate of $2W > 2B$. Redefine $T_N = N/2W$. Then the transform of (6.47) becomes
$$F(u) = \sum_{p=1}^{N} K_p(u) \sum_{n=-\infty}^{\infty} g_p(nT_N) e^{-j2\pi n u T_N} \Pi\left(\frac{u}{2W}\right).$$

Multiplying both sides by $\Pi(u/2B)$ leaves the result unaltered.
$$F(u) = \sum_{p=1}^{N} K_p(u) \sum_{n=-\infty}^{\infty} g_p(nT_N) e^{-j2\pi n u T_N} \Pi\left(\frac{u}{2B}\right). \tag{6.63}$$

In the time domain, this is equivalent to using the interpolation function set $\{\hat{k}_p(t)\}$ in place of $\{k_p(t)\}$ where
$$\hat{k}_p(t) = \int_{-B}^{B} K_p(u) e^{j2\pi u t} du$$
$$= k_p(t) * 2B\,\mathrm{sinc}(2Bt). \tag{6.64}$$

Inverse transforming (6.63) then gives us the oversampled version of (6.47).
$$f(t) = \sum_{p=1}^{N} \sum_{n=-\infty}^{\infty} g_p(nT_N) \hat{k}_p(t - nT_N). \tag{6.65}$$

As we shall demonstrate in Section 7.2.1, oversampling can reduce interpolation noise level due to noisy data. Thus, with all other factors equal, (6.65) should be used in lieu of (6.47) for interpolating oversampled signals.

6.4 Derivative Interpolation

Interpolation formulae for generating the derivative of a bandlimited signal can be obtained by direct differentiation of the cardinal series [900]. The result is
$$x^{(p)}(t) = \left(\frac{d}{dt}\right)^p x(t)$$
$$= (2B)^p \sum_{n=-\infty}^{\infty} x\left(\frac{n}{2B}\right) d_p(2Bt - n) \tag{6.66}$$

where

$$d_p(t) := \left(\frac{d}{dt}\right)^p \operatorname{sinc}(t) \qquad (6.67)$$

is the derivative kernel. From the derivative theorem of Fourier analysis, we can equivalently write

$$d_p(t) = \int_{-1/2}^{1/2} (j2\pi u)^p e^{j2\pi ut} du$$

$$= \frac{(-1)^p p!}{\pi t^{p+1}} \left[\sin(\pi t) \cos_{p/2}(\pi t) - \cos(\pi t) \sin_{\frac{p-1}{2}}(\pi t) \right] \qquad (6.68)$$

where the incomplete sine and cosine are defined, respectively, as

$$\cos_a(z) = \sum_{n=0}^{\lfloor a \rfloor} \frac{(-1)^n z^{2n}}{(2n)!} \qquad (6.69)$$

and

$$\sin_a(z) = \sum_{n=0}^{\lfloor a \rfloor} \frac{(-1)^n z^{2n+1}}{(2n+1)!}. \qquad (6.70)$$

The notation $\lfloor a \rfloor$ denotes "the greatest integer less than or equal to a". To allow for $p = 0$ in (6.68), we set $\sin_{-1/2}(t) = 0$. Then $d_0(t) = \operatorname{sinc}(t)$. In the evaluation of (6.68), we used the identity [544]

$$\int x^n e^{ax} dx = (-1)^n e^{ax} \sum_{k=0}^{n} (-1)^k \frac{(n-k)!}{k!} \frac{x^k}{a^{n-k+1}}.$$

6.4.1 Properties of the Derivative Kernel

This section will be devoted to exploring properties of the derivative kernel. For large t and even p, the $\cos_{p/2}(\pi t)$ term in (6.68) dominates. For odd p, $\sin_{(p-1)/2}(\pi t)$ dominates. This observation leads to the following asymptotic relation for $d_p(t)$ for large t.

$$\lim_{t \to \infty} d_p(t) = \begin{cases} (-1)^{p/2} \pi^p \operatorname{sinc}(t) & ; \ p \text{ even} \\ (-1)^{(p-1)/2} \pi^p \cos(\pi t)/(\pi t) & ; \ p \text{ odd}. \end{cases}$$

Convolution of $(2B)^{p+1} d_p(2Bt)$ with any bandlimited $x(t)$ yields $x^{(p)}(t)$. To show this, we write

$$(2B)^{p+1} \int_{-\infty}^{\infty} x(\tau) d_p(2B(t-\tau)) d\tau$$

$$= (-1)^p \int_{-\infty}^{\infty} x(\tau) \left(\frac{d}{d\tau}\right)^p \operatorname{sinc}(2B(t-\tau)) d\tau$$

$$= \int_{-B}^{B} X(u)(j2\pi u)^p e^{j2\pi ut} du$$

$$= x^{(p)}(t) \qquad (6.71)$$

where, in the second step, we have used the power theorem of Fourier analysis. This result is a generalization of that of Gallagher and Wise [478] who noted that the first derivative of a bandlimited signal can be achieved by a convolution with an appropriately scaled first–order spherical Bessel function $j_1(t) = -\frac{d}{dt}\,\text{sinc}(t/\pi)$.

Using $d_q(t)$ as the signal in (6.71) gives the recurrence relation

$$d_{p+q}(t) = \int_{-\infty}^{\infty} d_q(\tau) d_p(t-\tau)\, d\tau. \tag{6.72}$$

Thus, higher order kernels can be generated by convolution of lower ordered kernels.

A second obvious recurrence relation is

$$d_{p+q}(t) = \left(\frac{d}{dt}\right)^p d_q(t).$$

Using this expression with $q = 1$ and the relations

$$\frac{d}{dt}\cos_n(t) = -\sin_{n-1}(t)$$

$$\frac{d}{dt}\sin_n(t) = \cos_n(t)$$

gives, via (6.68), a third recurrence formula:

$$\frac{d}{dt} d_p(t) = d_{p+1}(t)$$

$$= \begin{cases} \frac{-(p+1)}{t} d_p(t) + \frac{(-1)^{p/2}\pi^p}{t}\cos(\pi t); & \text{even } p \\ \frac{-(p+1)}{t} d_p(t) - \frac{(-1)^{\frac{p-1}{2}}\pi^p}{t}\sin(\pi t); & \text{odd } p. \end{cases}$$

(6.73)

Alternate derivative interpolation can be achieved by recognizing that if $x(t)$ is bandlimited, so is $x^{(p)}(t)$. Therefore

$$x^{(p)}(t) = \sum_{m=-\infty}^{\infty} x^{(p)}\left(\frac{m}{2B}\right) \text{sinc}(2Bt - m).$$

Thus, the signal derivative is uniquely specified by its sample values which, from (6.66) can be computed by the discrete time convolution

$$x^{(p)}\left(\frac{m}{2B}\right) = (2B)^p \sum_{n=-\infty}^{\infty} x^{(p)}\left(\frac{n}{2B}\right) d_p(m-n) \tag{6.74}$$

where, from (6.68)[5]

$$d_p(m) = \begin{cases} \dfrac{-(-1)^{m+p} p!}{\pi m^{p+1}} \sin_{\frac{p-1}{2}}(\pi m); & m \neq 0 \\ (-1)^p \dfrac{\pi^p}{p+1} \delta[p-\text{even}]; & m = 0. \end{cases}$$

---

5. To derive the $m = 0$ case, it is easiest to use the integral in (4.66) with $t = m = 0$.

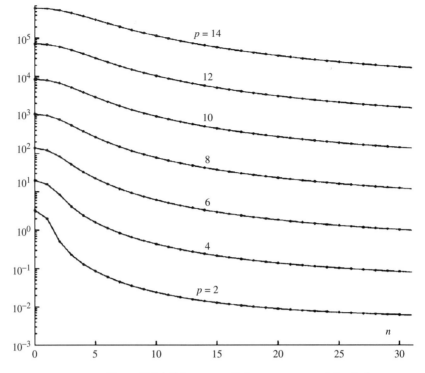

FIGURE 6.11. Plots of $|d_p(n)|$ for even p. Points are connected for clarity.

Note that the discrete derivative kernel is independent of the signal bandwidth. Plots of $|d_p(m)|$ are shown in Figures 6.11 and 6.12.

Using (6.70), the asymptotic behavior for $d_p(m)$ for large m is found to be

$$d_p(m) \longrightarrow \begin{cases} (-1)^{m+\frac{p}{2}} p\pi^{p-2}/m^2 \; ; & \text{even } p \\ \dfrac{(-1)^{m+\frac{p-1}{2}} \pi^{p-1}}{m} \; ; & \text{odd } p. \end{cases}$$

A recurrence relation for the discrete derivative kernel follows from the use of $d_q(n)$ as the signal in (6.74)

$$d_{p+q}(m) = \sum_{n=-\infty}^{\infty} d_q(n) d_p(m-n).$$

This is the discrete equivalent of (6.72).

A second recurrence relation immediately follows from (6.73) for $m \neq 0$:

$$d_{p+1}(m) = \begin{cases} \dfrac{-(p+1)}{m} d_p(m) + \dfrac{(-1)^{m+\frac{p}{2}} \pi^p}{m} \; ; & \text{even } p \\ \dfrac{-(p+1)}{m} d_p(m) & ; & \text{odd } p. \end{cases}$$

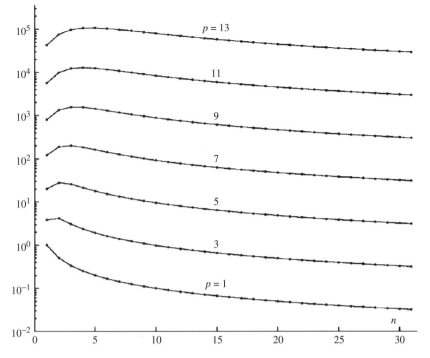

FIGURE 6.12. Plots of $|d_p(n)|$ for odd p. Points are connected for clarity.

The discrete derivative kernel is square summable. Since $d_p(m)$ is simply the mth Fourier coefficient of $(j2\pi u)^m$ for $|u| < 1/2$, we have

$$\sum_{m=-\infty}^{\infty} |d_p(m)|^2 = \int_{-\infty}^{\infty} |d_p(t)|^2 dt$$

$$= \int_{-1/2}^{1/2} |(j2\pi u)^p|^2 du$$

$$= \frac{\pi^{2p}}{2p+1}.$$

The sensitivity of derivative interpolation to additive sample noise is examined in Chapter 7. There, we show that the interpolation noise level increases significantly with p.

6.5 A Relation Between the Taylor and Cardinal Series

The discrete derivative kernel can be utilized to couple a bandlimited signal's Taylor series and sampling theorem expansion. If $x(t)$ is bandlimited, it is analytic everywhere. Thus, its Taylor series about $t = m/2B$,

$$x(t) = \sum_{p=0}^{\infty} \frac{(t - \frac{m}{2B})^p}{p!} x^{(p)}\left(\frac{m}{2B}\right),$$

converges for all t. Substituting (6.74) gives

$$x(t) = \sum_{p=0}^{\infty} \frac{(2Bt-m)^p}{p!} \sum_{n=-\infty}^{\infty} x\left(\frac{n}{2B}\right) d_p(m-n).$$

Since the series is absolutely convergent,[6] we can interchange the summation order

$$x(t) = \sum_{n=-\infty}^{\infty} x\left(\frac{n}{2B}\right) \sum_{p=0}^{\infty} \frac{(2Bt-m)^p d_p(m-n)}{p!}. \qquad (6.75)$$

The sum over p is recognized as the Taylor series expansion of $\text{sinc}(2Bt-n)$ about $t = m/2B$. Thus (6.75) reduces to the cardinal series.

6.6 Sampling Trigonometric Polynomials

A trigonometric polynomial is a bandlimited periodic function with a finite number of nonzero Fourier coefficients. A low pass trigonometric polynomial with period T can be written as

$$x(t) = \sum_{m=-N}^{N} c_m e^{-j2\pi mt/T}. \qquad (6.76)$$

This function is uniquely determined by $2N + 1$ coefficients. We therefore would expect that $2N + 1$ samples taken within a single period would suffice to uniquely specify $x(t)$. We will show that

$$x(t) = \sum_{q=1}^{P} x(qT_p) \, k\left(\frac{t}{T_p} - q\right) \qquad (6.77)$$

where $T_p = T/P$ is the sampling interval, P (assumed odd) is the number of samples per period and the interpolation function is

$$k(t) = \text{array}_P\left(\frac{t}{P}\right) \qquad (6.78)$$

We require that $P > 2N + 1$.

Proof. The cardinal series for $x(t)$ can be written as

$$x(t) = \sum_{p=-\infty}^{\infty} x(pT_p) \, \text{sinc}\left(\frac{t}{T_p} - p\right) \qquad (6.79)$$

where the sampling interval is

$$T_p = \frac{T}{P}; \quad P = 2M + 1 > 2N + 1. \qquad (6.80)$$

---
6. See Exercise 2.5.

We have assumed, for simplicity, that the odd number of samples taken in each period are the same. We can partition the sum in (6.79) as

$$\sum_{p=-\infty}^{\infty} = \ldots + \sum_{p=1+P}^{2P} + \sum_{p=1}^{P} + \sum_{p=1-P}^{0} + \ldots$$

$$= \sum_{n=-\infty}^{\infty} \sum_{p=1-nP}^{(1-n)P}$$

$$= \sum_{n=-\infty}^{\infty} \sum_{q=1}^{P}$$

where $q = p + nP$. Using this, and recognizing that $x\big((q-nP)T_p\big) = x(qT_p)$ reduces (6.79) to

$$x(t) = \sum_{q=1}^{P} x(qT_p)\, i_q(t)$$

where the interpolation function is

$$i_q(t) = \sum_{n=-\infty}^{\infty} \mathrm{sinc}\left(\frac{t+nT}{T_p} - q\right). \tag{6.81}$$

Using the Poisson sum formula

$$i_q(t) = \frac{1}{2M+1} \sum_{n=-M}^{M} e^{-j2\pi n\left(\frac{q}{P} - \frac{t}{T}\right)}$$

and the geometric series in (14.8) gives

$$i_q(t) = \frac{\sin \pi \left(\frac{t}{T_p} - q\right)/P}{\sin \pi \left(\frac{t}{T_p} - q\right)/P}$$

$$= \mathrm{array}_P\left(\frac{\frac{t}{T_p} - q}{P}\right)$$

and our proof is complete.

6.7 Sampling Theory for Bandpass Functions

A signal $x(t)$ is said to be bandpass with center frequency f_0 and bandwidth B if

$$X(u) \equiv 0;\quad 0 < |u| < f_L,\ f_U < |u| < \infty$$

where the upper and lower frequencies are $f_L = f_0 - B/2$ and $f_U = f_0 + B/2$ respectively. An example spectrum is shown in Figure 6.13. The signal is assumed to be real so that $X(u)$ is Hermetian.

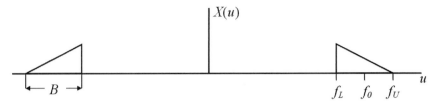

FIGURE 6.13. Parameters for the support of the Fourier transform of a bandpass signal.

We will discuss two techniques to characterize a bandpass function by its samples. The first requires preprocessing prior to sampling. The second uses samples taken directly from $x(t)$ at a rate of $2B$. A hybrid approach is left as an exercise.

6.7.1 Heterodyned Sampling

A bandpass signal can be heterodyned down to baseband by using the standard (coherent) upper sideband amplitude demodulation technique illustrated in Figure 6.14. The bandpass signal is first multiplied by a cosinusoid to obtain

$$y(t) = x(t)\cos(2\pi f_L t)$$

or, in the frequency domain

$$Y(u) = X(u) * \frac{1}{2}[\delta(u - f_L) + \delta(u + f_L)]$$
$$= \frac{1}{2}X(u - f_L) + \frac{1}{2}X(u + f_L).$$

The result is illustrated in Figure 6.15. The signal $y(t)$ is then low pass filtered to yield the baseband signal $z(t)$, which can be sampled by conventional means. If sampling is

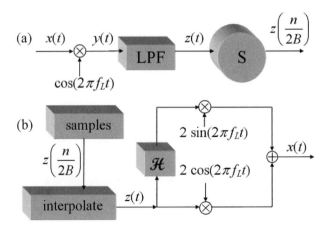

FIGURE 6.14. (a) Heterodyning a bandpass signal to baseband in order to apply conventional sampling. The encircled **S** is a sampler. (b) Restoration of the bandpass signal from the baseband signal's samples.

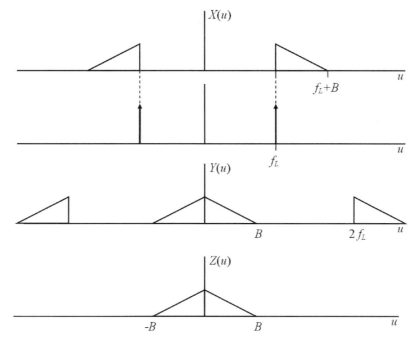

FIGURE 6.15. Illustration of heterodyning down to baseband.

performed at the Nyquist rate, then the baseband samples are

$$z\left(\frac{n}{2B}\right) = 2B \int_{-\infty}^{\infty} x(t) \cos(2\pi f_L t) \operatorname{sinc}(2Bt - n) dt.$$

Postprocessing is also required to regenerate $x(t)$ from samples of $z(t)$. We partition $Z(u)$ as

$$Z(u) = U(u) + L(u)$$

where

$$U(u) = Z(u)\,\mu(u)$$

and

$$L(u) = Z(u)\,\mu(-u).$$

Then, clearly

$$\frac{1}{2}X(u) = L(u+f_L) + U(u-f_L)$$
$$= Z(u+f_L)\mu(-u-f_L) + Z(u-f_L)\,\mu(u-f_L).$$

The inverse transform of the first term is the conjugate of the inverse transform of the second. Thus

$$\frac{1}{2}x(t) = 2\Re\, w(t)$$

where

$$w(t) = \{z(t)\exp(-j2\pi f_L t)\} * \left\{ \left(\frac{1}{2}\delta(t) - \frac{j}{2\pi t} \right) \exp(-j2\pi f_L t) \right\},$$

and $\Re$ is the real operator. Simplifying gives

$$x(t) = 2\, z(t) \cos(2\pi f_L t) + 2\, z_H(t) \sin(2\pi f_L t)$$

where the *Hilbert transform* of $z(t)$ is

$$z_H(t) = -\frac{1}{\pi} \int_{-\infty}^{\infty} \frac{z(\tau)\, d\tau}{t - \tau}.$$

6.7.2 Direct Bandpass Sampling

A bandpass signal can also be reconstructed by samples taken directly from the signal. With reference to Figure 6.11, assume that f_L is an integer multiple of B.

$$f_L = 2NB. \tag{6.82}$$

This relation can always be achieved by artificially increasing f_U, resulting in an equal incremental increase in B.

The reason for requiring (6.82) is made evident in Figure 6.16. When the bandpass signal is sampled at a rate of $2B$, the replicated spectra do not overlap with $X(u)$ (shown with solid lines). Therefore, $X(u)$ can be regained from the sample data with the use of a bandpass filter. Let's derive the specifics. Let

$$s(t) = \sum_{n=-\infty}^{\infty} x\left(\frac{n}{2B}\right) \delta\left(t - \frac{n}{2B}\right) = x(t)\, 2B\, \mathrm{comb}(2Bt) \tag{6.83}$$

so that

$$S(u) = X(u) * \mathrm{comb}(u/2B) = 2B \sum_{n=-\infty}^{\infty} X(u - 2nB).$$

The signal is regained with a bandpass filter.

$$X(u) = \frac{1}{2B} S(u) \left[\Pi\left(\frac{u + f_0}{B}\right) + \Pi\left(\frac{u - f_0}{B}\right) \right].$$

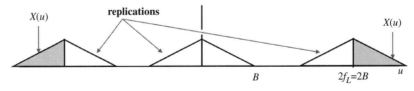

FIGURE 6.16. When f_L is an integer multiple of B, the spectral replications corresponding to a sampling rate of $2B$ do not overlap the original spectrum. Note, in this example, we had to artificially increase B to meet the integer multiplication criterion.

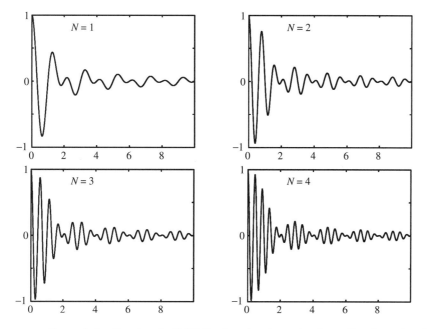

FIGURE 6.17. Interpolation functions in (6.85) for direct bandpass signal sampling for various N.

Inverse transforming gives

$$x(t) = s(t) * (\operatorname{sinc}(Bt) \cos(2\pi f_0 t)).$$

Substituting (6.83) and simplifying leaves

$$x(t) = \sum_{n=-\infty}^{\infty} x\left(\frac{n}{2B}\right) k(2Bt - n). \tag{6.84}$$

where the interpolation function is

$$k(t) = \operatorname{sinc}\left(\frac{t}{2}\right) \cos\left(\frac{\pi(2N+1)t}{2}\right). \tag{6.85}$$

Plots of $k(t)$ for various N are shown in Figure 6.17.

6.8 A Summary of Sampling Theorems for Directly Sampled Signals

A number of the sampling theorems discussed in this chapter can be written as

$$x(t) = \sum_{n \in S} x(t_n) k_n(t) \tag{6.86}$$

where S is a set of integers. A list of some applicable sampling theorems are in Table 6.1. The signal $x(t)$ is assumed to be (low pass or high pass) bandlimited with bandwidth B.

TABLE 6.1. Direct sample interpolation following the formula in (6.86). (a) The sum is over both n and p. (b) $2W = P/T$ where T is the signal's period and P is an odd integer. Parameter comments: (1) $B =$ signal's bandwidth $W > B$. (2) $r = B/W =$ sampling rate parameter, (3) $B_N = B/N$, (4) $\alpha_p =$ sample locations in recurrent nonuniform sampling

(Eq)/Section	Sample domain S	Sample locations t_n	Interpolation function $k_n(t)$
Cardinal Series (5.3)/5.1	$-\infty < n < \infty$	$\frac{n}{2B}$	$\text{sinc}(2Bt - n)$
Over-sampling (6.2)/6.2.1	$-\infty < n < \infty$	$\frac{n}{2W}$	$r\,\text{sinc}(2Bt - rn)$
Lost sample at $t = 0$ (6.94)/6.11	$n \neq 0$	$\frac{n}{2W}$	$r\,\text{sinc}(2Bt - rn)$ $+ \frac{r^2}{(1-r)}\text{sinc}(rn)\text{sinc}(2Bt)$
Recurrent nonuniform[a] (6.52)/6.3.3	$1 \leq p \leq P$ $-\infty < n < \infty$	$t_{np} = \frac{n}{2B_N} + \alpha_p$	$\text{sinc}(2B_N(t - t_{np}))$ $\times \prod_{\substack{q=1 \\ q \neq p}}^{N} \frac{\sin(2\pi B_N(t - t_{nq}))}{\sin(2\pi B_N(t_{np} - t_{nq}))}$
Bandpass signals (6.85)/6.7.2	$-\infty < n < \infty$	$\frac{n}{2B}$	$\text{sinc}\left(Bt - \frac{n}{2}\right)$ $\times \cos\left(\pi(2N+1)\left(Bt - \frac{n}{2}\right)\right)$
Trigonometric polynomials[b] (6.78)/6.6	$1 \leq n \leq P$	$\frac{n}{2W}$	$\text{array}_P\left(\frac{t-t_n}{P}\right)$

The sampling rate $2W$ exceeds the Nyquist rate. Note that, in each case, the function used for interpolation can itself be sampled and interpolated as

$$k_m(t) = \sum_{n \in S} k_m(t_n) \, k_n(t). \tag{6.87}$$

Excluded from this generalization are interpolations requiring samples of a signal's derivative or Hilbert transform. Derivative interpolation is likewise not included.

6.9 Lagrangian Interpolation

There exists a unique Nth order polynomial passing through any $N + 1$ samples. Lagrangian interpolation identifies that polynomial. Lagrangian interpolation, when applied to an infinite number of uniformly spaced samples, is equivalent to cardinal series interpolation.

In general, let $\{t_n\}$ denote a set of sample locations for a function $x(t)$. The corresponding Lagrangian interpolation from these samples is [1155]

$$y(t) = \sum_n x(t_n) k_n(t) \tag{6.88}$$

where

$$k_n(t) = \prod_{m \neq n} \frac{t - t_m}{t_n - t_m}.$$

Note that this function meets the interpolation function criterion

$$k_n(t_m) = \delta[n - m].$$

In other words, $y(t)$ passes through all the sample points.

We now show that if $\{t_n = nT | -\infty < n < \infty\}$, then (6.88) becomes the cardinal series. Under this assumption, the interpolation function clearly takes on the same form at every sample location. Thus

$$k_n(t) = k(t - nT).$$

Analysis of the $n = 0$ case therefore suffices. There,

$$k(t) = \prod_{m \neq 0} \left[1 - \left(\frac{t}{mT}\right)\right].$$

Separating the product into its positive and negative m portions followed by a multiplicative combination gives

$$k(t) = \prod_{m=1}^{\infty} \left[1 - \left(\frac{t}{mT}\right)^2\right].$$

Since [5]

$$\sin(z) = z \prod_{m=1}^{\infty} \left[1 - \left(\frac{z}{\pi m}\right)^2\right] \tag{6.89}$$

we conclude that

$$k(t) = \text{sinc}\left(\frac{t}{T}\right)$$

and our equivalence demonstration is complete.

6.10 Kramer's Generalization

The generalization of the sampling theorem by Kramer (1959) [772] can best be explained by a review of the conventional sampling theorem derivation in followed by a parallel generalized derivation.

Consider the inverse Fourier transform expression of a bandlimited function in (5.2). We can evaluate this expression without loss of information at the points $t = n/2B$ because the functions $\{\exp(-j\pi nu/B)| -\infty < n < \infty\}$ form a complete orthogonal basis set on the interval $-B < u < B$. Therefore, as explained in Section 2.4, the inner products expressed in (5.9) are sufficient for an orthogonal series expansion for $X(u)$ and therefore $x(t)$.

Consider, then, the generalized integral transform[7]

$$y(t) = \int_I Y(u) C(t; u) du \tag{6.90}$$

---

7. In this section, $y(t)$ and $Y(u)$ are *not* Fourier transform pairs.

where $C(t; u)$ is a given kernel, and I a given interval.[8] Assume that over the interval I, the functions $\{C(t_n; u) \mid -\infty < n < \infty\}$ form a complete orthonormal basis set which can be used to express $Y(u)$. Then $Y(u)$ can be expressed in an orthonormal expansion using samples of $y(t)$ as coefficients.

$$Y(u) = \sum_{n=-\infty}^{\infty} y(t_n)\, C^*(t_n; u).$$

Substituting into (6.90) gives a generalization of the cardinal series

$$y(t) = \sum_{n=-\infty}^{\infty} y(t_n)\, k_n(t) \tag{6.91}$$

where the nth interpolation function is

$$k_n(t) = \int_I C^*(t_n; u) C(t; u)\, du. \tag{6.92}$$

Equations (6.91) and (6.92) constitute Kramer's generalization.

6.11 Exercises

6.1. Fill in the details between the two equations in (6.2).

6.2. Let $x(t)$ have a bandwidth of B. Let $r = B/W \leq 1$. Consider the sinc squared interpolation:

$$y(t; A) = D \sum_{n=-\infty}^{\infty} x(nT)\, [A\, \mathrm{sinc}\,(A(t - nT))]^2$$

where D is a constant and $T = 1/2W$. Let C be such that

$$B \leq C \leq 2W - B.$$

(a) Find D such that

$$y(t; C) - y(t; B) = x(t).$$

(b) Find a filter $H(\frac{u}{2B})$ that gives $x(t)$ as an output when $y(t; C)$ is input.

6.3. (a) Let

$$k(2Bt) = \frac{1}{2B}\, e^{-at} \mu(t).$$

Restore the resulting generalized interpolation in (6.24) using a differentiator, a low pass filter and an amplifier.

(b) Same except

$$k(2Bt) = \frac{1}{2B}\, e^{-a|t|}.$$

Here, you are allowed an inverter, two amplifiers, two differentiators and a low pass filter for restoration.

---

8. For the specific case of Fourier series, $C(t; u) = \exp(j2\pi ut)/\sqrt{2B}$ and $I = \{u \mid -B < u < B\}$.

GENERALIZATIONS OF THE SAMPLING THEOREM

6.4. A signal's Hilbert transform can be obtained by passing the signal through a filter with frequency response

$$H(u) = -j\,\text{sgn}(u).$$

Let $f(t)$ be a bandlimited signal with bandwidth B and let $g(t)$ be its Hilbert transform. Find $f(t)$ from $\{g\left(\frac{n}{2B}\right) \mid -\infty < n < \infty\}$.

6.5. A bandlimited signal, $f(t)$, and its Hilbert transform are both sampled in phase at half their Nyquist rates. Generate the interpolation functions required to regain $f(t)$.

6.6. Generate an alternate method for restoring lost samples by evaluating (6.13) at the points $\{t = \frac{n}{2W} \mid n \notin \mathcal{M}\}$.

6.7. A signal's samples are

$$f(nT) = \begin{cases} \dfrac{4(-1)^{n/2}}{\pi(1-n^2)} & ;\ n \text{ even} \\ ? & ;\ n = \pm 1 \\ 0 & ;\ \text{otherwise.} \end{cases}$$

Given $r = 1/2$, find $f(\pm T)$.

6.8. Except for $n = 0$, a signal's samples are

$$f(nT) = \begin{cases} \dfrac{2}{n}(-1)^{n/2} & ;\ \text{even } n \neq 0 \\ \dfrac{4(-1)^{(n+1)/2}}{\pi n^2} & ;\ \text{odd } n \end{cases}.$$

The signal is known to be oversampled, but the value of r is uncertain. Find $f(0)$.

6.9. **Vandermonde determinants**

The determinant of the matrix

$$\begin{bmatrix} 1 & x_1 & x_1^2 & \ldots & x_1^{N-1} \\ 1 & x_2 & x_2^2 & \ldots & x_2^{N-1} \\ \vdots & \vdots & \vdots & \ldots & \vdots \\ 1 & x_N & x_N^2 & \ldots & x_N^{N-1} \end{bmatrix}$$

is called the *Vandermonde determinant* and is equal to

$$\Delta = \Pi_{1 \leq j < k \leq N}(x_k - x_j).$$

For example, for $N = 4$,

$$\Delta = (x_4 - x_3)(x_4 - x_2)(x_4 - x_1)$$
$$\times (x_3 - x_2)(x_2 - x_1)$$
$$\times (x_2 - x_1).$$

Use this result to derive (6.52) by using (6.51) in (6.50) with Cramer's rule.

6.10. Let $f(t)$ have bandwidth B. The signals $f(t-\alpha)$ and $f(t+\alpha)$ are sampled uniformly at a rate of B. Show that [1087]

$$f(t) = \frac{\cos(2\pi B\alpha) - \cos(2\pi Bt)}{2\pi B \, \sin(2\pi B\alpha)}$$

$$\times \sum_{n=-\infty}^{\infty} \frac{f(\frac{n}{B}+\alpha)}{B(t-\alpha)-n} - \frac{f(\frac{n}{B}-\alpha)}{B(t+\alpha)-n}.$$

6.11. (a) Derive the interpolation functions in (6.58) and (6.59).
 (b) Show that the formula in (6.60) not only interpolates the signal samples properly, but also interpolates the derivative samples.

6.12. Consider the case where the integrating detector bins in Section 6.2.4.1 are separated by gaps and, in lieu of (6.31), we are given

$$g\left(\frac{n}{T}\right) = \int_{t=(n-\frac{1}{2})\Upsilon}^{t=(n+\frac{1}{2})\Upsilon} f(t)dt.$$

where $\Upsilon < T$. Assuming $1/T$ equals or exceeds the Nyquist rate, craft a procedure to find the samples of $f(t)$.

6.13. Why can't we allow $M = N$ in (6.80) ?

6.14. Show that the Fourier coefficients of a trigonometric polynomial can be generated directly from the signal's samples by the matrix equation

$$\vec{c} = \mathbf{A}\,\vec{x}$$

where $\vec{x}$ contains the P signal samples, $\vec{c}$ contains the $2N+1$ Fourier coefficients and the nq^{th} element of $\mathbf{A}$ is

$$a_{nq} = \int_{-\frac{1}{2}}^{\frac{1}{2}} \text{array}_P\left(t-\frac{q}{P}\right) e^{j2\pi nt}\, dt$$

6.15. **Envelope detection**

In this problem, we consider the amplitude modulation process in Section 3.4. A bandlimited signal, $x(t)$, has bandwidth B is used to form the signal

$$z(t) = 2\,w(t)\,\cos(2\pi f_0 t)$$

where f_0 is the carrier frequency and

$$w(t) = x(t) + b.$$

The bias, b, is chosen such that $w(t) \geq 0$. Then, as illustrated in Figure 3.18, the signal $w(t)$ rides the envelope of the modulating sinusoid. The signal $z(t)$ touches $w(t)$ when $t = n/f_0$. We therefore have access to the samples $w\left(\frac{n}{f_0}\right)$ and therefore the samples

$$x\left(\frac{n}{f_0}\right) = w\left(\frac{n}{f_0}\right) - b.$$

(a) Are the maxima of $z(t)$ in Figure 3.18 equal to the samples $w\left(\frac{n}{f_0}\right)$? Hint: See Exercise 2.33.

(b) Under what conditions can we regain the baseband signal, $x(t)$, from the samples $x\left(\frac{n}{f_0}\right)$ using the sampling theorem?

(c) Form a signal by linearly connecting the points of $x\left(\frac{n}{f_0}\right)$. Can the baseband signal be recovered from this piecewise linear signal?

6.16. Let $v(t)$ denote a real baseband signal with a maximum frequency component of $B/2$. The signal

$$x(t) = v(t)\cos(2\pi f_0 t)$$

is bandpass. We showed that $x(t)$, when heterodyned to baseband, required a minimum sampling rate of $2B$. Show a technique whereby a down heterodyned version of $x(t)$ requires a sampling rate of half that much ($f_0 > B$).

6.17. **Implicit sampling** of a function $x(t)$ is illustrated in Figure 6.18. A sample is taken when $x(t)$ crosses a predetermined level. Assume that the levels are each separated by an interval of Δ and that one of the levels is at zero. Show that not all finite energy bandlimited signals are determined uniquely by their implicit samples for any finite value of Δ.   Hint: Assume an average sampling density of $2B$ is necessary to uniquely specify the signal and consider the function

$$y(t) = \text{sinc}^2(t) + \text{sinc}^2(t-a)$$

which is strictly positive when a is not an integer.

6.18. Does (6.85) satisfy the criterion for an interpolation function? If not, why?

6.19. What class of functions does Lagrangian interpolation always interpolate exactly using only N samples?

6.20. Derive a closed form expression for the interpolation function for recurrent nonuniform sampling using Lagrangian interpolation. Is it the same as (6.52)?

6.21. Does Lagrangian interpolation result in the expression in (6.2) for oversampled bandlimited signals?

6.22. An oversampled bandlimited function can be interpolated using the oversampled cardinal series as

$$x(t) = r\sum_{n=0}^{\infty} x\left(\frac{rn}{2B}\right)\text{sinc}(2Bt - rn) \qquad (6.93)$$

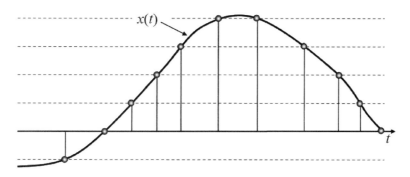

FIGURE 6.18. Implicit sampling.

where the sampling rate parameter, r, has been explicitly placed in the argument of $x(\cdot)$. For finite energy signals, this is true for $0 < r \leq 1$. Thus, as r changes in this range, $x(t)$ should not change and

$$\frac{d}{dr}x(t) = 0.$$

Show that, indeed, this is true. Hint: If $x(t)$ is bandlimited, so is $tx(t)$ and $tx'(t)$.

6.23. Show that for any oversampled finite energy bandlimited function that

$$x(t) = r \sum_{n \neq 0} x\left(\frac{n}{2W}\right) \left(\operatorname{sinc}(2Bt - rn) + \frac{r}{(1-r)} \operatorname{sinc}(rn) \operatorname{sinc}(2Bt) \right). \quad (6.94)$$

6.24. The lost sample formula in (6.11) works for finite energy signals. Does it work for $x(t) = \cos(2\pi Bt)$ when $r = 0.5$? In other words, can the sample at the origin, $x(0) = 1$ be expressed using the other samples?

6.25. A bandpass function with bandwidth B is directly sampled at a rate $2B$ where B is an integer multiple of f_L. The samples are interpolated using the conventional cardinal series. Outline the processing required to regain the original signal from this interpolation. Does it make a difference whether the integer multiple is odd or even?

6.26. For the filters in (6.61), derive the corresponding interpolation functions for $N = 3$.

6.27. **Complex Walsh functions**

For a given B, let $I = \{u | -B < u < B\}$. Let

$$C(t; u) = \{\operatorname{sgn}[\cos(2\pi ut)] + j \operatorname{sgn}[\sin(2\pi ut)]\} / \sqrt{2B}.$$

The set of functions, $\{C(\frac{n}{2B}; u) | -\infty < n < \infty\}$, known as *complex Walsh functions*, form a complete orthonormal basis set for finite energy functions on the interval I. Evaluate the interpolation functions, $k_n(t)$, corresponding to Kramer's generalization of the sampling theorem for this basis set.

6.12 Solutions for Selected Chapter 6 Exercises

6.2. (a) $\quad y(t; C) = D C^2 \sum_{n=-\infty}^{\infty} x(nT) \operatorname{sinc}^2 (C(t - nT))$

$$y(t; W) = D W^2 \sum_{n=-\infty}^{\infty} x(nT) \operatorname{sinc}^2 (W(t - nT))$$

$$x(t) = \sum_{n=-\infty}^{\infty} x(nT) k_r(t - n/2W)$$

where

$$k_r(t) = D \left[C^2 \operatorname{sinc}^2(Ct) - B^2 \operatorname{sinc}^2(Bt) \right]$$

$$\longleftrightarrow K_r(u) = D \left[C\Lambda(u/C) - B\Lambda(u/B) \right].$$

For $|u| \leq B$,

$$K_r(u) = D\left[C\left(1 - \frac{|u|}{C}\right) - B\left(1 - \frac{|u|}{B}\right)\right]$$

$$= T; \quad D = \frac{1}{2W(C-B)}$$

and $x(t)$ is recovered.

(b) Here

$$k_r(t) = D\,C^2 \operatorname{sinc}(Ct)$$

$$\longleftrightarrow K_r(u) = K\left(\frac{u}{2B}\right) = DC\Lambda\left(\frac{u}{C}\right)$$

and

$$H\left(\frac{u}{2B}\right) = \Pi\left(\frac{u}{2B}\right) / K\left(\frac{u}{2B}\right)$$

$$= \frac{1}{D(C-|u|)}\Pi\left(\frac{u}{2B}\right).$$

Notice, for $|u| \leq B$

$$\frac{1}{DC} \leq H\left(\frac{u}{2B}\right) \leq \frac{1}{D(C-B)}.$$

6.3. (a) $K\left(\frac{u}{2B}\right) = 1/(a + j2\pi u)$ and

$$H\left(\frac{u}{2B}\right) = (a + j2\pi u)\,\Pi\left(\frac{u}{2B}\right).$$

This inverse filter is shown in Figure 6.19.[9]

(b)

$$K\left(\frac{u}{2B}\right) = \frac{2a}{a^2 - (j2\pi u)^2}$$

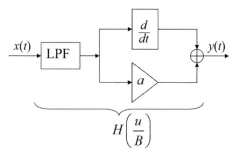

FIGURE 6.19. See Exercise 6.3a.

---

9. Note: the LPF should be used first to avoid differentiating discontinuities.

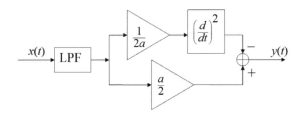

FIGURE 6.20. See Exercise 6.3b.

and

$$H\left(\frac{u}{2B}\right) = \left[\frac{a}{2} - \frac{1}{2a}(j2\pi u)^2\right]\Pi\left(\frac{u}{2B}\right).$$

The inverse filter is shown in Figure 6.20.

6.4. (b) $H(u) = -j\,\text{sgn}(u)$. Since $1/\text{sgn}(u) = \text{sgn}(u)$,

$$K(u) = \frac{j}{2B}\,\text{sgn}(u)\,\Pi\left(\frac{u}{2B}\right)$$

and

$$k(t) = \frac{j}{2B}\int_{-B}^{B}\text{sgn}(u)\,(j\sin(2\pi ut))\,du$$

$$= \frac{-1}{B}\int_{0}^{B}\sin(2\pi ut)\,du$$

$$= -\pi Bt\,\text{sinc}^2(Bt).$$

Thus, if $g(t)$ is the Hilbert transform of $f(t)$ with bandwidth B, then

$$f(t) = \frac{1}{\pi}\sum_{n=-\infty}^{\infty} g\left(\frac{n}{2B}\right)\frac{(-1)^n \cos(2\pi Bt) - 1}{2Bt - n}.$$

6.5. (a) For $M = 1$ and $\alpha = 0$, (6.55) becomes

$$\Delta(u) = j2\pi B.$$

Equation (6.56) becomes

$$K_1(u) = \frac{1}{j2\pi B^2}\left[\Pi\left(\frac{u}{B} - \frac{1}{2}\right) - \Pi\left(\frac{u}{B} + \frac{1}{2}\right)\right]$$

$$= \frac{1}{j2\pi B^2}\Pi\left(\frac{u}{B}\right) * \left(\delta\left(u - \frac{B}{2}\right) - \delta\left(u + \frac{B}{2}\right)\right).$$

Thus

$$k_1(t) = \frac{1}{\pi B}\,\text{sinc}(Bt)\,\sin(\pi Bt)$$

$$= \frac{\sin^2(\pi Bt)}{\pi^2 B^2 t}$$

$$= t\,\text{sinc}^2(Bt).$$

Equation (6.57) becomes

$$K_2(u) = \frac{1}{B^2}\left[-(u-B)\Pi\left(\frac{u}{B}-\frac{1}{2}\right) + (u+B)\Pi\left(\frac{u}{B}+\frac{1}{2}\right)\right]$$

$$= \frac{1}{B}\Lambda\left(\frac{u}{B}\right).$$

Thus

$$k_2(t) = \operatorname{sinc}^2(Bt)$$
$$= \frac{\sin^2(\pi Bt)}{(\pi Bt)^2}$$
$$= \operatorname{sinc}^2(Bt).$$

(b) The interpolation is

$$f(t) = \sum_{n=-\infty}^{\infty}\left[f\left(\frac{n}{B}\right)k_1\left(t-\frac{n}{B}\right) + f'\left(\frac{n}{B}\right)k_2\left(t-\frac{n}{B}\right)\right].$$

Since $k_1(p/B) = \delta[n]$ and $k_2(p/B) = 0$, the equation reduces to an identity at $t = m/B$. Differentiate

$$f'(t) = \sum_{n=-\infty}^{\infty}\left[f\left(\frac{n}{B}\right)k_1'\left(t-\frac{n}{B}\right) + f'\left(\frac{n}{B}\right)k_2'\left(t-\frac{n}{B}\right)\right].$$

Clearly

$$k_1'(t) = \frac{d}{dt}\operatorname{sinc}^2(Bt)$$
$$= 2B\operatorname{sinc}(Bt)d_1(Bt)$$

and

$$k_1'\left(\frac{n}{B}\right) = 0$$

since $d_1(0) = 0$ and $\operatorname{sinc}(n)$ is zero everywhere else. Note that

$$k_2(t) = t\,k_1(t),$$

Thus

$$k_2'(t) = t\,k_1'(t) + k_1(t)$$

and

$$k_2'\left(\frac{n}{B}\right) = \operatorname{sinc}^2(n)$$
$$= \delta[n].$$

The interpolation therefore also interpolates the derivative samples.

6.6. Doing so gives

$$\sum_{n \notin M} [\delta[n-m] - r\,\text{sinc}\,(r(n-m))]\, x\!\left(\frac{n}{2W}\right) = r \sum_{n \in M} x\!\left(\frac{n}{2W}\right) \text{sinc}\,(r(n-m))$$

or

$$\vec{\hat{g}} = \mathbf{S}\,\vec{x}$$

where $\vec{\hat{g}}$ has elements $\{\hat{g}\!\left(\frac{n}{2B}\right) \mid n \in M\}$ and

$$\hat{g}(nT) = \sum_{n \notin M} [\delta[n-m] - r\,\text{sinc}\,(r(n-m))]\, x\!\left(\frac{n}{2W}\right)$$

can be found from the known data.

6.7.

$$f(nT) = \begin{cases} \dfrac{4(-1)^{n/2}}{\pi(1-n^2)} & ;\ \text{even } n \\ 0 & ;\ \text{odd } n,\ n \neq \pm 1 \end{cases}$$

Therefore, for $T = 1/2W$,

$$g(\pm T) = \frac{1}{2} \sum_{\text{even } n} \frac{4(-1)^{n/2}}{\pi(1-n^2)} \text{sinc}\!\left(\frac{\pm 1 - n}{2}\right).$$

Note $g(T) = g(-T)$. Using the + gives

$$\text{sinc}\!\left(\frac{n-1}{2}\right) = \frac{2(-1)^{n/2}}{\pi(1-n)} \quad ; \text{even } n$$

and

$$g(\pm T) = \left(\frac{2}{\pi}\right)^2 \sum_{\text{even } n} \frac{1}{(1-n^2)(1-n)}$$

$$= \left(\frac{2}{\pi}\right)^2 \left[1 + 2\sum_{m=1}^{\infty} \frac{1}{(1-(2m)^2)^2}\right]$$

where we have let $2m = n$. Numerically,

$$g(\pm T) = \frac{1}{2}.$$

Now,

$$\mathbf{H} = \begin{bmatrix} -1/2 & 0 \\ 0 & 1/2 \end{bmatrix}.$$

Thus

$$f(\pm T) = 1.$$

Note: The function these samples were taken from is

$$f(t) = \text{sinc}\!\left(2Bt - \frac{1}{2}\right) + \text{sinc}\!\left(2Bt + \frac{1}{2}\right).$$

6.8. The samples are taken from the signal

$$f(t) = \frac{d}{dt}\operatorname{sinc}(t) = \frac{\cos(\pi t) - \operatorname{sinc}(t)}{t}.$$

Thus $f(0) = 0$. (Note $r = 1/2$). Using (6.8):

$$x(0) = -2 \sum_{\substack{\text{even } n \\ n \neq 0}} \frac{(-1)^{n/2}}{n} - \frac{4}{\pi} \sum_{\text{odd } n} \frac{(-1)^{\frac{n+1}{2}}}{n^2}$$

$$= 0$$

since both summands are odd.

6.13. We have implicitly assumed that $C_{\pm N} = 0$. If $P = 2N + 1$ then, in the spectral replication, the Dirac delta at $u = N/T$ would be aliased by the shifted Dirac delta originally at $u = -N/T$. We evaluate the Fourier coefficients in (6.76) with the familiar formula

$$c_n = \frac{1}{T}\int_{-T/2}^{T/2} x(\tau)\, e^{j2\pi n\tau/T}\, d\tau.$$

Substituting (6.77) followed by manipulation completes the problem.

6.16. Since $v(t)$ is real, $V(u)$ is Hermitian. Thus, as illustrated in Figure 6.21, $X(u)$ has a four fold symmetry. We can, therefore, heterodyne the center frequency f_0 (rather than the lower frequency f_L to the origin). Let

$$y(t) = x(t)\cos(2\pi f_0 t)$$
$$= v(t)\cos^2(2\pi f_0 t)$$
$$= \frac{1}{2}v(t) + \frac{1}{2}v(t)\cos(\pi f_0 t).$$

A lowpass filter gives

$$z(t) = \frac{1}{2}v(t)$$

which can be sampled at the Nyquist rate of B.

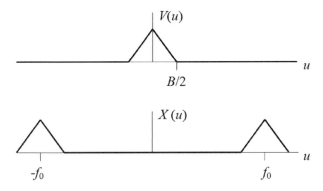

FIGURE 6.21. See Exercise 6.16.

6.17. Since $y(t) > 0$ and approaches zero for large $|t|$, no matter how small Δ, there exists some value of $|t|$ above which there will be no more samples. Thus, the number of samples is finite and the signal is not uniquely determined.

6.18. Clearly

$$k(n) = \mathrm{sinc}(n/2) \cos(\pi(2N+1)n/2).$$

Except for $n = 0$, $\mathrm{sinc}(n/2)$ is zero for even n. The cosine term is always zero for odd n. Therefore, (6.26) is satisfied.

6.19. Real $(N-1)^{st}$ order polynomials.

6.20. We wish to compute the interpolation function corresponding to α_p. The Lagrangian kernel in (6.88) can be partitioned as

$$k_p(t) = a_p(t) b_p(t)$$

where

$$a_p(t) = \prod_{m \neq 0} \frac{t - (mT_N + \alpha_p)}{\alpha_p - (mT_N + \alpha_p)}$$

is due to sample locations located distances mT from α_p and the contribution due to the remaining terms is

$$b_p(t) = \prod_{\substack{q=1 \\ q \neq p}} \prod_{m=-\infty}^{\infty} \frac{t - (mT_N + \alpha_p)}{\alpha_p - (mT_N + \alpha_p)}.$$

Using the product formula for $\sin(z)$ in (6.89) gives

$$a_p(t) = \mathrm{sinc}\left(2B_N(t - \alpha_p)\right).$$

After factoring out the zero term, the m product in $b_p(t)$ can be written as

$$\frac{t + \alpha_q}{\alpha_p - \alpha_q} \prod_{m \neq 0} \frac{1 - \dfrac{t - \alpha_q}{mT_N}}{1 - \dfrac{\alpha_p - \alpha_q}{mT_N}}.$$

Expressing both products as sincs and simplifying reveals the resulting interpolation function to be identical to that in (6.52).

6.21. No. Lagrangian interpolation would result in the conventional cardinal series. Recall that Lagrangian interpolation results in an interpolation where only the sample value contributes to the interpolation at that point. Equation (6.2), on the other hand, usually has every sample value contributing to the sample at $t = m/2W$.

GENERALIZATIONS OF THE SAMPLING THEOREM

6.22. Differentiating (6.93) gives

$$\frac{d}{dr}x(t) = \frac{d}{dr}\left(r\sum_{n=0}^{\infty} x\left(\frac{rn}{2B}\right)\text{sinc}(2Bt - rn)\right)$$

$$= \sum_{n=0}^{\infty} x\left(\frac{rn}{2B}\right)\text{sinc}(2Bt - rn)$$

$$+ r\sum_{n=0}^{\infty} \left(\frac{n}{2B}\right) x'\left(\frac{rn}{2B}\right)\text{sinc}(2Bt - rn) \qquad (6.95)$$

$$- r\sum_{n=0}^{\infty} n\, x\left(\frac{rn}{2B}\right)\text{sinc}'(2Bt - rn)$$

The first of the three terms is recognized as $x(t)/r$ and (6.95) becomes

$$\frac{d}{dr}x(t) = \frac{1}{r}x(t) + r\sum_{n=0}^{\infty} \left(\frac{n}{2B}\right) x'\left(\frac{rn}{2B}\right)\text{sinc}(2Bt - rn)$$

$$- r\sum_{n=0}^{\infty} n\, x\left(\frac{rn}{2B}\right)\text{sinc}'(2Bt - rn) \qquad (6.96)$$

For the second term, we use the hint and expand $tx'(t)$ into an oversampled cardinal series

$$tx'(t) = r\sum_{n=0}^{\infty} \left(\frac{rn}{2B}\right) x'\left(\frac{rn}{2B}\right)\text{sinc}(2Bt - rn).$$

Equation (6.96) thus becomes

$$\frac{d}{dr}x(t) = \frac{1}{r}x(t) + \frac{1}{r}t\, x'(t) - r\sum_{n=0}^{\infty} n\, x\left(\frac{rn}{2B}\right)\text{sinc}'(2Bt - rn). \qquad (6.97)$$

Using the hint and expanding $tx(t)$ into an oversampled cardinal series gives

$$tx(t) = r\sum_{n=0}^{\infty} \left(\frac{rn}{2B}\right) x\left(\frac{rn}{2B}\right)\text{sinc}(2Bt - rn).$$

Thus

$$\frac{d}{dt}tx(t) = r\sum_{n=0}^{\infty} (rn)\, x\left(\frac{rn}{2B}\right)\text{sinc}'(2Bt - rn)$$

and (6.97) becomes

$$\frac{d}{dr}x(t) = \frac{1}{r}x(t) + \frac{1}{r}t\, x'(t) - \frac{1}{r}\frac{d}{dt}tx(t)$$

$$= \frac{1}{r}\left[x(t) + t\, x'(t) - \left(x(t) + t\, x'(t)\right)\right]$$

$$= 0.$$

6.23. We write (6.2) as

$$x(t) = rx(0)\text{sinc}(2Bt) + r\sum_{n\neq 0} x(n/2B)\,\text{sinc}(2Bt - rn).$$

Substitute (6.11) and simplify. The same result can be obtained by filtering (6.12).

6.24.
$$x\left(\frac{n}{2W}\right) = \cos(\pi rn).$$

For $r = 0.5$, (6.11) becomes

$$x(0) = \sum_{n\neq 0} x\left(\frac{n}{2W}\right)\,\text{sinc}\left(\frac{n}{2}\right).$$

But $x\left(\frac{n}{2W}\right) = 0$ for odd values of n and $\text{sinc}\left(\frac{n}{2}\right) = 0$ for even values of $n \neq 0$. The series is thus equal to zero rather than the desired value of one. The lost sample restoration procedure clearly does not work in this case.

6.26. The matrix to solve is (for $\frac{B}{3} < u < B$)

$$\begin{bmatrix} S_1(u) \\ S_2(u) \\ S_3(u) \end{bmatrix} = \frac{2B}{3} \begin{bmatrix} 1 & 1 & 1 \\ j2\pi u & j2\pi(u - \frac{2B}{3}) & j2\pi(u - \frac{4B}{3}) \\ (j2\pi u)^2 & [j2\pi(u - \frac{2B}{3})]^2 & [j2\pi(u - \frac{4B}{3})]^2 \end{bmatrix}$$

$$\times \begin{bmatrix} F(u) \\ F(u - \frac{2B}{3}) \\ F(u - \frac{4B}{3}) \end{bmatrix}.$$

Using the third order Vandermonde determinant

$$\Delta(u) = j2\left(\frac{4\pi B}{3}\right)^3$$

and Cramer's rule, we have

$$F(u) = \frac{1}{2}\left(\frac{3}{2B}\right)^3 \left[\left(u - \frac{2B}{3}\right)\left(u - \frac{4B}{3}\right) S_1(u)\right.$$

$$+ j\left(\frac{1}{2\pi}\right)\left(\frac{3}{2B}\right)^2 (u - B)\, S_2(u)$$

$$\left. - \left(\frac{1}{8\pi^2}\right)\left(\frac{3}{2B}\right)^3 S_3(u)\right]; \quad \frac{B}{3} < u < B$$

$$F\left(u - \frac{2B}{3}\right) = -\left(\frac{3}{2B}\right)^3 u\left(u - \frac{4B}{3}\right) S_1(u)$$

$$- j\left(\frac{1}{\pi}\right)\left(\frac{3}{2B}\right)^3 \left(u - \frac{2B}{3}\right) S_2(u)$$

$$+ \left(\frac{1}{4\pi^2}\right)\left(\frac{3}{2B}\right)^3 S_3(u)\,;\quad \frac{B}{3} < u < B$$

GENERALIZATIONS OF THE SAMPLING THEOREM

$$F\left(u - \frac{4B}{3}\right) = \frac{1}{2}\left(\frac{3}{2B}\right)^3 u\left(u - \frac{2B}{3}\right) S_1(u)$$

$$+ j\left(\frac{1}{2\pi}\right)\left(\frac{3}{2B}\right)^3 \left(u - \frac{B}{3}\right) S_2(u)$$

$$- \left(\frac{1}{8\pi^2}\right)\left(\frac{3}{2B}\right)^3 S_3(u) \; ; \quad \frac{B}{3} < u < B.$$

To construct $F(u)$, we evaluate the equation

$$F(u) = K_1(u) S_1(u) + K_2(u) S_2(u) + K_3(u) S_3(u) \; ; -B < u < B.$$

Solving gives

$$K_1(u) = \frac{1}{2}\left(\frac{3}{2B}\right)^3 \left[\left(u - \frac{2B}{3}\right)\left(u - \frac{4B}{3}\right) \Pi\left(\frac{3u}{2B} - 1\right)\right.$$

$$- 2\left(u - \frac{2B}{3}\right)\left(u + \frac{2B}{3}\right) \Pi\left(\frac{3u}{2B}\right)$$

$$\left. + \left(u + \frac{2B}{3}\right)\left(u + \frac{4B}{3}\right) \Pi\left(\frac{3u}{2B} + 1\right)\right],$$

$$K_2(u) = j\left(\frac{1}{2\pi}\right)\left(\frac{3}{2B}\right)^2 \left[(u - B)\Pi\left(\frac{3u}{2B} - 1\right)\right.$$

$$\left. - 2u\Pi\left(\frac{3u}{2B}\right) + (u + B)\Pi\left(\frac{3u}{2B} + 1\right)\right]$$

and

$$K_3(u) = -\left(\frac{1}{8\pi^2}\right)\left(\frac{3}{2B}\right)^3 \left[\Pi\left(\frac{3u}{2B} - 1\right)\right.$$

$$\left. - 2\Pi\left(\frac{3u}{2B}\right) + \Pi\left(\frac{3u}{2B} + 1\right)\right].$$

With a bit of work we inverse Fourier transform to the interpolation functions in (4.57).

7

Noise and Error Effects

> As far as the laws of mathematics refer to reality, they are not certain; and as far as they are certain, they do not refer to reality.
> — Albert Einstein, [402b]
>
> In theory, theory and reality are the same. In reality, they're not.
> — unknown
>
> It is not certain that everything is uncertain.
> — Blaise Pascal (1623–1662) [1098]

7.1 Introduction

Exact interpolation using the cardinal series from unaliased samples assumes that (a) the values of the samples are known exactly, (b) the sample locations are known exactly (c) an infinite number of terms are used in the series, and (d) sampling is performed at a sufficiently fast rate. Deviation from these requirements results in interpolation error due to (a) data noise (b) jitter (c) truncation and (d) aliasing respectively. The perturbation to the interpolation from these sources of error is the subject of this chapter.

7.2 Effects of Additive Data Noise

If noise is superimposed on sample data, the corresponding interpolation will be perturbed. In this section, the nature of this perturbation is examined. The effect of data noise on continuous sampling interpolation is treated in Section 10.3.1.2. The multidimensional case is the topic of Section 8.10.2.

7.2.1 On Cardinal Series Interpolation

Suppose that the signal we sample is corrupted by real additive zero mean wide sense stationary noise, $\xi(t)$. Then, instead of sampling the deterministic bandlimited signal, $x(t)$, we would be sampling the signal

$$y(t) = x(t) + \xi(t). \tag{7.1}$$

From these samples, we form the series

$$z(t) = \sum_{n=-\infty}^{\infty} y\left(\frac{n}{2W}\right) \operatorname{sinc}(2Wt - n) \quad (7.2)$$

where the sampling rate, $2W$, equals or exceeds twice the bandwidth, B, of $x(t)$. Recall the sampling rate parameter

$$r = \frac{B}{W} \leq 1.$$

In general, $z(t)$ will equal $y(t)$ only at the sample point locations.

Substituting (7.1) into (7.2) reveals that

$$z(t) = x(t) + \eta(t) \quad (7.3)$$

where

$$\eta(t) = \sum_{n=-\infty}^{\infty} \xi\left(\frac{n}{2W}\right) \operatorname{sinc}(2Wt - n). \quad (7.4)$$

Therefore, $\eta(t)$ is the stochastic process generated by the samples of $\xi(t)$ alone and is independent of the signal. Note that, since $\xi(t)$ is zero mean, so is $\eta(t)$. Hence, expecting both sides of (7.3) leads us to the desirable conclusion that the expected value of the noisy interpolation is the noiseless signal.

$$\overline{z(t)} = x(t).$$

7.2.1.1 Interpolation Noise Level

A meaningful measure of the cardinal series' noise sensitivity is the *interpolation noise level* which, since $\xi(t)$ is zero mean, is, $\overline{\eta^2(t)}$. Towards this end, we will first find the autocorrelation for $\eta(t)$. From (7.4)

$$R_\eta(t - \tau) = E[\eta(t)\eta(\tau)]$$

$$= \sum_{n=-\infty}^{\infty} \sum_{m=-\infty}^{\infty} R_\xi\left(\frac{n-m}{2W}\right) \operatorname{sinc}(2Wt - n)\operatorname{sinc}(2W\tau - m)$$

$$= \sum_{k=-\infty}^{\infty} R_\xi\left(\frac{k}{2W}\right) \sum_{n=-\infty}^{\infty} \operatorname{sinc}(2W\tau - n + k)\operatorname{sinc}(2Wt - n) \quad (7.5)$$

where our assumption of the wide sense stationarity of $\eta(t)$ will shortly be justified and, in the second step, we have let $k = n - m$. The n sum in (7.5) is the cardinal series applied to $x(t) = \operatorname{sinc}(2W(\tau - t) + k)$. Then

$$R_\eta(t) = \sum_{k=-\infty}^{\infty} R_\xi\left(\frac{k}{2W}\right) \operatorname{sinc}(2Wt - k). \quad (7.6)$$

Thus, the interpolation noise autocorrelation is found from the cardinal series interpolation using sample values from the data noise autocorrelation.

To find the interpolation noise level, we simply evaluate (7.6) at $t = 0$. The remarkable result is

$$\overline{\eta^2} = \overline{\xi^2}. \tag{7.7}$$

That is, the cardinal series interpolation results in a noise level equal to that of the original signal before sampling [149].

7.2.1.2 Effects of Oversampling and Filtering

In many cases, one can reduce the interpolation noise level by oversampling and filtering. If, for example, we place $z(t)$ in (7.3) through a filter that is one for $|u| \leq B$ and zero otherwise, then the output is

$$z_r(t) = z(t) * 2B \, \text{sinc}(2Bt)$$
$$= x(t) + \eta_r(t)$$

where the stochastic process, $\eta_r(t)$, is defined as

$$\eta_r(t) = \eta(t) * 2B \, \text{sinc}(2Bt).$$

We now will show that

$$\overline{\eta_r^2} \leq \overline{\eta^2} = \overline{\xi^2}. \tag{7.8}$$

That is, filtering reduces or maintains the interpolation noise level since power due to the high frequency components of the noise is eliminated.

Using the appropriate form of (4.82) with $h(t) = 2B \, \text{sinc}(2Bt)$ gives

$$S_{\eta_r}(u) = S_\eta(u) \, \Pi\left(\frac{u}{2B}\right).$$

From (4.72),

$$\overline{\eta_r^2} = \int_{-B}^{B} S_\eta(u) du \tag{7.9}$$

whereas

$$\overline{\eta^2} = \int_{-W}^{W} S_\eta(u) du. \tag{7.10}$$

Since power spectral densities are nonnegative, comparison of (7.9) and (7.10) immediately reveals that the noise level is maintained or reduced as was advertised in (7.8).

We now investigate this reduction more specifically for two types of noise autocorrelations.

(a) Discrete White Noise
Here, we assume that

$$R_\xi\left(\frac{u}{2W}\right) = \overline{\xi^2}\delta[n].$$

Then, from (7.6):

$$R_\eta(t) = \overline{\xi^2}\operatorname{sinc}(2Wt).$$

Thus

$$S_\eta(u) = \frac{\overline{\xi^2}}{2W}\Pi\left(\frac{u}{2W}\right)$$

and, from (7.9)

$$\overline{\eta_r^2} = \frac{\overline{\xi^2}}{2W}\int_{-B}^{B}\Pi\left(\frac{u}{2W}\right)du$$
$$= r\overline{\xi^2}. \tag{7.11}$$

The noise level is reduced by the ratio of the Nyquist to the sampling rate.

(b) Laplace Autocorrelation

If the data noise has a Laplace autocorrelation with parameter λ as in (4.77), then the Fourier transform of (7.6) is

$$S_\eta(u) = \frac{\overline{\xi^2}}{2W}\sum_{n=-\infty}^{\infty} e^{\frac{-\lambda|n|}{2W}} e^{-j\pi nu/W}\Pi\left(\frac{u}{2W}\right)$$

$$= \frac{\overline{\xi^2}}{2W}\sum_{n=-\infty}^{\infty} e^{\frac{-\lambda|n|}{2W}}\cos(\pi nu/W)\Pi\left(\frac{u}{2W}\right)$$

where, in the second step, we have recalled that S_η is real. Continuing,

$$S_\eta(u) = \frac{\overline{\xi^2}}{2W}\left[1 + 2\sum_{n=1}^{\infty} e^{\frac{-\lambda n}{2W}}\cos(\pi nu/W)\right]\Pi\left(\frac{u}{2W}\right)$$

$$= \frac{\overline{\xi^2}}{2W}\left[1 + 2\Re\sum_{n=1}^{\infty} e^{\frac{-n(\lambda+j2\pi u)}{2W}}\right]\Pi\left(\frac{u}{2W}\right). \tag{7.12}$$

Applying the geometric series[1] to (7.12) and simplifying gives the (unfiltered) interpolation noise power spectral density [904]

$$S_\eta(u) = \frac{\overline{\xi^2}}{2W}\frac{\sinh\left(\frac{\lambda}{2W}\right)\Pi\left(\frac{\pi r}{2}\right)}{\cosh\left(\frac{\lambda}{2W}\right) - \cos(\pi u/W)}.$$

The power spectral density for $\eta_r(t)$ is the same, but is only nonzero over the interval $|u| \le B$. The filtered interpolation noise level, from (7.9), follows as

$$\overline{\eta_r^2} = \overline{\xi^2}\sinh\left(\frac{\lambda}{2W}\right)\int_0^r\left[\cosh\left(\frac{\lambda}{2W}\right) - \cos(\pi v)\right]^{-1}dv \tag{7.13}$$

---

1. See (14.6) in the Appendix.

where we have made the variable substitution $\nu = 2uT$ and have recognized the integrand is even. Since [544]

$$\int \frac{d\gamma}{a + b\cos(\gamma)} = \frac{2}{\sqrt{a^2 - b^2}} \arctan\left(\frac{\sqrt{a^2 - b^2}\tan(\gamma/2)}{a + b}\right) \; ; a^2 > b^2$$

equation (7.13) can be evaluated as

$$\overline{\eta_r^2} = \frac{2\overline{\xi^2}}{\pi} \arctan\left(\frac{\sinh\left(\frac{\lambda}{2W}\right)\tan(\frac{\pi r}{2})}{\cosh\left(\frac{\lambda}{2W}\right) - 1}\right). \tag{7.14}$$

Since the principle value of the arctan is strictly less than $\pi/2$, it is clear that the filtered interpolation noise level is less than the data noise level.

Note that for large λ, the Laplace autocorrelation approaches the autocorrelation of discrete white noise. This follows from

$$\lim_{\rho \to \infty} \frac{\sinh(\rho)}{\cosh(\rho) - 1} = 1.$$

The corresponding limiting case of (7.14) is thus the same as that for white noise in (7.11). Note also for $r = 1$ and η replacing η_r, that (7.14) reduces to (7.7).

Plots of $\overline{\eta_r^2}/\overline{\xi^2}$ are shown in Figure 7.1 as a function of r for various values of $\lambda/2W$. The higher the correlation between adjacent noise samples, the higher the filtered interpolation noise level.

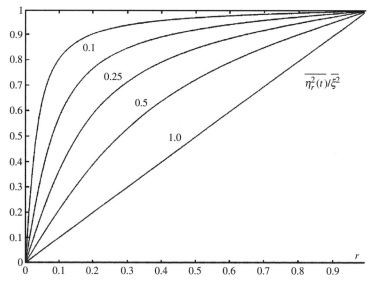

FIGURE 7.1. Plots of interpolation noise variance for additive data for various values of $\lambda/2W$ where λ is the parameter of the Laplace autocorrelation.

7.2.2 Interpolation Noise Variance for Directly Sampled Signals

The results of the previous section can be nicely generalized to the sampling theorems listed in Table 6.1, all of which can be written as

$$x(t) = \sum_{n \in S} x(t_n) k_n(t) \quad (7.15)$$

where S is a set of integers. Suppose the data were corrupted by real additive zero mean stationary noise $\xi(t)$. Then $x(t_n) + \xi(t_n)$ would appear in the summand of (7.15) rather than just $x(t_n)$. The result is clearly $x(t) + \eta(t)$ where the interpolation noise is

$$\eta(t) = \sum_{n \in S} \xi(t_n) k_n(t). \quad (7.16)$$

Our noisy interpolated signal is then

$$z(t) = x(t) + \eta(t). \quad (7.17)$$

Since $\xi(t)$ is zero mean, we have the desirable property that

$$\overline{z(t)} = x(t).$$

The second order statistics of $\eta(t)$ reveal the uncertainty of our estimate. Using (7.16), we have

$$R_\eta(t; \tau) = E[\eta(t)\eta(\tau)]$$
$$= \sum_{n \in S} \sum_{m \in S} R_\xi(t_n - t_m) k_n(t) k_m(\tau). \quad (7.18)$$

The interpolation noise variance follows as

$$\overline{\eta^2(t)} = R_\eta(t; t). \quad (7.19)$$

If there is oversampling, the signal, once interpolated, can be filtered to remove noise components not in the pass band. If the signal is low pass with bandwidth B, the result is

$$z_r(t) = z(t) * 2B \operatorname{sinc}(2Bt)$$
$$= x(t) + \eta_r(t) \quad (7.20)$$

where

$$\eta_r(t) = \eta(t) * 2B \operatorname{sinc}(2Bt).$$

Discrete White Noise. The expressions for the second order statistics simplify significantly if the noise samples $\{\xi(t_n) \mid n \in S\}$ are uncorrelated (white). Then

$$R_\xi(t_n - t_m) = \overline{\xi^2} \delta[n - m], \quad (7.21)$$

and (7.18) reduces to a single sum:

$$R_\eta(t; \tau) = \overline{\xi^2} \sum_{n \in S} k_n(t) k_n(\tau), \qquad (7.22)$$

and the *normalized interpolation noise variance* (NINV) becomes

$$\overline{\eta^2(t)}/\overline{\xi^2} = \sum_{n \in S} k_n^2(t). \qquad (7.23)$$

If the interpolated signal is filtered, the resulting NINV is

$$\overline{\eta_r^2(t)}/\overline{\xi^2} = \sum_{n \in S} \left[k_n^{(r)}(t)\right]^2 \qquad (7.24)$$

where, if the signal is low pass with bandwidth B,

$$k_n^{(r)}(t) = k_n(t) * 2B \operatorname{sinc}(2Bt). \qquad (7.25)$$

The general results of this section will now be applied to some specific cases.

7.2.2.1 Interpolation with Lost Samples

We here consider the NINV resulting from interpolation in the presence of lost samples [906]. We will demonstrate that the NINV increases when (a) the $r < 1$ sampling rate becomes close to that of Nyquist, (b) the number of lost samples increases and/or (c) the lost sample locations are "close" to each other. Analysis will be restricted to additive white noise as in (7.21).

(a) One Lost Sample

For one lost sample at the origin, we use the corresponding interpolation function in (6.12):

$$k_n(t) = \operatorname{sinc}(2Wt - n) + \frac{r}{1-r} \operatorname{sinc}(rn)\operatorname{sinc}(2Wt). \qquad (7.26)$$

Substituting into (7.23) gives

$$\begin{aligned}
\overline{\eta^2(t)}/\overline{\xi^2} &= \sum_{n \neq 0} \left[\operatorname{sinc}(2Wt - n) + \frac{r}{1-r} \operatorname{sinc}(rn)\operatorname{sinc}(2Wt)\right]^2 \\
&= \sum_{n=-\infty}^{\infty} \left[\operatorname{sinc}^2(2Wt - n) + \left(\frac{r}{1-r}\right)^2 \operatorname{sinc}^2(rn)\operatorname{sinc}^2(2Wt)\right. \\
&\quad \left. + \frac{2r}{1-r} \operatorname{sinc}(2Wt - n)\operatorname{sinc}(rn)\operatorname{sinc}(2Wt)\right] \\
&\quad - \frac{1}{(1-r)^2} \operatorname{sinc}^2(2Wt).
\end{aligned} \qquad (7.27)$$

Each of the three infinite sums can be evaluated in closed form. For the first term, we set $\tau = t$ in the cardinal series expansion

$$\operatorname{sinc}(2W(t - \tau)) = \sum_{n=-\infty}^{\infty} \operatorname{sinc}(2W\tau - n) \operatorname{sinc}(2Wt - n).$$

The cardinal series applied to

$$\text{sinc}(2Bt) = \sum_{n=-\infty}^{\infty} \text{sinc}(rn)\,\text{sinc}(2Wt - n) \quad (7.28)$$

lets us evaluate the third sum, and the series in (6.2) can similarly be applied to

$$\text{sinc}(2Bt) = r \sum_{n=-\infty}^{\infty} \text{sinc}(rn)\,\text{sinc}(2Bt - rn). \quad (7.29)$$

Setting $t = 0$ gives the second sum. Alternately, (7.29) is a low passed version of (7.28). Collecting terms and simplifying leaves

$$\overline{\eta^2(t)}/\overline{\xi^2} = 1 + \frac{2r}{1-r}\,\text{sinc}(2Wt)\,\text{sinc}(2Bt)$$

$$- \frac{1}{1-r}\,\text{sinc}^2(2Wt). \quad (7.30)$$

Note that, for large t, the NINV approaches one. This is consistent with (7.7) since, far removed from the origin, the effect of the lost sample is negligible.

The noise at the origin follows from (7.30) as

$$\overline{\eta^2(0)}/\overline{\xi^2} = \frac{r}{1-r}. \quad (7.31)$$

The result is monotonically increasing on $0 < r \leq 1$. Interestingly, for $r < 1/2$, the normalized interpolation noise level in (7.31) is less than one which is less than the noise level of the known sample data. Note, however, that we have yet to filter the high-frequency components of the discrete white noise.

For the filtered case for one lost sample, the interpolation function, from Table 6.1, is

$$k_n^{(r)}(t) = r\,\text{sinc}(2Bt - rn) + \frac{r^2}{1-r}\,\text{sinc}(rn)\,\text{sinc}(2Bt).$$

From (7.23), the corresponding NINV is:

$$\overline{\eta_r^2}/\overline{\xi^2} = r^2 \sum_{n \neq 0}\left[\text{sinc}(2Bt - rn) + \frac{r}{1-r}\,\text{sinc}(rn)\,\text{sinc}(2Bt)\right]^2.$$

Proceeding in a manner similar to that for the unfiltered case above, we obtain

$$\overline{\eta_r^2}/\overline{\xi^2} = \frac{r}{1-r}\left[1 - r\{1 - \text{sinc}^2(2Bt)\}\right]. \quad (7.32)$$

For large t, the noise level goes to the no lost sample filtered equivalent in (7.11). Note, in particular, from (7.31) that

$$\overline{\eta_r^2(0)} = \overline{\eta^2(0)}.$$

Hence, filtering the interpolation does not improve the uncertainty of the restoration of the lost sample. As we would expect from (7.7) and (7.11) respectively,

$$\overline{\eta^2(\pm\infty)} = \overline{\xi^2}$$

and

$$\overline{\eta_r^2(\pm\infty)} = r\overline{\xi^2}.$$

Plots of (7.30) and (7.32) are shown in Figure 7.2.

(b) Two Lost Samples
Let $M = 2$ and let the lost samples be located at the origin and at $x = k/2B$ for some specified positive integer k. The 2×2 matrix, $\mathbf{A}$, discussed in direct interpolation from M lost samples in Section 6.2.1 has elements

$$a_{11} = a_{22} = \frac{1-r}{\Delta}$$

and

$$a_{12} = a_{21} = \frac{r \operatorname{sinc}(rk)}{\Delta}$$

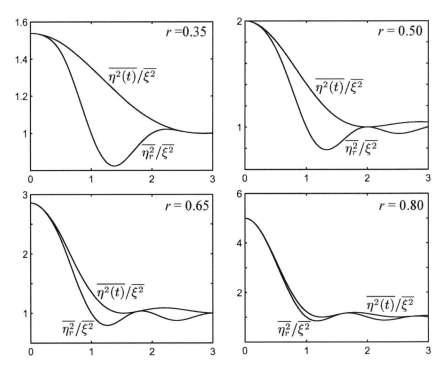

FIGURE 7.2. Plots of the normalized interpolation noise variance (NINV) for the case of a single lost sample vs. $t/2W$ for four different values of r. In all cases, the curve with the largest values are for $\overline{\eta^2/\xi^2}$ in (7.30) and the lower curves are the filtered case, $\overline{\eta_r^2/\xi^2}$, in (7.32). As expected, the NINV increases with r. In all cases, for both the filtered and unfiltered cases, the NINV asymptotically approaches one.

where

$$\Delta = (1-r)^2 - r^2 \operatorname{sinc}^2(rk).$$

The corresponding interpolation functions in (6.20) and (6.22) are substituted into (7.23) and (7.24), respectively. After straightforward yet tedious calculations, we obtain

$$\overline{\eta^2(t)}/\overline{\xi^2} = 1 - (\alpha^2 + \beta^2) + 2r\left[a_{11}(\alpha\tau + \beta\rho) + a_{12}(\alpha\rho + \beta\tau)\right]$$
$$+ r^2\left[(a_{11}^2 + a_{12}^2)(\alpha^2\lambda + 2\alpha\beta\gamma + \beta^2\lambda)\right.$$
$$\left. + 2a_{11}a_{12}(\alpha^2\gamma + 2\alpha\beta\lambda + \beta^2\gamma)\right] \quad (7.33)$$

and

$$\overline{\eta_r^2(t)}/\overline{\xi^2} = r - r^2(\alpha_1^2 + \beta_1^2)$$
$$+ 2r^3[a_{11}(\alpha_1\tau_1 + \beta_1\rho_1) + a_{12}(\alpha_1\rho_1 + \beta_1\tau_1)]$$
$$+ r^4[(a_{11}^2 + a_{12}^2)(\alpha_1^2\lambda + 2\alpha_1\beta_1\gamma + \beta_1^2\lambda)$$
$$+ 2a_{11}a_{12}(\alpha_1^2\gamma + 2\alpha_1\beta_1\lambda + \beta_1^2\gamma)] \quad (7.34)$$

where

$$\alpha = \operatorname{sinc}(2Wt - k), \quad \alpha_1 = \operatorname{sinc}(2Bt - rk)$$
$$\beta = \operatorname{sinc}(2Wt), \quad \beta_1 = \operatorname{sinc}(2Bt)$$

and

$$\rho = \sum_{p \neq 0,k} \operatorname{sinc}(rp)\operatorname{sinc}(2Wt - p)$$
$$= \beta_1 - \beta - \alpha \operatorname{sinc}(rk),$$
$$\rho_1 = \sum_{p \neq 0,k} \operatorname{sinc}(rp)\operatorname{sinc}(2Bt - rp)$$
$$= \left(\frac{1-r}{r}\right)\beta_1 - \alpha_1 \operatorname{sinc}(rk),$$
$$\tau = \sum_{p \neq 0,k} \operatorname{sinc}(r(p-k))\operatorname{sinc}(2Wt - p)$$
$$= \alpha_1 - \alpha - \beta \operatorname{sinc}(rk),$$
$$\tau_1 = \sum_{p \neq 0,k} \operatorname{sinc}(r(p-k))\operatorname{sinc}(2Bt - rp)$$
$$= \left(\frac{1-r}{r}\right)\alpha_1 - \beta_1 \operatorname{sinc}(rk),$$

$$\lambda = \sum_{p \neq 0, k} \operatorname{sinc}^2(rp)$$

$$= \sum_{p \neq 0, k} \operatorname{sinc}^2(r(p-k))$$

$$= \frac{1}{r} - 1 - \operatorname{sinc}^2(rk),$$

$$\gamma = \sum_{p \neq 0, k} \operatorname{sinc}(rp)\operatorname{sinc}(r(p-k))$$

$$= \left(\frac{1}{r} - 2\right) \operatorname{sinc}(rk).$$

Numerical examples of the NINV's in (7.33) and (7.34) are shown in Figures 7.3 and 7.4 for $k = 1$ and various r. The samples are lost at times $t = 0$ and 1. The asymptotic value of the unfiltered restoration approaches one and that of the filtered case, r. As the sampling rate becomes closer to the Nyquist rate, the problem of restoration of the two lost samples becomes more ill-conditioned.[2]

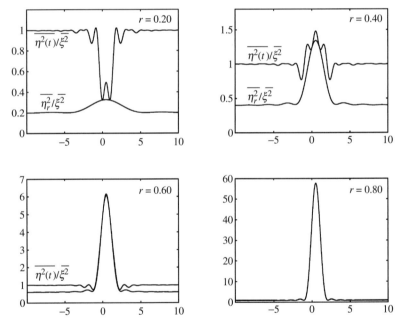

FIGURE 7.3. NINV in (7.33) and (7.34) for two adjacent lost samples (at $t = 0$ and $t = 1$) corresponding to, on the top row, $r = 0.2$ and 0.4 and, on the bottom row, $r = 0.6$ and 0.8. In all cases, the unfiltered NINV, $\overline{\eta^2(t)}/\overline{\xi^2}$ is the top curve, i.e., $\overline{\eta^2(t)}/\overline{\xi^2} \geq \overline{\eta_r^2(t)}/\overline{\xi^2}$ where $\overline{\eta_r^2(t)}/\overline{\xi^2}$ is the filtered NINV. The unfiltered NINV always asymptotically approaches one whereas the filtered NINV approaches r. For the $r = 0.8$ curve (bottom right), the filtered and unfiltered NINV's are graphically indistinguishable. The $r = 0.8$ curve is shown using a logarithm plot in the upper left of Figure 7.4.

---

2. See Appendix 14.5.

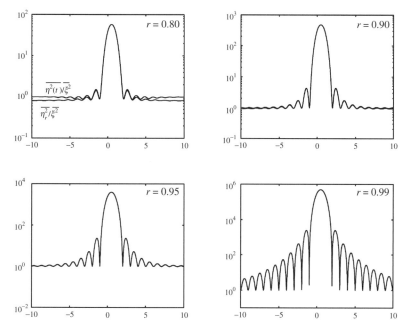

FIGURE 7.4. NINV in (7.33) and (7.34) for two adjacent lost samples (at $t = 0$ and $t = 1$) corresponding to, in the top row, $r = 0.8$ and 0.9 and, in the bottom row, $r = 0.95$ and 0.99. For all but the $r = 0.8$ case, the curves are graphically indistinguishable. The unfiltered NINV, $\overline{\eta^2(t)/\xi^2}$ always exceeds the unfiltered case, i.e., $\overline{\eta^2(t)/\xi^2} \geq \overline{\eta_r^2(t)/\xi^2}$ where $\overline{\eta_r^2(t)/\xi^2}$ is the filtered NINV. Since the NINV becomes larger as the sampling rate becomes closer to the Nyquist rate, whereat it becomes infinite, the restoration of the two lost samples becomes more ill-conditioned. As we move farther from the lost sample, $\overline{\eta^2(t)/\xi^2} \to 1$ and $\overline{\eta_r^2(t)/\xi^2} \to r$

For $r = 0.5$, the NINV in (7.33) and (7.34) is shown in Figure 7.5 for various k. For separation by $k = 1$ sample (upper left), the NINV for both the filtered and unfiltered cases is higher. For $k = 2, 3, 4$, the NINV's are commensurate. One might expect that, for large k, the NINV's would become the same as that as for a single lost sample. In essence, each point is sufficiently separated from the other so neither knows the other is there. This is illustrated in Figure 7.6 where NINV's for two lost samples are shown for $k = 4, 8$. The shapes of the curves are nearly identical.

The NINV at the filtered and unfiltered lost sample locations are identical. Examining the filtered and unfiltered NINV at the lost sample point locations yields

$$\overline{\eta^2(0)} = \overline{\eta^2\left(\frac{k}{2B}\right)} = \overline{\eta_r^2(0)} = \overline{\eta_r^2\left(\frac{k}{2B}\right)}$$
$$= r^2 \overline{\xi^2} \left[\left(a_{11}^2 + a_{12}^2\right) \lambda + 2 a_{11} a_{12} \gamma \right]. \tag{7.35}$$

As the separation, k, between the two lost samples increases, the NINV approaches that for a single lost sample. From (7.31), we have

$$\lim_{k \to \infty} \overline{\eta^2(0)} = \lim_{k \to \infty} \overline{\eta^2\left(\frac{k}{2B}\right)} = \lim_{k \to \infty} \overline{\eta_r^2(0)} = \lim_{k \to \infty} \overline{\eta_r^2\left(\frac{k}{2B}\right)} = \frac{r\overline{\xi^2}}{1-r} \tag{7.36}$$

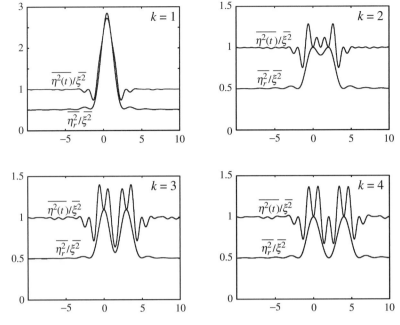

FIGURE 7.5. NINV in (7.33) and (7.34) for $r = 0.5$ and various k. The lost samples are at $(0, k)$. The unfiltered NINV, $\overline{\eta^2(t)}/\overline{\xi^2}$ is the top curve in all four plots. As we move farther from the lost sample, $\overline{\eta^2(t)}/\overline{\xi^2} \to 1$ and $\overline{\eta_r^2(t)}/\overline{\xi^2} \to r$.

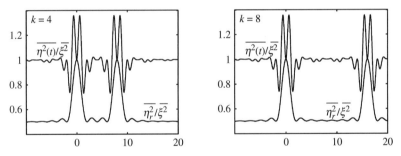

FIGURE 7.6. NINV in (7.33) and (7.34) for $r = 0.5$ and $k = 4$ and $k = 8$. The locations of the lost samples are sufficiently separated so that the NINV's around the lost samples are nearly the same. The unfiltered NINV, $\overline{\eta^2(t)}/\overline{\xi^2}$ is the top curve in both plots. As we move farther from the lost sample, $\overline{\eta^2(t)}/\overline{\xi^2} \to 1$ and $\overline{\eta_r^2(t)}/\overline{\xi^2} \to r$.

(c) A Sequence of Lost Samples

For an even greater number of lost samples, the obtaining of a closed form solution for the NINV using the previous methods becomes nearly intractable. Evaluation of the infinite series numerically becomes more attractive. Alternately, a concise matrix approach to the problem developed by Tseng [1414] can be used. We will not, however, review it here.

Numerically evaluated plots of the filtered NINV for three samples in a row are shown in Figure 7.8 for $r = 0.5$ and 0.8. The NINV at the lost sample locations is shown in Figure 7.9

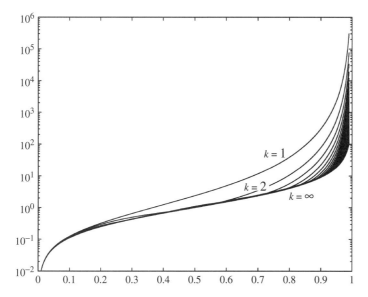

FIGURE 7.7. Log plot of the NINV for two lost samples at the sample point locations in (7.35). The top curve corresponds to lost sample separation $k = 1$. The next curve is for $k = 2$, etc. As $k \to \infty$, the curves, as predicted in (7.36), approach $r/(1 - r)$.

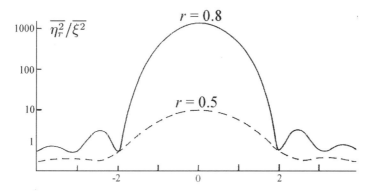

FIGURE 7.8. Filtered NINV for three lost samples at 0, $+1$, and -1 as a function of $2Wt$.

for M lost samples in a row. The noise level increases drastically with respect to the number of adjacent lost samples and sampling rate parameter. Correspondingly, the condition number of the $\mathbf{A} = [\mathbf{I} - \mathbf{S}]^{-1}$ matrix increases greatly with larger M and r.

7.2.2.2 Bandpass Functions

The NINV for bandpass function interpolation follows from Table 6.1 and (7.23) as

$$\overline{\eta^2(t)}/\overline{\xi^2} = \sum_{n=-\infty}^{\infty} \operatorname{sinc}^2\left(Bt - \frac{n}{2}\right) \cos^2\left(\pi(2N + 1)\left(Bt - \frac{n}{2}\right)\right). \qquad (7.37)$$

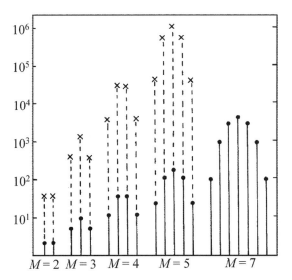

FIGURE 7.9. NINV of M lost samples in a row at the lost sample locations. The lower dot values in each case correspond to $r = 0.5$ and the upper ×'s to $r = 0.8$.

Expanding the bandpass function

$$x(t) = \operatorname{sinc} B(t - \tau) \cos(\pi(2N+1)B(t-\tau))$$

in the bandpass sampling theorem in (6.84) and (6.85) gives

$$\operatorname{sinc} B(t-\tau) \cos(\pi(2N+1)B(t-\tau))$$
$$= \sum_{n=-\infty}^{\infty} \operatorname{sinc}\left(B\tau - \frac{n}{2}\right) \cos\left(\pi(2N+1)\left(B\tau - \frac{n}{2}\right)\right)$$
$$\times \operatorname{sinc}\left(Bt - \frac{n}{2}\right) \cos\left(\pi(2N+1)\left(Bt - \frac{n}{2}\right)\right).$$

Evaluating this expression at $\tau = t$ reduces (7.37) to

$$\overline{\eta^2(t)}/\xi^2 = 1.$$

Therefore, as in the cardinal series, the interpolation noise variance is the same as the variance of the data noise.

7.2.3 On Papoulis' Generalization

Papoulis' generalization of the sampling theorem was presented in Section 6.3. Here, we will explore the effects of additive white noise superimposed on the sample data. We will expose the *ill-posedness* of a number of innocent appearing sampling theorems [285]. Interpolation is here defined to be ill-posed if the NINV cannot be bounded. Clearly, such sampling theorems should be avoided.

An example of an ill-posed sampling theorem is a special case of signal-derivative sampling. Shannon was the first to note that one could sample at half the Nyquist rate if at each sample location two samples were taken: one of the signal and one of the signal's derivative. Consider the seemingly innocent alteration of sampling at the Nyquist rate with interlaced signal and first derivative samples taken at each Nyquist interval. As we will demonstrate, restoration here is ill-posed. Indeed, subjecting the samples to sample-wise white noise renders the restoration unstable. Hence, one would wish to sample an odometer and speedometer simultaneously, rather than sequentially, to determine position.

Let $\{\xi_p(nT_N) \mid p = 1, 2, \ldots, N; -\infty < n < \infty\}$ denote a zero mean discrete stochastic noise sequence. If $g_p(nT_N) + \xi_p(nT_N)$ is used in (6.47) instead of $g_p(nT_N)$, the output is $f(t) + \eta(t)$ where

$$\eta(t) = \sum_{p=1}^{N} \sum_{n=-\infty}^{\infty} \xi_p(nT_N) k_p(t - nT_N). \tag{7.38}$$

We will assume that the discrete noise is stationary and white:

$$E[\xi_k(nT_N)\xi_p(mT_N)] = \overline{\xi_p^2}\, \delta[k-p]\, \delta[n-m]$$

where $\overline{\xi_p^2} = E[|\xi_p(nT_N)|^2]$ is the data noise variance of the pth sampled signal. The interpolation noise level then follows as

$$\overline{\eta^2(t)} = \sum_{p=1}^{N} \overline{\xi_p^2} \sum_{n=-\infty}^{\infty} |k_p(t - nT_N)|^2.$$

Clearly, $\overline{\eta^2(t)}$ is periodic with period T_N. Application of the Poisson sum formula yields

$$\overline{\eta^2(t)} = \frac{1}{T_N} \sum_{p=1}^{N} \overline{\xi_p^2} \sum_{n=-\infty}^{\infty} W_p(2nB_N)\, e^{j4\pi nB_N t} \tag{7.39}$$

where

$$|k_p(t)|^2 \longleftrightarrow W_p(u) = \int_{-B}^{B} K_p(\beta) K_p^*(\beta - u) d\beta.$$

Note that (7.39) is simply a Fourier series with coefficients

$$C_n = \frac{1}{T_N} \sum_{p=1}^{N} \overline{\xi_p^2}\, W_p(2nB_N). \tag{7.40}$$

We accordingly define the average interpolation noise variance by

$$C_0 = \frac{1}{T_N} \sum_{p=1}^{N} \overline{\xi_p^2}\, W_p(0)$$

$$= \frac{1}{T_N} \sum_{p=1}^{N} \overline{\xi_p^2} \int_{-B}^{B} |K_p(u)|^2\, du \tag{7.41}$$

or, using Parseval's theorem,

$$C_0 = \frac{1}{T_N} \sum_{p=1}^{N} \overline{\xi_p^2} \int_{-\infty}^{\infty} |k_p(t)|^2\, dt. \tag{7.42}$$

Thus the average interpolation noise variance is infinite if any one of the N interpolation functions has unbounded energy. Equivalently, if, for any $p = 1, 2, \ldots, N$,

$$\int_{-\infty}^{\infty} |k_p(t)|^2 \, dt = \int_{-B}^{B} |K_p(u)|^2 \, du = \infty \qquad (7.43)$$

then the restoration is ill-posed.

7.2.3.1 Examples

1. Derivative Sampling. Consider the $N = 1$ case corresponding to Mth-order derivative sampling

$$H_1(u) = (j2\pi u)^M$$

We can, in principle, regain all frequency components other than zero. Note, however, that (7.42) becomes

$$C_0 = \frac{\overline{\xi_1^2}}{(2\pi)^{2M} T_N} \int_{-B}^{B} u^{-2M} \, du$$

$$= \infty.$$

The corresponding sampling theorem is thus ill-posed.

2. Interlaced Signal-Derivative Sampling. A less obvious ill-posed sampling theorem occurs when we nonuniformly interlace Mth order derivative samples with signal samples. The sampling theorem for this problem was addressed in Section 4.2.3.2. The spectra of the interpolation functions $K_1(u)$ and $K_2(u)$ have real poles when $\Delta(u) = 0$ or $\Delta(u + B) = 0$ on the intervals $(0, B)$ and $(-B, 0)$ respectively. The former occurs when

$$u^M e^{-j2\pi(\alpha B + n)} = (u - B)^M; \, 0 \leq n < M$$

or

$$u = \frac{B}{2} \left[1 - j \cot\left(\frac{\pi(\alpha B + n)}{M} \right) \right].$$

One of these roots is real when (a) $\alpha = 0$ and M is even, or (b) $\alpha = \frac{1}{2B}$ and M is odd (corresponding to $n = M/2$ and $n = (M - 1)/2$, respectively.) In either case, the real pole generated by $\Delta(u)$ is at $B/2$ and that generated by $\Delta(u + B)$ is at $-B/2$. Clearly, application of (7.41) exposes these sampling theorems as ill-posed. Plots of the spectra of the interpolation functions are shown in Figure 7.10 for $M = 1$ for various values of α. Figure 7.11 illustrates the same process for $M = 2$.

7.2.3.2 Notes

1. Sample Contributions in the Ill-Posed Sampling Theorems. Insight into the ill-posedness of the sampling theorems can be gained by inspection of the interpolation functions. Consider, for example, $N = 1$ derivative sampling with $M = 1$. It follows that

$$k_1(t) = \frac{1}{4\pi B} \int_{-B}^{B} \frac{e^{j2\pi ut}}{ju} \, du$$

$$= \frac{1}{2\pi B} \text{Si}(2\pi B t)$$

NOISE AND ERROR EFFECTS   305

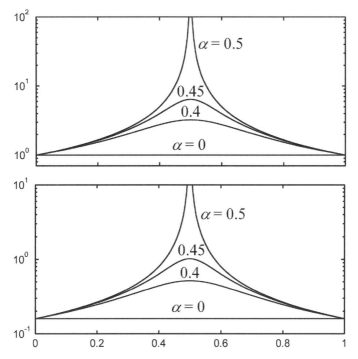

FIGURE 7.10. Illustration of the manner in which the interpolation functions in (6.56) and (6.57) approach poles for $M = 1$ as $\alpha \to 1/2B$. On top is a plot of $|K_1(u)|$ for different values of α. $|K_2(u)|$ is shown on the bottom. The bandwidth, B, is normalized to one.

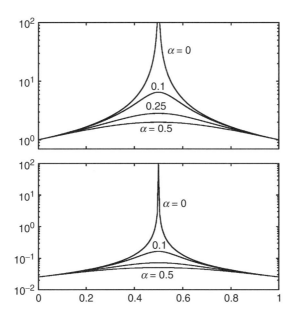

FIGURE 7.11. Same as Figure 7.10, except $M = 2$.

where Si(·) is the sine integral.[3] Since Si($\pm\infty$) = $\pm\pi/2$, interpolation at any point is affected significantly by every sample value, no matter how distant.

A similar contribution occurs for the ill-posed cases of interlaced signal-derivative sampling. We can invert (6.56) and (6.57). For $M = 2$ and $\alpha = 0$, the results are

$$k_1(t) = \frac{1}{2}\left(\sin(\pi Bt)\text{Si}(\pi Bt) + \cos(\pi Bt)\text{sinc}(Bt) + \text{sinc}^2(Bt)\right)$$

and

$$k_2(t) = \frac{1}{2}\frac{\sin(\pi Bt)\text{Si}(\pi Bt)}{(\pi B)^2}.$$

Again, the occurrence of the sine integrals makes possible equally significant contributions from all sample values, no matter how far removed from the point of interpolation. The weighted noise levels from each sample value thus add to a random variable with unbounded variance.

2. Effects of Oversampling. If we sample at a rate $2W > 2B$ and do not take advantage of the oversampling, the average interpolation noise level in (7.41) becomes

$$C_0 = \frac{\overline{\xi^2}}{T_N}\sum_{p=1}^{N}\int_{-W}^{W}|K_p(u)|^2\,du$$

where, now, $T_N = N/2W$. If, however, the filtered interpolation formula in (6.65) is used, then one can easily show that the average noise variance reduces to

$$C_0^{(r)} = \frac{\overline{\xi^2}}{T_N}\sum_{p=1}^{N}\int_{-B}^{B}|K_p(u)|^2\,du. \qquad (7.44)$$

Clearly

$$C_0^{(r)} \leq C_0.$$

Thus, as we have seen before, oversampling can reduce interpolation noise variance by allowing suppression of high frequency noise components.

As an example, consider the ill-posed interlaced signal derivative sampling theorem. If we sample at a rate greater than twice the Nyquist rate, the integral in (7.44) will not include the poles at $u = \pm W/2$ and the resulting sampling theorem becomes well-posed. At exactly twice the Nyquist rate, the integration limits in (7.44) are at the pole locations. Thus $C_0^{(r)} = \infty$. We can, however, discard the derivative samples and use the conventional (well-posed) sampling theorem to restore the signal. Thus we are confronted with the curious task of discarding the derivative samples to improve the interpolation noise level.

7.2.4 On Derivative Interpolation

Hamming [572] notes that "the estimation of derivatives from computed or tabulated values is dangerous." This is largely due to data uncertainty. We will show especially that, even

---
3. Defined in (2.77).

when the noise is bandlimited, NINV's for high order differentiation interpolation can be significantly high [900].

Recall from Section 6.4 on derivative interpolation that derivatives of bandlimited signals can be computed via

$$x^{(p)}(t) = (2W)^p \sum_{n=-\infty}^{\infty} x\left(\frac{n}{2W}\right) d_p(2Wt - n) \tag{7.45}$$

where the derivative kernel is

$$d_p(t) = \left(\frac{d}{dt}\right)^p \operatorname{sinc}(t).$$

Since $x^{(p)}(t)$ is bandlimited, it is unaltered by low-pass filtering. Passing (7.45) through a filter unity on $|u| < B$ and zero elsewhere gives

$$x^{(p)}(t) = r(2B)^p \sum_{n=-\infty}^{\infty} x\left(\frac{n}{2W}\right) d_p(2Bt - rn). \tag{7.46}$$

If the noise samples, $\xi(\frac{n}{2B})$, are added to the signal samples in (7.46), the result is $x^{(p)}(t) + \eta^{(p)}(t)$ where

$$\eta^{(p)}(t) = r(2B)^p \sum_{n=-\infty}^{\infty} \xi\left(\frac{n}{2W}\right) d_p(2Bt - rn).$$

Thus

$$R_{\eta_p}(t - \tau) = E\left[\eta^{(p)}(t)\eta^{(p)}(\tau)\right]$$

$$= r^2(2B)^{2p} \sum_{m=-\infty}^{\infty} R_\xi\left(\frac{m}{2W}\right) \sum_{n=-\infty}^{\infty} d_p(2B\tau - (n-m)r) \, d_p(2Bt - rn).$$

Since $x(t) = d_p(2B(\tau - t) + mr)$ is bandlimited, we can use (7.46) to evaluate the n sum above. Furthermore, since

$$d_p(t) = (-1)^p d_p(-t)$$

and

$$\left(\frac{d}{dt}\right)^p d_p(t) = d_{2p}(t)$$

it follows that

$$R_{\eta_p}(\tau) = (-1)^p r(2B)^{2p} \sum_{m=-\infty}^{\infty} R_\xi\left(\frac{m}{2W}\right) d_{2p}(2B\tau - rm). \tag{7.47}$$

Since $\eta_p(t)$ is zero mean, the interpolation noise is wide sense stationary. The corresponding interpolation noise variance is

$$\overline{\eta_p^2} = R_{\eta_p}(0)$$

$$= (-1)^p r(2B)^{2p} \sum_{m=-\infty}^{\infty} R_\xi\left(\frac{m}{2W}\right) d_{2p}(rm). \tag{7.48}$$

A spectral density description of the process can be obtained by first transforming (7.47).

$$R_{\eta_p}(t) \longleftrightarrow S_{\eta_p}(u)$$

$$= \frac{(2\pi u)^{2p}}{2W} \sum_{m=-\infty}^{\infty} R_{\xi}\left(\frac{m}{2W}\right) e^{-j\pi m u/W} \Pi\left(\frac{u}{2B}\right).$$

Application of the Poisson sum formula to the m sum gives

$$S_{\eta_p}(u) = (2\pi u)^{2p} \sum_{n=-\infty}^{\infty} S_{\xi}(u - 2nW) \Pi\left(\frac{u}{2B}\right)$$

where $R_{\xi}(t) \longleftrightarrow S_{\xi}(u)$ and $S_{\xi}(u)$ is the input noise power spectral density.

An alternate expression for the output noise level follows as

$$\overline{\eta_p^2} = \int_{-\infty}^{\infty} S_{\eta_p}(u) du$$

$$= (2\pi)^{2p} \sum_{n=-\infty}^{\infty} \int_{-B}^{B} u^{2p} S_{\xi}(u - 2nW) \, du. \qquad (7.49)$$

For the unfiltered case ($r = 1$), the integration interval in (7.49) is over $|u| \leq W$. Since $S_{\xi}(u) \geq 0$, filtering always results in a noise level equal to or better than the unfiltered case.

7.2.5 A Lower Bound on the NINV

Consider $\xi(t)$ input into the cascaded low-pass filter and p^{th}-order differentiator in Figure 7.12. Let $\theta_p(t)$ denote the output. Recall that, in general, the output spectral density $S_{\theta}(u)$ due to a spectral density input $S_i(u)$ into a system with transfer function $H(u)$ is:

$$S_{\theta}(u) = |H(u)|^2 S_i(u).$$

Thus

$$S_{\theta}(u) = (2\pi u)^{2p} S_{\xi}(u) \Pi\left(\frac{u}{2B}\right)$$

and

$$\overline{\theta_p^2} = (2\pi)^{2p} \int_{-B}^{B} u^{2p} S_{\xi}(u) \, du. \qquad (7.50)$$

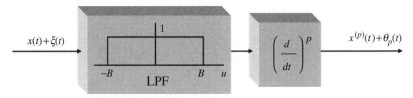

FIGURE 7.12. Cascaded low-pass filter and p^{th} order differentiator. The output noise level, $\overline{\theta_p^2}$, is a lower bound for p^{th} order derivative interpolation from the input samples.

Compare this to (7.49). Since $S_\xi(u) \geq 0$, it follows that $\overline{\theta_p^2}$ is a lower bound for the output noise level

$$\overline{\eta_p^2} \geq \overline{\theta_p^2}.$$

Equality is achieved when $\xi(t)$ has band-limited spectral density (say over the interval $|u| \leq \Omega$) and the sampling rate is sufficiently high to avoid aliasing (i.e., $2W - \Omega > B$). A lower sampling rate would result in aliasing and a higher output noise level.

For finite $\overline{\xi^2}$, $S_\xi(u) \to 0$ as $|u| \to \infty$. We see from (7.49) that $\overline{\eta_p^2} \to \overline{\theta_p^2}$ as $2W \to \infty$. Hence, the bound can be approached arbitrarily closely by an appropriate increase in sampling rate. Note that we can guarantee from (7.49) that $\overline{\eta_p^2}$ strictly decreases with r if $S_\xi(u)$ strictly decreases with $u > 0$. This spectral density property is applicable to a Laplace autocorrelation. It is, however, not applicable to triangular autocorrelation.

7.2.5.1 Examples

(a) Triangular Autocorrelation. Consider the triangle autocorrelation parameterized by $a > 0$.

$$R_\xi(\tau) = \overline{\xi^2} \Lambda\left(\frac{\tau}{a}\right). \tag{7.51}$$

Substituting into (7.48) gives the normalized error.

$$\frac{\overline{\eta_p^2}}{\overline{\xi^2}} = (-1)^p r(2B)^{2p} \left[d_{2p}(0) + 2\sum_{m=1}^{N}\left(1 - \frac{m}{T}\right) d_{2p}(rm) \right] \tag{7.52}$$

where

$$T = 2Wa$$

and $N = \lfloor T \rfloor$. Plots of (7.52) are shown in Figure 7.13 for $2B = 1$ and $a = 0.5$ and 0.1.

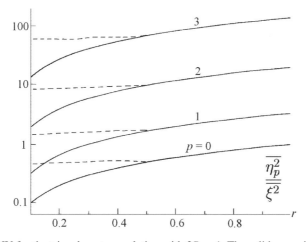

FIGURE 7.13. NINV for the triangle autocorrelation with $2B = 1$. The solid curve is for $a = 0.1$ and the dashed curve for $a = 0.5$. The curves are identical for $r > 1/2$ where the noise samples are white. For $a = 0.1$, the noise samples are white for $r > 0.1$.

Of specific interest is the case where sampling is performed such that $T < 1$. The noise samples are then white. That is,

$$R_\xi\left(\frac{n}{2W}\right) = \overline{\xi^2}\,\delta[n]. \tag{7.53}$$

Since

$$\begin{aligned}d_{2p}(0) &= \int_{-\frac{1}{2}}^{\frac{1}{2}} (j2\pi u)^{2p}\,du \\ &= \frac{(-1)^p \pi^{2p}}{2p+1},\end{aligned} \tag{7.54}$$

we have for white samples

$$\overline{\eta_p^2}/\overline{\xi^2} = \frac{r(2\pi B)^{2p}}{2p+1}. \tag{7.55}$$

For the conventional sampling theorem, $p = 0$ and the noise level is improved by a factor of r. This result is also in (7.11).

Since $2B = 1$ in Figure 7.13, the plots there are equivalent to (7.55) for $r > a$. Note that (7.55) is independent of the a parameter-thus the merging of the $a = 0.1$ and 0.5 plots at $r = 0.5$. For the domain shown, all of the $a = 0.1$ samples are white.

Note also that (a) the NINV increases dramatically with the order of differentiation and (b) there can exist a point whereupon a further increase of sampling rate results in an insignificant improvement in the interpolation noise level.

Since $\Lambda(t) \longleftrightarrow \text{sinc}^2 u$, from (7.50), the normalized lower bound for the triangle autocorrelation is

$$\overline{\theta_p^2}/\overline{\xi^2} = 2a(2\pi)^{2p} \int_0^B u^{2p}\,\text{sinc}^2(au)\,du. \tag{7.56}$$

To place this in more palatable form, we rewrite it as

$$\overline{\theta_p^2}/\overline{\xi^2} = \frac{(2\pi)^{2p}}{\pi^2 a} \int_0^B u^{2p-2}[1 - \cos(2\pi au)]\,du. \tag{7.57}$$

For even index

$$\begin{aligned}d_{2q}(t) &= \int_{-\frac{1}{2}}^{\frac{1}{2}} (j2\pi v)^{2q} e^{-j2\pi vt}\,dv \\ &= 2(2\pi)^{2q}(-1)^q \int_0^{\frac{1}{2}} v^{2q}\cos(2\pi vt)\,dv.\end{aligned}$$

For $u = 2Wv$, it follows from (7.57) that for $p > 1$

$$\overline{\theta_p^2}/\overline{\xi^2} = \frac{2(2B)^{2p}(-1)^p}{a}[d_{2p-2}(0) - d_{2p-2}(2aB)].$$

The $p = 0$ case follows immediately from (7.56) using integration by parts. Thus, using (7.54),

$$\overline{\theta_p^2}/\overline{\xi^2} = \begin{cases} \dfrac{4^p B^{2p-1}}{a}[\dfrac{\pi^{2p-2}}{2p-1} + (-1)^p d_{2p-2}(2aB)] \; ; \; p > 0 \\ \dfrac{2}{\pi}[\text{Si}(2\pi aB) - \sin(\pi aB)\,\text{sinc}(aB)] \quad ; \; p = 0. \end{cases}$$

Lower bounds for each of the plots in Figure 7.13 are graphically indistinguishable from the $r = 0.1$ values.

(b) Laplace Autocorrelation. A second tractable solution results from the Laplace autocorrelation parameterized by λ.

$$R_\xi(\tau) = \overline{\xi^2}\, e^{-\lambda|\tau|}. \tag{7.58}$$

is considered here. Rewriting (7.48) as

$$\overline{\eta_p^2} = \frac{(2\pi)^{2p}}{2W} \int_{-B}^{B} u^{2p} \sum_{n=-\infty}^{\infty} R_\xi\left(\frac{n}{2W}\right) e^{-j\pi nu/W}\, du,$$

we can show in a manner similar similar to that in Section 7.2.1 that the NINV can be written as

$$\overline{\eta_p^2}/\overline{\xi^2} = (2\pi W)^{2p} \sinh\left(\frac{\lambda}{2W}\right) \int_0^r \frac{u^{2p}\, du}{\cosh(\frac{\lambda}{2W}) - \cos(\pi u)}. \tag{7.59}$$

The well-behaved (strictly increasing) integrand in (7.59) provides for straightforward digital integration. Sample plots of (7.59) are shown in Figure 7.14 for $2B = 1$. As with the previous example, the NINV increases significantly with p.

Two special cases of (7.59) are worthy of note.

a) The $p = 0$ case as plotted in Figure 7.13 simply corresponds to conventional sampling theorem interpolation followed by filtering. For this case, (7.59) becomes the integral in (7.13) whose solution is (7.14).

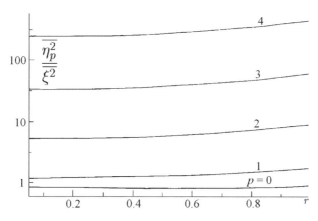

FIGURE 7.14. NINV for Laplace autocorrelation with $\lambda = 2B = 1$.

b) If

$$\frac{\lambda}{2W} \gg 1 \qquad (7.60)$$

then (7.59) approaches

$$\overline{\eta_p^2/\xi^2} = (2\pi W)^{2p} \int_0^r u^{2p} du$$

$$= \frac{r(2\pi B)^{2p}}{2p+1} \qquad (7.61)$$

which is the same result as the discrete white noise case in (7.54). Indeed, when (7.60) is applicable, the noise samples are very nearly white and the plots for white samples in Figure 7.13 can be used as excellent approximations.

Here, as in the previous example, one must be cautioned on comparing equally parameterized interpolation noise levels for differing p. If we are dealing with temporal functions, the units of $\overline{\eta_p^2/\xi^2}$ are (seconds)$^{-2p}$.

For a lower bound for the Laplace autocorrelation, we transform (7.58) and substitute into (7.50). The result is

$$\overline{\theta_p^2/\xi^2} = \frac{4(2\pi)^{2p}}{\lambda} \int_0^B \frac{v^{2p} dv}{1+(\frac{2\pi v}{\lambda})^2}. \qquad (7.62)$$

Setting $u = 2\pi v/\lambda$ gives

$$\overline{\theta_p^2/\xi^2} = \frac{2}{\pi} \lambda^{2p} \int_0^\varepsilon \frac{u^{2p} du}{1+u^2} \qquad (7.63)$$

where $\varepsilon = 2\pi B/\lambda$. The $p = 0$ case follows immediately. For $p > 0$, consider first the case where $\varepsilon < 1$. With $z = -u^2$, the denominator in (7.63) can be expanded via the geometric series in (14.6).

$$\sum_{m=0}^\infty z^m = (1-z)^{-1}; \ |z| < 1. \qquad (7.64)$$

The resulting integral is evaluated to give

$$\overline{\theta_p^2/\xi^2} = \frac{2}{\pi} \lambda^{2p} \sum_{m=0}^\infty (-1)^m \frac{\varepsilon^{2m+2p+1}}{2m+2p+1}.$$

Set $n = m + p$ and recall the Taylor series

$$\arctan z = \sum_{n=0}^\infty \frac{(-1)^n z^{2n+1}}{2n+1}. \qquad (7.65)$$

Thus, for $\varepsilon < 1$,

$$\overline{\theta_p^2/\xi^2} = \frac{2}{\pi} (-1)^p \lambda^{2p} \left(\arctan \varepsilon - \sum_{n=0}^{p-1} (-1)^n \frac{\varepsilon^{2n+1}}{2n+1} \right).$$

For $\varepsilon > 1$ we rewrite (7.63) as

$$\overline{\theta_p^2}/\overline{\xi^2} = \frac{2}{\pi}\lambda^{2p}\left[\int_0^{1-} + \int_{1+}^{\varepsilon}\right]\frac{u^{2p}\,du}{1+u^2}$$

$$= \frac{2}{\pi}\lambda^{2p}\left[(-1)^p\left(\frac{\pi}{4} - \sum_{n=0}^{p-1}\frac{(-1)^n}{2n+1}\right) + \int_{1+}^{\varepsilon}\frac{u^{2p}\,du}{1+u^2}\right].$$

From (7.64), it follows that

$$-\sum_{n=0}^{\infty} z^{-n} = \frac{z}{1-z}; \quad |z| > 1.$$

Again, with $z = -u^2$, we obtain

$$\overline{\theta_p^2}/\overline{\xi^2} = \frac{2}{\pi}\lambda^{2p}\left[(-1)^p\left(\frac{\pi}{4} - \sum_{n=0}^{p-1}\frac{(-1)^n}{2n+1}\right) + \sum_{m=0}^{\infty}(-1)^m \frac{\varepsilon^{2p-2m-1}-1}{2p-2m-1}\right].$$

Set $n = m - p$ in the m sum and use (7.65). Recognizing $\pi/2 - \arctan(1/\varepsilon) = \arctan\varepsilon$ for $\varepsilon > 0$ again yields (7.66). Placing these results in recursive form gives

$$\frac{\overline{\theta_p^2}}{\overline{\xi^2}} = \begin{cases} \frac{2}{\pi}\arctan\varepsilon & ; p = 0 \\ \frac{2\lambda^2}{\pi}[\varepsilon - \arctan\varepsilon] & ; p = 1 \\ \frac{2\lambda(2\pi B)^{2p-1}}{\pi(2p-1)} - \lambda^2 \frac{\overline{\theta_{p-1}^2}}{\overline{\xi^2}} & ; p > 1 \end{cases} \quad (7.66)$$

where

$$\varepsilon = \frac{2\pi B}{\lambda}.$$

Graphically, these bounds are also indistinguishable from the corresponding smallest values on the plots in Figure 7.14.

7.3 Jitter

Jitter occurs when samples are taken near to but not exactly at the desired sample locations. We here consider only the case of direct uniform sampling [1079]. Instead of the sample set $x(n/2W)$, we have the sample set

$$\left\{x\left(\frac{n}{2W} - \sigma_n\right)\Big| -\infty < n < \infty\right\}$$

where σ_n is the jitter offset of the nth sample. We assume that the jitter offsets are unknown. If they are known, then sampling theorems for irregularly spaced samples can be used [951].

In this section, we will show that cardinal series interpolation of jittered samples yields a biased estimate of the original signal. Although the bias can be corrected with an inverse filter, the resulting interpolation noise variance is increased.

7.4 Filtered Cardinal Series Interpolation

We know that

$$x(t) = r \sum_{n=-\infty}^{\infty} x\left(\frac{n}{2W}\right) \operatorname{sinc}(2Bt - rn)$$

and thus may be motivated to estimate $x(t)$ directly from the jittered data as

$$y(t) = r \sum_{n=-\infty}^{\infty} x\left(\frac{n}{2W} - \sigma_n\right) \operatorname{sinc}(2Bt - rn). \tag{7.67}$$

The interpolation error is obtained by subtracting these two expressions.

$$\eta(t) = y(t) - x(t)$$

$$= r \sum_{n=-\infty}^{\infty} \xi_n \operatorname{sinc}(2Bt - rn) \tag{7.68}$$

where

$$\xi_n = x\left(\frac{n}{2W} - \sigma_n\right) - x\left(\frac{n}{2W}\right). \tag{7.69}$$

If the jitter deviations $\{\sigma_n\}$ are identically distributed random variables, then the interpolation in (7.67), does not give an unbiased estimate of $x(t)$. That is

$$\overline{y(t)} = \widetilde{x(t)} \neq x(t) \tag{7.70}$$

where, for any function $v(t)$, we define

$$\widetilde{v(t)} = v(t) * f_\sigma(t) \tag{7.71}$$

and $f_\sigma(t)$ is the probability density function describing each σ_n. To show this, we first note that if $x(t)$ is bandlimited with bandwidth B, then so is $\widetilde{x(t)}$. Hence

$$\widetilde{x(t)} = r \sum_{n=-\infty}^{\infty} \widetilde{x\left(\frac{n}{2W}\right)} \operatorname{sinc}(2Bt - rn).$$

The expectation of a single jittered sample is

$$\overline{x\left(\frac{n}{2W} - \sigma_n\right)} = \int_{-\infty}^{\infty} f_\sigma(\tau) x\left(\frac{n}{2W} - \tau\right) d\tau$$

$$= \widetilde{x\left(\frac{n}{2W}\right)}. \tag{7.72}$$

Substituting this into the expected version of (7.67) substantiates our claim in (7.70).

7.4.1 Unbiased Interpolation from Jittered Samples

Equation (7.70) reveals that cardinal series interpolation from jittered samples results in a biased estimate of the original signal. Motivated by our analysis in the previous section and

Section 6.2.4 on reconstruction from a filtered signal's samples, we propose the interpolation formula

$$z(t) = \sum_{n=-\infty}^{\infty} x\left(\frac{n}{2W} - \sigma_n\right) k\left(t - \frac{n}{2W}\right) \qquad (7.73)$$

where

$$k(t) \longleftrightarrow \frac{1}{2W} \frac{\Pi\left(\frac{u}{2B}\right)}{\Phi_\sigma(u)} \qquad (7.74)$$

and the jitter offset's characteristic function, $\Phi_\sigma(u)$, is the Fourier transform of its probability density function.

$$f_\sigma(t) \longleftrightarrow \Phi_\sigma(u).$$

We claim that $z(t)$ is an unbiased estimate of $x(t)$.

$$\overline{z(t)} = x(t). \qquad (7.75)$$

Furthermore, the interpolation noise variance is given by

$$\text{var } z(t) = E\left[\{z(t) - \overline{z(t)}\}^2\right]$$

$$= \sum_{n=-\infty}^{\infty} \left[\widetilde{x^2\left(\frac{n}{2W}\right)} - \widetilde{x\left(\frac{n}{2W}\right)}^2\right] k^2\left(t - \frac{n}{2W}\right) \qquad (7.76)$$

when the jitter offsets are independent. As jitter becomes less and less pronounced,

$$f_\sigma(t) \to \delta(t).$$

From (7.72), we thus expect to see the interpolation noise variance in (7.76) correspondingly approach zero.

Proof. Expecting (7.73) and substituting (7.72) gives

$$\overline{z(t)} = \sum_{n=-\infty}^{\infty} \widetilde{x\left(\frac{n}{2W}\right)} k\left(t - \frac{n}{2W}\right).$$

Fourier transforming and using (7.74) gives

$$\overline{z(t)} \longleftrightarrow \frac{1}{2W} \sum_{n=-\infty}^{\infty} \widetilde{x\left(\frac{n}{2W}\right)} e^{-j\pi nu/W} \frac{\Pi\left(\frac{u}{2B}\right)}{\Phi_\sigma(u)}$$

$$= X(u)\Phi_\sigma(u) \frac{\Pi\left(\frac{u}{2B}\right)}{\Phi_\sigma(u)}$$

$$= X(u) \qquad (7.77)$$

where we have recognized from (7.71) that

$$\widetilde{x(t)} \longleftrightarrow X(u) \Phi_\sigma(u).$$

Inverse transforming (7.77) completes our proof of (7.75).

To show (7.76), we first compute the autocorrelation

$$R_z(t;\tau) = \sum_{n=-\infty}^{\infty}\sum_{m=-\infty}^{\infty} \overline{x\left(\frac{n}{2W}-\sigma_n\right)x\left(\frac{m}{2W}-\sigma_m\right)} k\left(t-\frac{n}{2W}\right) k\left(\tau-\frac{m}{2W}\right). \quad (7.78)$$

Since the σ_n's are independent

$$\overline{x\left(\frac{n}{2W}-\sigma_n\right)x\left(\frac{m}{2W}-\sigma_m\right)} = \begin{cases} \widetilde{x^2\left(\frac{n}{2W}\right)} & ; n=m \\ \widetilde{x\left(\frac{n}{2W}\right)}\widetilde{x\left(\frac{m}{2W}\right)} & ; n\neq m. \end{cases}$$

Substituting into (7.78) gives

$$R_z(t;\tau) = \sum_{n=-\infty}^{\infty} \widetilde{x^2\left(\frac{n}{2W}\right)} k\left(t-\frac{n}{2W}\right) k\left(\tau-\frac{n}{2W}\right)$$

$$+ \sum_{n=-\infty}^{\infty}\sum_{m=-\infty}^{\infty} \widetilde{x\left(\frac{n}{2W}\right)}\widetilde{x\left(\frac{m}{2W}\right)} k\left(t-\frac{n}{2W}\right) k\left(\tau-\frac{m}{2W}\right)$$

$$- \sum_{n=-\infty}^{\infty} \widetilde{x\left(\frac{n}{2W}\right)}^2 k\left(t-\frac{n}{2W}\right) k\left(\tau-\frac{n}{2W}\right)$$

$$= \sum_{n=-\infty}^{\infty} \left[\widetilde{x^2\left(\frac{n}{2W}\right)} - \widetilde{x\left(\frac{n}{2W}\right)}^2\right] k\left(t-\frac{n}{2W}\right) k\left(\tau-\frac{n}{2W}\right) + x(t)x(\tau).$$

Using the relationship

$$\text{var } z(t) = R_z(t;t) - x^2(t)$$

gives our desired result.

7.4.2 Effects of Jitter In Stochastic Bandlimited Signal Interpolation

Our analysis in this section will show that the use of an inverse filter to obtain an unbiased interpolated estimate from jittered samples will increase the variance of the estimate. Instead of a deterministic signal, $x(t)$, we consider the analysis of the previous section as applied to a wide sense stationary stochastic signal, $\chi(t)$, with mean $\overline{\chi}$ and autocorrelation $R_\chi(\tau)$.

We assume that jitter locations (which we will express in vector form as $\vec{\sigma}$) are independent of $\chi(t)$. Thus, the joint probability density function for $\vec{\sigma}$ and $\chi(t)$ can be expressed as the product of the probability density of $\vec{\sigma}$ with that of $\chi(t)$. Thus, the expectation of any function $w[\vec{\sigma};\chi(t)]$ can be written

$$Ew[\vec{\sigma};\chi(t)] = E_\chi E_{\vec{\sigma}} w[\vec{\sigma};\chi(t)] \quad (7.79)$$

where E_χ and $E_{\vec{\sigma}}$ denote expectation with respect to $\chi(t)$ and $\vec{\sigma}$ respectively. Thus, if (7.73) is used to interpolate jittered samples from the stochastic process $\chi(t)$, we conclude from (7.75) that

$$E_{\vec{\sigma}} z(t) = \chi(t).$$

In accordance with (7.79) we expectate both sides with respect to $\chi(t)$ and conclude that

$$Ez(t) = \overline{\chi(t)}. \tag{7.80}$$

If $v(t)$ is any stochastic signal with constant expectation $\overline{v}$, we conclude from (7.71) that

$$E\widetilde{v(t)} = \int_{-\infty}^{\infty} \overline{v(\tau)} f_\sigma(t-\tau) d\tau = \overline{v}.$$

Reinterpret (7.76) as

$$E_{\vec{\sigma}}\left[z(t) - \overline{z(t)}\right]^2 = \sum_{n=-\infty}^{\infty} \left[\widetilde{\chi^2\left(\frac{n}{2W}\right)} - \widetilde{\chi\left(\frac{n}{2W}\right)}^2\right] k^2\left(t - \frac{n}{2W}\right).$$

Expectation of both sides with respect to χ therefore gives

$$\operatorname{var} z(t) = \operatorname{var}(\chi(t)) \sum_{n=-\infty}^{\infty} k^2\left(t - \frac{n}{2W}\right)$$

where

$$\operatorname{var} \chi(t) = \overline{\chi^2} - \overline{\chi}^2$$

is a constant. Applying the Poisson sum formula gives

$$\operatorname{var} z(t) = 2W \operatorname{var}[\chi(t)] \sum_{n=-\infty}^{\infty} \wp(2nW) e^{-j4\pi nWt}$$

where

$$\wp(u) = K(u) \star K(u)$$

and $\star$ denotes deterministic autocorrelation.[4] Since $K(u) = 0$ for $|u| > B$, we conclude that $\wp(u) = 0$ for $|u| > 2B$. Thus

$$\wp(2nW) = \wp(0)\delta[n].$$

Since

$$\wp(0) = \frac{1}{(2W)^2} \int_{-B}^{B} |\Phi_\sigma(u)|^{-2} du$$

we conclude that

$$\frac{\operatorname{var} z(t)}{\operatorname{var} \chi(t)} = \frac{1}{2W} \int_{-B}^{B} |\Phi_\sigma(u)|^{-2} du. \tag{7.81}$$

This and equation (7.80) are our desired results.

---

4. See Table 2.3.

7.4.2.1 NINV of Unbiased Restoration

Using the filtered cardinal series in (7.67) on the stochastic signal $\chi(t)$ results in a now unbiased estimate

$$\overline{y(t)} = \overline{\chi}$$

where, now

$$\frac{\text{var } y(t)}{\text{var } \chi(t)} = r. \qquad (7.82)$$

Since density functions are non-negative, the characteristic function obeys the inequality

$$|\Phi_\sigma(u)| \leq \Phi_\sigma(0) = 1. \qquad (7.83)$$

Thus, (7.81) always equals or exceeds r. The price of an unbiased estimate is a higher NINV. This price can be measured in the ratio of (7.81) to (7.82):

$$\rho = \frac{\text{var } z(t)}{\text{var } y(t)} = 2 \int_0^{1/2} |\Phi_\sigma(2Bv)|^{-2} \, dv \geq 1.$$

7.4.2.2 Examples

In the following specific examples of unbiased restoration from jittered samples, s is the standard deviation of the jitter density. We will also find useful the parameter

$$\kappa = 2\pi B s.$$

In all cases, we will see ρ increase with increasing κ.

1. **Gaussian Jitter.** If

$$f_\sigma(t) = \frac{1}{\sqrt{2\pi} s} \exp\left(\frac{-t^2}{2s^2}\right)$$

then

$$\Phi_\sigma(u) = \exp\left(-2(\pi s u)^2\right).$$

Thus

$$\rho = 2 \int_0^{1/2} e^{(2\kappa v)^2} \, dv.$$

A plot is shown in Figure 7.15 versus κ.

2. **Laplace Jitter.** Let

$$f_\sigma(t) = \frac{1}{\sqrt{2} s} \exp\left(\frac{-\sqrt{2}|t|}{s}\right).$$

Then

$$\Phi_\sigma(u) = \frac{2/s^2}{(2/s^2) + (2\pi u)^2}$$

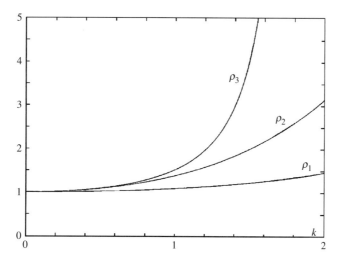

FIGURE 7.15. A plot of ρ for Gaussian (ρ_1), Laplace (ρ_2) and uniform (ρ_3) jitter as a function of $\kappa = 2\pi Bs$.

and we obtain the closed form solution

$$\rho = 1 + \frac{1}{3}\kappa^2 + \frac{1}{2\sigma}\kappa^4.$$

A plot is shown in Figure 7.15.

3 **Uniform Jitter.** Uniform jitter is characterized by the density

$$f_\sigma(t) = \frac{1}{\varphi} \Pi(t/\varphi) \longleftrightarrow \operatorname{sinc}(\varphi u)$$

where $\varphi = \sqrt{12}s$. The density function is thus zero for

$$|t| > \frac{\varphi}{2} = \frac{\sqrt{12}\, s}{2}.$$

Therefore

$$\rho = 2 \int_0^{1/2} \frac{dv}{\operatorname{sinc}^2(\frac{\sqrt{12}\,\kappa v}{\pi})}.$$

A plot is shown in Figure 7.15. Note that there is a singularity in the integrand when any zero of the sinc lies within the interval of integration. This occurs when

$$\varphi \geq \frac{1}{B}$$

or, in other words, when the temporal locations of two adjacent samples have a finite probability of interchanging their position in time. The value of ρ in such cases is unbounded. This case reveals that obtaining an unbiased estimate by inverse filtering can be an unstable undertaking.

7.5 Truncation Error

The cardinal series requires an infinite number of terms to exactly interpolate a bandlimited signal from its samples. In practice, only a finite number of terms can be used. We will examine the resulting *truncation error* for both deterministic and stochastic bandlimited signals.

7.5.1 An Error Bound

We can write the truncated cardinal series approximation of $x(t)$ as

$$x_N(t) = \sum_{n=-N}^{N} x\left(\frac{n}{2B}\right) \text{sinc}(2Bt - n).$$

The error resulting from using a finite number of terms is referred to as truncation error:

$$e_N(t) = |x(t) - x_N(t)|^2. \tag{7.84}$$

We will show that [1079]

$$e_N(t) \leq 2B(E - E_N)[\text{sinc}(2Bt + N) - \text{sinc}(2Bt - N)](-1)^n \frac{\sin(2\pi Bt)}{\pi}$$

$$= (E - E_N)\frac{4NB/\pi^2}{N^2 - (2Bt)^2}\sin^2(2\pi Bt) \; ; \; |t| < \frac{N}{2B} \tag{7.85}$$

where E is the energy of $x(t)$ and E_N is the energy of $x_N(t)$. Using Parseval's theorem in (5.34) applied to $x_N(t)$ gives,

$$E_N = \frac{1}{2B} \sum_{n=-N}^{N} \left|x\left(\frac{n}{2B}\right)\right|^2.$$

From (7.85), as we would expect,

a) since $E_\infty = E$, the error bound tends to zero as $N \to \infty$.
b) for $|n| \leq N$, the error $e_N(\frac{n}{2B})$ is zero.

A plot of

$$\frac{e_N(\tau/2B)}{2B(E - E_N)} = \frac{2N \sin^2(\pi t)}{\pi^2 (N^2 - t^2)}$$

is shown in Figure 7.16 for $N = 10$. Also shown is the curve's envelope

$$\frac{2N/\pi^2}{N^2 - \tau^2}.$$

The envelope for $N = 50$ is shown in Figure 7.17. Note that, at $\tau = N - 1$, the envelope has a value, for large N, of about $1/\pi^2 \approx 0.10$.

NOISE AND ERROR EFFECTS   321

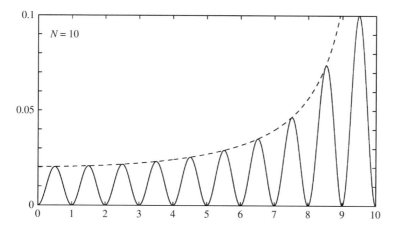

FIGURE 7.16. Truncation error bound and it's envelope for $N = 10$.

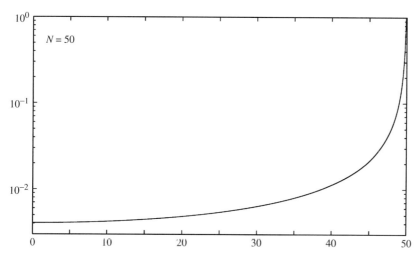

FIGURE 7.17. Truncation error envelope for $N = 50$.

Proof. Using the cardinal series expansion in (3.4), we have

$$e_N(t) = \frac{1}{\pi^2}\sin^2(2\pi Bt)\left|\sum_{|n|>N} \frac{(-1)^n x(\frac{n}{2B})}{2Bt - n}\right|^2$$

applying Schwarz's inequality[5] gives

$$e_N(t) \leq \frac{\sin^2(2\pi Bt)}{2B\pi^2}(E - E_N)\sum_{|n|>N}\left(t - \frac{n}{2B}\right)^{-2} \quad (7.86)$$

---

5. See Appendix 14.1.

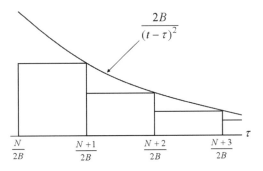

FIGURE 7.18. The left side of the inequality in (7.88) is the area under the rectangles. The right side of the inequality is the area under the $2B(t - \tau)^{-2}$ curve over the interval $(N/2B, \infty)$.

where we have recognized that

$$\frac{1}{2B} \sum_{|n|>N} \left| x\left(\frac{n}{2B}\right) \right|^2 = E - E_N.$$

We can write (7.86) as

$$e_N(t) \leq \frac{\sin^2(2\pi Bt)}{2B\pi^2}(E - E_N) \left[\sum_{n=-\infty}^{-N-1} + \sum_{n=N+1}^{\infty} \right] \left(t - \frac{n}{2B} \right)^{-2}. \quad (7.87)$$

Motivated by Figure 7.18, we can write, for $t < N/2B$,

$$\sum_{n=N+1}^{\infty} \left(t - \frac{n}{2B} \right)^{-2} \leq 2B \int_{N/2B}^{\infty} (t - \tau)^{-2} \, d\tau$$

$$= 2B \left(\frac{N}{2B} - t \right)^{-1}. \quad (7.88)$$

Similarly, for $t > N/2B$,

$$\sum_{n=N+1}^{\infty} \left(t + \frac{n}{2B} \right)^{-2} \leq 2B \left(\frac{N}{2B} + t \right)^{-1}.$$

Substitution of these two inequalities into (7.87) followed by simplification yields our desired result in (7.85).

7.6 Exercises

7.1. According to (7.11), the filtered interpolation noise level can be made arbitrarily small by simply increasing the sampling rate. Why is this result intuitively not satisfying as $r \to \infty$? Explain why we cannot make the noise level arbitrarily small in practice.

7.2. The interpolation function for trigonometric polynomials has infinite energy. Thus, application of either (7.23) or (7.24) results in an infinite NINV classifying the interpolation as ill-posed. Explain why this reasoning is faulty. Derive the filtered and unfiltered NINV using (6.77) assuming additive discrete white noise on the P data points.

7.3. The stochastic process in (7.38) is wide sense *cyclostationary* [1076, 1329]. Define the stochastic process

$$\Psi(t) = \eta(t - \Theta)$$

where Θ is a uniform random variable on the interval $(0, T_N)$ and $\eta(t)$ is given by (7.38).

(a) Show that $\Psi(t)$ is wide sense stationary.

(b) Evaluate $\overline{\Psi^2}$ in terms of the Fourier coefficients in (7.40).

7.4. In both Figures 7.13 and 7.14, colored noise always produces a worse NINV than white noise. Is this always true? If not, specify a noise for which the statement is not true.

7.5. Using the result of Exercise 2.23, show that the interpolation noise variance due to jitter in (7.68) can be bounded as

$$\overline{\eta^2(t)} \leq (2\pi B)^3 E \sigma_{max}/(3\pi)$$

where the jitter deviation is bounded by

$$\sigma_{max} \geq |\sigma_n|$$

and we have assumed sampling at the Nyquist rate.

7.6. Using (7.76), let

$$E_z = \int_{-\infty}^{\infty} \text{var } z(t) \, dt.$$

Similarly, let

$$E_y = \int_{-\infty}^{\infty} \text{var } y(t) \, dt.$$

Show that $E_z \geq E_y$.

7.7. Show that the interpolation noise in (7.68) is not zero mean.

7.8. Compute the NINV in (7.9) when the additive noise has a Cauchy autocorrelation with parameter γ.

$$R_\xi(\tau) = \frac{\overline{\xi^2}}{(\tau/\gamma)^2 + 1}.$$

7.9. Consider restoring a single lost sample at the origin when $r \leq 1/2$. We could use (6.11) or delete every other sample and use the filtered cardinal series in (6.2). Contrast the corresponding NINV's.

7.10. A bandlimited signal, $x(t)$, with nonzero finite energy, can be characterized by a finite number of samples if there exists a T such that $x(t) \equiv 0$ for $|t| > T$. Show, however, that no such class of signals exists.

7.11. Show that

$$\int_{-\infty}^{\infty} e_N(t) \, dt = E - E_N$$

where $e_N(t)$ is the truncation error in (7.84).

7.7 Solutions for Selected Chapter 7 Exercises

7.1. The explanation is unreasonable because the noise level on a bandlimited signal could, in the limit, be reduced to zero. The resolution is that any physical continuous noise has a finite correlation length. Thus, as the samples are taken closer and closer, the noise eventually must become correlated and the white noise assumption is violated. For continuous white noise, all noise samples are uncorrelated. But the noise level is infinite.

7.3. $\overline{\Psi^2} = c_0$.

7.4. Consider a power spectral density that is nonzero only over the interval $B < |u| < W$. Interpolating and filtering would yield a zero NINV.

7.6. Clearly

$$E_z = \sum_{n=-\infty}^{\infty} \left[\widetilde{x^2(n/2W)} - \widetilde{x(n/2W)}^2\right] \int_{-\infty}^{\infty} k^2(t)\, dt.$$

From (7.74) and Parseval's theorem,

$$\int_{-\infty}^{\infty} k^2(t)\, dt = \frac{1}{(2W)^2} \int_{-B}^{B} |\Phi_\theta(u)|^{-2} du.$$

The case of $y(t)$ can be obtained by $k(t) = r\,\mathrm{sinc}(2Bt)$. Since

$$r^2 \int_{-\infty}^{\infty} \mathrm{sinc}^2(2Bt)\, dt = \frac{1}{(2W)^2} \int_{-B}^{B} du$$

$$= \frac{r}{2W} \leq \int_{-\infty}^{\infty} k^2(t)\, dt,$$

We conclude that $E_z \geq E_y$.

7.7. Clearly

$$\overline{\eta(t)} = \widetilde{x(t)} - x(t) \neq 0.$$

7.8. The power spectral density of the noise follows from the transform of (7.6);

$$S_\eta(u) = 2W\,\gamma^2\,\overline{\xi^2} \left[\frac{1}{(2W\gamma)^2} + 2\sum_{n=1}^{\infty} \frac{\cos(\pi nu/W)}{n^2 + (2W\gamma)^2}\right] \Pi\left(\frac{u}{2W}\right).$$

Since [544]

$$\sum_{k=1}^{\infty} \frac{\cos(kx)}{k^2 + a^2} = \frac{\pi}{2a} \frac{\cosh(a(\pi - x))}{\sinh(a\pi)} - \frac{1}{2a^2}\quad ;\ 0 < x \leq 2\pi,$$

we conclude that

$$S_\eta(u) = \gamma\pi\overline{\xi^2}\, \frac{\cosh(2\pi\gamma(W-u))}{\sinh(2\pi\gamma W)} \Pi\left(\frac{u}{2W}\right).$$

Substituting into (7.9) and evaluating gives

$$\overline{\eta_r^2}/\overline{\xi^2} = \sinh(2\pi\gamma B)\,\coth(2\pi\gamma W)$$
$$= \sinh(2\pi r\gamma W)\,\coth(2\pi\gamma W).$$

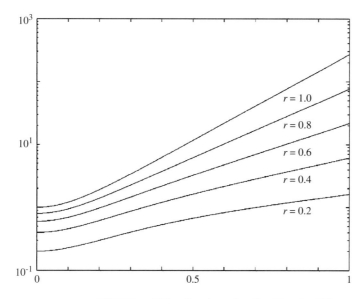

FIGURE 7.19. NINV for additive Couchy noise. See Exercise 7.8.

A semilog plot of the NINV is shown in Figure 7.19 as a function of γW for various values of r.

7.9. By deleting every other sample, the sampling rate parameter becomes $2r$. The NINV for a single sample is $r/(1-r)$. Since

$$2r < r/(1-r) \; ; \; r < 1/2,$$

we conclude that (6.11) yields a smaller NINV.

7.10. Since $x(t)$ is bandlimited, it is analytic. Thus, its Taylor series expanded about any given point converges everywhere. We choose an interval for which $x(t)$ is identically zero. The resulting Taylor series, however, converges to $x(t) \equiv 0$ everywhere.

8

Multidimensional Signal Analysis

> Most people never think about dimensions, except in geometry and architecture classes, for the simple reason that our dimensional realm never changes. Since birth we have lined in the four dimensions of length, width, height, and time ... The laws of physics God designed for physical life require that all the matter and energy of the universe be situated on (and confined to) the manifold, or 'envelope' of these four dimensions.
>
> Hugh Ross [1193].

> We must admit with humility that, while number is purely a product of our minds, space has a reality outside our minds, so that we cannot completely prescribe its properties a priori.
>
> Karl Friedrich Gauss (1777–1855), Letter to Bessel, 1830.

> But my Lord has shewn me the intestines of all my countrymen in the Land of Two Dimensions by taking me with him into the Land of Three. What therefore more easy than now to take his servant on a second journey into the blessed region of the Fourth Dimension, where I shall look down with him once more upon this land of Three Dimensions, and see the inside of every three dimensioned house, the secrets of the solid earth, the treasures of the mines in Spaceland, and the intestines of every solid living creature, even of the noble and adorable Spheres.
>
> Edwin A. Abbott in **Flatland** (1884) [3].

8.1 Introduction

N dimensional signals are characterized as values in an N dimensional space. Each point in the space is assigned a value, possibly complex. Each dimension in the space can be discrete, continuous, or on a time scale.[1] A black and white movie can be modelled as a three dimensional signal. A color picture can be modelled as three signals in two dimensions, one each, for example, for red, green and blue.

This chapter explores Fourier characterization of different types of multidimensional signals and corresponding applications. Some signal characterizations are straightforward extensions of their one dimensional counterparts. Others, even in two dimensions, have properties not found in one dimensional signals. We are fortunate to be able to visualize structures in two, three, and sometimes four dimensions. It assists in the intuitive generalization of properties to higher dimensions.

---

1. See Chapter 12 on time scales.

Fourier characterization of multidimensional signals allows straightforward modelling of reconstruction of images from their tomographic projections. Doing so is the foundation of certain medical and industrial imaging, including CAT (for *computed axial tomography*) scans.

Multidimensional Fourier series are based on models found in nature in periodically replicated crystal Bravais lattices [987, 1188]. As is one dimension, the Fourier series components can be found from sampling the Fourier transform of a single period of the periodic signal.

The multidimensional cosine transform, a relative of the Fourier transform, is used in image compression such as JPG images.

Multidimensional signals can be filtered. The McClellan transform is a powerful method for the design of multidimensional filters, including generalization of the large catalog of zero phase one dimensional FIR filters into higher dimensions.

As in one dimension, the multidimensional sampling theorem is the Fourier dual of the Fourier series. Unlike one dimension, sampling can be performed at the Nyquist density with a resulting dependency among sample values. This property can be used to reduce the sampling density of certain images below that of Nyquist, or to restore lost samples from those remaining.

Multidimensional signal and image analysis is also the topic of Chapter 9 on time frequency representations, and Chapter 11 where POCS is applied signals in higher dimensions.

8.2 Notation

A continuous multidimensional function is denoted $x\left(\vec{t}\right)$ where the coordinates of the N dimensional space be expressed by the vector

$$\vec{t} := [t_1 \ t_2 \ t_3 \ldots t_n \ldots t_N]^T. \tag{8.1}$$

and $-\infty < t_n < \infty$ for $1 \leq n \leq N$. The superscript T denotes transposition. The ℓ_m vector norm of $\vec{t}$ is

$$||\vec{t}||_m := \left[\sum_{n=1}^{N} |t_n|^m\right]^{1/m}. \tag{8.2}$$

The ℓ_2 norm of this vector, $\vec{t}$, is thus determined by the expression[2]

$$||\vec{t}||_2^2 = \vec{t}^T \vec{t} = \sum_{n=1}^{N} t_n^2.$$

Geometrically, the ℓ_2 vector norm is the Euclidean length of the vector. The ℓ_1 norm, also dubbed the *city block metric*, is

$$||\vec{t}||_1 = \sum_{n=1}^{N} |t_n|.$$

---

2. In the absence of a subscript for $||\cdot||$, we assume the ℓ_2 norm. That is
$$||\cdot|| := ||\cdot||_2.$$

The ℓ_∞ norm is determined solely by the dominate term in the vector.

$$||\vec{t}||_\infty = \max_{n=1}^{N} |t_n|.$$

For discrete signals, a multidimensional signal is denoted using square brackets as in $x[\vec{m}]$ where

$$\vec{m} = [m_1 \ m_2 \ m_3 \ \ldots \ m_n \ \ldots m_N]^T.$$

and $m_n = \ldots, -2, -1, 0, 1, 2, \ldots$ for $1 \le n \le N$.

Multidimensional signals can be mixed, with a continuous domain in one dimension and a discrete or time scale domain in another. An example is a black and white celluloid movie where a sequence of continuous two dimensional images is presented in discrete time steps.

8.3 Visualizing Higher Dimensions

Conceptual extrapolation to higher dimensions is often straightforward. A two dimensional circle becomes a sphere in three dimensions and, in N dimensions, a hypersphere. The equation describing the locus of points on an N dimensional hypersphere is

$$||\vec{t}||_2 = r$$

where r is the radius of the sphere. If the points internal to the hypersphere are included, we have an N dimensional hyperball that includes all points satisfying $||\vec{t}||_2 \le r$. Similarly, a two dimensional box becomes a three dimensional cube and, in N dimensions, a hypercube. The locus of points on the hypercube are described by all the points satisfying $||\vec{t}||_\infty \le r^2$.

The mathematical extensions to higher dimensions are often straightforward, but can mislead the intuition. As an example, consider a unit radius circle embedded in a square. The Euclidean distance from the middle of the circle to a point on the circle is one. The distance from the center to a vertex of the square is $\sqrt{2}$. In three dimensions, we have a unit radius sphere imbedded in a box. The distance from the center to the surface of the sphere is one. The distance to a vertex of the box is $\sqrt{3}$. Thus, in general, for N dimensions, a unit radius hypersphere embedded in a hypercube has one unit to the surface of the sphere and $\sqrt{N}$ units to the vertex of the box. Hence, for $N = 10,000$ dimensions, we traverse one unit to arrive at the hypersphere surface and 99 more units to arise at the vertex of the hypercube. Geometric visualization of such properties in higher dimensions is beyond our ability. Hilbert spaces[3] can have an infinite number of dimensions.

8.3.1 N Dimensional Tic Tac Toe

Higher dimensions can be pictured in lower dimensions. In Figure 8.1, a three dimensional tic tac toe cube on the left is shown as three matrices of two dimensions on the right. The object of the the game is to colinearly align four X's or O's in three dimensions. The 5×5 matrix of 5×5 matrices in Figure 8.2 is a two dimensional representation of a tic

---

3. See Section 11.3.1.

MULTIDIMENSIONAL SIGNAL ANALYSIS

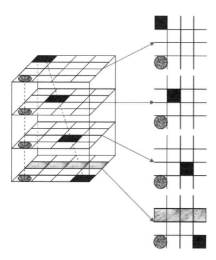

FIGURE 8.1. Tic tac toe in three dimensions can be characterized in two dimensions. Four in a row in any direction wins. Three winning combinations are shown.

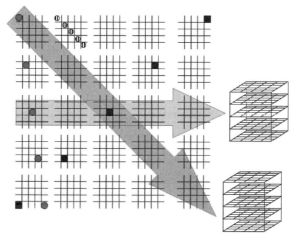

FIGURE 8.2. Tic tac toe in four dimensions can be characterized in two dimensions. Five in a row in any direction wins. Three winning combinations of five in a row are shown.

tac toe game in four dimensions. Each linear sequence of five matrices represents a three dimensional cube. A line through any three dimensional cube wins. The four dimensional game can be played in two dimensions on a sheet of paper. As is seen in Figure 8.3, this is also true of the game in five dimensions. There is a 6×6 matrix of 6×6 matrices of 6×6 matrices. The visualization of winning tic tac toe lines clearly becomes more difficult as the dimensions increase.

8.3.2 Vectorization

The tic tac toe example aids in understanding *vectorization*: the reduction of higher dimensional linear mappings into lower dimensions. Consider the two dimensional

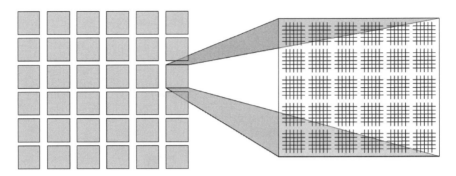

FIGURE 8.3. Tic tac toe in five dimensions can also be characterized in two dimensions.

superposition sum

$$y[n_1, n_2] = \sum_{k_1=1}^{K_1} \sum_{k_2=1}^{K_2} h[n_1, n_2; k_1, k_2] x[k_1, k_2]; \; 1 \leq n_l \leq N_1 \text{ and } 1 \leq n_2 \leq N_2. \quad (8.3)$$

Here, $y[n_1, n_2]$ is the system output of a linear system with point spread function $h[n_1, n_2; k_1, k_2]$ corresponding to an input $x[k_1, k_2]$. Inversion of this mapping, i.e., finding the input from the output, can be achieved through mapping of (8.3) to a one dimensional mapping. We unwrap the $K_1 \times K_2$ matrix row by row and let $\ell = (K_1 - 1)k_2 + k_1$. Similarly, let $m = (N_1 - 1)n_2 + n_1$. The superposition sum in (8.3) then becomes

$$y[m] = \sum_{\ell=1}^{L} h[m, \ell] x[\ell]; \; 1 \leq m \leq M \quad (8.4)$$

where $M = N_1 N_2$ and $L = K_1 K_2$. The two dimensional mapping in (8.3) is revealed as a sequence of simultaneous equations that, in the case of inversion, can be solved using conventional tools for solving simultaneous equations in one dimension.

8.4 Continuous Time Multidimensional Fourier Analysis

The Fourier transform of $x(\vec{t})$ is defined by the integral

$$X(\vec{u}) := \int_{\vec{t}} x(\vec{t}) \exp\left(-j2\pi \vec{u}^T \vec{t}\right) d\vec{t} \quad (8.5)$$

where

$$\int_{\vec{t}} := \int_{t_1=-\infty}^{\infty} \int_{t_2=-\infty}^{\infty} \cdots \int_{t_N=-\infty}^{\infty} \quad (8.6)$$

and

$$d\vec{t} := dt_1 \, dt_2 \cdots dt_N. \quad (8.7)$$

A short hand notation for a Fourier transform pair in (8.5), as was the case in one dimension, is

$$x(\vec{t}) \longleftrightarrow X(\vec{u}).$$

The inverse of the N dimensional Fourier transform is

$$x(\vec{t}) = \int_{\vec{u}} X(\vec{u}) \exp(j2\pi \vec{u}^T \vec{t}) d\vec{u} \qquad (8.8)$$

Table 8.1 contains a list of multidimensional Fourier transforms theorems. The Fourier transform in $N = 2$ dimensions can be written

$$X(u_1, u_2) = \int_{-\infty}^{\infty} x(t_1, t_2) e^{-j2\pi(u_1 t_1 + u_2 t_2)} dt_1 dt_2 \qquad (8.9)$$

TABLE 8.1. Properties of the multidimensional Fourier transform

(a) Transform

$$x(\vec{t}) \longleftrightarrow X(\vec{u}) = \int_{\vec{t}} x(\vec{t}) \exp(-j2\pi \vec{u}^T \vec{t}) d\vec{t}$$

(b) Linearity

$$\sum_k a_k x_k(\vec{t}) \longleftrightarrow \sum_k a_k X_k(\vec{u})$$

(c) Inversion

$$x(\vec{t}) = \int_{\vec{u}} X(\vec{u}) \exp(j2\pi \vec{u}^T \vec{t}) d\vec{u} \longleftrightarrow X(\vec{u})$$

(d) Shift

$$x(\vec{t} - \vec{\tau}) \longleftrightarrow X(\vec{u}) \exp(-j2\pi \vec{u}^T \vec{\tau})$$

(e) Separability

$$\prod_{n=1}^{N} x_n(t_n) \longleftrightarrow \prod_{n=1}^{N} X_n(u_n)$$

(f) Rotation, Scale and Transposition

$$x(\mathbf{A}\vec{t}) \longleftrightarrow \frac{X(\mathbf{A}^{-T}\vec{u})}{|\det \mathbf{A}|}$$

(g) Convolution

$$x(\vec{t}) * h(\vec{t}) = \int_{\vec{\tau}} x(\vec{\tau}) h(\vec{t} - \vec{\tau}) d\tau \longleftrightarrow X(\vec{u}) H(\vec{u})$$

(h) Modulation

$$x(\vec{t}) h(\vec{t}) \longleftrightarrow X(\vec{u}) * H(\vec{u})$$

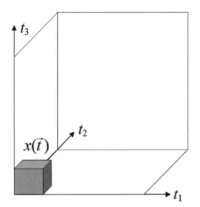
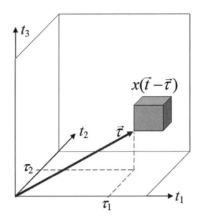

FIGURE 8.4. Illustration of a shift operation in $N = 3$ dimensions. The function is shown on the left. The function, shifted by the vector $\vec{\tau}$, is shown on the right.

8.4.1 Linearity

The Fourier transform in (8.5) is linear.[4]

8.4.2 The Shift Theorem

A function $x(\vec{t})$ is shifted by the fixed vector $\vec{\tau}$ to form the shifted function $x(\vec{t} - \vec{\tau})$. The process can be interpreted as repositioning the origin to the point $\vec{t} = \vec{\tau}$. This is illustrated in Figure 8.4 in $N = 3$ dimensions. On the left, a function $x(\vec{t})$ is one inside the cube and zero outside. The cube is shifted by the vector $\vec{\tau}$ on the right. The cube has been shifted so that it's previous value at the origin is now positioned at the position $\vec{t} = \vec{\tau}$.

If the cube, starting as shown at the left in Figure 8.4, moves with component velocities given by the vector $\vec{v} = [v_1 \ v_2 \ v_3]^T$, then the position of the cube at time s is the shift $x(\vec{t} - \vec{v}s)$.

The shift theorem in Table 8.1 is a straightforward generalization of the one dimensional counterpart.

$$x(\vec{t} - \vec{\tau}) \leftrightarrow \int_{\vec{\xi}} x(\vec{\xi} - \vec{\tau}) \exp\left(-j2\pi \vec{u}^T \vec{\xi}\right) d\vec{\xi}$$

$$= \int_{\vec{t}} x(\vec{t}) \exp\left(-j2\pi \vec{u}^T (\vec{t} + \vec{\tau})\right) d\vec{t}$$

$$= X(\vec{u}) \exp\left(-j2\pi \vec{u}^T \vec{\tau}\right)$$

8.4.3 Multidimensional Convolution

The multidimensional convolution integral is

$$y(\vec{t}) = x(\vec{t}) * h(\vec{t})$$
$$= \int_{\vec{\tau}} x(\vec{\tau}) h(\vec{t} - \vec{\tau}) d\vec{\tau}. \tag{8.10}$$

---

4. See Exercise 8.5.

In one dimension, the kernel h is referred to as the system *impulse response* or, in optics, as the system *line spread function*. In higher dimensions, h is also termed the system *point spread function* or *Green's function*. In one dimension, systems that can be characterized by convolution integrals are said to be *time invariant*. In higher dimensions, they are *shift invariant* or, in the case of optics, *isoplanatic*.

Proof of the convolution theorem in Table 8.1 follows.

$$Y(\vec{u}) = \int_{\vec{t}} y(\vec{t})e^{-j2\pi\vec{u}^T\vec{t}}d\vec{t}$$

$$= \int_{\vec{t}} \left[\int_{\vec{\tau}} x(\vec{\tau})h(\vec{t}-\vec{\tau})d\vec{\tau} \right] e^{-j2\pi\vec{u}^T\vec{t}}d\vec{t}$$

$$= \int_{\vec{\tau}} x(\vec{\tau}) \left[\int_{\vec{t}} h(\vec{t}-\vec{\tau})e^{-j2\pi\vec{u}^T\vec{t}}d\vec{t} \right] d\vec{\tau}$$

$$= \int_{\vec{\tau}} x(\vec{\tau}) \left[H(\vec{u})e^{-j2\pi\vec{u}^T\vec{\tau}} \right] d\vec{\tau}.$$

$$= \left[\int_{\vec{\tau}} x(\vec{\tau})e^{-j2\pi\vec{u}^T\vec{\tau}}d\vec{\tau} \right] H(\vec{u})$$

$$= X(\vec{u})H(\vec{u})$$

Between the third and fourth lines we have applied the shift theorem.

8.4.3.1 The Mechanics of Two Dimensional Convolution

In one dimension, convolution mechanics consists of the operations of "flip and shift". Similar mechanics hold in higher dimensions.

In two dimensions, the convolution integral in (8.10) is

$$y(t_1, t_2) = \int_{t_1=\infty}^{\infty} \int_{t_2=\infty}^{\infty} x(\tau_1, \tau_2)h(t_1-\tau_1, t_2-\tau_2) \, d\tau_1 d\tau_2 \qquad (8.11)$$

For two dimensions, the mechanics of convolution are illustrated in Figure 8.5. In Figures 8.5(1) and 8.5(2) are pictured the functions, $x(t_1, t_2)$ and $h(t_1, t_2)$, to be convolved. Since the integral in (8.11) is with respect to τ_1 and τ_2, we show, in Figure 8.5(3), the function $h(\tau_1, \tau_2)$. For a fixed (t_1, t_2), we shift in Figure 8.5(4) to form the function

$$k(\tau_1, \tau_2) = h(\tau_1-t_1, \tau_2-t_2).$$

The transposition of k in both variables, as shown in Figure 8.5(5), gives

$$k(-\tau_1, -\tau_2) = h(t_1-\tau_1, t_2-\tau_2).$$

Note that the argument of h here is exactly to the form required in the convolution integral in (8.11). Therefore, in lieu of the one dimensional "flip and shift", two dimensional convolution requires a "transposition and shift" of the impulse response. The "transposition and shift" manipulates the impulse response in Figure 8.5(3) into that in Figure 8.5(5). Evaluation of the two dimensional convolution integral in (8.11) is now straightforward. The function $h(t_1-\tau_1, t_2-\tau_2)$ is multiplied by $x(\tau_1, \tau_2)$ and the product integrated. The result is the output $y(t_1, t_2)$ for the value of (t_1, t_2) used in the shift. For our running example, the output is equal to the overlapping area, shown as black, in Figure 8.5(6).

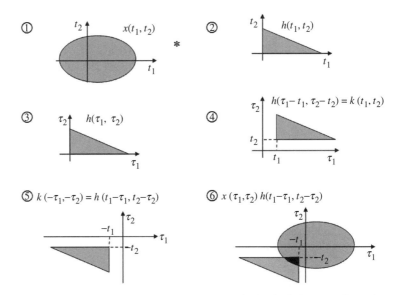

FIGURE 8.5. Convolution mechanics. The functions are one in the shaded areas and are otherwise zero.

8.4.4 Separability

A function $x(\vec{t})$ is said to be separable if it can be written as the product of one dimensional functions:

$$x(\vec{t}) = \prod_{n=1}^{N} x_n(t_n) \qquad (8.12)$$

The separability theorem states that the corresponding multidimensional Fourier transform is the product of the one dimensional Fourier transforms of the one dimensional signals.

$$x(\vec{t}) \leftrightarrow X(\vec{u}) = \prod_{n=1}^{N} X_n(u_n)$$

where

$$x_n(t_n) \leftrightarrow X_n(u_n) = \int_{-\infty}^{\infty} x_n(t_n) \exp(-j2\pi u_n t_n) dt_n$$

is a one dimensional transform pair.

Proof. Substitute (8.12) into (8.5) and separate integrals.

Example 8.4.1. *A Sum of Separable Functions.* The separability theorem allows evaluation of certain multidimensional transforms using one dimensional transforms. Consider the $N = 2$ dimensional example in Figure 8.6 where a function is one inside the two symmetrically spaced squares and is zero outside. We can write

$$x(t_1, t_2) = \Pi\left(\frac{t_1}{c} - \frac{1}{2}\right) \Pi\left(\frac{t_2}{c} - \frac{1}{2}\right) + \Pi\left(\frac{t_1}{c} + \frac{1}{2}\right) \Pi\left(\frac{t_2}{c} + \frac{1}{2}\right) \qquad (8.13)$$

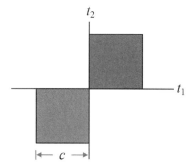

FIGURE 8.6. A two dimensional signal is equal to one inside the two squares and is otherwise zero.

From the separability theorem

$$\Pi\left(\frac{t_1}{c} \pm \frac{1}{2}\right)\Pi\left(\frac{t_2}{c} \pm \frac{1}{2}\right) \leftrightarrow c^2\text{sinc}(cu_1)\text{sinc}(cu_2)e^{\pm j\pi c(u_1+u_2)} \quad (8.14)$$

Superimposing the + and − components of (8.14) and using the linearity property of Fourier transformation gives us the transform of (8.13):

$$x(t_1, t_2) \leftrightarrow 2c^2\text{sinc}(cu_1)\text{sinc}(cu_2)\cos(\pi c(u_1+u_2)) \quad (8.15)$$

The above example is a case where a function is the difference of two separable functions. One can establish that, for any two dimensional function, we can always find one dimensional functions such that

$$x(t_1, t_2) = \sum_{m=-\infty}^{\infty} x_m^{(1)}(t_1) x_m^{(2)}(t_2)$$

That is, any two dimensional function can be expressed as the superposition of separable functions. More generally, in N dimensions,

$$x(\vec{t}) = \sum_{m=-\infty}^{\infty} \prod_{n=1}^{N} x_m^{(n)}(t_n). \quad (8.16)$$

The choices of the one dimensional functions are not unique.

Example 8.4.2. Fourier Series. As a two dimensional example of (8.16), consider the set of functions, $x(t_1, t_2)$, that are zero outside the interval $0 \leq t_1 \leq T$. For a fixed t_2, we can write the Fourier series

$$x(t_1, t_2) = \sum_{m=-\infty}^{\infty} \alpha_m(t_2) e^{j2\pi m t_1/T}; \quad 0 \leq t_1 \leq T$$

where

$$\alpha_n(t_2) = \frac{1}{T}\int_{t_1=0}^{T} x(t_1, t_2) e^{-j2\pi m t_1/T} dt_1.$$

Comparing with (8.16),

$$x_m^{(1)}(t_1) = e^{j2\pi m t_1/T}; \quad 0 \leq t_1 \leq T,$$

and

$$x_m^{(2)}(t_2) = \alpha_m(t_2).$$

Application of the linearity and separability properties of the multidimensional Fourier transform to (8.16) gives

$$x(\vec{t}) \leftrightarrow \sum_{m=-\infty}^{\infty} \prod_{n=1}^{N} X_m^{(n)}(u_n) \qquad (8.17)$$

where

$$x_m^{(n)}(t) \longleftrightarrow X_m^{(n)}(u).$$

8.4.5 Rotation, Scale and Transposition

For any nonsingular matrix $\mathbf{A}$, the function $x(\mathbf{A}\vec{t})$ can be envisioned as a rotation, scaling, and/or transposition[5] of the function $x(\vec{t})$. All operations are performed about a stationary origin, since $\vec{t} = \vec{0} \to \mathbf{A}\vec{t} = \vec{0}$.

We can illustrate simple linear distortion with the unit square at the right side of Figure 8.7. The (t_1, t_2) axis on the right are distorted to $\mathbf{A}\vec{t}$ on the left, where

$$\mathbf{A} = \begin{bmatrix} a_{11} & a_{12} \\ a_{21} & a_{22} \end{bmatrix}.$$

The $t_1 = constant$ axis is now along the line $a_{11}t_1 + a_{12}t_2 = constant$. Likewise, the $t_2 = constant$ axis is now along the line $a_{21}t_1 + a_{22}t_2 = constant$. This is illustrated in Figure 8.7. The unit square on the left is thus distorted into the parallelogram on the right. The more general mapping of two dimensional functions for the matrix $\mathbf{A}$ is evident in Figure 8.8. Each small square on the left is mapped to a parallelogram on the right.

The examples in Figures 8.7 and 8.8 assume the elements of $\mathbf{A}$ are positive.

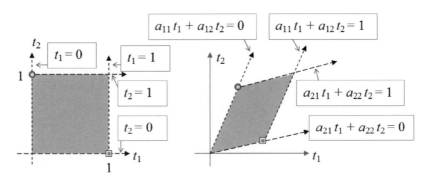

FIGURE 8.7. Illustration in two dimensions of the effect of the coordinate transfer of $\vec{t}$ to $\mathbf{A}\vec{t}$.

---

5. In multidimensions, transposition occurs when the function is reversed in one or more axes. For $x(t_1, t_2, t_3)$, for example, $x(t_1, -t_2, t_3)$, $x(-t_1, t_2, -t_3)$ and $x(-t_1, -t_2, -t_3)$ are all referred to as transpositions.

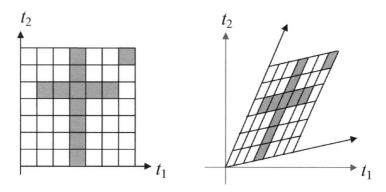

FIGURE 8.8. Mapping of a two dimensional function from coordinates $\vec{t}$ to coordinates $\mathbf{A}\vec{t}$.

8.4.5.1 Transposition

Negative elements in the **A** matrix impose transposition. For the matrix,

$$\mathbf{A} = \begin{bmatrix} -1 & 0 \\ 0 & 1 \end{bmatrix},$$

the t_1 axis is reversed corresponding to a horizontally flipped coordinate system.

8.4.5.2 Scale

Scaling the unit square, shown at the left of Figure 8.7, is performed by $\mathbf{S}\vec{t}$ where the *scaling matrix* is

$$\mathbf{S} = \begin{bmatrix} \frac{1}{M_1} & 0 \\ 0 & \frac{1}{M_2} \end{bmatrix}. \tag{8.18}$$

The horizontal scaling replaces t_1 by t_1/M_1. The coordinate $(t_1, t_2) = (1, 0)$ is therefore replaced by $(t_1, t_2) = (M_1, 0)$ thereby effecting a magnification of M_1 in the horizontal direction.

More generally, in N dimensions, multidimensional scaling is performed using

$$\mathbf{S} = \text{diag} \begin{bmatrix} \frac{1}{M_1} & \frac{1}{M_2} & \cdots & \frac{1}{M_N} \end{bmatrix} \tag{8.19}$$

where diag denotes a diagonal matrix.

8.4.5.3 Rotation

To rotate a function counterclockwise by an angle of θ, we use the *rotation matrix*

$$\mathbf{R} = \begin{bmatrix} \cos(\theta) & \sin(\theta) \\ -\sin(\theta) & \cos(\theta) \end{bmatrix} \tag{8.20}$$

Note that $\mathbf{R}^{-1} = \mathbf{R}^T$.

8.4.5.4 Sequential Combination of Operations

Care must be taken in sequential application of rotation, scale, transposition and shift operations. One might suppose that a rotation followed by a scale operation would be accomplished by $\mathbf{SR}\vec{t}$. The proper operation is, however, $\mathbf{RS}\vec{t}$. We illustrate with an example.

Example 8.4.3. Sequential Operations. The mechanics of sequential application of linear distortions and shifts is illustrated in Figure 8.10. Beginning with the function $x(\vec{t}) = \Pi(t_1)\Pi(t_2)$ on the left, we desire to generate the function $z(\vec{t})$ on the right. The shaded areas correspond to values of one while unshaded areas are zero.

In the first operation, we rotate the function $45°$ to form $w(\vec{t}) = x(\mathbf{R}\vec{t})$ where the rotation matrix, $\mathbf{R}$, uses $\theta = 45°$. Next, to form $y(\vec{t})$, the w function is scaled and

$$y(\vec{t}) = w(\mathbf{S}\vec{t})$$
$$= x(\mathbf{RS}\vec{t}).$$

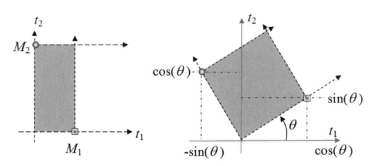

FIGURE 8.9. Illustration in two dimensions of (left) scaling by magnification factors M_1 and M_2, and (right) counterclockwise rotation by an angle of θ.

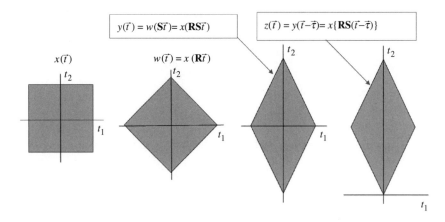

FIGURE 8.10. Steps for generating shifted diamond on the right from the square on the left.

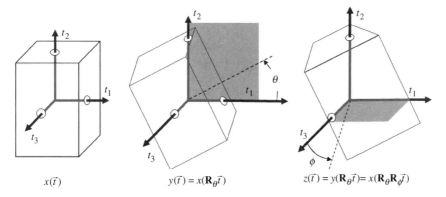

FIGURE 8.11. Illustration of rotation in three dimensions.

Thus, for rotation followed by scale, the proper sequence of matrix operations is **RS**. Lastly, we form

$$\begin{aligned} z(\vec{t}) &= y(\vec{t} - \vec{\tau}) \\ &= w\left(\mathbf{S}(\vec{t} - \vec{\tau})\right) \\ &= x\left(\mathbf{RS}(\vec{t} - \vec{\tau})\right). \end{aligned}$$

8.4.5.5 Rotation in Higher Dimensions

The rotation matrix in (8.20) is for two dimensions. Sequential application of two dimensional rotations in planes, however, can be applied to effect rotations in higher dimensions. An illustration for extension to three dimensions is shown in Figure 8.11. A three dimensional function, $x(\vec{t}) = x(t_1, t_2, t_3)$, is shown in the left hand of the figure. We first rotate counterclockwise in the (t_1, t_2) plane to form $y(\vec{t}) = x(\mathbf{R}_\theta \vec{t})$ where, keeping the t_3 coordinate fixed and, in accordance to the rotation matrix in (8.20),

$$\mathbf{R}_\theta = \begin{bmatrix} \cos(\theta) & \sin(\theta) & 0 \\ -\sin(\theta) & \cos(\theta) & 0 \\ 0 & 0 & 1 \end{bmatrix}.$$

Next, we rotate $y(\vec{t})$ in Figure 8.11 in the (t_3, t_1) plane to form

$$z(\vec{t}) = y(\mathbf{R}_\phi \vec{t})$$

where

$$\mathbf{R}_\phi = \begin{bmatrix} -\sin(\phi) & 0 & \cos(\phi) \\ 0 & 1 & 0 \\ \cos(\phi) & 0 & \sin(\phi) \end{bmatrix}.$$

Therefore, we conclude that $z(\vec{t}) = x(\mathbf{R}\vec{t})$ where

$$\mathbf{R} = \mathbf{R}_\theta \mathbf{R}_\phi = \begin{bmatrix} -\cos(\theta)\sin(\phi) & \sin(\theta) & \cos(\theta)\cos(\phi) \\ \sin(\theta)\sin(\phi) & \cos(\theta) & -\sin(\theta)\cos(\phi) \\ \cos(\phi) & 0 & \sin(\phi) \end{bmatrix}. \qquad (8.21)$$

8.4.5.6 Effects of Rotation, Scale and Transposition on Fourier Transformation

To determine the effects of rotation and scale on a function's Fourier transform, we evaluate

$$x(\mathbf{A}\vec{t}) \longleftrightarrow \int_{\vec{t}} x(\mathbf{A}\vec{t}) \exp(-j2\pi \vec{u}^T \vec{t}) d\vec{t}$$

$$= \frac{1}{|\det \mathbf{A}|} \int_{\vec{\tau}} x(\vec{\tau}) \exp(-j2\pi (\mathbf{A}^{-T}\vec{u})^T \vec{\tau}) d\vec{\tau}$$

$$= \frac{X(\mathbf{A}^{-T}\vec{u})}{|\det \mathbf{A}|} \tag{8.22}$$

where we have made the substitution $\vec{\tau} = \mathbf{A}\vec{t}$ and $|\det \mathbf{A}|$ is the transformation Jacobian. We assume $\mathbf{A}$ is not singular. The superscript $-T$ denotes inverse transposition, i.e., $\mathbf{A}^{-T} = (\mathbf{A}^{-1})^T = (\mathbf{A}^T)^{-1}$.

Example 8.4.4. Scaling. For the scaling matrix in (8.18), we conclude

$$x\left(\frac{t_1}{M_1}, \frac{t_2}{M_2}\right) \longleftrightarrow M_1 M_2 \, X(M_1 u_1, M_2 u_2)$$

More generally, if $\mathbf{S}$ is the diagonal matrix in (8.19), we have

$$x(\mathbf{A}\vec{t}) \longleftrightarrow M_1 M_2 \ldots M_N X(M_1 u_1, M_2 u_2, \ldots, M_N u_N)$$

Example 8.4.5. Rotation. For the rotation matrix in (8.20),

$$\mathbf{R}^{-1} = \mathbf{R}^T. \tag{8.23}$$

Therefore,

$$x(\mathbf{R}\vec{t}) \leftrightarrow X(\mathbf{R}\vec{u}) \tag{8.24}$$

and, in two dimensions, rotating a function simply rotates its transform by the same angle. This is also true in higher dimensions.

The rotated square in Figure 8.10 is $w(\vec{t}) = x(\mathbf{R}\vec{t})$ where the rotation angle is 45°. Then, if $x(\vec{t}) = \Pi(t_1)\Pi(t_2) \leftrightarrow \text{sinc}(u_1)\text{sinc}(u_2)$,

$$w(\vec{t}) = x(\mathbf{R}\vec{t})$$
$$\longleftrightarrow X(\mathbf{R}^{-1}\vec{u})$$
$$= X(\mathbf{R}^T \vec{u})$$
$$= X\left(\frac{u_1 - u_2}{\sqrt{2}}, \frac{u_1 + u_2}{\sqrt{2}}\right)$$
$$= \text{sinc}\left(\frac{u_1 - u_2}{\sqrt{2}}\right) \text{sinc}\left(\frac{u_1 + u_2}{\sqrt{2}}\right)$$

Example 8.4.6. Sequential Operation. The rotated square is shown on the right of Figure 8.9, call it $z(\vec{t})$, can be obtained from $x(\vec{t}) = \Pi(t_1)\Pi(t_2)$ by (1) first translating to the first quadrant to form $y(\vec{t}) = x(\vec{t} - \vec{\tau})$ where $\vec{\tau} = [\frac{1}{2} \; \frac{1}{2}]^T$, and then (2) rotating to form $z(\vec{t}) = y(\mathbf{R}\vec{t}) = x(\mathbf{R}\vec{t} - \vec{\tau})$. Applying the shift theorem

$$y(\vec{t}) \leftrightarrow X(\vec{u})e^{-j2\pi \vec{u}^T \vec{\tau}}.$$

Step (2) gives

$$z(\vec{t}) \leftrightarrow Y(\mathbf{R}^T \vec{u})$$
$$= X(\mathbf{R}^T \vec{u})e^{-j2\pi (\mathbf{R}^T \vec{u})^T \vec{\tau}}.$$

Since

$$\mathbf{R}^T \vec{u} = \begin{bmatrix} u_1 \cos(\theta) - u_2 \sin(\theta) \\ u_1 \sin(\theta) - u_2 \cos(\theta) \end{bmatrix},$$

and

$$\left(\mathbf{R}^T \vec{u}\right)^T \vec{\tau} = [u_1 \cos(\theta) - u_2 \sin(\theta)] \tau_1 + [u_1 \sin(\theta) + u_2 \cos(\theta)] \tau_2$$

then, since $X(\vec{u}) = \mathrm{sinc}(u_1)\mathrm{sinc}(u_2)$, we conclude

$$Z(\vec{u}) = \mathrm{sinc}(u_1 \cos(\theta) - u_2 \sin(\theta)) \, \mathrm{sinc}(u_1 \sin(\theta) - u_2 \cos(\theta))$$
$$\times \exp(-j\pi \{(u_1 (\cos(\theta) + \sin(\theta)) + u_2 (\cos(\theta) - \sin(\theta))\}) \quad (8.25)$$

The results of the above examples illustrate visualization of the effects of rotation and scaling on a function's transform. Rotation in the time domain corresponds to an equivalent rotation in the transform domain whereas, for scaling, a compression in one dimension in the time domain is countered by an expansion along the corresponding frequency domain dimension.

8.4.6 Fourier Transformation of Circularly Symmetric Functions

A function is said to be *circularly symmetric* or, in higher dimensions, *spherically symmetric* if,

$$x(\vec{t}) = f(r) \quad (8.26)$$

where

$$r = ||\vec{t}|| = \sqrt{\vec{t}^T \vec{t}}. \quad (8.27)$$

That is, the multidimensional function varies only with the radial variable, r. An illustration in two dimensions is in Figure 8.12.

The two dimensional polar equivalent of the multidimensional Fourier transform in (8.9) can be obtained by the variable substitutions

$$t_1 = r\cos(\theta); \; t_2 = r\sin(\theta)$$
$$dt_1 dt_2 = r\, dr\, d\theta$$

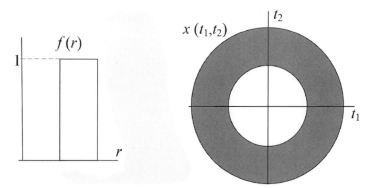

FIGURE 8.12. On the left is the one dimensional function $f(r) = \Pi\left(r - \frac{3}{2}\right)$. The annulus $x(t_1, t_2) = f(\|\vec{t}\|)$ is shown on the right.

and the polar frequency coordinates, (ρ, ϕ), related by

$$u_1 = \rho\cos(\phi); \quad u_2 = \rho\sin(\phi). \tag{8.28}$$

We will interchangeably write $r = \|\vec{t}\|$ and $\rho = \|\vec{u}\|$. It follows that

$$x(\vec{t}) = f(r) \longleftrightarrow \int_{\theta=0}^{2\pi}\int_{r=0}^{\infty} f(r)\exp\left(j2\pi r\rho\cos(\theta - \phi)\right) r\, dr\, d\theta$$

where we have used a trigonometric identity. Note that the integrand in θ is a periodic function of θ with period of 2π. Since the integral over any period of a periodic function independent of the starting point of the integration,

$$x(\vec{t}) \longleftrightarrow \int_{\theta=0}^{2\pi}\int_{r=0}^{\infty} f(r)\exp\left(j2\pi r\rho\cos(\theta)\right) r\, dr\, d\theta. \tag{8.29}$$

From this expression, we see the two dimensional Fourier transform of a circularly symmetric function in also circularly symmetric, i.e., (8.29) is not a function of ϕ. Hence, we can write the two dimensional Fourier transform property

$$x(\vec{t}) = f(r) \longleftrightarrow X(\vec{u}) = F(\rho). \tag{8.30}$$

A useful Bessel function identity here is[6]

$$J_n(z) = \frac{1}{2\pi}\int_{-\pi}^{\pi} e^{jnt} e^{jz\sin(t)} dt. \tag{8.31}$$

Application of the $n = 0$ case to (8.29) gives our desired result for the transform for two dimensional circularly symmetric functions in (8.30) as

$$f(r) \longleftrightarrow F(\rho) = 2\pi \int_{r=0}^{\infty} r f(r) J_0(2\pi r\rho) dr. \tag{8.32}$$

---

6. This integral is that used to find the Fourier series coefficients of a periodic function, in this case $e^{jz\sin(t)}$ with period 2π.

This relation is named a *Fourier-Bessel* or (zeroth order) *Hankel* transforms. The transform is commonly written as

$$\mathcal{H}f(r) = F(\rho) \tag{8.33}$$

where $\mathcal{H}$ is the Hankel transform operator.

In summary, given a circularly symmetric function, $f(r)$, we can compute the two dimensional Fourier transform using (8.9) and evaluating

$$\int_{-\infty}^{\infty} \int_{-\infty}^{\infty} f\left(\sqrt{t_1^2 + t_2^2}\right) e^{-j2\pi(u_1 t_1 + u_2 t_2)} dt_1 dt_2$$

or we can use the Hankel transform in (8.32). The results will be the same. Use of the Hankel transform is usually more straightforward. An exception is the Gaussian in Exercise 8.14(a).

Hankel transform theorems are the topic of Exercise 8.13.

8.4.6.1 The Two Dimensional Fourier Transform of a Circle

Consider the unit radius circle

$$f(r) = \Pi\left(\frac{r}{2}\right). \tag{8.34}$$

Substituting into (8.32) gives

$$F(\rho) = 2\pi \int_0^1 r J_0(2\pi r \rho) dr.$$

Since

$$\frac{d}{d\alpha} \alpha J_1(\alpha) = \alpha J_0(\alpha)$$

we conclude that

$$F(\rho) = 2 \operatorname{jinc}(\rho). \tag{8.35}$$

A plot of $F(\rho)$ is in Figure 8.13.

8.4.6.2 Generalization to Higher Dimensions

In N dimensions, spherically symmetric functions still obey (8.26) and (8.27). The N dimensional Fourier transform of $x(\vec{t}) = f(r)$ is spherically symmetric with radial variable $\rho = ||\vec{u}||$. Hence, in N dimensions,

$$x(\vec{t}) = f(r) \longleftrightarrow X(\vec{u}) = F(\rho).$$

In lieu of computing the N dimensional transform in (8.5), we can evaluate the Hankel transform generalized to N dimensions [149].

$$F(\rho) = \frac{2\pi}{\rho^{\frac{N}{2}-1}} \int_0^\infty f(r) \, r^{\frac{N}{2}} J_{\frac{N}{2}-1}(2\pi \rho r) dr. \tag{8.36}$$

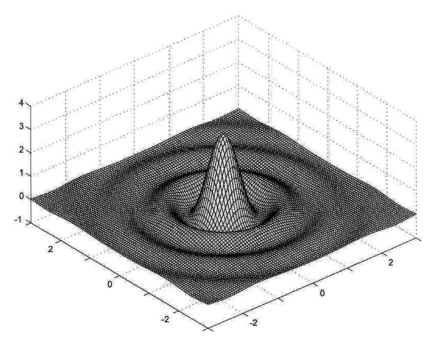

FIGURE 8.13. A plot of 2 jinc(ρ). An optically generated jinc is shown in Figure 13.19.

8.4.6.3 Polar Representation of Two Dimensional Functions that are Not Circularly Symmetric

Any two dimensional function can be expressed in polar coordinates as $x(r, \theta)$. The two dimensional function is periodic in θ with a period of 2π and can therefore be represented as a Fourier series

$$x(r, \theta) = \sum_{n=-\infty}^{\infty} c_n(r) e^{jn\theta} \tag{8.37}$$

where

$$c_n(r) = \frac{1}{2\pi} \int_0^{2\pi} x(r, \theta) e^{-jn\theta} d\theta \tag{8.38}$$

are referred to as *circular harmonics*.

As in the circularly symmetric case, we express the two dimensional Fourier transform in (8.5) in polar form. The result is

$$X(\rho, \phi) = \int_{\theta=0}^{2\pi} \int_{r=0}^{\infty} x(r, \theta) \exp\left(j2\pi r\rho \cos(\theta - \phi)\right) r dr d\theta. \tag{8.39}$$

Substituting the circular harmonic expansion in (8.37) and using the Bessel function identity in (8.31), we obtain after some manipulation,

$$X(\rho, \phi) = \sum_{n=-\infty}^{\infty} j^n e^{jn\phi} \mathcal{H}_n[c_n(r)] \tag{8.40}$$

where the nth order Hankel transform is defined by

$$\mathcal{H}_n[f(r)] = 2\pi \int_0^\infty rf(r) \, J_n(2\pi r\rho) dr \tag{8.41}$$

Note, in particular, that (8.40) is a polar Fourier series of the spectrum of $X(\rho, \phi)$ with circular harmonics $(-j)^n \mathcal{H}_n[c_n(r)]$. The nth order Hankel transform of an image's nth circular harmonic therefore gives the nth circular harmonic of the image's two dimensional Fourier transform.

The special case of the zeroth order Hankel in (8.33) is $\mathcal{H}_0 = \mathcal{H}$. Indeed, the circularly symmetric Fourier-Bessel transform in (8.33) is a special case of the nonsymmetric transform in (8.40). This follows from the observation that, for circularly symmetric functions, the spherical harmonics are such that

$$c_n(r) = f(r)\delta[n].$$

There are no circular harmonics except for the zero frequency component $n = 0$.

8.5 Characterization of Signals from their Tomographic Projections

In tomography, information of an image's *tomographic projections* is used to reconstruct the image. The mathematics for doing this from planar projections using the inverse Radon transform is the topic of this section.

8.5.1 The Abel Transform and Its Inverse

An illustration of the tomographic projection of a circularly symmetric object, $f(r)$, is in Figure 8.14. For each fixed t_2, the area of the one dimensional slice as a function of t_2 is computed. The result is the tomographic projection, $p(t_2)$. Specifically,

$$p(t_2) = \int_{-\infty}^\infty f(r) \, dt_1$$
$$= 2\int_0^\infty f(r) \, dt_1$$

where, in the second step, we have recognized that the integrand is an even function. To change the integration to a function of r, we make the variable substitution

$$t_1 = \sqrt{r^2 - t_2^2} \text{ for } r \geq t_2 \quad \rightarrow \quad dt_1 = \frac{rdr}{\sqrt{r^2 - t_2^2}}.$$

Then, abandoning the subscript on t_2, we have

$$p(t) = 2\int_{r=t}^\infty \frac{f(r)dr}{\sqrt{r^2 - t^2}} \tag{8.42}$$

This, the *Abel transform*, is simply the tomographic projection of a circularly symmetric function. The inverse problem is finding the function, $f(r)$, from its tomographic projection, $p(t)$. This is the *inverse Abel transform*.

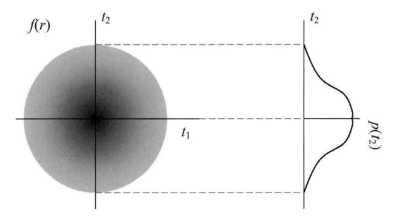

FIGURE 8.14. The Abel transform, $p(t_2)$, is the projection of a circularly symmetric function, $f(r)$.

To invert the Abel transform in (8.42), we will first massage the forward Abel transform into a convolution integral [148]. The first step is to explicitly place the the integration limits in the integrand as a unit step

$$p(t) = 2 \int_{-\infty}^{\infty} \frac{f(r)\mu(r-t)dr}{\sqrt{r^2 - t^2}}.$$

Make the variable substitutions

$$\rho = r^2 \to r = \sqrt{\rho} \to dr = \frac{d\rho}{2\sqrt{\rho}}$$

$$\tau = t^2 \to t = \sqrt{\tau} \tag{8.43}$$

Since $\mu\left(\sqrt{\rho} - \sqrt{\tau}\right) = \mu\left(\rho - \tau\right)$, the result of the substitution is

$$p\left(\sqrt{\tau}\right) = 2 \int_{-\infty}^{\infty} \frac{f\left(\sqrt{\rho}\right) \sqrt{\rho} \mu\left(\rho - \tau\right)}{\sqrt{\rho - \tau}} \frac{d\rho}{2\sqrt{\rho}}$$

$$= \int_{-\infty}^{\infty} \frac{\mu(\rho - \tau)}{\sqrt{\rho - \tau}} f\left(\sqrt{\rho}\right) d\rho.$$

We have therefore expressed the Abel transform as the convolution

$$p\left(\sqrt{\tau}\right) = \frac{\mu(-\tau)}{\sqrt{-\tau}} * f\left(\sqrt{\tau}\right). \tag{8.44}$$

We will find it useful to use the variable ρ in lieu of τ in this convolution.

$$p\left(\sqrt{\rho}\right) = \frac{\mu(-\rho)}{\sqrt{-\rho}} * f\left(\sqrt{\rho}\right). \tag{8.45}$$

Let $\mathcal{F}$ denote the Fourier transform operator that maps the variable ρ into the frequency variable, u. From the Fourier transforms listed in Table 2.4, we find

$$\rho^{-1/2} \mu(\rho) \longleftrightarrow (j2u)^{-1/2}.$$

Imposing the transpose theorem (see Table 2.3), we have

$$(-\rho)^{-1/2}\mu(-\rho) \leftrightarrow (-j2u)^{-1/2} \tag{8.46}$$

and, applying the convolution theorem, (8.45) becomes

$$\mathcal{F}\left[p\left(\sqrt{\rho}\right)\right] = (-j2u)^{-1/2}\mathcal{F}\left[f\left(\sqrt{\rho}\right)\right].$$

Multiplying both sides by $j2\pi u$ and applying the derivative theorem gives

$$\mathcal{F}\left[\frac{d}{d\rho}p\left(\sqrt{\rho}\right)\right] = -\pi\sqrt{-j2u}\,\mathcal{F}\left[f\left(\sqrt{\rho}\right)\right].$$

Solving gives

$$\mathcal{F}\left[f\left(\sqrt{\rho}\right)\right] = -\frac{1}{\pi}(-j2u)^{-1/2}\mathcal{F}\left[\frac{d}{d\rho}p\left(\sqrt{\rho}\right)\right].$$

Inverse transforming and again using (8.46) gives

$$f\left(\sqrt{\rho}\right) = -\frac{1}{\pi}\frac{\mu(-\rho)}{\sqrt{-\rho}} * \frac{d}{d\rho}p\left(\sqrt{\rho}\right). \tag{8.47}$$

From the chain rule of differentiation,

$$\frac{d}{d\rho}p\left(\sqrt{\rho}\right) = \frac{1}{2\sqrt{\rho}}\,p'\left(\sqrt{\rho}\right),$$

and the convolution can be written as

$$f\left(\sqrt{\rho}\right) = -\frac{1}{2\pi}\int_{-\infty}^{\infty}\frac{\mu(\tau-\rho)}{\sqrt{\tau(\tau-\rho)}}\,p'\left(\sqrt{\tau}\right)d\tau$$

To make this expression in terms of its original variables, we apply the substitutions in (8.43). Since $d\tau = 2t\,dt$ and $\mu(t^2 - \rho^2) = \mu(t - \rho)$, we obtain the inverse Abel transform.

$$f(r) = -\frac{1}{\pi}\int_{t=r}^{\infty}\frac{p'(t)dt}{\sqrt{t^2 - r^2}}. \tag{8.48}$$

This is the inverse Abel transform. Equivalently, integrating by parts gives

$$f(r) = -\frac{1}{\pi}\int_{t=r}^{\infty}\sqrt{t^2 - r^2}\,\frac{d}{dt}\left(\frac{p'(t)}{t}\right)dt. \tag{8.49}$$

8.5.1.1 The Abel Transform in Higher Dimensions

To extend the Abel transform to higher dimensions, define the radial variable $r_N = ||\vec{t}_N||$ where $\vec{t}_N = [t_1\ t_2\ t_3\ \ldots\ t_N]^T$. The projection of the function $f_N(r_N)$ to one dimension is

$$f_1(r_1) = \int_{t_2}\int_{t_3}\int_{t_4}\ldots\int_{t_N} f_N(r_N)dt_N\ldots dt_4 dt_3 dt_2. \tag{8.50}$$

This integration can be done in stages, the nth of which is

$$f_{n-1}(r_{n-1}) = \int_{t_n} f_n(r_n)\, dt_n$$

$$= \int_{t_n} f_n\left(\sqrt{r_{n-1}^2 + t_n^2}\right) dt_n$$

Clearly, $f_{n-1}(r_{n-1})$ is the Abel transform of $f_n(r_n)$. Multidimensional Abel transforms can therefore be performed as sequential one dimensional Abel transforms. The same is true for the inverse Abel transform.

8.5.2 The Central Slice Theorem

The *central slice theorem* says that the one dimensional Fourier transform of a tomographic projection of a two dimensional function is equal to a slice of the function's two dimensional Fourier transform along a line. Consequently, knowledge of all projections at all angles suffices to fill in all values in the frequency plane. Since the Fourier transform of the image is known, the image is thus totally defined by its projections. This property, illustrated in Figure 8.15, will now be derived.

The projection, $p(t_1)$, in Figure 8.15 is obtained by vertical line integrals of the image $x(\vec{t})$. Thus

$$p(t_1) = \int_{-\infty}^{\infty} x(t_1, t_2)\, dt_2.$$

The one dimensional Fourier transform of the projection is

$$P(u_1) = \int_{-\infty}^{\infty} p(t_1) e^{-j2\pi u_1 t_1}\, dt_1$$

$$= \int_{-\infty}^{\infty}\int_{-\infty}^{\infty} x(t_1, t_2) e^{-j2\pi u_1 t_1}\, dt_1\, dt_2. \tag{8.51}$$

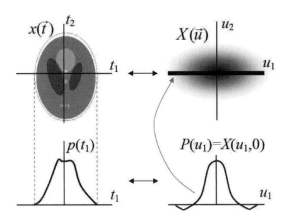

FIGURE 8.15. Illustration of the central slice theorem.

The two dimensional Fourier transform of the image is

$$X(u_1, u_2) = \int_{-\infty}^{\infty} \int_{-\infty}^{\infty} x(t_1, t_2) e^{-j2\pi(u_1 t_1 + u_2 t_2)} dt_1 dt_2.$$

Comparing with (8.51) yields the central slice theorem.

$$X(u_1, 0) = P(u_1).$$

Thus, as illustrated in Figure 8.15, the Fourier transform of the projection is equal to the slice of X along the u_1 axis.

Information about X along the u_2 axis is obtained by Fourier transforming a horizontal projection of the image. To find X along a 45° line, we Fourier transform the image projection at a 45° angle. The value of X at any point (u_1, u_2) can therefore be obtained by Fourier transforming an appropriate projection. The central slice theorem clearly illustrates sufficient information exists in the tomographic projections of an image to reconstruct the image.

8.5.3 The Radon Transform and Its Inverse

The central slice theorem convinces us that an image can be reconstructed from its projections. The inverse Radon transfom gives us the algorithm to do so.

Geometry illustrating the Radon transform is in Figure 8.16. As was the case in Figure 8.15, we preform a projection except the image is now rotated clockwise by an angle of θ. Recall, from (8.24), rotation of an image in two dimensions is equivalent to rotating its Fourier transform by the same angle. Let the projection parallel to t_2 be

$$p_\theta(t_1) = \int_{-\infty}^{\infty} x(\mathbf{R}_\theta \vec{t}) dt_2. \tag{8.52}$$

This mapping of the two dimensional image, $x(\vec{t})$, to the two dimensional function, $p_\theta(t)$ is the Radon transform. An image and its Radon transform are shown in Figure 8.17. The reconstruction problem is finding the two dimensional image from its projections. This is the inverse Radon transform.

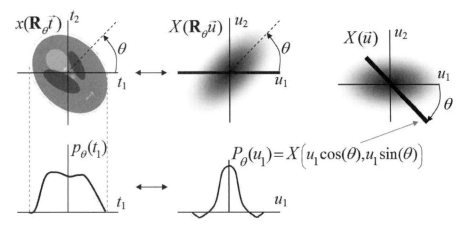

FIGURE 8.16. Illustration of the geometry behind the Radon transform.

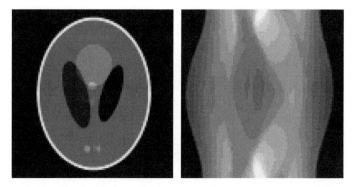

FIGURE 8.17. An image (left - the Shepp-Logan phantom) and its Radon transform.

The Fourier transform, $P_\theta(u_1)$, is, by the central slice theorem, equal to the horizontal slice on the Fourier transform plane. Rotating back to the original coordinate system, shown on the rightmost figure in Figure 8.16, we find that

$$P_\theta(u_1) = X(u_1 \cos(\theta), u_1 \sin(\theta)). \tag{8.53}$$

Note

$$\begin{aligned} P_{\phi+\pi}(\rho) &= X(\rho \cos(\theta + \pi), \rho \sin(\theta + \pi)) \\ &= X(-\rho \cos(\theta), -\rho \sin(\theta)) \\ &= P_\phi(-\rho). \end{aligned} \tag{8.54}$$

Consider, then, the inverse Fourier transform using the polar coordinates in (8.28).

$$x(t_1, t_2) = \int_{\phi=0}^{2\pi} \int_{\rho=0}^{\infty} X(\rho \cos(\phi), \rho \sin(\phi)) e^{-j2\pi\rho(t_1 \cos(\phi) + t_2 \sin(\phi))} \rho \, d\rho \, d\phi.$$

Substitute (8.53),

$$\begin{aligned} x(t_1, t_2) &= \int_{\phi=0}^{2\pi} \int_{\rho=0}^{\infty} P_\phi(\rho) e^{-j2\pi\rho(t_1 \cos(\phi) + t_2 \sin(\phi))} \rho \, d\rho \, d\phi \\ &= \left[\int_{\phi=0}^{\pi} + \int_{\phi=\pi}^{2\pi} \right] \int_{\rho=0}^{\infty} P_\phi(\rho) e^{-j2\pi\rho(t_1 \cos(\phi) + t_2 \sin(\phi))} \rho \, d\rho \, d\phi. \end{aligned} \tag{8.55}$$

Using (8.55) applied to the integral from $\phi = \pi$ to 2π gives

$$\int_{\phi=\pi}^{2\pi} \int_{\rho=0}^{\infty} P_\phi(\rho) e^{-j2\pi\rho(t_1 \cos(\phi) + t_2 \sin(\phi))} \rho \, d\rho \, d\phi$$

$$= \int_{\phi=0}^{\pi} \int_{\rho=0}^{\infty} P_\phi(-\rho) e^{-j2\pi\rho(t_1 \cos(\phi+\pi) + t_2 \sin(\phi+\pi))} \rho \, d\rho \, d\phi$$

$$= \int_{\phi=0}^{\pi} \int_{\rho=0}^{\infty} P_\phi(-\rho) e^{j2\pi\rho(t_1 \cos(\phi)+t_2 \sin(\phi))} \rho \, d\rho \, d\phi$$

$$= \int_{\phi=0}^{\pi} \int_{\rho=-\infty}^{0} P_\phi(\rho) e^{-j2\pi\rho(t_1 \cos(\phi)+t_2 \sin(\phi))} (-\rho) \, d\rho \, d\phi.$$

Then (8.55) becomes

$$x(t_1, t_2) = \int_{\phi=0}^{\pi} \int_{\rho=0}^{\infty} P_\phi(\rho) e^{-j2\pi\rho(t_1 \cos(\phi)+t_2 \sin(\phi))} \rho \, d\rho \, d\phi$$

$$+ \int_{\phi=0}^{\pi} \int_{\rho=-\infty}^{0} P_\phi(\rho) e^{-j2\pi\rho(t_1 \cos(\phi)+t_2 \sin(\phi))} (-\rho) \, d\rho \, d\phi \qquad (8.56)$$

$$= \int_{\phi=0}^{\pi} \left[\int_{\rho=-\infty}^{\infty} |\rho| P_\phi(\rho) e^{-j2\pi\rho(t_1 \cos(\phi)+t_2 \sin(\phi))} d\rho \right] d\phi$$

Define the filtered projection by the Fourier transform pair

$$\hat{p}(t) \leftrightarrow |u| P_\phi(u) \qquad (8.57)$$

where

$$p_\phi(t) \leftrightarrow P_\phi(u).$$

The function $\hat{p}(t)$ is therefore obtained by passing the projection through a filter with a frequency response of $|u|$. It is therefore appropriately dubbed a *filtered projection*. Equation (8.56) can be expressed in terms of the filtered projection.

$$x(t_1, t_2) = \int_{\phi=0}^{\pi} \hat{p}(t_1 \cos(\phi) + t_2 \sin(\phi)) \, d\phi. \qquad (8.58)$$

This is the inverse Radon transform.

Since

$$|u| = \frac{\text{sgn}(u)(j2\pi u)}{j2\pi},$$

and

$$\frac{j}{\pi t} \leftrightarrow \text{sgn}(u),$$

the filtered projection in (8.57) can then be expressed as the convolution

$$\hat{p}_\phi(t) = \frac{1}{2\pi^2 t} * \frac{dp_\phi(t)}{dt}.$$

Equivalently, in terms of the Hilbert transform, the filtered projection can be written as

$$\hat{p}_\phi(t) = -\frac{1}{2\pi} \mathcal{H}\left(\frac{dp_\phi(t)}{dt}\right) = -\frac{1}{2\pi} \frac{d}{dt} \mathcal{H} p_\phi(t).$$

The inverse Radon transform in (8.58) requires knowledge of projections at all angles. In practice, only a finite number of projections is known and we approximate (8.58) by N projections.

$$x(t_1, t_2) \approx \sum_{n=0}^{N-1} \hat{p}(t_1 \cos(\phi_n) + t_2 \sin(\phi_n)) \Delta\phi, \qquad (8.59)$$

where $\phi_n = n\Delta\phi$ and $\Delta\phi = \pi/N$. Each projection is filtered and smeared back into two dimensions along the angle where the original projection was obtained. This process is referred to as *back projection*. All of the back projected filtered projections are summed to obtain an approximation to the original image. The reconstruction process is appropriately termed *filtered back projection*.

Tomography variations and modelling are developed extensively beyond the introductory treatment given here [711].

8.6 Fourier Series

Multidimensional periodic functions can be expanded in a multidimensional generalization of the Fourier series. First, we must establish some formality for characterizing multidimensional periodicity.

8.6.1 Multidimensional Periodicity

The geometry of multidimensional periodicity in N dimensions can be dictated by N *periodicity vectors*, $\{\vec{p}_n | 1 \leq n \leq N\}$, that map the set of all N dimensional vector of integers, $\vec{m}$, to a grid of points. This is illustrated in two dimensions in Figure 8.18 for the vectors $\vec{p}_1 = [1, 1]^T$ and $\vec{p}_2 = [2, 0]^T$. The integer vector $\vec{m}_1 = [2, -1]^T$, marked $(2, -1)$ in Figure 8.18, corresponds to the point $\vec{t} = 2\vec{p}_1 - \vec{p}_2$. Likewise, each of the points in Figure 8.18 is assigned a vector of integers in accordance to the multiplicity of times each vector contributes to the point.

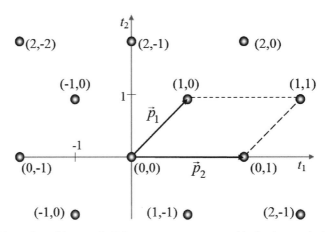

FIGURE 8.18. Illustration of how periodicity vectors generate a grid of points such that each point is assigned a vector of integers.

The matrix comprised of the columns equal to the periodicity vector,

$$\mathbf{P} = [\vec{p}_1 | \vec{p}_2 | \cdots | \vec{p}_N],$$

is referred to as the *periodicity matrix*. For the example in Figure 8.18, the periodicity matrix is

$$\mathbf{P} = \begin{bmatrix} 1 & 2 \\ 1 & 0 \end{bmatrix}. \tag{8.60}$$

Requiring that the $\vec{p}_n$'s not be collinear is equivalent to requiring that $\mathbf{P}$ not be singular (det $\mathbf{P} \neq 0$). The set of all points can be written as

$$\{\mathbf{P}\vec{m} | \vec{m} = \text{all vectors of integers}\}.$$

A given replication geometry does not have a unique periodicity matrix. Besides the periodicity vectors shown in Figure 8.18, any of the periodicity vector pairs in Figure 8.19 will generate the same replication geometries. The $\mathbf{P}$ vector's in this figure are

$$\mathbf{P}_1 = \begin{bmatrix} 1 & -1 \\ 1 & 1 \end{bmatrix}; \quad \mathbf{P}_2 = \begin{bmatrix} 2 & -1 \\ 0 & 1 \end{bmatrix}$$

$$\mathbf{P}_3 = \begin{bmatrix} 1 & 0 \\ 1 & 2 \end{bmatrix}; \quad \mathbf{P}_4 = \begin{bmatrix} 2 & 3 \\ 0 & 1 \end{bmatrix}. \tag{8.61}$$

The area, T, of the multidimensional parallelogram defined by periodicity vectors is $T = |\det \mathbf{P}|$. For the example in Figure 8.18, evaluating the parallelogram, and computing the magnitude of the determinate of (8.60), both yield an area of $T = 2$. This is also the area of all the parallelograms in Figure 8.19 as defined by the periodicity matrices. The determinants of the four matrices in (8.61) is likewise, in each case, equal to 2.

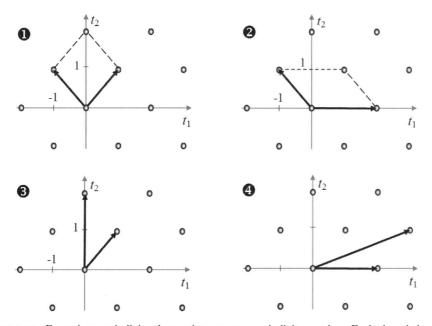

FIGURE 8.19. For a given periodicity, there exist numerous periodicity matrices. Each, though, has the same determinant.

Definition 8.6.1. Multidimensional Periodicity.

An N dimensional function, $S(\vec{u})$, is said to be periodic if there exists a set of N non-collinear vectors, $\{\vec{p}_n | 1 \leq n \leq N\}$, such that, for all $\vec{u}$,

$$S(\vec{u}) = S(\vec{u} - \vec{p}_1) = S(\vec{u} - \vec{p}_2) = \cdots = S(\vec{u} - \vec{p}_N). \tag{8.62}$$

Alternately, if

$$S(\vec{u}) = S(\vec{u} - \mathbf{P}\vec{m})$$

for all vectors of integers, $\vec{m}$, the function is periodic.

A periodic $N = 2$ signal with the periodicity matrix in (8.60) is shown in Figure 8.20.

The concept of a *period* is not as straightforward as in one dimension since we now must deal with geometrical forms rather than with simple intervals. A *tile* is any region that, when replicated in accordance with the periodicity matrix, will fill the entire space without gaps. Parallelograms defined by the periodicity vectors always generate a tile. The parallelogram example, illustrated in Figure 8.21 for the two dimensional periodic replication of crosses in Figure 8.20, generalizes straightforwardly into higher dimensions. Rectangular and hexagonal tiles for the same periodicity are shown in Figures 8.22 and 8.23. Tiles, as illustrated in Figure 8.24, can even be disjoint. For a given periodicity matrix, the choices for tiles shapes is endless. As we will shortly show, however, all tiles share the property of equal area.[7]

Note that there exists one replication point per tile. The number of periods per unit area, T, is thus equal to the area of a cell which can be calculated from the parallelogram configuration as

$$T = |\det \mathbf{P}|$$

The periodicity in Figure 8.21, for example, has an area of 2 units.

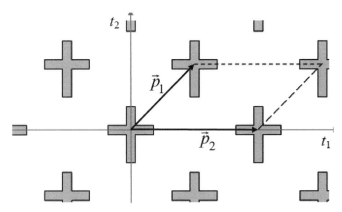

FIGURE 8.20. A periodic function in two dimensions with the periodicity matrix in (8.60). The function is one inside the shaded area and is otherwise zero.

---

7. In this chapter, area means, in general, multidimensional area, e.g., for $N = 3$, volume.

MULTIDIMENSIONAL SIGNAL ANALYSIS 355

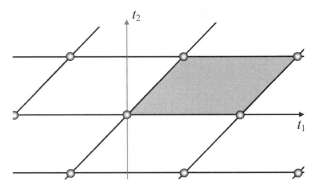

FIGURE 8.21. Completing the parallelogram using the periodicity vectors as legs always generates a tile. The shaded parallelogram is a tile for the periodic replication of crosses in Figure 8.20.

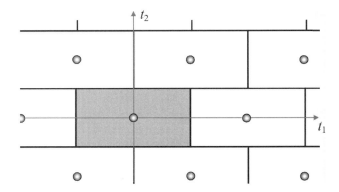

FIGURE 8.22. The shaded rectangle is a tile for the periodic replication of crosses in Figure 8.20.

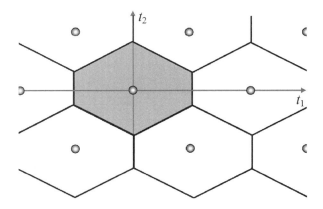

FIGURE 8.23. The shaded hexagon is a tile for the periodic replication of crosses in Figure 8.20.

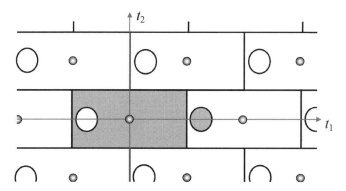

FIGURE 8.24. The shaded area, a rectangle minus the circle plus the circle in the adjacent rectangle, is a tile for the periodic replication of crosses in Figure 8.20.

8.6.2 The Multidimensional Fourier Series Expansion

In this section, we derive the Fourier series for a periodic function, $S(\vec{u})$, with given periodicity matrix $\mathbf{P}$. The periodic function will be assumed to exist in the $\vec{u}$ domain for reasons that will become obvious shortly. First, we establish the following.

Multidimensional Comb Lemma

$$\sum_{\vec{m}} \delta(\vec{t} - \mathbf{Q}\vec{m}) \leftrightarrow |\det \mathbf{P}| \sum_{\vec{m}} \delta(\vec{u} - \mathbf{P}\vec{m}) \qquad (8.63)$$

where

$$\mathbf{Q}^T = \mathbf{P}^{-1} \qquad (8.64)$$

and the multidimensional Dirac delta is

$$\delta(\vec{t}) = \delta(t_1)\delta(t_2)\cdots\delta(t_N)$$

Proof. If we define

$$\lambda(\vec{t}) = \sum_{\vec{m}} \delta(\vec{t} - \mathbf{Q}\vec{m})$$

then

$$\lambda(\mathbf{Q}\vec{t}) = \frac{\sum_{\vec{m}} \delta(\vec{t} - \vec{m})}{|\det \mathbf{Q}|} = |\det \mathbf{P}| \prod_{n=1}^{N} \text{comb}(t_n)$$

where we have used the identities

$$|\det \mathbf{P}| = \frac{1}{|\det \mathbf{Q}|}$$

and

$$\delta(\mathbf{A}\vec{t}) = \frac{\delta(\vec{t})}{|\det \mathbf{A}|}. \qquad (8.65)$$

Using the last entry in Table 2.4 and the separability theorem, we conclude that

$$\lambda(\mathbf{Q}\vec{t}) \leftrightarrow |\det \mathbf{P}| \prod_{n=1}^{N} \text{comb}(u_n) = |\det \mathbf{P}| \sum_{\vec{m}} \delta(\vec{u} - \vec{m})$$

Using the rotation and scale theorem, we find that

$$\lambda(\vec{t}) \leftrightarrow \sum_{\vec{m}} \delta(\mathbf{Q}^T \vec{u} - \vec{m})$$

from which (8.63) immediately follows.

Multidimensional Fourier Series

Using (8.63), we can now show that a periodic function, $S(\vec{u})$ with periodicity matric $\mathbf{P}$ can be expressed via the Fourier series expansion

$$S(\vec{u}) = \sum_{\vec{m}} c[\vec{m}] \exp(-j2\pi \vec{u}^T \mathbf{Q}\vec{m}) \qquad (8.66)$$

where the Fourier coefficients are

$$c[\vec{m}] = |\det \mathbf{Q}| \int_{\vec{u} \in \mathcal{C}} S(\vec{u}) \exp(j2\pi \vec{u}^T \mathbf{Q}\vec{m}) d\vec{u} \qquad (8.67)$$

and $\mathcal{C}$ is any tile.

Proof. Let $X(\vec{u}) = S(\vec{u})/|\det \mathbf{P}|$ over any tile and be zero elsewhere. This function $X(\vec{u})|\det \mathbf{P}|$ is a period of $S(\vec{u})$. We can therefore replicate it to form $S(\vec{u})$.

$$S(\vec{u}) = |\det \mathbf{P}| X(\vec{u}) * \sum_{\vec{m}} \delta(\vec{u} - \mathbf{P}\vec{m}) \qquad (8.68)$$

Using the modulation theorem in Table 8.1, and the multidimensional comb lemma in (8.63), we conclude

$$s(\vec{t}) = x(\vec{t}) \sum_{\vec{m}} \delta(\vec{t} - \mathbf{Q}\vec{m}) = \sum_{\vec{m}} x(\mathbf{Q}\vec{m}) \delta(\vec{t} - \mathbf{Q}\vec{m})$$

where

$$x(\vec{t}) \leftrightarrow X(\vec{u}) \qquad (8.69)$$

and

$$s(\vec{t}) \leftrightarrow S(\vec{u}).$$

Transforming gives

$$S(\vec{u}) = \sum_{\vec{m}} x(\mathbf{Q}\vec{m}) \exp(-j2\pi \vec{u}^T \mathbf{Q}\vec{m}) \qquad (8.70)$$

and, after noting that $c[\vec{m}] = x(\mathbf{Q}\vec{m})$, our proof is complete.

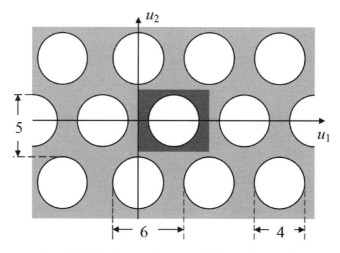

FIGURE 8.25. A periodic function equal to one within the circles and zero without.

Example 8.6.1. Consider the periodic function, $S(\vec{u})$, shown in Figure 8.25.
The function is zero in the circles and one outside. A periodicity matrix for this function is

$$\mathbf{P} = \begin{bmatrix} 6 & 3 \\ 0 & 5 \end{bmatrix}$$

We choose the rectangular periodicity cells shown with the dashed lines. To find the Fourier coefficients, we choose the cell centered at (3, 0). Correspondingly,

$$|\det \mathbf{P}|X(\vec{u}) = \Pi\left(\frac{u_1 - 3}{6}\right) \Pi\left(\frac{u_2}{5}\right) - \Pi\left(\frac{\sqrt{(u_1 - 3)^2 + u_2^2}}{4}\right).$$

Therefore

$$|\det \mathbf{P}|x(\vec{t}) = [30 \operatorname{sinc}(6t_1) \operatorname{sinc}(5t_2) - 8 \operatorname{jinc}(2\sqrt{t_1^2 + t_2^2})] \exp(-j6\pi t_1)$$

and, since

$$\mathbf{Q}\vec{m} = \left[\begin{array}{c} \frac{m_1}{6} \\ -\frac{m_1}{10} + \frac{m_2}{5} \end{array} \right]$$

we conclude that our desired Fourier series coefficients are

$$c[m_1, m_2] = (-1)^{m_1} \left\{ \delta[m_1] \operatorname{sinc}\left(\frac{m_1}{2} - m_2\right) \right.$$

$$\left. - \frac{4}{15}\operatorname{jinc}\left(2\sqrt{\left(\frac{m_1}{6}\right)^2 + \left(\frac{m_1}{10} - \frac{m_2}{5}\right)^2}\right) \right\}.$$

The resulting Fourier series using these coefficients is

$$S(u_1, u_2) = \sum_{m_1=-\infty}^{\infty} \sum_{m_2=-\infty}^{\infty} c[m_1, m_2] \exp\left[-j2\pi \left(\frac{m_1 u_1}{6} - \frac{m_1 u_2}{10} + \frac{m_2 u_2}{5}\right)\right]$$

8.6.3 Multidimensional Discrete Fourier Transforms

The DTFT of an M dimensional discrete signal, $x[\vec{n}]$, is

$$X(\vec{f}) = \sum_{\vec{n}} x[\vec{n}] e^{-j2\pi \vec{f}^T \vec{n}}$$

This is recognized as a Fourier series with an identity matrix as the periodicity matrix. The inverse is thus

$$x[\vec{n}] = \int_{box} X(\vec{f}) e^{j2\pi \vec{f}^T \vec{n}} d\vec{f}$$

where[8]

$$\int_{box} := \int_{f_1=-\frac{1}{2}}^{\frac{1}{2}} \int_{f_2=-\frac{1}{2}}^{\frac{1}{2}} \cdots \int_{f_M=-\frac{1}{2}}^{\frac{1}{2}}.$$

Let $x[\vec{n}]$ be defined in a rectangular box with length N_m in the mth dimension, i.e., $\{0 \leq n_m \leq N_m;\ 1 \leq m \leq M\}$. The DFT of the multidimensional signal follows as

$$X[\vec{k}] = \sum_{\vec{n} \in box} x[\vec{n}] e^{-j2\pi \vec{k}^T \mathbf{N}^{-1} \vec{n}}$$

where the $M \times M$ diagonal matrix is

$$\mathbf{N} := \text{diag}[N_1\ N_2 \cdots N_M].$$

and

$$\sum_{\vec{n} \in box} = \sum_{n_1=0}^{N_1-1} \sum_{n_2=0}^{N_2-1} \cdots \sum_{n_M=0}^{N_M-1}.$$

Equivalently

$$X[\vec{k}] = \sum_{\vec{n}} x[\vec{n}] \prod_{m=1}^{M} e^{-j2\pi k_m n_m / N_M}.$$

The inverse DFT follows as

$$x[\vec{n}] = \frac{1}{\det \mathbf{N}} \sum_{\vec{n} \in box} X[\vec{k}] e^{-j2\pi \vec{k}^T \mathbf{N}^{-1} \vec{n}}.$$

In the special case of $M = $ two dimensions, the DFT and inverse DFT are

$$X[k_1, k_2] = \sum_{n_1=0}^{N_1-1} \sum_{n_2=0}^{N_2-1} x[n_1, n_2] e^{-j2\pi \left(\frac{n_1 k_1}{N_1} + \frac{n_2 k_2}{N_2}\right)}. \tag{8.71}$$

---

8. Alternately, integration can be over any unit interval tile cube.

and

$$x[n_1, n_2] = \frac{1}{N_1 N_2} \sum_{k_1=0}^{N_1-1} \sum_{k_2=0}^{N_2-1} X[k_1, k_2] e^{j2\pi \left(\frac{n_1 k_1}{N_1} + \frac{n_2 k_2}{N_2}\right)}.$$

8.6.3.1 Evaluation

The FFT (fast Fourier transform) [160, 322, 323, 625, 1111] is an efficient method for evaluating the one dimensional DFT. It can be used to efficiently evaluate higher order DFT's. For two dimensions, we can take the FFT of each of the rows of $x[n_1, n_2]$ to form the $N_1 \times N_2$ matrix

$$\tilde{x}[k_1, n_2] = \sum_{n_1=0}^{N_1-1} x[n_1, n_2] e^{-j2\pi n_1 k_1 / N_1}.$$

Evaluating the FFT of each column of $\tilde{x}[k_1, n_2]$ gives the 2D DFT in (8.71). More generally, sequential application of a one dimensional DFT or FFT to a multidimensional function one dimension at a time will result in a multidimensional DFT.

8.6.3.2 The Role of Phase in Image Characterization

In this section, we explore the role of the magnitude and phase of a 2D DFT in the characterization of an image. We will show that, the phase of the DFT of an image is more important than the magnitude of the DFT.

Two images are shown at the top of Figure 8.26. We will call them *Josh* and *Family*. Both images are Fourier transformed and the magnitude of the transform is set to one. The inverse transform of the phase only functions is shown in the second row of Figure 8.26.[9] The nature of the original images is still present. Indeed, the phase images are akin to high pass filtered versions of the original images. This is because the magnitude of the Fourier transform of the images have much of their mass around the low frequencies. By setting the magnitude to one, the high frequencies are essentially amplified as they would be by a high pass filter.

In the third row of Figure 8.26 is the inverse Fourier transform when the transform phase is set to zero.[10] Close examination reveals no relationship to the original images.

In the bottom row, the magnitudes and phases are swapped. On the left, the magnitude of the Fourier transform of *Family* and the phase of the transform of *Josh* was mixed and inverse transformed. The *Josh* picture clearly emerges. On the right is the mixture of the phase of the Fourier transform of *family* with the magnitude of the transform of *Josh*. The *Family* picture emerges.

The images in Figure 8.26 clearly demonstrate the dominate importance of the phase of the Fourier transform over magnitude in preservation of the original image.

8.7 Discrete Cosine Transform-Based Image Coding

The two dimensional discrete cosine transform (DCT) is a common tool in image compression. The two dimensional DCT [116] on an $N \times N$ array is the straightforward

---

9. The inverse transform is normalized [0, 1] to allow display as an image.
10. No normalization is applied here. When a pixel value exceeded one (white), it is displayed as one.

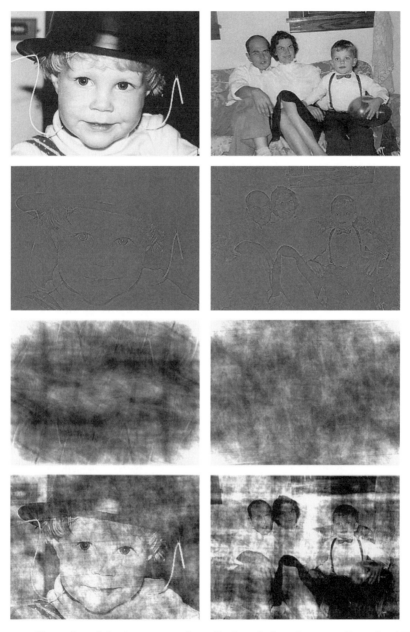

FIGURE 8.26. Illustration of the importance of amplitude and phase in a two dimensional Fourier transform. *Josh* is shown in the upper left and *Family* in the upper right. The phase of the Fourier transform *Josh* using the magnitude of the Fourier transform of *Family* is shown at the bottom left. On the bottom right is the result of using the phase of *Family* and the magnitude of *Josh*. This illustrates the phase of transform has more of an effect on the image than the magnitude.

generalization of the one dimensional DCT case of (2.121) in Section 2.7.

$$X_{\cos}[k_1, k_2] = \frac{1}{4} C[k_1]C[k_2] \sum_{n_1=0}^{N-1} \sum_{n_2=0}^{N-1} x[n_1, n_2]$$
$$\times \cos\left(\frac{\pi(2n_1 + 1)k_1}{2N}\right) \cos\left(\frac{\pi(2n_2 + 1)k_2}{2N}\right); \quad (8.72)$$
$$0 \leq k_1 < N, \ 0 \leq k_2 < N$$

where, for convenience, we rewrite (2.122)

$$C[k] = \begin{cases} 2^{-\frac{1}{2}} & ; k = 0 \\ 1 & ; \text{otherwise.} \end{cases} \quad (8.73)$$

The inverse 2-D DCT is

$$x[n_1, n_2] = \frac{1}{4} \sum_{k_1=0}^{N-1} \sum_{k_2=0}^{N-1} C[k_1]C[k_2] X_{\cos}[k_1, k_2]$$
$$\times \cos\left(\frac{\pi(2n_1 + 1)k_1}{2N}\right) \cos\left(\frac{\pi(2n_2 + 1)k_2}{2N}\right); \quad (8.74)$$
$$0 \leq n_1 < N, \ 0 \leq n_2 < N.$$

8.7.1 DCT Basis Functions

The 2-D DCT is used in image compression, including JPEG[11] (a.k.a. JPG) image compression. Images are separated into 8×8 square blocks of pixels. An example 8×8 block is shown in the gray scale image in Figure 8.27. Each of these blocks is subjected to an $N = 8$ DCT using (8.72). The basis functions for the DCT are the orthogonal basis functions

$$\phi_{k_1, k_2}[n_1, n_2] = \cos\left(\frac{\pi(2n_1 + 1)k_1}{2N}\right) \cos\left(\frac{\pi(2n_2 + 1)k_2}{2N}\right);$$
$$0 \leq n_1 < N, \ 0 \leq n_2 < N,$$
$$0 \leq k_1 < N, \ 0 \leq k_2 < N.$$

These 64 functions are shown in Figure 8.28.

8.7.2 The DCT in Image Compression

DCT coding is effective for image encoding. If the image across the 8×8 block, for example, is constant, then only one entry in the DCT is nonzero.

---

11. Joint Photographic Experts Group.

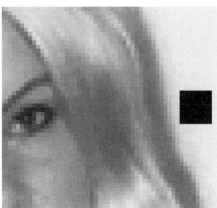

FIGURE 8.27. An 8 × 8 block of pixels in this 345 × 363 pixel is shown on the left as black square. A zoomed version of the image is shown on the right.

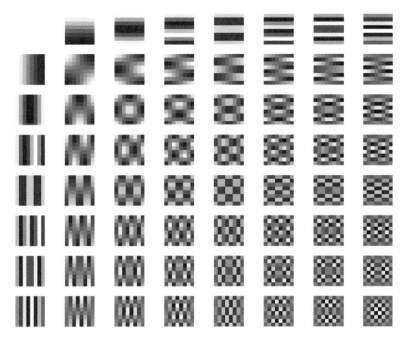

FIGURE 8.28. The 64 orthogonal basis functions used in DCT image encoding.

Here is an example of lossy image compression for gray scale images. Consider the 8×8 block from a one byte per pixel image[12]

$$\mathbf{X} = \begin{bmatrix} 9 & 11 & 58 & 37 & 26 & 39 & 188 & 212 \\ 13 & 15 & 69 & 45 & 23 & 36 & 185 & 215 \\ 50 & 54 & 87 & 37 & 38 & 52 & 196 & 216 \\ 53 & 56 & 81 & 37 & 45 & 59 & 198 & 216 \\ 62 & 60 & 38 & 48 & 116 & 129 & 210 & 216 \\ 50 & 49 & 32 & 71 & 151 & 159 & 213 & 216 \\ 24 & 24 & 47 & 127 & 188 & 192 & 215 & 216 \\ 34 & 34 & 65 & 141 & 193 & 196 & 213 & 214 \end{bmatrix}. \quad (8.75)$$

The gray level block for this matrix is shown on the left in Figure 8.29. The typical DCT encoding process consists of the following steps.

(a) **Subtraction.** 128 is subtracted from each of the elements in the block in (8.75). The result is

$$\mathbf{X}_- = \begin{bmatrix} -119 & -117 & -70 & -91 & -102 & -89 & 60 & 84 \\ -115 & -113 & -59 & -83 & -105 & -92 & 57 & 87 \\ -78 & -74 & -41 & -91 & -90 & -76 & 68 & 88 \\ -75 & -72 & -47 & -91 & -83 & -69 & 70 & 88 \\ -66 & -68 & -90 & -80 & -12 & 1 & 82 & 88 \\ -78 & -79 & -96 & -57 & 23 & 31 & 85 & 88 \\ -104 & -104 & -81 & -1 & 60 & 64 & 87 & 88 \\ -94 & -94 & -63 & 13 & 65 & 68 & 85 & 86 \end{bmatrix}. \quad (8.76)$$

(b) **Perform DCT.** The DCT of $\mathbf{X}_-$, rounded to the nearest integer, is

$$DCT(\mathbf{X}_-) = \begin{bmatrix} -199 & -177 & 5 & -10 & -1 & -2 & 9 & 17 \\ -487 & 65 & -42 & -31 & 10 & 1 & 12 & -5 \\ 153 & 143 & -77 & -3 & 11 & -2 & 15 & -8 \\ -74 & -134 & -9 & 42 & -1 & -13 & 8 & 17 \\ 10 & -37 & 22 & 9 & -4 & -3 & -6 & 2 \\ 40 & 46 & 16 & -17 & 0 & 6 & -7 & -6 \\ -48 & 1 & 9 & -10 & -1 & 2 & -2 & -1 \\ 64 & 33 & -15 & -6 & 2 & 3 & 5 & -2 \end{bmatrix}. \quad (8.77)$$

(c) **Quantization.** The image compression occurs at this, the quantization step. A fixed quantization matrix $\mathbf{Q}$ is used. An example quantization matrix is

$$\mathbf{Q} = \begin{bmatrix} 16 & 11 & 10 & 16 & 24 & 40 & 51 & 61 \\ 12 & 12 & 14 & 19 & 26 & 58 & 60 & 55 \\ 14 & 13 & 16 & 24 & 40 & 57 & 69 & 56 \\ 14 & 17 & 22 & 29 & 51 & 87 & 80 & 62 \\ 18 & 22 & 37 & 56 & 68 & 109 & 103 & 77 \\ 24 & 35 & 55 & 64 & 81 & 104 & 113 & 92 \\ 49 & 64 & 78 & 87 & 103 & 121 & 120 & 101 \\ 72 & 92 & 95 & 98 & 112 & 100 & 103 & 99 \end{bmatrix}. \quad (8.78)$$

Each matrix element in (8.77) is divided by the corresponding matrix element in the quantization matrix. The larger the element in the quantization matrix, the more

---

12. One byte = 8 bits suffices to specify 2^8 gray levels from black (0) to white (255).

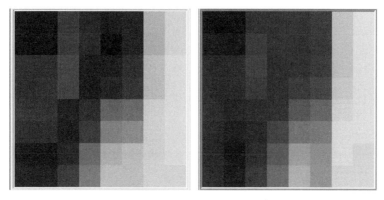

FIGURE 8.29. Example of DCT encoding of a gray scale image. On the left is the original 8×8 block and, on the right, the result of encoding.

quantization and, hence, the more information loss. If each element of $\mathbf{Q}$ is one, there is no quantization.

The result of quantization in our running example, rounded, is

$$\mathbf{D} = [DCT(\mathbf{X}_-)]./\mathbf{Q} = \begin{bmatrix} -12 & -16 & 1 & -1 & 0 & 0 & 0 & 0 \\ -41 & 5 & -3 & -2 & 0 & 0 & 0 & 0 \\ 11 & 11 & -5 & 0 & 0 & 0 & 0 & 0 \\ -5 & -8 & 0 & 1 & 0 & 0 & 0 & 0 \\ 1 & -2 & 1 & 0 & 0 & 0 & 0 & 0 \\ 2 & 1 & 0 & 0 & 0 & 0 & 0 & 0 \\ -1 & 0 & 0 & 0 & 0 & 0 & 0 & 0 \\ 1 & 0 & 0 & 0 & 0 & 0 & 0 & 0 \end{bmatrix} \quad (8.79)$$

where the notation "./" denotes element by element matrix division. It is this matrix in which the image block is encoded. Due, in part, to the preponderance of zeros, the information in this matrix requires less transmission time than that of the original image block, $\mathbf{X}$. The cost of this property is information loss in the original image block. The block $\mathbf{X}$ in (8.75) cannot be reconstructed with precision from $\mathbf{D}$, but, as outlined in the following steps, can be done so approximately.

(d) **Coding, Transmission, Receiving and Decoding.** The matrix is encoded and transmitted. Huffman coding, for example, encodes each number as a binary string according to their frequency of occurrence. The coded signal is received and decoded at the receiver end to reconstruct the $\mathbf{D}$ matrix in (8.79)[116].

(e) **Inverse quantization.** At the receiver, each element of the $\mathbf{D}$ matrix is multiplied by the corresponding element in the quantization matrix, $\mathbf{Q}$. For our running example, this inverse quantization results in

$$\mathbf{B} = \mathbf{Q}.*\mathbf{D} = \begin{bmatrix} -192 & -176 & 10 & -16 & 0 & 0 & 0 & 0 \\ -492 & 60 & -42 & -38 & 0 & 0 & 0 & 0 \\ 154 & 143 & -80 & 0 & 0 & 0 & 0 & 0 \\ -70 & -136 & 0 & 29 & 0 & 0 & 0 & 0 \\ 18 & -44 & 37 & 0 & 0 & 0 & 0 & 0 \\ 48 & 35 & 0 & 0 & 0 & 0 & 0 & 0 \\ -49 & 0 & 0 & 0 & 0 & 0 & 0 & 0 \\ 72 & 0 & 0 & 0 & 0 & 0 & 0 & 0 \end{bmatrix}. \quad (8.80)$$

where the notation ".*" denotes element by element matrix multiplication.

(f) **Inverse DCT.** The received matrix is subjected to an inverse DCT. Since 128 was subtracted from each pixel in Step 1, 128 is now added to each pixel. For our example, the result is

$$\mathbf{Y} = DCT^{-1}\mathbf{B} = \begin{bmatrix} 13 & 15 & 53 & 50 & 51 & 63 & 193 & 219 \\ 26 & 31 & 64 & 52 & 48 & 61 & 192 & 218 \\ 44 & 49 & 73 & 51 & 50 & 67 & 194 & 217 \\ 53 & 54 & 68 & 50 & 66 & 87 & 203 & 217 \\ 53 & 44 & 54 & 55 & 96 & 116 & 213 & 216 \\ 47 & 32 & 47 & 73 & 134 & 143 & 219 & 216 \\ 42 & 27 & 53 & 100 & 168 & 158 & 217 & 215 \\ 40 & 28 & 62 & 120 & 187 & 163 & 213 & 214 \end{bmatrix} \quad (8.81)$$

This is our final decoded image block which should be close to the original image block **X** in (8.75). The ratio of the absolute error as a percent of the full pixel range of 255 is

$$\%\mathbf{E} = \frac{100|\mathbf{Y} - \mathbf{X}|}{255} = \begin{bmatrix} 2 & 2 & 2 & 5 & 10 & 9 & 2 & 3 \\ 5 & 6 & 2 & 3 & 10 & 10 & 3 & 1 \\ 2 & 2 & 5 & 5 & 5 & 6 & 1 & 0 \\ 0 & 1 & 5 & 5 & 8 & 11 & 2 & 0 \\ 4 & 6 & 6 & 3 & 8 & 5 & 1 & 0 \\ 1 & 7 & 6 & 1 & 7 & 6 & 2 & 0 \\ 7 & 1 & 2 & 11 & 8 & 13 & 1 & 0 \\ 2 & 2 & 1 & 8 & 2 & 13 & 0 & 0 \end{bmatrix} \quad (8.82)$$

The maximum error is thus 13% of the full gray scale range.[13] More important than numerical values is the perception of the decoded image to the eye. The matrix **Y** is shown in gray scale form on the right in Figure 8.29 next to the original image block, **X**. Although the reproduction is not perfect, it is close.

8.8 McClellan Transformation for Filter Design

There are a wealth of designs available for one dimensional FIR filter design [650, 991, 1054]. The McClellan transform [387, 970, 971, 980, 976] allows extension and variation in these designs to higher dimensions [270, 277, 833, 1003, 1034, 1035, 1145, 1146, 1301, 1413].

The DFT of the impulse response, $h[n]$, of an FIR filter can be written[14]

$$H(f) = \sum_{n=-N}^{N} h[n] e^{-j2\pi fn}.$$

We restrict attention to zero phase filters, *i.e* filters where the frequency response is real. This imposes Hermetian symmetry on the impulse response.

$$h[n] = h^*[-n].$$

---

13. A more commonly used measure of goodness of one byte image restoration is the *peak signal-to-noise ratio* (PSNR) in (11.35) of Section 11.4.6. The higher the PSNR, the better the match. For our example, *PSNR* = 25 dB.

14. This impulse response is not causal but can be easily made so by a shift. It is used in this form to avoid unnecessarily complicated notation.

If we further assume the impulse response is real, then the frequency response becomes

$$H(f) = \sum_{m=-N}^{-1} h[m]e^{-j2\pi fm} + h[0] + \sum_{n=1}^{N} h[n]e^{-j2\pi fn}$$

$$= \sum_{n=1}^{N} h[-n]e^{j2\pi fn} + h[0] + \sum_{n=1}^{N} h[n]e^{-j2\pi fn}$$

$$= h[0] + \sum_{n=1}^{N} h[n]\left(e^{j2\pi fm} + e^{-j2\pi fn}\right)$$

$$= h[0] + 2\sum_{n=1}^{N} h[n]\cos(2\pi fn).$$

Equivalently, we can write the frequency response of the zero phase filter as

$$H(f) = \sum_{n=0}^{N} a[n]\cos(2\pi fn) \tag{8.83}$$

where

$$a[n] = \begin{cases} h[0] & ; n = 0 \\ 2h[n] & ; 1 \leq n \leq N \end{cases}$$

In the McClellan transform, we dub $H(f)$ the *prototype filter*. The choice of the $a[n]$'s and the constraint on the size of N dictate the frequency response of the filter.

The McClellan transform maps one dimensional filter designs into higher dimensions. It makes use of Chebyshev polynomials of the first kind defined in (2.84) which, for reference, we rewrite here

$$T_n(\cos\theta) = \cos n\theta. \tag{8.84}$$

The frequency response in (8.83) can then be written as

$$H(f) = \sum_{n=0}^{N} a[n]\, T_n\left(\cos(2\pi f)\right) \tag{8.85}$$

We now introduce the *transformation function*, $\Phi(f_1, f_2)$, for the McClellan transform. The transformation function is typically realized with a small, e.g., 3×3, impulse response. The McClellan transform of the one dimensional to two dimensional filter with frequency response, $H(f_1, f_2)$, is

$$H(f_1, f_2) = \sum_{n=0}^{N} a[n]\, T_n\left(\Phi(f_1, f_2)\right). \tag{8.86}$$

The transformation is therefore

$$\cos(2\pi f) = \Phi(f_1, f_2). \tag{8.87}$$

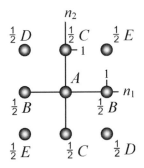

FIGURE 8.30. The impulse response for a 3 × 3 transformation function.

A 3 × 3 transformation function impulse response, pictured in Figure 8.30, is

$$\phi[n_1, n_2] = A\,\delta[n_1]\delta[n_1] + \frac{1}{2}B\left(\delta[n_1 - 1] + \delta[n_1 + 1]\right)\delta[n_2]$$
$$+ \frac{1}{2}C\,\delta[n_1]\left(\delta[n_2 - 1] + \delta[n_1 + 1]\right)$$
$$+ \frac{1}{2}D\left(\delta[n_1 - 1]\delta[n_2 + 1] + \delta[n_1 + 1]\delta[n_2 - 1]\right)$$
$$+ \frac{1}{2}E\left(\delta[n_1 + 1]\delta[n_2 + 1] + \delta[n_1 - 1]\delta[n_2 - 1]\right).$$

The transformation function follows as

$$\Phi(f_1, f_2) = A + B\cos(2\pi f_1) + C\cos(2\pi f_2)$$
$$+ D\cos\left(2\pi(f_1 - f_2)\right) + E\cos\left(2\pi(f_1 + f_2)\right). \tag{8.88}$$

The choice of the parameters, $\{A, B, C, D, E\}$ dictate the shape of the transformation function. Parameters initially proposed by McClellan [970] are

$$A = -\frac{1}{2},\ B = C = \frac{1}{2},\ D = E = \frac{1}{4}. \tag{8.89}$$

Then

$$\Phi(f_1, f_2) = \frac{1}{2}\left(-1 + \cos(2\pi f_1) + \cos(2\pi f_2) + \cos(2\pi f_1)\cos(2\pi f_2)\right). \tag{8.90}$$

As with the 2-D DTFT of any signal, this function is rectangularly periodic with unit periodicity in both dimensions. The contour plot of (8.90), shown in Figure 8.31 on the $\left(-\frac{1}{2}, \frac{1}{2}\right) \times \left(-\frac{1}{2}, \frac{1}{2}\right)$ square, is nearly circular near the origin and grows more and more square away from the origin. This structure proves to be very important in the McClellan transform design of numerous 2-D filters.

An important characteristic of multidimensional signals is the *conservation of contours* property. The contour plot of Φ in Figure 8.31 is the same as the contour plots of Φ^2, e^Φ and $\Phi^2 + e^\Phi$. The values assigned to the contours change, but the contour shapes remain the same. The contour corresponding to $\Phi = \frac{1}{2}$, for example, will be the contour for $\Phi^2 = \frac{1}{4}$. Thus, the contours of $\Phi(f_1, f_2)$ will generally[15] be the same as for the function

---

15. There are counter examples. $g(\Phi) = 1$, for example, has no defined contours.

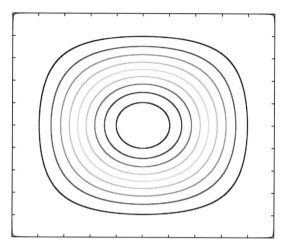

FIGURE 8.31. Contour plot of the transformation function in (8.90).

$g(\Phi(f_1,f_2))$. A key observation in the McClellan transform in (8.86) is that the contours of the transformation function, $\Phi(f_1,f_2)$, are the same as the contours of the frequency response, $H(f_1,f_2)$.

Thus, the key points about the McClellan transform are

(a) The contours of the transformation function, Φ, are the same as the contours of the filter, $H(f_1,f_2)$.
(b) The values along these contours are determined solely by the coefficients, $a[n]$.

Example 8.8.1. For proper choices of $a[n]$ coefficients, the transformation function, Φ, with the contours in Figure 8.31 can be used to design 2-D

(a) low pass filters equal to one inside a specified contour and zero outside,
(b) high pass filters with a frequency response equal to the complement of the low pass filter,
(c) bandpass filters equal to one between two of the contours and zero otherwise, and
(d) band stop filters, equal to the complement of the bandpass filter.

A remarkable property of the transformation function in (8.90) is

$$\Phi(f_1, 0) = \cos(2\pi f_1). \tag{8.91}$$

With attention to (8.85) and (8.83), along the f_1 axis, the McClellan transform in (8.86) then becomes

$$H(f_1, 0) = \sum_{n=0}^{N} a[n]\, T_n\left(\Phi(f_1, 0)\right)$$

$$= \sum_{n=0}^{N} a[n]\, T_n\left(\cos(2\pi f_1)\right)$$

$$= \sum_{n=0}^{N} a[n]\, \cos(2\pi n f_1) \tag{8.92}$$

$$= H(f_1)$$

Thus, the slice of $H(f_1, f_2)$ along the f_1 axis is the prototype filter, $H(f_1)$. The values along this slice will follow the contours of the transformation function in Figure 8.31. Hence, if we design $H(f)$ to be a low pass filter and use the coefficients $a[n]$ in the McClellan transform, then, as described above, $H(f_1, f_2)$ will be a low pass filter. The same is true for high pass, bandpass and band stop filters.

An example of a low pass filter from the McClellan transform is shown in Figure 8.32. A simple Fourier series for $X(f) = \Pi(f/\alpha)$ on the interval $|f| \leq \frac{1}{2}$ is used. The corresponding Fourier series coefficients are $h[n] = \alpha \, \text{sinc}(\alpha n)$. The 1-D prototype filter is shown in Figure 2.1 for $\alpha = \frac{1}{3}$ and $N = 10, 25, 50$ and 100. The corresponding McClellan transform filters are shown in Figure 8.32.

A 2-D bandpass filter designed using the McClellan transform is shown in Figure 8.33. Cutoff frequencies are $\alpha_1 = 0.2$ and $\alpha_2 = 0.4$. Again, a simple Fourier series filter with Fourier coefficients of $h[n] = \alpha_2 \, \text{sinc}(\alpha_2 n) - \alpha_1 \, \text{sinc}(\alpha_1 n)$ are used. Filters corresponding to $N = 10$ and 25 are shown.

The coefficients in (8.89) giving rise to the transformation function contours correspond to only one case where prototype filters can be applied. There are numerous other possibilities, many much more subtle.

Example 8.8.2. Two Dimensional Hilbert Transform. We seek a McClellan transform that generates a frequency response

$$H(f_1, f_2) = \text{sgn}(f_1)\text{sgn}(f_2); \ |f_1| \leq \frac{1}{2}; \ |f_q| \leq \frac{1}{2}$$

$$= \begin{cases} 1 \ ; \ (f_1, f_2) \in \text{quadrants I and II in the unit square} \\ -1 \ ; \ (f_1, f_2) \in \text{quadrants II and IV in the unit square} \end{cases} \quad (8.93)$$

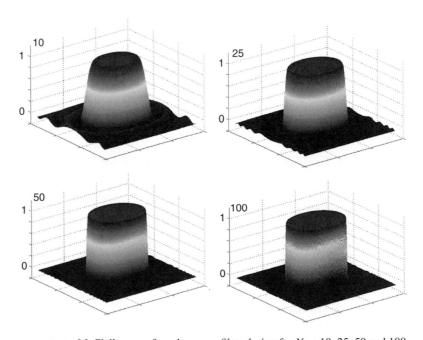

FIGURE 8.32. McClellan transform low pass filter design for $N = 10, 25, 50$ and 100.

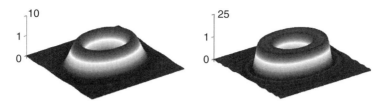

FIGURE 8.33. McClellan transform bandpass filter design for $N = 10$ and 25.

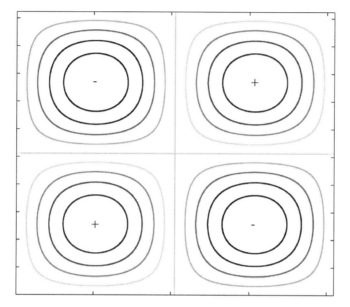

FIGURE 8.34. Contour plot of the 3×3 prototype with transformation function $\Phi(f_1, f_2) = \sin(2\pi f_1)\sin(2\pi f_2)$.

We choose a transformation function corresponding to $A = B = C = 0$, $D = -E = \frac{1}{2}$. From (8.88), we have

$$H(f_1, f_2) = \frac{1}{2}\cos(2\pi(f_1 - f_2)) - \frac{1}{2}\cos(2\pi(f_1 + f_2))$$
$$= \sin(2\pi f_1)\sin(2\pi f_2)$$

The contour plot of this transformation function is shown if Figure 8.34. It is positive in quadrants I and III and negative in quadrants II and IV. To find an appropriate prototype filter, we examine the McClellan transform which, from (8.87), is

$$\cos(2\pi f) = \sin(2\pi f_1)\sin(2\pi f_2 t). \tag{8.94}$$

On the interval $[-\frac{1}{2}, \frac{1}{2}]$, the cosine term is positive of $|f| < \frac{1}{2}$ and -1 otherwise. Therefore, the prototype filter desired is one for $|f| < \frac{1}{4}$ and -1 otherwise. Using a Fourier series, let $h[n] = 0$ for $n = 0$ and $2\,\text{sinc}(n/2)$ otherwise. The McClellan transform using these parameters is shown in Figure 8.35 for $N = 10, 25, 50$ and 100.

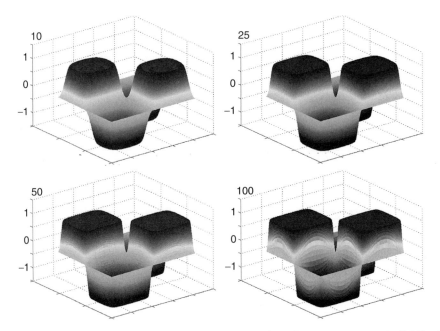

FIGURE 8.35. Synthesis of the two dimensional Hilbert transform frequency response in (8.93) using a McClellan transform for $N = 10, 25, 50$ and 100.

8.8.1 Modular Implementation of the McClellan Transform

The McClellan representation in (8.86) contains Chebyshev polynomials that obey the recurrence relation in (2.85) which, for convenience, we repeat her.

$$T_{n+1}(t) = 2tT_n(t) - T_{n-1}(t). \tag{8.95}$$

This recursion relationship can be used to characterize the sum in (8.86) in the modular form shown in Figure 8.36. The first two terms in the sum, $T_0(\Phi) = 1$ and $T_1(\Phi) = \Phi$, are multiplied, respectively, by $h[0]$ and $2h[1]$ and are fed to the summing bus at the bottom. The term $T_2(\Phi) = 2\Phi T_1(\Phi) - T_0(\Phi)$ is formed from the first two stages and has a tap of $2h[3]$. $T_3(\Phi)$ through $T_5(\Phi)$ are similarly formed.

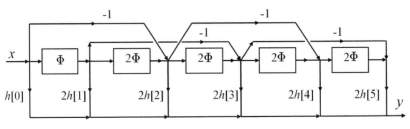

FIGURE 8.36. Modular implementation of the McClellan transform based on the recurrence relationship of the Chebyshev polynomials in (8.95).

8.8.2 Implementation Issues

There are alternative methods to perform filtering. For a low pass filter, for example, applying a two dimensional FFT to an image, multiplying the low pass filter, and performing an inverse FFT, can also achieve the filtering operation. For certain implementations, polynomial synthesis of filters using transformation functions, such as the McClellan transform, can be very sensitive to small computational perturbations [387].

8.9 The Multidimensional Sampling Theorem

The sampling theorem in higher dimensions was first proposed by Peterson and Middleton in 1962 [1112].

We have established in one dimension that the Shannon sampling theorem is the Fourier dual of a Fourier series.[16] The same is true in higher dimensions. Indeed, we can glean important insight into the multidimensional sampling theorem from our development of the Fourier series in Section 8.6.

The sampling theorem will be developed in N dimensions and illustrated in two dimensions in Figures 8.37 and 8.38. In the left column in Figure 8.37 is pictured functions in $\vec{t}$ and the right their corresponding Fourier transforms are shown in the right column as a function of $\vec{u}$. For the left column, row (1) times row (2) equals row (3). This means that on the right column, row (3) is the convolution of rows (1) and (2). Row (3) at the bottom of Figure 8.37 is repeated on the top of 8.38. In the right column, row (3) is multiplied by row (4) to obtain row (5). This means that, in the left column, row (5) is the convolution of rows (3) and (4). Let's look at these operations more closely to see how the multidimensional sampling theorem emerges.

In row (1) of Figure 8.37, a function $x(\vec{t})$ has a Fourier transform $X(\vec{u})$ that is zero outside a tile $\mathcal{C}$. The tile shape is arbitrary. All that is required is existence of a periodicity matrix, $\mathbf{P}$, such that replication of the tile fills the entire $\vec{u}$ space. In row (2), in the $\vec{t}$ space on the left, Dirac deltas are placed periodically in accordance with the matrix $\mathbf{Q} = \mathbf{P}^{-T}$ In the placement of the Dirac deltas, the $\mathbf{Q}$ matrix acts as a periodicity matrix. We repeat the results of the multidimensional comb lemma in (8.63) which relates this array of Dirac deltas on the left in row (2) to its Fourier transform on the right.

$$\sum_{\vec{m}} \delta(\vec{t} - \mathbf{Q}\vec{m}) \leftrightarrow |\det \mathbf{P}| \sum_{\vec{m}} \delta(\vec{u} - \mathbf{P}\vec{m}). \qquad (8.96)$$

Let the function $s(\vec{t})$ be the product of the signal with the array of Dirac deltas.

$$\begin{aligned} s(\vec{t}) &= x(\vec{t}) \sum_{\vec{m}} \delta(\vec{t} - \mathbf{Q}\vec{m}) \\ &= \sum_{\vec{m}} x(\mathbf{Q}\vec{m}) \delta(\vec{t} - \mathbf{Q}\vec{m}). \end{aligned} \qquad (8.97)$$

---

16. See Section 5.3.2.

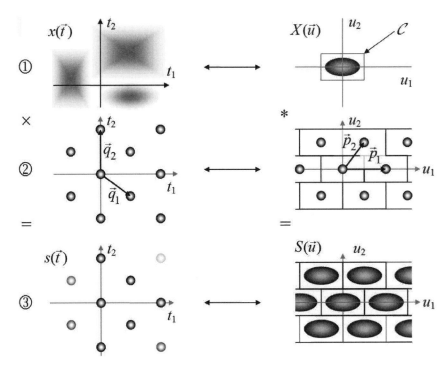

FIGURE 8.37. Illustration of multidimensional sampling. Continued in Figure 8.38.

The function $s(\vec{t})$ therefore is a function only of the sample points of $x(\vec{t})$. Its Fourier transform, using (8.96), is

$$S(\vec{u}) = |\det \mathbf{P}| X(\vec{u}) * \sum_{\vec{m}} \delta(\vec{u} - \mathbf{P}\vec{m})$$

$$= |\det \mathbf{P}| \sum_{\vec{m}} X(\vec{u} - \mathbf{P}\vec{m})$$

Therefore, the signal's spectrum, $X(\vec{u})$, is replicated in the $\vec{u}$ space on tiles with a periodicity matrix **P**. This is illustrated in row (3) of Figure 8.37. We have assumed that $X(\vec{u})$ is zero outside of the tile. If this were not the case, the sampling would be aliased. Thus, from $S(\vec{u})$ we can regain the original signal spectrum by

$$X(\vec{u}) = \frac{S(\vec{u})\Pi_\mathcal{C}(\vec{u})}{|\det \mathbf{P}|} \qquad (8.98)$$

where

$$\Pi_\mathcal{C}(\vec{u}) = \begin{cases} 1\,; & \vec{u} \in \mathcal{C} \\ 0\,; & \text{otherwise.} \end{cases}$$

The function $S(\vec{u})$ is clearly periodic with periodicity matrix **P**. Its Fourier series is obtained by the Fourier transform of (8.97).

$$S(\vec{u}) = \sum_{\vec{m}} x(\mathbf{Q}\vec{m}) e^{-j2\pi \vec{u}^T \mathbf{Q}\vec{m}}.$$

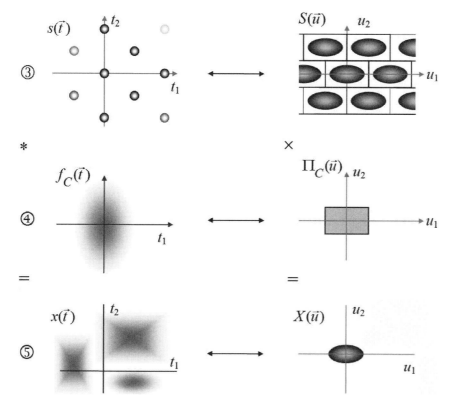

FIGURE 8.38. Illustration of restoration from multidimensional samples. The top row, (3), is the same as the bottom row in Figure 8.37.

Thus, from (8.98),

$$X(\vec{u}) = \frac{1}{|\det \mathbf{P}|} \sum_{\vec{m}} x(\mathbf{Q}\vec{m}) e^{-j2\pi \vec{u}^T \mathbf{Q}\vec{m}} \Pi_C(\vec{u}).$$

Inverse Fourier transformation gives the desired expression for the multidimensional sampling theorem.

$$x(\vec{t}) = \sum_{\vec{m}} x(\mathbf{Q}\vec{m}) f_C(\vec{t} - \mathbf{Q}\vec{m}) \qquad (8.99)$$

where the interpolation function, $f_C(\vec{t})$, is defined by the transform pair

$$f_C(\vec{t}) \longleftrightarrow \frac{\Pi_C(\vec{u})}{|\det \mathbf{P}|}. \qquad (8.100)$$

Equivalently

$$f_C(\vec{t}) = |\det \mathbf{Q}| \int_{\vec{u} \in C} e^{-j2\pi \vec{u}^T \vec{t}} d\vec{t}. \qquad (8.101)$$

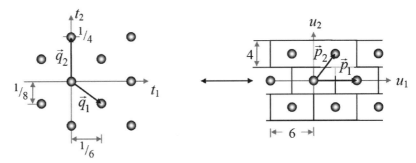

FIGURE 8.39. An example of a sampling matrix, **Q** and its corresponding periodicity matrix, $\mathbf{P} = \mathbf{Q}^{-T}$.

The **Q** matrix is called the *sampling matrix* and dictates the geometry of the uniform sampling. We decompose **Q** into vector columns.

$$\mathbf{Q} = [\vec{q}_1 | \vec{q}_2 | \ldots | \vec{q}_N]. \tag{8.102}$$

As with the periodicity matrix, each component vector is a basis vector for sampling. An example is shown in Figure 8.39 for $N = 2$ with

$$\mathbf{Q} = \frac{1}{24} \begin{bmatrix} 4 & 0 \\ -3 & 6 \end{bmatrix}; \quad \mathbf{P} = \begin{bmatrix} 6 & 3 \\ 0 & 4 \end{bmatrix}.$$

Consider the parallelogram defined by the column vectors of **Q**. Visualize replication to fill the entire space. Clearly, there is a one to one correspondence of a sample to a parallelogram. Since the area of the parallelogram is $|\det \mathbf{Q}|$, we conclude that the *sampling density* corresponding to **Q** is

$$\text{SD} = \frac{1}{|\det \mathbf{Q}|} = |\det \mathbf{P}| \text{ samples/unit area}.$$

The sampling geometry in Figure 8.39, for example, has a density of 24 samples per unit area.

We call the region $\mathcal{A}$, over which a spectrum $X(\vec{u})$ is not identically zero, the spectral *support*. In order for $x(\vec{t})$ to follow our definition of bandlimitedness, $\mathcal{A}$ must be totally contained within an N dimensional sphere of finite radius.

8.9.1 The Nyquist Density

The lowest rate at which a temporal bandlimited signal can be uniformly sampled without aliasing is the *Nyquist rate*. This measure can be generalized to higher dimensions. For a given spectral support, $\mathcal{A}$, the *Nyquist density* is that density resulting from maximally packed unaliased replication of the signal's spectrum. Equivalently, for a given spectral support, the Nyquist density corresponds to the replication tile with minimum area.

For some spectral supports, numerous sampling geometries can achieve the Nyquist density. An example is shown in Figure 8.40. A rectangular spectral support is shown in the upper left of the figure. Three different tilings are shown that achieve the Nyquist density. Periodicity vectors are shown as bold arrows.

In other cases, there is a single sampling geometry to achieve the Nyquist density. An example is when the support is an equalateral hexagon as shown in Figure 8.41.

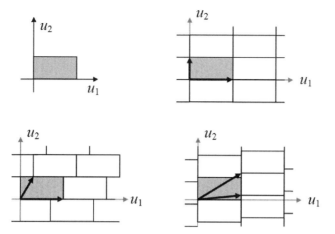

FIGURE 8.40. An example of a rectangular support (upper left) and three different sets of periodicity vectors that will achieve the Nyquist density.

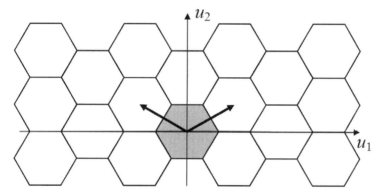

FIGURE 8.41. The support of maximally packed circles is a hexagon. The periodicity vectors are shown as bold arrows.

8.9.1.1 Circular and Rectangular Spectral Support

Any imaging system that uses lenses with circular pupils will generate images the spectra of which have a circular spectral region of support if the monochromatic illumination is either coherent or incoherent [493, 518]. Consider, then, the case where the support of an $N = 2$ dimensional spectrum is a circle of radius W. If we limit ourselves to rectangular sampling (i.e., restrict $\mathbf{Q}$ to be a diagonal matrix), then the closest we can pack circles is in Figure 8.42.

The periodicity and sampling matrices are, respectively,

$$\mathbf{P} = \begin{bmatrix} 2W & 0 \\ 0 & 2W \end{bmatrix}$$

$$\mathbf{Q} = \begin{bmatrix} T & 0 \\ 0 & T \end{bmatrix}$$

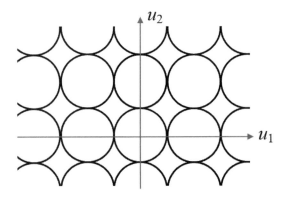

FIGURE 8.42. A rectangular array of circles.

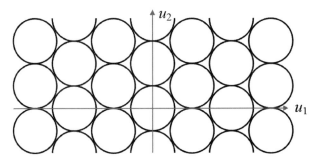

FIGURE 8.43. Maximally packed circles, shown here, achieve the Nyquist density. The hexagonal tile for this optimal periodic replication is shown at the right of Figure 8.45.

where

$$T = \frac{1}{2W}.$$

The tile, $\mathcal{C}$, corresponding to rectangular sampling is the square shown on the left of Figure 8.45.

Rectangular packed circles, however, do not achieve the Nyquist density. This, rather, is achieved by the packing shown in Figure 8.43.

The corresponding periodicity matrix is

$$\mathbf{P} = \begin{bmatrix} W & -W \\ \sqrt{3}W & \sqrt{3}W \end{bmatrix}. \qquad (8.103)$$

The sampling matrix for (8.103) follows as

$$\mathbf{Q} = \begin{bmatrix} T & -T \\ \dfrac{T}{\sqrt{3}} & \dfrac{T}{\sqrt{3}} \end{bmatrix} \qquad (8.104)$$

The tiles for the hexagonal geometries for the periodicity matrix and the sampling matrix in (8.104) are shown in Figure 8.44. Note the rotation of the hexagons. On the left hand

MULTIDIMENSIONAL SIGNAL ANALYSIS   379

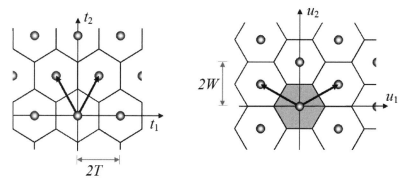

FIGURE 8.44. An optimal sampling strategy achieving the Nyquist density when an image's spectral support is a circle with radius W. The hexagonal tiling on the right is achieved with the periodicity matrix in (8.103). This dictates the hexagonal sampling shown on the left. The corresponding sampling matrix is given in (8.104).

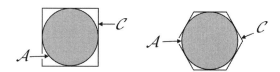

FIGURE 8.45. On the left is the tile $\mathcal{C}$ corresponding to the circular support $\mathcal{A}$ when the circles are packed rectangularly as shown at the left in Figure 8.42. On the right, maximally packed circles, as shown on the right in Figure 8.42, require a hexagonal tile.

side of Figure 8.44, the pointed ends of the hexagons point up and down. On the right, they point left to right.

8.9.1.2 Sampling Density Comparisons

What is the savings in sampling density of the hexagonal geometry in over the rectangular sampling geometry? For hexagonal sampling, the (Nyquist) sampling density is equal to the area of the hexagon shaded on the right of Figure 8.45 which in turn, is equal to the determinant magnitude of (8.103).

$$\mathrm{SD}_{nyq} = 2\sqrt{3}W^2. \qquad (8.105)$$

The corresponding density for rectangular sampling is clearly

$$\mathrm{SD}_{rect} = 4W^2.$$

The ratio of these densities is

$$r_2 = \frac{\mathrm{SD}_{nyq}}{\mathrm{SD}_{rect}} = 0.866$$

where the subscript denotes $N = 2$ dimensions. Thus, use of hexagonal sampling reduces the sampling density by 13.4% over rectangular sampling.

This result has been extended to higher dimensions for an N dimensional hypersphere of radius W [387, 984]. The minimum rectangular sampling density is $\mathrm{SD}_{rect} = (2W)^N$.

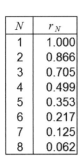
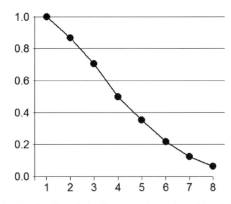

FIGURE 8.46. For maximally packed spheres of radius W in N dimensions, the table of the left is the ratio of area of the multidimensional tiles to $(2W)^N$. This is the tile area corresponding to when the spheres are packed rectangularly. On the right is a plot of these numbers.

The Nyquist density, SD_{nyq}, can be evaluated from the geometry of maximally packed spheres. The ratio

$$r_N = \frac{SD_{nyq}}{SD_{rect}}$$

is plotted in Figure 8.46 up to eight dimensions.

8.9.2 Generalized Interpolation Functions

In this section, we extend some of the results of Section 6.2 on generalized interpolation functions to higher dimensions.

8.9.2.1 Tightening the Integration Region

If the spectral support of a signal, when replicated, has regions that are identically zero, then there exists alternate interpolation functions than $f_C(\vec{t}\,)$ in (8.101). Furthermore, we will later show forms of these alternate interpolation functions, in comparison to $f_C(\vec{t}\,)$, can have greater immunity to noise in the interpolation process.

In the sampling mechanics illustrated in Figure 8.38, a single spectrum is extracted from the replicated spectra by multiplying the replicated spectra by a function constant within a tile and zero otherwise. The choice of a tile, however, may not be necessary if the replications leave regions that are identically zero. Tighter integration regions may be possible.

To illustrate interpolation functions with tighter integration regions, consider a spectral replication in accordance to the rotated rectangle tile on the left of Figure 8.47. A single tile is shown in the left. The region covered by the tile is denoted by C. Within the tile on the right is the spectral support of the replicated spectra. It is shown shaded and is assumed to be identically zero outside of a region A. Also shown by a dashed line is the boundary of a region, B, that encloses the spectral support (i.e., $A \in B$), but excludes higher order spectra. Although not shown in Figure 8.47, the B region can extend beyond the tile region. Multiplication of the replicated spectra by a function equal to one inside B and zero outside will yield the isolated spectrum. Clearly, both A and C are special cases of B.

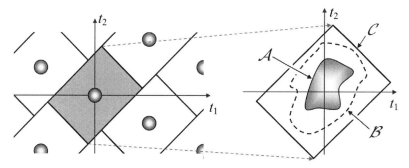

FIGURE 8.47. Illustration of three regions in a spectral replication. $\mathcal{B}$ is any region that isolates the spectral support. Special cases are the tile region, $\mathcal{C}$ and the spectral support region, $\mathcal{A}$.

Thus, in lieu of the multidimensional sampling theorem in (8.99), we have

$$x(\vec{t}) = \sum_{\vec{m}} x(\mathbf{Q}\vec{m}) f_\mathcal{B}(\vec{t} - \mathbf{Q}\vec{m}) \tag{8.106}$$

where the interpolation function with a tighter integration region is

$$f_\mathcal{B}(\vec{t}) = |\det \mathbf{Q}| \int_{\vec{u} \in \mathcal{B}} e^{-j2\pi \vec{u}^T \vec{t}} d\vec{t} \leftrightarrow \frac{\Pi_\mathcal{B}(\vec{u})}{|\det \mathbf{P}|}. \tag{8.107}$$

Note that, in one dimension, $\mathcal{A} = \mathcal{B} = \mathcal{C}$ at the Nyquist rate. In higher dimensions, such an equality is not assured at the Nyquist density.

8.9.2.2 Allowing Slower Roll Off

In one dimension, we were able to allow the spectrum of the interpolation function to be any desired function over those intervals where the signal's spectrum was identically zero. This allowed use of interpolation functions whose spectra had a slower roll off (see Figure 6.2). An analogous freedom occurs in higher dimensions. We can write, for example

$$x(\vec{t}) = \sum_{\vec{m}} x(\mathbf{Q}\vec{m}) f_\mathcal{B}^+(\vec{t} - \mathbf{Q}\vec{m}) \tag{8.108}$$

where

$$f_\mathcal{B}^+(\vec{t}) = f_\mathcal{B}(\vec{t}) + \int_{\mathcal{B} \cap \bar{\mathcal{A}}} \Re(\vec{u}) \exp(j2\pi \vec{u}^T \vec{t}) d\vec{u} \tag{8.109}$$

where $\Re(\vec{u})$ is any convenient function and $\cap$ denotes intersection. $\mathcal{B} \cap \bar{\mathcal{A}}$ denotes any region, $\mathcal{B}$, with the spectral support, $\mathcal{A}$ region excluded. Thus, integration is over all points in $\mathcal{B}$ not contained in $\mathcal{A}$. Equivalently,

$$f_\mathcal{B}^+(\vec{t}) \leftrightarrow |\det \mathbf{P}| \, \Pi_\mathcal{A}(\vec{u}) + [\, \Pi_\mathcal{B}(\vec{u}) - \Pi_\mathcal{A}(\vec{u})] \Re(\vec{u}).$$

8.10 Restoring Lost Samples

In this section, we will show that an arbitrarily large but finite number of lost samples can be regained from those remaining for certain band-limited signals even when sampling is performed at the Nyquist density [911].

8.10.1 Restoration Formulae

Let $\mathcal{M}$ denote a set of M integer vectors corresponding to the M lost-sample locations in an N- dimensional bandlimited signal sampled in accordance with a sampling matrix, $\mathbf{Q}$. As an example, consider Figure 8.48 where the sampling matrix is

$$\mathbf{Q} = \begin{bmatrix} 1 & -1 \\ 1 & 1 \end{bmatrix}.$$

A total of $M = 3$ lost samples are shown by hollow dots. It follows that

$$\mathcal{M} = \left\{ \begin{bmatrix} 0 \\ 0 \end{bmatrix}, \begin{bmatrix} 0 \\ 1 \end{bmatrix}, \begin{bmatrix} 1 \\ -2 \end{bmatrix} \right\}.$$

8.10.1.1 Lost Sample Restoration Theorem

If $x(\vec{t})$ is a bandlimited signal and the sampling matrix, $\mathbf{Q}$, is chosen to ensure that there is no aliasing, then the missing samples can be regained from solution of the M equations

$$\sum_{\vec{n} \notin \mathcal{M}} x(\mathbf{Q}\vec{n}) \left[\delta[\vec{k} - \vec{n}] - f_B(\mathbf{Q}(\vec{k} - \vec{n})) \right]$$
$$= \sum_{\vec{n} \in \mathcal{M}} x(\mathbf{Q}\vec{n}) f_B(\mathbf{Q}(\vec{k} - \vec{n})); \quad \vec{k} \in \mathcal{M} \tag{8.110}$$

assuming that the solution is not singular. The left-hand side of (8.110) contains the unknown samples. The right-hand side can be found from the known data.

Proof. We can write (8.107) as

$$x(\vec{t}) = \left[\sum_{\vec{n} \notin \mathcal{M}} + \sum_{\vec{n} \in \mathcal{M}} \right] x(\mathbf{Q}\vec{n}) f_B(\vec{t} - \mathbf{Q}\vec{m})$$

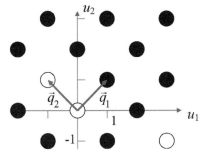

FIGURE 8.48. Illustration of the three lost samples in the set $\mathcal{M} = \{\vec{m}_1, \vec{m}_2, \vec{m}_3\}$. The location of the three lost samples is $\mathbf{Q}\vec{m}_k$; $k = 1, 2, 3$.

This expression can be evaluated at M points, and we can solve for the sample set $\{x(\mathbf{Q}\vec{n})|\vec{n} \in \mathcal{M}\}$. Let these M points be the $\vec{t} = \mathbf{Q}\vec{k}$, where $\vec{k} \in \mathcal{M}$:

$$x(\mathbf{Q}\vec{k}) = \left[\sum_{\vec{n} \notin \mathcal{M}} + \sum_{\vec{n} \in \mathcal{M}}\right] x(\mathbf{Q}\vec{n}) f_{\mathcal{B}}\{\mathbf{Q}(\vec{k}-\vec{n})\} \, ; \quad \vec{k} \in \mathcal{M}$$

Rearranging gives (8.110).

8.10.1.2 Restoration of a Single Lost Sample

For a single lost sample at the origin, if $f_{\mathcal{B}}(\vec{0}) \neq 1$, then

$$x(\vec{0}) = \left[1 - f_{\mathcal{B}}(\vec{0})\right]^{-1} \sum_{\vec{n} \neq \vec{0}} x(\mathbf{Q}\vec{n}) f_{\mathcal{B}}(-\mathbf{Q}\vec{n}) \qquad (8.111)$$

This follows from (8.110) for $M = 1$ and $\mathcal{M}$ containing only the origin. Note that the signal's interpolation can then be written directly void of the sample at the origin.

$$x(\vec{t}) = \sum_{\vec{n} \neq \vec{0}} x(\mathbf{Q}\vec{n}) \left(f_{\mathcal{B}}(\vec{t} - \mathbf{Q}\vec{n}) + \left[1 - f_{\mathcal{B}}(\vec{0})\right]^{-1} f_{\mathcal{B}}(-\mathbf{Q}\vec{n}) f_{\mathcal{B}}(\vec{t}) \right)$$

8.10.1.2.1 A Sufficient Condition for Singularity

A sufficient condition for (8.110) to be singular is when the integration region, $\mathcal{B}$, is over a tile, $\mathcal{C}$.

Proof. On a tile, the functions $\{\exp(j2\pi \vec{u}^T \mathbf{Q}\vec{n})\}$ form an orthogonal basis set. From (8.99) with $\mathcal{B} = \mathcal{C}$ we have

$$f_{\mathcal{B}}(\mathbf{Q}\vec{n}) = |\det \mathbf{Q}| \int_{\mathcal{C}} \exp(j2\pi \vec{u}^T \mathbf{Q}\vec{n}) d\vec{u} = \delta[\vec{n}].$$

The left-hand side of (8.110) is thus zero and the resulting set of equations singular.

8.10.2 Noise Sensitivity

Our purpose here is investigation of the lost sample restoration algorithm's performance when inaccurate data are used. In general, the algorithm becomes more unstable when (1) M increases and/or (2) the area corresponding to $\mathcal{B}$ increases with respect to that of the area of the tile $\mathcal{C}$. Indeed, restoration is no longer possible when $\mathcal{B} = \mathcal{C}$.

The restoration algorithm in (8.110) is linear. Let $\xi(\vec{t})$ denote a zero mean stochastic process. If $x(\vec{t})$ is uncorrelated with $\xi(\vec{t})$, then the use of $\{x(\mathbf{Q}\vec{n}) + \xi(\mathbf{Q}\vec{n})|\vec{n} \notin \mathcal{M}\}$ in (8.110) instead of $\{x(\mathbf{Q}\vec{n})|\vec{n} \notin \mathcal{M}\}$ will result in $\{x(\mathbf{Q}\vec{n}) + \eta(\mathbf{Q}\vec{n})|\vec{n} \in \mathcal{M}\}$, where $\{\eta(\mathbf{Q}\vec{n})|\vec{n} \in \mathcal{M}\}$ is the response to $\{\xi(\mathbf{Q}\vec{n})|\vec{n} \notin \mathcal{M}\}$ alone.

$$\sum_{\vec{n} \notin \mathcal{M}} \eta(\mathbf{Q}\vec{n})\{\delta[\vec{k}-\vec{n}] - f_{\mathcal{B}}\{\mathbf{Q}(\vec{k}-\vec{n})\}\}$$
$$= \sum_{\vec{n} \in \mathcal{M}} \xi(\mathbf{Q}\vec{n}) f_{\mathcal{B}}\{\mathbf{Q}(\vec{k}-\vec{n})\} \vec{k} \in \mathcal{M}. \qquad (8.112)$$

The restoration noise, η, depends linearly on the data noise, ξ. Thus the cross correlation between these two processes and the autocorrelation of η can be determined from a given input data noise autocorrelation.

Our treatment will be limited to the case when a single sample is lost and the data noise is white i.e.,

$$E[\xi(\mathbf{Q}\vec{n})\xi^*(\mathbf{Q}\vec{m})] = \overline{\xi^2}\delta[\vec{n} - \vec{m}] \tag{8.113}$$

where $\overline{\xi^2}$ is the data noise level (variance) and E denotes expectation. With no loss in generality, we place the lost sample at the origin, and (8.112) becomes

$$\eta(\vec{0}) = [1 - f_B(\vec{0})]^{-1} \sum_{\vec{n} \neq \vec{0}} \xi(\mathbf{Q}\vec{n}) f_B(-\mathbf{Q}\vec{n}).$$

Taking the square of the magnitude, expectating, and using (8.113) gives

$$\frac{\overline{\eta^2(\vec{0})}}{\overline{\xi^2}} = [1 - f_B(\vec{0})]^{-2} \sum_{\vec{n} \neq \vec{0}} |f_B(-\mathbf{Q}\vec{n})|^2 \tag{8.114}$$

where the restoration noise level is then

$$\overline{\eta^2(\vec{0})} = E[|\eta^2(\vec{0})|].$$

The sum in (8.114) can be evaluated through (8.111) with $x(\vec{t}) = f^*(-\vec{t})$. Since $F_B(\vec{u})$ is real, $f^*(-\vec{t}) = f(\vec{t})$. The result is

$$\frac{\overline{\eta^2(\vec{0})}}{\overline{\xi^2}} = \frac{f_B(\vec{0})}{1 - f_B(\vec{0})}. \tag{8.115}$$

This result has a fascinating geometrical interpretation. From (8.107)

$$f_B(\vec{0}) = |\det \mathbf{Q}| \int_B d\vec{u}.$$

But, with an illustration in Figure 8.47,

$$B = \int_B d\vec{u} = \text{area of integration}, \mathcal{B}.$$

and

$$C = \int_C d\vec{u} = |\det \mathbf{P}| = \text{area of a tile}.$$

Thus (8.115) can be written as

$$\frac{\overline{\eta^2(\vec{0})}}{\overline{\xi^2}} = \left(\frac{C}{B} - 1\right)^{-1}. \tag{8.116}$$

The restoration noise level is thus directly determined by the area of the integration region for $f_\mathcal{B}(\vec{t})$ and the area of a tile. Equation (8.116) is a strictly increasing function of $\mathcal{B}$. Thus, for minimum restoration noise level, we choose $\mathcal{B} = \mathcal{A}$ = the region of support of the transform of the signal $x(\vec{t})$.

For Nyquist density sampling in one dimension, $\mathcal{A} = \mathcal{B} = \mathcal{C}$. In this case oversampling is required to restore lost samples. For higher dimensions, the restoration capability is dependent on the region of support of the signal's spectrum. If the support is in the shape of a tile (e.g., rectangular, hexagonal), then restoration is not possible at the Nyquist density.

8.10.2.1 Filtering

Discrete white noise has a uniform spectral density and thus significant high-frequency energy. Once lost data have been restored, the data noise level can be reduced by filtering the result through $\mathcal{B}$ assuming that $\mathcal{B} < \mathcal{C}$. The noise level at the lost sample location remains the same. The noise level at locations far removed from the lost-sample locations will asymptotically be the same as that for the filtered noisy samples if no data were lost. If $\xi(\mathbf{Q}\vec{n})$ is zero mean and stationary, then after filtering, the process $\Psi(\mathbf{Q}\vec{n})$ is also stationary. If the data noise is white as in (8.113), its spectral density is uniform in $\mathcal{C}$. Thus if we filter the noise through $\mathcal{B}$, the resulting normalized noise level is

$$\frac{\overline{\Psi^2}}{\overline{\xi^2}} = \frac{\mathcal{B}}{\mathcal{C}}. \tag{8.117}$$

To minimize, we clearly would choose $\mathcal{B} = \mathcal{A}$.

For a single lost sample in discrete white noise, the ratio of the restoration noise level to that of data far removed is, after filtering through $\mathcal{B}$,

$$\frac{\overline{\eta^2(\vec{0})}}{\overline{\Psi^2}} = \left(1 - \frac{\mathcal{B}}{\mathcal{C}}\right)^{-1} \tag{8.118}$$

where we have used (8.116) and (8.117). To minimize, we again would choose $\mathcal{B} = \mathcal{A}$. Note that (8.118) exceeds both one and (8.116). Indeed, elementary manipulation reveals

$$\frac{\overline{\eta^2(\vec{0})}}{\overline{\Psi^2}} = \frac{\overline{\eta^2(\vec{0})}}{\overline{\xi^2}} + 1.$$

8.10.2.2 Deleting Samples from Optical Images

The Nyquist sampling density for images whose spectra have circular support is achieved when the circles in the frequency domain are densely packed as is shown in Figure 8.43.

Note, as is shown at the right in Figure 8.45, the area of $\mathcal{A}$ is less than that of $\mathcal{C}$. Thus, in the absence of noise, an arbitrary number of lost image samples can be restored from those (infinite number) remaining. For $\mathcal{B} = \mathcal{A}$, the interpolation function here is

$$f_\mathcal{A}(t_1, t_2) = 2W^2 |\det \mathbf{Q}| \operatorname{jinc}(Wr) = \frac{\operatorname{jinc}(Wr)}{\sqrt{3}}.$$

We can numerically illustrate the effects of discrete white noise on restoring a lost sample from an image that has a spectrum with circular support. Suboptimal rectangular sampling is considered first, followed by the optimal hexagonal case. Both cases are extended to higher dimensions.

Rectangular Sampling

If limited to rectangular sampling, the restoration noise level from (8.116) follows as

$$\frac{\overline{\eta^2(\vec{0})}}{\overline{\xi^2}} = \left(\frac{4}{\pi} - 1\right)^{-1} \approx 3.66. \tag{8.119}$$

After filtering through the $\mathcal{A}$ circle, the ratio of the restoration noise level to the data noise level at points far removed from the origin is

$$\frac{\overline{\eta^2(\vec{0})}}{\overline{\Psi^2}} = \left(1 - \frac{\pi}{4}\right)^{-1} \approx 4.66 \tag{8.120}$$

where we have used (8.118) with $B = A = \pi W^2$. The lost-sample noise is thus 6.7 dB above the filtered data noise at infinity.

The results can easily be extended to higher dimensions. Assume that the spectrum has support within an N-dimensional hypersphere of radius W. The sphere area (volume) is [1513]

$$A_{sphere}(N) = \begin{cases} \dfrac{2^N \pi^{\frac{N-1}{2}} \left(\frac{N-1}{2}\right)! W^N}{N!} & ; \text{odd } N \\[8pt] \dfrac{\pi^{\frac{N}{2}} W^N}{\left(\frac{N}{2}\right)!} & ; \text{even } N. \end{cases} \tag{8.121}$$

For rectangular sampling, $C = (2W)^N$. The corresponding plots of $\overline{\eta^2(\vec{0})}/\overline{\xi^2}$ and $\overline{\eta^2(\vec{0})}/\overline{\Psi^2}$ are shown as solid lines in Figure 8.49.

A single hexagonal tile is shown on the right of Figure 8.45 for minimum density sampling. If the radius of the inscribed circle is W, the area of the hexagon is

$$C = 2\sqrt{3} W^2$$

Thus, from (8.116) for $B = A = \pi W^2$

$$\frac{\overline{\eta^2(\vec{0})}}{\overline{\xi^2}} = \left[\frac{2\sqrt{3}}{\pi} - 1\right]^{-1} \approx 9.74$$

and, similarly, from (8.118)

$$\frac{\overline{\eta^2(\vec{0})}}{\overline{\Psi^2}} = \left[1 - \frac{\pi}{2\sqrt{3}}\right]^{-1} \approx 10.74$$

As one would expect, these values ($\approx$ 10dB) are greater than those of the corresponding rectangular sampling cases in (8.119) and (8.120).

In higher dimensions, Nyquist sampling corresponds to densely packing hyperspheres in the frequency domain. We use the table in Figure 8.46 in conjunction with (8.121) to generate the restoration noise level plots in Figure 8.49 for Nyquist density sampling when the signal's spectrum support is a hypersphere. The plots are shown with broken lines and, as we would expect, exceed the corresponding rectangular sampling results.

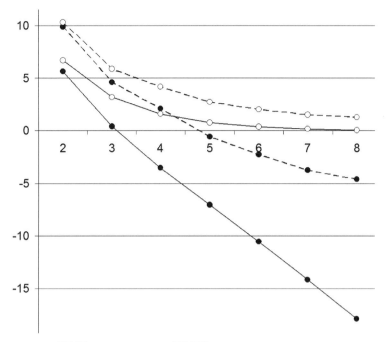

FIGURE 8.49. Plots of $\overline{\eta^2}/\overline{\xi^2}$ (filled circles) and $\overline{\eta^2}/\overline{\Psi^2}$ (open circles) in dB [$10\log_{10}(\cdot)$]. The solid lines are for minimum density rectangular sampling and the dashed for Nyquist (hexagonal) sampling.

8.11 Periodic Sample Decimation and Restoration

In the previous section, we showed that if gaps exist in the replication of spectra, then an arbitrarily large but finite number of lost samples can be regained from those remaining. In this section, we show that under the same circumstances, certain periodic sample decimation can be reversed from knowledge of the remaining samples. Thus, an infinite number of lost samples can be restored in certain scenarios. This procedure is applicable in certain cases even at the Nyquist density. The overall sampling density can thus be reduced to below the Nyquist density.

Fundamentally, we will distinguish between the Nyquist and *minimum sampling* densities which, in one dimension, are the same. The Nyquist density is the area of a tile when the replicated spectra are most densely packed. The minimum sampling density, we will show, is equal to the area of the support of a signal's spectrum. With reference to the right side of Figure 8.45, we showed, for example, that the Nyquist density for a signal with a circular spectrum is, $2\sqrt{3}W^2$. We will show that the minimum sampling density is the area of the circle, πW^2, a sampling density reduction of over 9%. In practice, this reduction can be achieved by the decimation procedure described in this section.

8.11.1 Preliminaries

Before a discussion of the procedure to restore decimated samples, we need to establish a formality for decimation notation. Consider sampling geometry at the top of Figure 8.50 with sampling matrix

$$\mathbf{Q} = \begin{bmatrix} T & -T \\ T & T \end{bmatrix}.$$

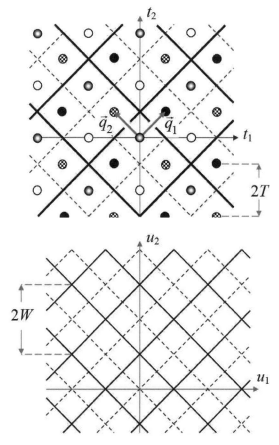

FIGURE 8.50. (Top) Samples are divided into four subgroups. (Bottom) A tile for all the samples is the large square diamond. Each diamond contains four subtiles. A subtile is a tile for any one sample subgroup.

As shown, the samples are divided into four groups labelled one through four. The sampling matrix for the group of solid dots is clearly

$$\mathbf{D} = \begin{bmatrix} 2T & -2T \\ 2T & 2T \end{bmatrix}.$$

Each of the other subgroups has the same sampling matrix, but with a different offset vector. Rather than the origin, replication is centered about the offset vector. Using the four samples in the bold square diamond, these offset vectors are

$$\mathbf{Q}\vec{e}_1 = \begin{bmatrix} T \\ -T \end{bmatrix}; \quad \mathbf{Q}\vec{e}_2 = \begin{bmatrix} 0 \\ 0 \end{bmatrix}$$

$$\mathbf{Q}\vec{e}_3 = \begin{bmatrix} -T \\ -T \end{bmatrix}; \quad \mathbf{Q}\vec{e}_4 = \begin{bmatrix} 0 \\ -2T \end{bmatrix}.$$

Correspondingly, we define the offset matrix

$$\mathbf{E} = [\,\vec{e}_1 \mid \vec{e}_2 \mid \vec{e}_3 \mid \vec{e}_4\,] = \begin{bmatrix} 0 & 0 & -1 & -1 \\ -1 & 0 & 0 & -1 \end{bmatrix}.$$

Let's examine the frequency domain periodicity generated by these sampling geometries. The periodicity matrix for all of the samples is

$$\mathbf{P} = \mathbf{Q}^{-T} = \begin{bmatrix} W & -W \\ W & W \end{bmatrix}.$$

The large square diamonds shown at the bottom of Figure 8.50 are corresponding tile. (Each contains four smaller square diamonds.) For any one of the sample subgroups, the periodicity matrix is

$$\mathbf{P}_D = \mathbf{D}^{-T} = \begin{bmatrix} \dfrac{W}{2} & -\dfrac{W}{2} \\ \dfrac{W}{2} & \dfrac{W}{2} \end{bmatrix}.$$

The smaller square diamonds at the bottom of Figure 8.50 outlined by broken lines can serve as periodicity *subtiles* for a sample subgroup. A tile, $\mathcal{C}$, and subtile, $\mathcal{C}_D$, for this example are shown in Figure 8.51.

We can generalize our observations. For a given $\mathbf{Q}$, we have

$$\mathbf{D} = \mathbf{Q}\mathbf{M}$$

where $\mathbf{M}$ is a nonsingular matrix of integers. For our previous example

$$\mathbf{M} = \begin{bmatrix} 2 & 0 \\ 0 & 2 \end{bmatrix}.$$

The $\mathbf{E}$ matrix of offset vectors is obtained through examination of any period of sample subgroups. There are

$$L = |\det \mathbf{M}|$$

subgroups and L subtiles per tile.

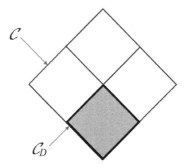

FIGURE 8.51. A diamond shaped tile divided into four subtiles.

The spectrum of all of the samples is

$$s(\vec{t}) = \sum_{\vec{n}} x(\mathbf{Q}\vec{n})\delta(\vec{t} - \mathbf{Q}\vec{n})$$

$$\leftrightarrow S(\vec{u}) = \sum_{\vec{n}} x(\mathbf{Q}\vec{n}) \exp(-j2\pi \vec{u}^T \mathbf{Q}\vec{n}).$$

The samples can be recovered via

$$x(\mathbf{Q}\vec{n}) = |\det \mathbf{M}| \int_{\mathcal{C}} S(\vec{u}) \exp(j2\pi \vec{u}^T \mathbf{Q}\vec{n}) d\vec{u}$$

where $\mathcal{C}$ is any tile.

The *ith* sample subgroup has a spectrum of

$$s_i(\vec{t}) = \sum_{\vec{n}} x(\mathbf{D}\vec{n} - \mathbf{Q}\vec{e}_i)\delta(\vec{t} - \mathbf{D}\vec{n} + \mathbf{Q}\vec{e}_i) \longleftrightarrow$$

$$S_i(\vec{u}) = \exp(-j2\pi \vec{u}^T \mathbf{Q}\vec{e}_i) \sum_{\vec{n}} x(\mathbf{D}\vec{n} - \mathbf{Q}\vec{e}_i) \exp(-j2\pi \vec{u}^T \mathbf{D}\vec{n});$$

$$1 \leq i \leq L. \tag{8.122}$$

A sample subgroup can be obtained from

$$x(\mathbf{D}\vec{n} - \mathbf{Q}\vec{e}_i) = |\det \mathbf{D}| \int_{\mathcal{C}_D} S_i(\vec{u}) \exp\left(j2\pi \vec{u}^T (\mathbf{D}\vec{n} - \mathbf{Q}\vec{e}_i)\right) d\vec{u} \tag{8.123}$$

where $\mathcal{C}_D$ is any subtile. Clearly

$$S(\vec{u}) = \sum_{i=1}^{L} S_i(\vec{u}). \tag{8.124}$$

8.11.2 First Order Decimated Sample Restoration

Consider unaliased replication of a spectrum in such a manner that gaps occur. As we have seen, this can even occur at Nyquist densities. We will show that if a subtile is totally subsumed in a gap contained within a tile, then any sample subgroup can be expressed as a linear combination of those sample subgroups remaining. Thus, a sample subgroup can be lost and the signal can still be interpolated from those samples remaining.

Proof. If a subtile, $\mathcal{C}_D$, lies in a gap, then $S(\vec{u})$ is identically zero there. Thus, for $\vec{u}$ within the subtile we have from (8.124)

$$S(\vec{u}) = 0 = S_L(\vec{u}) + \sum_{i=1}^{L-1} S_i(\vec{u})$$

or

$$S_L(\vec{u}) = -\sum_{i=1}^{L-1} S_i(\vec{u}).$$

Substituting into (8.123) and simplifying gives

$$x(\mathbf{D}\vec{n} - \mathbf{Q}\vec{e}_L) = -|\det \mathbf{D}| \sum_{i=1}^{L-1} \int_{C_D} S_i(\vec{u}) \exp\left(j2\pi \vec{u}^T (\mathbf{D}\vec{n} - \mathbf{Q}\vec{e}_L)\right) d\vec{u}.$$

Substituting (8.122) and simplifying gives

$$x(\mathbf{D}\vec{n} - \mathbf{Q}\vec{e}_L) = -\sum_{i=1}^{L-1} \sum_{\vec{m}} x(\mathbf{D}\vec{n} - \mathbf{Q}\vec{e}_i) f\left(\mathbf{D}(\vec{n} - \vec{m}) + \mathbf{Q}(\vec{e}_i - \vec{e}_L)\right) \quad (8.125)$$

where

$$f(\vec{t}) = -|\det \mathbf{D}| \int_{C_D} \exp(j2\pi \vec{u}^T \vec{t}) d\vec{u} \quad (8.126)$$

is the interpolation kernel. Equation (8.125) shows the manner by which the Lth sample subgroup can be recovered from the remaining $L - 1$ subgroups.

Example 8.11.1. We reconsider the minimum regular rectangular sampling density of $N = 2$ dimensional signals whose spectrum has a circular support. The spectral replication of Figure 8.42 is redrawn in Figure 8.52 with a bold outlined tilted rectangle shown as a tile. We divide the tile into eight identical square diamond subtiles. One of the tiles clearly falls into the gap among the replications. Using the periodicity matrix in (8.102), we find such subtiles can be generated by using

$$\mathbf{M} = \begin{bmatrix} 2 & -2 \\ 2 & 2 \end{bmatrix}.$$

A sample grouping that will achieve these subtiles is shown in Figure 8.53. The deleted samples are shown as solid dots.

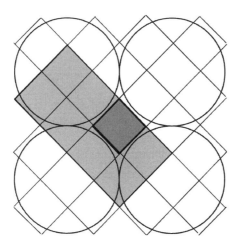

FIGURE 8.52. A tile for the replicated circles shown is the bold outlined tilted rectangle. The rectangle can be divided into eight square diamond subtiles one of which, shown shaded, lies within a gap among the circles.

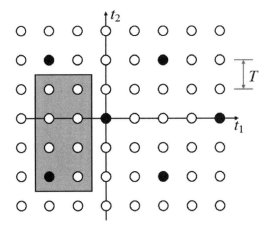

FIGURE 8.53. The decimation shown with solid dots achieves the periodicity subtile structure shown in Figure 8.52. Therefore, the samples at the locations of the solid dots can be deleted and restored as a linear combination of those remaining.

The interpolation function for this example follows from (8.126) using the square diamond in Figure 8.52 as the integration period, $\mathcal{C}_D$. The result is

$$f(t_1, t_2) = -\text{sinc}\left(\frac{t_1 + t_2}{4T}\right) \text{sinc}\left(\frac{t_1 - t_2}{4T}\right) e^{\frac{j\pi(t_1 + t_2)}{T}}$$

This decimation geometry reduces the sampling rate to $\frac{7}{8}(2W)^2 = 3.50W^2$ which is still higher than the Nyquist density of $2\sqrt{3}W^2 \approx 3.46W^2$.

8.11.3 Sampling Below the Nyquist Density

If there are gaps among spectral replications at the Nyquist density, then first order sample decimations can always be applied thereby reducing the overall sampling density below that of Nyquist [284, 288, 916]. We offer two examples.

8.11.3.1 The Square Doughnut

A spectral support is shown in Figure 8.54 that is zero inside the small square and zero outside the large square.

The Nyquist density is clearly $(2W)^2$. The large square, however, can be divided into a three by three array of smaller squares, the center one of which falls within an identically zero region. We may therefore decimate every ninth sample as shown in Figure 8.55.

The decimated samples can be recovered using (8.126) with $\mathcal{C}_D$ as the small square in Figure 8.54. The required interpolation function follows as

$$f(t_1, t_2) = -\text{sinc}\left(\frac{2Wt_1}{3}\right) \text{sinc}\left(\frac{2Wt_2}{3}\right).$$

The resulting sampling density is 8/9 that of Nyquist and, since the sampling density is equal to the spectral support of the square doughnut, the minimum sampling density has been achieved.

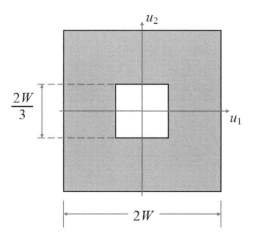

FIGURE 8.54. A "square doughnut" spectral support for which a sub-Nyquist sampling density is possible. Indeed, we show how to achieve the minimum sampling density.

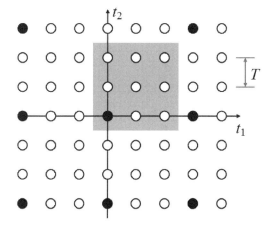

FIGURE 8.55. If a signal has the spectral support shown in Figure 8.54, then every ninth sample can be decimated as shown here. The composite sampling density is the minimum sampling density and is sub-Nyquist.

8.11.3.2 Sub-Nyquist Sampling of Optical Images

We return to the example of maximally packed circles as illustrated in Figure 8.43. A tile for this replication follows from choosing midpoints at four gaps and forming the parallelogram shown in Figure 8.56.

The tile is divided into four congruent parallelogram subtiles. The upper right parallelogram is divided into four smaller parallelograms. The process is repeated one final time. A small parallelogram, shown in black, is totally contained in a gap among the circles. Using this as a subtile, we see that we can reduce the overall sampling density to $\frac{63}{64}$ that of Nyquist. A sample decimation procedure to achieve this is shown in Figure 8.57.

8.11.4 Higher Order Decimation

First order decimation restoration can be straightforwardly generalized. If $M < L$ nonoverlapping subtiles lie in a gap within one tile, then M sample subgroups can be

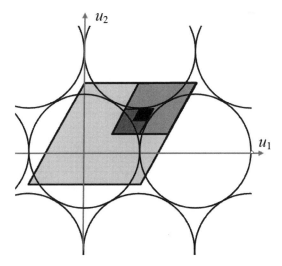

FIGURE 8.56. The small black parallelogram subtile lies totally in the gap among maximally packed circles. Since the subtile is $\frac{1}{64}^{th}$ the area of the larger parallelogram tile, the (Nyquist) sampling density can be reduced by a factor of $\frac{63}{64}$. A corresponding sample decimation strategy is shown in Figure 8.57

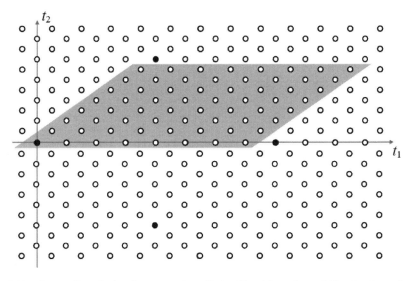

FIGURE 8.57. A sampling decimation procedure that will achieve the subtile structure shown in Figure 8.56. The solid dots represent the decimation locations.

represented as linear combinations of the remaining $L - M$. The analysis procedure is identical to that in the previous section, except that one solves for the M sample subgroups' spectra corresponding to M functional equations. The analysis is similar to a multidimensional extension of Papoulis' generalization as presented in Section 6.3. Details are given elsewhere [284, 287, 288].

Higher order decimation can be used to establish that the minimum sampling density for a bandlimited signal is equal to the area of its support. In Figure 8.58, for example, we have replication of circular support resulting from rectangular sampling. Each large

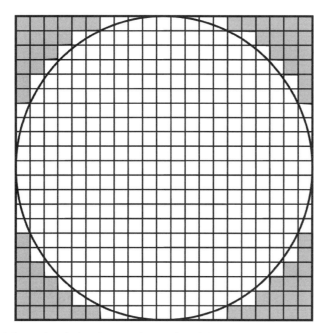

FIGURE 8.58. Higher order decimation can be used to illustrate that the sampling density for a bandlimited image can be reduced to the area of its spectral support and, therefore, the minimum sampling density.

square tile is divided into 441 square subtiles. In each tile, 68 of these subtiles (shown shaded) lie in a gap. Thus, if the circular radius is W, the sampling rate is reduced from $(2W)^2$ to $\frac{373}{441}(2W)^2 = 3.38W^2$, which is below the Nyquist density of $3.46W^2$. Ultimately, by increasing the number of subtiles, the sampling density can be reduced to $3.46W^2$, the area of the circular support and, for the circular spectral support, the minimum sampling density.

8.12 Raster Sampling

Raster sampling of a two dimensional signal, $x(t_1, t_2)$, is illustrated in Figure 8.59. The most familiar application of raster scanning is in analog (CRT) television images. The set of functional slices, $\{x(nT, t_2) | \infty < n < \infty\}$, serve as the raster sample set. The problem is, from this set, to reconstruct the original signal.

The sampled function of one dimensional vertical slices can be expressed via

$$s(t_1, t_2) = x(t_1, t_2) \times \frac{1}{T} \text{comb}\left(\frac{t}{T}\right)$$

$$= \sum_{n=-\infty}^{\infty} x(nT, t_2)\delta(t_1 - nT) \quad (8.127)$$

where T is the sampling interval. From the separability theorem, in two dimensions,

$$f(t_1) = f(t_1) \times 1 \leftrightarrow F(u_1)\delta(u_2).$$

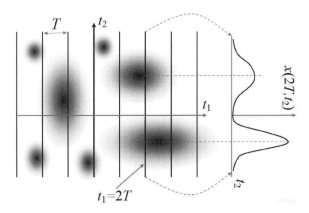

FIGURE 8.59. Illustration of raster sampling. Shown on the right is the one dimensional sample, $x(3T, t_2)$, taken from the slice of $x(t_1, t_2)$ along the line $t_1 = 2T$.

Thus, in two dimensions,

$$\text{comb}(t_1) \leftrightarrow \text{comb}(u_1)\delta(u_2)$$

and the Fourier transform of (8.127) gives

$$S(u_1, u_2) = X(u_1, u_2) * [\text{comb}(Tu_1)\delta(u_2)]$$

$$= \frac{1}{T} \sum_{n=-\infty}^{\infty} X\left(u_1 - \frac{n}{T}, u_2\right).$$

This is illustrated in Figure 8.60. The spectrum on the left is replicated in one dimension at intervals or $\frac{1}{T}$. Let the width of the spectrum, as illustrated in Figure 8.60, be B_2. In order for there to be no aliasing in the raster sampled signal, the signal spectra must not overlap. To do so, we make the familiar requirement

$$T \leq \frac{1}{2B_2}.$$

The Nyquist rate for raster sampling is thus

$$2B_2 = \frac{1}{2T}.$$

Note, in Figure 8.60, the spectral support need not be finite in the u_2 direction.

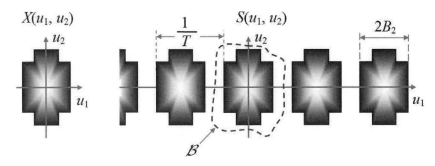

FIGURE 8.60. Spectral replication as a consequence of raster sampling.

Assuming no aliasing, to interpolate the original signal from the raster samples, the sampled signal must be run through a filter that isolates the zeroth order spectrum in the replication. This can be, for example, the region $\mathcal{B}$ shown in Figure 8.60. Then

$$X(u_1, u_2) = S(u_1, u_2) \times T\Pi_\mathcal{B}(u_1, u_2).$$

Choices for $\Pi_\mathcal{B}(u_1, u_2)$ are $\Pi(T\vec{u})$ and $\mathcal{B} = \mathcal{A}$ = the spectral support. The interpolation follows as

$$x(t_1, t_2) = \sum_{n=-\infty}^{\infty} x(nT, t_2) f_\mathcal{B}(t_1 - nT, t_2)$$

where the interpolation function is found by

$$f_\mathcal{B}(t_1, t_2) \leftrightarrow T\Pi_\mathcal{B}(u_1, u_2).$$

8.12.1 Bandwidth Equivalence of Line Samples

Here we establish the interesting property that, if $x(t_1, t_2)$ is bandlimited, all of its one dimensional parallel slices have the same bandwidth. As a consequence, every slice in a raster scan has the same bandwidth. We can establish this by inspection of Figure 8.61 where the support of a bandlimited signal, $X(u_1, u_2)$, is shown in the bottom right corner. Its inverse, $x(t_1, t_2)$, is in the upper left. The function, $\chi(t_1, u_2)$ shown in the lower left corner. It can be computed as the one dimensional inverse Fourier transform of each horizontal slice of $X(u_1, u_2)$.

$$\chi(t_1, u_2) = \int_{u_1=-\infty}^{\infty} X(u_1, u_2) e^{j2\pi u_2 t_2} du_1.$$

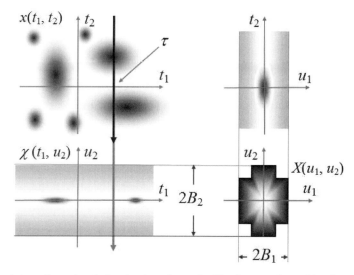

FIGURE 8.61. A two dimensional signal, $x(t_1, t_2)$, can be Fourier transformed by first transforming each vertical slice to form $\chi(t_1, u_2)$ in the bottom left corner. Each horizontal slice of $\chi(t_1, u_2)$ can be Fourier transformed to form the two dimensional Fourier transform, $X(u_1, u_2)$, in the bottom right hand corner. This perspective helps establish that all parallel one dimensional slices of two dimensional functions have the same bandwidth.

Since $X(u_1, u_2) = 0$ for $|u_2| > B_2$, it follows that $\chi(t_1, u_2)$ is also zero for $|u_2| > B_2$. The χ function can also be determined by the one dimensional Fourier transformation of each vertical slice of $x(t_1, t_2)$

$$\chi(t_1, u_2) = \int_{t_2=-\infty}^{\infty} x(t_1, t_2) e^{-j2\pi u_2 t_2} dt_2,$$

For a fixed slice $t_1 = \tau$, this is a one dimensional Fourier transform.

$$x(\tau, t) \longleftrightarrow \chi(\tau, u).$$

Clearly, the slice $x(\tau, t_2)$ is bandlimited with bandwidth B_2. This is true for any choice of τ. We thus conclude *any two parallel slices of a two dimensional bandlimited function have the same spectral support* (e.g., bandwidth.)

8.13 Exercises

8.1. What is a one dimensional sphere? What is a one dimensional cube?

8.2. [§] In Figure 8.62, the distance from point A to B on the unit square is $\sqrt{2}$. If we travel, as shown on the left, horizontally to the southeast corner, and then north, we travel a total of two units. If the trip is broken as shown in the middle figure, yet all steps are either horizontal or vertical, the distance is still two units. This is even true when, as shown on the right, the steps are made smaller. The distance travelled from A to B is still two units. Take this to the limit as the step size approaches zero. We then have the diagonal connecting A to B and the distance travelled is still two, not $\sqrt{2}$. Expose the faulty reasoning here.

8.3. The superposition integral in two dimensions mapping the contents in a $\Lambda_1 \times \Lambda_2$ region to a $T_1 \times T_2$ region is

$$y(t_1, t_2) = \int_{\tau_1=0}^{\Lambda_1} \int_{\tau_2=0}^{\Lambda_2} h(t_1, t_2; \tau_1, \tau_2) x(\tau_1, \tau_2) d\tau_1 d\tau_2;$$

$$0 \le t_1 \le T_1 \text{ and } 0 \le t_1 \le T_1.$$

This is a continuous version of the superposition sum in (8.3). For the discrete case, the two dimensional mapping can be expressed as the one dimensional mapping in (8.4). Explore a similar dimensional reduction for the continuous case.

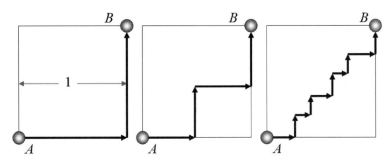

FIGURE 8.62. A flawed illustration that $\sqrt{2} = 2$.

8.4. § In the game of three dimensional tic tac toe in Figure 8.1, is it possible to have a tie game? If so, illustrate one.

8.5. Show that the multidimensional Fourier transform in (8.5) is linear.

8.6. Equation (8.23) states the inverse of a rotation matrix is the transposition of the rotation matrix. Verify this for the rotation matrix in (8.21).

8.7. For a given $x(\vec{t})$, interpret $x(\mathbf{A}\vec{t})$ when $\mathbf{A}$ is singular.

8.8. **Pitch, roll and yaw.** The pitch, roll and yaw of an aircraft is illustrated in Figure 8.63. The roll is the angular displacement of the aircraft on the (t_1, t_2) plane rotated about the t_3 axis. Pitch and yaw are analogously defined.

(a) Evaluate the three rotation matrices for pitch, roll and yaw angles of, respectively, θ_p, θ_r and θ_y.

(b) Compute the composite rotation matrix. Scaling matrices generally do not commute. Do these matrices commute?

(c) If $x(t_1, t_2, t_3) = 1$ inside the plane and zero outside at time zero with no rotation, compute the function at time s assuming speed c with $\theta_p = \theta_r = \theta_y = 45°$.

8.9. On the (t_1, t_2) plane, form the line connecting $t_2 = 2$ on the t_2 axis to the point $t_1 = 1$ on the t_1 axis. Let the right triangle defined by this segment and portions of the two positive axes form a triangle in the first quadrant. Let $x(\vec{t})$ be equal to one inside this triangle and zero outside. Also, let $\vec{\tau} = [\ 1\ 2\ ]^T$ and $\mathbf{R}_\theta$ be a rotation matrix for $\theta = \pi/4$. Let $\mathbf{S}$ be a scaling matrix with diagonal elements $[\ \frac{1}{2}\ \frac{2}{3}\ ]^T$. Provide detailed sketches of

(a) $x(\vec{t})$

(b) $x(\vec{t} - \vec{\tau})$

(c) $x(\mathbf{R}_\theta \vec{t})$

(d) $x(\mathbf{S}\vec{t})$

(e) $x(\mathbf{SR}_\theta \vec{t})$

(f) $x(\mathbf{R}_\theta \mathbf{S}\vec{t})$

(g) $x(\mathbf{SR}_\theta (\vec{t} - \vec{\tau}))$

(h) $x(\mathbf{SR}_\theta \vec{t} - \vec{\tau})$

(i) $x(\mathbf{S}(\mathbf{R}_\theta \vec{t} - \vec{\tau}))$

(j) $x(\mathbf{R}_\theta \mathbf{S}(\vec{t} - \vec{\tau}))$

(k) $x(\mathbf{R}_\theta \mathbf{S}\vec{t} - \vec{\tau})$

(l) $x(\mathbf{R}_\theta (\mathbf{S}\vec{t} - \vec{\tau}))$

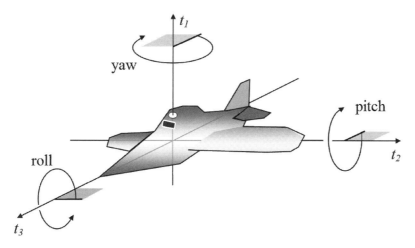

FIGURE 8.63. Illustration of the angular variable definitions for pitch, roll and yaw.

8.10. Consider three points on the (t_1, t_2) plane.

$$\vec{t}_1 = \begin{bmatrix} 0 \\ 0 \end{bmatrix}; \quad \vec{t}_2 = \begin{bmatrix} 0 \\ 2 \end{bmatrix}; \quad \vec{t}_3 = \begin{bmatrix} 1 \\ 0 \end{bmatrix}.$$

Using a "+" for $\vec{t}_1$, a "o" for $\vec{t}_2$ and an "×" for $\vec{t}_3$, plot the location of the following points.

(a) $\vec{t}$

(b) $\vec{t} - \vec{\tau}$

(c) $\mathbf{R}_\theta \vec{t}$

(d) $\mathbf{S}\vec{t}$

(e) $\mathbf{S}\mathbf{R}_\theta \vec{t}$

(f) $\mathbf{R}_\theta \mathbf{S}\vec{t}$

(g) $\mathbf{S}\mathbf{R}_\theta \left(\vec{t} - \vec{\tau}\right)$

(h) $\mathbf{S}\mathbf{R}_\theta \vec{t} - \vec{\tau}$

(i) $\mathbf{S}\left(\mathbf{R}_\theta \vec{t} - \vec{\tau}\right)$

(j) $\mathbf{R}_\theta \mathbf{S} \left(\vec{t} - \vec{\tau}\right)$

(k) $\mathbf{R}_\theta \mathbf{S}\vec{t} - \vec{\tau}$

(l) $\mathbf{R}_\theta \left(\mathbf{S}\vec{t} - \vec{\tau}\right)$

The matrices and vectors are defined in Exercise 8.9. The transformations here correspond to those in Exercise 8.9.

8.11. Evaluate the areas of the triangles
   (a) in Exercise 8.10.
   (b) in Exercise 8.9.

8.12. Compute the Fourier transform of the two dimensional functions shown in Figure 8.64. All are one inside and zero outside the curves shown. The ellipses in (c) and (d) are identical.

8.13. Let $f(r) \leftrightarrow F(\rho)$ denote Hankel transform pairs. Prove the following theorems.
   (a) Linearity.
   (b) Scaling

$$f(ar) \longleftrightarrow \frac{1}{|a|^2} F\left(\frac{\rho}{a}\right).$$

   (c) Inversion

$$f(r) = 2\pi \int_0^\infty \rho F(\rho) J_0(2\pi r \rho) d\rho. \tag{8.128}$$

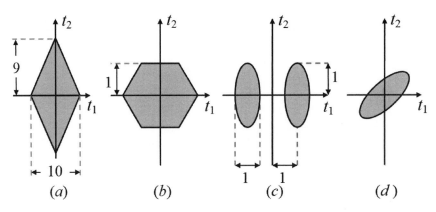

FIGURE 8.64. The ellipse in (d) is rotated 45° and had the same dimension as the ellipses in (c).

8.14. Compute the Hankel transform of
   (a) $f(r) = \exp(-\pi r^2)$.
   (b) the function $f(r)$ illustrated in Figure 8.12.
8.15. Show that the Hankel transform in N dimensions in (8.36) reduces to the one dimensional Fourier transform when $N = 1$ and the two dimensional Hankel transform when $N = 2$. Hint: For the $N = 1$ case, consider using a lower order spherical Bessel function.
8.16. What is the relation between the Abel transform and the Hankel transform of a circularly symmetric function, $f(r)$? HINT: Apply the central slice theorem.
8.17. Consider the three dimensional probability density function, $f_{V_1,V_2,V_3}(v_1, v_2, v_3)$, describing the equilibrium velocity of particles in an ideal gas. The particle velocity in any one dimension is not dependent on the velocity in any perpendicular direction. Thus, V_1, V_2, and V_1 and independent random variables and the density function is separable.

$$f_{V_1,V_2,V_3}(v_1, v_2, v_3) = f_{V_1}(v_1) f_{V_2}(v_2) f_{V_2}(v_2).$$

The marginal densities will be identical so

$$f_{V_1}(v) = f_{V_2}(v) = f_{V_3}(v) = f_V(v).$$

Since the statistics of the probability density function are independent of direction, the probability density function will also be spherically symmetric so that the density can be written as

$$f_{V_1,V_2,V_3}(\rho) = f_V(v_1) f_V(v_2) f_V(v_3)$$

where

$$\rho^2 = v_1^2 + v_2^2 + v_3^2.$$

There only exists one function type, $f_V(v)$, that can satisfy both the required separability and spherical symmetric properties for this problem. What is it?
8.18. Show that the magnitudes of the circular harmonics of $x(r, \theta)$ are the same as those for the rotated image $x(r, \theta - \phi)$.
8.19. **Fractional derivatives.** The inverse Abel transform touches on the idea of fractional derivatives. The derivative theorem states

$$\left(\frac{d}{dt}\right)^n x(t) \leftrightarrow (j2\pi u)^n X(u).$$

If n is not an integer, we have a fractional derivative. Express the fractional derivative

$$\left(\frac{d}{dt}\right)^{1/2} x(t)$$

as a convolution integral.
8.20. Derive the inverse Abel transform expression in (8.48) using (8.49).

8.21. (a) If a projection, $p(t)$, is zero for $|t| > T$, show that the inverse Abel transform in (8.49) can be written as

$$f(r) = \frac{1}{\pi} \int_{t=r}^{\infty} \sqrt{t^2 - r^2} \, \frac{d}{dt}\left(\frac{p'(t)}{t}\right) dt - \frac{p'(T)}{\pi T}\sqrt{T^2 - r^2}. \qquad (8.129)$$

(b) Evaluate the inverse Abel transform of the Parzen window in Table 9.2.

8.22. In the Radon transform, the function $p_\theta(t_1)$ in (8.52) is referred to as a *sinogram*. To show why, evaluate the Radon transform of a single point.

$$x(t_1, t_2) = \delta(t_1 - \tau_1)\delta(t_2 - \tau_2).$$

8.23. Compute the Fourier series coefficients for the periodic functions
   (a) $S(\vec{u}) = \sum_{\vec{n}} G(\vec{u} - \mathbf{P}\vec{n})$ where $G(\vec{u})$ is a given function that is not a period of $S(\vec{u})$
   (b) $S(u_1, u_2) = \cos[2\pi(\sin u_1 + \sin u_2)]$

8.24. What is the frequency response, $H(f_1, f_2)$, for the transformation function in (8.8.2) when the prototype filter, on the interval $|f| \leq \frac{1}{2}$, is

   (a) $H(f) = \Pi(4f)$, (c) $H(f) = \Lambda(4f)$,
   (b) $H(f) = \Lambda(2f)$, (d) $H(f) = \Lambda(3f)$.

8.25. A sampling geometry is denoted by $\mathbf{Q}$. Show that $\mathbf{QM}$ produces the same geometry when $\mathbf{M}$ is a matrix of integers and $|\det \mathbf{M}| = 1$.

8.26. Replace the multidimensional sample data, $x(\mathbf{Q}\vec{m})$, by the noisy data $x(\mathbf{Q}\vec{m}) + \xi(\mathbf{Q}\vec{m})$. The interpolation result is $x(\vec{t}) + \Psi(\vec{t})$. Assume that $\xi(\mathbf{Q}\vec{m})$ is zero mean and wide sense stationary with variance $\overline{\xi^2}$.
   (a) Show for $B = C$, that $\overline{\Psi^2(\vec{t})} = \overline{\xi^2}$. Thus the interpolation noise level is the same as the data noise level.
   (b) Show that when the data noise is discrete white noise, that the interpolation noise level is given by (8.117).

8.27. A signal $x(t_1, t_2)$ is known to have a spectrum that is identically equal to zero outside of an equilateral triangle. If a side of the triangle is B, what is a sampling matrix that we can use to minimize the sampling density? Draw the replicated spectra resulting from this sampling geometry.

8.28. A signal $x(t_1, t_2)$ is known to have a spectrum that is identically zero outside of a circle of finite radius in the (u_1, u_2) plane. The signal is sampled at its Nyquist density corresponding to this circle and the sample at the origin is lost. If all of the known samples are zero, what is the value of the lost sample?

8.29. For what dimension is the volume of a hypersphere in (8.121), for fixed radius, W, maximum? What happens to the volume of a hypersphere for fixed W as $N \to \infty$?

8.30. For a fixed $x(\vec{t})$ and $\mathbf{Q}$, numerically investigate the convergence rate of (8.111) for various B's.

8.31. Instead of $x(\mathbf{D}\vec{m} - \mathbf{Q}\vec{e}_i)$ on the right side of (8.125), suppose we had $x(\mathbf{D}\vec{m} - \mathbf{Q}\vec{e}_i) + \xi(\mathbf{D}\vec{m} - \mathbf{Q}\vec{e}_i)$ where ξ is zero mean discrete white noise with variance $\overline{\xi^2}$. Thus

$$E[\xi(\mathbf{D}\vec{m} - \mathbf{Q}\vec{e}_i)\xi(\mathbf{D}\vec{n} - \mathbf{Q}\vec{e}_i)] = \overline{\xi^2}\delta[\vec{n} - \vec{m}]\delta[i - j].$$

Let the response to this noisy data be

$$x(\mathbf{D}\vec{m} - \mathbf{Q}\vec{e}_L) + \eta(\vec{n}).$$

Clearly, $\eta(\vec{n})$ is zero mean and has the same variance,

$$\mathrm{E}[|\eta(\vec{n})|^2] = \overline{\eta^2}$$

for all $\vec{n}$. Find a closed form expression for the NINV, $\overline{\eta^2}/\overline{\xi^2}$.

8.32. Consider the parallelogram periodicity cell for maximally packed circles pictured in Figure 8.65. The parallelogram can be divided into two equilateral triangles. One triangle is positioned, as shown, with its three vertices centered in gaps. The triangle is divided into four smaller equilateral triangles. The center triangle is again divided. Note that the small equilateral triangle falls completely in a gap.
  (a) Using this observation, describe a second order decimation scheme to reduce sampling below the Nyquist rate. (Note: as shown, the periodicity cell can be oriented as to contain these two gaps.)
  (b) What is the overall density of this decimation scheme?
  (c) For your choice of subgroup decimation, derive the required interpolation functions.

8.33. Assume that functions in Figure 8.64 are functions of (u_1, u_2) rather than (t_1, t_2). We wish to raster sample $N = 2$ dimensional signals that have spectra with supports illustrated in Figure 8.64(a), (b) and (d). We have the freedom to choose the sense of sampling. All sample slices, for example, can be horizontal, all can be at 45° or 30°, etc.
  (a) In each case, specify a sampling sense that produces the minimum number of line samples per unit interval. Also, give this minimum rate.
  (b) Sketch the resulting spectral replication.

FIGURE 8.65. Geometry for Exercise 8.32.

8.34. Multidimensional system theory is a straightforward generalization of the one dimensional systems theory presented in Section 3. The response to an input $x(t_1, t_2)$ is $\mathcal{S}[x(t_1, t_2)] = y(t_1, t_2)$. Classify each system below as isoplanatic and/or linear. If linear, specify its point-spread function, $h(t_1, t_2; \tau_1, \tau_2)$. If linear and isoplanatic, specify $h(t_1, t_2)$.

(a) Magnifier.
$$y(t_1, t_2) = \frac{1}{M^2} x\left(\frac{t_1}{M}, \frac{t_2}{M}\right); 0 \neq M \neq 1.$$

(b) Let $f(z)$ be a nonlinear function.
$$y(t_1, t_2) = f[x(t_1, t_2)].$$

(c) Rotation by an angle θ.
$$y(t_1, t_2) = x(t_1 \cos\theta + t_2 \sin\theta, t_2 \cos\theta - t_1 \sin\theta).$$

(d) Translation.
$$y(t_1, t_2) = x(t_1 - \xi_1, t_2 - \xi_2).$$

(e) Fourier transform.
$$y(t_1, t_2) = \int_{\tau_1=-\infty}^{\infty} \int_{\tau_2=-\infty}^{\infty} x(\tau_1, \tau_2) e^{-j2\pi(t_1\tau_1 + t_2\tau_2)} d\tau_1 d\tau_2.$$

8.14 Solutions for Selected Chapter 8 Exercises

8.1. In one dimension, both a cube and a sphere are line segments.

8.2. The distance of two is the ℓ_1 norm between A and B. The diagonal measure of $\sqrt{2}$ is the ℓ_2 or Euclidian norm.

8.4. An example of a tie game in three dimensional tic tac toe is shown in Figure 8.66.

8.8. From (8.21), $\mathbf{R} = \mathbf{R}_\theta \mathbf{R}_\phi$. Hence, $\mathbf{R}^T = \mathbf{R}_\phi^T \mathbf{R}_\theta^T$ and

$$\begin{aligned}
\mathbf{R}\mathbf{R}^T &= \left[\mathbf{R}_\theta \mathbf{R}_\phi\right]\left[\mathbf{R}^T \mathbf{R}_\phi^T \mathbf{R}_\theta^T\right] \\
&= \mathbf{R}_\theta \left[\mathbf{R}_\phi \mathbf{R}_\phi^T\right] \mathbf{R}_\theta^T \\
&= \mathbf{R}_\theta \mathbf{I} \mathbf{R}_\theta^T \\
&= \mathbf{I}
\end{aligned}$$

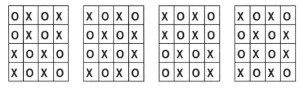

FIGURE 8.66. A tie game.

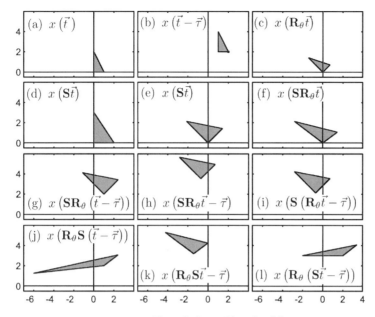

FIGURE 8.67. The solution to Exercise 8.9.

verifying

$$\mathbf{R}^T = \mathbf{R}^{-1}.$$

8.9. See Figure 8.67.
8.10 See Figure 8.68.
8.11 In two dimensions, the area of a parallelogram formed by vectors $\vec{t}_1$ and $\vec{t}_2$ is the determinate

$$\left| \begin{bmatrix} \vec{t}_1 \vdots \vec{t}_2 \end{bmatrix} \right|.$$

For two square matrices, $\mathbf{A}$ and $\mathbf{B}$, we have the identity

$$|\mathbf{AB}| = |\mathbf{A}||\mathbf{B}|.$$

Since

$$|\mathbf{S}| = \frac{1}{3},$$
$$|\mathbf{R}_\theta| = 1,$$

and the area of the original triangle in is one, it is straightforward to show in Exercise 8.10 that each transformation not containing $\mathbf{S}$ (cases (a) through (c)) resulted in triangles with unit area, whereas all transformations containing $\mathbf{R}_\theta$ (cases (d) through (l)) resulted in triangles with an area of $\frac{1}{3}$. The areas of the triangles in Exercise 8.9 are the reciprocal. In cases (a) through (c), the area is one. In cases (d) through (l), the area is

$$|\mathbf{S}^{-1}| = \frac{1}{|\mathbf{S}|} = 3.$$

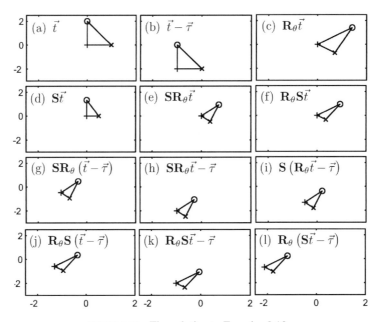

FIGURE 8.68. The solution to Exercise 8.10.

8.13. (b) The scaling theorem for the Hankel transform follows directly from the scaling theorem applied to the two dimensional Fourier transform.
(c) The most obvious approach to derive the inverse Hankel transform is substitution of the inversion formula into the Hankel transform and manipulation of the mathematics to obtain an identity. Although correct, the approach requires use of Bessel function identities. A more straightforward approach begins with writing the inverse Fourier transform in two dimensions using (8.8).

$$x(t_1, t_2) = \int_{-\infty}^{\infty} \int_{-\infty}^{\infty} X(u_1, u_2) e^{j2\pi(u_1 t_1 + u_2 t_2)} du_1 du_2.$$

Following steps analogous to Section 8.4.6 immediately results in the inversion formula for the Hankel transform given in (8.128).

8.14. (a) In two dimensions, we know from the separability theorem that

$$e^{-\pi r^2} = e^{-\pi(t_1^2 + t_2^2)} = e^{-\pi t_1^2} e^{-\pi t_2^2} \longleftrightarrow e^{-\pi u_1^2} e^{-\pi u_2^2} = e^{-\pi \rho^2}.$$

Thus, the Hankel transform of $e^{-\pi r^2}$ is $e^{-\pi \rho^2}$.
(b) The function $f(r)$ in Figure 8.12 is the difference between two circles. Its Hankel transform will thus be the difference between two jinc's.

8.16. The Abel transform is the projection of $f(r)$. From the central slice theorem, we know that the Fourier transform of a tomographic projection is a slice of the two dimensional Fourier transform. Since the two dimensional Fourier transform of $f(r)$ is circularly symmetric and all slices are identical, we conclude that the one dimensional Fourier transform of the Abel transform is the Hankel transform of $f(r)$. Equivalently, the inverse Hankel transform of the one dimensional

Fourier transform of the Abel transform returns the original function. In other words[149]

$$f(y) = \int_0^\infty \left\{ \int_{x=-\infty}^\infty \left[\left(\int_{\xi=x}^\infty \frac{2\xi f(\xi)}{(\xi^2 - x^2)^{\frac{1}{2}}} d\xi \right) e^{-j2\pi\rho x} dx \right] \rho \, J_0(2\pi\rho y) \right\} d\rho.$$

8.17. Since, for c and C constants,

$$Ce^{-cv_1^2} e^{-cv_2^2} e^{-cv_3^2} = Ce^{-c\rho^2},$$

our answer is the Gaussian random variable. Details of the constants are determined by the mass of the particle, m, and temperature, T. The result is the Maxwell-Boltzmann velocity distribution for ideal gases.

$$f_{V_1,V_2,V_3}(v_1, v_2, v_3) = \sqrt{\left(\frac{1}{2\pi mkT}\right)^3} \, e^{\frac{-m(v_1^2+v_2^2+v_3^2)}{2kT}}.$$

where k is Boltzmann's constant.[17]

8.19.

$$\left(\frac{d}{dt}\right)^{1/2} x(t) \leftrightarrow (j2\pi u)^{1/2} X(u)$$

$$= \frac{j}{(-j2\pi u)^{1/2}} (j2\pi u) X(u)$$

$$= \frac{j}{\sqrt{\pi}} (-j2u)^{-1/2} (j2\pi u) X(u)$$

Using (8.46),

$$\left(\frac{d}{dt}\right)^{1/2} x(t) = \frac{1}{\sqrt{\pi}} \left(t^{-1/2} \mu(-t)\right) * \frac{d}{dt} x(t)$$

8.20. Integrate the inverse Abel transform in (8.48) by parts. Using standard notation

$$u = \frac{p'(t)}{t} \qquad dv = \frac{t}{\sqrt{t^2 - r^2}} dt$$

$$du = \frac{d}{dt}\left(\frac{p'(t)}{t}\right) dt \quad v = \sqrt{t^2 - r^2}.$$

Equation (8.49) follows directly.

8.21. (b) The inverse Abel transform of the Parzen window is, for $T = 1$ [1515],

$$f(r) = \begin{cases} f_1(r) \; ; \; 0 \leq r \leq \frac{1}{2} \\ f_2(r) \; ; \; \frac{1}{2} \leq r \leq 1 \end{cases}$$

---

17. See Section 4.4.3.1.

where

$$f_1(r) = \frac{9}{\pi}\left(\frac{b}{2} - r^2 \ln\left(\frac{\frac{1}{2}-b}{r}\right)\right) + \frac{6}{\pi}\left(\frac{9b}{4} - \frac{3a}{2} + c\ln\left(\frac{1+a}{\frac{1}{2}+b}\right)\right),$$

$$f_2(r) = \frac{6}{\pi}\left(\frac{-3a}{2} + c\ln\left(\frac{1+a}{4}\right)\right),$$

and

$$a = \sqrt{1-r^2},$$

$$b = \sqrt{\frac{1}{4} - r^2},$$

$$c = 1 + \frac{r^2}{2}.$$

8.24. In all figures in this exercise, $N = 25$.
(a) For $H(f) = \Pi(4f)$, see Figure 8.69.
(b) For $H(f) = \Lambda(2f)$, see Figure 8.70.
(c) For $H(f) = \Lambda(4f)$, see Figure 8.71.
(d) For $H(f) = \Lambda(3f)$, a plot of $-H(f_1, f_2)$ is shown in Figure 8.72.

8.29. A plot of the volume of a unit radius hypersphere given by (8.121) is shown in Figure 8.73. The maximum of 5.263789014 occurs at $N = 5$ dimensions. Clearly, as the dimension N increases, the volume approaches zero.

8.33. (a) The minimum raster sampling rate occurs when sampling with parallel lines with a slope of 9/5 (or −9/5). The resulting minimum sampling rate is equal to the shortest distance between two parallel sides of the parallelogram which can be shown, after a bit of work, to be $1/T = 2\sqrt{56}$.
(b) Use horizontal lines at a sampling interval $T = 1/2$.
(c) Sample with lines of unit slope separated by $T = 1$ intervals. Corresponding spectral replications are shown in Figure 8.74.

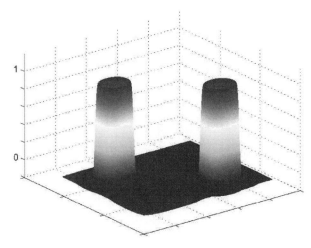

FIGURE 8.69. Solution to Exercise 8.24 (a). A Fourier series was used with $h[n] = \alpha \, \text{sinc}(\alpha n)$ for $\alpha = \frac{1}{4}$.

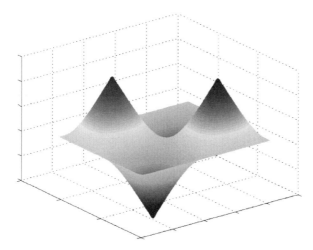

FIGURE 8.70. Solution to Exercise 8.24 (b). A Fourier series was used with $h[n] = \alpha \, \text{sinc}^2(\alpha n)$ for $\alpha = \frac{1}{2}$.

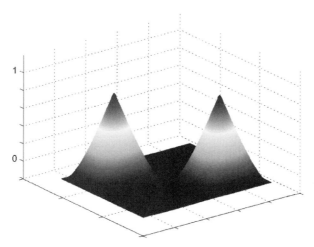

FIGURE 8.71. Solution to Exercise 8.24 (c). A Fourier series was used with $h[n] = \alpha \, \text{sinc}^2(\alpha n)$ for $\alpha = \frac{1}{4}$.

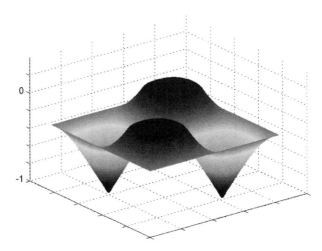

FIGURE 8.72. Solution to Exercise 8.24 (d) The negative of $H(f_1, f_2)$ is plotted. A Fourier series was used with $h[n] = \alpha \operatorname{sinc}^2(\alpha n)$ for $\alpha = \frac{1}{3}$.

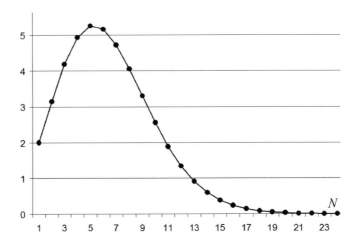

FIGURE 8.73. The volume of a unit radius hypersphere as a function of dimension, N.

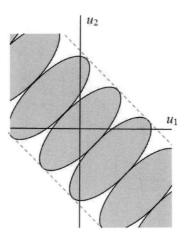

FIGURE 8.74. See the solution to Exercise 8.33.

9

Time-Frequency Representations

God exists since mathematics is consistent, and the Devil exists since we cannot prove it.
Andre Weil (1906–1998) [423]

Medicine makes people ill, mathematics makes them sad, and theology makes them sinful.
Martin Luther (1483–1546) [1287]

The good Christian should beware of mathematicians, and all those who make empty prophecies. The danger already exists that the mathematicians have made a covenant with the devil to darken the spirit and confine man in the bonds of Hell.[1]
Saint Augustine [38]

9.1 Introduction

The Fourier transform is not particularly conducive in the illustration of the evolution of frequency with respect to time. A representation of the temporal evolution of the spectral content of a signal is referred to as a *time-frequency representation* (TFR). The TFR, in essence, attempts to measure the instantaneous spectrum of a dynamic signal at each point in time.

Musical scores, in their most fundamental interpretation, are TFR's. The fundamental frequency of the note is represented by the vertical location of the note on the staff. Time progresses as we read notes from left to right. The musical score shown in Figure 9.1 is an example. Temporal assignment is given by the note types. The 120 next to the quarter note indicates the piece should be played at 120 beats per minute. Thus, the duration of a quarter note is one half second.

The frequency of the A above middle C is, by international standards, 440 Hertz. Adjacent notes notes have a ratio of $2^{1/12}$. The note, A#, for example, has a frequency of $440 \times 2^{1/12} = 466.1637615$ Hertz. Middle C, nine half tones (a.k.a. semitones or chromatic steps) below A, has a frequency of $440 \times 2^{-9/12} = 261.6255653$ Hertz. The interval of an octave doubles the frequency. The frequency of an octave above A is twelve half tones, or, $440 \times 2^{12/12} = 880$ Hertz.[2] The frequency spacings in the time-frequency representation of musical scores such as Figure 9.1 are thus logarithmic. This is made more clear in the alternate representation of the musical score in Figure 9.2 where time is on

---

1. During the fourth century, mathematicians were often equated with astrologers.
2. For more details, see Section 13.1.

FIGURE 9.1. A simple musical score is a TFR.

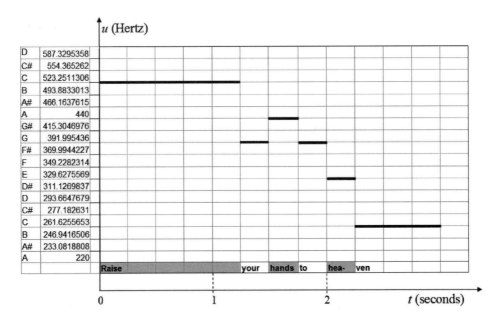

FIGURE 9.2. A representation of the musical score in Figure 9.1 as a function of time and frequency.

the horizontal axis and frequency on the vertical. At every point in time where there is no rest, a frequency is assigned. To make chords, numerous frequencies can be assigned to a point in time. Further discussion of the technical theory of western harmony is in Section 13.1.

If each of the notes in the musical score were represented by a sinusoid, the corresponding signal, $x(t)$, will consist of a sequence of sinusoids of the form

$$x(t) = \begin{cases} \sin(2\pi \times 523.2511306t) \; ; \; 0 \leq t < 1.25 \\ \sin(2\pi \times 391.995436t) \; ; \; 1.25 \leq t < 1.5 \\ \sin(2\pi \times 440t) \; \; \; \; \; \; \; \; \; \; \; \; \; \; \; ; \; 1.5 \leq t < 1.75 \\ \sin(2\pi \times 391.995436t) \; ; \; 1.75 \leq t < 2.0 \\ \sin(2\pi \times 329.6275569t) \; ; \; 2.0 \leq t < 2.25 \\ \sin(2\pi \times 261.6255653t) \; ; \; 2.25 \leq t < 2.75. \end{cases} \quad (9.1)$$

How do we transform a one dimensional temporal waveform of the type in (9.1) to a TFR of the type in Figure 9.2? Doing so is the fundamental problem addressed in TFR synthesis. The signals are generally much more complex than sinusoids and consist of numerous frequency components at each point in time.

9.2 Short Time Fourier Transforms and Spectrograms

TFR's can be computed at every instance in time or at specified intervals of time. For the case of specified intervals, the TFR is said to be *decimated*[3] *in time*. Decimation is appropriate when the TFR is either not changing quickly in time, or the TFR is unable to track rapid changes.

One way to generate an undecimated TFR display of a temporal signal, $x(t)$, is by placing the signal into a bank of bandpass filters, each filter being tuned to a different frequency. It is in this spirit that the *short time Fourier transform* is formulated. The short time Fourier transform, $S(t; u)$, of a signal, $x(t)$, is defined as

$$S(t; u) = \int_{-\infty}^{\infty} w(\tau) x(t - \tau) e^{-j2\pi u \tau} d\tau \qquad (9.2)$$

where $w(t)$ is a *sliding window*.

The discrete time equivalent of the short time Fourier transform is

$$S[n; f] = \sum_{-\infty}^{\infty} w[k] x[n - k] e^{-j2\pi f k} \qquad (9.3)$$

where $w[n]$ is the discrete time window and f is the discrete time frequency. Typically, the window is assumed to be of finite duration and is zero, say, for $|n| > L$.

9.3 Filter Banks

A filter bank [451, 805, 977, 1348, 1429, 1430] can be used to generate short time Fourier transforms for specific frequency values of the form $S(t, f_n)$ for some set of f_n.

A linearly calibrated filter bank using ideal bandpass filters is illustrated in Figure 9.3. The center frequencies are separated as $f_n = f_{n+1} + 2B$ and are of width $2B$. We can show this is a sampled version of the short time Fourier transform in (9.2). For an input $x(t) \longleftrightarrow X(u)$, the output of the nth filter is

$$S(t, f_n) = \int_{u=f_n-B}^{f_n+B} X(u) e^{j2\pi u t} du$$

or, equivalently

$$S(t, f_n) = \int_{-\infty}^{\infty} X(u) \Pi\left(\frac{u - f_n}{2B}\right) e^{j2\pi u t} du.$$

Using the convolution theorem

$$S(t, f_n) = x(t) * 2B \operatorname{sinc}(2Bt) e^{-j2\pi f_n t},$$

---

3. The *deci* prefix in *decimate* has long been abandoned to mean one tenth, as it does in the measure decibel (dB) meaning one tenth of a bel. The term has its origin in the Roman empire where, in a regiment of soldiers to be disciplined, one soldier in ten was chosen for punishment. In battle, soldiers had the choice between engaging in spirited battle or being disciplined after. The Greek historian Polybius of Megalopolis describes the procedure [1133] "The tribune assembles the legion, and brings up those guilty of leaving the ranks, reproaches them sharply, and finally chooses by lots sometimes five, sometimes eight, sometimes twenty of the offenders, so adjusting the number thus chosen that they form as near as possible the tenth part of those guilty of cowardice. Those on whom the lot falls are bastinadoed [beating of the soles of the feet with a stick as a form of torture or punishment] mercilessly."

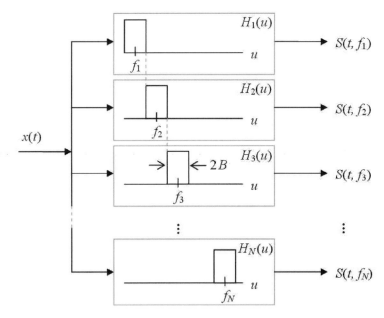

FIGURE 9.3. A filter bank for generating linearly spaced frequency samples the short time Fourier transform.

or

$$S(t, f_n) = \int_{-\infty}^{\infty} 2B \, \text{sinc}(2Bt) x(t - \tau) \, e^{-j2\pi f_n t}. \tag{9.4}$$

Comparing with (9.2), we see this is the short time Fourier transform of $x(t)$ sampled at $u = f_n = f_0 + 2nB$ with window

$$w(t) = 2B \, \text{sinc}(2Bt).$$

Our choice of block orthogonal[4] filters assures that the sum of the signals $S(t, f_n)$ over n will give the signal $x(t)$ over the band of frequencies considered.

Filter banks need not be linearly calibrated. If, for example, we desire 10 equally spaced samples per decade, we would generate samples of the type

$$S(t, f_n) = \int_{u=f_n}^{10^{1/10} f_n} X(u) e^{j2\pi ut} du \tag{9.5}$$

in reference to a fundamental frequency, f_0. Equivalently, $f_n = f_0 10^{n/10}$. Likewise, if we would like twelve notes per octave as is the case in music,[5] then for some reference frequency, f_0,

$$S(t, f_n) = \int_{u=f_n}^{2^{1/12} f_n} X(u) e^{j2\pi ut} du. \tag{9.6}$$

Thus $f_n = f_0 2^{n/12}$.

---

4. Two functions $\Theta(u)$ and $\Phi(u)$ are *block orthogonal* if $\Theta(u)\Phi(u) = 0$.
5. See Section 13.1.

TIME-FREQUENCY REPRESENTATIONS

TABLE 9.1. Some commonly used windows. $\hat{t} = t/T$ and I_0 denotes a modified Bessel function of the first kind. (See Exercize 2.18.) The number β is a parameter of the Kaiser window

boxcar	$\Pi(\hat{t})$
Bartlett	$(1-\|\hat{t}\|)\Pi\left(\frac{\hat{t}}{2}\right)$
Parzen	$(1 - 6\hat{t}^2 + 6\|\hat{t}\|^3)\Pi\left(\frac{\hat{t}}{2}\right) + 2[1-\|\hat{t}\|^3]\Pi\left(\|2\hat{t} - \frac{3}{2}\|\right)$
Hanning[9]	$\cos^2(\pi\hat{t}/2)\Pi(\frac{\hat{t}}{2})$
Hamming	$[0.54 + 0.46\cos(\pi\hat{t})]\Pi\left(\frac{\hat{t}}{2}\right)$
Blackman	$\left[0.42 + 0.5\cos + 0.08\cos(2\pi\hat{t})\right]\Pi\left(\frac{\hat{t}}{2}\right)$
Lanczos	$\operatorname{sinc}(\hat{t})\Pi\left(\frac{\hat{t}}{2}\right)$
Welch	$(1 - \|\hat{t}\|^2)\Pi\left(\frac{\hat{t}}{2}\right)$
Kaiser	$\frac{I_0(\pi\beta(1-\hat{t})^{\frac{1}{2}})}{I_0(\pi\beta)}\Pi\left(\frac{\hat{t}}{2}\right)$

The filter banks in (9.5) and (9.6), however, are not special cases of the short time Fourier transform in (9.2).[6]

9.3.1 Commonly Used Windows

Typically, $w(t)$ is of finite duration

$$w(t) = w(t)\,\Pi\left(\frac{t}{2T}\right). \tag{9.7}$$

We will also assume a normalization of $w(0) = 1$. Some commonly used windows are listed in Table 9.1.

Windows are chosen for their leakage-resolution tradeoff.[7] This is illustrated in Figure 9.4 for the case of a continuous time window, $w(t)$. A log plot of the normalized Fourier transform of a typical window is shown. Ideally, we would $\frac{W(u)}{W(0)} = \delta(u)$. Since, however, $w(t)$ is of finite duration, this is not possible. The best we can do is design the window to minimize the size of its main lobe, Δ. To avoid the leaching of adjacent frequencies, we also desire to keep the height of the first side lobe, δ, as small as possible. For a fixed window length, however, decreasing Δ increases δ and visa versa. The boxcar window, for example, has superb resolution but terrible leakage. For the windowing approach, we are left with choosing a tradeoff.[8]

9.3.2 Spectrograms

We refer to the function $|S(t; u)|^2$ as the *spectrogram*.

$$|S(t;u)|^2 = \left|\int_{-\infty}^{\infty} w(\tau)x(t-\tau)e^{-j2\pi u\tau}d\tau\right|^2 \tag{9.8}$$

---

6. They are, however, constant Q filter banks. See Exercises 9.3 and 9.4
7. In optics, windowing is referred to as apodization. See Section 13.2.5.
8. The same tradeoff is encountered in beamforming with a finite aperture. See Section 13.2.5.1.
9. Also known as the Hann window

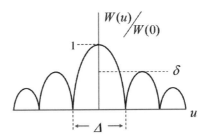

FIGURE 9.4. Illustration of the leakage-resolution tradeoff in a window.

Note that the spectrogram can be written as

$$| S(t; u) |^2 = \left| \int_{-\infty}^{\infty} w(\tau) x(t-\tau) e^{-j2\pi u(\tau-\varphi)} d\tau \right|^2 \tag{9.9}$$

where φ is an arbitrary phase. The arbitrariness of the choice of φ allows for simplification in some computational architectures for the spectrogram.

Similarly, for discrete time signals, the spectrogram may be defined by

$$| S[n; f] |^2 = \left| \sum_{-\infty}^{\infty} x[k] w[n-k] e^{-j2\pi f(k-\phi)} \right|^2 \tag{9.10}$$

where ϕ is arbitrary.

9.3.3 The Mechanics of Short Time Fourier Transformation

The mechanics of the short time Fourier transform are pictured in Figure 9.5. Illustrated is a window, $w(\tau)$, and a signal $x(\tau)$. Motivated by (9.2), we picture the functions $w(\tau)$ and $x(t-\tau)$ for a specific value of t. The product of these two functions, zero outside of the window, is Fourier transformed. This transform is the spectral representation used for $x(t)$ at time $t = \tau$. As t increases, the function slides to the right and the spectrum is generated for each point in time. In a relative sense, the signal can alternately be considered stationary and the window moving. This observation makes clear why the short time Fourier transform is also referred to as a *sliding window Fourier transform*.

9.3.3.1 The Time Resolution Versus Frequency Resolution Trade Off

The traditional scaling theorem of Fourier analysis states that coordinate compression in the time domain corresponds to inverse expansion in the frequency domain and visa versa. In the short time Fourier transform, this theorem unfortunately manifests itself as a time resolution versus frequency resolution trade off.[10] We can have good resolution in time but not frequency with the spectrogram, or good frequency resolution at the cost of poorer time resolution.

---

10. In the context of this section, the time resolution versus frequency resolution trade off differs from Heisenberg's uncertainty principle with which it is sometimes confused. See Section 13.3.

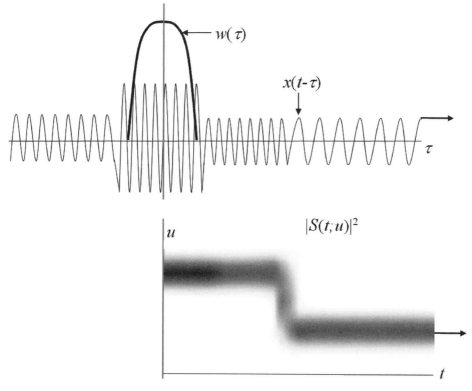

FIGURE 9.5. Mechanics for generating the short time Fourier transformation and spectrogram of a signal.

Visualize a sinusoid of one frequency abruptly changing to a sinusoid of another frequency. Far from the transition point, we best estimate the frequency of either sinusoid with the use of a long window corresponding to a large value of T. A long window will contain a large number of cycles and will Fourier transform into a very sharp and narrow frequency band. Using a long window, however, does not allow us to readily identify the point of transition. In its sliding, the long window contains this portion of the signal for a significantly long period of time. The preponderance of the signal within the window during the attempt to measure the signal, in effect, clouds identification from the short time Fourier transform of the exact time the frequency transition occurs. The answer, of course, is to shorten the duration of the window. This, however, increases the time resolution but at the cost of the frequency resolution. This, as illustrated in Figure 9.6, is the time resolution versus frequency resolution trade off. This limitation is one of the motivations for investigation of other time-frequency distributions such as Cohen's *generalized time-frequency representations* (GTRF's) [309, 310, 312].

9.3.4 Computational Architectures

There are a number of useful computational architectures for short time Fourier transforms. In most cases, the approach can be used effectively on both discrete time and continuous time signals.

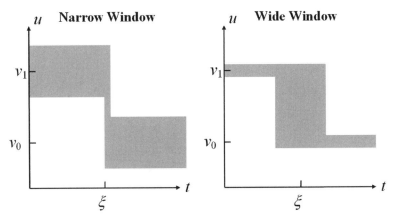

FIGURE 9.6. Illustration of the time resolution versus frequency resolution trade off for the short time Fourier transform and the spectrogram. A sinusoid with frequency v_1 makes a transition to a sinusoid of frequency v_0 at time $t = \xi$. For a window of short duration, as shown on the left, the frequency resolution will be poor since, for good frequency resolution, a large number of cycles must be present within the window. The thick horizontal swaths centered around v_1 and v_0 indicate a high uncertainty of the frequency values. If the window is narrow, on the other hand, the time resolution is high. The narrow window contains the transition point, $t = \xi$, for only a short period of time. Time resolution, indicated by the short vertical swath centered at $t = \xi$ on the left hand figure, is therefore high for a narrow window. The converse is true for a window of long duration. This is shown on the right side. Wide windows contain a larger number of cycles thereby allowing high frequency resolution. The horizontal swaths centered about v_1 and v_0 in the right hand figure, are therefore narrow. Time resolution, on the other hand, is poor. The transition at $t = \xi$ is contained in the wide window for a long period of time. Therefore, there is a large vertical swath on the right centered at $t = \xi$.

9.3.4.1 Modulated Inputs

Generation of the short time Fourier transform using a modulated version of the signal is best introduced by rewriting (9.2) as

$$S(t; u) = \{[x(t)e^{j2\pi ut}] * w(t)\}e^{-j2\pi u\tau} \quad (9.11)$$

where $*$ denotes convolution. The straightforward implementation of (9.11) is shown in Figure 9.7. For purposes of discussion, assume that $x(t)$ is a signal with significant spectral content at the frequency f. The Fourier transform of $x(t)\exp(j2\pi u\tau)$ translates the u component of the spectrum of $x(t)$ to be shifted to the origin. The window filter acts as a low pass filter to extract this low frequency component which, by multiplication by $\exp(j2\pi u\tau)$, is modulated back into a bandpass signal.

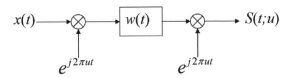

FIGURE 9.7. A system to evaluate the short time Fourier transform of a continuous time signal at frequency u. The impulse response of the LTI filter is the window, $w(t)$.

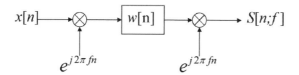

FIGURE 9.8. A system to evaluate the short time Fourier transform of a discrete time signal at frequency f. The impulse response of the LTI filter is the window, $w[N]$ which, for finite duration windows, can be implemented with an FIR filter.

In order to generate a TFR across the frequency spectrum, a number of processors of the type in Figure 9.7 can be used, each with the multiplicative oscillator tuned to a different frequency.

The discrete implementation, shown in Figure 9.8, is identical in concept. We rewrite (9.3) as

$$S[n; f] = \{(x[n]e^{j2\pi fn}) * w[n]\}e^{-j2\pi fn}. \qquad (9.12)$$

When the window filter, $w[n]$, is of finite duration, it can be implemented using a standard finite impulse response (FIR) filter [1149].

The spectrogram is achieved by simply performing the magnitude squared of the short time Fourier transform. Thus, since $|\exp(j\theta)| = 1$, the output sinusoid multiplication is not needed in Figures 9.7 and 9.8. Furthermore, the phase of the multiplicative complex exponential is no longer important when computing the spectrogram.

9.3.4.2 Window Design Using Truncated IIR Filters

Convolution of rectangles. A truncated IIR filter is an IIR filter configured with additional circuitry to require the composite impulse response to be of finite extent [919]. An example is shown in Figure 9.9. The first stage of the filter, from $x[n]$ to $v[n]$, is an IIR filter with an impulse response of a unit step. The second filter, from $v[n]$ to $y[n]$, is an FIR filter with impulse response $\delta[n] - \delta[n - (L+1)]$. The composite impulse response is a convolution of the two component impulse responses, and is equal to one for $0 \leq n \leq L$ and is otherwise zero-a shifted boxcar window of length $L + 1$.

When filters are cascaded, their impulse responses are convolved. Since the triangular Bartlett window is the convolution of two rectangles, cascading two of the filters in Figure 9.9 and a multiplier of $\frac{1}{L+1}$ will implement a Bartlett window of length $2L + 1$. Similarly, the Parzen window, equal to the auto-convolution of four rectangles, can be implemented by cascading four of the boxcar windows shown in Figure 9.9. Generating a spectrogram at frequency f using a Parzen window is shown in Figure 9.10.

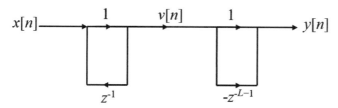

FIGURE 9.9. A boxcar window implemented as a truncated IIR filter. Cascading two of these windows gives a Bartlett window. Cascading four of them gives a Parzen window. (See Figure 9.10.)

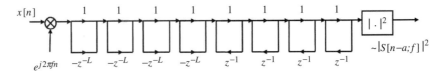

FIGURE 9.10. Using the architecture in Figure 9.5 with four cascaded boxcar windows depicted in Figure 9.9 generates a spectrogram of a signal $x[n]$ at frequency f with a Parzen window of length $2a = 2(L-3)$. (The output is proportional to the spectrogram because the window was not normalized.)

TABLE 9.2. Cosine component decomposition of some commonly used causal windows, $w[n]$, of length $L+1$

Type	Function	q_0	q_1	q_2
boxcar	1	1		
Hanning	$\frac{1}{2}\left[1 - \cos\left(\frac{2\pi n}{L}\right)\right]$	$\frac{1}{2}$	$-\frac{1}{2}$	
Hamming	$0.54 - 0.46\cos\left(\frac{2\pi n}{L}\right)$	0.54	-0.46	
Blackman	$0.42 - 0.5\cos\left(\frac{2\pi n}{L}\right) + 0.08\cos\left(\frac{4\pi n}{L}\right)$	0.42	-0.5	0.08

Truncated Sinusoid Impulse Responses. Sinusoid impulse responses can also be used in truncated IIR filters. If a causal sine wave is added to the same wave shifted by half a period, only the first hump of the original sine wave remains. A number of such filters can be placed in parallel to implement windows with multiple cosine components. In principal, any window can be thus implemented since any even function of finite duration can be characterized by a Fourier series of cosines. In practice, the number of cosine components should be small. Three such windows are the Hanning, Hamming and Blackman windows. These are summarized in Table 9.2. The windows each have a duration of L and have been shifted into positive time territory to allow for causal implementation.

Each of the filters in Table 9.2 can be expressed as a special case of the following causal window[11] of length $L+1$.

$$w[n] = \begin{cases} \beta_0 + \sum_{q=1}^{Q} \beta_q \cos\left(\frac{2\pi nq}{L}\right) & ; 0 \le n \le L \\ 0 & ; \text{otherwise} \end{cases} \quad (9.13)$$

The term with the constant β_0 can be implemented using a truncated IIR boxcar window of the type in Figure 9.9. In order to synthesize the other terms in (9.13), we will use filters with a marginally stable impulse response of

$$h[n] = \cos(n\theta)\mu[n]. \quad (9.14)$$

---

11. A discrete time causal window is zero for $n < 0$. Any window can be made causal by shifting it into positive territory.

The z transform of this cosine filter is

$$H_z(z) = \sum_{n=0}^{\infty} h[n]z^{-n}$$

$$= \sum_{n=0}^{\infty} \cos(n\theta)z^{-n}$$

$$= \Re \sum_{n=0}^{\infty} e^{jn\theta} z^{-n}$$

$$= \Re \sum_{n=0}^{\infty} \left(e^{j\theta} z^{-1}\right)^n$$

$$= \Re \frac{1}{1 - e^{j\theta}z^{-1}} \frac{1 - e^{-j\theta}z^{-1}}{1 - e^{-j\theta}z^{-1}}$$

$$= \Re \frac{1 - e^{-j\theta}z^{-1}}{1 - 2z^{-1}\cos(\theta) + z^{-2}}$$

$$= \frac{1 - \cos(\theta)z^{-1}}{1 - 2z^{-1}\cos(\theta) + z^{-2}}.$$

Review the general transfer function in (2.136) and the IIR schematic in Figure 2.29. We can implement the impulse response in (9.14) as shown on the left hand side of Figure 9.11. The left hand side filter, from $x[n]$ to $v[n]$, has the impulse response of (9.14).

The second portion of the filter in Figure 9.11, from $v[n]$ to $y[n]$, subtracts $v[n-L]$ from $v[n]$. To effect truncation of the sinusoid impulse response, we set

$$\theta = \frac{2\pi q}{L}.$$

The composite impulse response and z transform transfer function of the filter in Figure 9.11 is then

$$h_q[n] = k[n] - k[n-L]$$

$$= \begin{cases} \cos\left(\frac{2\pi qn}{L}\right) & ; 0 \leq n < L \\ 0 & ; \text{otherwise} \end{cases} \quad (9.15)$$

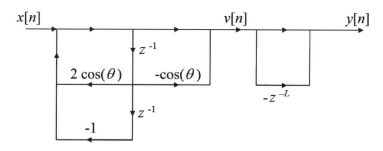

FIGURE 9.11. A truncated IIR filter for implementing a finite duration sinusoid impulse response.

and

$$H_q(z) = \sum_{n=0}^{\infty} h_q[n]z^{-n}$$

$$= \frac{z^{-1} \cos\left(\frac{2\pi q}{L}\right)(1 - z^{-2L})}{1 - 2z^{-1}\cos\left(\frac{2\pi q}{L}\right) + z^{-2}}. \tag{9.16}$$

Thus, filters of the type in (9.15) can be connected in parallel to achieve cosine series type windows. An example is shown in Figure 9.12 where the short time Fourier transform is performed at frequency f. It follows the architecture in Figure 9.8 with a truncated IIR filter inserted. The filter implements windows of the type in (9.13) for $Q \leq 2$, including, for appropriate choices of the β's in Table 9.2, the Blackman window. For the Hanning and Hamming windows, only $Q = 1$ is required. They can therefore be implemented by deleting the bottom (β_2) filter component and appropriate choices of β_0 and β_1.

The parallel connection of $-z^{-L}$ and a unit gain, common to all of the filters, can be factored from the filters. Only one such circuit is needed and is in the upper right of the

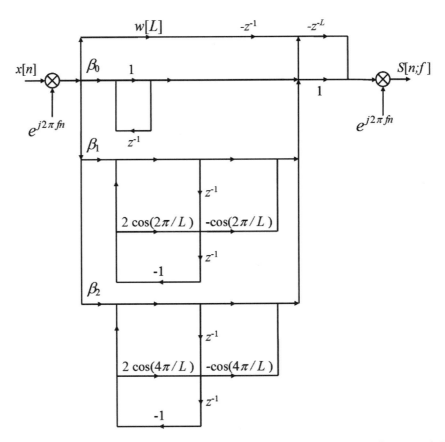

FIGURE 9.12. Generating the short time Fourier transform of a signal, $x[n]$ at the frequency f. For appropriate choices of the β's, any of the filters in Table 9.2 can be implemented. If a β is zero, the stage can be omitted.

circuit in Figure 9.12. Additional circuitry is added for end point correction of the window. In (9.15), the cosine filter is implemented for $0 \leq n < L$ rather that the desired $0 \leq n \leq L$. The window at the end point, $w[L]$, is therefore not included in the implementation and in added explicitly in Figure 9.12. For the Hanning window, $w[L] = 0$ and this portion of the circuit is not required.

9.3.4.3 Modulated Windows

The short time Fourier transform in (9.3) can also be written as

$$S[n; f] = \sum_{k=-L}^{L} x[n-k]\omega[k; u]$$

$$= x[n] * \omega[n; f] \tag{9.17}$$

where the *modulated window* is

$$\omega[n; f] = w[n]e^{-j2\pi nf}.$$

First, consider the case where the window is a causal boxcar window of length L. Then

$$\omega[n; f] = \begin{cases} e^{-j2\pi nf} & ; \ 0 \leq n < L \\ 0 & ; \ \text{otherwise.} \end{cases} \tag{9.18}$$

We can implement this as a truncated IIR filter. The transfer function of the impulse response of

$$k[n] = e^{-j2\pi fn}\mu[n]$$

has a z transform of

$$K_z(z) = \frac{1}{1 - z^{-1}e^{j2\pi f}}.$$

This can be implemented by the simple IIR filter shown on the left of Figure 9.13, from $x[n]$ to $v[n]$. To cancel the impulse response at times L and after, the right stage is used. The impulse response is delayed L units and multiplied by $e^{-j2\pi Lf}$. After time L, the composite impulse response of the filter is the difference

$$k[n] - e^{-j2\pi Lf}k[n-L] = e^{-j2\pi fn}\mu[n] - e^{-j2\pi Lf}\left(e^{-j2\pi f(n-L)}\mu[n-L]\right)$$

$$= e^{-j2\pi fn}(\mu[n] - \mu[n-L])$$

$$= \begin{cases} e^{-j2\pi fn} & ; \ 0 \leq n < L \\ 0 & ; \ \text{otherwise} \end{cases}$$

$$= \omega[n; f]$$

and we have the modulated boxcar window in (9.18).

More generally, the modulated window using a cosine series is

$$\omega[k; f] = \begin{cases} \sum_{q=0}^{Q} \beta_q \cos(2\pi v_q n)e^{-j2\pi nf} & ; \ 0 \leq n \leq \lfloor T \rfloor \\ 0 & ; \ \text{otherwise} \end{cases} \tag{9.19}$$

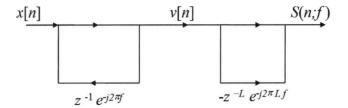

FIGURE 9.13. Modulated boxcar window implementation.

where T is an arbitrary rational time and the v_q's arbitrary frequencies. To emulate the cosine series in (9.13), we would choose $T = L$ and $v_q = q/L$. For $0 \leq n \leq \lfloor T \rfloor$, (9.19) can be written as

$$\omega[k;f] = \beta_0 e^{-j2\pi nf} + \frac{1}{2}\sum_{q=1}^{Q} \beta_q \left[e^{-j2\pi n(f-v_q)} + e^{-j2\pi n(f+v_q)} \right]. \quad (9.20)$$

The β_0 term can be implemented as in Figure 9.13. Each other q term can be implemented using two filters of the type in Figure 9.13 connected in parallel. Details are left as an exercise.

9.4 Generalized Time-Frequency Representations

Improvement of the performance of the spectrogram has a long history dating back to the work of Gabor [474]. Leon Cohen [309, 310, 312] introduced the analytically appealing *generalized time-frequency representation* (GTFR) of a temporal signal. Forms of the GTFR have been applied to bioengineering [271, 272, 273, 392, 985], biomedical signal processing [118, 132, 632, 975, 1195, 1498], fault diagnosis [1529], feature extraction [844], geophysics [265], music [1120, 1121], neurophysiology, [98], nondestructive testing [1090], parameter estimation [1191], radar [495], signal detection [804], sonar [495, 1125], speech [816, 1019, 1587], tomography [1523], ultrasound [923, 1029, 1030, 1194], and vibration analysis [276, 852, 1021, 1058, 1589].

The GTFR of a signal, $x(t)$, is [281, 405, 406, 554, 555, 766, 805, 853, 1019, 1067, 1464, 1480, 1522, 1584, 1585]

$$C(t, u) = \int_{-\infty}^{\infty} \int_{-\infty}^{\infty} \hat{\phi}(t - \xi, \tau) x\left(\xi - \frac{\tau}{2}\right) x^*\left(\xi + \frac{\tau}{2}\right) e^{-j2\pi u\tau} d\xi d\tau. \quad (9.21)$$

where $\hat{\phi}(t, \tau)$ is the kernel of the GTFR and u is the frequency variable. The choice of the kernel dictates the performance of the GTFR, akin to the effect of a window in the spectrogram. Constraints are placed on the kernel in order to enhance various characteristics of the GTFR [64, 65, 292, 293, 299, 309, 310, 312, 1587].

The discrete equivalent of (9.21) is

$$C[n, u] = \sum_{k=-\infty}^{\infty} \sum_{m=-\infty}^{\infty} \hat{\phi}[n - m, k] x\left[m - \frac{|k|}{2}\right] x^*\left[m + \frac{|k|}{2}\right] e^{-j2\pi uk}. \quad (9.22)$$

TABLE 9.3. Fourier transform relationships among the various GTFR kernel forms

	t	f
τ	$\hat{\phi}(t,\tau) \rightarrow$	$\phi(f,\tau)$
	$\downarrow$	$\downarrow$
u	$\hat{\Phi}(t,u) \rightarrow$	$\Phi(f,u)$

We can define Fourier transforms on the kernel. The variables t and f denote time and frequency in a Fourier transform relation. So do τ and u.[12] The kernel forms are

$$\phi(f,\tau) = \int_{t=-\infty}^{\infty} \hat{\phi}(t,\tau) e^{-j2\pi ft} dt, \qquad (9.23)$$

$$\hat{\Phi}(t,u) = \int_{\tau=-\infty}^{\infty} \hat{\phi}(t,\tau) e^{-j2\pi u\tau} d\tau,$$

and

$$\Phi(f,u) = \int_{-\infty}^{\infty}\int_{-\infty}^{\infty} \hat{\phi}(t,\tau) e^{-j2\pi(u\tau+ft)} dt d\tau$$

$$= \int_{-\infty}^{\infty} \phi(f,\tau) e^{-j2\pi u\tau} d\tau$$

$$= \int_{-\infty}^{\infty} \hat{\Phi}(t,u) e^{-j2\pi ft} dt.$$

Note that $\hat{\phi}(t,\tau)$ and $\Phi(f,u)$ are two dimensional Fourier transform pairs.[13]

The various forms of the kernel are summarized in Table 9.3. Alternate albeit equivalent expressions for the GTFR using these different kernel forms are[14]

$$C(t,u) = \int_{-\infty}^{\infty}\int_{-\infty}^{\infty}\int_{-\infty}^{\infty} \phi(f,\tau)$$
$$\times x\left(\xi - \frac{\tau}{2}\right) x^*\left(\xi + \frac{\tau}{2}\right) e^{j2\pi(ft-\tau u - f\xi)} d\xi d\tau df \qquad (9.24)$$

and[15]

$$C(t,u) = \int_{-\infty}^{\infty}\int_{-\infty}^{\infty} \Phi(f, u-v) X\left(v - \frac{f}{2}\right) X^*\left(v + \frac{f}{2}\right) e^{j2\pi ft} dv df. \qquad (9.25)$$

9.4.1 GTFR Mechanics

A geometrical interpretation of the continuous time GTFR in (9.21) is shown in Figure 9.14. The signal $x\left(\xi - \frac{\tau}{2}\right)$ on the (ξ,τ) plane is the signal $x(t)$ "smeared" along the line $\xi = \frac{\tau}{2}$.

---

12. This notation temporarily departs from our convention of using f as a frequency variable for discrete time and u for continuous time.
13. Multidimensional Fourier transforms are discussed Section 8.4.
14. See Exercise 9.12b.
15. See Exercise 9.12b.

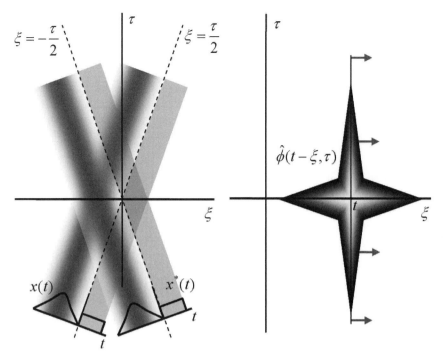

FIGURE 9.14. Illustration of the mechanics of computing a GTFR using (9.21).

Similarly, $x^*\left(\xi + \frac{|\tau|}{2}\right)$ is $x^*(t)$ "smeared" along the line $\xi = -\frac{\tau}{2}$. The two smears are multiplied to give $x\left(\xi - \frac{\tau}{2}\right) x^*\left(\xi + \frac{\tau}{2}\right)$ on the (ξ, τ). This is shown on the left in Figure 9.14.

For a fixed t, the kernel $\hat{\phi}(\xi - t, \tau)$ can be visualized on the (ξ, τ) plane. The notation dictates that, as t increases, the kernel moves along the ξ axis and, at time t, is centered on $\xi = t$. This is pictured on the right side of Figure 9.14.

To implement the GTFR in (9.21), the product of the smeared signals on the left of Figure 9.14 is multiplied, at time t, by the kernel on the right. This product is projected onto a one dimensional signal on the τ axis by integrating over all ξ. This projection at time t is

$$c(t, \tau) = \int_{\xi=-\infty}^{\infty} \hat{\phi}(t - \xi, \tau) x\left(\xi - \frac{\tau}{2}\right) x^*\left(\xi + \frac{\tau}{2}\right) d\xi.$$

The Fourier transform of this function with respect to τ gives the signal's GTFR in (9.21).

9.4.2 Kernel Properties

Using the kernels summarized in Table 9.3, we can straightforwardly state some of the commonly used constraints imposed on the GTFR and their corresponding interpretation as kernel constraints. These constraints are used in Section 11.4.10 to synthesize GTFR kernels under different design criteria. Oh et al. have shown no GTFR can simultaneously obey all constraints[16] [1037, 1043]. Therefore, use of the GTFR requires the adopting of performance trade offs.

---

16. This is the topic of Section 11.4.10

9.4.3 Marginals

Cohen reasoned that a valid TFR should have meaningful marginals. If we integrate over frequency, we should obtain the signal power.

$$\int_{-\infty}^{\infty} C(t,u)du = |x(t)|^2. \tag{9.26}$$

Integrating over time should result in the signal's power spectral density.

$$\int_{-\infty}^{\infty} C(t,u)dt = S_x(u) \tag{9.27}$$

where

$$R_x(t) \longleftrightarrow S_x(u) \tag{9.28}$$

and the signal's autocorrelation is

$$R_x(\tau) = \int_{-\infty}^{\infty} x\left(\xi - \frac{\tau}{2}\right) x^*\left(\xi + \frac{\tau}{2}\right) d\xi. \tag{9.29}$$

Substituting (9.29) and (9.28) into (9.27) gives

$$\int_{-\infty}^{\infty} C(t,u)dt = \int_{\tau=-\infty}^{\infty} \left[\int_{\xi=-\infty}^{\infty} x\left(\xi - \frac{\tau}{2}\right) x^*\left(\xi + \frac{\tau}{2}\right) d\xi\right] e^{-j2\pi u\tau} d\tau. \tag{9.30}$$

Equations 9.26 and 9.30 are known, respectively, as the *time marginal* and *frequency marginal* constraints of the GTFR. These constraints are satisfied if the GTFR kernel has certain properties. Establishing these properties is our next task.

For the time marginal constraint, we integrate (9.21) over u and obtain

$$\int_{-\infty}^{\infty} C(t,u)du = \int_{-\infty}^{\infty}\int_{-\infty}^{\infty} \hat{\phi}(t-\xi,\tau) x\left(\xi - \frac{\tau}{2}\right) x^*\left(\xi + \frac{\tau}{2}\right) \delta(\tau) d\xi d\tau$$

$$= \int_{\xi=-\infty}^{\infty} \hat{\phi}(t-\xi,0)|x(\xi)|^2 d\xi. \tag{9.31}$$

The time marginal constraint in (9.26) is therefore satisfied if the kernel has the property

$$\hat{\phi}(t,0) = \delta(t). \tag{9.32}$$

For the frequency marginal, we similarly have

$$\int_{-\infty}^{\infty} C(t,u)dt = \int_{\tau=-\infty}^{\infty}\int_{\xi=-\infty}^{\infty} \left[\int_{t=-\infty}^{\infty} \hat{\phi}(t-\xi,\tau)dt\right]$$
$$\times x\left(\xi - \frac{\tau}{2}\right) x^*\left(\xi + \frac{\tau}{2}\right) e^{-j2\pi u\tau} d\xi d\tau. \tag{9.33}$$

The frequency marginal in (9.30) is then satisfied if

$$\int_{t=-\infty}^{\infty} \hat{\phi}(t,\tau)dt = 1. \tag{9.34}$$

9.4.3.1 Kernel Constraints

There are many constraints that can be imposed on the kernel of the GTFR. We begin with restatement of the time and frequency marginal constraints.

(a) **Time Marginal Constraint**
The time marginal constraint in (9.26) is satisfied when the kernel satisfies (9.32). From the Fourier transform in (9.23), this is true when

$$\phi(f, 0) = 1. \tag{9.35}$$

(b) **Frequency Marginal Constraint**
Evaluating (9.23) at $f = 0$ reveals that satisfying the frequency marginal constraint in (9.34) is equivalent to requiring that

$$\phi(0, \tau) = 1. \tag{9.36}$$

(c) **Time Resolution Constraint**
The requirement that the input on the interval $-T \leq t - \xi \leq T$ contribute to the GTFR only on the same interval can be cast as a *cone constraint* [1587]. This requires that $\hat{\phi}(t, \tau)$ be identically zero outside of the cone shown on the left in Figure 9.15. In other words,

$$\hat{\phi}(t, \tau) = \hat{\phi}(t, \tau) \Pi\left(\frac{t}{\tau}\right) \Pi\left(\frac{\tau}{2T}\right). \tag{9.37}$$

(d) **Frequency Resolution Constraint**
Comparing (9.25) with (9.21) immediately suggests a frequency resolution constraint that is the dual of the cone constraint in (9.37).

$$\Phi(f, u) = \Phi(f, u) \Pi\left(\frac{f}{u}\right) \Pi\left(\frac{f}{2B}\right) \tag{9.38}$$

where B is the frequency dual of T. This constraint is shown on the right hand side of Figure 9.15. For finite B and T, the constraints in (9.37) and (9.38) cannot

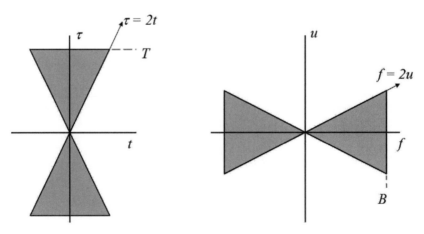

FIGURE 9.15. LEFT: The signal, $x(t)$, on the interval $-T \leq t - \xi \leq T$ should contribute to the GTFR only on the same interval. When the GTFR kernel $\hat{\phi}(t, \tau)$ lies within the shaded cone area, this constraint is satisfied. RIGHT: The dual *bow tie* constraint applied in the frequency domain. The cone kernel GTFR is discussed in Section 9.4.4.4.

be satisfied simultaneously. We are therefore motivated to formulate the following relaxed version of (9.38)

$$0 \leq \Phi(f, u) \leq \alpha(f, u) \tag{9.39}$$

where $\alpha(f, u)$ is a given positive function such as

$$\alpha(f, u) = \alpha(0, 0)\exp\left(-\beta|f| + \gamma|u|\right) \tag{9.40}$$

where β and γ are given constants. The bandwidth, B, can here be interpreted as the point 3dB from the maximum, or $B = \frac{\ln 2}{\beta}$. Note, also, that the interference constraint in (9.45) can be imposed by the bound in (9.39), possibly in a relaxed form.

(e) **Interference Suppression Constraint**

The GTFR contains the product of the signal with itself. When the signal is sinusoidal, there will be undesirable beat frequencies. A proper choice of GTFR kernel can reduce these beat frequencies. To see this, consider a signal consisting of two sinusoids of frequencies f_1 and f_2.

$$x(t) = \exp(j2\pi f_1 t) + \exp(j2\pi f_2 t). \tag{9.41}$$

The GTFR of this signal is[17]

$$\begin{aligned} C(t, u) = & \Phi(0, u - f_1) + \Phi(0, u - f_2) \\ & + \Phi\left(f_1 - f_2, u - \frac{f_1 + f_2}{2}\right) e^{j2\pi(f_1 - f_2)t} \\ & + \Phi\left(f_2 - f_1, u - \frac{f_1 + f_2}{2}\right) e^{j2\pi(f_2 - f_1)t}. \end{aligned} \tag{9.42}$$

The first two terms are those desired. The first, $\Phi(0, u - f_1)$, is a function centered at $u = f_1$ for all time, t, and the desired TFR for $\exp(j2\pi f_1 t)$. Likewise for $\Phi(0, u - f_2)$. The two remaining terms are at the beat frequency $(f_1 + f_2)/2$ and requires suppression. Assuming $\Phi(f, u) = \Phi^*(-f, u)$, the beat frequency is suppressed when

$$\Phi\left(f_1 - f_2, u - \frac{f_1 + f_2}{2}\right) = 0. \tag{9.43}$$

With attention focused on the u variable of $\Phi(f, u)$, this constraint is met if

$$\Phi(f, u) = \Theta(u)\delta(f)$$

where $\Theta(u)$ is an arbitrary one dimensional function and $\delta(f)$ is the Dirac delta. This is equivalent to requiring that $\hat{\phi}$ not be a function of t.

$$\hat{\phi}(t, \tau) = \theta(\tau). \tag{9.44}$$

A relaxed interference constraint is

$$\Phi(f, u) = \Phi(f, u)\Pi\left(\frac{f}{2\Delta}\right) \tag{9.45}$$

where Δ is the *interference bandwidth*.

---

17. See Exercise 9.13.

(f) **Realness Constraint**
A sufficient condition for $C(t, u)$ to be real is that the kernel be conjugately symmetric.

$$\phi(f, \tau) = \phi^*(-f, -\tau). \tag{9.46}$$

This is equivalent to requiring that $\hat{\Phi}(t, u)$ is real.

$$\Re\hat{\Phi}(t, u) = \hat{\Phi}(t, u) \tag{9.47}$$

where $\Re$ denotes *the real part of*.

(g) **Time Symmetry Constraint**
At a given point temporal point, past and future time are symmetrically treated if

$$\phi(f, \tau) = \phi^*(f, -\tau). \tag{9.48}$$

Note that, assuming differentiability, it follows that

$$\left.\frac{\partial \phi(f, \tau)}{\partial \tau}\right|_{\tau=0} = 0.$$

This and (9.35) constitute the *instantaneous frequency* constraint.

(h) **Frequency Symmetry Constraint**
Similarly, for frequency symmetry, we impose the constraint

$$\phi(f, \tau) = \phi^*(-f, \tau). \tag{9.49}$$

Again, assuming differentiability, this requires that

$$\left.\frac{\partial \phi(f, \tau)}{\partial f}\right|_{f=0} = 0. \tag{9.50}$$

Note that imposition of any two of the previous three constraints imposes the third. Equation (9.50) and (9.36) constitute the *group delay* constraint.

(i) **Non-negativity Constraint**
We may wish to require that the kernel is positive in the sense that

$$\hat{\phi}(t, \tau) = \Re\hat{\phi}(-t, \tau)\mu[\Re\hat{\phi}(-t, \tau)] \tag{9.51}$$

where $\mu(\cdot)$ is the unit step. In other words, the real part of $\hat{\phi}(f, \tau)$ is non-negative.

(j) **Finite Area Constraint**
A property that in useful in the iterative synthesis procedure of the GTFR kernel in Section 11.4.10.2. is

$$\Phi(0, 0) = \gamma > 0 \tag{9.52}$$

or, equivalently,

$$\int_{-\infty}^{\infty}\int_{-\infty}^{\infty} \hat{\phi}(t, \tau) dt d\tau = \gamma.$$

TIME-FREQUENCY REPRESENTATIONS 431

9.4.4 Example GTFR's

The literature on GTFR's is vast [312]. Some special cases are summarized in Table 9.4. We here present a select few examples.

9.4.4.1 The Spectrogram

Although not obvious by inspection of the equation, the spectrogram in (9.8) is a GTFR with a diamond shaped kernel. To show this, we expand (9.8) into

$$|S(t,u)|^2 = \int_{\alpha=-\infty}^{\infty} w(\alpha)x(t-\alpha)e^{-j2\pi\alpha u}d\alpha$$
$$\times \int_{\beta=-\infty}^{\infty} w(\beta)x^*(t-\beta)e^{-j2\pi\beta u}d\beta$$
$$= \int_{\alpha=-\infty}^{\infty}\int_{\beta=-\infty}^{\infty} w(\alpha)w(\beta)$$
$$\times x(t-\alpha)x^*(t-\beta)e^{-j2\pi(\alpha-\beta)u}d\alpha d\beta$$

where we assume the window is real. Fixing α and setting $\tau = \alpha - \beta$ in the β integral produces

$$|S(t,u)|^2 = \int_{\alpha=-\infty}^{\infty}\left[\int_{\tau=-\infty}^{\infty} w(\alpha)w(\alpha-\tau)e^{j2\pi u\tau}d\tau\right]d\alpha.$$

TABLE 9.4. Examples of some special cases of Cohen's GTFR. Others include [312] the Margenau-Hill distribution [879] ($\phi(f,\tau) = \cos(\pi uf)$), the Kirkwood-Rihaczek distribution [741, 1178] ($\phi(f,\tau) = \exp(j\pi f\tau)$) the Choi-Williams distribution [299] ($\phi(f,\tau) = \exp(-(2\pi f\tau)^2/\sigma)$) and the Page distribution [1070] ($\phi(f,\tau) = \exp(j2\pi f|\tau|)$).

GTFR	Kernel: $\phi(f,\tau)$	$C(t,u) =$
Cohen's GTFR [309]	$\phi(f,\tau)$	$\int_{-\infty}^{\infty}\int_{-\infty}^{\infty}\int_{-\infty}^{\infty}\phi(f,\tau)$ $\times \exp(j2\pi(ft-\tau u-f\xi))$ $\times x\left(\xi-\frac{\tau}{2}\right)x^*\left(\xi+\frac{\tau}{2}\right)d\xi d\tau df$
Wigner distribution [1491]	1	$\int_{-\infty}^{\infty} x\left(t-\frac{\tau}{2}\right)x^*\left(t+\frac{\tau}{2}\right)$ $\times \exp(-j2\pi u\tau)d\tau$
Spectrogram See Section 9.2	$\int_{-\infty}^{\infty} w\left(\xi+\frac{\tau}{2}\right)w^*\left(\xi-\frac{\tau}{2}\right)$ $\times \exp(-j2\pi\xi f)d\xi$	$\left|\int_{-\infty}^{\infty} x(\tau)w(t-\tau)\right.$ $\left.\times \exp(-j2\pi\tau u)d\tau\right|^2$
Born-Jordon [309]	$\text{sinc}(f\tau)$	$\int_{-\infty}^{\infty} \frac{1}{|\tau|}\exp(-j2\pi\tau u)$ $\times \left[\int_{-\infty}^{\infty} x\left(\xi-\frac{\tau}{2}\right)x^*\left(\xi+\frac{\tau}{2}\right)d\xi\right]d\tau$
Cone Kernel [1587]	$g(\tau)|\tau|\text{sinc}(2af\tau)$	$\int_{-\infty}^{\infty} g(\tau)\exp(-j2\pi\tau u)$ $\times \left[\int_{t-\frac{|\tau|}{2}}^{t+\frac{|\tau|}{2}} x\left(\xi+\frac{\tau}{2}\right)x^*\left(\xi-\frac{\tau}{2}\right)d\xi\right]d\tau$

Reversing integration order and setting

$$\xi = \alpha - \frac{\tau}{2}$$

gives, assuming an even window,

$$|S(t,u)|^2 = \int_{\tau=-\infty}^{\infty}\int_{\xi=-\infty}^{\infty} w\left((t-\xi)-\frac{\tau}{2}\right) w\left((t-\xi)+\frac{\tau}{2}\right)$$
$$\times\, x\left(\xi-\frac{\tau}{2}\right) x^*\left(\xi+\frac{\tau}{2}\right) e^{-j2\pi u\tau}\,d\xi\,d\tau.$$

Comparing with the GTFR in (9.21) reveals the spectrogram is a GTFR with a kernel

$$\hat{\phi}(t,\tau) = w\left(t-\frac{\tau}{2}\right) w\left(t+\frac{\tau}{2}\right).$$

If $w(t)$ is zero for $-T \le t \le T$, then $w(t) = w(t)\Pi\left(\frac{t}{2T}\right)$ and

$$\hat{\phi}(t,\tau) = \hat{\phi}(t,\tau)\,d(t,\tau)$$

where

$$d(t,\tau) = \Pi\left(\frac{t-\frac{\tau}{2}}{2T}\right)\Pi\left(\frac{t+\frac{\tau}{2}}{2T}\right).$$

This support is one inside the diamond shown on the left in Figure 9.16 and zero otherwise. The spectrogram therefore has a diamond shaped kernel.

Two chirps with linear increasing and decreasing frequencies have a spectrogram as shown in Figure 9.17. The spectrogram of on-off sinusoids of various frequencies is illustrated in Figure 9.18.

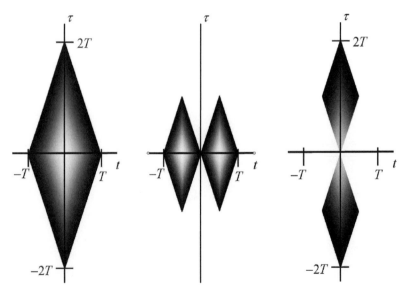

FIGURE 9.16. The diamond kernel on the left is that of a spectrogram using a window zero outside the interval $-T \le t \le T$. The two diamonds in the middle are generated using spectrograms of the same signal on $(-T, 0)$ and $(0, T)$. For judicious choice of windows, the subtraction of the two shorter duration spectrograms from the longer results in the double diamond kernel GTFR on the right.

TIME-FREQUENCY REPRESENTATIONS 433

FIGURE 9.17. The spectrogram of two chirps. A waterfall plot is on the top and a gray level plot on the bottom.
Source: [1043] S. Oh, R.J. Marks II and L.E. Altas. "Kernel synthesis for generalized time-frequency distributions using the method of alternating projections onto convex sets." *IEEE Transactions on Signal Processing*, vol. 42, No.7, July 1994. pp. 1653–1661.

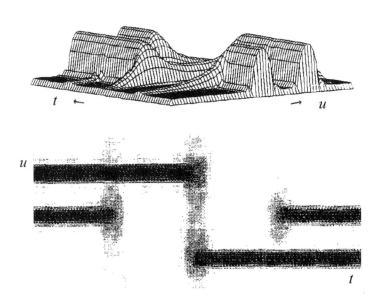

FIGURE 9.18. The spectrogram of simultaneous switching tones. A waterfall plot is on the top and a gray level plot on the bottom.
Source: [1043] S. Oh, R.J. Marks II and L.E. Altas. "Kernel synthesis for generalized time-frequency distributions using the method of alternating projections onto convex sets." *IEEE Transactions on Signal Processing*, vol. 42, No.7, July 1994. pp. 1653–1661.

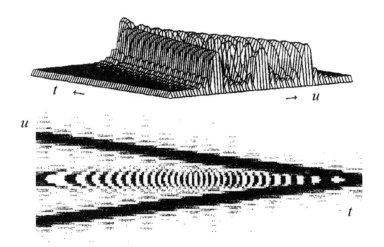

FIGURE 9.19. The Wigner distribution of two chirps. A waterfall plot is on the top and a gray level plot on the bottom. Compare with the spectrogram in Figure 9.18.
Source: [1043] S. Oh, R.J. Marks II and L.E. Atlas. "Kernel synthesis for generalized time-frequency distributions using the method of alternating projections onto convex sets." *IEEE Transactions on Signal Processing*, vol. 42, No.7, July 1994. pp. 1653–1661.

9.4.4.2 The Wigner Distribution

As defined in Table 9.4, the Wigner distribution is a special case of Cohen's GTFR. It has better resolution than the spectrogram but, as illustrated in Figures 9.19 and 9.20, at the cost of cross frequency interference.

9.4.4.3 Kernel Synthesis Using Spectrogram Superposition

Easily computed spectrograms can be superimposed to synthesize GTFR's with various kernels.[18] We illustrate synthesis of the double diamond kernel [1038] shown on the right in Figure 9.16. The diamond kernel, shown on the left in Figure 9.16, is the support of a spectrogram over $-T \leq t \leq T$. Next, a spectrogram is generated on $-T \leq t \leq 0$. This has a support in the small diamond on the left in the middle of Figure 9.16. A spectrogram on $0 \leq t \leq T$ is the small triangle on the right. If the windows are chosen properly, the shorter duration spectrograms can be subtracted from the spectrogram over the longer interval to form the double diamond kernel on the right.

9.4.4.4 The Cone Shaped Kernel

The continuous time GTFR with a cone shaped kernel [338, 339, 492, 704, 923, 1038, 1042, 1043, 1126, 1451, 1587] has been applied to bioengineering [271, 272, 273, 392, 985], biomedical signal processing [632, 975], geophysics [265], nondestructive testing [1090], sonar [1125], speech [1587], ultrasound [923, 1194], and vibration analysis [276]. It is defined as

$$C(t; u) = \int_{\tau=-T}^{T} \int_{\xi=t-\frac{|\tau|}{2}}^{t+\frac{|\tau|}{2}} \varphi(\tau) x\left(\xi - \frac{\tau}{2}\right) x^*\left(\xi + \frac{\tau}{2}\right) e^{-j2\pi u\tau} d\xi d\tau \qquad (9.53)$$

---

18. Doing so is an unpublished idea of Seho Oh.

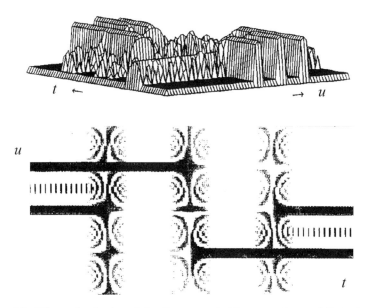

FIGURE 9.20. The Wigner distribution of simultaneous switching tones. A waterfall plot is on the top and a gray level plot on the bottom. There is high interference. Compare with the spectrogram in Figure 9.18.
Source: [1043] S. Oh, R.J. Marks II and L.E. Atlas. "Kernel synthesis for generalized time-frequency distributions using the method of alternating projections onto convex sets." *IEEE Transactions on Signal Processing*, vol. 42, No.7, July 1994. pp. 1653–1661.

where $\varphi(t)$ is a window that is zero outside of the interval $-T \leq t \leq T$. The kernel is said to be cone shaped because the integration in (9.53) is over the cone shown on the left of Figure 9.15.

The discrete time equivalent is

$$C[n,f] = \sum_{k=-L}^{L} \sum_{m=n-\frac{|k|}{2}}^{n+\frac{|k|}{2}} \varphi[k] x\left[m - \frac{k}{2}\right] x^*\left[m + \frac{k}{2}\right] e^{-j2\pi kf} \qquad (9.54)$$

where $\varphi[n]$ is a discrete window that is zero outside of the interval $-L \leq k \leq L$. We will assume the window is real and even

$$\varphi[k] = \varphi[-k]. \qquad (9.55)$$

Examples of spectrograms versus a cone shaped kernel GTFR are shown in Figures 9.21 and 9.22. The time and frequency resolutions are much better, but beat frequency artifacts appear [312, 1042].

9.4.4.5 Cone Kernel GTFR Implementation Using Short Time Fourier Transforms.

One advantage of the cone kernel GTFR[19] is its ease in computing for undecimated time.[20] For the discrete case, the cone kernel GTFR can be computed as illustrated in Figure 9.23 for

---

19. Also called the ZAM GTFR using the initials of the last name of the three authors who initially proposed the cone kernel GTFR [276, 729, 975, 1042, 1587].
20. The temporal resolution was so good, decimation of the cone kernel GTFR in Figure 9.22 resulted in the missing of transitions.

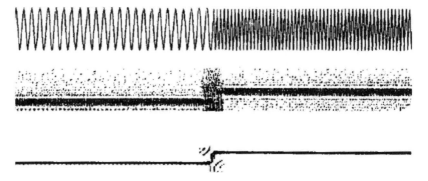

FIGURE 9.21. The spectrogram and cone kernel GTFR for a frequency transition. The cone kernel GTFR is on the bottom and the spectrogram is immediately above. The frequency transition for the cone kernel GTFR is remarkably better, but artifacts are apparent in the transition. For the spectrogram, better frequency resolution corresponds to thinner horizontal lines. This can be achieved by making the window longer. Doing so, though, will make the interval of frequency transition longer. This illustrates the frequency versus time resolution tradeoff in spectrograms.
Source: [1587] Yunxin Zhao, L.E. Atlas and R.J. Marks II. "The use of cone-shape kernels for generalized time- frequency representations of nonstationary signals". *IEEE Transactions on Acoustics, Speech and Signal Processing*, vol. 38, pp. 1084–1091 (1990).

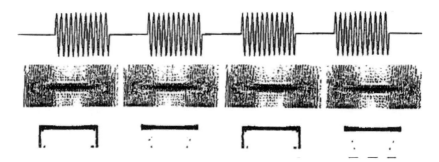

FIGURE 9.22. The spectrogram and cone kernel GTFR for switched frequency pulses. The cone kernel GTFR is on the bottom and the spectrogram is immediately above. The frequency transition for the cone kernel GTFR is remarkably better. The GTFR is decimated. The frequency transition apparent in the GTFR of the first and third pulses are not in the second and fourth. The transition times in the second and fourth frequency pulses skipped due to decimation.
Source: [1587] Yunxin Zhao, L.E. Atlas and R.J. Marks II. "The use of cone-shape kernels for generalized time- frequency representations of nonstationary signals". *IEEE Transactions on Acoustics, Speech and Signal Processing*, vol. 38, pp. 1084–1091 (1990).

real signals. This nonlinear filter produces the GTFR at a single frequency. The short time Fourier transform can be generated straightforwardly using an FFT or using the methods described in Section 9.2.

To show the validity of Figure 9.23, we begin by defining the difference

$$\Delta C[n,f] = C[n,f] - C[n-1,f].$$

If we can evaluate $\Delta C[n,f]$, the GTFR can be easily updated using the recursion

$$C[n,f] = C[n-1,f] + \Delta C[n,f]. \tag{9.56}$$

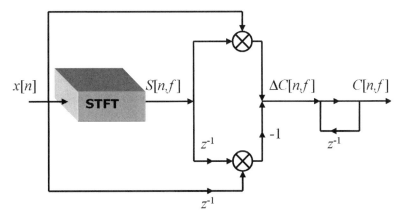

FIGURE 9.23. Evaluation of the GTFR for a cone shaped kernel for frequency f.

Using (9.54),

$$\Delta C[n,f] = \sum_{k=-L}^{L} \left[\sum_{m=n-\frac{|k|}{2}}^{n+\frac{|k|}{2}} - \sum_{m=n-1-\frac{|k|}{2}}^{n-1+\frac{|k|}{2}} \right] \varphi[k]$$

$$\times x\left[m - \frac{k}{2}\right] x^*\left[m + \frac{k}{2}\right] e^{-j2\pi kf}. \quad (9.57)$$

Since

$$\sum_{m=P}^{Q} (z[m] - z[m-1]) = z[Q] - z[P-1]$$

the sum in (9.57) becomes

$$\Delta C[n,f] = \sum_{k=-L}^{L} \varphi[k] \left\{ x\left[n + \frac{|k|}{2} - \frac{k}{2}\right] x^*\left[n + \frac{|k|}{2} + \frac{k}{2}\right] \right.$$

$$\left. - x\left[(n-1) - \frac{|k|}{2} - \frac{k}{2}\right] x^*\left[(n-1) - \frac{|k|}{2} + \frac{k}{2}\right] \right\} e^{-j2\pi kf}.$$

Breaking the sum into $k = 0$, $k > 0$ and $k < 0$ components allows simplification.

$$\Delta C[n,f] = \varphi[0] \left(|x[n]|^2 - |x[n-1]|^2 \right)$$

$$+ \sum_{k=1}^{L} \varphi[k] \left\{ x[n]x^*[n+k] - x[(n-1)-k]x^*[n-1] \right\} e^{-j2\pi kf}$$

$$+ \sum_{k=-L}^{-1} \varphi[k] \left\{ x[n-k]x^*[n] - x[n-1]x^*[(n-1)+k] \right\} e^{-j2\pi kf}. \quad (9.58)$$

In the third line, replace k by $-k$. Using (9.55), the second and third lines of (9.58) are then recognized as complex conjugates of each other. Thus, (9.58) becomes

$$\Delta C[n,f] = \varphi[0]\left(|x[n]|^2 - |x[n-1]|^2\right)$$

$$+ 2\Re x[n] \sum_{k=1}^{L} \varphi[k] x^*[n-k] e^{-j2\pi kf}$$

$$- 2\Re x^*[n-1] \sum_{k=1}^{L} \varphi[k] x^*[(n-1)-k] e^{-j2\pi kf}. \tag{9.59}$$

The sums here are recognized as short time Fourier transforms that can be generated using FFT's or the filtering methods discussed in Section 9.2.

9.4.4.6 Cone Kernel GTFR Implementation for Real Signals.

We will now demonstrate the architecture in Figure 9.23 for computing the cone kernel GTFR when the signal and the window, $\varphi[k]$, are real. In such a case, (9.59) becomes

$$\Delta C[n,f] = \varphi[0]\left(|x[n]|^2 - |x[n-1]|^2\right)$$

$$+ 2x[n] \sum_{k=1}^{L} \varphi[k] x[n-k] \cos(2\pi kf)$$

$$- 2x[n-1] \sum_{k=1}^{L} \varphi[k] x[(n-1)-k] \cos(2\pi kf)$$

or, equivalently,

$$\Delta C[n,f] = x[n]\, S[n,f] - x[n-1]\, S[n-1,f] \tag{9.60}$$

where the short time Fourier transform is

$$S[n,f] = \sum_{k=0}^{L} \hat{\varphi}[k] x[n-k] \cos(2\pi kf)$$

and the window, $\hat{\varphi}$, is

$$\hat{\varphi}[k] = \begin{cases} \varphi[0] & ; k = 0 \\ 2\varphi[k] & ; 1 \leq k \leq L \\ 0 & ; \text{otherwise.} \end{cases}$$

The cone kernel GTFR is implemented in Figure 9.23 using (9.56) where (9.60) generates $\Delta C[n,f]$.

9.4.4.7 Kernel Synthesis Using POCS

Signal synthesis using alternating projection onto convex sets (POCS) is the topic of Chapter 11. Kernel synthesis for Cohen's GTFR using POCS is the topic of Section 11.4.10.2.

9.5 Exercises

9.1. Let $x(t)$ have units of volts. What are the units of the short time Fourier transform? Of the GTFR?

9.2. Compute the short time Fourier transform of the signal

$$x(t) = \cos(2\pi f - t)\mu(-t) + \cos(2\pi f + t)\mu(t).$$

Explore the time resolution versus frequency resolution tradeoff as a function of the window parameter, $2T$.

9.3. **Filter Banks**

(a) Show that, unlike the linearly calibrated filter bank in (9.4), the nonlinearly calibrated filter banks in (9.5) and (9.6) do *not* reduce to the sampled short time Fourier transform in (9.2).

(b) If windows are allowed to vary for different values of frequency, we can generalize (9.2) to

$$S(t; u) = \int_{-\infty}^{\infty} w(\tau, u) x(t - \tau) e^{-j2\pi u \tau} d\tau. \qquad (9.61)$$

Show the nonlinearly calibrated filter banks in (9.5) and (9.6) reduce to frequency samples from this generalized form of the sampled short time Fourier transform. What are the sampled windows?

(c) The sum of the filter outputs from the filter bank in (9.4) give the signal $x(t)$ over the band of frequencies considered. Is the same true of (9.5) and (9.6)?

(d) Is the generalized spectrogram $|S(t; u)|^2$ using $S(t; u)$ in (9.61) a GTFR?

9.4. **Constant Q Filter Banks.** The Q or *quality factor* of a bandpass filter is

$$Q = \frac{u_0}{2B}.$$

where u_0 is the resonant frequency and the bandwidth is $2B$. If all of the filter Q's in a filter bank are identical, the filter bank is called *constant Q* [382, 1425].

(a) Show that the linearly calibrated filter bank in (9.4) is not constant Q. Assume the resonant frequency is the arithmetic mean (i.e., the average) of the upper and lower limits of the frequency band.

(b) Show that the nonlinearly calibrated filter banks in (9.5) and (9.6) are constant Q filter banks. Assume the resonant frequency is the geometric mean (i.e., the root of the product) of the upper and lower limits of the frequency band. What is the constant Q of these two filter banks?

9.5. **Chirp Fourier Transform.** A signal, $x[n]$, is multiplied by $e^{j\pi n^2 v}$ and is passed through a linear time-invariant filter the output of which is multiplied by $e^{j\pi n^2 v}$. This is shown in Figure 9.24.

(a) When[21]

$$h[n] = e^{j\pi n^2 v} \qquad (9.62)$$

---

[21]. Signals like $\cos(2\pi f_0 t)$ have a constant frequency, f_0. Thus, $\cos(2\pi v t^2)$ can be thought of as having a linearly increasing frequency vt. For this reason, signals like (9.62) are called *chirps* in accordance to its sound in the audio band. Since the frequency in (9.62) increases linearly, this is a *linear chirp*.

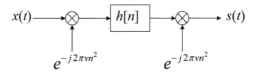

FIGURE 9.24. See Exercise 9.5.

show that

$$s[n] = \left[\left(x[n]e^{-j\pi n^2 v}\right) * e^{j\pi n^2 v}\right] e^{-j\pi n^2 v}$$

reduces to a Fourier transform.

(b) For real time implementation, the filter in Figure 9.24 must be causal. Redefine

$$h[n] = \begin{cases} e^{j\pi n^2 v} & ; n \geq 0 \\ 0 & ; n < 0. \end{cases}$$

Does the system still perform a Fourier transform?

(c) Let $x[n]$ be zero outside of the interval $0 \leq n \leq L - 1$. Using the impulse response in (9.62) and $v = 1/L$, demonstrate how the system in Figure 9.24 can compute the discrete Fourier transform of $x[n]$.

$$X[k] = \sum_{n=0}^{L-1} x[n] e^{-j2\pi kn/L}. \qquad (9.63)$$

9.6. Alter the discrete time processor shown in Figure 9.7 as follows. Replace the input sinusoid by $\exp(j2\pi[\{n - \phi\}u + \varphi])$ where ϕ and φ are arbitrary and the output sinusoid multiplication stage by a modulus squared operation. Show that the result is the spectrogram of the input signal.

9.7. Consider the impulse response of the LTI system implementing Goertzel's algorithm in Section 3.5 which we repeat here.

$$h_G[n] = e^{j2\pi(n+1)k/N}. \qquad (9.64)$$

(a) Evaluate the transfer function for this impulse response.
(b) Design an IIR filter to implement this transfer function.

9.8. We have not allowed for the required causality of the filters in Figure 9.8. The $w[n]$ filter at time n, for example, requires information about the input data L units into the future to compute the current output. For real time implementation, the fastest short time Fourier transform that we can compute is $S[n - L; u]$. Alter the architecture in Figure 9.8 so that the required filter is causal and the output of the processor is $S[n - L; u]$.

9.9. (a) Derive the following bound for the spectrogram in (9.9).

$$|S(t; u)|^2 \leq E_w \int_{\tau=t-T}^{t+T} |x(\tau)|^2 d\tau \qquad (9.65)$$

where E_w is the energy of $w(t)$. Use Schwarz's inequality.[22]

(b) Derive the dual expression for (9.65) for the discrete time spectrogram in (9.10).

---

22. See Appendix 14.1.

9.10. The window in Figures 9.11, 9.12, and 9.13 are marginally stable. Implement the windows and note any stability issues. Will the marginal stability impact implementation? If so, what are some variations that will address these problems?

9.11. Draw the modulated window implementation of a $Q = 2$ filter in (9.20) for $T = L$ and $v_q = q/L$. Simplify the circuit including factoring out filter components common to each stage. Make certain end point correction is addressed.

9.12. There are numerous alternate forms in which the GTFR can be written.
  (a) Show that the alternate GTFR expression in (9.24) reduces to (9.21).
  (b) Derive the frequency domain representation of the GTFR in (9.25) from the definition in (9.21).

9.13. Show that the GTFR of the two sinusoids in (9.41) is (9.43).

9.14. Derive a similar expression for computing the continuous time cone kernel GTFR that was derived in Section 9.4.4.4 for the discrete case. Specifically, when $\varphi(\tau)$ is real and even, show the derivative of the continuous time cone kernel GTFR in (9.53) is

$$\frac{dC(t,u)}{dt} = 2\Re\, x(t) \int_{-T}^{T} \tilde{\varphi}(\tau) x^*(t+\tau) e^{-j2\pi u \tau} d\tau \qquad (9.66)$$

where

$$\tilde{\varphi}(\tau) = \varphi(\tau)\,\mathrm{sgn}(\tau). \qquad (9.67)$$

The integral in (9.66) is recognized as a continuous short time Fourier transform. The cone kernel GTFR can be obtained by running $\frac{dC(t,u)}{dt}$ through an analog integrator. Note that, when $x(t)$ is real, (9.66) becomes

$$\frac{dC(t,u)}{dt} = 2\, x(t) \int_{-T}^{T} \tilde{\varphi}(\tau) x^*(t+\tau) \cos(2\pi u \tau) d\tau.$$

9.15. **The ambiguity function**, used in radar and sonar [282, 291, 1510], is a measure of a signal to simultaneously measure the range and velocity of a moving target [39, 894, 1207, 1229, 1372]. It is given by the Fourier transform

$$\chi(v,\tau) = \int_{-\infty}^{\infty} f(t) f(t-\tau) e^{-j2\pi v t} dt.$$

  (a) Perform an inverse Fourier transform of the ambiguity function in v followed by a forward Fourier transform on τ. How does this relate to the GTFR?
  (b) Compute the ambiguity function of a rectangular pulse of duration $2T$.

$$f(t) = \Pi\left(\frac{t}{2T}\right). \qquad (9.68)$$

  (c) Provide a sketch in the (v, τ) plane of the locus of points where the solution in (b) is identically zero.
  (d) Repeat (b) and (c) for the double pulse

$$f(t) = \Pi\left(\frac{t}{6T}\right) - \Pi\left(\frac{t}{2T}\right).$$

9.6 Solutions for Selected Chapter 9 Exercises

9.7. (a) The transfer function is

$$H(z) = \sum_{n=0}^{\infty} h_G[n]\, z^{-n}$$

$$= \sum_{n=0}^{\infty} e^{j2\pi(n+1)k/N} z^{-n}$$

$$= e^{j2\pi k/N} \sum_{n=0}^{\infty} \left(z^{-1} e^{j2\pi k/N}\right)^n.$$

Using the geometric series,

$$H(z) = e^{j2\pi k/N} \frac{1}{1 - z^{-1} e^{j2\pi k/N}}$$

$$= e^{j2\pi k/N} \frac{1}{1 - z^{-1} e^{j2\pi k/N}} \frac{1 - z^{-1} e^{-j2\pi k/N}}{1 - z^{-1} e^{-j2\pi k/N}} \qquad (9.69)$$

$$= e^{j2\pi k/N} \frac{1 - z^{-1} e^{-j2\pi k/N}}{1 - 2\cos(2\pi k/N) + z^{-2}}.$$

(b) See Figure 9.25.

9.9. (a)

$$|S(t; u)|^2 = \left| \int_{-\infty}^{\infty} w(\tau) x(t - \tau) e^{-j2\pi u(\tau - \theta)} d\tau \right|^2$$

$$= E_w \int_{\tau = t - T}^{t + T} |x(\tau)|^2 d\tau.$$

9.12.

$$C(t, u) = \int_{-\infty}^{\infty} \int_{-\infty}^{\infty} \int_{-\infty}^{\infty} \phi(f, \tau) e^{j2\pi(ft - \tau u - f\xi)}$$

$$\times x\left(\xi + \frac{\tau}{2}\right) x^*\left(\xi - \frac{\tau}{2}\right) d\xi d\tau df$$

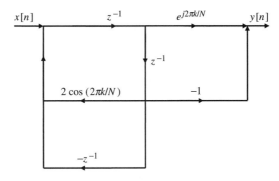

FIGURE 9.25. IIR filter implementing the Goertzel's algorithm transfer function in (9.69) in Exercise 9.7.

$$= \int_{-\infty}^{\infty} \int_{-\infty}^{\infty} \left[\int_{-\infty}^{\infty} \phi(f,\tau) e^{j2\pi f(t-\xi)} df \right] \quad (9.70)$$
$$\times x\left(\xi + \frac{\tau}{2}\right) x^*\left(\xi - \frac{\tau}{2}\right) e^{-j2\pi \tau u} d\xi d\tau$$

The inverse Fourier transform of (9.23) is

$$\hat{\phi}(t,\tau) = \int_{t=-\infty}^{\infty} \phi(f,\tau) e^{j2\pi ft} df.$$

Application to (9.71) gives the desired result in (9.21)

9.13. The GTFR in (9.25) can be written as

$$C(t,u) = \int_{f=-\infty}^{\infty} \left[\Phi(f,u) \overset{u}{*} X\left(u + \frac{f}{2}\right) X^*\left(u - \frac{f}{2}\right) \right] e^{j2\pi ft} df \quad (9.71)$$

where

$$\overset{u}{*}$$

indicates convolution with respect to u. Given $x(t)$ in (9.41), we have

$$X(u) = \delta(u - f_1) + \delta(u - f_2)$$

and

$$X\left(u + \frac{f}{2}\right) X^*\left(u - \frac{f}{2}\right) = \delta\left(u - \frac{f}{2} + f_1\right) \delta\left(u + \frac{f}{2} - f_1\right)$$
$$+ \delta\left(u - \frac{f}{2} + f_1\right) \delta\left(u + \frac{f}{2} - f_2\right)$$
$$+ \delta\left(u - \frac{f}{2} + f_2\right) \delta\left(u + \frac{f}{2} - f_1\right)$$
$$+ \delta\left(u - \frac{f}{2} + f_2\right) \delta\left(u + \frac{f}{2} - f_2\right). \quad (9.72)$$

Since

$$\Phi(f,u) \overset{u}{*} \delta(u-\theta_1)\delta(u-\theta_2) = \Phi(f, u-\theta_1)\delta(\theta_2 - \theta_1),$$

substituting (9.72) into (9.71) gives

$$C(t,u) = \int_f \Phi\left(f, u + \frac{f}{2} - f_1\right) \delta(f) e^{j2\pi ft} df$$
$$+ \int_f \Phi\left(f, u + \frac{f}{2} - f_1\right) \delta(f - (f_1 - f_2)) e^{j2\pi ft} df$$
$$+ \int_f \Phi\left(f, u + \frac{f}{2} - f_2\right) \delta(f - (f_2 - f_1)) e^{j2\pi ft} df$$
$$+ \int_f \Phi\left(f, u + \frac{f}{2} - f_2\right) \delta(f) e^{j2\pi ft} df.$$

Equation (9.42) follows immediately from the Dirac delta's sifting property.

9.14. Using Leibniz's rule,[23] the derivative of the cone kernel GTFR in (9.53) is

$$\frac{dC(t,u)}{dt} = \int_{\tau=-T}^{T} \frac{d}{dt} \left\{ \int_{\xi=t-\frac{|\tau|}{2}}^{t+\frac{|\tau|}{2}} \varphi(\tau) x\left(\xi - \frac{\tau}{2}\right) x^*\left(\xi + \frac{\tau}{2}\right) d\xi \right\}$$
$$\times e^{-j2\pi u\tau} d\tau$$
$$= \int_{-T}^{T} \varphi(\tau) \left[x\left(t + \frac{|\tau|}{2} - \frac{\tau}{2}\right) x^*\left(t + \frac{|\tau|}{2} + \frac{\tau}{2}\right) \right.$$
$$\left. - x\left(t - \frac{|\tau|}{2} - \frac{\tau}{2}\right) x^*\left(t - \frac{|\tau|}{2} + \frac{\tau}{2}\right) \right] e^{-j2\pi u\tau} d\tau.$$

Breaking the integration into $\tau < 0$ and $\tau > 0$ sections gives

$$\frac{dC(t,u)}{dt} = \int_{\tau=0}^{T} \varphi(\tau) \left\{ x(t) x^*(t+\tau) - x(t-\tau) x^*(t) \right\} e^{-j2\pi u\tau} d\tau$$
$$+ \int_{\tau=-T}^{0} \varphi(\tau) \left\{ x(t-\tau) x^*(t) - x(t) x^*(t+\tau) d\tau \right\} e^{-j2\pi u\tau} d\tau$$

The integrands are similar and the integral can be recombined, using (9.67), into

$$\frac{dC(t,u)}{dt} = \int_{-T}^{T} \tilde{\varphi}(\tau) \left\{ x(t) x^*(t+\tau) - x(t-\tau) x^*(t) \right\} e^{-j2\pi u\tau} d\tau.$$

Equivalently

$$\frac{dC(t,u)}{dt} = x(t) \int_{-T}^{T} \tilde{\varphi}(\tau) x^*(t+\tau) e^{-j2\pi u\tau} d\tau$$
$$- x^*(t) \int_{-T}^{T} \tilde{\varphi}(\tau) x(t-\tau) e^{-j2\pi u\tau} d\tau.$$

In the second integral, substitute $-\tau$ for τ. If $\varphi(\tau)$ is even, $\tilde{\varphi}(\tau)$ is odd and

$$\frac{dC(t,u)}{dt} = x(t) \int_{-T}^{T} \tilde{\varphi}(\tau) x^*(t+\tau) e^{-j2\pi u\tau} d\tau$$
$$+ x^*(t) \int_{-T}^{T} \tilde{\varphi}(\tau) x(t+\tau) e^{j2\pi u\tau} d\tau.$$

Since $\tilde{\varphi}(\tau)$ is real, the terms are complex conjugates of each other and

$$\frac{dC(t,u)}{dt} = 2\Re\, x(t) \int_{-T}^{T} \tilde{\varphi}(\tau) x^*(t+\tau) e^{-j2\pi u\tau} d\tau$$

which is the desired result in (9.66).

---

23. See Appendix 14.2.

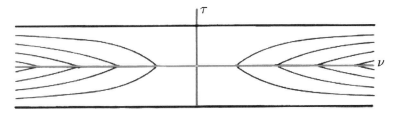

FIGURE 9.26. The zero locus plot for the ambiguity function of a single rectangular pulse. See the solution to Exercise 9.15(b).
Source: [886] R.J. Marks II, J.F. Walkup, and T.F. Krile. Ambiguity function display: An improved coherent processor. *Applied Optics Vol. 16*, pages 746–750, 1977.

9.15. The ambiguity function
   (b) The ambiguity function for the pulse in (9.68) is

$$\chi(\nu, \tau) = (2T - |\tau|) \operatorname{sinc}(\nu(2T - |\tau|)) e^{-j2\pi\nu\tau} \Pi\left(\frac{\tau}{4T}\right).$$

This is zero when the argument of the sinc is a nonzero integer, n and $|\tau| \leq T$. Thus

$$\nu = \frac{n}{2T - |\tau|} \; ; \; |\tau| \leq T.$$

The corresponding zero locus plot is shown in Figure 9.26.

   (c) For the double pulse, a bit of work gives the ambiguity function

$$\chi(\nu,\tau) = \begin{cases} 4(2T-|\tau|)\operatorname{sinc}(\nu(2T-|\tau|))\cos(4\pi T\nu)e^{-j2\pi\nu\tau} & ; |\tau| \leq 2T \\ -(2T-|\tau|)\operatorname{sinc}(\nu(2T-|\tau|))e^{-j\pi\nu\tau} & ; 2T \leq |\tau| \leq 4T \\ (6T-|\tau|)\operatorname{sinc}(\nu(6T-|\tau|))e^{-j\pi\nu\tau} & ; 4T \leq |\tau| \leq 6T \\ 0 & ; |\tau| \geq 6T. \end{cases}$$

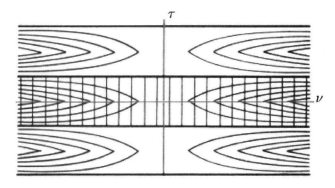

FIGURE 9.27. The zero locus plot for the ambiguity function of a double rectangular pulse. See the solution to Exercise 9.15(c).
Source: [886] R.J. Marks II, J.F. Walkup, and T.F. Krile. Ambiguity function display: An improved coherent processor. *Applied Optics Vol. 16*, pages 746–750, 1977.

The zero locus follows from when either the sinc or the cos is zero. We use $m \in \mathbb{Z}$ to parameterize the cosine and $n \neq 0$ for the sincs. The result is

$$\nu = \begin{cases} \dfrac{2m+1}{8T} & ; \quad |\tau| \leq 2T \\[6pt] \dfrac{n}{|\tau|-2T} & ; \quad |\tau| \leq 4T \\[6pt] \dfrac{n}{6T-|\tau|} & ; \ 4T \leq |\tau| \leq 6T \end{cases}$$

The corresponding zero locus plot is shown in Figure 9.27.

10

Signal Recovery

> There exists, if I am not mistaken, an entire world which is the totality of mathematical truths, to which we have access only with our mind, just as a world of physical reality exists, the one like the other independent of ourselves, both of divine creation,
> Charles Hermite (1822–1901)

> One merit of mathematics few will deny: it says more in fewer words than any other science. The formula, $e^{i\pi} = -1$ expressed a world of thought, of truth, of poetry, and of the religious spirit 'God eternally geometrizes.'
> David Eugene Smith [1189]

> The imaginary number is a fine and wonderful recourse of the divine spirit, almost an amphibian between being and not being.
> Gottfried Whilhem Leibniz (1646–1716) [468]

> He who can properly define and divide is to be considered a god.
> Plato [468]

10.1 Introduction

The literature on the recovery of signals and images is vast (e.g., [23, 110, 112, 257, 391, 439, 791, 795, 933, 934, 937, 945, 956, 1104, 1324, 1494, 1495, 1551]). In this Chapter, the specific problem of recovering lost signal intervals from the remaining known portion of the signal is considered. Signal recovery is also a topic of Chapter 11 on POCS.

10.2 Continuous Sampling

To this point, sampling has been discrete. Bandlimited signals, we will show, can also be recovered from continuous samples. Our definition of continuous sampling is best presented by illustration. A signal, $f(t)$, is shown in Figure 10.1a, along with some possible continuous samples. Regaining $f(t)$ from knowledge of $g_e(t) = f(t)\Pi(t/T)$ in Figure 10.1b is the *extrapolation problem* which has applications in a number of fields. In optics, for example, extrapolation in the frequency domain is termed *super resolution* [2, 40, 367, 444, 500, 523, 641, 720, 864, 1016, 1099, 1117].

Reconstructing $f(t)$ from its tails [i.e., $g_i(t) = f(t)\{1 - \Pi(t/T)\}$] is the *interval interpolation problem*. *Prediction*, shown in Figure 10.1d, is the problem of recovering a signal with knowledge of that signal only for negative time.

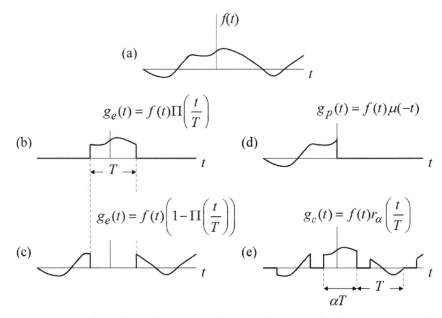

FIGURE 10.1. Illustration of continuous sampling (a) the original signal; (b) extrapolation; (c) interpolation; (d) prediction; (e) periodic continuous sampling.

Lastly, illustrated in Figure 10.1e, is *periodic continuous sampling*. Here, the signal is known in sections periodically spaced at intervals of T. The *duty cycle* is α. Reconstruction of $f(t)$ from this data includes a number of important reconstruction problems as special cases.

(a) By keeping αT constant, we can approach the extrapolation problem by letting T go to ∞.
(b) Redefine the origin in Figure 10.1e to be centered in a zero interval. Under the same assumption as (a), we can similarly approach the interpolation problem.
(c) Redefine the origin as in (b). Then the interpolation problem can be solved by discarding data to make it periodically sampled.
(d) Keep T constant and let $\alpha \to 0$. The result is reconstructing $f(t)$ from discrete samples as discussed in Chapter 5. Indeed, this model has been used to derive the sampling theorem [246].

Figures 10.1b-e all illustrate continuously sampled versions of $f(t)$. In this chapter, we will present techniques by which the signal can be reconstructed assuming that $f(t)$ is bandlimited. *In the absence of noise*, any finite energy bandlimited signal can be reconstructed from any continuous sample. Such signals are *analytic* (or *entire*) everywhere [125]. Thus, if we know the signal within an arbitrarily small neighborhood centered at $t = \tau$, we can compute the value of the function and all its derivatives at τ and generate a Taylor series about $t = \tau$ that converges everywhere.

$$f(t) = \sum_{n=-\infty}^{\infty} \frac{(t-\tau)^n f^{(n)}(\tau)}{n!}. \tag{10.1}$$

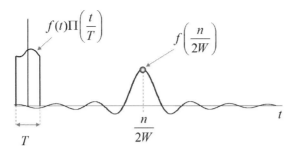

FIGURE 10.2. Illustration that a sample value far removed from an interval has an effect on the signal in that interval.

In practice, of course, this can only be done to an approximation. We could, for example, empirically determine $f(\tau)$ and $f^{(1)}(\tau)$ and maybe even $f^{(2)}(\tau)$. But higher order derivatives will eventually become critically muddled by measurement inexactitude. Thus, in our example using two derivatives, we could at best fit a quadratic to the signal at $t = \tau$. In the absence of uncertainty, however, restoration can be performed from any continuously sampled bandlimited signal.

Intuitively, one should not be surprised that a known portion of a bandlimited signal can be extended at least near to where the signal is known. Bandlimited signals are smooth. The extension of such a signal must be similarly smooth. Simply continuing the curve will in general yield a good estimate "near" to where the signal is known. If, for example, we are given $f(t)\Pi(t/T)$ in Figure 10.2, and told that it is a portion of a bandlimited signal, we then know that the signal is smooth. We could, with pencil in hand, extend the signal a bit beyond the known portion $|t| \leq T/2$ and have confidence that our estimation of the extension of known portion of the signal was fairly accurate. If we were asked to pencil in the signal far removed from the known signal, however, we would not be able to do so with the information we have.

For the extrapolation problem, Pask [1099] explained the relationship between the known interval and those samples far removed from the interval using the cardinal series interpolation. In Figure 10.2, the sample $f(n/2W)$ is assumed to be outside of the known interval. From the sampling theorem, the interpolation contribution of this sample is a sinc function whose tails will intersect the interval $|t| \leq T$ and thus make a contribution to the known portion of the signal. Thus, the known portion of the signal contains information about the unknown part of the signal. Note, however, as we go farther and farther away from the known portion of the signal, the contribution from the wiggles of the sinc becomes less and less.

Unlike the Taylor series treatment that examines a signal and its derivatives at a single point, most of the restoration algorithms in this chapter make use of all of the known continuous portion of the signal. Like the Taylor series treatment, the sensitivity of these algorithms to inexactitude must be considered. Indeed, a number of the algorithms are *ill-posed*. This means that a small amount of noise on the known data can render the reconstruction unstable. In such cases, further *a priori* information about $f(t)$ must be included in the algorithm i.e., the original problem statement is too vague. If we assume only that $f(t)$ is bandlimited with finite energy, the extrapolation and prediction problems are ill-posed whereas restoration from periodic continuous samples and interpolation problems are not.

10.3 Interpolation From Periodic Continuous Samples

The given data for the periodic continuously sampled signal in Figure 10.1e is

$$g_c(t) = f(t)r_\alpha\left(\frac{t}{T}\right) \tag{10.2}$$

where the duty cycle, α, lies between zero and one and the pulse train is

$$r_\alpha(t) = \sum_{n=-\infty}^{\infty} \Pi\left(\frac{t-n}{\alpha}\right). \tag{10.3}$$

Given that $f(t)$ is bandlimited, the problem is to find $f(t)$ given $g_c(t)$ and the signal's bandwidth, B.

The degradation process described by (10.2) is illustrated by the top three functions in Figure 10.3. The corresponding operation in the frequency domain, shown in the bottom three functions in Figure 10.3 is

$$G_c(u) = F(u) * TR_\alpha(Tu) \tag{10.4}$$

where the upper case letters denote the Fourier transforms of the corresponding functions in (10.2) and the asterisk denotes convolution. Expanding (10.3) in a Fourier series followed by transformation gives

$$TR_\alpha(Tu) = \sum_{n=-\infty}^{\infty} C_n \delta(u - nT) \tag{10.5}$$

where

$$C_n = \alpha \operatorname{sinc}(\alpha n).$$

Then the spectrum of the degradation in (10.4) can be written:

$$G_c(u) = \sum_{n=-\infty}^{\infty} C_n F(u - nT).$$

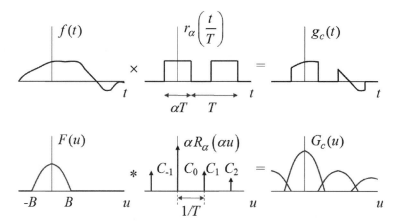

FIGURE 10.3. Illustration of the degradation of $f(t)$ to $g(t)$ (top row) in t and (bottom row) in the frequency domain.

10.3.1 The Restoration Algorithm

Clearly, if the sampling rate $1/T$ exceeds $2B$, the replicated spectra do not overlap and $F(u)$ can be regained from $G(u)$ by a simple low pass filter. We are interested in restoration when the data is aliased. If one of the spectra overlaps the right half zero order spectrum as in Figure 10.4a, we have first order aliasing. If two overlap, as in Figure 10.4b, we have second order aliasing, etc. In general, the order of aliasing is

$$M = \lfloor 2BT \rfloor.$$

Consider Figure 10.5 in which $2M+1$ shifted versions of $G_c(u)$ are shown, i.e., the set

$$\left\{ G_c\left(u - \frac{m}{T}\right) \Big| -M \leq m \leq M \right\}.$$

The interfering component spectra in each shifted G_c are shown not overlapping for presentation clarity. We now simply need to weight the mth shifted G_c by a coefficient b_m so that

$$\sum_{m=-M}^{M} b_m G_c\left(u - \frac{m}{T}\right) \Pi\left(\frac{u}{2B}\right) = F(u). \tag{10.6}$$

With attention again to Figure 10.5, this is equivalent to summing the weights of the component spectra in each column to give zero for the interfering spectra and one for the zero order spectrum. That is, find the b_m's which satisfy

$$\sum_{m=-M}^{M} b_m C_{n-m} = \delta[n]; \quad -M \leq n \leq M. \tag{10.7}$$

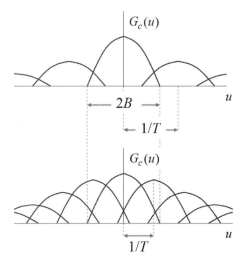

FIGURE 10.4. Illustration of (top) first order aliasing and (bottom) second order aliasing.

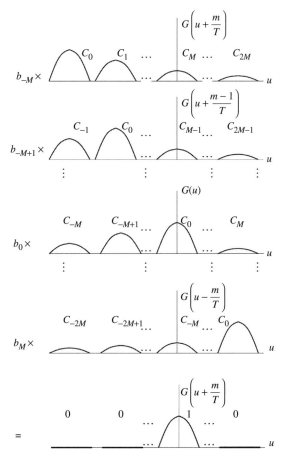

FIGURE 10.5. Illustration of the methodology of restoring Mth order aliased data by summing $2M+1$ shifted and weighted versions for the degraded spectrum.

Viewing this as a matrix operation,

$$\begin{bmatrix} C_0 & C_{-1} & \cdots & C_{-M} & \cdots & C_{-2M} \\ C_1 & C_0 & \cdots & C_{-M+1} & \cdots & C_{-2M+1} \\ \vdots & \vdots & & \vdots & \vdots & \vdots \\ C_M & C_{M-1} & \cdots & C_0 & \cdots & C_{-M} \\ \vdots & \vdots & & \vdots & \vdots & \vdots \\ C_{2M} & C_{2M-1} & \cdots & C_M & \cdots & C_0 \end{bmatrix} \begin{bmatrix} b_{-M} \\ b_{-M} \\ \vdots \\ b_0 \\ \vdots \\ b_{-M} \end{bmatrix} = \begin{bmatrix} 0 \\ 0 \\ \vdots \\ 1 \\ \vdots \\ 0 \end{bmatrix}, \qquad (10.8)$$

it is clear the b_m's can be solved for by solution of a *Toeplitz*[1] set of equations.

Inverse transforming (10.6) gives the time domain restoration formula

$$f(t) = \left[g_c(t) \Theta_M \left(\frac{t}{T} \right) \right] * 2B\text{sinc}(2Bt) \qquad (10.9)$$

---

1. The nm^{th} element of the matrix is a function only of $n - m$.

where $\Theta_M(t)$ is the trigonometric polynomial

$$\Theta_M(t) = \sum_{m=-M}^{M} b_m \, e^{j2\pi mt}. \tag{10.10}$$

Note, however, that since

$$g_c(t) = g_c(t) r_\alpha \left(\frac{t}{T}\right)$$

we only require knowledge of $\Theta_M(t)$ where r_α is unity. Thus we define the periodic function

$$\psi_M(t) = \Theta_M(t) r_\alpha(t). \tag{10.11}$$

In lieu of (10.9), the restoration algorithm now becomes [902]

$$f(t) = \left[g_c(t) \psi_M \left(\frac{t}{T}\right) \right] * 2B \, \text{sinc}(2Bt). \tag{10.12}$$

This is illustrated in Figure 10.8.

Expanding $\psi_M(t)$ in a Fourier series gives

$$\psi_M(t) = \sum_{n=-\infty}^{\infty} d_n \, e^{j2\pi nt}$$

The coefficients are

$$d_n = \int_{-\frac{1}{2}}^{\frac{1}{2}} \psi_M(t) \, e^{-j2\pi nt} dt$$

$$= \int_{-\frac{\alpha}{2}}^{\frac{\alpha}{2}} \Theta_M(t) \, e^{j2\pi nt} dt$$

$$= \alpha \sum_{m=-M}^{M} b_m \, \text{sinc}(\alpha(n-m))$$

where, in the last step we have used (10.10). From (10.7), we conclude that

$$d_n = \begin{cases} \alpha \sum_{m=-M}^{M} b_m \, \text{sinc}(\alpha(n-m)) \, ; \, |n| \geq M \\ \delta[n] \qquad\qquad\qquad\qquad\qquad ; \, |n| \leq M. \end{cases}$$

Note that the d_n's are also the weights of the remaining spectra after restoration. Plots of $\psi_M(t)$ for $\alpha = 0.5$ are shown in Figure 10.6. Plots of $\psi_2(t)$ for various duty cycles are shown in Figure 10.7.

This is our desired result.

10.3.1.1 Trigonometric Polynomials

If $f(t)$ is the trigonometric polynomial,

$$f(t) = \sum_{n=-N}^{N} \beta_n \, e^{j2\pi nt/T}$$

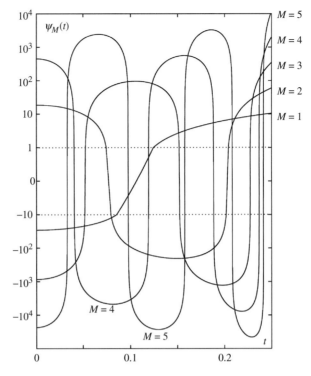

FIGURE 10.6. Plots of $\psi_M(t) = \psi_M(-t)$ for $\alpha = 0.5$ and $M = 1, 2, 3, 4$, and 5. The vertical scale is linear for $|\psi_M(t)| < 1$ and logarithmic otherwise.

then, for $|t| < T/2$, the restoration algorithm in (10.12) becomes

$$f(t) = \left[f(t) \, \Pi\left(\frac{t}{\alpha T}\right) \psi_M\left(\frac{t}{T}\right) \right] * \frac{1}{T} h\left(\frac{t}{T}\right) \tag{10.13}$$

where

$$h(t) = (2N+1) \, \text{array}_{2N+1}(t)$$

$$\longleftrightarrow H(u) = \sum_{n=-N}^{N} \delta(u-n).$$

This restoration process is illustrated in Figure 10.9. Note that $H(uT)$ acts as a sampler in the frequency domain. Plots of $h(t)$ are shown in Figure 10.10 for various N.

Proof. We can write (10.12) as

$$f(t) = 2B \sum_{n=-\infty}^{\infty} \int_{(n-\frac{\alpha}{2})T}^{(n+\frac{\alpha}{2})T} g_c(\tau) \, \psi_M\left(\frac{\tau}{T}\right) \text{sinc}\,(2B(t-\tau)) \, d\tau.$$

Since both g_c and ψ_M are periodic, setting $\xi = \tau - nT$ gives

$$f(t) = \int_{-\alpha T/2}^{\alpha T/2} g_c(\xi) \, \psi_M\left(\frac{\xi}{T}\right) k(t-\xi) \, d\xi$$

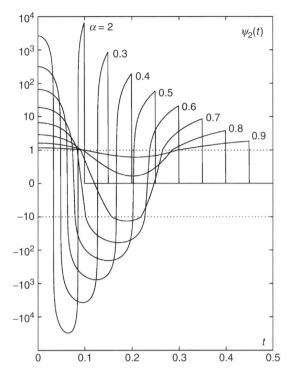

FIGURE 10.7. Plots of $\psi_2(t)$ for various α. The vertical scale is linear for $|\psi_2(t)| < 1$ and logarithmic otherwise.

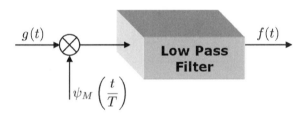

FIGURE 10.8. Restoration of continuously sampled signals. The periodic function $\psi_M(t)$ is parameterized by the duty cycle and the severity of aliasing. The low-pass filter has the same bandwidth, B, as the restored signal.

where

$$k(t) = 2B \sum_{n=-\infty}^{\infty} \text{sinc}\,(2B(t-nT)).$$

Recognizing that $B = N/T$, we can evaluate this sum in the same manner we evaluated (6.81). The result is

$$k(t) = \frac{\sin\left(\pi(2N+1)\frac{t}{T}\right)}{T \sin\left(\frac{\pi t}{T}\right)} = \frac{2N+1}{T}\,\text{array}_{2N+1}\left(\frac{t}{T}\right)$$

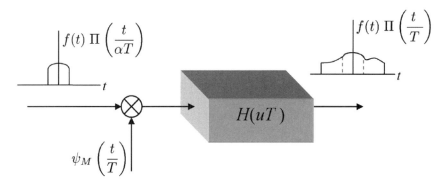

FIGURE 10.9. Restoration of a period of a trigonometric polynomial known only over the subperiod of $|t| < \frac{1}{2}$.

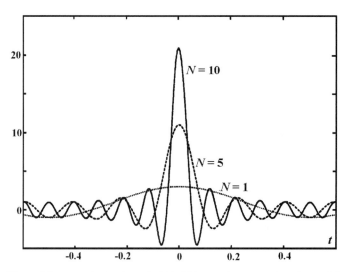

FIGURE 10.10. Plots of the convolution kernels $h(t)$. When a trigonometric polynomial of period T is known for $|t| < \frac{T}{2}$, the convolution of this function with the known portion of the polynomial (after weighting with $\psi_M(t)$) results in a whole period of the trigonometric polynomial.

and (10.13) results. Note that we must have the strict inequality $M > 2N$ since, at $M = 2N$, we have first order aliasing due to the Dirac delta nature of the spectrum of $f(t)$.

10.3.1.2 Noise Sensitivity

Here, we explore the performance of the restoration algorithm in (10.13) in the presence of additive wide sense stationary zero mean noise, $\xi(t)$ [903]. Because of linearity, an input of $g_c(t) + \xi(t)$ into the restoration algorithm will yield an output of $f(t) + \eta(t)$, where $\eta(t)$ is the algorithm response to $\xi(t)$ alone. Using (10.12), it follows that

$$\eta(t) = \left[\xi(t) \, \psi_M\left(\frac{t}{T}\right) \right] * 2B \, \text{sinc}(2Bt).$$

The interpolation noise, $\eta(t)$, is also zero mean although it is not stationary.

SIGNAL RECOVERY

The restoration noise level follows as

$$\overline{\eta^2(t)} = E\left[\eta^2(t)\right]$$

$$= E\left[\int_{\tau=-\infty}^{\infty} \xi(\tau)\,\psi_M\left(\frac{\tau}{T}\right) 2B\mathrm{sinc}\,(2B(t-\tau))\,d\tau \right.$$

$$\left. \times \int_{\lambda=-\infty}^{\infty} \xi(\lambda)\,\psi_M\left(\frac{\lambda}{T}\right) 2B\mathrm{sinc}\,(2B(t-\lambda))\,d\lambda\right]$$

$$= 4B^2 \int_{-\infty}^{\infty}\int_{-\infty}^{\infty} R_\xi(\tau-\lambda)\psi_M\left(\frac{\tau}{T}\right)\psi_M\left(\frac{\lambda}{T}\right)$$

$$\times\,\mathrm{sinc}\,(2B(t-\tau))\,\mathrm{sinc}\,(2B(t-\lambda))\,d\tau d\lambda.$$

By straightforward integral manipulation we obtain

$$\overline{\eta^2(t)} = \int_{-\infty}^{\infty} R_\xi(\gamma) h(t;\gamma)\,d\gamma \qquad (10.14)$$

where

$$h(t;\gamma) = 2B\,\psi_M\left(\frac{\gamma}{T}\right)\mathrm{sinc}\,(2B(t-\gamma)) \star 2B\,\psi_M\left(\frac{\gamma}{T}\right)\mathrm{sinc}\,(2B(t-\gamma)). \qquad (10.15)$$

The $\star$ denotes deterministic autocorrelation[2] with respect to γ. The output noise level in (10.14) is an even periodic function with period T.

White Noise. For continuous white noise

$$R_\xi(\tau) = \overline{\xi^2}\delta(\tau). \qquad (10.16)$$

Equation (10.14) becomes

$$\frac{\overline{\eta^2(t)}}{\overline{\xi^2}} = h(t;0)$$

$$= (2B)^2 \mathrm{sinc}^2(2Bt) * \psi_M^2\left(\frac{t}{T}\right). \qquad (10.17)$$

From (10.11)

$$\psi_M^2(t) = r_\alpha(t)\Theta_M^2(t)$$

where, from (10.10)

$$\Theta_M^2(t) = \sum_{k=-M}^{M}\sum_{r=-M}^{M} b_k b_r\, e^{j2\pi(k+r)t}. \qquad (10.18)$$

Fourier transforming both sides of (10.17) gives

$$\frac{\overline{\eta^2(t)}}{\overline{\xi^2}} \longleftrightarrow 2B\Lambda\left(\frac{u}{2B}\right)\mathcal{F}\,\psi_M^2\left(\frac{t}{T}\right)$$

$$= 2B\Lambda\left(\frac{u}{2B}\right)\left[TR_\alpha(Tu) * \mathcal{F}\Theta_M^2\left(\frac{t}{T}\right)\right] \qquad (10.19)$$

---

2. See Table 2.3 for the definition of deterministic correlation.

where $\mathcal{F}$ denotes the Fourier transform operator. Substituting (10.5) and the transform of (10.18) into (10.19) gives

$$\overline{\frac{\eta^2(t)}{\xi^2}} = 2B\Lambda\left(\frac{u}{2B}\right) \sum_{p=-\infty}^{\infty} c_p \sum_{k=-M}^{M} b_k \sum_{r=-M}^{M} b_r \delta\left(u + \frac{(k+r-p)}{T}\right)$$

$$= 2B\,\Lambda\left(\frac{u}{2B}\right) \sum_{k=-M}^{M} b_k \sum_{r=-M}^{M} b_r \sum_{q=-M}^{M} c_{k+r-q}\delta\left(u - \frac{q}{T}\right)$$

where $q = k + r - p$ and we have recognized that the finite extent of the triangle function lets through only $2M + 1$ of the Dirac delta functions. Evaluating $\Lambda(u/2B)$ at $u = q/T$ and inverse transforming gives the desired result:

$$\overline{\frac{\eta^2(t)}{\xi^2}} = \sum_{k=-M}^{M} b_k \sum_{r=-M}^{M} b_r \sum_{q=-M}^{M} \left(2B - \frac{|q|}{T}\right) c_{k+r-q}\, e^{-j2\pi qt/T}. \qquad (10.20)$$

An illustration of the restoration noise level for various duty cycles for first degree aliasing is shown in Figure 10.12. The effects of variation of the aliasing order are illustrated in Figure 10.11.

Colored Noise. With the aim of placing (10.15) in more tractable form for colored noise, we Fourier transform with respect to γ using the correlation theorem of Fourier analysis

$$H(t; \nu) = \int_{-\infty}^{\infty} h(t; \gamma)\, e^{-j2\pi \nu \gamma}\, d\gamma$$

$$= \left| T\, \Psi_M(T\nu) * \left[e^{-j2\pi \nu t}\, \Pi\left(\frac{\nu}{2B}\right) \right] \right|^2 \qquad (10.21)$$

where $\Psi_M(\nu)$ is the Fourier transform of $\psi_M(\gamma)$ and convolution is with respect to ν. Clearly

$$\Psi_M(\nu) = \sum_{n=-\infty}^{\infty} d_n \delta(\nu + n).$$

Thus (10.21) becomes

$$H(t; \nu) = \left| \sum_{n=-\infty}^{\infty} d_n \exp\left(-j2\pi(\nu + \frac{n}{T})t\right) \Pi\left(\frac{\nu + n/T}{2B}\right) \right|^2$$

$$= \left| \sum_{n=-\infty}^{\infty} d_n \exp\left(\frac{-j2\pi nt}{T}\right) \Pi\left(\frac{\nu + n/T}{2B}\right) \right|^2. \qquad (10.22)$$

Since

$$\Pi\left(\frac{\nu + \frac{n}{T}}{2B}\right) \Pi\left(\frac{\nu + \frac{m}{T}}{2B}\right) = \Pi\left(\frac{m-n}{4BT}\right) \Pi\left(\frac{\nu + \frac{m+n}{2T}}{2B - \frac{|n-m|}{T}}\right),$$

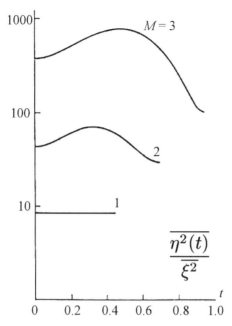

FIGURE 10.11. NINV for additive white noise for various orders of aliasing M. The values of T corresponding to $M = 1, 2, 3$ are $T = 0.9, 1.4, 1.9$, respectively. Because of symmetry, plots are needed only for $0 < x < \frac{T}{2}$. ($2B = 2$ and $\alpha = 0.6$.)

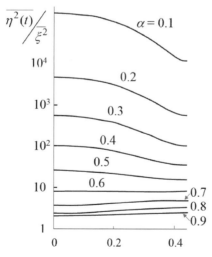

FIGURE 10.12. NINV for additive white noise for various duty cycles. $2B = 2$ and $T = 0.9$, giving M = first-order aliasing.

substituting into (10.22) and further recognizing from (10.7) that $d_m = \delta[m]$ for $|m| < M$ gives

$$H(t; \nu) = \Pi\left(\frac{\nu}{2B}\right)$$

$$+ \sum_{n=-M}^{M} \sum_{\substack{|m|>M \\ |n-m|\leq M}} d_n d_m \Pi\left(\frac{\nu + \frac{m+n}{2T}}{2B - \frac{|m-n|}{T}}\right) \exp\left(\frac{-j2\pi(n-m)t}{T}\right)$$

$$= \Pi\left(\frac{\nu}{2B}\right)$$

$$+ \sum_{n=-M}^{M} \sum_{\substack{|m|\leq M \\ |n-m|>M}} d_n d_{n-m} \Pi\left(\frac{\nu + \frac{2n-m}{2T}}{2B - \frac{|m|}{T}}\right) \exp\left(\frac{-j2\pi mt}{T}\right). \quad (10.23)$$

Using the power theorem, (10.14) can be written as

$$\overline{\eta^2(t)} = \int_{-\infty}^{\infty} S_\xi(u) H(t; u)\, du \quad (10.24)$$

where the power spectral density, $S_\xi(u)$, is the Fourier transform of $R_\xi(t)$. Define the odd indefinite integral

$$I_\xi(u; t) = \int_0^u S_\xi(v) H(t; v)\, dv.$$

Then substituting (10.23) into (10.24) and recognizing that $\eta^2(t)$ is real gives the Fourier series

$$\overline{\eta^2(t)} = 2 I_\xi(B; t)$$

$$+ \sum_{n=-M}^{M} \sum_{\substack{|m|\leq M \\ |n-m|>M}} d_n d_{n-m} \cos\left(\frac{2\pi mt}{T}\right)$$

$$\times \left[I_\xi\left(\frac{m - |m| - 2n}{2T} + B; t\right) \right.$$

$$\left. - I_\xi\left(\frac{m + |m| - 2n}{2T} - B; t\right) \right]. \quad (10.25)$$

For white noise, as in (10.16), $I_\xi(u; t) = \overline{\xi^2}$. The equivalent result in (10.20), however, is in closed form.

Laplace autocorrelation. For an example application of (10.25), consider the Laplace autocorrelation

$$R_\xi(\tau) = \overline{\xi^2}\, e^{-\lambda|\tau|}$$

where λ is a specified positive parameter. Then

$$I_\xi(u; t) = \overline{\xi^2}\, \frac{\arctan(2\pi u/\lambda)}{\pi}.$$

In the numerical examples to follow, B is set to one. Figure 10.13 shows the dependence of output noise level on the duty cycle α for first-order aliasing. The dependence of the Laplace parameter is shown in Figure 10.14 for a fixed duty cycle. As λ increases, adjacent points of the input noise become less correlated and the interpolation noise level decreases. Dependence of the output noise level on order of aliasing is illustrated in Figure 10.15.

10.3.2 Observations

10.3.2.1 Comparison with the NINV of the Cardinal Series

For certain combinations of the parameters T, $2B$, and α, the continuously sampled signal can be discretely sampled uniformly at or in excess of the Nyquist rate. Let this rate be

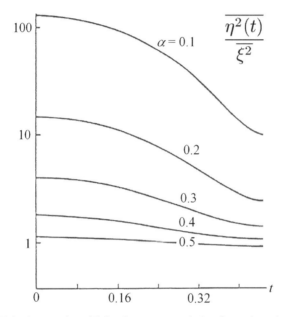

FIGURE 10.13. NINV for input noise with Laplace autocorrelation for various duty cycles α. $2B = 2$ and $T = 0.9$, giving M = first-degree aliasing. The Laplace parameter is $\lambda = 2$.

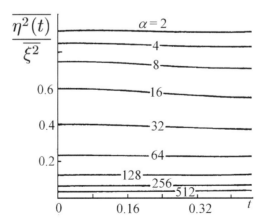

FIGURE 10.14. NINV for additive input noise with Laplace autocorrelation for various Laplace parameters λ. $2B = 2$ and $T = 0.9$, giving M = first-degree aliasing. The duty cycle is $\alpha = 0.6$.

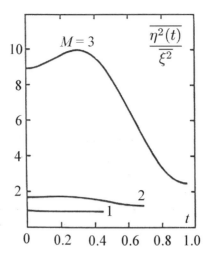

FIGURE 10.15. NINV for additive input noise with Laplace autocorrelation for various degrees of aliasing. The T values corresponding to $M = 1, 2, 3$ are $T = 0.9, 1.4, 1.9$, respectively. Because of symmetry, plots are needed only for $0 < x < T/2$. ($2B = 2$ and $\alpha = 0.6$, and $\lambda = 2$).

denoted by $2W > 2B$. The result is the same as if we had discretely sampled the original signal at a rate of $2W$.

Let T, $2B$, and α be such that this uniform sampling can be performed. Assume that, as in the previous section, each sample point is perturbed by additive Laplace autocorrelation noise with parameter α. When the noisy samples are interpolated and passed through a filter unity on $|u| < B$ and zero otherwise, the resulting NINV is given by (7.14). One would expect that the periodic continuous sample restoration would yield a lower noise level since more data are used in the recovery. As the results in Table 10.1 indicate, this is indeed the case.

10.3.2.2 In the Limit as an Extrapolation Algorithm

Keeping αT constant and letting T tend to ∞ alters our algorithm to an extrapolation algorithm if $f(t) \to 0$ as $t \to \infty$. As a consequence, the degree of aliasing, M, becomes unbounded. It is clear from Figures 10.13 and 10.17 that the noise sensitivity in this case increases enormously. This is our first empirical observation that the extrapolation problem is ill-posed.

TABLE 10.1. Comparison of noise levels for some cases in which the signal can be restored using either the continuously sampled signal-restoration algorithm or the conventional sampling theorem (followed by filtering). The former, in each case, has a lower level. In each case, the Laplace parameter is $\lambda = 2$

Example	T	$2B$	α	$2W$	min $\overline{\eta^2(t)}/\overline{\xi^2}$	max $\overline{\eta^2(t)}/\overline{\xi^2}$	$\overline{\eta_0^2}/\overline{\xi^2}$
(a)	1	1.5	0.90	5	0.7460	0.7478	0.7647
(b)	1	1.5	0.70	3	0.7768	0.7842	0.8020
(c)	5	0.3	0.96	1	0.2810	0.2822	0.3754

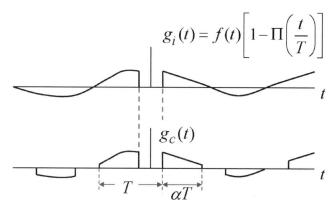

FIGURE 10.16. Forming a continuously sampled signal, $g_c(t)$, from a known interpolation problem signal, $g_i(t)$.

10.3.2.3 Application to Interval Interpolation

Use of the periodic continuous sample restoration algorithm for interval interpolation is shown in Figure 10.16. The known data, shown in the top figure, consists of the signal's tails. By selectively throwing away portions of the known data, we can form the continuously sampled signal shown. Our algorithm can be applied and the signal restored.

The unknown interval of length $(1 - \alpha)T$ must stay fixed. Note, however, that we have freedom in our choice of T. If we choose T to be small, then we have a small duty cycle and, as is illustrated in Figure 10.12, a correspondingly large restoration noise level. If we choose T to be large, then the order of aliasing increases and, as witnessed by Figures 10.11 and 10.14, the restoration noise level is also large. These observations suggest that there might exist some intermediate value of T that has optimal restoration noise properties [908].

An example where this is the case is pictured in Figure 10.17, where the normalized interpolation noise variance (NINV) at the origin is plotted versus T for the additive white noise in (4.74). As T increases, the restoration noise level decreases until T is sufficiently large to increase the order of aliasing. Then, as shown, the noise level makes a quantum leap and begins decreasing again until the next order of aliasing is reached. (Values of M are given at the top of the plot.) Note that in this case, the relative minima increase with T and, for minimum restoration noise level, the best choice for T is $1 - \epsilon$ where $0 < \epsilon \ll 1$.

As is shown in Figure 10.18, the relative minima can also increase with T. Here, the noise has a Laplace autocorrelation with parameter $\lambda = 2$. All other parameters are the same. Increasing λ to 10 again yields decreasing minima as shown in Figure 10.19. Note that, in any case, the interpolation noise level is finite. By this measure, the interpolation problem is thus well posed.

10.4 Interpolation of Discrete Periodic Nonuniform Decimation

Discrete periodic nonuniform decimation occurs then a discrete time signal, $f[n]$, is periodically set to zero over a specified interval. The problem is the discrete time version of continuous sampling presented in Section 10.2. Let the period of the decimation be P and

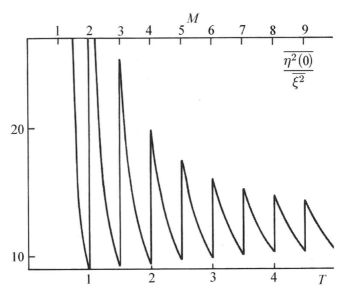

FIGURE 10.17. Continuous sample restoration of the interval interpolation problem yields this $\overline{\eta^2(0)}/\overline{\xi^2}$ curve when the data are perturbed by white noise. The optimum choice of T is a bit below one ($2B = 2, (1 - \alpha)T = 0.4$).

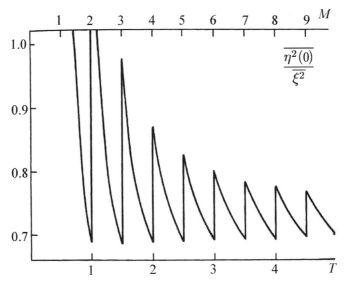

FIGURE 10.18. Same as Figure 10.17, except the noise has a Laplace autocorrelation with parameter $\lambda = 2$. The best choice of T is a bit below 1.5.

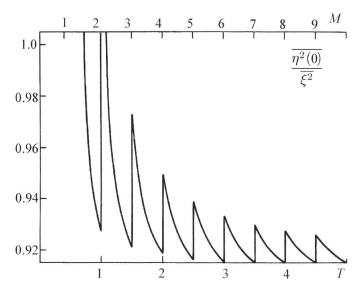

FIGURE 10.19. Same as Figure 10.18, except $\lambda = 10$. The minima here decrease.

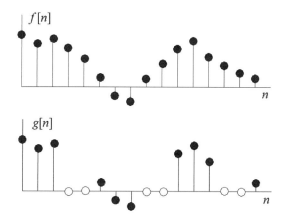

FIGURE 10.20. Illustration of *discrete periodic nonuniform decimation*. Here, the period is $P = 5$. Each of the first $N = 3$ values are kept and the remaining $P - N = 2$ are discarded. This is the discrete time version of continuous sampling illustrated in Figure 10.1.

let $N < P$. Within a period, the first N values are known and the remaining $P - N$ are set to zero. This periodic decimation is repeated for all values of n. The periodically nonuniformly decimated signal is denoted by $g[n]$. Given that the original signal, $f[n]$, is bandlimited, the problem is, given $g[n]$ and the bandwidth B, to find, if possible, $f[n]$. This is illustrated in Figure 10.20 for $P = 5$ and $N = 3$.

Analysis in this section is restricted to the case where the first N of P values are known in each period. The discrete periodic nonuniform decimation problem, though, is more general than this. In a period of $P = 5$, for example, the identity of the first, third and fifth values may be known and the others not. Interpolation in the case of these more general problems follows straightforwardly from the analysis to follow.

10.4.1 Problem Description

To generate the decimated signal, $g[n]$, define the discrete rectangular pulse train

$$r_P^N[n] = \sum_{p=-\infty}^{\infty} \Pi[pP \leq n < pP + N] \tag{10.26}$$

where $\Pi[n_- \leq n < n_+] = 1$ for $n_- \leq n < n_+$ and is otherwise zero. Then

$$g[n] = f[n] r_P^N[n]. \tag{10.27}$$

Our task is, when possible, to find $f[n]$ given its bandwidth and $g[n]$.

Taking the DTFT[3] of both sides of (10.27) gives the circular convolution[4]

$$G(v) = F(v) * \left(R_P^N(v) \Pi(v) \right) \tag{10.28}$$

where $R_P^N(v)$ is the DTFT of $r_P^N[n]$. From (10.26)

$$R_P^N(v) = \sum_{p=-\infty}^{\infty} \sum_{n=pP}^{pP+N-1} e^{-j2\pi nv}.$$

In the n sum, let $m = n - pP$. Then

$$R_P^N(v) = \sum_{p=-\infty}^{\infty} e^{-j2\pi pPv} \sum_{m=0}^{N-1} e^{-j2\pi mv}. \tag{10.29}$$

The p sum is recognized as the Fourier series of $\text{comb}(Pv)$ and, using a geometric series, we can show

$$\sum_{m=0}^{N-1} e^{-j2\pi mv} = N e^{-j\pi(N-1)v} \text{array}_N(v). \tag{10.30}$$

Equation (10.29) therefore becomes

$$R_P^N(v) = N \text{comb}(Pv) e^{-j\pi(N-1)v} \text{array}_N(v)$$

$$= \sum_{p=-\infty}^{\infty} a_p \delta\left(v - \frac{p}{P}\right)$$

where

$$a_p = \frac{N}{P} e^{-j\pi(N-1)p/P} \text{array}_N\left(\frac{p}{P}\right). \tag{10.31}$$

---

3. See Equation (2.112).
4. See Table 2.6.

10.4.1.1 Degree of Aliasing

The function $f[n]$ is assumed to have bandwidth $B < \frac{1}{2}$ so that[5]

$$F(v) = F(v) r_{2B}(v).$$

The convolution replicates this periodic spectrum in a possibly aliased fashion. An example is shown in Figure 10.21 for $P = 3$. Since one spectrum from the right overlaps the zeroth order spectrum, the degree of aliasing is $M = 1$. In general, the lowest frequency component of the Mth spectrum is at $v = \frac{M}{P} - B$. We wish to determine the largest value of M that infringes on the interval of the zeroth order spectrum, i.e., the largest value of M such that $\frac{M}{P} - B \leq B$. The degree of aliasing follows as

$$M = \lfloor 2PB \rfloor \tag{10.32}$$

where $\lfloor x \rfloor$ denotes the largest integer not exceeding x.

10.4.1.2 Interpolation

For Mth order aliasing, as is the case for continuous sampling, a total of $2M$ overlapping spectra must be eliminated from their overlap of the zeroth order spectrum. We desire coefficients $\{\beta_q | - M \leq q \leq M\}$ such that superimposing $2M + 1$ versions of $G(u)$ with various shifts eliminates the aliasing spectra and reconstructs the zeroth order spectrum exactly. We therefore seek the coefficients that solve

$$\left[\sum_{q=-M}^{M} \beta_q G\left(v - \frac{q}{P}\right) \right] \Pi\left(\frac{v}{2B}\right) = F(v); \quad |v| \leq B. \tag{10.33}$$

This is illustrated in Figure 10.22 for the spectra aliasing the zeroth order spectra. If these overlaps are cleared, then, due to the periodicity of DTFT spectra, all aliasing spectra will

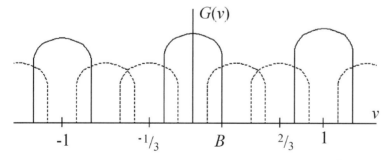

FIGURE 10.21. Illustration of $M = $ 1st order aliasing corresponding to $B = \frac{1}{4}$ and $P = 3$. The spectrum, $F(u)$, is shown with solid lines and the aliasing spectra with broken lines. From (10.32), $M = \frac{3}{2} = 1$.

---

5. Recall that

$$r_\alpha(t) = \sum_{n=-\infty}^{\infty} \Pi\left(\frac{t-n}{\alpha}\right).$$

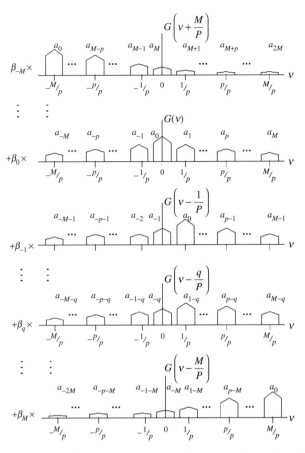

FIGURE 10.22. Weighting each of $2M+1$ shifted versions of the aliased signal in the frequency domain can, in certain instances, be used to remove the aliased spectra. Compare with Figure 10.5 which is for the continuous case.

be eliminated. From Figure 10.22, the β_q coefficients are solutions to the following system of $2M+1$ linear equations.

$$\sum_{q=-M}^{M} \beta_q a_{p-q} = \begin{cases} 1 \,;\, p = 0 \\ 0 \,;\, 1 \leq |p| \leq M. \end{cases} \tag{10.34}$$

In matrix-vector notation (10.34) is

$$\mathbf{A}[N, P]\vec{\beta} = \vec{\delta} \tag{10.35}$$

where $\mathbf{A}$ is a $(2M+1) \times (2M+1)$ matrix with elements

$$\begin{aligned}(\mathbf{A}[N, P])_{nm} &= a_{n-m} \\ &= \frac{N}{P} e^{-j\pi(N-1)p/P} \text{array}_N\left(\frac{p}{P}\right), \end{aligned} \tag{10.36}$$

$\vec{\beta}$ is a vector of the β_q's and $\vec{\delta}$ is a matrix of zeros except with a single "1" in the middle. The vector $\vec{\beta}$ is therefore equal to the middle column of $\mathbf{A}^{-1}$.

Assuming these equations do not contain colinear terms, the values of $\{\beta_q | -M \leq q \leq M\}$ can be solved numerically. Since

$$f[n] = \int_{-\frac{1}{2}}^{\frac{1}{2}} F(v) e^{j2\pi vn} dv,$$

$$= \int_{-B}^{B} F(v) e^{j2\pi vn} dv, \qquad (10.37)$$

we have, from (10.33),

$$f[n] = \int_{-B}^{B} F(v) e^{j2\pi vn} dv$$

$$= \int_{-\frac{1}{2}}^{\frac{1}{2}} \left[\sum_{q=-M}^{M} \beta_q G\left(v - \frac{q}{P}\right) \Pi\left(\frac{v}{2B}\right) \right] e^{j2\pi vn} dv \qquad (10.38)$$

$$= \sum_{q=-M}^{M} \beta_q \int_{-\frac{1}{2}}^{\frac{1}{2}} \left[G\left(v - \frac{q}{P}\right) \Pi\left(\frac{v}{2B}\right) \right] e^{j2\pi vn} dv.$$

Since

$$f[n] = \int_{-\frac{1}{2}}^{\frac{1}{2}} [X(v) H(v)] e^{j2\pi vn} dv$$

$$= x[n] * h[n]$$

where $*$ denotes discrete convolution, $x[n] \leftrightarrow X(v)$, and $h[n] \leftrightarrow H(v)$. And since

$$2B \, \text{sinc}(2Bn) \leftrightarrow \Pi\left(\frac{v}{2B}\right),$$

and

$$g[n] e^{j2\pi np/P} \leftrightarrow G\left(v - \frac{q}{P}\right), \qquad (10.39)$$

(10.38) becomes

$$f[n] = \left[\Theta_M\left(\frac{n}{P}\right) g[n] \right] * 2B \, \text{sinc}(2Bn) \qquad (10.40)$$

where we define the trigonometric polynomial

$$\Theta_M(v) = \sum_{q=-M}^{M} \beta_q e^{j2\pi qv}.$$

Since, however, $g[n] = g[n] r_P^N[n]$, knowledge of $\Theta_M\left(\frac{n}{P}\right)$ is required only when $r_P^N[n] = 1$. Therefore, define the periodic function

$$\Psi_M\left(\frac{n}{P}\right) = \Theta_M\left(\frac{n}{P}\right) r_P^N[n]$$

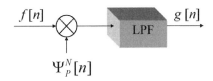

FIGURE 10.23. Signal flow diagram for restoring a periodic nonuniform decimated discrete signal. The bandwidth of the low pass filter is B.

and (10.40) becomes

$$f[n] = \left[g[n]\Psi_M\left(\frac{n}{P}\right)\right] * 2B\,\text{sinc}(2Bn). \tag{10.41}$$

This procedure for regaining $f[n]$ from $g[n]$ is illustrated in Figure 10.23.

10.4.1.3 The A[N, P] Matrix

The ability to solve the set of equations in (10.35) is dependent on the condition[6] of the matrix $\mathbf{A}[N, P]$. Here are some special cases

- $\mathbf{A}[1, P]$ has a zero determinate and is therefore singular. Thus, interpolation is not possible when $N = 1$. To show this, we note that, since $\text{array}_1(x) = 1$, each element of the $\mathbf{A}[1, P]$ matrix in (10.36) is

$$a_{nm} = \frac{N}{P}.$$

- $\mathbf{A}[N, P]$ is singular when[7] $B \geq \frac{1}{2}$.
- Here are the numerically computed condition numbers for $\mathbf{A}[N, P]$ when $M = 1$.

$P \Rightarrow$	2	3	4	5	6	7	8
$N = 1$	∞	∞	∞	∞	∞	∞	∞
2	∞	∞	∞	∞	∞	>10^{16}	>10^{16}
3	–	∞	4	13.1	34.0	73.7	141
4	–	–	∞	2.50	6.00	13.0	25.3
5	–	–	–	1	2.00	4.00	7.58
6	–	–	–	–	1	1.75	3.09
7	–	–	–	–	–	1	1.60
8	–	–	–	–	–	–	1

A plot of the condition number for $M = 1$ for P up to 40 is shown in Figure 10.24. A similar plot for second order aliasing ($M = 2$) is shown in Figure 10.25.

10.4.2 The Periodic Functions, $\Psi_M(\nu)$

The periodic functions, $\Psi_M(\nu)$, needed to interpolate the periodic nonuniformly decimated signal in (10.41) cannot always be computed. The $\mathbf{A}[N, P]$ matrix must have a sufficiently

---

6. See Appendix 14.5.
7. See Exercise 10.4.

SIGNAL RECOVERY   471

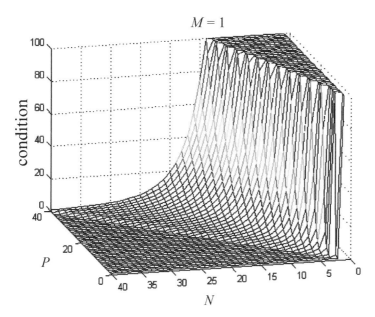

FIGURE 10.24. A plot of matrix condition number for first order ($M = 1$) aliasing for various N and P. (Values for $N < P$ are shown set to zero).

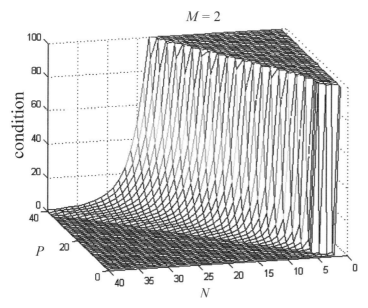

FIGURE 10.25. A plot of matrix condition number for second order ($M = 2$) aliasing for various N and P. (Values for $N < P$ are shown set to zero.)

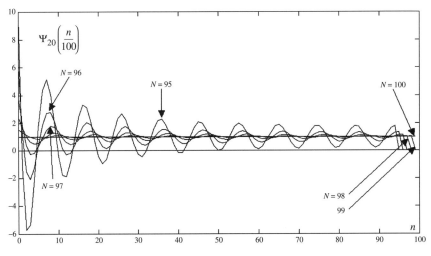

FIGURE 10.26. Plots of $\Psi_M\left(\frac{n}{P}\right)$ for $M = 20$, $P = 100$ and $95 \leq N \leq 100$.

small condition number.[8] An example of the resulting periodic functions are shown in Figure 10.26 for $(M, P) = (20, 100)$.

10.4.3 Quadrature Version

If $f[n]$ is real, its DTFT is conjugately symmetric. That is, $F(\nu) = F^*(-\nu)$. Therefore, if $f[n]$ is bandlimited, knowledge of $F(\nu)$ is only required on the interval $0 \leq \nu \leq B$. With reference to Figure 10.22, only the aliasing for positive ν must be removed. There are M aliasing spectra to the right of the zeroth order spectra. In addition, there are spectra to the left of the zeroth order spectrum. The maximum frequency component of the $-m$th spectrum is $\frac{-m}{P} + B$. Let $-M_2$ be the minimum value of $-m$ for which this value is positive. Solving $\frac{-m}{P} + B > 0$ gives

$$M_2 = \lfloor BP \rfloor.$$

Note, for M even, $2M_2 = M$. Instead of the equation in (10.33), we desire to solve

$$\sum_{q=-M_2}^{M} \hat{\beta}_q G\left(\nu - \frac{q}{P}\right) = F(\nu); \quad 0 \leq \nu \leq B. \tag{10.42}$$

To find the $\hat{\beta}$ coefficients, instead of (10.34), we solve the following set of $M + M_2 + 1$ linear equations.

$$\sum_{q=-M_2}^{M} \hat{\beta}_q a_{p-q} = \begin{cases} 1; & p = 0 \\ 0; & -M_2 \leq p \leq -1 \text{ and } 1 \leq p \leq M. \end{cases} \tag{10.43}$$

---

8. See (14.10).

10.5 Prolate Spheroidal Wave Functions

A set of orthogonal functions which prove useful in the extrapolation and interval interpolation problems are the *prolate spheroidal wave functions* (PSWF's). Their use in these problems was initially reported by Slepian and Pollak [1285] in the first of a classic series of papers [1280, 1281, 1282, 1284, 1285, 1286].

The PSWF's can be defined as the solution of the integral equation

$$\lambda_n \psi_n(t) = 2B \int_{-T/2}^{T/2} \psi_n(\tau) \, \text{sinc}(2B(t - \tau)) \, d\tau \tag{10.44}$$

where $0 \leq n < \infty$ and the λ_n's are the eigenvalues. Equivalently

$$\lambda_n \psi_n(t) = \left[\psi_n(t) \, \Pi\left(\frac{t}{T}\right) \right] * 2B\text{sinc}(2Bt). \tag{10.45}$$

The PSWF's can thus be viewed as the eigenfunctions of low pass filtering signals of finite support.

Although not explicitly stated in the notation, both $\psi_n(t)$ and λ_n are continuous functions of the time-bandwidth product

$$c = 2BT. \tag{10.46}$$

Plots of some PSWF's are shown in Figure 10.27

10.5.1 Properties

Here we present without proof some significant properties of the PSWF's and their eigenvalues [1076].

(a) The eigenvalues of the PSWF's are real. Note from (10.45), that energy in $\psi_n(t)$ is reduced first by truncation and then by filtering. Thus, each λ_n has a magnitude less

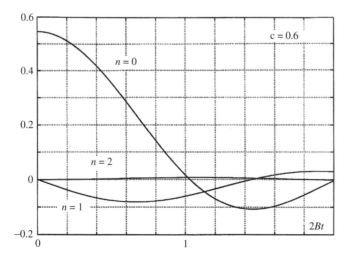

FIGURE 10.27. PSWF's $\psi_0(t)$, $\psi_1(t)$, and $\psi_2(t)$ vs. $2Bt$ for $c = 0.6$.

than unity. We will choose them to be positive and will index them in decreasing order.

$$1 > \lambda_0 > \lambda_1 > \cdots > 0. \qquad (10.47)$$

(b) From (10.45), the PSWF's are clearly bandlimited and thus are not affected by low pass filtering.

$$\psi_n(t) = \psi_n(t) * 2B\text{sinc}(2Bt). \qquad (10.48)$$

(c) For a given c, the PSWF's are orthonormal on $(-\infty, \infty)$.

$$\int_{-\infty}^{\infty} \psi_n(\tau)\psi_m(\tau)d\tau = \delta[n-m]. \qquad (10.49)$$

Furthermore, they form a complete basis set for finite energy bandlimited signals. Thus, if $f(t)$ is bandlimited, then

$$f(t) = \sum_{n=0}^{\infty} a_n \psi_n(t) \qquad (10.50)$$

where

$$a_n = \int_{-\infty}^{\infty} f(t)\psi_n(t)dt. \qquad (10.51)$$

Like the sampling theorem, convergence of (10.50) is uniform [478].

(d) For a given c, the PSWF'S are orthogonal on the interval $|t| \leq T/2$.

$$\int_{-T/2}^{T/2} \psi_n(\tau)\psi_m(\tau)\,d\tau = \lambda_n\,\delta[n-m]. \qquad (10.52)$$

Furthermore, the PSWF's are a complete basis set for finite energy functions on the interval $|t| \leq T/2$. Specifically

$$h(t) = \sum_{n=-N}^{N} b_n\psi_n(t); \; |t| \leq \frac{T}{2} \qquad (10.53)$$

where

$$\lambda_n b_n = \int_{-T/2}^{T/2} h(t)\psi_n(t)dt.$$

Like the Fourier series, convergence of (10.53) is assured in the mean square sense.

(e) The PSWF's are also eigenfunctions of a sort for the Fourier transform. Specifically

$$\psi_n(t) \longleftrightarrow \sqrt{\frac{T}{2B\lambda_n}}\,\psi_n\left(\frac{Tu}{2B}\right)\Pi\left(\frac{u}{2B}\right). \qquad (10.54)$$

Similarly, for the truncated PSWF,

$$\psi_n(t)\,\Pi\left(\frac{t}{T}\right) \longleftrightarrow \sqrt{\frac{T\lambda_n}{2B}}\,\psi_n\left(\frac{Tu}{2B}\right). \qquad (10.55)$$

This follows from (10.54) and duality.

(f) The PSWF's are difficult to deal with numerically. For our purposes, they will prove to be primarily an analytic tool. Their detail structure, shown in Figure 10.27, is of secondary interest.

10.5.2 Application to Extrapolation

Let $f(t)$ be bandlimited with known bandwidth B. The extrapolation problem is to regain $f(t)$ from

$$g_e(t) = f(t)\, \Pi\left(\frac{t}{T}\right). \tag{10.56}$$

A solution to this problem is obtained by expanding $g(t)$ into a PSWF series:

$$g_e(t) = \sum_{n=0}^{\infty} a_n \psi_n(t)\, \Pi\left(\frac{t}{T}\right) \tag{10.57}$$

where

$$\lambda_n a_n = \int_{-T/2}^{T/2} g_e(t)\, \psi_n(t)\, dt. \tag{10.58}$$

Also, since $f(t)$ is bandlimited, it can be expanded as in (10.50). The coefficients of an orthogonal function expansion are unique. Thus, the coefficients in (10.50) and (10.58) are the same. The significant point is that these coefficients can be determined only with knowledge of $g_e(t)$ via (10.58). Then $f(t)$ can be found from (10.50) and our extrapolation is complete.

This result should bother our intuition. For example, telephone conversation waveforms can be considered bandlimited. Our result says that the entirety of a phone conversation can be determined if we know only a word or two in the middle. This, of course, is an unacceptable conclusion.

The resolution of this apparent paradox between mathematics and intuition lies in the fact that our analysis has been to this point deterministic. In practice, the known portion of the signal will be accompanied by some type of noise. To understand how noise affects the algorithm, we must examine the structure of the eigenvalues shown in Figure 10.28. Fix c. For n below a certain number, the eigenvalues are essentially one. Above that threshold, they are close to zero. A typical plot of λ_n versus n is shown in Figure 10.29.

Consider, then, the evaluation of the coefficients in (10.58) when either the integral computation and/or $g_e(t)$ is accompanied by a small degree of inexactitude. If n is above the threshold, then division by $\lambda_n \ll 1$ will greatly magnify this error. Thus, the a_n coefficients can be only computed reliably up to that threshold which we will call S.

To get a feeling for the value of S, consider again, $g_e(t)$. If we sample this known portion at the Nyquist rate, $2B$, over a time interval of duration T, then the total number of non-zero samples is about $2BT$. This is the time-bandwidth product discussed in Section 5.4.3. It is also roughly the number of discrete values required to specify $g_e(t)$ to a "good" approximation. We can show empirically that this is the threshold we seek. The value

$$S = 2BT$$

has also been called the *Shannon number* [1407] or time-bandwidth product. Note that $S = c$. Thus, we conclude that in most any practical situation, $g_e(t)$ can be represented by roughly S numbers, be they samples or PSWF coefficients. In very high signal-to-noise ratio situations, however, it is possible to add a few degrees of freedom to a truncated signal.

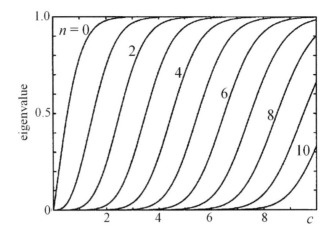

FIGURE 10.28. Eigenvalues, λ_n, of the PSWF integral equation.

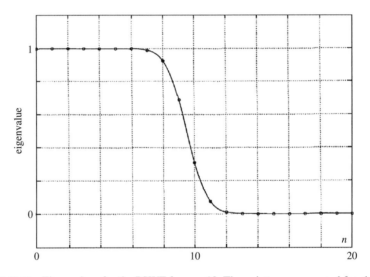

FIGURE 10.29. Eigenvalues for the PSWF for $c = 10$. The points are connected for clarity.

10.5.3 Application to Interval Interpolation

Mathematically, the interpolation problem is similar to extrapolation with a significant difference- interval interpolation is well posed. As is shown in Figure 10.1c, our given data here is

$$g_i(t) = f(t)\left[1 - \Pi\left(\frac{t}{T}\right)\right]. \tag{10.59}$$

From (10.44) and (10.48), we conclude that

$$(1 - \lambda_n)\psi_n(t) = 2B \int_{\tau > \frac{|T|}{2}} \psi_n(\tau)\operatorname{sinc}(2B(t - \tau))d\tau. \tag{10.60}$$

It follows that the PSWF's are complete for $g_i(t)$ when $f(t)$ is bandlimited. Equation (10.50) applies and the expansion coefficients can be found with knowledge only of $g_i(t)$:

$$a_n = (1 - \lambda_n)^{-1} \int_{\tau > \frac{|T|}{2}} g_i(\tau) \psi_n(\tau) d\tau.$$

Here, we are dividing the integral by a small number when n is less than S. The difference here is that the number is finite. Thus, finite data error yields a finite amount of interval interpolation error and the problem is well posed.

10.6 The Papoulis-Gerchberg Algorithm

The *Papoulis-Gerchberg algorithm* (PGA) [500, 1084] is an ingenious technique for restoring any continuously sampled bandlimited signal without directly using PSWF's. Indeed, the algorithm requires only the operations of filtering and truncation [263, 440, 469, 924, 1144, 1218, 1375, 1518, 1527].

The PGA was first discovered by Papoulis [1084] but was first published in an archival journal, independently, by Gerchberg [500]. DeSantis and Gori [367], also independently, published Papoulis' algorithm proof shortly after Gerchberg's paper appeared.

The PGA is applicable to each of the continuous sampling problems illustrated in Figure 10.1 when $f(t)$ is bandlimited [263, 440, 1218]. The algorithm is most easily proved for the cases of extrapolation and interval interpolation. Since most of the work performed has been on the extrapolation problem, this will be our main focus. As before, our results will be ill-posed. Problems-not algorithms-are well or ill-posed.

10.6.1 The Basic Algorithm

The PGA is illustrated in Figure 10.30 for the case of extrapolation. The known portion of the signal is $g_e(t)$ as shown in Figure 10.1b. We know, secondly, that our signal to be restored, $f(t)$, is bandlimited with bandwidth B. The PGA iterates back and forth between the time and frequency domains reinforcing these criteria.

Beginning with $g_e(t)$, the first step in the algorithm is Fourier transformation. Since $g_e(t)$ is of finite extent, its transform will be identically zero nowhere. This is contrary to our knowledge that the signal to be restored is bandlimited. Thus, we make the spectrum that of a bandlimited function in step 2 by multiplying by $\Pi(u/2B)$. Step 3 is inverse transformation back to the time domain. This signal, shown at the bottom of Figure 10.30, is clearly bandlimited. It is not, however, equal to $g_e(t)$ on the interval $|t| < T/2$. To impose this criterion, we first set the signal to zero on $|t| \leq T/2$ in step 4 by multiplying by $1 - \Pi(u/2B)$. Then, in step 5, the known portion of the signal, $g_e(t)$, is inserted in the dead space. We will call this signal $f_1(t)$.

In general, $f_1(t)$ will be a discontinuous function which, in turn, cannot be bandlimited. Thus, we need to reimpose the criterion of bandlimitedness. Thus, in step 6, we begin the set of the same operations again.

Denote the results of the Nth iteration by $f_N(t)$. We will show in the next section that

$$\lim_{N \to \infty} f_N(t) = f(t).$$

Thus, our extrapolation is performed.

Note that, by simple alteration, the algorithm can be applied to the interval interpolation, prediction and, indeed, to any continuously sampled bandlimited signal.

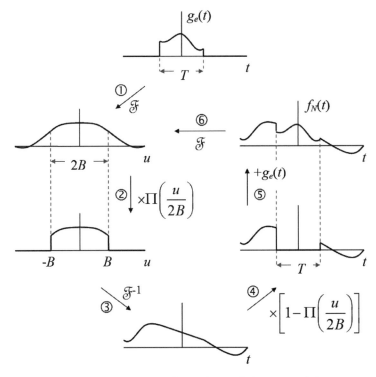

FIGURE 10.30. Illustration of the PGA applied to extrapolation.

Let's operationally compress the PGA of Figure 10.30. Steps 1, 2 and 3 are simply a low pass filtering operation. We define the low pass filter operator by

$$\mathbf{B}_B h(t) = h(t) * 2B\, \mathrm{sinc}(2Bt).$$

The operation of discarding the center can be modelled by $1 - \mathbf{D}_T$ where the duration limiting operator is defined by

$$\mathbf{D}_T h(t) = h(t)\, \Pi\left(\frac{t}{T}\right).$$

With this notation, the PGA can be written as

$$f_{N+1}(t) = g_e(t) + [1 - \mathbf{D}_T]\mathbf{B}_B f_N(t) \qquad (10.61)$$

with initialization

$$f_0(t) = g_e(t). \qquad (10.62)$$

The PGA using operators is illustrated in Figure 10.31.

A numerical example of the PGA for extrapolation is shown in Figure 10.32 for the case of a sinc. The function $\mathrm{sinc}(t)$ for $|t| \leq 0.1$ is the known portion of the signal. "Good" convergence takes place in only eight iterations. We must remark, however, that sincs extrapolate well. Furthermore, the only noise in the unknown signal is computational round off error.

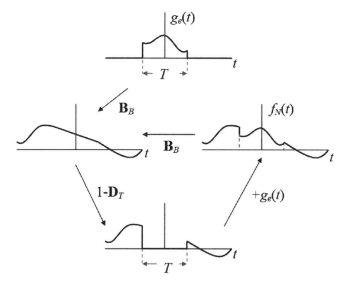

FIGURE 10.31. An equivalent illustration of the PGA algorithm.

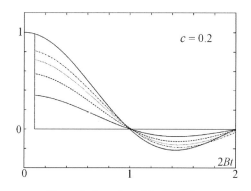

FIGURE 10.32. Numerical results of the PGA for extrapolation.

10.6.1.1 An Alternate Derivation of the PGA Using Operators

An alternate derivation of the PGA makes use of the identity

$$\mathbf{B}_B f(t) = f(t).$$

This follows from the bandlimitedness of $f(t)$. Thus

$$g_e(t) = [1 - (1 - \mathbf{D}_T)\mathbf{B}_B] f(t). \tag{10.63}$$

If the operator in square brackets can be inverted, then we have

$$f(t) = [1 - (1 - \mathbf{D}_T)\mathbf{B}_B]^{-1} g_e(t). \tag{10.64}$$

For inversion, we generalize the geometric series in (14.6) and write:

$$f(t) = \sum_{n=0}^{\infty} [(1 - \mathbf{D}_T)\mathbf{B}_B]^n g_e(t).$$

In the spirit of truncation, define

$$f_N(t) = \sum_{n=0}^{N} [(1 - \mathbf{D}_T)\mathbf{B}_B]^n g_e(t). \tag{10.65}$$

Equations (10.61) and (10.62) follow as a direct consequence.

10.6.1.2 Application to Interpolation

The PGA can be used in principle for restoration of any bandlimited signal that is continuously sampled in the absence of noise. The algorithm for interval interpolation is, for example

$$f_{N+1}(t) = g_i(t) + \mathbf{D}_T \mathbf{B}_B f_N(t) \tag{10.66}$$

where

$$\begin{aligned} g_e(t) &= (1 - \mathbf{D}_T) f(t) \\ &= f_0(t). \end{aligned} \tag{10.67}$$

The interpolation equivalents of (10.64) and (10.65) are, respectively,

$$f(t) = [(1 - \mathbf{D}_T)\mathbf{B}_B]^{-1} g_i(t) \tag{10.68}$$

and

$$f_N(t) = \sum_{n=0}^{N} [\mathbf{D}_T \mathbf{B}_B]^n g_i(t). \tag{10.69}$$

10.6.2 Proof of the PGA using PSWF's

Here, we offer a proof of the PGA using PSWF's for the case of extrapolation. [367, 907, 1085]

We begin by noting that (10.44) and (10.48) can be respectively written as

$$\mathbf{B}_B \mathbf{D}_T \psi_n(t) = \lambda_n \psi_n(t)$$

and

$$\mathbf{B}_B \psi_n(t) = \psi_n(t).$$

Also, from (10.57) we can write the known portion of the signal as

$$g_e(t) = \sum_{n=0}^{\infty} a_n \mathbf{D}_T \psi_n(t). \tag{10.70}$$

From (10.61) and (10.62), the first iteration of the PGA can be written:

$$f_1(t) = g_e(t) + (1 - \mathbf{D}_T)\mathbf{B}_B + \sum_{n=0}^{\infty} a_n \mathbf{D}_T \psi_n(t)$$

$$= g_e(t) + (1 - \mathbf{D}_T) \sum_{n=0}^{\infty} a_n \lambda_n \psi_n(t). \qquad (10.71)$$

Expanding $g_e(t)$ via (10.70) and applying the second iteration gives:

$$f_2(t) = g_e(t) + (1 - \mathbf{D}_T) \sum_{n=0}^{\infty} a_n (2\lambda_n - \lambda_n^2) \psi_n(t).$$

Repeating

$$f_3(t) = g_e(t) + (1 - \mathbf{D}_T) \sum_{n=0}^{\infty} a_n (3\lambda_n - 3\lambda_n^2 + \lambda_n^3) \psi_n(t)$$

$$f_4(t) = g_e(t) + (1 - \mathbf{D}_T) \sum_{n=0}^{\infty} a_n (4\lambda_n - 6\lambda_n^2 + 4\lambda_n^3 - \lambda_n^4) \psi_n(t).$$

This is a sufficient number of iterations to recognize that the coefficients of the eigenvalues are binomial. We can show by induction that

$$f_N(t) = g_e(t) + (1 - \mathbf{D}_T) \sum_{n=0}^{\infty} a_n \left[1 - \sum_{k=0}^{N} \binom{N}{k} (-\lambda_n)^k \right] \psi_n(t)$$

or, using the binomial series,[9]

$$f_N(t) = g_e(t) + (1 - \mathbf{D}_T) \sum_{n=0}^{\infty} a_n [1 - (1 - \lambda_n)^N] \psi_n(t). \qquad (10.72)$$

Proof. From (10.71), the result is true for $N = 1$. We will assume (10.72) and show the corresponding equation for $N + 1$ follows. An additional iteration gives

$$g_e(t) + (1 - \mathbf{D}_T)\mathbf{B}_B f_{N+1}(t)$$

$$= g_e(t) + (1 - \mathbf{D}_T) \sum_{n=0}^{\infty} a_n [\lambda_n + (1 - \lambda_n)\{1 - (1 - \lambda_n)\}^N] \psi_n(t).$$

After some algebra, this relationship becomes (10.72) for $N + 1$. The validity of (10.72) is thus proved.

Since, from (10.47),

$$0 < \lambda_n < 1 \qquad (10.73)$$

---

9. See Equation 14.4.

we conclude that
$$\lim_{N\to\infty} [1 - (1-\lambda_n)^N] = 1$$
and thus, in the limit,
$$\lim_{N\to\infty} f_N(t) = g_e(t) + (1-\mathbf{D}_T)\sum_{n=0}^{\infty} a_n \psi_n(t)$$
$$= \Pi(t/T)f(t) + [1 - \Pi(t/T)]f(t)$$

where we have used (10.56) and (10.50). This completes our proof of the PGA.

10.6.3 Remarks

A fundamental problem of the PGA is the ill-posed nature of the extrapolation problem. We note, however, that extrapolation is ill-posed in the sense that the restoration noise level cannot be bounded. This makes sense. The uncertainty of restoration should in general increase as we remove ourselves farther and farther from the known portion of the signal. Clearly, the uncertainty of restoration with knowledge of a signal only on a interval of length T will be enormous at a distance of, say, $10^{10} \times T$ away. Thus, extrapolation is ill-posed in a global manner. As has been mentioned, however, one might expect "good" results near to where the (smooth) signal is known.

A problem is ill-posed because insufficient information about the restored signal has been provided. To *regularize* such problems, either additional information about the signal must be provided, or the class of allowable solutions restricted.

The PGA, as is shown in Section 11.4.3, is a special case of the POCS paradigm. POCS affords great flexibility for introducing additional constraints in the restoration algorithm.

10.7 Exercises

10.1. § **Slepian's Paradox**
   (a) Why can a causal signal never be bandlimited?
   (b) David Slepian presents a scenario whereby a causal signal must also be bandlimited. He wrote [1283]

   "[A] pair of solid copper wires will not propagate electromagnetic waves at optical frequencies, and so the signals I receive over such a pair must be bandlimited. In fact, it makes little physical sense to talk of energy received over wires at frequencies higher than some finite cutoff W, say 10^{20} Hertz. It would seem, then, that signals must be bandlimited.

   "On the other hand, however, signals of limited bandwidth W are finite Fourier transforms ... and irrefutable mathematical arguments show them to be extremely smooth. They possess derivatives of all orders. Indeed, such integrals are entire [analytic] functions of t, completely predictable from any little piece, and they cannot vanish on any t interval unless they vanish everywhere. Such signals cannot start or stop, but must go on forever. Surely real signals start and stop, and so they cannot be bandlimited!"

   Can you resolve Slepian's paradox?

10.2. The periodic continuously sampled signal at the bottom of Figure 10.16 can be written:

$$g_c(t) = f(t)\left[1 - r_{1-\alpha}\left(\frac{t}{T}\right)\right].$$

Assume $f(t)$ has bandwidth B. Specify how to find the trigonometric polynomial, $\Theta_M(t)$, such that

$$f(t) = [g_c(t)\psi_M(t)] * 2B\,\text{sinc}(2Bt)$$

where

$$\psi_M(t) = \Theta_M(t)[1 - r_{1-\alpha}(t)].$$

10.3. (a) If $f(t)$ is real, $F(u)$ is conjugately symmetric.

$$F(u) = F^*(-u)$$

Thus, knowledge of $F(u)$ for $u \geq 0$ is sufficient to uniquely specify $f(t)$. Write the formula for $f(t)$ in terms of $F(u)$ for $u > 0$.

(b) Consider, then, first order aliasing for periodic continuous sampling of a bandlimited signal. As shown in Figure 10.33, we add two weighted versions of $G(u)$ to rid ourselves of the positive frequency overlap. Specify $F(u)$ for positive u in terms of $G(u)$ and $G\left(u - \frac{1}{T}\right)$.

(c) Find $f(t)$ directly from $g(t)$.

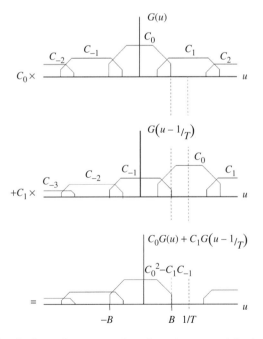

FIGURE 10.33. Removing the first order spectrum by subtracting two weighted and shifted versions of the degraded spectrum.

10.4. (a) Derive the relationship in (10.30).
   (b) Coefficient periodicity.
      (i) Show that the coefficient a_p in (10.31) is periodic with period P.
      (ii) Are there cases where the periodicity of a_p will render the set of equations in (10.34) singular? Use this with the inequality $M \leq 2BP$ to show $\mathbf{A}$ is singular when $B > \frac{1}{2}$.

10.5. Generalize the analysis for restoration of $f[n]$ from $g[n] = f[n]p_P[n]$ where $p_P[n]$ is a given periodic function with period P [910].

10.6. Are there cases where the $\mathbf{A}$ matrix is singular and the set of equations in (10.43) are not?

10.7. For $M = 1$, the $\mathbf{A}$ matrix is

$$\mathbf{A} = \begin{bmatrix} a_0 & a_1 & a_2 \\ a_{-1} & a_0 & a_1 \\ a_{-2} & a_{-1} & a_0 \end{bmatrix}$$

Alternate shifting can be used. We can, for example, use[10]

$$\mathbf{A} = \begin{bmatrix} a_{52} & a_{53} & a_{54} \\ a_{-5} & a_{-4} & a_{-3} \\ a_7 & a_6 & a_5 \end{bmatrix}.$$

For (a) fixed but arbitrary M, and (b) for the case of no two equal rows of coefficients, can such alternate shifts be used to generate nonsingular $\mathbf{A}$ matrices for some shifts and singular for others?

10.8. Find the PSWF expansion of $2B \operatorname{sinc}(2B(t - \tau))$.

10.9. The cardinal series can be viewed as an orthogonal expansion using $\{\operatorname{sinc}(2Wt - n) \mid -\infty < n < \infty\}$ as the basis function set.
   (a) Show these functions form an orthogonal set.
   (b) Express the PSWF expansion coefficients, a_n, in terms of the samples $f(n/2B)$ assuming $f(t)$ is bandlimited. The result is an infinite sum. Is it well-posed?
   (c) Similarly, express, $f(n/2B)$ as a weighted sum of PSWF coefficients, a_n. Is it well-posed?

10.10. (a) Show that (10.52) follows as a consequence of (10.49) and (10.44).
   (b) Show that (10.55) follows as a consequence of (10.49) and (10.54).

10.11. **The Extrapolation Matrix**
Consider the discrete implementation solution of the operator equation in (10.63). The discretized operators become matrices. For an arbitrary signal, $z(t)$, we have

$$w(t) := \mathbf{B}_B z(t) = 2B \operatorname{sinc}(2Bt) * z(t) = 2B \int_{-\infty}^{\infty} z(\tau) \operatorname{sinc}(2B(t - \tau)) d\tau$$

which, for a sample interval of Δ, can be discretized as

$$\vec{w} = \hat{\mathbf{B}}_B \vec{z}$$

where $(\vec{w})_n = w(n\Delta)$, $(\vec{z})_n = z(n\Delta)$, and the matrix $\hat{\mathbf{B}}_B$ contains elements

$$\left(\hat{\mathbf{B}}_B\right)_{nm} = 2B\Delta \operatorname{sinc}(2B(n-m)\Delta).$$

---

10. Note the indices of the third row *decrease* as we move to the right.

Likewise, the operator $\mathbf{D}_T$ in (10.63) can be discretized as a matrix $\hat{\mathbf{D}}_T$ of all zeros except for ones on the diagonal at locations corresponding to where the signal is passed and zeros otherwise. The discretized version of (10.63) then becomes

$$\vec{g}_e = \left[\mathbf{I} - \left(\mathbf{I} - \hat{\mathbf{D}}_T\hat{\mathbf{B}}_B\right)\right]\vec{f}.$$

We can then solve for $\vec{f}$ and write

$$\vec{f} = \hat{\mathbf{E}}\,\vec{g}_e$$

where the *extrapolation matrix* [899, 897, 907, 1205, 1289] is the matrix inverse

$$\hat{\mathbf{E}} = [\mathbf{I} - (\mathbf{I} - \hat{\mathbf{D}}_T)\hat{\mathbf{B}}_B]^{-1}.$$

(a) Simulate application of the performance of the extrapolation matrix for various bandlimited signals.
(b) Try the alternate expression $\mathbf{B}_B = \mathbf{IDFT} \times \hat{\mathbf{D}}_{2B} \times \mathbf{DFT}$ where **IDFT** and **DFT** correspond to inverse discrete Fourier transform and discrete Fourier transform matrices.
(c) Repeat (a) and (b), but for interpolation.
(d) Evaluate the condition[11] of the extrapolation and interpolation matrices in (a), (b) and (c). Comment on the relationship between performance and the condition number.

10.12. Clearly, if $\psi_n(t)$ is a solution of (10.44), then so is $const \times \psi_n(t)$. Is there an ambiguity in the definition of the PSWF? If not, how have we removed it?

10.13. **Parseval's Theorem for PSWF's**
(a) For the expansion in (10.50), show that the energy of $f(t)$ is

$$E = \sum_{n=0}^{\infty} |a_n|^2.$$

(b) Derive a similar expression for the energy of $h(t)$ in (10.53).

10.14. We pass a signal, $x(t) = x(t)\,\Pi(t/T)$, through a low pass filter with bandwidth B. The output is $y(t)$. Assuming $x(t)$ has unit energy, what input will yield an output with the maximum energy? What is this energy?

10.15. Using the PSWF's, prove the PGA as applied to interval interpolation.

10.16. For the extrapolation problem, $g_e(t)$ in (10.56) can be expressed in terms of a Fourier series:

$$g_e(t) = \sum_{n=-\infty}^{\infty} g_n \exp\left(\frac{j2\pi nt}{T}\right) \Pi\left(\frac{t}{T}\right).$$

(a) Find the a_n's in (10.58) directly as an infinite weighted sum of the Fourier coefficients, g_n. Is this restoration well-posed or ill-posed?
(b) Conversely, find the g_n's. Comment again on the posedness.

10.17. Show that (10.61) follows as a consequence of (10.65).

---

11. See (14.10).

10.18. (a) Assume we know apriori that the signal to be restored is non-negative, i.e., $f(t) \geq 0$. Incorporate this constraint into the PGA for extrapolation.
(b) Suppose $f(t) = R_\xi(t)$ is an autocorrelation. Incorporate this information into the PGA for extrapolation.

10.19. The geometric series, $\sum_{n=0}^{\infty} z^n = 1/(1-z)$, converges if $|z| < 1$. Similarly, a sufficient condition for the equivalent equation:

$$(1 - \mathbf{H})^{-1} h(t) = \sum_{n=0}^{\infty} \mathbf{H}^n h(t)$$

is that

$$\| \mathbf{H} \| < 1.$$

The operator norm can be defined by

$$\| \mathbf{H} \| = \sup_{\| h(t) \| = 1} \| \mathbf{H} h(t) \|$$

where the L_2 norm is defined by

$$\| y(t) \|^2 = \int_{-\infty}^{\infty} | y(t) |^2 \, dt$$

and sup denotes 'supremum.'
(a) For the extrapolation algorithm, from (10.64), $\mathbf{H} = (1 - \mathbf{D}_T) \mathbf{B}_B$. Compute $\| \mathbf{H} \|$.
(b) For interval interpolation, from (10.66), $\mathbf{H} = \mathbf{D}_T \mathbf{B}_B$. Compute $\| \mathbf{H} \|$.

10.20. We apply the PGA to the restoration of a single lost sample at the origin of an oversampled signal. The given signal, $g(t)$, is shown in Figure 10.34. It can be written as

$$g(t) = r \sum_{n \neq 0} f(nT) \text{sinc}(2Wt - rn)$$

where $T = 1/2W$ and $f(t)$ has bandwidth of $B < W$. From $g(t)$ we wish to find $f(0)$. As is shown in Figure 10.34, we first Fourier transform this sequence and multiply by $\Pi(u/2B)$ to form $S_N(u)$. The inverse Fourier transform of this expression is evaluated at the origin and used as an estimate for the lost sample. After the Nth iteration, we have

$$s_N(t) = T f_N(0) + g(t).$$

The iteration is repeated. Evaluate $f_\infty(0)$ and compare your answer with (6.11).

10.21. Generalize the analysis in Section 10.4 for restoration of $f[n]$ from $g[n] = f[n] p_P[n]$ where $p_P[n]$ is a given periodic function with period P [910].

10.8 Solutions for Selected Chapter 10 Exercises

10.1. Slepian's Paradox
(a) A causal signal cannot be bandlimited. Since all bandlimited signals are analytic, they can be expressed using the Taylor series expansion in (10.1) about

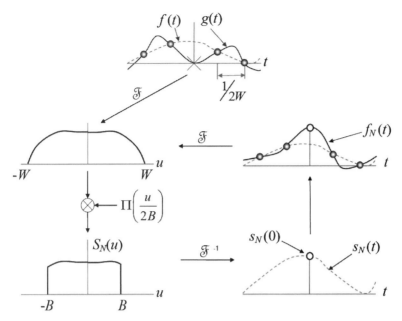

FIGURE 10.34. Application of the PGA to the restoration of a single lost sample at the origin of an oversampled signal.

any point in the signal. For a causal signal, we choose a negative value of time where the signal and all of its derivatives are zero. The Taylor series converges to identically zero. This, indeed, is the only signal that is both bandlimited and causal.

(b) The resolution of the paradox is more philosophical than mathematical. Slepian writes "My solution to this dilemma will certainly not please all of you: it rests on matters I do not fully understand myself" [1283]. Basically, models are often inappropriately equated to reality.[12] Finite precision measurements also play a role. Slepian notes there are innumerable causal and bandlimited signals so close that we are unable to empirically measure the difference. From this point of view, the answer to (a) is "yes". There are signals that are both bandlimited and causal in the sense that, empirically, we can't tell the difference.

10.2.
$$g_c(t) = f(t)\left[1 - r_{1-\alpha}\left(\frac{t}{T}\right)\right]$$

$$1 - r_{1-\alpha}\left(\frac{t}{T}\right) = \sum_{n=-\infty}^{\infty} c_n e^{-j2\pi nt/T}$$

$$\hat{c}_n = \begin{cases} 1-\alpha & ; \quad n=0 \\ -(1-\alpha)\,\text{sinc}\,((1-\alpha)n) & ; \quad n \neq 0 \end{cases}$$

---

12. See, for example, Slepian's quote at the beginning of Chapter 2.

The b_n coefficients are solutions of

$$\sum_{m=-M}^{M} \hat{b}_m \, \hat{c}_{n-m} = \delta[n] \; ; \; |n| \leq M.$$

and

$$\hat{\theta}_M(t) = \sum_{m=-M}^{M} \hat{b}_m \, e^{-j2\pi mt}$$

$$\hat{\psi}_M(t) = \hat{\theta}_M(t) \left(1 - r_{1-\alpha}\left(\frac{t}{T}\right)\right).$$

10.3. (a) $f(t) = 2\Re\left[\int_0^\infty F(u) \, e^{-j2\pi ut} \, du\right]$. If $f(t)$ is bandlimited, the upper integration limit is B.

(b) From Figure 10.33, we have

$$F(u) = \frac{c_0 \, G_1(u) - c_1 \, G(u - 1/T)}{c_0^2 - c_{-1} c_1}.$$

(c) $f(t) = \dfrac{2}{(c_0^2 - c_{-1} c_1)} \Re\left[g(t)\left(c_0 - c_1 \, e^{j2\pi t/T}\right)\right]$

$\quad * [B \, \text{sinc}(Bt) \, e^{j\pi Bt}]$

$= \dfrac{2}{(c_0^2 - c_{-1} c_1)} \left[\left\{g(t)\left[c_0 \, \cos \pi Bt \right.\right.\right.$

$\qquad \left.\left.\left. - c_1 \, \cos\left(\pi\left(B - \dfrac{2}{T}\right)t\right)\right]\right\} * B\text{sinc}Bt\right] \cos \pi Bt$

$+ \dfrac{2}{(c_0^2 - c_{-1} c_1)} \left[\left\{g(t)\left[c_0 \, \sin \pi Bt \right.\right.\right.$

$\qquad \left.\left.\left. - c_1 \, \sin\left(\pi\left(B - \dfrac{2}{T}\right)t\right)\right]\right\} * B\text{sinc}Bt\right] \sin \pi Bt.$

(d) See Figure 10.35.

10.4. (a) Write

$$S = \sum_{m=0}^{N-1} e^{-j2\pi mv} = \sum_{m=0}^{N-1} \varphi^m$$

where

$$\varphi = e^{-j2\pi v}.$$

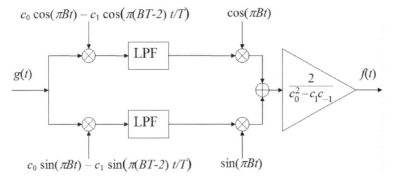

FIGURE 10.35. Restoration of first order aliased data from the shift method illustrated in Figure 10.33, using the conjugate symmetry property of the spectrum of a real signal. The filters are one for $|u| < B/2$ and zero elsewhere.

This is a geometric series in (14.5). Thus

$$S = \frac{1 - \varphi^N}{1 - \varphi}$$

$$= \varphi^{\frac{N-1}{2}} \frac{\varphi^{-N/2} - \varphi^{N/2}}{\varphi^{-\frac{1}{2}} - \varphi^{\frac{1}{2}}}$$

$$= e^{-j\pi(N-1)v} \frac{\sin(\pi N v)}{\sin(\pi v)} \qquad (10.74)$$

from which (10.30) readily follows.

(b) (i) From (10.31), it follows that

$$a_{p+\ell P} = \frac{N}{P} e^{-j\pi(N-1)(p+\ell P)/P} \text{array}_N\left(\frac{p+\ell P}{P}\right). \qquad (10.75)$$

Since

$$e^{-j\pi(N-1)(p+\ell P)/P} = e^{-j\pi(N-1)p/P} \times e^{-j\pi(N-1)\ell}$$

$$= (-1)^{(N-1)\ell} e^{-j\pi(N-1)p/P}, \qquad (10.76)$$

and

$$\text{array}_N\left(\frac{p+\ell P}{P}\right) = \frac{\sin(\pi N(p+\ell P)/P)}{N \sin(\pi(p+\ell P)/P)}$$

$$= \frac{(-1)^{\ell N} \sin(\pi N(p+\ell P)/P)}{(-1)^\ell N \sin(p+\ell P)/P}$$

$$= (-1)^{(N-1)\ell} \text{array}_N\left(\frac{p}{P}\right), \qquad (10.77)$$

it follows, after substitution of (10.76) and (10.77) into (10.75), that

$$a_{p+\ell P} = a_p$$

and periodicity is established.

(ii) Yes. Consider the first row of the matrix **A**. It is of length $2M + 1$ and is a window of the periodic function a_p. As we go down row by row, the periodic sequence a_p is shifted one interval. If sufficient shifts occur to result in duplication of the first row, the matrix is singular. This occurs if there are enough matrix rows to allow P shifts. Since there are $2M + 1$ rows allowing for a maximum of $2M$ shifts (the first row is zero shifts), the singularity thus occurs when

$$2M \geq P.$$

Combining this inequality with $2BP \geq M$, we conclude that the **A** matrix is always singular when $2B > \frac{1}{2}$.

10.8. $2B \operatorname{sinc}(2B(t - \tau)) = \sum_{n=0}^{\infty} a_n \psi_n(t)$

where

$$a_n = 2B \int_{-\infty}^{\infty} \operatorname{sinc}(2B(t - \tau)) \psi_n(\tau) d\tau$$

$$= \psi_n(t).$$

Thus

$$2B \operatorname{sinc}(2B(t - \tau)) = \sum_{n=0}^{\infty} \psi_n(t)\psi_n(\tau).$$

10.9. (a) From the power theorem

$$\int_{-\infty}^{\infty} \operatorname{sinc}(2Bt - n) \operatorname{sinc}(2Bt - m) \, dt$$

$$= \int_{-\infty}^{\infty} \left[\frac{1}{2B} e^{-j\pi nu/B} \Pi\left(\frac{u}{2B}\right) \right] \left[\frac{1}{2B} e^{-j\pi mu/B} \Pi\left(\frac{u}{2B}\right) \right] du$$

$$= \frac{1}{2B} \delta[n - m].$$

(b) Clearly

$$f(t) = \sum_{-\infty}^{\infty} f\left(\frac{n}{2B}\right) \operatorname{sinc}(2Bt - n)$$

$$= \sum_{n=0}^{\infty} a_n \psi_n(t).$$

Multiply both sides by $\psi_n(t)$ and integrate. Using (10.49) and (6.52) gives:

$$a_m = \frac{1}{2B} \sum_{n=0}^{\infty} f\left(\frac{n}{2B}\right) \psi_m\left(\frac{n}{2B}\right).$$

SIGNAL RECOVERY   491

(c) Here, multiply by sinc($2Bt - m$) and integrate

$$f\left(\frac{m}{2B}\right) = 2B \sum_{n=0}^{\infty} a_n \psi_n\left(\frac{m}{2B}\right).$$

Both are well–posed.

10.10. (a) Since $\psi_n(t)$ in bandlimited, we know that $\psi_n(t) = \psi_n(t) * 2B \,\text{sinc}(2Bt)$. Thus

$$\int_{-\frac{T}{2}}^{\frac{T}{2}} \psi_n(t) \psi_m(t)\, dt$$

$$= 2B \int_{-\frac{T}{2}}^{\frac{T}{2}} \psi_n(t) \int_{-\infty}^{\infty} \psi_m(\tau) \,\text{sinc}\,(2B(t-\tau))\, d\tau\, dt$$

$$= 2B \int_{-\infty}^{\infty} \psi_m(\tau) \int_{-\frac{T}{2}}^{\frac{T}{2}} \psi_n(t) \,\text{sinc}\,(2B(t-\tau))\, dt\, d\tau$$

$$= \lambda_n \int_{-\infty}^{\infty} \psi_n(\tau) \psi_m(\tau)\, d\tau$$

$$= \lambda_n\, \delta[n-m].$$

10.12. The ambiguity is removed by requiring each PSWF to have unit energy.

10.13. (a) $\displaystyle\int_{-\infty}^{\infty} |f(t)|^2\, dt = \int_{-\infty}^{\infty} \left(\sum_{n=0}^{\infty} a_n \psi_n(t)\right)^2 dt$

$$= \sum_{n=0}^{\infty} \sum_{m=0}^{\infty} a_n a_m \int_{-\infty}^{\infty} \psi_n(t)\psi_m(t)dt$$

$$= \sum_{n=0}^{\infty} \sum_{m=0}^{\infty} a_n a_m \delta[n-m]$$

$$= \sum_{n=0}^{\infty} a_n^2.$$

(b) $\displaystyle\int_{-\infty}^{\infty} |h(t)|^2\, dt = \sum_{n=0}^{\infty} \lambda_n\, |b_n|^2.$

10.14. $$x(t) = x(t)\, \Pi\left(\frac{t}{T}\right)$$

$$= \sum_{n=0}^{\infty} b_n\, \psi_n(t)$$

$$y(t) = x(t) * 2B\,\text{sinc}(2Bt)$$

$$= 2B \int_{-\infty}^{\infty} x(\tau)\, \text{sinc}\,(2B(t-\tau))\, d\tau$$

$$= 2B \sum_{n=0}^{\infty} b_n \int_{-\infty}^{\infty} \psi_n(\tau) \, \text{sinc}\,((2B(t-\tau)) \, d\tau$$

$$= \sum_{n=0}^{\infty} b_n \, \lambda_n \, \psi_n(t).$$

From (6.23), the energies of x and y are

$$E_x = \sum_{n=0}^{\infty} \lambda_n \, |b_n|^2 = \sum_{n=0}^{\infty} (\sqrt{\lambda_n} \, b_n)^2$$

and

$$E_y = \sum_{n=0}^{\infty} (\lambda_n b_n)^2 = \sum_{n=0}^{\infty} \lambda_n (\sqrt{\lambda_n} b_n)^2.$$

We wish to maximize E_y subject to $E_x = 1$. Since λ_0 is the largest eigenvalue, we choose $b_n = \sqrt{\lambda_0} \, \delta[n]$ and

$$x(t) = \frac{\psi_0(t) \, \Pi\,(t/T)}{\sqrt{\lambda_0}}.$$

The output thus has energy $E_y = \lambda_0$.

10.15. The results are the same as for extrapolation except that, λ_n becomes $(1 - \lambda_n)$ and $(1 - \mathbf{D}_T)$ replaces $\mathbf{D}_T$. Thus, instead of (10.72), we have

$$f_N(t) = g_i(t) + \mathbf{D}_T \sum_{n=0}^{\infty} a_n (1 - \lambda_n^N) \psi_N(t).$$

As with the extrapolation case, the validity of this equation can be proven by induction. Convergence again follows due to (10.73).

10.16. $g_e(t) = \sum_{n=0}^{\infty} g_n \, e^{j2\pi nt/T}$

$$= \sum_{n=0}^{\infty} a_n \, \psi_n(t) \, ; \quad |t| \leq T/2.$$

(a) Multiplying both sides by $\psi_m(t)$ and integrate over $|t| \leq T/2$.

$$\lambda_m \, a_m = \sum_{n=0}^{\infty} g_n \int_{-\infty}^{\infty} \psi_m(t) \, e^{-j2\pi nt/T} \, dt$$

or, using (10.55),

$$a_m = \frac{1}{\lambda_m} \sum_{n=-\infty}^{\infty} g_n \sqrt{\frac{T \lambda_m}{2B}} \, \Psi_m\left(\frac{n}{2B}\right)$$

$$= \sqrt{\frac{T}{2B\lambda_m}} \sum_{n=-\infty}^{\infty} g_n \, \psi_m\left(\frac{n}{2B}\right).$$

The restoration is ill–posed due to the $1/\sqrt{\lambda_m}$ coefficient.

(b) Using the top equation, multiply both sides by $e^{j2\pi nt/T}$ and integrate over $|t| \le T/2$.

$$Tg_m = \sum_{n=0}^{\infty} a_n \int_{-\frac{T}{2}}^{\frac{T}{2}} \psi_m(t)\, e^{j2\pi nt/T}\, dt.$$

Again, using (10.55)

$$g_m = \frac{1}{T} \sum_{n=0}^{\infty} a_n \sqrt{\frac{T\lambda_n}{2B}}\, \psi_m\left(\frac{-n}{2B}\right)$$

$$= \frac{1}{\sqrt{S}} \sum_{n=0}^{\infty} \sqrt{\lambda_n}\, a_n\, \psi_m\left(\frac{-n}{2B}\right).$$

The result is well–posed.

10.18. (a) Instead of (10.61), we have

$$f_{N+1}(t) = g_e(t) + \mathbf{R}(1 - \mathbf{D_T})\mathbf{B_B}\, f_N(t)$$

where the nonlinear half wave rectifier operator is defined by

$$\mathbf{R}\, h(t) = h(t)\, \mu[h(t)].$$

(b) Here we know $S_\xi(u) \ge 0$. Since

$$\mathbf{B_B} = \mathbf{F}^{-1}\, \mathbf{D_B}\, \mathbf{F}$$

our altered algorithm could be:

$$f_{N+1}(t) = g_e(t) + (1 - \mathbf{D_T})\, \mathbf{F}^{-1}\, \mathbf{R}\, \mathbf{D_B}\, \mathbf{F}\, f_N(t).$$

This is Howard's *minimum negativity constraint* restoration algorithm. See Section 11.4.4.

10.19. (a) Let $h(t)$ have finite energy. Then $\mathbf{B_B}\, h(t) = f(t)$ is bandlimited and can be written as

$$f(t) = \sum_{n=0}^{\infty} a_n\, \psi_n(t).$$

(Note that maximum energy occurs when $h(t)$ is chosen to be bandlimited.) Thus,

$$\|\mathbf{H}\| = \sup_{\|f(t)\|=1} \|\mathbf{H} f(t)\|$$

$$\|\mathbf{H}\| = \sup_{\|f(t)\|=1} \|(1 - \mathbf{D_T}) f(t)\|$$

where $f(t)$ is bandlimited. Now

$$(1 - \mathbf{D_T}) f(t) = \sum_{n=0}^{\infty} a_n\, (1 - \mathbf{D_T})\, \psi_n(t)$$

and

$$\| (1 - \mathbf{D_T})f(t) \|^2 = \sum_{n=0}^{\infty} \sum_{m=0}^{\infty} a_n a_m^* \int_{|t| \geq \frac{T}{2}} \psi_n(t)\, \psi_m(t)\, dt$$

$$= \sum_{n=0}^{\infty} (1 - \lambda_n)\, |a_n|^2$$

where we have used (5.41). Since, from Exercise 7.7a,

$$\sum_{n=0}^{\infty} |a_n|^2 = 1$$

we choose $a_\infty = 1$ (since $\lambda_\infty = 0$) and $\| \mathbf{H} \| = 1$.

(b) Same as above, except

$$\| \mathbf{H} \| = \sup_{\|f(t)\|=1} \| \mathbf{D_T} f(t) \|.$$

Since

$$\| \mathbf{D_T} f(t) \|^2 = \sum_{n=0}^{\infty} \lambda_n\, |a_n|^2,$$

we choose $|a_n| = 1$ since λ_0 is max and $\| \mathbf{H} \| = \sqrt{\lambda_0} < 1$.

10.20. Clearly

$$x_N(0) = \int_{-\infty}^{\infty} X_N(u)\, du$$

$$= T \int_{-\infty}^{\infty} \int_{-\infty}^{\infty} \left[x_{N-1}(0)\delta(t) + \sum_{n=0}^{\infty} x(nT)\, \delta(t - \tau) \right] e^{-j2\pi u t}\, dt\, du$$

$$= T \int_{-\infty}^{\infty} \left[x_{N-1}(0) \sum_{n=0}^{\infty} x(nT)\, e^{-j2\pi n u t} \right] du$$

$$= r\, x_{N-1}(0) + r \sum_{n=0}^{\infty} x(nT)\, \text{sinc}(rn)$$

where $r = 2BT$. Letting $N \longrightarrow \infty$ gives

$$x_\infty(0) = r x_\infty(0) + r \sum_{n=0}^{\infty} x(nT)\, \text{sinc}(rn).$$

Solving for $x_\infty(0)$ therefore results in (6.11). Note also that $x_N(t) \longrightarrow x(t)$.

11

Signal and Image Synthesis: Alternating Projections Onto Convex Sets

> I had a feeling once about Mathematics-that I saw it all. Depth beyond depth was revealed to me-the Byss and Abyss. I saw-as one might see the transit of Venus or even the Lord Mayor's Show-a quantity passing through infinity and changing its sign from plus to minus. I saw exactly why it happened and why the tergiversation was inevitable but it was after dinner and I let it go.
> — Sir Winston Leonard Spencer Churchill (1874–1965) [424]

11.1 Introduction

Alternating projections onto convex sets (POCS)[1] [319, 918, 1324, 1333] is a powerful tool for signal and image restoration and synthesis.[2] The desirable properties of a reconstructed signal may be defined by a convex set of constraint parameters. Iteratively projecting onto these convex constraint sets can result in a signal which contains all desired properties. Convex signal sets are frequently encountered in practice and include the sets of bandlimited signals, duration limited signals, causal signals, signals that are the same (e.g., zero) on some given interval, bounded signals, signals of a given area and complex signals with a specified phase.

POCS was initially introduced by Bregman [156] and Gubin *et al.* [558] and was later popularized by Youla & Webb [1550] and Sezan & Stark [1253]. POCS has been applied to such topics as acoustics [300, 1381], beamforming [426], bioinformatics [484], cellular radio control [1148], communications systems [29, 769, 1433], deconvolution and extrapolation [718, 907, 1216], diffraction [421], geophysics [4], image compression [1091, 1473], image processing [311, 321, 470, 471, 672, 736, 834, 1065, 1069, 1093, 1473, 1535, 1547, 1596], holography [880, 1381], interpolation [358, 559, 1266], neural networks [1254, 1543, 909, 913, 1039], pattern recognition [1444, 1588], optimization [598, 1359, 1435], radiotherapy [298, 814, 1385], remote sensing [1223], robotics [740], sampling theory [399, 1334, 1542], signal recovery [320, 737, 1104, 1428, 1594], speech processing [1450], superresolution [399, 633, 654, 834, 1393, 1521], television [736, 786], time-frequency analysis [1037, 1043], tomography [1103, 713, 1212, 1213, 1275, 916, 1322, 1060, 1040], video processing [560, 786, 1092], and watermarking [19, 1470].

---

1. The *alternating* term is implicit in the POCS paradigm, but traditionally not included in the acronym.
2. Portions of this chapter follow closely the development by Marks [918].

11.2 Geometical POCS

Although signal processing applications of POCS use sets of signals, POCS is best visualized viewing the operations on sets of points. In this section, POCS is introduced geometrically in two and three dimensions. Such visualization of POCS is invaluable in application of the theory.

11.2.1 Geometrical Convex Sets

A set, A, is convex if for every vector $\vec{x}_1 \in A$ and every $\vec{x}_2 \in A$, it follows that $\alpha \vec{x}_1 + (1 - \alpha)\vec{x}_2 \in A$ for all $0 \leq \alpha \leq 1$. In other words, as illustrated in Figure 11.1, the line segment connecting $\vec{x}_1$ and $\vec{x}_2$ is totally subsumed in A. If any portion of the cord connecting two points lies outside of the set, the set is not convex. This is illustrated in Figures 11.2 and 11.3. Examples of geometrical convex sets include balls, boxes, lines, line segments, cones and planes.

Closed convex sets are those that contain their boundaries. In two dimensions, for example, the set of points

$$A_1 = \left\{ (t_1, t_2) \mid t_1^2 + t_2^2 < 1 \right\} \tag{11.1}$$

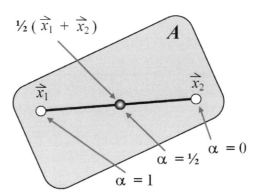

FIGURE 11.1. If a set A is convex, the line segment formed by connecting any two points, $\vec{x}_1$ and $\vec{x}_2$, in the set is totally subsumed within the set. Thus, if $\vec{x}_1 \in A$ and $\vec{x}_2 \in A$, then $\forall \alpha \ni 0 \leq \alpha \leq 1$, we require that $\alpha \vec{x}_1 + (1 - \alpha)\vec{x}_2 \in A$ for A to be convex.

FIGURE 11.2. Examples of three convex sets of points in three dimensions: a cube, a sphere and a solid cylinder. In all three cases, the line connecting any two points in the object is totally subsumed in the object.

SIGNAL AND IMAGE SYNTHESIS: ALTERNATING PROJECTIONS ONTO CONVEX SETS

FIGURE 11.3. Examples of three sets of points in three dimensions that are *not* convex: a cross, a hollow cylinder and a crescent. In each case, there are at least two points in the set that, when connected with a line, lie outside of the set.

is not closed since the points on the circle $t_1^2 + t_2^2 = 1$ are not in the set. The set,

$$A = \left\{ (t_1, t_2) | \ t_1^2 + t_2^2 \leq 1 \right\}$$

is a closed convex set. A is referred to as the *closure* of A_1. Henceforth, all convex sets will be considered closed.

Strictly convex sets are those that do not contain any flat boundaries. A ball is strictly convex whereas a box is not. Neither is a line.[3]

11.2.2 Projecting onto a Convex Set

The *projection* onto a convex set is illustrated in Figure 11.4. For a given $\vec{x} \notin A$, the projection onto A is the unique vector $\vec{y} \in A$ such that the distance between $\vec{x}$ and $\vec{y}$ is minimum. If $\vec{x} \in A$, then the projection onto A is $\vec{x}$. In other words, the projection and the vector are the same. Projections are *idempotent* in that two projection operators are the same as one. If the projection of $\vec{x}$ onto A is denoted $P_A\vec{x}$, then the idempotent property can be written as

$$P_A = P_A^2 \tag{11.2}$$

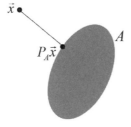

FIGURE 11.4. The oval, A, is convex. The projection of $\vec{x}$ onto A is the unique element in A closest to $\vec{x}$, and is denoted $P_A\vec{x}$.

---

3. Rigorously, a set A is strictly convex if for any distinct $\vec{x}_1, \vec{x}_2 \in A$, $(\vec{x}_1 + \vec{x}_2)/2$ is an interior point of A. A vector $\vec{x}$ is called an *interior point* of closed set A if $\vec{x} \in A$ and $\vec{x} \notin \text{Closure}[E - A]$ where E is the universal set. In other words, an interior point does not lie on the boundary of a closed convex set.

11.2.3 POCS

There are three outcomes in the application of POCS. Each depends on the various ways that the convex sets intersect.

1. The remarkable primary result of POCS is that, given two or more convex sets with nonempty intersection, alternately projecting among the sets will converge to a point included in the intersection [1324, 1550]. This is illustrated in Figure 11.5. Two convex sets are shown. The oval shaped collection of points is A and the line segment is B. The intersection of the two convex sets, $A \cap B$, is the portion of the line segment lying within the oval. POCS will always converge to an element of $A \cap B$. Start with the point $\vec{y}_0$. This point is first projected onto B, then onto A, onto B, etc. This repeated iteration converges to the point $\vec{y}_\infty$ which, as advertised, lies on $A \cap B$. Starting at different initializations results in convergence to different points on $A \cap B$. If an initialization point is chosen on $A \cap B$, POCS will "converge" to the initialization point. As another example, starting at the point $\vec{x}_0$ in Figure 11.5 converges to the point $\vec{x}_\infty \in A \cap B$.

The result extends to three or more convex sets with nonempty intersection. For N convex sets, $\{A_n | 1 \leq n \leq N\}$, define the intersection

$$\Theta = \bigcap_{n=1}^{N} A_n.$$

If Θ is not empty, then repeated alternated projections among the N sets, in any order, will result in a limit point in Θ.

This is illustrated in Figure 11.6 for three convex sets that intersect in a single point. Thus, no matter where the iteration begins in the plane, alternating projections among the three convex sets will converge to that point.

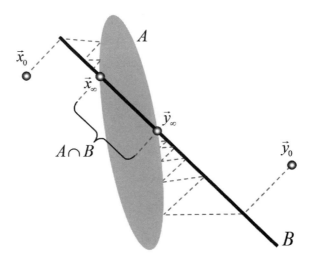

FIGURE 11.5. Alternating projection between two or more convex sets with nonempty intersection results in convergence to a fixed point in the intersection. The oval, A, and the line segment B, are convex. Initializing the iteration at $\vec{y}_0$ results in convergence to $\vec{y}_\infty \in A \cap B$. Similarly, the vector $\vec{x}_0$ results in convergence to $\vec{x}_\infty \in A \cap B$.

SIGNAL AND IMAGE SYNTHESIS: ALTERNATING PROJECTIONS ONTO CONVEX SETS   499

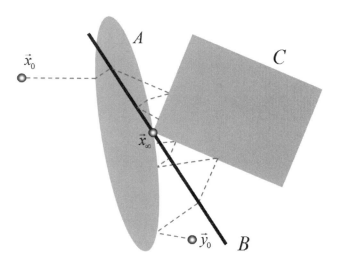

FIGURE 11.6. Shown are three convex sets. A in an oval, B is a line segment and C is a rectangle. The three sets intersect in a single point. Alternating projections will converge to the point $\vec{x}_\infty$ independent of initialization. The initialization point $\vec{y}_0$ converges to $\vec{x}_\infty$. So does initialization at $\vec{x}_0$.

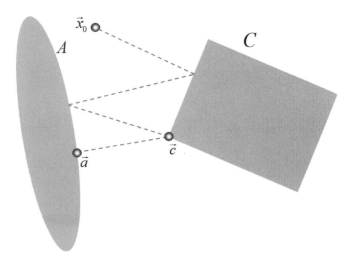

FIGURE 11.7. If two convex sets, A and C, do not intersect, POCS converges to a limit cycle, here between the points $\vec{a}$ and $\vec{c}$. The point $\vec{c} \in C$ is the point in C closest to the set A and $\vec{a} \in A$ is the point in A closest to the set C. POCS will converge to this limit cycle for all points of initialization.

2. If two convex sets do not intersect, convergence is to a *limit cycle* that is a mean square solution to the problem. Specifically, the cycle is between points in each set that are closest in the mean square sense to the other set [509]. This is illustrated in Figure 11.7.
3. Conventional sequential alternating POCS breaks down in the important case where three or more convex sets do not intersect [1552]. POCS, rather, converges to *greedy limit cycles* that are dependent on the ordering of the projections and do not display any desirable optimality properties. This is illustrated in Figure 11.8. When the projection order is $A \rightarrow B \rightarrow C$, the limit cycle is $\vec{a} \in A$, $\vec{b} \in B$,

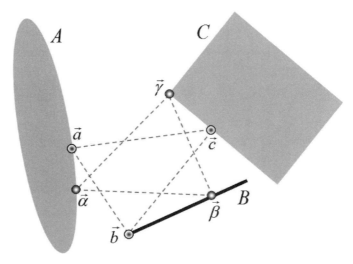

FIGURE 11.8. If three or more convex sets do not intersect, POCS converges to greedy limit cycles. As illustrated here, the limit cycles can differ for different orders of set projection.

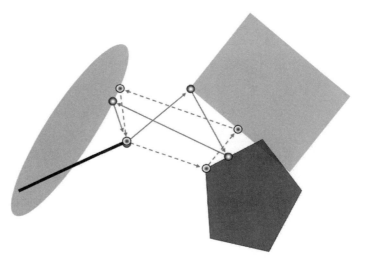

FIGURE 11.9. Four convex sets with a nonempty intersection and two possible limit cycles.

and $\vec{c} \in C$. When the projection order is $C \to B \to A$, the limit cycle is $\vec{\gamma} \in C$, $\vec{\beta} \in B$, and $\vec{\alpha} \in A$.

This greedy limit cycle can also occur when some of the convex sets intersect and others don't. This is illustrated in Figure 11.9.[4]

The POCS paradigm, however, is useful when three or more convex sets do not intersect if, rather than *sequential* alternating projections, one uses *simultaneous*

---

4. Oh *et.al.* [1041, 917] have applied Zadeh's ideas on *fuzzy* convex sets [1562] to the case where the convex sets do not intersect. The idea is to find a point that is 'near' to each of the sets. The term 'near' is referred to as a *fuzzy linguistic variable*.

weighted projections. For a set of N convex sets, $S = \{A_n | 1 \leq n \leq N\}$, the *simultaneous weighted projections*[5] of an arbitrary point, $\vec{x}$, is defined as

$$P_S \vec{x} := \sum_{n=1}^{N} w_n P_n \vec{x}$$

where P_n is the projection of $\vec{x}$ onto the convex set A_n and the w_n's are weights of the importance of each projection. These N projection results are summed in a weighted fashion to give the next element of the iteration. The iteration is repeated until convergence, i.e.,

$$\vec{x}_{i+1} = P_S \vec{x}_i.$$

where i is the iteration index. Repeated applications of sequential weighted projections converges to a point that minimizes the weighted sum of square of weighted distances [1552]. Specifically, the iteration converges to $\vec{x}_\infty$ such that

$$d_S = \sum_{n=1}^{N} w_n \|\vec{x}_\infty - P_n \vec{x}_\infty\|^2$$

is minimum where $\|\vec{y} - \vec{x}\|$ is the distance between $\vec{y}$ and $\vec{x}$. This is illustrated in Figure 11.10.

11.3 Convex Sets of Signals

The geometrical view introduced in the previous section allows powerful interpretation of POCS applied in a Hilbert space, $\mathcal{H}$. The concepts can be applied to functions in both continuous and discrete time.[6]

---

5. The sequential weighted projections is not, itself, a projection. It is not, for example, idempotent since, with reference to (11.2), $P_S \neq P_S^2$.

6. i.e., from an L_2 to an ℓ_2 space. One advantage of discrete time is the guaranteed strong convergence of POCS. Denote the nth element of a sequence x by $x[n]$ and let $o^{(k)}[n]$ be the kth POCS iteration. Let $o[n]$ be the point of convergence and $x[n]$ is any point (including the origin) in ℓ_2, then

$$\lim_{k \to \infty} \|x[n] - o^{(k)}[n]\|^2 = \|x[n] - o[n]\|^2$$

for all $x[n]$. The square of the ℓ_2 norm of a sequence $x[n]$, is

$$\|x[n]\|^2 = \sum_{n=-\infty}^{\infty} |x[n]|^2.$$

In continuous time, only weak convergence can generally be assured [1550]. Specifically

$$\lim_{n \to \infty} \int_{-\infty}^{\infty} x(t) o^{(n)}(t) dt = \int_{-\infty}^{\infty} x(t) o(t) dt$$

for all $x(t)$ in L_2. Strong convergence subsumes weak convergence.

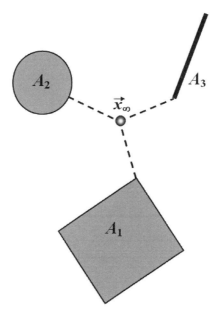

FIGURE 11.10. If three or more convex sets do not intersect, simultaneous weighted projections can be used to find a representative solution. In this example, all weights are the same. The solution corresponding to the three convex sets is $\vec{x}_\infty$. The weighted sum of distance squared is minimized. Here, all of the distances are the same.

11.3.1 The Hilbert Space

An image, $x(t)$, is in $\mathcal{H}$ if[7]

$$\|x(t)\| < \infty$$

where the norm of $x(t)$ is

$$\|x(t)\| = \sqrt{\int_{-\infty}^{\infty} |x(t)|^2 dt}. \qquad (11.3)$$

The *energy* of $x(t)$ is (see (2.1))

$$E = \|x(t)\|^2. \qquad (11.4)$$

The space thus consists of all finite energy signals. Geometrically, $\vec{x} = x(t)$ can be visualized as a point in the Hilbert space a distance of $\|x(t)\|$ from the origin, $x(t) \equiv 0$. Similarly, the (mean square) distance between two points $x(t)$ and $y(t)$ is $\|x(t) - y(t)\|$.

Two objects, $w(t)$ and $z(t)$, are said to be *orthogonal* if their inner product is zero, i.e.,

$$\int_{-\infty}^{\infty} w(t) z^*(t) dt = 0 \qquad (11.5)$$

where the superscript asterisks denotes complex conjugation.

---

7. Signals in a Hilbert space are also required to be Lebesgue measurable.

The concept applies directly to a Hilbert space of finite energy discrete time signals.[8] The signal norm is now

$$\|x[n]\| = \sqrt{\sum_{n=-\infty}^{\infty} |x[n]|^2}.$$

The discrete time signal's energy[9] is $E = \|x[n]\|^2$. The mean square distance between two points $x[n]$ and $y[n]$ is $\|x[n] - y[n]\|$. Two objects, $w[n]$ and $z[n]$, are orthogonal if their discrete time inner product is zero.

$$\sum_{-\infty}^{\infty} w[n]z^*[n]dt = 0. \quad (11.6)$$

Some examples of signal sets follow.

11.3.1.1 Convex Sets

Analogous to the definition given for sets of vectors, a set of signals, A, is convex if, for $0 \leq \alpha \leq 1$, the signal

$$\alpha u_1(t) + (1-\alpha)u_2(t) \in A \quad (11.7)$$

when $u_1(t), u_2(t) \in A$.

11.3.1.2 Subspaces

Subspaces (*a.k.a.* linear manifolds) can be visualized as hyperplanes in the space. A set of signals, S, is a subspace if

$$\beta u_1(t) + \gamma u_2(t) \in S \quad (11.8)$$

when $u_1(t), u_2(t) \in S$, and β, γ are any finite real numbers. Subspaces are convex.[10] The origin, $u(t) \equiv 0$, is an element of all subspaces.[11]

11.3.1.3 Null Spaces

We denote a *null space* corresponding to a subspace S, as $S^\perp$. The null space consists of all signals orthogonal to all elements in S.

$$S^\perp = \left\{ x(t) \,\middle|\, \int x(t)u^*(t)dt = 0 \text{ for all } u(t) \in S \right\}.$$

In three dimensions, the null space of a line is the plane (subspace) perpendicular to the line. Note that

$$\left(S^\perp\right)^\perp = S.$$

---

8. i.e., the space of all ℓ_2 signals.
9. See (2.2).
10. See Exercise 11.3.
11. If $u(t) \in S$, then, with $u_1(t) = u(t) = -u_2(t)$ and $\beta = \gamma = 1$, we have, from (11.8) that $0 = u(t) - u(t) \in S$.

11.3.1.4 Linear Varieties

A *linear variety* [854] is any hyperplane in the space. It need not go through the origin. All subspaces are linear varieties. A linear variety, L, can always be defined as

$$L = \left\{ u(t) = u_S(t) + d(t) \mid u_S(t) \in S \right\} \tag{11.9}$$

where S is a subspace and $d(t) \notin S$ is a fixed displacement vector. The formation of a linear variety from a subspace is illustrated in Figure 11.11.

All linear varieties are convex.

11.3.1.5 Cones

A set C containing is a cone if $x(t) \in C$ implies $\alpha x(t) \in C$ for all $\alpha > 0$. A convex cone is a cone that also obeys the properties of a convex set. Orthants[12] and line segments drawn from the origin to infinity in any direction are examples of convex cones.

11.3.1.6 Convex Hulls and Convex Cone Hulls

The *convex hull* of a collection of points is the smallest convex set containing the points. This is illustrated in two dimensions in Figure 11.12. If the plane were a board in which nails were driven, the convex hull would be the shape of a rubber band placed around the nails. Interior points in the hull do not contribute to its shape.

Similarly, a *convex cone hull* of a collection of points is the smallest convex cone containing the points. This is illustrated in Figure 11.13. Each edge contains a point in the set. As is the case for the convex hull, there can be additional interior points inside the cone that do not contribute to the shape of the cone.

11.3.2 Some Commonly Used Convex Sets of Signals

A number of commonly used signal classes are convex. In this section, examples of these sets and their projection operators are given.

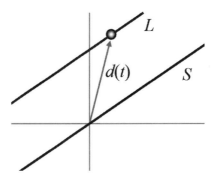

FIGURE 11.11. The subspace, S, is shown here as a line. The function $d(t)$, assumed not to lie in the subspace, S, is a displacement vector. The tail of the vector, $d(t)$, is at the origin ($u(t) \equiv 0$). The set of all points in S added to $d(t)$ forms the linear variety, L.

---

12. In two dimensions, an orthant is a quadrant. In three, an octant.

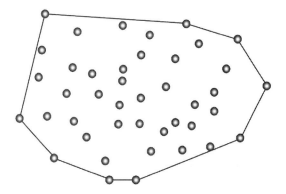

FIGURE 11.12. Illustration of the convex hull of a set of points.

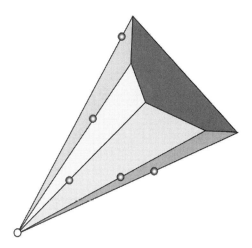

FIGURE 11.13. Illustration of the convex cone hull of a set of points. The origin, at the lower left, is marked with a hollow dot. Points interior to the cone are not shown.

Projection operators will be denoted by a P. The notation

$$x(t) = P_K\, y(t)$$

is read "$x(t)$ is the projection of $y(t)$ onto the convex set A_K". Note, if $y(t) \in A_K$, then $P_K\, y(t) = y(t)$. Most projections are intuitively straightforward. In many instances, the projection is obtained simply by forcing the signal to conform with the constraint in the most obvious way.

11.3.2.1 Matrix equations

To illustrate subspaces in matrix relationships, consider the matrix $\mathbf{S}$ and the degradation[13]

$$\vec{i} = \mathbf{S}\vec{o} \qquad (11.10)$$

---

13. The vector $\vec{o}$ is for *object* and $\vec{i}$ for *image*.

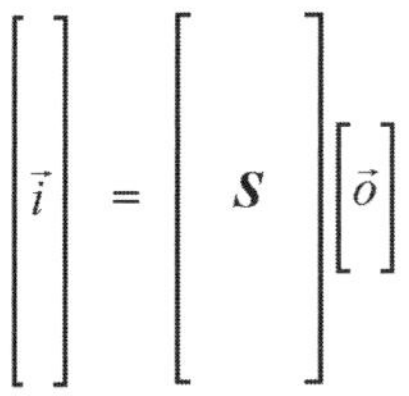

$$\begin{bmatrix} \\ \vec{i} \\ \\ \end{bmatrix} = \begin{bmatrix} \\ \mathbf{S} \\ \\ \end{bmatrix} \begin{bmatrix} \vec{o} \end{bmatrix}$$

FIGURE 11.14. Illustration of a matrix, **S**, that is not full rank.

The vector, $\vec{i}$, is a linear combination of the columns of $\mathbf{S} \in \mathbb{R}^{M \times N}$. Such a vector is said to lie in the *column space* of **S**. Denote the column space by $C_\mathbf{S}$. We assume the $M \times N$, $M > N$ matrix, **S**, as illustrated in Figure 11.14, is not full rank. Our goal is to reconstruct the object, $\vec{o}$, from the image, $\vec{i}$. If the image is corrupted by noise or other uncertainty, there may not be an object that satisfies the matrix equation in (11.10).

From (11.10) we can write

$$\mathbf{S}^T \vec{i} = \mathbf{S}^T \mathbf{S} \vec{o}$$

Formally solving for $\vec{o}$ gives the *pseudo inverse* solution[14]

$$\vec{o}_{PI} = [\mathbf{S}^T \mathbf{S}]^{-1} \mathbf{S}^T \vec{i} \qquad (11.11)$$

The vector $\vec{o}_{PI}$ will be equal to $\vec{o}$ in (11.10) if $\vec{i}$ is a linear combination of the rows of **S**. Otherwise, it will not be. For an arbitrary $\vec{i}$ that is not a linear combination of the columns of **S**, the vector $\vec{o}_{PI}$ when substituted into the matrix equation in (11.10) cannot generate $\vec{i}$. It, rather, generates the vector in the column space $C_\mathbf{S}$ of the matrix **S** that is closest to $\vec{i}$ in the mean square sense.

The *projection matrix* follows from substituting the pseudo inverse in (11.11) into the matrix equation in (11.10). The result is $\mathbf{S}\vec{o}_{PI} = \mathbf{P}_\mathbf{S} \vec{i}$ where the projection matrix is

$$\mathbf{P}_\mathbf{S} = \mathbf{S}[\mathbf{S}^T \mathbf{S}]^{-1} \mathbf{S}^T. \qquad (11.12)$$

As illustrated in Figure 11.15, the projection matrix projects an arbitrary vector $\vec{i}$ onto the column space of **S**. As is necessary for a projection operator, the projection matrix is idempotent, i.e., projecting twice is the same as projecting once.

$$\mathbf{P}_\mathbf{S}^2 = \mathbf{P}_\mathbf{S}.$$

Consider the signal space illustrated in Figure 11.15. The pseudo-inverse solution of (11.10) lies on the subspace $C_\mathbf{S}$ onto which $\mathbf{P}_\mathbf{S}$ projects. Denote the null space, a.k.a. the orthogonal complement, of $C_\mathbf{S}$ by $C_{\perp \mathbf{S}}$. The projection operator onto this space is simply

$$\mathbf{P}_{\perp \mathbf{S}} = \mathbf{1} - \mathbf{P}_\mathbf{S}$$

where **1** is an identity matrix. If the projection onto the orthogonal complement is

$$\vec{i}_\perp = \mathbf{P}_{\perp \mathbf{S}} \vec{i},$$

---

14. The solution, $\vec{o}_{PI}$, is also referred to as the *minimum norm solution*.

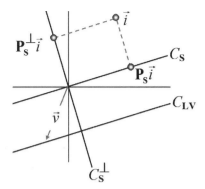

FIGURE 11.15. Geometrical illustration of the subspaces, linear varieties and projections in the matrix equation in (11.10).

we are assured that

$$\vec{i} = (\mathbf{P}_{\perp\mathbf{S}} + \mathbf{P}_{\mathbf{S}})\vec{i}$$

Define the linear variety parallel to the column space of $\mathbf{S}$ as

$$C_{\mathrm{LV}} = \{\vec{u} \mid \vec{u} = \vec{v} + \vec{i} \text{ for all } \vec{i} \in C_{\mathbf{S}}\}$$

where $\vec{v}$ is a given vector not collinear with $\vec{i}$. The projection onto the linear variety for an arbitrary vector, $\vec{z}$, is

$$P_{\mathrm{LV}}\vec{z} = \mathbf{P}_{\mathbf{S}}\vec{z} + \mathbf{P}_{\perp\mathbf{S}}\vec{v}.$$

11.3.2.2 Bandlimited Signal

The set of bandlimited signals with bandwidth B is

$$A_{\mathrm{BL}} = \{x(t) \mid ; X(u) = 0 \text{ for } |u| > B\}. \tag{11.13}$$

where $x(t) \leftrightarrow X(u)$. The set, A_{BL}, is a subspace since $\beta x_1(t) + \gamma x_2(t)$ is bandlimited when both $x_1(t)$ and $x_2(t)$ are bandlimited. The projection of a signal, $y(t)$, onto A_{BL}, is obtained by placing $y(t)$ through a low pass filter.

$$\begin{aligned} x(t) &= P_{\mathrm{BL}}\, y(t) \\ &= \int_{-B}^{B} Y(u)e^{j2\pi t}d\omega \end{aligned} \tag{11.14}$$

where $y(t) \leftrightarrow Y(u)$ are Fourier transform pairs.

11.3.2.3 Duration Limited Signals

The convex set of duration limited signals is, for a given $T > 0$,

$$A_{\mathrm{TL}} = \left\{x(t) \,\middle|\, x(t) = 0 \text{ for } |t| > \frac{T}{2}\right\}. \tag{11.15}$$

Since the weighted sum of any two duration limited signals is duration limited, the set A_{TL} is recognized as a subspace. To project an arbitrary $y(t)$ onto this set, the portion of $y(t)$ for $|t| > T$ is simply set to zero.

$$x(t) = P_{TL}\, y(t)$$
$$= \begin{cases} y(t)\,; & |t| \leq \frac{T}{2} \\ 0 \,; & |t| > \frac{T}{2}. \end{cases} \tag{11.16}$$

11.3.2.4 Real Transform Positivity

The set of signals with real and nonnegative Fourier transform is

$$A_{POS} = \{x(t) \mid \Re\, X(u) \geq 0\} \tag{11.17}$$

where $\Re$ denotes 'the real part of' and $x(t) \leftrightarrow X(u)$. The set A_{POS} is a cone.

To project $y(t)$ onto this set, the Fourier transform, $Y(u)$, is first evaluated. The projection follows as

$$x(t) = P_{POS}\, y(t)$$
$$= \int_{-\infty}^{\infty} \left[\Re Y(u)\mu\left(\Re Y(u)\right) + j\Im Y(u)\right] e^{j2\pi u t}\, du$$

where $\Im$ denotes 'the imaginary part of'. In other words, the Fourier transform is made to be real and nonnegative and is then inverse transformed to find the projection onto A_{POS}.

11.3.2.5 Constant Area Signals

1. **For continuous signals.** Over an interval $\mathcal{I}$, the set of signals with given area, ρ, are

$$A_{CA} = \left\{ x(t) \,\middle|\, \int_{\mathcal{I}} x(t)\, dt = \rho \right\}. \tag{11.18}$$

Let the area of a signal on the interval $\mathcal{I}$ be $\rho_y = \int_{\mathcal{I}} y(t)\, dt$ and the interval $I = \int_{\mathcal{I}} dt$. Then the projection of a signal $y(t)$ onto A_{CA} is[15]

$$P_{CA} y(t) = \begin{cases} y(t) & ;\, t \notin \mathcal{I} \\ y(t) - \frac{1}{I}\left(\rho_y - \rho\right) & ;\, t \in \mathcal{I} \end{cases}. \tag{11.19}$$

As illustrated in Figure 11.16, the projection simply corresponds to raising or lowering the function on $\mathcal{I}$ until the desired area is achieved.

2. **For discrete signals.** The discrete equivalent of the set is

$$A_{CA} = \left\{ \vec{x} \,\middle|\, \sum_{\mathcal{I}} x[n] = \rho \right\} \tag{11.20}$$

and the corresponding projection is

$$(P_{CA}\vec{y})_n = \begin{cases} y_n & ;\, n \notin \mathcal{I} \\ y_n - \frac{1}{I}(\rho_y - \rho) & ;\, n \in \mathcal{I} \end{cases}$$

---

[15]. See Exercise 11.4(h) for a generalization of this convex set from a plane to a convex slab, and the corresponding projection.

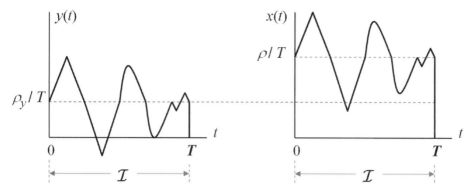

FIGURE 11.16. On the left is the signal $y(t)$ over the interval $\mathcal{I} = \{t | 0 \leq t \leq T\}$. The area of $y(t)$ on this interval is ρ_y. To project onto the set of signals with area ρ on $\mathcal{I}$, the function $y(t)$ on the interval $\mathcal{I}$ is simply raised or lowered to the point where the area is ρ. This is illustrated on the right.

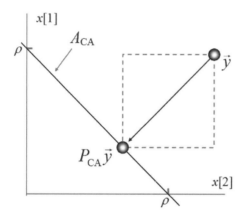

FIGURE 11.17. Two dimensional example of projecting onto a signal with specified area over a given interval.

where

$$\rho_y = \sum_{\mathcal{I}} y[n]$$

and

$$I = \sum_{\mathcal{I}} 1.$$

To illustrate this set and its projection, consider the two points, $x[1]$ and $x[2]$ in Figure 11.17. The set of all points where $x[1] + x[2] = \rho$ is the linear variety A_{CA} corresponding to a 45° line intersecting each axis at ρ. A vector, $\vec{y}$, lies off axis. The projection onto the line is shown. The projection is surrounded by a square. Each dimensionis reduced by the same amount to achieve the projection. This extends to higher dimensions and to continuous functions. To project onto the constant area subspace, the function on the interval $\mathcal{I}$ is simply raised or lowered to the point where the desired area, ρ, is achieved.

510 HANDBOOK OF FOURIER ANALYSIS AND ITS APPLICATIONS

3. **Standard simplex.** When $\rho = 1$, the intersection of A_{CA} and the first orthant (i.e., the set of nonnegative signals) is called a *standard simplex*. All discrete probability density functions lie on a standard simplex.

4. **Weighted areas for discrete signals.** A generalization of the constant area convex set is the constant weighted area convex set defined as

$$A_{CA} = \{\vec{x} \mid \vec{w}^T \vec{x} = \rho\}. \tag{11.21}$$

With no loss of generality, we require the weighting vector, $\vec{w}$, have unit norm. Thus

$$\|\vec{w}\|^2 = \vec{w}^T \vec{w} = 1.$$

This set, A_{CA}, is a linear variety. The projection onto this set is

$$P_{CA}\vec{y} = \vec{y} - \vec{w}\vec{w}^T \vec{y} + \rho \vec{w}. \tag{11.22}$$

Note that right multiplying (11.22) by $\vec{w}^T$ gives

$$\vec{w}^T P_{CA} \vec{y} = \rho.$$

The projection in (11.22) is illustrated in two dimensions, $x[1]$ and $x[2]$, in Figure 11.18. Shown is the unit vector, $\vec{w}$. The corresponding null space consists of all vectors orthogonal to $\vec{w}$.

$$A_{\perp \vec{w}} = \{\vec{x} \mid \vec{w}^T \vec{x} = 0\}.$$

The vector $\rho \vec{w}$ is perpendicular to the null space. The translation of each element in $A_{\perp \vec{w}}$ by $\rho \vec{w}$ is the linear variety, A_{CA}.

To project onto the linear variety, A_{CA}, of weighted signals with constant area, we will first project onto the null space, $A_{\perp \vec{w}}$, and then add the perpendicular offset vector, $\rho \vec{w}$. The projection of an arbitrary vector, $\vec{y}$, onto the line defined by the

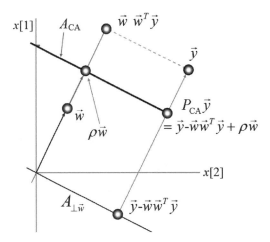

FIGURE 11.18. Two dimensional example of projecting onto a signal with a weighted area over a given interval.

unit vector, $\vec{w}$, is $\vec{w}\left(\vec{w}^T \vec{y}\right)$. This vector, subtracted from $\vec{y}$, gives the projection of $\vec{y}$ onto the null space.

$$P_{\perp \vec{w}} \vec{y} = \vec{y} - \vec{w}\left(\vec{w}^T \vec{y}\right).$$

Adding the offset vector gives the desired projection in Figure 11.22.

5. **Weighted areas for continuous signals.** The continuous version of the constant area signals in (11.21) is

$$A_{CA} = \left\{ x(t) \,\Big|\, \int_{-\infty}^{\infty} w(t) x(t) dt = \rho \right\} \quad (11.23)$$

where the weighting function has a unit norm

$$\|w(t)\|^2 = \int_{-\infty}^{\infty} |w(t)|^2 dt = 1.$$

The projection of a signal, $y(t)$, onto this linear variety is

$$P_{CA} y(t) = y(t) - w(t) \int_{-\infty}^{\infty} w(\tau) x(\tau) d\tau + \rho w(t).$$

11.3.2.6 Bounded Energy Signals

The set of signals with bounded energy, for a given bound η, is

$$A_{BE} = \{ x(t) \mid E \leq \eta \}$$

where E is defined in (11.4). The set A_{BE} is a ball. If the signal is outside of the ball, we project onto the ball. If the energy of the signal is less than η, the signal is in the ball and we leave it as is. The projection follows as

$$x(t) = P_{BE}\, y(t)$$

$$= \begin{cases} y(t) & ; \|y(t)\|^2 \leq \eta \\ \sqrt{\eta}\, y(t)/\|y(t)\| & ; \|y(t)\|^2 > \eta. \end{cases}$$

11.3.2.7 Constant Phase Signals

The convex set of signals whose Fourier transform has a specified phase, $\phi(u)$, is

$$A_{CP} = \{ x(t) \mid \angle X(u) = \phi(u) \}$$

where $X(u) = |X(u)| e^{j \angle X(u)}$.

The projection onto this convex set is illustrated by geometry shown in Figure 11.19.

$$x(t) = P_{CP} y(t)$$

$$\leftrightarrow X(u) = \begin{cases} |Y(u)| \cos(\phi(u) - \angle Y(u))\, e^{j\phi(u)} & ; |\phi(u) - \angle Y(u)| \leq \frac{\pi}{2} \\ 0 & ; \text{otherwise.} \end{cases}$$

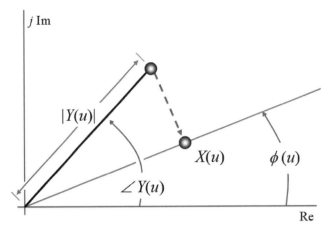

FIGURE 11.19. Geometry for projection onto the convex set of signals whose Fourier transform has a specified phase, $\phi(u)$.

11.3.2.8 Bounded Signals

1. **Signal Bound.** For a given real signal, $a(t)$, the set of bounded signals

$$A_{BS} = \{x(t) \mid \Re x(t) \leq a(t)\}$$

is convex. If $a(t) \equiv$ constant, then A_{BS} is a cube.

The choice of projection is obvious. The signal is set to $\Re x(t) = z(t)$ if $\Re y(t)$ is too large. Otherwise, the signal remains as is.

$$x(t) = P_{BS} \, y(t)$$
$$= \begin{cases} y(t) & ; \Re y(t) \leq a(t) \\ a(t) + j \Im y(t) & ; \Re y(t) > a(t). \end{cases}$$

2. **Two Bounds.** The upper and lower bounds, $a_\ell(t)$ and $a_u(t)$, can be different. The set

$$A_{BS} = \{x(t) \mid a_\ell(t) \leq \Re x(t) \leq a_u(t)\}$$

is convex.

3. **Bounded Derivatives.** For a fixed n, the set of signals with bounded derivatives is convex.

$$A_{DB} = \left\{ x(t) \, \Big| \, \left(\frac{d}{dt}\right)^n x(t) \leq a_n(t) \right\}.$$

where $a_n(t)$ is the bound. The discrete equivalent for the set of bounded first derivative is the bounded first difference. If the bound is constant, then for the first difference,

$$A_{DB} = \{x[n] \mid |x[n+1] - x[n]| \leq \epsilon \}. \tag{11.24}$$

The set for two points is shown in Figure 11.20. The set for three points is shown in Figure 11.21.

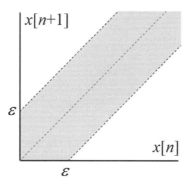

FIGURE 11.20. The convex set, A_{DB} in (11.24) for two points.

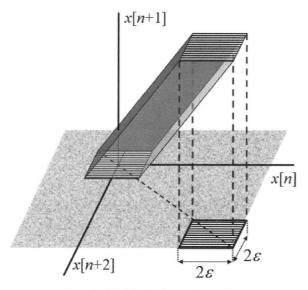

FIGURE 11.21. The convex set, A_{DB} in (11.24) for three points. The set corresponds to $|x[n+1] - x[n]| \leq \epsilon$ and $|x[n+2] - x[n+1]| \leq \epsilon$.

11.3.2.9 Signals With Identical Middles

Let $a(x)$ be a given signal on the interval $|t| \leq \frac{T}{2}$. Define

$$A_{\mathrm{IM}} = \left\{ x(t) \, \bigg| \, x(t) = a(x) \text{ for } |t| \leq \frac{T}{2}; \text{ anything for } |t| > \frac{T}{2} \right\} \quad (11.25)$$

This set is a linear variety. The parallel subspace corresponds to $a(t) = 0$ and the displacement vector is $d(t) = a(t)$. To project onto A_{IM}, one merely inserts the desired signal, $a(t)$, into the appropriate interval.

$$\begin{aligned} x(t) &= P_{\mathrm{IM}} \, y(t) \\ &= \begin{cases} a(t) \, ; & |t| \leq \frac{T}{2} \\ y(t) \, ; & |t| > \frac{T}{2} \end{cases} \end{aligned}$$

11.4 Example Applications of POCS

A number of commonly used reconstruction and synthesis algorithms are special cases of POCS. In this section, we look at some specific examples.

11.4.1 Von Neumann's Alternating Projection Theorem

When all of the convex sets are linear varieties, POCS is equivalent to Von Neumann's alternating projection theorem [1025]. The solution of simultaneous linear equations in the next section and the Papoulis-Gerchberg algorithm in Section 11.4.3 are two examples.

11.4.2 Solution of Simultaneous Equations

POCS can be used to solve the N simultaneous equations

$$z_n = \vec{s}_n^T \vec{x}; \quad 1 \leq n \leq N. \tag{11.26}$$

These equations describe N distinct convex sets (planes) as defined in (11.21). We restate here using the notation in (11.26).

$$A_n = \left\{ \vec{x} \mid z_n = \vec{s}_n^T \vec{x} \right\} \tag{11.27}$$

Assuming the set of equations is not singular, these planes intersect in a single point, $\vec{x}$, that solves the set of simultaneous equations in (11.26). Using (11.22) with the unit vector

$$\vec{w} = \frac{\vec{s}_n}{\|\vec{s}_n\|}$$

gives the projection of an arbitrary vector, $\vec{y}$, onto the nth plane as

$$P_n \vec{y} = \vec{y} - \frac{\vec{s}_n \vec{s}_n^T}{\|\vec{s}_n\|^2} \vec{y} - z_n \frac{\vec{s}_n}{\|\vec{s}_n\|}.$$

11.4.3 The Papoulis-Gerchberg Algorithm

D.C. Youla [1549] was the first to recognize the Papoulis-Gerchberg algorithm, described in Section 10.6, is a special case of POCS between a subspace and a linear variety. The two convex sets are

- The set of all signals equal to $g_e(t)$ on the interval $|t| \leq T/2$.[16]

$$A_{IM} = \left\{ f(t) \mid f(t) = g_e(t); \quad |t| \leq T/2 \right\}$$

- The set of all bandlimited signals, A_{BL}, with a bandwidth B or less. This set is defined in (11.13).

The alternating projections in the Papoulis-Gerchberg algorithm are performed between the subspace A_{BL} and the linear variety A_{IM}.

---

16. Equivalent to the definition in (11.25) with middle $a(x) = g_e(t)$.

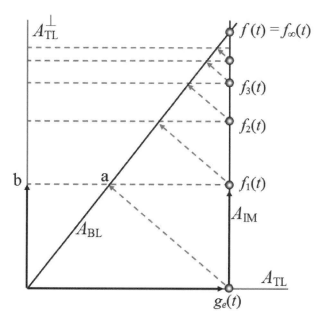

FIGURE 11.22. D.C. Youla's [1550, 1551] illustration of the convergence of the PGA algorithm in Figure 10.31 in a Hilbert space.

A geometrical illustration of the PGA is shown in Figure 11.22 for the case of extrapolation [1550, 1551]. The subspace on the horizontal axis consists of all duration limited signals, A_{TL}. The signal $x(t) \in A_{TL}$ if it is zero for $|t| \leq \frac{T}{2}$. Using the notation in Section 10.6,

$$\mathbf{D}_T x(t) = x(t) \rightarrow x(t) \in A_{TL}.$$

Thus, as shown, our known signal, $g_e(t)$, lies in this space.

The null space of A_{TL}, denoted $A_{TL}^{\perp}$, consists of all signals equal to zero for $t > \frac{T}{2}$. Thus

$$(1 - \mathbf{D}_T) x(t) = x(t) \rightarrow x(t) \in A_{TL}^{\perp}.$$

The spaces are clearly perpendicular. If $x(t) \in A_{TL}$ and $u(t) \in A_{TL}^{\perp}$, then $\int x(t)u^*(t)dt = 0$. The null space is shown as a vertical line in Figure 11.22.

The space of bandlimited signals with bandwidths not exceeding B is shown as a line in labelled A_{BL} Figure 11.22. Then

$$\mathbf{B}_B x(t) = x(t) \rightarrow x(t) \in A_{BL}.$$

The only signal common to A_{BL}, A_{TL}, and $A_{TL}^{\perp}$ is $x(t) \equiv 0$. The three lines in Figure 11.22 thus intersect at the origin.

For the PGA, the known portion of the signal, $g_e(t)$, is zero outside of the interval $|t| \leq \frac{T}{2}$ and therefore, as shown in Figure 11.22, lies in the subspace A_{TL}. In Figure 11.11, we form the linear variety, L, by adding the signal $d(t)$ to every point in the subspace S. Likewise, in Figure 11.22, we form the linear variety A_{IM} by adding $g_e(t)$ to every point in the subspace $A_{TL}^{\perp}$. This linear variety is set of signals with an identical middle, the identical middle being $g_e(t)$ on $t \leq \frac{T}{2}$.

With our space established, we can now follow the POCS interpretation of the PGA in Figure 10.31. Beginning with $g_e(t)$ at the top in Figure 10.31, we perform a low pass filter. This is equivalent, in Figure 11.22, of projecting the point $g_e(t)$ onto the subspace A_{BL} to point 'a'. The next step in is setting the middle of the signal to zero as is illustrated at the bottom of Figure 10.31. The result is equivalent in Figure 11.22 to projecting the point 'a' onto the subspace of signals with no middles-namely $A_{\text{TL}}^{\perp}$. The point 'a' is thus projected and becomes the point 'b'. The next step in Figure 10.31 is to add $g_e(t)$ to the signal without middle to form $f_N(t)$ for $N =$ the first iteration. In Figure 11.22, we add the vectors $g_e(t)$ and 'a' to obtain $f_1(t)$. Thus completes the first iteration of the PGA. Subsequent iterations, as shown in Figure 11.22, produce $f_2(t), f_3(t)$, etc. Because the bandlimited signal, $f(t)$, is analytic, the known portion of the signal, $g_e(t)$, uniquely determines $f(t)$. The subspace of bandlimited signals and the linear variety of identical middles, A_{IM}, intersect at a single point. The iteration continues to the fixed point $f_\infty(t) = f(t)$. As is always the case with extrapolation, the assumption of no noise can be fatal to the restoration process since the problem is ill-posed.

11.4.4 Howard's Minimum-Negativity-Constraint Algorithm

Howard [626, 624] proposed a procedure for extrapolation of an interferometric signal known in a specified interval when the spectrum of the signal was known (or desired) to be real and non-negative. The technique was applied to experimental interferometric data and performs well.

As shown by Cheung et al. [289], Howard's procedure was a special case of POCS. Iteration is between the cone of signals with non-negative Fourier transforms defined by A_{POS} in (11.17) and the set of signals with identical middles, A_{IM}, as defined in (11.25). As with the Papoulis-Gerchberg algorithm, the middle is equal to the known portion of the signal.

The geometrical interpretation of Howard's minimum-negativity-constraint algorithm is similar to that of the Papoulis-Gerchberg algorithm, except that the subspace A_{BL} is replaced by a cone corresponding to A_{POS}.

11.4.5 Associative Memory

POCS can be used to construct an associative memory, also called a content addressable memory [912, 913, 914, 1039, 1333]. In a nutshell, conventional memory supplies an address wherein information is stored. As associative memory, on the other hand, supplies a portion of the information in the memory and requires construction of the remainder of the information. One might, for example, be given a picture of a nose and desire, from a data base, to find the corresponding face.

Let the data to be stored in the associative memory be encoded as N library vectors of length $M > N$.

$$\{\vec{f}_n \mid 1 \leq n \leq N\}.$$

We stack these vectors into the *library matrix*

$$\mathbf{S} = \left[\vec{f}_1 | \vec{f}_2 | \cdots | \vec{f}_N \right]. \tag{11.28}$$

11.4.5.1 Template Matching

A straightforward and effective method to implement an associative memory is *template matching*. Let a vector, $\vec{y}$ be equal to a library vector over P elements. For example, if $\vec{y}$ is associated with library vector $\vec{f}_1$, we might have the following.

$$\vec{f}_1 = \begin{bmatrix} 1 & 2 & 3 & 4 & 5 & 6 & 7 & 8 & 7 & 6 & 5 & 4 & 3 & 2 & 1 \end{bmatrix}^T$$

$$\vec{y} = \begin{bmatrix} - & - & 3 & - & - & 6 & 7 & - & - & - & 5 & 4 & - & 2 & - \end{bmatrix}^T$$

where the dash denotes unknown values. We do not know that the values in $\vec{y}$ come from $\vec{f}_1$ and, given the library matrix, wish to determine the unknown numbers. Let χ (for *clamped*) denote the set of indices where $\vec{y}$ is known, i.e., $(\vec{y})_m$ is known for all $m \in \chi$. A simple and effective method of template matching amounts to comparison of each of the clamped elements to the corresponding elements of each of the library vectors and find the smallest mean square error. That is, we wish to minimize over all N library vectors

$$\sum_{m \in \chi} \left| (\vec{f}_n)_m - (\vec{y})_m \right|^2.$$

In more conventional notation, the best matching library vector, $\vec{f}_\ell$ has index

$$\ell = \arg \min_{1 \leq n \leq N} \sum_{m \in \chi} \left((\vec{f}_n)_m - (\vec{y})_m \right)^2.$$

11.4.5.2 POCS Based Associative Memories

POCS can be used to implement associative memories. We will establish the convex sets required for an associative memory. The library matrix in (11.28) is not full rank and we can form the projection matrix in (11.12). This matrix projects any vector onto the subspace defined by the library vectors. This is our first convex set. To initiate the associative search, we require a portion of the information to be restored. The known portion of the signal forms an *identical middles* linear variety. This is our second convex set. Alternating back and forth will converge to the associative recall if the two linear manifolds converge at a single point. The iteration is similar to the PGA except the associative memory subspace replaces the bandlimited signal set.

11.4.5.3 POCS Associative Memory Examples

To illustrate the POCS associative memory, we use the $N = 56$ images shown in Figures 11.23 and 11.24. Each image contains 225×226 gray level pixels normalized[17] from $-\frac{1}{2}$ to $\frac{1}{2}$.

The average of all 56 images, shown on the left in Figure 11.25, resembles a severely blurred face.

---

17. All associative memory examples in this section are displayed as one byte gray level images. The $-\frac{1}{2}$ to $\frac{1}{2}$ normalization is used to form the image library matrix in order to improve convergence. If the normalization $(0,1)$ is used, all library vectors lie in the first orthant and no two vectors, unless they lie on planes defining the orthant, can be orthogonal. If the library vectors are bipolar, they can be orthogonal or nearly orthogonal. POCS converges more quickly for near orthogonal subspaces. Indeed, for two orthogonal lines, POCS converges in a single step. For this reason, we have chosen the bipolar $-\frac{1}{2}$ to $\frac{1}{2}$ library.

FIGURE 11.23. Continued in next Figure.

An example of convergence of the POCS associative memory using the image of Fourier is shown in Figures 11.26 and 11.27. Four different initializations on each of the four rows corresponding to 1%, 4%, 9% and 16% of the original image. The columns show the results of 2, 8, 32, 128 and 512 POCS iterations terminating on the identical middles linear variety. The visually identified trend is numerically verified in Figure 11.28 where the error, $\|\vec{f}_1 - \vec{y}_p\|^2$, is shown where $\vec{f}_1$ is the original image of Fourier, $\vec{y}_p$ is the pth POCS iteration on the linear variety of identical middles.

Noise imposed on the known portion of the image can also effect convergence. This is shown in Figures 11.33 and 11.34 where zero mean white Gaussian noise is added to each pixel. The standard deviations of the noise are $\sigma = 0.15, 0.3, 0.45$ and 0.60. In order to keep the pixels within the $(-\frac{1}{2}, \frac{1}{2})$ interval, values above $\frac{1}{2}$ are set to $\frac{1}{2}$ and values below $-\frac{1}{2}$ are set to $-\frac{1}{2}$. This was in both the initialization and during each iteration. Doing so is equivalent to adding a third convex set to the POCS recovery-the set of all signals bounded by $\pm\frac{1}{2}$). In all cases, the iteration converges to an image clearly recognizable as Fourier.

The larger the library, the slower the convergence. This is illustrated in Figures 11.29 and 11.30 where convergence of Fourier is shown for library sizes 56, 28, 14 and 7 library vectors. The error plot is shown in Figure 11.31.

Convergence is shown in Figures 11.32 and 11.33 when the known portion of the signal is corrupted by additive discrete white Gaussian noise with standard deviation σ. The additive noise has standard deviations of $\sigma = 0.15, 0.30, 0.45$ and 0.60. When the signal

SIGNAL AND IMAGE SYNTHESIS: ALTERNATING PROJECTIONS ONTO CONVEX SETS 519

FIGURE 11.24. Pictures of great mathematicians serve as elements of the associative member library. Each image's size is 226 × 225 pixels.

FIGURE 11.25. (LEFT) The average of all of the mathematician pictures in Figure 11.24. (RIGHT) The picture of Fourier from the library. This image will be used to illustrate convergence of the associative memory.

FIGURE 11.26. Continued in the next figure.

plus noise exceeds $\frac{1}{2}$, it is clamped to $\frac{1}{2}$. Values below $-\frac{1}{2}$ are clamped to $-\frac{1}{2}$. The higher the noise level, the greater the steady state error of the restored image in comparison to the original Fourier picture. The error of the restoration for the different noise initializations as a function of iteration is shown in Figures 11.32 and 11.33. The POCS iteration will not converge to the exact image of Fourier. The error in Figure 11.34 does thus not approach zero. Nevertheless, the image of Fourier in Figures 11.32 and 11.33 clearly emerges from the POCS associative memory.

It is fun to initiate the POCS iteration with elements not in the library. An example, shown in Figure 11.35, is initialized with Euler's head and Jacobi's mouth. Fifty iterations result in an image that favors Jacobi. Termination is shown both at the linear variety of identical middles (eye open) and the library subspace.

Another initialization with a portion of a picture not in the library is shown in Figure 11.36.

FIGURE 11.27. Associative memory convergence using Fourier using the library of pictures in Figure 11.23 and 11.24. The greater the percentage (1%, 4%, 9%, 16%) of the known image, the faster the convergence. An error curve is shown in Figure 11.28.

11.4.5.4 POCS Associative Memory Convergence

Here we analyze the performance of the POCS associative memory. Let $\vec{y}$ be the initialization of the iteration where

$$\vec{y} = \begin{bmatrix} \vec{y}_\chi \\ \vec{0} \end{bmatrix}. \tag{11.29}$$

The vector component, $\vec{y}_\chi$, contains known values of the vector and $\vec{0}$ is a vector of zeros. Any vector $\vec{v}$ can be written as

$$\vec{v} = \begin{bmatrix} \vec{v}_\chi \\ \vec{v}_{\overline{\chi}} \end{bmatrix}. \tag{11.30}$$

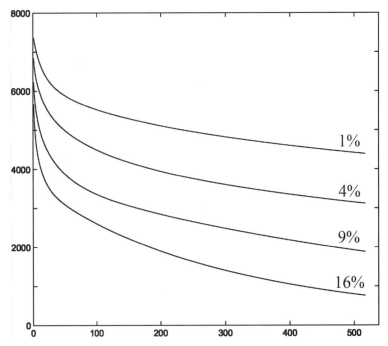

FIGURE 11.28. Plots of the error $\|\vec{f}_1 - \vec{y}_p\|^2$ where $\vec{f}_1$ is the original image of Fourier and $\vec{y}_p$ is the pth POCS iteration on the linear variety of identical middles. The library of pictures in Figure 11.23 and 11.24 was used. The columns show the results of 2, 8, 32, 128 and 512 POCS iteration. The greater the percentage (1%, 4%, 9%, 16%) of the known image, the faster the convergence.

For an arbitrary vector of $\vec{v}$, the projection onto the linear variety of identical middles corresponding to $\vec{y}_\chi$ is

$$P_\chi \vec{v} = \begin{bmatrix} \vec{y}_\chi \\ \vec{v}_{\overline{\chi}} \end{bmatrix}.$$

The projection of a vector $\vec{v}$ onto the library subspace is

$$P_S \vec{v} = \mathbf{T}\vec{v}.$$

The POCS iteration between the two iteration can then be written as

$$\vec{y}_{r+1} = P_\chi \mathbf{T} \vec{y}_r. \tag{11.31}$$

with initialization, from (11.29), of $\vec{y}_0 = \vec{y}$. Note, however, that the lower portion of the vector $\mathbf{T}\vec{y}_r$ is always discarded in the next projection. To sharpen our analysis, we divide the $N \times N$ projection matrix into submatrices.

$$\mathbf{T} = \begin{bmatrix} \mathbf{T}_2 & \mathbf{T}_1 \\ \mathbf{T}_3 & \mathbf{T}_4 \end{bmatrix}.$$

If $\vec{y}_\chi$ is of length $N - P$, then $\mathbf{T}_2$ is $(N - P) \times (N - P)$ and $\mathbf{T}_4$ is $P \times P$. The top of $\vec{y}_r$ is always clamped to $\vec{y}_\chi$. Thus

$$\vec{y}_r = \begin{bmatrix} \vec{y}_\chi \\ \vec{\phi}_r \end{bmatrix}.$$

where $\vec{\phi}_r$ (ϕ for *floating*) denotes the only portion of the vector that changes from iteration to iteration. Equation (11.31) can be written as

$$\begin{bmatrix} \vec{y}_\chi \\ \vec{\phi}_{r+1} \end{bmatrix} = P_\chi \begin{bmatrix} \mathbf{T}_2 & \mathbf{T}_1 \\ \mathbf{T}_3 & \mathbf{T}_4 \end{bmatrix} \begin{bmatrix} \vec{y}_\chi \\ \vec{\phi}_r \end{bmatrix}.$$

Solving for the floating component gives the iteration

$$\vec{\phi}_{r+1} = \mathbf{T}_3 \vec{y}_\chi + \mathbf{T}_4 \vec{\phi}_r. \tag{11.32}$$

There are at least two interesting uses of this equation.

1. **Linear Convergence.** Let the $P \times P$ matrix $\mathbf{T}_4$ have P eigenvalues, λ_p, and P eigenvectors, $\vec{\psi}_p$, where $1 \le p \le P$. Thus

$$\lambda_p \vec{\psi}_p = \mathbf{T}_4 \vec{\psi}_p.$$

Expand the result of the rth floating vector as

$$\vec{\phi}_r = \sum_{p=1}^{P} \alpha_r[p] \vec{\psi}_p$$

where $\alpha_r[p] = \vec{\phi}_r^T \vec{\psi}_p$. The iteration in (11.32) is then

$$\sum_{p=1}^{P} \alpha_{r+1}[p] \vec{\psi}_p = \mathbf{T}_3 \vec{y}_\chi + \sum_{p=1}^{P} \lambda_p \alpha_r[p] \vec{\psi}_p.$$

Premultiplying by $\vec{\psi}_q^T$ and invoking the orthonormality of the eigenvectors

$$\vec{\psi}_q^T \vec{\psi}_p = \delta[p-q]$$

gives

$$\alpha_{r+1}[q] = \beta_q + \lambda_q \alpha_r[q]$$

where

$$\beta_q := \vec{\psi}_q^T \mathbf{T}_3 \vec{y}_\chi.$$

This difference equation has a solution of

$$\alpha_r[q] = \frac{1 - \lambda_q^r}{1 - \lambda_q} \beta_q + \lambda_q^r \alpha_0[q]$$

where $\alpha_0[q]$ is the initialization. Since we have assumed zero initial conditions, $\alpha_0[q] = 0$ and

$$\alpha_r[q] = \frac{1 - \lambda_q^r}{1 - \lambda_q} \beta_q.$$

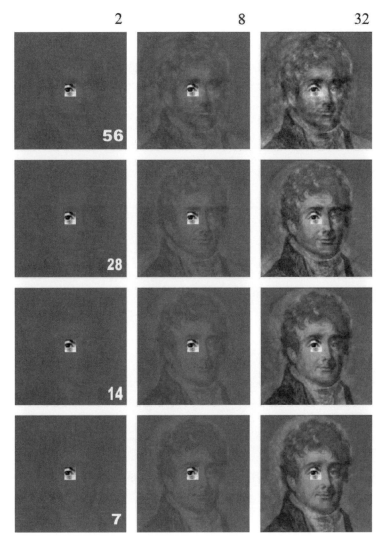

FIGURE 11.29. Continued in the next figure.

If the magnitude of all eigenvalues is less than one, we have $\lim_{r\to\infty} \lambda_q^r = 0$ and

$$\alpha_\infty[q] = \frac{1}{1-\lambda_q}\,\beta_q.$$

The error in the rth iteration is then

$$\epsilon_r[q] = \alpha_\infty[q] - \alpha_r[q]$$
$$= \lambda_q^r \beta_q/(1-\lambda_q).$$

We note that

$$\frac{\epsilon_{r+1}[q]}{\epsilon_r[q]} = \lambda_q. \qquad (11.33)$$

FIGURE 11.30. Convergence of the POCS associative memory as a function of the library size. The convergence of Fourier for 56, 28, 14 and 7 library vectors is shown. In each case, the Fourier is included in the library. A plot of error is shown in the next figure (Figure 11.31). The larger the library, the slower the convergence.

When the ratio of the error in two sequential iterations is a constant, as it is here, the iteration is said to *linearly converge*. As a consequence, (11.33) has a solution

$$\epsilon_r[q] = \epsilon_r[0]\lambda_q^r.$$

We here see that the convergence of POCS is dominated by the largest eigenvalue of $\mathbf{T}_4$. The closer the magnitude of this eigenvalue is to one, the longer it takes for λ_q^r to go to zero and for the iteration to converge.

To illustrate, consider the error curve in Figure 11.31. The ratio of successive errors, shown in Figure 11.37, approaches a constant. The largest eigenvalue is

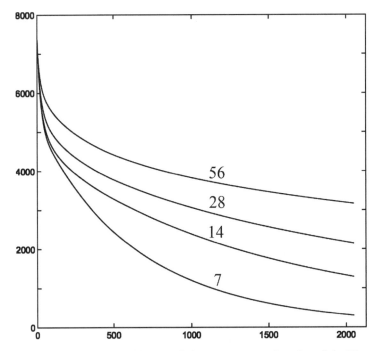

FIGURE 11.31. Convergence of the POCS associative memory as a function of the library size. The error is shown as a function of iteration for 56, 28, 14 and 7 library vectors. Images of the convergence are shown in the previous figure (Figure 11.30).

here dominating and the contributions of the smaller eigenvalues have decayed to insignificance.

2. **Noniterative Solution.** The POCS iteration in (11.32) can be solved by solution of P linear equations. Assuming convergence, we let $r \to \infty$ in (11.32). Then [721, 907, 1205]

$$\vec{\phi}_\infty = \mathbf{T}_3 \vec{y}_\chi + \mathbf{T}_4 \vec{\phi}_\infty.$$

and, assuming $\mathbf{I} - \mathbf{T}_4$ is not singular, the iteration will converge to

$$\vec{\phi}_\infty = (\mathbf{I} - \mathbf{T}_4)^{-1} \mathbf{T}_3 \vec{y}_\chi.$$

11.4.5.5 POCS Associative Memory Capacity

The maximum number of vectors that can be placed in the POCS associative memory is, in principle, $M - 1$ non-colinear library vectors when the length of the library vector is M. The associative memory's performance, however, degrades as the number of library vectors is increased.

The number of elements in the library vector can be made greater than M thereby increasing the capacity. This can be done by adding extra elements to each library vector consisting of, say, randomly generated values or nonlinear combinations of other elements [913].

11.4.5.6 Heteroassociative Memory

A *heteroassociative memory* is one wherein a stimulus vector, say $\vec{s}_n$, is configured to produce a response vector, $\vec{r}_n$. These stimulus and response vectors can be stacked into an library vector

$$\vec{f}_n = \begin{bmatrix} \vec{s}_n \\ \vec{r}_n \end{bmatrix}. \tag{11.34}$$

A number of stimulus-response vector pairs can thus be construed as a library to which the POCS associative memory can be applied. The input vector corresponding to the stimulus is clamped and the floating portion, corresponding to the response vector, is reconstructed using POCS.

11.4.6 Recovery of Lost Image Blocks

POCS can be used for block loss recovery. The block loss recovery problem is illustrated in Figure 11.38. A compressed image is transmitted over a lossy digital channel. The received image has missing blocks. Such might be characteristic, for example, of JPEG image compression where the image is partitioned into 8×8 blocks of pixels and coded with a discrete cosine transform (DCT). An intermediate corruption of transmission would result in loss of a whole block. The block loss recovery problem is to fill in the loss blocks using the known portions of the image without retransmitting the image.

Yang *et al.* propose a projection based spatially-adaptive reconstruction of images [1533, 1534]. In Yang *et al.*'s algorithm, convex sets are developed for de-blocking or capturing the smoothness properties of the desired image [1533, 1534]. We closely follow the POCS approach of Park *et al.* [1091, 1092, 1093] who have applied POCS to the block loss recovery problem with remarkable success.

Three convex sets are used. Assume an $N \times N$ block of pixels are lost.

1. **The convex hull of adjacent blocks.** The first convex set is the convex cone hull of surrounding images. For an $N \times N$ block of missing pixels, there are $8N$ touching $N \times N$ blocks of known pixels. This is illustrated by the left hand figure in Figure 11.39 where four such adjacent blocks are shown. When the top left block shifts one pixel to the right, we obtain a new $N \times N$, albeit with $N \times (N-2)$ common pixels. We expect the missing block to resemble, in texture and grey level, these known images. The $8N$ points form the foundation for a convex hull that serves as the first convex set in the POCS reconstruction of the missing block.
2. **Range constraint.** The reconstructed pixels are required to lie within a range of x_{min} and x_{max}. In a gray scale image with black assigned zero and white one, any realizable restoration must lie between zero and one. Signals satisfying this constraint form the convex set of bounded signals.
3. **Smoothness constraint.** The smoothness constraint eliminates abrupt gray level jumps between adjacent pixels. More about this shortly.

Rather than reconstructing the entire missing block at once, Park *et al.* opted to reconstruct the block one column (or one row) at a time [815, 1416]. The horizontal case is illustrated by the illustration on the right in Figure 11.39. Using the known pixels, the lost pixels in, say, the left hand column, are reconstructed using POCS. Once found, these pixels are fixed and the block is moved a pixel to the right to expose another column of lost pixels.

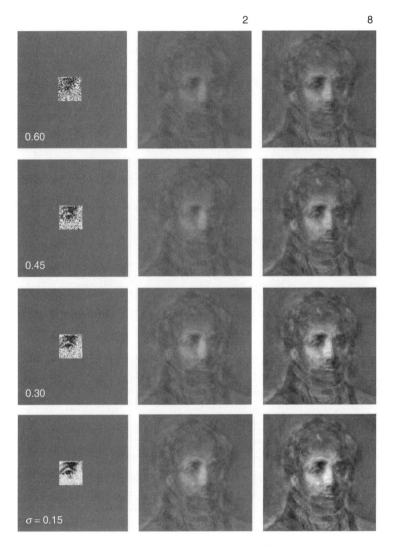

FIGURE 11.32. Continued in the next figure.

The blocks on the left and right eventually meet in the middle and the reconstruction process is complete.[18]

The smoothness constraint requires the difference between a reconstructed pixel and its adjacent pixel be bounded. Thus, if a known pixel has a value of x, the reconstructed pixel can take on only values in the range $x \pm \alpha$ where α is the maximum allowed deviation. This convex set corresponds to a box of linear dimension 2α centered about the known pixel values. The projection is straightforward. If the reconstructed pixel exceeds $x + \alpha$, it is set

---

18. Should the blocks be moved in from the left and right wherein pixel columns are reconstructed as shown on the right side of Figure 11.39, or from the top and bottom wherein rows of pixels are reconstructed? Park *et al.* [1091, 1092, 1093] reasoned the blocks should slide along the same direction of any line like structure. Thus, prior to reconstructing a block, the structure around the missing block is examined. If there is line like texture running horizontally, the reconstruction shown on the right side of Figure 11.39 is used. If the lines run vertically, the blocks are moved in from the top and the bottom.

FIGURE 11.33. Associative memory convergence of Fourier using the library of pictures in Figure 11.23 and 11.24. The lower the noise level, the lower the error. An error curve is shown in Figure 11.34.

to $x + \alpha$. If $x - \alpha$ exceeds the reconstructed pixel's value, it is set to $x - \alpha$. Otherwise, it remains as is.

Projections in the POCS don't always converge to a single point in this application, but the difference in solutions is typically negligible.

To illustrate the effectiveness of the approach, the *Healthy Girl* image in Figure 11.40 was subjected to the loss of 8 × 8 pixel blocks as is shown in Figure 11.41. The reconstruction using POCS is shown in Figure 11.42. A close-up of the reconstruction process along the girl's shoulder is shown in Figure 11.43. Although the restoration is quite good, the process does have flaws as the close-up of the POCS restoration in Figure 11.44 reveals.

A second example is given in Figure 11.45 (original *Masquerade* image), Figure 11.46 (every forth 8 × 8 block removed) and Figure 11.47 (the POCS restoration).

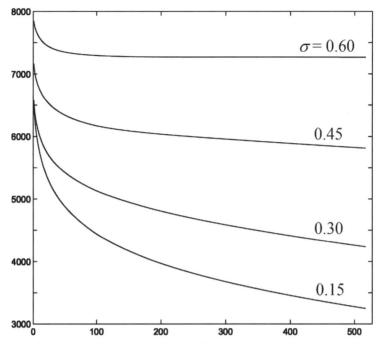

FIGURE 11.34. Plots of the error $\|\vec{f}_1 - \vec{y}_p\|^2$ where $\vec{f}_1$ is the original image of Fourier and $\vec{y}_p$ is the pth POCS iteration on the linear variety of identical middles. The library of pictures in Figures 11.23 and 11.24 was used. The columns show the results of 1, 25, 50, 100 and 300 POCS iteration. The lower the noise level, the closer POCS converges to the original picture of Fourier. (Note that the error range is not shown full scale.)

FIGURE 11.35. Initialization with portions of elements in library of mathematicians Jacobi and Euler shown in the last row in Figure 11.24. From left to right is (a) an initialization of POCS using portions of both faces, (b) the POCS convergence on the linear variety, and (c) the projection of this image on the library subspace. Fifty iterations are used.

An alternate lost block scenario is shown in Figure 11.48 where 16 × 16 pixel blocks of the original image in Figure 11.40 are lost. The restoration, shown in Figure 11.49 is, visually, not as good as the 8 × 8 block size loss restored in Figure 11.42 even though the total number of missing pixels in each case is approximately the same.

A smaller 16 × 16 block loss POCS is shown in Figure 11.50.

Lastly, the *Healthy Girl* image in Figure 11.40 looses every other row of eight pixels as shown in Figure 11.51. The restoration is shown in Figure 11.52.

There exist numerous other methods to restore missing blocks of pixels [20, 14, 20, 592, 1361, 1273]. A review is given by Wang and Zhu [1467]. A common metric for the goodness

SIGNAL AND IMAGE SYNTHESIS: ALTERNATING PROJECTIONS ONTO CONVEX SETS 531

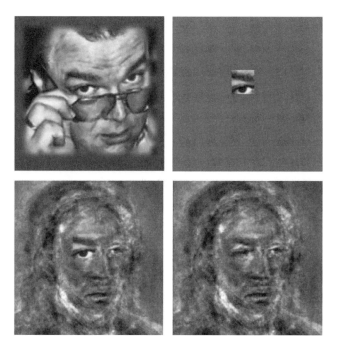

FIGURE 11.36. Initialization with an element not in the library of the great mathematicians in Figure 11.24. In the top row from left to right is (a) an image not in the library and (b) an eyeball from the image used as the known portion of the iteration. In the second row, we see (c) the result of the POCS iteration ending on the linear variety of all images with the eyeball (open eye), and (d) projection of this image on the library subspace (closed eye). One hundred iterations are used.

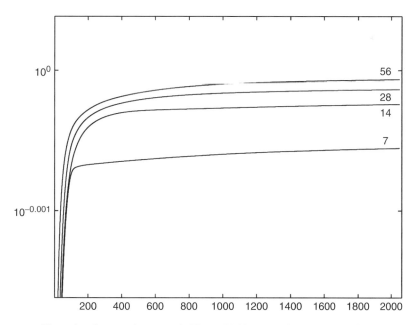

FIGURE 11.37. The ratio of successive errors in Figure 11.31 approaches a constant thereby illustrating linear convergence.

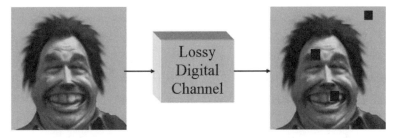

FIGURE 11.38. Illustration of the block loss recovery problem.

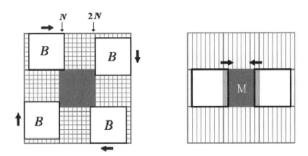

FIGURE 11.39. (left) For an $N \times N$ block of missing pixels, there are $8N$ blocks of $N \times N$ pixel blocks touching the missing block. Here, four are shown. (Right) Reconstructing the missing block of pixels one column at a time.)
Source: [1093] Jiho Park, Dong Chul Park, R.J. Marks II and M.A. El-Sharkawi, "Recovery of Image Blocks using the Method of Alternating Projections." *IEEE Transactions on Image Processing*, March 2005.

of restoration of a one byte (256 gray level) image of dimension $N_1 \times N_2$ pixels is the *peak signal-to-noise ratio* [504] defined as

$$PSNR = 10 \, \log \left(\frac{N_1 \cdot N_2 \cdot 255^2}{\sum_{m=1}^{N_1} \sum_{n=1}^{N_2} |f(m,n) - g(m,n)|^2} \right) \quad (11.35)$$

where f is the original and g is the given degradation. Using this metric, Park *et al.* [1091, 1092, 1093] found the POCS restoration performed better than the other proposed techniques [20, 14, 20, 592, 1361, 1273].

11.4.7 Subpixel Resolution

When the aperture of a photoreceptor is too large, high frequency information is lost. This is illustrated in Figure 11.53 where an object is imaged through an integrating square aperture. The aperture integrates, or sums, the object over the square aperture and returns a single number. The subpixel restoration problem is construction of the original image from a number of summed images.

11.4.7.1 Formulation as a Set of Linear Equations

The subpixel resolution problem can be interpreted as the solution to a large number of simultaneous equations. Visualize the original image as a long column vector. An summed image is the sum of a subset of these values and can be viewed as the inner product of a row vector of ones and zeros and the original image vector. The row vectors for a number of different summed images, all containing either ones or zeros, can be stacked into a matrix. The image vector, multiplied by this matrix of ones and zeros, gives a vector of summed image values. The corresponding set of linear equations can be solved to find the original image.

11.4.7.2 Subpixel Resolution using POCS

POCS can also be used to generate the subpixel resolution from summed images. The set of all objects that has a sum over an aperture as shown in Figure 11.53 is a convex set. Indeed, it is a constant area linear variety. To project onto this convex set, the set of all pixels in an image are equally adjusted so the resultant sum is equal to the value measures.

As shown in Figure 11.54, placing the aperture in a plurality of locations, taking a measurement of the sum, and projecting the image onto the convex set corresponding to that sum, allows for reconstruction of the original object. On the top left is shown a 274×401 pixel object and, on the top right, this image viewed though a sequence of 8×8 pixel apertures. To generate this low resolution image, each 8×8 pixel block is averaged. Even with this low resolution imaging device, generation of high resolution pictures is possible if the 8×8 imaging aperture is moved to other places on the image. To illustrate, we choose a random location in the 274×401 image to place the center of the 8×8 pixel aperture. At this location, we read a single number equal to the average of the values of the 64 pixels. All images that have this average value over these pixels form a discrete version of the constant area convex set in (11.18). The projection onto this set, from (11.19), is performed by adding to (or subtracting from) each of the 64 pixels the same number so that the resulting average is equal to the measured average.

The effects of repeatedly choosing 8×8 aperture locations in a uniformly random manner and projecting the image onto the reading from this aperture is shown in the bottom four images in Figure 11.54. In addition to the constant area convex sets, the images are constrained to fall in the interval zero (black) to one (white).

11.4.8 Reconstruction of Images from Tomographic Projections

In Section 8.5.3 in Chapter 8, the inverse Radon transform is shown as a methodology to reconstruct images from their tomographic projections. We will show doing so is also possible using POCS. Indeed, the convex sets are identical to those used for subpixel resolution in Section 11.4.7. Consider the projection, $p_\theta(t)$, of an image in (8.52) in Section 8.5.3. An example is shown in Figure 11.55. The sample of the projection at $t = \tau$, as shown, is equal to the sum of pixels along a line. As in the case of subpixel resolution, we have a measure equal to the sum of pixels over a specified region of the image. Given a number of such sums corresponding to numerous lines at different angles, the same mathematics used in subpixel resolution can be used to reconstruct the image.

Illustration of convergence is shown in Figures 11.56 and 11.57. Projections are taken along randomly chosen lines.[19] Results are shown for various numbers of lines.

---

19. Due to the *Bertrand paradox* [1077], we must be careful how we define *random*. (See Exercise 11.6.) A pixel is chosen at random within the image. An angle is then chosen on the interval $(0, \pi)$ to determine the slope of the line.

FIGURE 11.40. Original *Healthy Girl* image.
Source: [1093] Jiho Park, Dong Chul Park, R.J. Marks II and M.A. El-Sharkawi, "Recovery of Image Blocks using the Method of Alternating Projections." *IEEE Transactions on Image Processing*, March 2005.

FIGURE 11.41. The *Healthy Girl* image in Figure 11.40 subject to 8 × 8 pixel block loss.
Source: [1093] Jiho Park, Dong Chul Park, R.J. Marks II and M.A. El-Sharkawi, "Recovery of Image Blocks using the Method of Alternating Projections." *IEEE Transactions on Image Processing*, March 2005.

FIGURE 11.42. POCS restoration of the degraded *Healthy Girl* image in Figure 11.41. Compare to the original image in Figure 11.40.
Source: [1093] Jiho Park, Dong Chul Park, R.J. Marks II and M.A. El-Sharkawi, "Recovery of Image Blocks using the Method of Alternating Projections." *IEEE Transactions on Image Processing*, March 2005.

11.4.9 Correcting Quantization Error for Oversampled Bandlimited Signals

A bandlimited signal with bandwidth B is sampled at a rate of $2W > 2B$ samples per second. The signal can be recovered from the samples using (6.2) which we repeat here

$$x(t) = r \sum_{n=-\infty}^{\infty} x\left(\frac{n}{2W}\right) \operatorname{sinc}(2Bt - rn) \, ; \, r = \frac{B}{W}.$$

Consider the case where the samples of $x(t)$ are quantized as illustrated in Figure 11.59. Each sample is rounded to the nearest quantization level. The quantization levels are separated by an interval of 2Δ. Denote the quantization of the sample $x\left(\frac{n}{2W}\right)$ by q_n.

POCS can be applied to correction of quantization error [1333]. Define the convex set

$$C_n = \left\{ x(t) \,\bigg|\, Q\left[x\left(\frac{n}{2W}\right)\right] = q_n \text{ and } x(t) \text{ is bandlimited with bandwidth } B < W \right\}$$

where the quantization operator, Q, rounds the value on which it operates to the nearest quantization level. The convex sets on which the solution lies is (a) the set of all C_n's and (b) the set of bandlimited signals with bandwidth $B < W$ in (11.13).

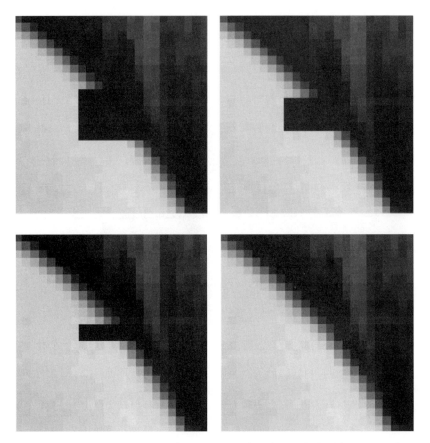

FIGURE 11.43. A close-up of four stages of the narrowing of the 8 × 8 missing block restoration process along the *Healthy Girl*'s shoulder.
Source: [1093] Jiho Park, Dong Chul Park, R.J. Marks II and M.A. El-Sharkawi, "Recovery of Image Blocks using the Method of Alternating Projections." *IEEE Transactions on Image Processing*, March 2005.

FIGURE 11.44. A close-up of the restored *Healthy Girl* in Figure 11.42 shows minor flaws on the edges.
Source: [1093] Jiho Park, Dong Chul Park, R.J. Marks II and M.A. El-Sharkawi, "Recovery of Image Blocks using the Method of Alternating Projections." *IEEE Transactions on Image Processing*, March 2005.

Let $y(t)$ be a bandlimited signal[20] with bandwidth W. If, at $t = \frac{n}{2W}$, the signal $y(t)$ quantizes to q_n, then $y(t) \in C_n$. Otherwise, we need to project it onto C_n. If the sample is too large, we need to change it to the largest value that quantizes to q_n. If too small, we choose the smallest value that quantizes to q_n. This is accomplished by the following projection onto C_n.

$$P_n y(t) = \begin{cases} y(t) & ; Q\left[y\left(\frac{n}{2W}\right)\right] = q_n \\ y(t) + \left(q_k + \Delta - y\left(\frac{n}{2W}\right)\right) \operatorname{sinc}(2Wt - n) & ; Q\left[y\left(\frac{n}{2W}\right)\right] > q_n \\ y(t) + \left(q_k - \Delta - y\left(\frac{n}{2W}\right)\right) \operatorname{sinc}(2Wt - n) & ; Q\left[y\left(\frac{n}{2W}\right)\right] > q_n \end{cases}$$

The projection onto bandlimited signals in (11.14) can be written

$$P_{\text{BL}}\, y(t) = y(t) * 2B \operatorname{sinc}(2Bt).$$

It follows that

$$P_{\text{BL}}\, P_n y(t) = \begin{cases} z(t) & ; Q\left[y\left(\frac{n}{2W}\right)\right] = q_n \\ z(t) + r\left(q_k + \Delta - y\left(\frac{n}{2W}\right)\right) \operatorname{sinc}(2Bt - rn) & ; Q\left[y\left(\frac{n}{2W}\right)\right] > q_n \\ (t) + r\left(q_k - \Delta - y\left(\frac{n}{2W}\right)\right) \operatorname{sinc}(2Bt - rn) & ; Q\left[y\left(\frac{n}{2W}\right)\right] > q_n \end{cases} \quad (11.36)$$

where

$$z(t) = P_{\text{BL}}\, y(t) = y(t) * 2B \operatorname{sinc}(2Bt).$$

Repeated application of the projection sequence in (11.36) will converge to a signal that satisfies the convex constraints. Clearly, if the quantization level is too large, a number of signals will satisfy these constraints.

11.4.10 Kernel Synthesis for GTFR's

Cohen's generalized time-frequency distribution (GTFR), introduced in Section 9.4, requires choice of a two dimensional kernel. The kernel directly effects many performance attributes of the GTFR such as time resolution, frequency resolution, realness, and conformity to time and frequency marginals. These are detailed in Section 9.4.2. A number of different kernels may suffice for a given performance constraint (high frequency resolution, for example). Interestingly, many sets of kernels satisfying commonly used performance constraints are convex. In this Section, following the work of Oh et al. [1043], we describe a method whereby kernels can be designed that satisfy two or more of these constraints. If there exists a non-empty intersection among the constraint sets, POCS guarantees convergence to a kernel that satisfies all of the constraints. If the constraints can be partitioned into two sets, each with a nonempty intersection, then POCS guarantees convergence to a kernel that satisfies the inconsistent constraints with minimum mean square error. We apply kernels synthesized using POCS to generation of some example GTFR's and compare their performance to the spectrogram, Wigner distribution and cone kernel GTFR.

11.4.10.1 GTFR Constraints as Convex Sets

The generalized time-frequency representation is introduced in Section 9.4. Properties of the GTFR can be controlled by imposition of constraints on the GTFR's kernel. These propertis

---

20. If $y(t)$ is not bandlimited, we feed it through a low pass filter with bandwidth W.

FIGURE 11.45. Original *Masquerade* image.
Source: [1093] Jiho Park, Dong Chul Park, R.J. Marks II and M.A. El-Sharkawi, "Recovery of Image Blocks using the Method of Alternating Projections." *IEEE Transactions on Image Processing*, March 2005.

FIGURE 11.46. Every fourth 8×8 block of pixels is removed from the *Masquerade* image in Figure 11.45.
Source: [1093] Jiho Park, Dong Chul Park, R.J. Marks II and M.A. El-Sharkawi, "Recovery of Image Blocks using the Method of Alternating Projections." *IEEE Transactions on Image Processing*, March 2005.

FIGURE 11.47. The POCS restoration of the degraded *Masquerade* image in Figure 11.46. Compare with the original image in Figure 11.45.
Source: [1093] Jiho Park, Dong Chul Park, R.J. Marks II and M.A. El-Sharkawi, "Recovery of Image Blocks using the Method of Alternating Projections." *IEEE Transactions on Image Processing*, March 2005.

are listed in Section 9.4.3.1. We will now show that each of these constraints corresponds to a convex set and that GTFR kernels can be synthesized using POCS [1037, 1043].

1. **Time and Frequency Resolution Projections**
   For the time resolution constraint in Section 9.4.3.1, the signal outside of the cone, shown on the left in Figure 9.15, simply set to zero.

$$P_{C_1}\hat{\phi}(t,\tau) = \hat{\phi}(t,\tau)\Pi\left(\frac{t}{\tau}\right)\Pi\left(\frac{\tau}{2T}\right).$$

Similarly, for frequency resolution constraint, the area outside of the bow tie, shown on the right hand side shown on the left in Figure 9.15, is set to zero.

$$P_{C_{3,1}}\Phi(f,u) = \Phi(f;u)\Pi\left(\frac{f}{u}\right)\Pi\left(\frac{f}{2B}\right). \quad (11.37)$$

A relaxed version of frequency resolution is

$$P_{C_{3,1}}\Phi(f,u) = \begin{cases} 0 & ; \ \Phi(f,u) < 0 \\ \Phi(f,u) & ; \ 0 \leq \Phi(f,u) \leq \alpha(f,u) \\ \alpha(f,u) & ; \ \Phi(f,u) > \alpha(f,u). \end{cases}$$

FIGURE 11.48. The *Healthy Girl* image in Figure 11.40 subject to 16 × 16 pixel block loss.
Source: [1093] Jiho Park, Dong Chul Park, R.J. Marks II and M.A. El-Sharkawi, "Recovery of Image Blocks using the Method of Alternating Projections." *IEEE Transactions on Image Processing*, March 2005.

FIGURE 11.49. The POCS restoration of the degraded *Healthy Girl* image in Figure 11.48. Compare with the original image in Figure 11.40.
Source: [1093] Jiho Park, Dong Chul Park, R.J. Marks II and M.A. El-Sharkawi, "Recovery of Image Blocks using the Method of Alternating Projections." *IEEE Transactions on Image Processing*, March 2005.

FIGURE 11.50. Original (left), degraded with 16 × 16 pixel block loss (middle) and POCS restored image.
Source: [1093] Jiho Park, Dong Chul Park, R.J. Marks II and M.A. El-Sharkawi, "Recovery of Image Blocks using the Method of Alternating Projections." *IEEE Transactions on Image Processing*, March 2005.

2. **Realness and Symmetry Constraint**
   The realness constraint in Section 9.4.3.1 can be imposed by the projection operator

   $$P_{C_6}\hat{\Phi}(t,u) = \Re\hat{\Phi}(t,u)$$

   or, equivalently, in the (f,τ) plane,

   $$P_{C_6}\phi(f,\tau) = \frac{1}{2}\left[\phi(f,\tau) + \phi^*(-f,-\tau)\right].$$

   Similarly, for the symmetry constraints in (9.48) and (9.49), the respective projection operators can be written as

   $$P_{C_7}\phi(f,\tau) = \frac{1}{2}\left[\phi(f,\tau) + \phi^*(f,-\tau)\right]$$

   and

   $$P_{C_8}\phi(f,\tau) = \frac{1}{2}\left[\phi(f,\tau) + \phi^*(-f,\tau)\right]. \tag{11.38}$$

3. **Relaxed Interference Projection**
   Motivated by (9.45), the projection operator corresponding to the relaxed interference term is

   $$P_{C_2}\Phi(f,u) = \Phi(f,u)\Pi\left(\frac{f}{2\Delta}\right)$$

   Note that if Δ is large enough and B is small enough, the frequency resolution projection in (11.37) subsumes this projection.

4. **Non-negative Projection**
   The non-negativity constraint can be imposed by the projection

   $$P_{C_9}\hat{\phi}(t,\tau) = \hat{\phi}(t,\tau)\mu[\hat{\phi}(t,\tau)].$$

5. **Marginal and Cone Constraint**
   In some cases, projections can be best described by the intersection of two or more convex constraints. Combining the time resolution constraint and the

FIGURE 11.51. The *Healthy Girl* image in Figure 11.40 looses every other row of 8 pixels.
Source: [1093] Jiho Park, Dong Chul Park, R.J. Marks II and M.A. El-Sharkawi, "Recovery of Image Blocks using the Method of Alternating Projections." *IEEE Transactions on Image Processing*, March 2005.

frequency marginal constraint in Section 9.4.3.1, we can write the projection on the intersection of the three sets as the convex set operator

$$P_{C_1 \cap C_4 \cap C_5} \hat{\phi}(t, \tau) = \begin{cases} \delta(t) & ; \tau = 0 \\ \left[\hat{\phi}(t, \tau) + \varphi(\tau) \right] \Pi(\frac{t}{\tau}) \Pi(\frac{\tau}{2T}) & ; \text{otherwise} \end{cases}$$

where $\cap$ denotes intersection, $\delta(\cdot)$ is Dirac delta function and

$$\varphi(\tau) = \frac{1}{|\tau|} \left[1 - \int_{-\frac{|\tau|}{2}}^{\frac{|\tau|}{2}} \hat{\phi}(t, \tau) dt \right].$$

Combining the time resolution and the finite area constraints, we have the projection

$$P_{C_1 \cap C_{10}} \hat{\phi}(t, \tau) = \left[\hat{\phi}(t, \tau) + \varphi \right] \Pi\left(\frac{t}{\tau}\right) \Pi\left(\frac{\tau}{2T}\right)$$

where

$$\varphi = \frac{1}{T^2} \left[1 - \int_{\tau=-T}^{T} \int_{t=-\frac{|\tau|}{2}}^{\frac{|\tau|}{2}} \hat{\phi}(t, \tau) dt d\tau \right].$$

FIGURE 11.52. Restoration of the degradation of the *Healthy Girl* image in Figure 11.51. Compare with the original image in Figure 11.40.
Source: [1093] Jiho Park, Dong Chul Park, R.J. Marks II and M.A. El-Sharkawi, "Recovery of Image Blocks using the Method of Alternating Projections." *IEEE Transactions on Image Processing*, March 2005.

11.4.10.2 GTFR Kernel Synthesis Using POCS

The use of POCS in the design of GTFR kernels is now evident. We choose from a menu of convex constraints that we desire our GTFR to obey. By alternately projecting between the corresponding convex sets, we hope to synthesize a corresponding kernel. If the convex sets intersect, a kernel meeting all constraints will be generated. If the constraint sets do not intersect, the iteration may break into a limit cycle. The result may or may not be acceptable depending on the magnitude of the mean square error.

We present two preliminary examples. Both are presented on a 128×128 grid. The kernels in both examples used both the cone and bow tie constraints. The value of T in each case corresponded to truncating the grid so that the cone was a peak to peak height of 64. Both examples resulted in a kernel that was positive and symmetric.

Example 11.4.1. uses, in addition, both marginal constraints. We take the alternating projection between the set $C_{3,1} \cap C_{3,2}$ and $C_1 \cap C_4 \cap C_5 \cap C_9$. The resulting kernel is pictured in Figure 11.60. It resembles a truncated Born-Jordan kernel [309, 310, 312] which has a $\frac{1}{|\tau|}$ taper within the cone. Indeed, for $B = \infty$ and $T = \infty$, the Born-Jordan kernel satisfies all the constraints. Specifically,

$$\hat{\phi}(t, \tau) = \frac{1}{|\tau|} \Pi\left(\frac{t}{\tau}\right)$$

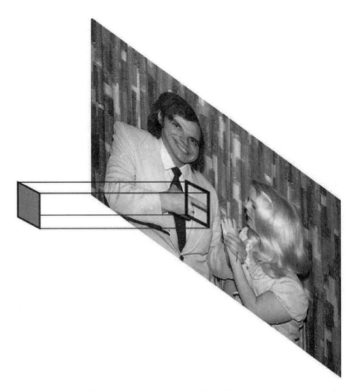

FIGURE 11.53. The square imaging aperture integrates the object in the picture and returns a single number equal to the sum of all the pixels within the aperture. As the aperture moves, this number changes.

satisfies the cone constraint. Furthermore, the marginal constraints are met as are the symmetry constraints of (9.46), (9.48) and (9.49), the realness constraint of (9.47) and the nonnegativity constraint in (9.51). Furthermore,

$$\Phi(f, u) = \frac{1}{|f|} \Pi\left(\frac{u}{f}\right)$$

satisfies the untruncated bow tie constraint.

The iterative POCS synthesis of this kernel does not converge. This empirically proves that, for the kernel dimensions used, *there does not exist a kernel that satisfies all of the constraints.*

Application of this kernel to two converging linear chirps resulted in the 25dB waterfall and 30dB gray level display in Figure 11.61 and the two tone signal with transition is in Figure 11.62. From floor to peak is 25dB.

Example 11.4.2. removes the marginal constraints. We take the alternating projection between $C_{3,1} \cap C_{3,2}$ and $C_1 \cap C_9 \cap C_{10}$. This resulted in the kernel in Figure 11.63. The outcome of the POCS design, smoothed with a Hanning window, was applied to the same linear chirp and two tone signal problem. The result of linear chirp and two tone signal are shown in Figures 11.64 and 11.65 respectively.

FIGURE 11.54. Subpixel resolution using POCS and the constant area convex set. After a half million projections, the image at the bottom left resembles the original image at the upper left.

11.4.11 Application to Conformal Radiotherapy

The concept of fundamental radiotherapy is illustrated in Figure 11.66. We follow closely the development of Lee *et al.* [814].

On the left in Figure 11.66, a tumor is shown imbedded in healthy tissue. The tumor is to be destroyed using externally applied radiation. The radiation, though, destroys both the tumor and the healthy tissue. Therefore, as illustrated on the right hand side of Figure 11.66, beams of radiation are introduced through the tumor at different angles. Thus, the tumor is zapped with every beam and the surrounding healthy tissue is exposed less. Radiotherapy is made more complex by the existence of critical organs in the vicinity of the tumor, e.g., a tumor around the spinal cord. The critical organs are less resilient to radiation and their exposure must be minimized.

Conformal radiotherapy allows beams to be sculpted. Thus, beam cross sections can be made more intense in the middle than the edges, or more intense on one side than the other. Given the locations of critical organs and tumor, the beam design problem in conformal radiotherapy is determination of beam profiles to best capture such features.

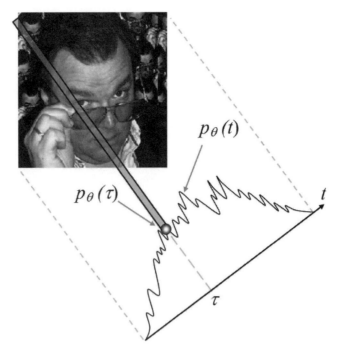

FIGURE 11.55. The function $p_\theta(t)$ is a projection of the image. For discrete images, the sample of the projection, $p_\theta(\tau)$, is equal to the sum of the pixels along a line.

Synthesis of beam profiles for a given dose prescription is a central problem in radiotherapy. Care must be taken in the beam design to expose the tumor volume at a high level, to avoid significant irradiation of critical organs, and to minimize exposure of all other tissue. POCS, we will show, is a viable approach to beam design.

Note that the conformal radiotherapy problem is closely related to tomographic reconstruction from projections.[21] For tomography, an image is reconstructed from its projections. In conformal radiotherapy, the image of the tumor and surrounding tissue is given, and the beam projections must be constructed.

The POCS method for synthesizing pencil beam cross sections to produce desired dose prescriptions consists of the following steps.

1. Specify convex constraints the prescribed dose must satisfy. These constraints can originate from either dose prescription, physics or mathematical concerns.
2. Establish the projection onto each of these convex sets.
3. Alternately project among the constraint sets. Iterate until convergence is achieved. The result is a beam profile synthesis for the dose prescription.

Design will be described and examples given for all beams lying in a plane. Extension to the three dimensions is straightforward. Also, the physics of dose synthesis is a continuous problem. We will discretize it to allow application of digital analysis. Let $b(x, \theta)$, the beam intensity function, correspond to the intensity discretized at angle θ crossing the

---

21. See Sections 8.5 and 11.4.8.

axis perpendicular to θ at the beam element position x. Assume there are Q linear beam arrays at fixed equal angular intervals,

$$b_k(x) := b\left(x, \frac{2\pi k}{Q}\right); \quad 1 \leq k \leq Q.$$

The kth array is discretized into N pencil beam elements that can be expressed by the vector

$$\vec{b}_k = [\, b_{k1} \; b_{k2} \cdots b_{kN} \,]^T.$$

The same spatial interval between adjacent pencil beam elements is assumed. The dose domain is discretized into M pixels. The dose contribution to the point m from the nth beam element in the kth beam array (b_{kn}) is $(a_{mn})_k$. The corresponding dose computation matrix, $\mathbf{A} \in \mathbb{R}^{M \times N}$, is defined for M tissue points. The dose from the kth beam array is

$$\vec{d}_k = [\, d_{k1} \; d_{k2} \cdots d_{kM} \,]^T$$
$$= \mathbf{A}_k \vec{b}_k. \qquad (11.39)$$

The contributions from all Q beam arrays must be summed to give the total dose delivered to each pixel. Physically, the matrix $\mathbf{A}_k = (a_{mn})_k$ is the discretized kernel, $\mathbf{A}(\gamma, x, 2\pi k/Q)$ that specifies the dose to point γ from the pencil beams in the kth linear array. The vector, γ, is discretized into the M pixels. This geometry is illustrated in Figure 11.67.

For Q discretized beam arrays, there are Q dose vectors,

$$\left\{ \vec{d}_k \mid 1 \leq k \leq Q \right\}.$$

These Q vectors are stacked to form the parent dose vector

$$\vec{d} = \left[\, \vec{d}_1^T \; \vec{d}_2^T \cdots \vec{d}_Q^T \,\right] \qquad (11.40)$$

where $\vec{d} \in \mathbb{R}^{MN \times 1}$. This is the space in which the dose constraint sets, all convex, are defined.

The total dose vector at M tissue points, t, is the sum of dose vectors from every incident beam and can be computed from the parent dose vector in (11.40).

$$\vec{t} = [\, t_1 \; t_2 \cdots t_M \,]^T = \sum_{k=1}^{Q} \vec{d}_k = \sum_{k=1}^{Q} \mathbf{A}_k \vec{b}_k$$

11.4.11.1 Convex Constraint Sets

To synthesize the beam elements, the following constraint sets are used. Each is convex. The projection operation is given for each set.

1. **Beam dose constraint set.** Given the dose computation matrix, $\mathbf{A}_k$, and the beam vector, $\vec{b}_k$, for the kth beam vector, the resulting dose vector is given by (11.39). The dimension of the dose vector exceeds the number of beam elements. In other words, the dose vector has a larger number of degrees of freedom than the beam vector. The matrix, $\mathbf{A}_k$, is thus not full rank. We therefore use the pseudo inverse[22]

---

22. See Section 11.3.2.1.

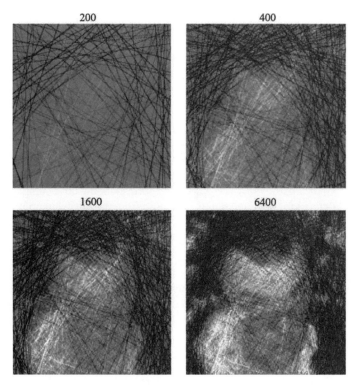

FIGURE 11.56. Illustration of convergence of reconstruction of an image from tomographic projections along a line. Each figure is labelled with the number of line projections. The original image is shown on the bottom right in Figure 11.57. (Continued in Figure 11.57.)

to find the $\vec{b}_k$'s. The corresponding convex set can be expressed as

$$C_B = \left\{ \vec{d} \mid \vec{d}_k = \mathbf{A}_k \vec{b}_k,\ 1 \leq k \leq Q \right\}.$$

The projection onto this convex set is

$$P_B \vec{d} = \left[\left[P_B \vec{d} \right]_1^T \left[P_B \vec{d} \right]_2^T \cdots \left[P_B \vec{d} \right]_Q^T \right]^T. \tag{11.41}$$

where the projection of the kth component is

$$[P_B \vec{d}\,]_k = \mathbf{T}_k \vec{d},$$

and $\mathbf{T}_k$ is the projection matrix

$$\mathbf{T}_k = \mathbf{A}_k \left(\mathbf{A}_k^T \mathbf{A}_k \right)^{-1} \mathbf{A}_k^T. \tag{11.42}$$

2. **Target dose constraint set.** This constraint set requires the delivered dose to match the prescribed dose in the target volume. Let T denote a subset of numbers from 1 to M corresponding to indices of the target volume. Let the prescribed dose vector be

$$(\vec{p})_k = \begin{cases} p_k\ ; k \in T \\ 0\ ; \text{otherwise}. \end{cases}$$

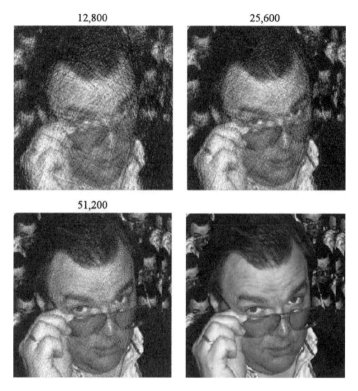

FIGURE 11.57. (Continued from Figure 11.56.)

These values are determined by sampling the continuous prescribed dose. In our simulations, each p_k is set to one. The target dose constraint set is then defined by

$$C_T = \left\{ \vec{d} \ \middle| \ \sum_{i=1}^{Q} d_{ij} = \begin{cases} p_j & ; j \in T \\ \text{don't care} & ; \text{otherwise.} \end{cases} \right\}.$$

The projection onto C_T is[23]

$$P_T \vec{d} = \left[\left[P_T \vec{d} \right]_1^T \left[P_T \vec{d} \right]_2^T \ldots \left[P_T \vec{d} \right]_Q^T \right]^T \quad (11.43)$$

where

$$[P_T \vec{d}\,]_k = \vec{d}_k + \frac{1}{Q} \mathbf{I}_T \left(\vec{p} - \sum_{i=1}^{Q} \vec{d}_i \right).$$

The diagonal matrix, $\mathbf{I}_T$, serves as a spatial discriminator and is given by

$$(\mathbf{I}_T)_{jj} = \begin{cases} 1 & ; j \in T \\ 0 & ; \text{otherwise.} \end{cases}$$

Thus only the projection components intersecting the target dose are affected

---

23. This is the projection onto constant area in Section 11.3.2.5.

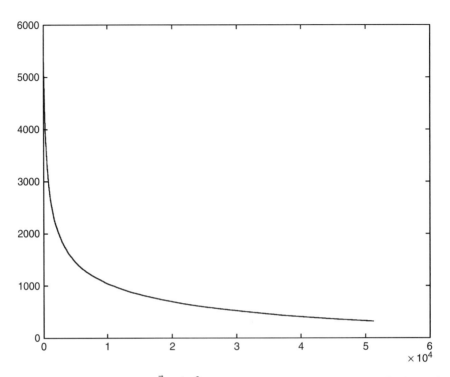

FIGURE 11.58. A plots of the error $\|\vec{f} - \vec{y}_p\|^2$ versus number of tomographic line projections, where $\vec{f}$ is the original image and $\vec{y}_p$ is the pth POCS iteration. As is the case with the associative memory, the convergence is linear.

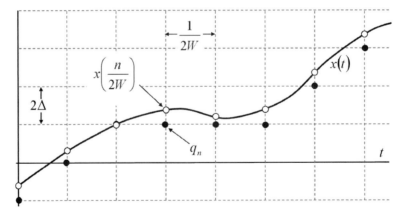

FIGURE 11.59. Illustration of quantization and quantization error for samples of an oversampled bandlimited signal.

SIGNAL AND IMAGE SYNTHESIS: ALTERNATING PROJECTIONS ONTO CONVEX SETS

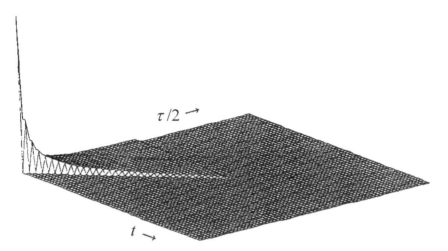

FIGURE 11.60. One quadrant of the symmetric cone kernel in the (t, τ) plane synthesized using POCS.
Source: [1043] S. Oh, R.J. Marks II and L.E. Atlas. "Kernel synthesis for generalized time-frequency distributions using the method of alternating projections onto convex sets." *IEEE Transactions on Signal Processing*, vol. 42, No. 7, July 1994. pp. 1653–1661.

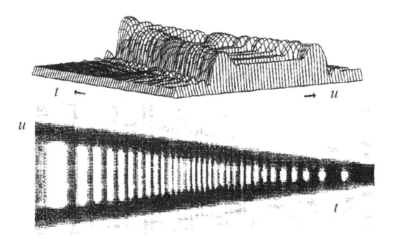

FIGURE 11.61. Waterfall and gray-level display of two linearly converging chirps using the POCS designed kernel in Figure 11.60. Compare with the spectrogram and Wigner distribution of the same signal in Figures 9.17 and 9.19.
Source: [1043] S. Oh, R.J. Marks II and L.E. Atlas. "Kernel synthesis for generalized time-frequency distributions using the method of alternating projections onto convex sets." *IEEE Transactions on Signal Processing*, vol. 42, No. 7, July 1994. pp. 1653–1661.

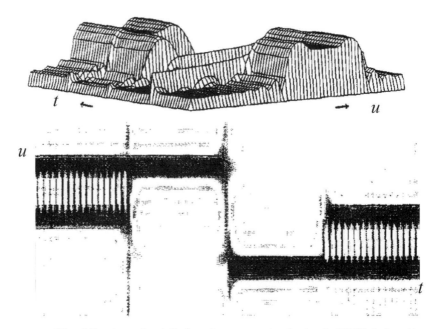

FIGURE 11.62. Waterfall and gray-level display of a two tone signal using the POCS designed kernel in Figure 11.60. Compare with the spectrogram and Wigner distribution of the same signal in Figures 9.18 and 9.20.
Source: [1043] S. Oh, R.J. Marks II and L.E. Atlas. "Kernel synthesis for generalized time-frequency distributions using the method of alternating projections onto convex sets." *IEEE Transactions on Signal Processing*, vol. 42, No. 7, July 1994. pp. 1653–1661.

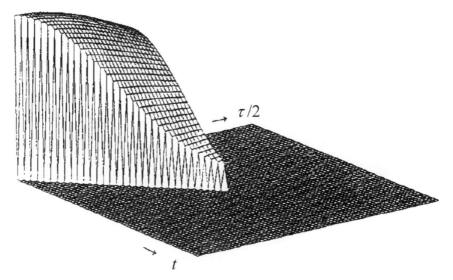

FIGURE 11.63. One quadrant of the symmetric cone kernel in the (t, τ) plane synthesized using POCS without use of the marginal constraints.
Source: [1043] S. Oh, R.J. Marks II and L.E. Atlas. "Kernel synthesis for generalized time-frequency distributions using the method of alternating projections onto convex sets." *IEEE Transactions on Signal Processing*, vol. 42, No. 7, July 1994. pp. 1653–1661.

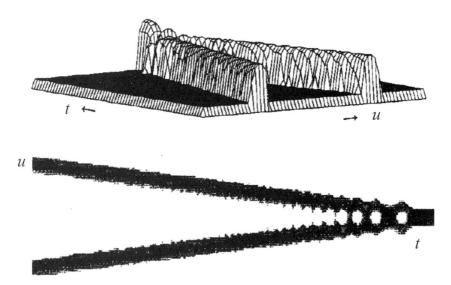

FIGURE 11.64. Waterfall and gray-level display of two linearly converging chirps using the POCS designed kernel in Figure 11.63 where marginals are not a required constraint. Compare with more constrained POCS kernel in Figure 11.61 and the spectrogram and Wigner distribution of the same signal in Figures 9.17 and 9.19.
Source: [1043] S. Oh, R.J. Marks II and L.E. Atlas. "Kernel synthesis for generalized time-frequency distributions using the method of alternating projections onto convex sets." *IEEE Transactions on Signal Processing*, vol. 42, No. 7, July 1994. pp. 1653–1661.

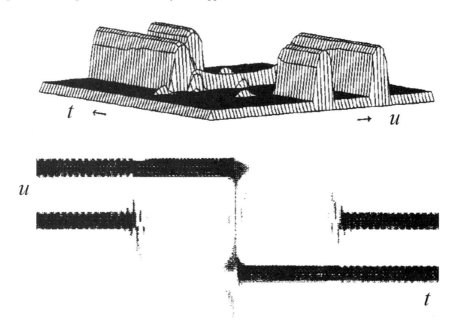

FIGURE 11.65. Waterfall and gray-level display of a two tone signal using the POCS designed kernel in Figure 11.63 where marginals are not a required constraint. Compare with more constrained POCS kernel in Figure 11.62 and the spectrogram and Wigner distribution of the same signal in Figures 9.18 and 9.20.
Source: [1043] S. Oh, R.J. Marks II and L.E. Atlas. "Kernel synthesis for generalized time-frequency distributions using the method of alternating projections onto convex sets." *IEEE Transactions on Signal Processing*, vol. 42, No. 7, July 1994. pp. 1653–1661.

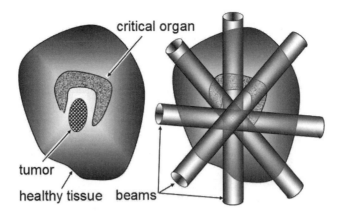

FIGURE 11.66. Illustration of radiotherapy for a tumor.

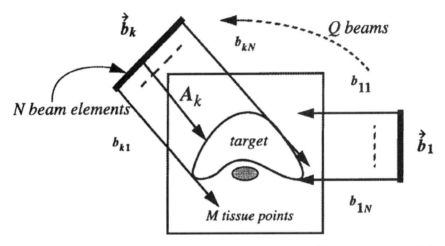

FIGURE 11.67. Geometry of the dose computation plane. There are Q beam positions, each beam containing N beam elements. The dose computation matrix corresponding to the kth beam is $\mathbf{A}_k$ whose dimension is $M \times N$, where M is the number of tissue points.
Source: [814] S. Lee, P.S. Cho, R.J. Marks II and S. Oh. "Conformal Radiotherapy Computation by the Method of Alternating Projection onto Convex Sets." *Phys. Med. Biol.*, vol. 42, July 1997, pp. 1065–1086.

3. **Organ dose constraint set.** This constraint set controls the dose in risk where the dose must be kept low. Let S denote a subset of numbers from 1 to M corresponding to the indices of the critical organ region. The organ dose constraint set is

$$C_o = \left\{ \vec{d} \ \middle| \ 0 \leq \sum_{i=1}^{Q} \sum_{j \in S} d_{ij} \leq E_1, \ \vec{d} \in C_B \right\}$$

where E_1 is the upper limit of allowable integrated dose in the critical region. To show that C_o is convex, let $\vec{f}, \vec{g} \in C_o$. Then, from the definition of the set, C_o,

$$0 \leq \alpha \sum_{i=1}^{Q} \vec{c}^T \vec{f}_i \leq \alpha E_1$$

and

$$0 \leq (1-\alpha) \sum_{i=1}^{Q} \vec{c}^T \vec{g}_i \leq (1-\alpha) E_1.$$

Adding gives

$$0 \leq \alpha \sum_{i=1}^{Q} \vec{c}^T \vec{f}_i + (1-\alpha) \sum_{i=1}^{Q} \vec{c}^T \vec{g}_i \leq \alpha E_1.$$

Since the convexity of C_B is already proved, the set, C_o is convex.
The projection operator for the set C_o is

$$P_o \vec{d} = \left[\left[P_o \vec{d} \right]_1^T \left[P_o \vec{d} \right]_2^T \cdots \left[P_o \vec{d} \right]_Q^T \right]^T \tag{11.44}$$

where

$$[P_o \vec{d}]_k = \begin{cases} \mathbf{T}_k \left(\vec{d}_k + \dfrac{E_1 - \sum_{j=1}^{Q} \vec{r}_j^T \mathbf{T}_j \vec{d}_j}{\sum_{j=1}^{Q} \vec{r}_j^T \mathbf{T}_j \vec{d}_j} \vec{r} \right) & ; \sum_{j=1}^{Q} \vec{r}_j^T \mathbf{T}_j \vec{d}_j > E_1 \\ \mathbf{T}_k \left(\vec{d}_k - \dfrac{\sum_{j=1}^{Q} \vec{r}_j^T \mathbf{T}_j \vec{d}_j}{\sum_{j=1}^{Q} \vec{r}_j^T \mathbf{T}_j \vec{d}_j} \vec{r} \right) & ; \sum_{j=1}^{Q} \vec{r}_j^T \mathbf{T}_j \vec{d}_j < E_1 \\ \mathbf{T}_k \vec{d}_k & ; \text{otherwise} \end{cases}$$

and

$$(\vec{r})_k = \begin{cases} 1 \, ; \, k \in S \\ 0 \, ; \, \text{otherwise} \end{cases}$$

4. **Non-negative beam constraint.** Physics dictates that each beam element have a non-negative value. This set requires adherence to this property. The corresponding convex non-negative beam constraint sets, one for each beam, are[24]

$$C_k = \{ \vec{d} \mid \vec{d}_k = \mathbf{A}_k \vec{b}_k, \, \vec{b}_k \geq 0 \}; \; 1 \leq k \leq Q.$$

The set C_k can be expressed as the intersection of a number of convex component sets.

$$C_k = \bigcap_{n=1}^{N} C_{kn},$$

where the component sets are

$$C_{kn} = \{ \vec{d} \mid \vec{d}_k = \mathbf{A} \vec{b}_k, \, (\vec{b}_k)_n \geq 0 \}; \; 1 \leq n \leq N.$$

---

24. The notation $\vec{b}_k > 0$ means each element of $\vec{b}_k$ is non-negative.

For a given $\vec{d}$, the sign of each $(\vec{b}_k)_n$ can be found from the pseudo inverse

$$\vec{b}_k = \left(\mathbf{A}_k^T \mathbf{A}_k\right)^{-1} \mathbf{A}_k^T \vec{d}_k.$$

The projection onto C_{kn} is

$$P_{kn}\vec{d} = \begin{cases} \mathbf{T}_{kn}\vec{d}_k \; ; & (\vec{b}_k)_n < 0 \\ \vec{d}_k & ; \text{ otherwise} \end{cases}$$

and the matrix $\mathbf{A}_{kn} \in \mathbb{R}^{M \times (N-1)}$ is formed by removing the nth column of $\mathbf{A}_k$. In lieu of projecting onto each C_k, projection is performed sequentially onto each C_{kn} set. These projection operators can be concatenated into the single operator

$$\Theta_+ = \prod_{k=1}^{Q} \prod_{n=1}^{N} P_{kn}.$$

Although Θ_+ is composed of a string of projection operators, it, itself is not a projection operator. It does not, for example, directly project onto the set

$$C_+ = \bigcap_{k=1}^{Q} C_k.$$

After a projection operator component of Θ_+ is performed, the constraint corresponding to the previous projection operator component may no longer be satisfied. Thus, after Θ_+ is applied, some of the beam elements can still be negative.

Using these convex constraint sets and corresponding projections, the dose can be synthesized using POCS. Let ℓ be the POCS iteration counter. Let $b_i[\ell]$ be the ℓth beam and $d_i[\ell]$ be the ith vector in $d[\ell]$ at the ℓth iteration. Then $d[\ell+1]$ is obtained by the recursion

$$d[\ell+1] = \Theta_+ P_B P_T P_o d[\ell].$$

POCS will also converge if some projections are used more than others [1324]. Lee *et al.* [814] found faster convergence using

$$d[\ell+1] = \Theta_+ (P_B P_T)^L P_o d[\ell].$$

where $(P_B P_T)^L$ denotes L repeated projections.

Once the iteration has converged, the beam vector, $b_i[\ell]$, can be uniquely determined using the minimum mean square error solution

$$b_i[\ell] = \left(\mathbf{A}_i^T \mathbf{A}_i\right)^{-1} \mathbf{A}_i^T d[\ell]; \; 1 \leq i \leq Q.$$

Residual negative beam weights due to constraint set non-intersection or early iteration truncation are set to zero. Lee *et al.* [814] found that if there were any negative beam weights, they were relatively small in magnitude.

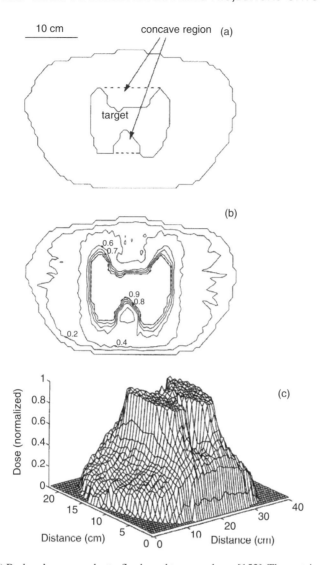

FIGURE 11.68. (a) Brahme's concave butterfly shaped tumor volume [152]. The matrix size of the dose distribution plane is 35 × 63. (b) and (c) Results of POCS dose synthesis applied to Brahmes butterfly shaped tumor after 15 iterations. 31 beams, each containing 55 beam elements were used. (b) The isodose contours in full curves and the tumor contour in broken curves. (c) A 3D plot of the relative dose versus the transaxial coordinates.

Source: [814] S. Lee, P.S. Cho, R.J. Marks II and S. Oh. "Conformal Radiotherapy Computation by the Method of Alternating Projection onto Convex Sets." *Phys. Med. Biol.*, vol. 42, July 1997, pp. 1065–1086.

11.4.11.2 Example Applications

Here are three example simulations of POCS applied to the conformal therapy problem.

1. The first example, a target with concave regions, is the Brahme's butterfly shaped tumor volume [152]. This target shape, illustrated in Figure 11.68(a), presents a potential challenge for treatment planning. The difficulty arises because of the concave normal tissue regions above and below the tumor.

Without intensity modulation, uniform delivery of the prescribed dose to the target can be achieved only at the expense of undesirable escalation of dose to the concave regions. The maximum width and height of the external contour are 40 cm and 22 cm, respectively, and corresponds to 63 by 35 tissue elements. Thirty-one spaced beams spaced angularly uniformly, each containing 55 elements, are used. The results of inverse planning after 15 iterations using the POCS algorithm are shown in Figures 11.68(b) and (c). Sharp dose gradients around the target border are observed. The organ dose constraint specified in terms of integrated limiting dose, E_1, to the concave regions was set to 50% of the maximum value. The intensity profiles of the incident beams after 15 iterations is shown in Figure 11.69.

2. The second example is adopted from Bortfeld *et al.* [140]. As shown in Figure 11.70(a), the irradiation volume consists of a horseshoe-shaped target and a dose limiting organ within the concave region. The computation results after 15 iterations using 31 beams with 55 elements each and a 20% organ dose constraint are shown in Figures 11.70(b) and 11.70(c). The dose falls off sharply outside the target and dips further near the critical organ. The corresponding beam intensity modulation profiles are shown in Figure 11.71.

3. Our final example contains two organs that are large relative to the target, as shown in Figure 11.72(a). The external contour width, height, dose matrix size, and the number of beam elements are the same as those used in the previous example. The results after 30 iterations using 31 beams with a 30% organ dose constraint are shown in Figures 11.72(b) and 11.72(c). The corresponding synthesized beam profiles are shown in Figure 11.73. The complex target organ geometry demonstrates the difficulty in achieving acceptable treatment planning. Although we have applied a 30% value to both organ A and B, these values can be varied independently according to the clinical requirement.

11.5 Generalizations

11.5.1 Iteration Relaxation

In certain instances, POCSc can converge painfully slowly. An example is shown in Figure 11.74. One technique to accelerate convergence is relaxing the projection operation by using a relaxed projection with parameter λ [1550].

$$P_{\text{relaxed}} = \lambda P + (I - \lambda)I. \qquad (11.45)$$

11.5.2 Contractive and Nonexpansive Operators

An operator, $\mathcal{O}$, is said to be *contractive* if, for all $\vec{w} \neq \vec{z}$,

$$\|\mathcal{O}\vec{w} - \mathcal{O}\vec{z}\| < \|\vec{w} - \vec{z}\|. \qquad (11.46)$$

In other words, operating on the two vectors place them closer together. This is illustrated in Figure 11.75. A useful property of contractive operators [1020] is, for any initialization, the iteration

$$\vec{o}^{(N+1)} = \mathcal{O}\vec{o}^{(N)} \qquad (11.47)$$

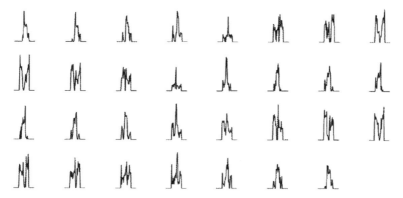

FIGURE 11.69. Beam intensity profiles synthesized for Brahmes geometry. The upper left profile corresponds to the beam which is incident horizontally from the right side of Figure 11.68(a). The gantry rotates in a counterclockwise direction. Each beam consists of 55 elements spanning the width of the external contour.
Source: [814] S. Lee, P.S. Cho, R.J. Marks II and S. Oh. "Conformal Radiotherapy Computation by the Method of Alternating Projection onto Convex Sets." *Phys. Med. Biol.*, vol. 42, July 1997, pp. 1065–1086.

converges to a unique fixed point

$$\vec{o}^{(\infty)} = \mathcal{O}\vec{o}^{(\infty)}.$$

A POCS projection is, however, not contractive.[25] Projection operators, rather, are *nonexpansive*. The $\mathcal{O}$ operator is nonexpansive if equality is allowed

$$\|\mathcal{O}\vec{w} - \mathcal{O}\vec{z}\| \leq \|\vec{w} - \vec{z}\|.$$

For nonexpansive operators, the iteration in (11.47) can converge to a number of fixed points.

A relaxed nonexpansive operator, as in (11.45), however, is contractive [509]. Applications of contractive operators do not have the elegant geometrical interpretation of POCS.

11.5.2.1 Contractive and Nonexpansive Functions

A function, $f(t)$, is contractive if

$$\left|\frac{df(t)}{dt}\right| < 1. \qquad (11.48)$$

and a nonexpansive if

$$\left|\frac{df(t)}{dt}\right| \leq 1. \qquad (11.49)$$

---

25. If, for example, both $\vec{w}$ and $\vec{z}$ are in the convex set, and $\mathcal{O}$ is a projection operator, then $\|\mathcal{O}\vec{w} - \mathcal{O}\vec{z}\| = \|\vec{w} - \vec{z}\|$ and (11.46) is violated.

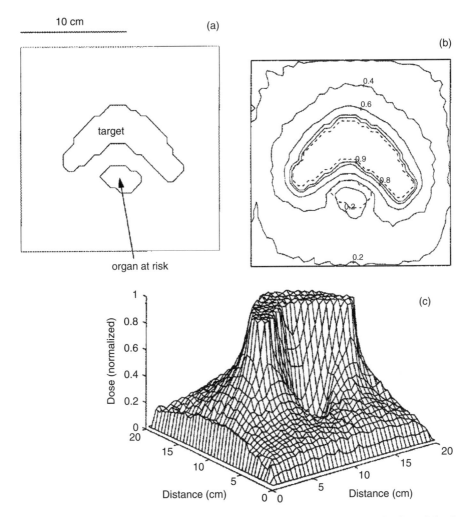

FIGURE 11.70. (a) Bortfelds horseshoe-shaped target with an organ at risk. Matrix size of the dose distribution plane is 39 × 39. (b) and (c) Results of the POCS dose synthesis applied to Bortfelds geometry after 15 iterations. Thirty one beams, each containing 55 beam elements were used. The value of 20% dose constraint was imposed on the critical structure. (b) The isodose contours in full curve and the tumor-organ contours in broken curve. (c) A 3D plot of the relative dose versus the transaxial coordinates.
Source: [814] S. Lee, P.S. Cho, R.J. Marks II and S. Oh. "Conformal Radiotherapy Computation by the Method of Alternating Projection onto Convex Sets." *Phys. Med. Biol.*, vol. 42, July 1997, pp. 1065–1086.

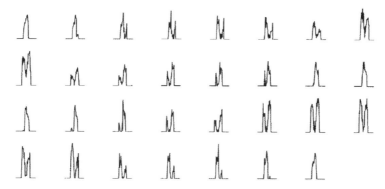

FIGURE 11.71. Beam intensity profiles obtained for Bortfelds geometry. The upper left profile corresponds to the beam which is incident horizontally from the right side of Figure 11.70(a). The gantry rotates in a counterclockwise direction. Each beam consists of 55 elements spanning the width of the square external contour.
Source: [814] S. Lee, P.S. Cho, R.J. Marks II and S. Oh. "Conformal Radiotherapy Computation by the Method of Alternating Projection onto Convex Sets." *Phys. Med. Biol.*, vol. 42, July 1997, pp. 1065–1086.

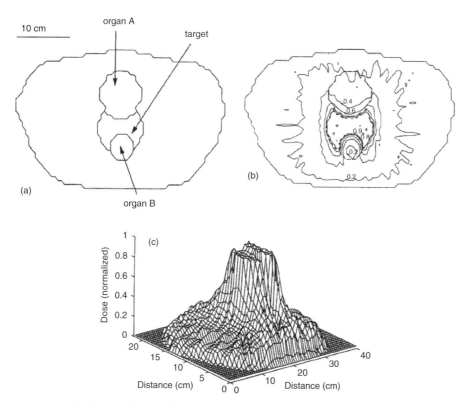

FIGURE 11.72. (a) A tumor that is adjacent to two relatively large organs at risk. The matrix size of the dose distribution plane is 35×63. (b) and (c) Results of the POCS dose synthesis applied to the geometry described in (a). Thirty one beams and 30 iterations were used. (b) The isodose contours in full curve and the tumor-organ contours in broken curve. (c) A 3D plot of the relative dose versus the transaxial coordinates.
Source: [814] S. Lee, P.S. Cho, R.J. Marks II and S. Oh. "Conformal Radiotherapy Computation by the Method of Alternating Projection onto Convex Sets." *Phys. Med. Biol.*, vol. 42, July 1997, pp. 1065–1086.

FIGURE 11.73. Beam intensity profiles obtained after 30 iterations for the geometry described in Figure 11.72(a). The upper left profile corresponds to the beam which is incident horizontally from the right side of the Figure 11.72 (a). The gantry rotates in a counterclockwise direction. Each beam consists of 55 elements spanning the width of the external contour.
Source: [814] S. Lee, P.S. Cho, R.J. Marks II and S. Oh. "Conformal Radiotherapy Computation by the Method of Alternating Projection onto Convex Sets." *Phys. Med. Biol.*, vol. 42, July 1997, pp. 1065–1086.

FIGURE 11.74. A geometrical example of slowly converging POCS. The two linear varities intersection, far to the right, is the ultimate fixed point of the iteration.

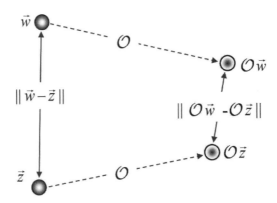

FIGURE 11.75. An operator, $\mathcal{O}$, is contractive if, after application to two vectors, the points are closer together than originally. (See (11.46).)

Proof. For the contractive case, let $\tau > t$. Then

$$|f(\tau) - f(t)| = \left| \int_\tau^t \frac{df(\xi)}{d\xi} d\xi \right| \leq \int_\tau^t \left| \frac{df(\xi)}{d\xi} \right| d\xi.$$

Using (11.48),

$$|f(\tau) - f(t)| < \int_\tau^t d\xi = \tau - t = |\tau - t|$$

A nearly identical proof follows for the nonexpansive case in (11.49).
If a function $f(t)$ is contractive, the iteration

$$t_{n+1} = f(t_n) \qquad (11.50)$$

converges to a unique point, t_∞, for all initializations.

An example is $f(t) = e^{-t}$ for $t > 0$. The iteration in (11.50) converges to $t_\infty = 0.56714329$ which is the solution to the transcendental equation

$$t_\infty = e^{-t_\infty}. \qquad (11.51)$$

On a calculator, repeated alternate punching of the e^x key and the CHS (change sign) key will thus always converge to $t_\infty = 0.56714329$.

11.6 Exercises

11.1. Consider the following interesting iteration. Choose any number, spell it and count the letters. The count is the next number. The process is repeated. The iteration will always end at the number four.[26] For example, 100 is spelled "one hundred" and contains 10 letters. "Ten" has three letters. "Three" has five letters. "Five" has four letters. "Four" has four letters and the iteration has thus converged to a fixed point. Is this iteration contractive?

11.2. For the POCS solution of simultaneous linear equations in Section 11.4.2.
   (a) Explain the convergence of POCS when the planes do not intersect at a single point. How does the initialization effect the fixed point of POCS in such situations?
   (b) Use the POCS iteration to invert a large matrix. Plot the error.
   (c) Should we expect the convergence of the iteration to be linear?

11.3. Show that the subspace defined in (11.8) and the linear variety in (11.9) are convex sets as defined in (11.7).

11.4. Are the following set of signals convex?
   (a) The set of signals that are monotonically increasing on a given closed interval, $\mathcal{I}$.
   (b) The set of even signals.
   (c) The set of odd signals.

---

[26]. This has never been proved but there is, as of yet, no counter example using concise descriptions of numbers using common English words.

(d) The set of continuous signals, $x(t)$, where, for a fixed n and ϵ, the signal obeys

$$\left|\left(\frac{d}{dt}\right)^n x(t)\right| \leq \epsilon.$$

(e) The set of discrete time signals, $x[n]$, where, for a fixed ϵ, $|x[n+1] - x[n]| \leq \epsilon$.
(f) The set of discrete time signals, $x[n]$, where, on a fixed interval, $\mathcal{I}$, and for a given function, $g[n]$, and constant ρ,

$$\sum_{n \in \mathcal{I}} g[n]\, x[n] = \rho.$$

(g) The set of complex functions where, for a fixed $|x|$, $x = |x|e^{j\angle x}$, where $-\frac{\pi}{2} \leq \angle x \leq \frac{\pi}{2}$.
(h) The set of real signals for which, for a fixed a and b,

$$\rho_1 \leq \int_a^b x(t)\,dt \leq \rho_2.$$

(i) The set of conjugately symmetric signals, $x(t) = x^*(-t)$.

11.5. Establish the projections for those signal sets in Exercise 11.4 that are convex.

11.6. Consider measurements from an experiment represented as points on the continuous x axis. As illustrated in Figure 11.76, our POCS problem is to estimate the probability density function, $f_X(x)$, from which these measurements were made [1333]. Two convex sets for the density function are positivity and unit area. In addition, the area of the density over any randomly chosen interval, T, must be equal to the percentage of points in the interval. For a given T, the set of all functions whose area is equal to the percentage of points in the interval is a convex set.

(a) If we repeatedly choose random intervals and project onto the convex set corresponding to the interval then, while also projecting on unit area functions and nonnegative functions, POCS will converge to the unique function obeying all of the constraints. What is that function?

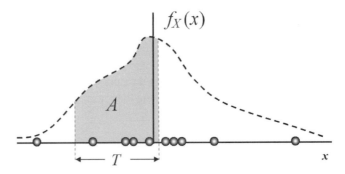

FIGURE 11.76. An unknown probability density function, $f_X(x)$, generates outcomes shown by the dots on the x axis. Given these realizations, our task is estimation of the probability density function. Over any arbitrary interval, T, the area of the density should be equal to the percentage of points in the interval. In this figure, there are 10 points. The interval T contains 4 points. Thus the area of the density function over T should be 0.4. The set of all functions with an area equal to 0.4 on T is a convex set.

(b) The result in (a) is unacceptable. Identify one or more convex smoothness constraints that remedy this problem.
(c) Describe how your new convex constraint set or sets in (b) intersect with the old ones.
(d) Show, by computer emulation, the result of your additional constraint(s).

11.7. §**Bertrand's Paradox.** Choosing random lines in Section 11.4.8 for tomographic reconstruction using POCS requires definition of the term *random*. (See Footnote 19 in Section 11.4.8.) Consider the circle in the NW corner of Figure 11.77. Inscribed is an equilateral triangle. Inscribed in the triangle is a smaller shaded circle. If the larger circle has radius r, then each side of the triangle has a length of $\sqrt{3}\,r$ and the diameter of the small shaded circle is r. Our task is to choose a line to randomly intersect the large circle and find the probability of the cord inside the large circle has a length not less than $\sqrt{3}\,r$. Let the length of the cord be ℓ. We then seek to evaluate

$$p = \Pr\left[\ell \geq \sqrt{3}\,r\right].$$

Paradoxically, we will now show there are at least three solutions and

$$\frac{1}{2} = p = \frac{1}{3} = p = \frac{1}{4}.$$

Here are the three solutions.

1. **Northeast Solution.** With no loss of generality, we can, after the random cord is chosen, rotate the circle to view the cord horizontally. Consider, then,

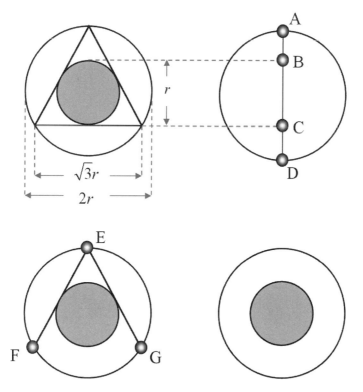

FIGURE 11.77. Illustration of the Bertrand paradox.

the circle shown in the NE corner of Figure 11.77. The cord, once rotated, will perpendicularly intersect $\overline{AD}$. The probability the cord exceeds $\sqrt{3}\,r$ is equal to the probability the cord has a midpoint on the line segment joining points B and C. Thus

$$p = \frac{\text{length of } \overline{BC}}{\text{length of } \overline{AD}} = \frac{1}{2}.$$

2. **Southwest Solution.** Let, as shown in the SW corner of Figure 11.77, the cord enters the circle at point E. The probability of the cord exceeds $\sqrt{3}\,r$ is then equal to the probability the other end of the cord exits the circle in the arc $\overline{FG}$. This is given by the arc length of $\overline{FG}$ divided by the circle's circumference. Then

$$p = \frac{\text{arc length of } \overline{FG}}{2\pi r} = \frac{1}{3}.$$

3. **Southeast Solution.** Lastly, we note, in order for the cord length to exceed $\sqrt{3}\,r$, the midpoint of the randomly chosen cord must lie within the shaded circle. This is illustrated in the SE corner of Figure Figure 11.77. Thus

$$p = \frac{\text{area of small shaded circle}}{\text{area of the large circle}} = \frac{\pi \left(\frac{r}{2}\right)^2}{\pi r^2} = \frac{1}{4}.$$

This analysis giving three different solutions for the same problem is clearly flawed.
(a) Resolve the Bertrand paradox.
(b) In Section 11.4.8, the POCS iteration will converge for many procedures for randomly choosing projection lines. For POCS to work, we need only to assure each pixel is visited sufficiently in the iteration process, but the frequency of visitation need not be the same for all pixels [1551]. Some algorithms for choosing random lines, however, will converge faster in some portions of the image than in others. What is a methodology for choosing random lines to assure the same statistical rate of convergence for each pixel?

11.8. **The Gerchberg-Saxton Algorithm.** A complex valued image, $x(t_1, t_2)$, and its Fourier transform, $X(u_1, u_2)$, are known in magnitude only. We can iteratively transform back and forth and, at each stage, retain the phase only and impose the magnitude. Thus, in the nth iteration we Fourier transform x_n to X_n and form

$$\hat{X}_n = |X|\,\exp(j\angle X_n)$$

where $|X|$ is the known transform magnitude. We then inverse transform $\hat{X}_n$ to x_{n+1} and form the image $\hat{x}_{n+1} = |x|\exp(j\angle x_{n+1})$ where $|x|$ is the known magnitude of the image. The iteration is repeated until desired convergence. This procedure is known as the Gerchberg-Saxton algorithm [275, 489, 501, 502, 587, 642, 1201, 1278, 1350, 1487, 1514, 1531, 1570].
(a) Is this an implementation of POCS?
(b) Choose two images and use the Gerchberg-Saxton algorithm to construct phases so that the two images are nearly Fourier transform pairs.

11.9. Let $f(t)$ be a contractive function and $g(t)$ be a nonexpansive function.
   (a) Show that both $f(g(t))$ and $g(f(t))$ are contractive functions.
   (b) Let $f(t) = e^{-t}$ and $g(t) = 2\cos^2(t)$. Find the solutions to $t_\infty = f(g(t_\infty))$ and $t_\infty = g(f(t_\infty))$.
   (c) Solve the transcendental equation

$$t_\infty = e^{-e^{-e^{-e^{-(3+2\cos^2(t_\infty))}}}}$$

11.10. Evaluate
   (a) The limit of $x(t) = e^{-e^{-e^{\cdots e^{-t}}}}$.
   (b) The limit of $x(t) = \cos(\cos(\cos(\cdots \cos(t))))$.

11.11. The unique solution to $\tau = f(t) = e^{-t}$ is the same as the solution to $t = f^{-1}(\tau) = -\ln t$. Application of $t_n = f(t_n)$ converges to t_∞ that satisfies (11.51).
   (a) Does $\tau_n = f^{-1}(\tau_n)$ converge to the unique solution $\tau_\infty = f^{-1}(\tau_\infty) = 0.56714329$?
   (b) Can we generalize? If $\tau = f(t)$ is contractive, when is $t = f^{-1}(\tau)$ contractive?

11.7 Solutions for Selected Chapter 11 Exercises

11.1. Let the operator for the iteration be Θ. Thus, for example, $\Theta(10) = 3$ and $\Theta(3) = 5$. A single counterexample shows this iteration is not contractive. Since

$$\|\Theta(12) - \Theta(2)\| = |6 - 3| = 3 \not< |12 - 2| = 10,$$

the iteration is not contractive.

11.4. All of these sets are convex except for (g) which is not.
   (b) The projection of $y(t)$ onto the set of even signals is

$$P_E\, y(t) = \frac{1}{2}(x(t) + x(-t)).$$

   (c) The projection of $y(t)$ onto the set of odd signals is

$$P_O\, y(t) = \frac{1}{2}(x(t) - x(-t)).$$

   (h) This set is an example of a *convex slab* equal to the region between two parallel planes. Let

$$\rho_y = \int_a^b y(t)\,dt.$$

The projection of real $y(t)$ onto the convex set on the interval $a \le t \le b$ is

$$P_{12}y(t) = \begin{cases} y(t) & ;\ \rho_1 \le \rho_y \le \rho_2 \\ y(t) + \frac{1}{b-a}(\rho_1 - \rho_y) & ;\ \rho_y < \rho_1 \\ y(t) + \frac{1}{b-a}(\rho_2 - \rho_y) & ;\ \rho_y > \rho_2. \end{cases}$$

11.6. For a more detailed treatment, including comparison to other methods to construct probability density functions from data, see Stark *et al.* [1333].
   (a) The probability function will be equally weighted probability masses (Dirac deltas) at all of the point locations.
11.8. **The Gerchberg-Saxton Algorithm.** See Figure 11.78.
   (a) This is not an implementation of POCS. The set of all images with a fixed magnitude, $|x|$, is not convex. Neither is the set of images whose Fourier transforms have magnitude $|X|$.
   (b) An example of the Gerchberg-Saxton algorithm is shown in Figure 11.78 for a 512×512 image *Sad Man*. The Gerchberg-Saxton is to synthesize phases in the

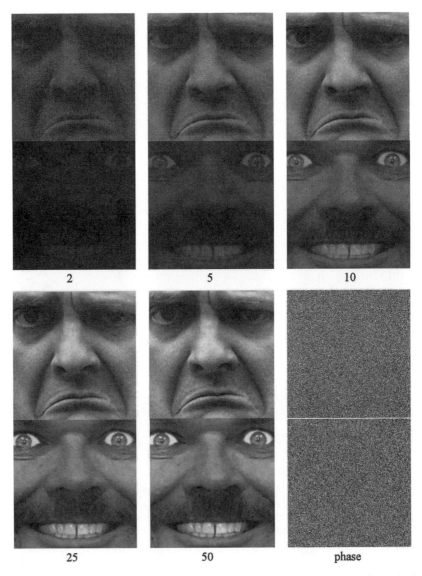

FIGURE 11.78. Example implementation of the Gerchberg-Saxton algorithm. Phase is synthesized in both the spatial and Fourier domains so that an image with magnitude of *Sad Man* has a Fourier transform with magnitude *Glad Man*. See Problem 11.8.

spatial and Fourier domains such that the Fourier transform of *Sad Man* is *Glad Man*. The synthesis is shown for various iterations. The result for 50 iterations is visually indistinguishable from the original images. The synthesized phase in both domains is shown at the bottom left. The maximum pixel value in each image is set to one for display purposes. For the phase plots, white corresponds to $-\pi$ and black to π.

11.9. (a) Apply the chain rule of differentiation. For $f(g(t))$,

$$\left|\frac{d}{dt}f(g(t))\right| = \left|\frac{dg(t)}{dt}f'(g(t))\right|$$
$$= \left|\frac{dg(t)}{dt}\right| \left|f'(g(t))\right| < 1.$$

(b) Applying the iteration in (11.50), we numerically converge to

$$t_\infty = f(g(t_\infty)) \to t_\infty = 0.1408$$

and

$$t_\infty = g(f(t_\infty)) \to t_\infty = 1.9606.$$

(c) The function is contractive. Applying the iteration in (11.50), we numerically converge to

$$t_\infty = e^{-e^{-e^{-e^{-(3+2\cos^2(t_\infty))}}}} \to t_\infty = 0.6884.$$

11.10. Both of these are contractive operators. The limit therefore can be numerically found through repeated iteration.
(a) The limit $e^{-e^{\cdots e^{-t}}} = 0.5671433$ for all t. Note that $\xi = 0.5671433$ is the solution for $\xi = \exp(-\xi)$.
(b) The limit $\cos(\cos(\cos(\cdots\cos(t)))) = 0.738513$ for all t. Likewise, $\xi = 0.738513$ is the solution to $\xi = \cos(\xi)$.

11.11. The inverse of a contractive function is not contractive. Set $\tau = f(t)$ and $t = f^{-1}(\tau)$.
(a) The iteration $\tau_{n+1} = -\ln(\tau_n)$ does not converge.
(b) Since

$$\frac{d\tau}{dt} = \frac{1}{\frac{dt}{d\tau}},$$

we see that, if

$$\left|\frac{d\tau}{dt}\right| < 1,$$

then

$$\left|\frac{dt}{d\tau}\right| > 1.$$

The inverse of a contractive function can therefore never be contractive.

12

Mathematical Morphology and Fourier Analysis on Time Scales

> God made the integers, all else is the work of man.
> Leopold Kronecker [87]

> The great advances in science usually result from new tools rather than from new doctrines.
> Freeman Dyson

> Reason's last step is the recognition that there are an infinite number of things which are beyond it.
> Blaise Pascal, (1623–1662) [1098]

12.1 Introduction

Mathematical morphology, used extensively in image processing, tracks the support domains for the operation of convolution and deconvolution. Morphology is also important in the modelling of signals on *time scales*. Time scale theory provides a generalization tent under which the operations of discrete and continuous time signal and Fourier analysis rest as special cases. The time scale paradigm provides modelling under which a rich class of hybrid signals and systems can be analyzed.

12.2 Mathematical Morphology Fundamentals

We begin with introductory material on mathematical morphology which is foundational to the development of time scale theory.

The support of convolution is related to the operation of dilation in mathematical morphology. Mathematical morphology is most commonly associated with image processing. Applications of morphology was initially applied to binary black and white images by Matheron [966]. The field is richly developed [506, 578]. Here, we outline the fundamentals.

12.2.1 Minkowski Arithmetic

In N dimensions, let $\mathbb{X}$ and $\mathbb{H}$ denote a set of vectors or, in the degenerate case of one dimension, a set of real numbers.

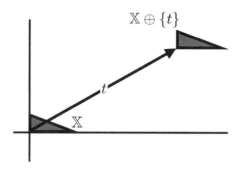

FIGURE 12.1. Illustration of set translation achieved by Minkowski addition, or dilation, of the triangular set, $\mathbb{T}$, with a single point, t.

Definition 12.2.1. Minkowski Addition and Dilation. The *Minkowski addition*, a.k.a. dilation, of $\mathbb{X}$ and $\mathbb{H}$, denoted $\mathbb{X} \oplus \mathbb{H}$, is

$$\mathbb{X} \oplus \mathbb{H} := \{ t \mid t = \xi + \tau \; \forall \, \xi \in \mathbb{H}, \; \tau \in \mathbb{X} \}.$$

Thus, the Minkowski addition of two sets is the set of the pairwise sums of all of the vectors in the two sets. Let

$$\mathbb{Y} = \mathbb{X} \oplus \mathbb{H}.$$

Example 12.2.1. If $\mathbb{H}$ consists of a single point, $\mathbb{H} = \{t\}$, then, as shown in Figure 12.1, Minkowski addition is equivalent to a translation of the set. Each vector in $\mathbb{X}$ has t added to it.

Example 12.2.2. If $\mathbb{X} = \{0, 2, 6\}$ and $\mathbb{H} = \{2, 4, 6\}$, then $\mathbb{Y} = \{2, 4, 6, 8, 10, 12\}$. If $\mathbb{X} = \{1, [2, 4], [10, 11]\}$ and $\mathbb{H} = \{[1, 2]\}$, then $\mathbb{Y} = \{[2, 6], [11, 13]\}$.

Minkowski addition is also called *dilation*. To see why, consider the two dimensional Minkowski addition example shown in Figure 12.2. Two sets of vectors, $\mathbb{X}$ and $\mathbb{Y}$, are shown in the left hand figure. To perform the Minkowski sum, we wish to add all of the vectors in the two sets. The boundary of the Minkowski addition set, as shown in the middle of Figure 12.2, is determined by the translation of the set $\mathbb{H}$ around $\mathbb{X}$. The result, shown on the

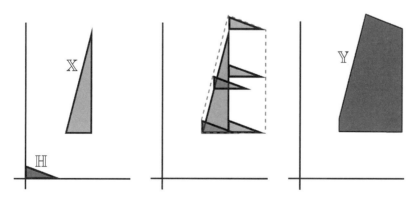

FIGURE 12.2. Illustration of Minkowski addition, also called dilation.

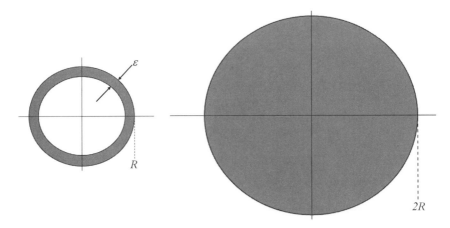

FIGURE 12.3. The dilation of the annulus on the left with itself gives the circle shown on the right.

right hand side, is the Minkowski sum of the two sets. The reason for dubbing this operation *dilation* is now apparent. The set $\mathbb{X}$ is enlarged, or *dilated*, by the operation of Minkowski addition with the set $\mathbb{H}$.

Example 12.2.3. A centered annulus of radius R and thickness ϵ is pictured on the left in Figure 12.3. The dilation of this annulus with itself is the circle of radius $2R$ shown on the right in Figure 12.3.

12.2.2 Relation of Convolution Support to the Operation of Dilation

Dilation dictates functional support in convolution, i.e., the support of the convolution of two signals is the dilation of the supports.

Definition 12.2.2. The *support* of a signal, $x(\vec{t})$, is the region in the signal's domain where the signal is not identically zero.

Let $x(\vec{t})$ be identically zero outside a region $\mathbb{X}$ and $h(\vec{t})$ be zero outside a region $\mathbb{H}$. Thus, $\mathbb{X}$ and $\mathbb{H}$ denote the support of $x(\vec{t})$ and $h(\vec{t})$. Then the convolution

$$y(\vec{t}) = x(\vec{t}) * h(\vec{t}) \tag{12.1}$$

is zero outside of the set

$$\mathbb{Y} = \mathbb{X} \oplus \mathbb{H}. \tag{12.2}$$

Example 12.2.4. If $x(t)$ is zero outside of the interval $\mathbb{X} = \{[1, 5], [11, 12]\}$ and $h(t)$ is identically zero outside the interval $\mathbb{H} = \{[5, 6]\}$, then $y(t) = x(t) * h(t)$ will be zero outside of the interval

$$\mathbb{Y} = \mathbb{X} \oplus \mathbb{H} = \{[6, 11], [16, 18]\}.$$

Example 12.2.5. Let

$$x(t) = \Pi(t) + \delta(t - 8).$$

Thus, the support of $x(t)$ is $\mathbb{X} = \{[-\frac{1}{2}, \frac{1}{2}], 8\}$. If

$$h(t) = \Pi(t-5) + \delta(t-10),$$

then the corresponding support is $\mathbb{H} = \{[4.5, 5.5], 10\}$. Thus

$$y(t) = x(t) * h(t) = \Lambda(t-5) + \Pi(t-10) + \Pi(t-13) + \delta(t-18)$$

has support $\mathbb{Y} = \{[4, 6], [9.5, 10.5], [12.5, 13.5], 18\}$. As promised, the support of the convolution is the dilation of the supports.

$$\mathbb{Y} = \mathbb{X} \oplus \mathbb{H}.$$

Equation 12.2 does not say the support of the convolution in (12.1) is $\mathbb{Y}$. As is illustrated in the following example, it is, rather, subsumed in $\mathbb{Y}$.

Example 12.2.6. With reference to Exercise 5.9 and its solution, consider the signal $x(t)$ with a Fourier transform

$$x(t) \longleftrightarrow X(u) = |\text{sinc}(u)|.$$

The signal $x(t)$ is identically zero nowhere on the interval $\mathbb{X} = \{t | -\infty < t < \infty\}$. Yet, with $h = x$ (and therefore $\mathbb{H} = \mathbb{X}$), we conclude

$$x(t) * h(t) \longleftrightarrow \text{sinc}^2(u)$$

so that

$$x(t) * h(t) = \Lambda(t)$$

which is zero outside of the interval $(-1 < t < 1)$. This interval is of course subsumed within but not equal to

$$\mathbb{Y} = \mathbb{X} \oplus \mathbb{H} = \{t | -\infty < t < \infty\}.$$

12.2.3 Other Morphological Operations

Definition 12.2.3. The negative of a set of vectors, $\hat{\mathbb{H}}$, is defined as

$$\hat{\mathbb{H}} = \{k \mid k = -h \; \forall \, h \in \mathbb{H}\}.$$

An example of a set and its negative is pictured in Figure 12.4.

Definition 12.2.4. The *Minkowski subtraction* of the set $\mathbb{H}$ from the set $\mathbb{Y}$, denoted $\mathbb{Y} \ominus \mathbb{H}$, is

$$\mathbb{Y} \ominus \mathbb{H} := \{t \mid \hat{\mathbb{H}} + t \subset \mathbb{Y}\}.$$

Definition 12.2.5. The *erosion* of the set $\mathbb{Y}$ by the set $\mathbb{H}$, is $\mathbb{Y} \ominus \hat{\mathbb{H}}$

Since $\hat{\hat{\mathbb{H}}} = \mathbb{H}$, we see from the previous two definitions the erosion of $\mathbb{Y}$ by $\mathbb{H}$ is

$$\mathbb{Y} \ominus \hat{\mathbb{H}} = \{t \mid \mathbb{H} + t \subset \mathbb{Y}\}.$$

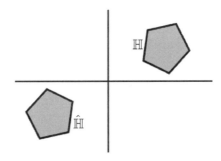

FIGURE 12.4. Example of a set of vectors, $\mathbb{H}$, and it's negative, $\hat{\mathbb{H}}$.

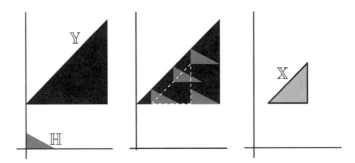

FIGURE 12.5. Illustration of the erosion of $\mathbb{X} = \mathbb{Y} \ominus \hat{\mathbb{H}}$.

Example 12.2.7. If $\mathbb{Y} = \{1, 3, 4, 6\}$ and $\mathbb{H} = \{0, 2\}$, the erosion is $\mathbb{Y} \ominus \hat{\mathbb{H}} = \{1, 4\}$.

Example 12.2.8. If $\mathbb{Y} = \{[1, 4], 6\}$ and $\mathbb{H} = \{[0, 1], 2\}$, the erosion is $\mathbb{Y} \ominus \hat{\mathbb{H}} = \{[1, 2], 3\}$.

Example 12.2.9. A two dimensional erosion, $\mathbb{X} = \mathbb{Y} \ominus \hat{\mathbb{H}}$, is illustrated in Figure 12.5. The two sets, $\mathbb{Y}$ and $\mathbb{H}$, are shown in the left hand figure. The set of all translations of $\mathbb{H}$ totally subsumed in $\mathbb{Y}$ is illustrated in the middle figure. The set of all translations by t that achieve this is outlined by the dashed white triangle in the middle figure. This triangle is drawn again on the right hand figure, and represents the erosion of $\mathbb{Y}$ by $\mathbb{H}$.

The reason for the terminology *erosion* is now evident. The large triangle has been eroded to the smaller triangle.

Example 12.2.10. In Figure 12.3, the erosion of the large circle on the right by the annulus on the left results in a circle of radius R.

Definition 12.2.6. The *complement* of a set $\mathbb{X}$ within the allowable domain is

$$\mathbb{X}^c = \{t | t \notin \mathbb{X}\}.$$

Definition 12.2.7. The *opening* of a set $\mathbb{Y}$ by a set $\mathbb{H}$, denoted $\mathbb{Y} \circ \mathbb{H}$, is

$$\mathbb{Y} \circ \mathbb{H} := \left(\mathbb{Y} \ominus \hat{\mathbb{H}} \right) \oplus \mathbb{H}. \tag{12.3}$$

FIGURE 12.6. Examples of morphological operations on the binary image, *Jack*. The white region corresponds to set membership so that erosion, for example, makes the white region larger. Dilation makes the white region smaller.

Definition 12.2.8. The *closing* of a set $\mathbb{X}$ by a set $\mathbb{H}$, denoted $\mathbb{X} \bullet \mathbb{H}$, is

$$\mathbb{X} \bullet \mathbb{H} := \left(\mathbb{X} \oplus \hat{\mathbb{H}} \right) \ominus \mathbb{H}. \tag{12.4}$$

In image processing applications, the set $\mathbb{H}$ is dubbed the *structuring element*.

Example 12.2.11. Examples of Minkowski arithmetic on an image are shown in Figure 12.6 on the image *Jack*. White corresponds to vectors in the set. Erosion rids the image of small isolated white regions. In the $\mathbb{Y} \ominus \hat{\mathbb{H}}$ image, for example, the teeth are nearly eliminated. Erosion also increases isolated black points such as on the forehead in the $\mathbb{Y} \ominus \hat{\mathbb{H}}$ image. These enlarged black points are removed by a subsequent dilation as can be seen in the

FIGURE 12.7. The image *Three Dolls* used to illustrate edge detection in Figure 12.8.

$\mathbb{Y} \circ \mathbb{H}$ image. The converse happens in the dilated $\mathbb{Y} \oplus \mathbb{H}$ image. The small white point on the cheek is enlarged and small black regions, such as the dimple, are removed. Eroding the image to form $\mathbb{Y} \bullet \mathbb{H}$ makes the enlarged white space small again. The dimple, lost through dilation, remains lost.

Definition 12.2.9. The *set subtraction* of $\mathbb{T}$ from $\mathbb{Y}$ is defined by
$$\mathbb{Y} - \mathbb{T} = \{t \mid t \in \mathbb{Y}, \ t \notin \mathbb{T}\}.$$

Example 12.2.12. Using the definition of set subtraction, we can generate set boundaries. In binary images, the result is edge detection. Consider the image *Three Dolls* in Figure 12.7. The subtraction of the image, $\mathbb{Y}$, from its dilation is the set $(\mathbb{Y} \oplus H) - \mathbb{Y}$. This forms the *outside boundary* of the set $\mathbb{Y}$ as shown in Figure 12.8. The difference $\mathbb{Y} - (\mathbb{Y} \ominus \hat{\mathbb{H}})$, also shown in Figure 12.8, forms the *inside boundary*.

12.2.4 Minkowski Algebra

The Minkowski operations have many useful and insightful algebraic properties [506].

Theorem 12.2.1. *Dilation is commutative since*
$$\mathbb{X} \oplus \mathbb{H} = \mathbb{H} \oplus \mathbb{X}.$$

Since
$$(\mathbb{X} \oplus \mathbb{H}) \oplus \mathbb{T} = \mathbb{X} \oplus (\mathbb{H} \oplus \mathbb{T}),$$
dilation is also associative.

Theorem 12.2.2. The dilation duality property is
$$\mathbb{X} \oplus \hat{\mathbb{H}} = \left(\mathbb{X}^c \ominus \hat{\mathbb{H}}\right)^c. \tag{12.5}$$

FIGURE 12.8. Examples of edge detection. The structuring element, $\mathbb{H}$, is 2×2. The original binary image is in Figure 12.7.

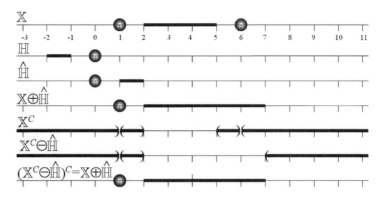

FIGURE 12.9. Illustration of the dilation duality property in (12.5).

Proof. Clearly,

$$t \in \left(\mathbb{X}^c \ominus \hat{\mathbb{H}}\right)^c$$

iff

$$t \notin \left(\mathbb{X}^c \ominus \hat{\mathbb{H}}\right)^c.$$

This is true iff $\exists\, \xi \in \hat{\mathbb{H}} \ni t \notin \mathbb{X}^c + \xi$ which, in turn, is true iff $\exists\, \xi \in \hat{\mathbb{H}} \ni t \in \mathbb{X}^c + \xi$. This is true iff $t \in \mathbb{X} \oplus \hat{\mathbb{H}}$.

Example 12.2.13. Let $\mathbb{X} = \{1, [2, 5], 6\}$ and $\mathbb{H} = \{[-2, -1], 0\}$. These sets, and subsequent operations on these sets, are shown in Figure 12.9. It follows that $\hat{\mathbb{H}} = \{0, [1, 2]\}$ and

$$\mathbb{X} \oplus \hat{\mathbb{H}} = \{1, [2, 7]\}. \tag{12.6}$$

For the right side of (12.5), we have

$$\mathbb{X}^c = \{(-\infty, 1), (1, 2), (5, 6), (6, \infty)\}$$

from which we conclude

$$\mathbb{X}^c \ominus \hat{\mathbb{H}} = \{(-\infty, 1), (1, 2), (7, \infty)\}.$$

Taking the complement, $\left(\mathbb{X}^c \ominus \hat{\mathbb{H}}\right)^c$, gives a result identical with (12.6).

Example 12.2.14. The dilation duality in Theorem 12.2.2 is illustrated in Figure 12.11 in two dimensions. On the top left, we have a binary image which we denote by $\mathbb{X}$. A pixel is in the set if it is white. The complement of the set, $\mathbb{X}^c$, is shown in the upper right hand corner. It is the negative of $\mathbb{X}^c$. The erosion of the negative by a 2×2 structuring element, $\mathbb{H}$, is shown on the bottom left. When negated, this image, shown on the bottom right, is equivalent to the dilation of the original image, $\mathbb{X}$, with the structuring element, $\hat{\mathbb{H}}$.

Theorem 12.2.3. *The erosion duality property is*

$$\mathbb{X} \ominus \hat{\mathbb{H}} = \left(\mathbb{X}^c \oplus \hat{\mathbb{H}}\right)^c. \tag{12.7}$$

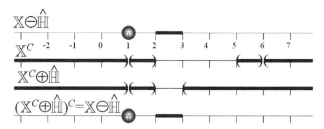

FIGURE 12.10. The sets $\mathbb{X}$, $\mathbb{H}$ and $\hat{\mathbb{H}}$ are shown in the top three lines of Figure 12.9. This figure illustrates the erosion duality property in (12.7).

Proof. Take the complement of (12.5) to give

$$\mathbb{X}^c \oplus \hat{\mathbb{H}} = \left[(\mathbb{X}^c)^c \ominus \hat{\mathbb{H}}\right]^c = \left(\mathbb{X} \oplus \hat{\mathbb{H}}\right)^c.$$

Complementing both sides gives (12.7).

Example 12.2.15. To illustrate the dilation duality property, we use the same sets, $\mathbb{X}$ and $\mathbb{H}$, used in Example 12.2.13. These sets on the real line and $\hat{\mathbb{H}}$ are illustrated in the top three lines of Figure 12.9. The illustration of the dilation duality property is in Figure 12.10 where, in the top line, we have the dilation

$$\mathbb{X} \ominus \hat{\mathbb{H}} = \{1, [2, 3]\}. \tag{12.8}$$

Also, $\mathbb{X}^c \oplus \hat{\mathbb{H}} = \{(-\infty, 1), (1, 2), (3, \infty)\}$. The complement of this set, shown at the bottom of Figure 12.10, is the same as (12.8).

Example 12.2.16. The erosion duality in Theorem 12.2.3 is illustrated in Figure 12.12. The original image, $\mathbb{X}$, and its complement, $\mathbb{X}^c$, are shown in the upper left and upper right of Figure 12.11. The dilation with $\hat{\mathbb{H}}$, a 2×2 pixel array, is shown on the left in Figure 12.12. The complement of this set, shown on the right is, according to the erosion property in (12.7), the erosion of the image, $\mathbb{X}$, with the structuring element, $\mathbb{H}$.

Theorem 12.2.4. *The dilation subset property: If* $\mathbb{H} \subset \mathbb{K}$, *then the following dilation subset relations hold. For all* $\mathbb{X}$,

$$\mathbb{X} \oplus \mathbb{H} \subset \mathbb{X} \oplus \mathbb{K}. \tag{12.9}$$

Proof. Follows immediately from the definition of Minkowski addition.

Example 12.2.17. The dilation of subset property in Theorem 12.2.4 is illustrated in Figure 12.13.

Theorem 12.2.5. *The erosion subset property:*[1] *If* $\mathbb{H} \subset \mathbb{K}$, *then the following erosion subset relationship holds. For all fixed* $\mathbb{X}$,

$$\mathbb{X} \ominus \hat{\mathbb{H}} \supset \mathbb{X} \ominus \hat{\mathbb{K}}. \tag{12.10}$$

---

1. If erosion order is reversed, we have $\mathbb{H} \ominus \hat{\mathbb{X}} \subset \mathbb{K} \ominus \hat{\mathbb{X}}$

FIGURE 12.11. Two dimensional illustration of the dilation duality theorem in (12.5). The original *Joshua* image is shown in the upper left.

Proof. Since $\mathbb{H} \subset \mathbb{K}$, for a vector t, we have $\mathbb{K} \oplus \{t\} \subset \mathbb{X}$ implying $\mathbb{H} \oplus \{t\} \subset \mathbb{X}$. Thus $t \in \mathbb{X} \ominus \hat{\mathbb{K}}$ implies $t \in \mathbb{X} \ominus \hat{\mathbb{H}}$.

Example 12.2.18. Let $\mathbb{X} = \{0, 2, 3, 4, 5, 8\}$ and $\mathbb{K} = \{0, 2, 3\}$. Then $\mathbb{X} \ominus \hat{\mathbb{K}} = \{0, 2\}$. If $\mathbb{H} = \{0, 2\} \subset \mathbb{K}$, then $\mathbb{X} \ominus \hat{\mathbb{H}} = \{0, 2, 3\}$ and (12.10) is satisfied.

Example 12.2.19. The erosion subset property is illustrated for a binary image in Figure 12.15.

Theorem 12.2.6. *Distributive properties.*

$$\mathbb{X} \oplus (\mathbb{H} \cup \mathbb{K}) = (\mathbb{X} \oplus \mathbb{H}) \cup (\mathbb{X} \oplus \mathbb{K})$$
$$\mathbb{X} \oplus (\mathbb{H} \cap \mathbb{K}) \subset (\mathbb{X} \oplus \mathbb{H}) \cap (\mathbb{X} \oplus \mathbb{K})$$
$$\mathbb{Y} \ominus (\mathbb{H} \cup \mathbb{K}) = (\mathbb{Y} \ominus \mathbb{H}) \cap (\mathbb{Y} \ominus \mathbb{K})$$
$$(\mathbb{H} \cap \mathbb{K}) \ominus \mathbb{Y} = (\mathbb{H} \ominus \mathbb{Y}) \cap (\mathbb{K} \ominus \mathbb{Y})$$

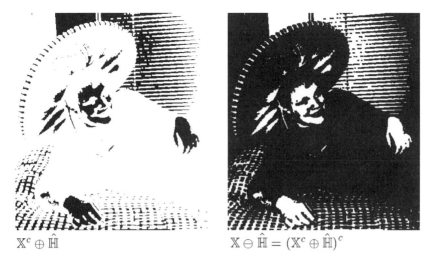

FIGURE 12.12. Illustration of the erosion duality theorem in (12.7). The pictures corresponding to $\mathbb{X}$ and $\mathbb{X}^c$ are shown in the top row of Figure 12.11.

Proof. See Exercise 12.2.

Theorem 12.2.7. *Opening and closing dualities.*

i. $(\mathbb{X} \bullet \mathbb{H})^c = \mathbb{X}^c \circ \mathbb{H}$.
ii. $(\mathbb{X} \circ \mathbb{H})^c = \mathbb{X}^c \bullet \mathbb{H}$.

Proof. i.

$$(\mathbb{X} \bullet \mathbb{H})^c = \left[(\mathbb{X} \oplus \hat{\mathbb{H}}) \ominus \mathbb{H}\right]^c$$

$$= \left[\mathbb{X} \oplus \hat{\mathbb{H}}\right]^c \oplus \mathbb{H}$$

$$= \left[\mathbb{X}^c \ominus \hat{\mathbb{H}}\right] \oplus \mathbb{H}$$

$$= \mathbb{X}^c \circ \mathbb{H}$$

ii.

$$(\mathbb{X} \circ \mathbb{H})^c = \left[(\mathbb{X}^c)^c \circ \mathbb{H}\right]^c$$

$$= \left[(\mathbb{X}^c \bullet \mathbb{H})^c\right]^c$$

$$= \mathbb{X}^c \bullet \mathbb{H}$$

Definition 12.2.10. An operation, $\diamond$, is *antiextensive* if, for all $\mathbb{Y}$ and $\mathbb{H}$,

$$\mathbb{Y} \diamond \mathbb{H} \subset \mathbb{Y}.$$

If for all $\mathbb{Y}$ and $\mathbb{H}$,

$$\mathbb{Y} \diamond \mathbb{H} \supset \mathbb{Y}.$$

the operation is said to be *extensive*.

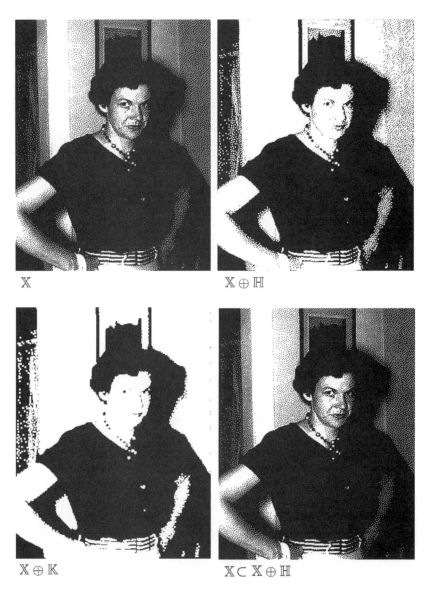

FIGURE 12.13. Illustration of the dilation subset property on a binary image. The original *Lenore* image, $\mathbb{X}$, is shown in the upper left hand corner. Dilation by a 2×2 structuring element, $\mathbb{H}$, is shown at the upper right. Dilation using a 4×4 set, $\mathbb{K} \supset \mathbb{H}$, is shown at the bottom left. The dilation subset property $\mathbb{X} \oplus \mathbb{H} \subset \mathbb{X} \oplus \mathbb{K}$ is illustrated in the bottom right image of Figure 12.14 where all of the black, corresponding to $\mathbb{X} \oplus \mathbb{H}$, is contained in the gray, corresponding to $\mathbb{X} \oplus \mathbb{K}$. (Continued in Figure 12.14.)

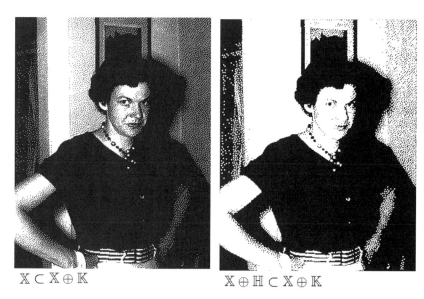

$$\mathbb{X} \subset \mathbb{X} \oplus \mathbb{K} \qquad\qquad \mathbb{X} \oplus \mathbb{H} \subset \mathbb{X} \oplus \mathbb{K}$$

FIGURE 12.14.   Continuation of Figure 12.13.

Theorem 12.2.8. *Opening is antiextensive since*

$$\mathbb{X} \circ \mathbb{H} \subset \mathbb{X}. \qquad\qquad (12.11)$$

Closing is extensive since

$$\mathbb{Y} \subset \mathbb{Y} \bullet H.$$

Proof. See Exercise 12.2.

Theorem 12.2.9. *Opening and closing are idempotent.*

  i. *Opening is idempotent in the sense that*

$$\mathbb{X} \circ \mathbb{H} = (\mathbb{X} \circ \mathbb{H}) \circ \mathbb{H}.$$

  ii. *Closing is idempotent in the sense that*

$$\mathbb{Y} \bullet \mathbb{H} = (\mathbb{Y} \bullet \mathbb{H}) \bullet \mathbb{H}.$$

Proof. See Exercise 12.2.

12.3 Fourier and Signal Analysis on Time Scales

Time scales, introduced by Hilger [607], are a unification theory under which continuous and discrete time signals and systems are subsumed. Discrete and continuous time Fourier transforms and convolutions, for example, are subsumed in the generalized time scale Fourier transform and convolution integral. The literature on time scales is vast [9, 10, 128, 129, 340, 341, 352, 497, 548, 593, 860, 922, 1165, 1208, 1209, 1360]. We limit treatment here to Fourier analysis on time scales[2] [921].

---

2. The contributions of Drs. Ian A. Gravagne and John M. Davis to this section are gratefully acknowledged.

FIGURE 12.15. Illustration of the erosion subset property on a binary image. The original *Marilee* image, $\mathbb{X}$, is shown at the upper left. It's erosion by a 2×2 structuring element, $\mathbb{K}$, is in the upper right. Erosion by a 4×4 element is at the bottom left. The erosion subset property is evident in the three remaining gray images where two images are shown using two different gray levels. $\mathbb{X} \ominus \hat{\mathbb{H}} \subset \mathbb{X}$, $\mathbb{X} \ominus \hat{\mathbb{K}} \subset \mathbb{X}$, and $\mathbb{X} \ominus \hat{\mathbb{H}} \subset \mathbb{X} \ominus \hat{\mathbb{K}}$ are represented. (Continued in Figure 12.16.)

A time scale, $\mathbb{T}$, consists of a set of closed intervals including individual point locations on the real line, $\mathbb{R}$. Some examples are shown in Figure 12.17. $\mathbb{N}$ denotes the natural numbers, $\{1, 2, 3, \cdots\}$. From top to bottom,

(a) $\mathbb{R}$ consists of the entire real line. Continuous time signals are on $\mathbb{R}$.
(b) $h\mathbb{Z}$ consists of points equally spaced at intervals of h. Discrete time signals are on $\mathbb{Z}$.
(c) $h\mathbb{Z}_n$ contains the origin and points separated by an interval h beginning at nh. The time scale shown here is $\mathbb{Z}_3$.

MATHEMATICAL MORPHOLOGY AND FOURIER ANALYSIS ON TIME SCALES 585

$\mathbb{X} \ominus \hat{\mathbb{K}} \subset \mathbb{X}$ $\qquad$ $\mathbb{X} \ominus \hat{\mathbb{H}} \subset \mathbb{X} \ominus \hat{\mathbb{K}}$

FIGURE 12.16. Continuation of Figure 12.15.

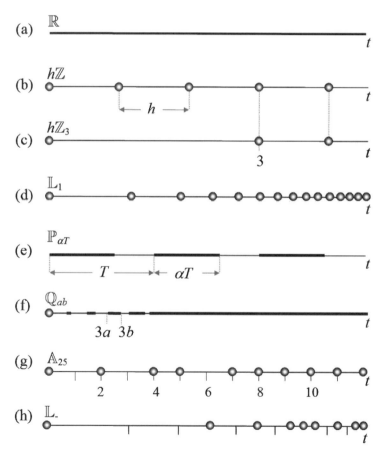

FIGURE 12.17. Some different time scales. The $\mathbb{R}$ and $\mathbb{Z}$ time scales correspond, respectively, to continuous and discrete time.

(d) $\mathbb{L}_k$, for a specified k, consists of all points $\{t_n \mid t_n = \log(k(n-1)+1)\}$ for $n \in \mathbb{N}$. For example

$$\mathbb{L}_1 = \{\log(1), \log(2), \log(3), \cdots\}$$

is shown in Figure 12.17. The log base is arbitrary but fixed.

(e) $\mathbb{P}_{\alpha T}$ consists of intervals of length αT separated by a distance of T.

(f) $\mathbb{Q}_{ab}$ contains the origin and the union of all of the intervals $na \leq t \leq nb$ for $n \in \mathbb{N}$. There will be a time, η, where all $t \geq \eta$ are in $\mathbb{Q}_{ab}$, i.e. $\mathbb{Q}_{ab}$ becomes continuous with no breaks for $t \geq \eta$. This occurs when the intervals begin to overlap. The nth and the $(n+1)^{\text{st}}$ intervals overlap when $nb \geq (n+1)a$ or $n > \frac{a}{b-a}$. Then

$$\eta = \left(1 + \left\lfloor \frac{a}{b-a} \right\rfloor\right) a. \tag{12.12}$$

For example

$$\mathbb{Q}_{10,13} = \{t \mid t = 0 \text{ and } 10 \leq t \leq 13 \text{ and } 20 \leq t \leq 26$$
$$\text{and } 30 \leq t \leq 39 \text{ and } t \geq 40\}.$$

Thus, as required by (12.12), $\eta = \left(1 + \left\lfloor \frac{10}{13-10} \right\rfloor\right) \times 10 = 40$.

(g) $\mathbb{A}_{\xi\eta}$ consists of the origin and all $t = n\xi + m\eta$ for all integers $n, m \in \mathbb{N}$. Shown in Figure 12.17 is

$$\mathbb{A}_{2,5} = \{t \mid 0, 2, 4, 5, 7, 8, 9, 10, 11, \cdots\}.$$

(h) $\mathbb{L}_-$ is $\mathbb{L}_1$ with all of the logs of prime numbers removed.

(i) A *discrete time scale*, $\mathbb{D}$, consists of a set of discrete points, $\mathbb{D} = \{t_n\}$. The time scales $h\mathbb{Z}$, $h\mathbb{Z}_n$, $\mathbb{L}_k$ and $\mathbb{A}_{\xi\eta}$ are discrete time scales.

The time scales $\mathbb{R}^+$ and $\mathbb{Z}^+$ are the time scales $\mathbb{R}$ and $\mathbb{Z}$ for $t \geq 0$. Time scales with no negative components are called *causal time scales*. The time scale $\mathbb{D}^+$ denotes a discrete time scale with no negative components. The time scales $\mathbb{R}$ and $\mathbb{R}^+$ are the time scales used for continuous time signals and systems, while $\mathbb{Z}$ and $\mathbb{Z}^+$ are the time scales of discrete time signals and systems.

12.3.1 Background

Here are some foundational definitions for time scales. A complete introduction to time scales is in Bohner and Peterson [128].

The *graininess*, $\mu(t)$, is the distance between a point $t \in \mathbb{T}$ and the closest point in $\mathbb{T}$ following t.[3] For $\mathbb{T} = h\mathbb{Z}$, it follows that $\mu(t) = h$. For $\mathbb{T} = \mathbb{R}$, $\mu(t) = dt$.

12.3.1.1 The Hilger Derivative

The *Hilger derivative* of a function $x(t)$ on a time scale $\mathbb{T}$ is

$$x^\Delta(t) = \frac{x(t + \mu(t)) - x(t)}{\mu(t)}.$$

---

[3]. The function $\mu(t)$ is also used in this text as the unit step function. The distinction between the graininess and the unit step function will be clear in the context of use.

At a right dense[4] point, the Hilger derivative is interpreted in the limiting sense and $x^\Delta(t) = \frac{d}{dt}x(t)$. For $\mathbb{T} = \mathbb{Z}$, the Hilger derivative is the forward difference

$$\Delta x[n] = x[n+1] - x[n] \tag{12.13}$$

where $x[n] = x(n)$.

12.3.1.2 Hilger Integration

Integration on a time scale is most easily expressed as an anti-derivative. Let $y(t) = x^\Delta(t)$. Then, for $a, b \in \mathbb{T}$ Hilger integration has the property

$$\int_a^b y(t)\Delta t = x(b) - x(a).$$

Let $C \in \mathbb{T}$ denote all points in $\mathbb{T}$ that are right dense and $D \in \mathbb{T}$ be those that are not. Clearly $\mathbb{T} = C + D$. Then

$$\int_{t \in \mathbb{T}} y(t)\Delta t = \int_{-\infty}^{\infty} y_\mathbb{R}(t)dt.$$

where $y_\mathbb{R}(t) = y_C(t) + y_D(t)$ and

$$y_D(t) = \sum_{\tau \in D} y^\mu(t)\delta(t - \tau),$$

and

$$y_C(t) = \begin{cases} y(t) \; ; \; t \in C \\ 0 \; \; \; ; \; t \notin C. \end{cases}$$

We have adopted the notation, for $\mu(t) > 0$,

$$y^\mu(t) := y(t)\mu(t).$$

Let $\mathbb{D}$ be a discrete time scale. Then, for a function $x(t)$ on $\mathbb{D}$,

$$\int_{t \in \mathbb{D}} x(t)\Delta t = \sum_{t_n \in \mathbb{D}} x^\mu(t_n).$$

12.3.2 Fourier Transforms on a Time Scale

The Fourier transform of a signal, $x(t)$, on a time scale, $\mathbb{T}$, is [608, 609]

$$X(u) = \int_{t \in \mathbb{T}} x(t)e^{-j2\pi ut}\Delta t. \tag{12.14}$$

Fourier transforms on continuous and discrete time are special cases. For the continuous time scale, $\mathbb{R}$, (12.14) becomes the continuous time Fourier transform (CTFT) in (2.10). Discrete time signals are on the time scale $\mathbb{Z}$. The time scale Fourier transform in (12.14) then becomes the discrete time Fourier transform (DTFT) in (2.112).

---

4. A time scale is *right dense* at t when $\mu(t) = dt$.

For a discrete time scale, $\mathbb{D}$, the Fourier transform in (12.14) can be written as

$$X(u) = \sum_{t_n \in \mathbb{D}} x^\mu(t_n) e^{-j2\pi u t_n}. \tag{12.15}$$

We will find the following rectangle function definition useful.

Definition 12.3.1. For a discrete causal time scale, $\mathbb{D}^+$, define $\Pi_N(t)$ to be equal to one for the first $N+1$ points in $\mathbb{D}^+$ (i.e. values of t_n for $0 \leq n \leq N$) and zero otherwise.

The following two examples use the discrete time scale $\mathbb{L}_1$ using $\log_{10}$.

Example 12.3.1. Figure 12.18 shows the Fourier transform of the signal corresponding to $x^\mu(t) := x(t)\mu(t) = \cos(2\pi f t)\Pi_{20,000}(t)$ for $f=4$. As is shown in the upper right, the magnitude of the Fourier transform peaks at the frequency $u=4$. An analogous example for $x(t) = \cos(2\pi f t)\Pi_{20,000}(t)$ is shown in Figure 12.19. In both figures, the real and imaginary portions of $X(u)$ are shown in the bottom right and bottom left of the figure.

Example 12.3.2. Shown in Figure 12.20 are the magnitudes of the time scale Fourier transform magnitude of $x(t) = \cos(2\pi f t)\Pi_K(t)$ for $f=10$ on $\mathbb{L}_1$ for $K = \{100, 1000, 10000, 100000\}$. All plots are on shown the same scale. A similar figure for $x^\mu(t) := x(t)\mu(t) = \cos(2\pi f t)\Pi_K(t)$ for $f=10$ is shown in Figure 12.21. Here, each plot range is different.

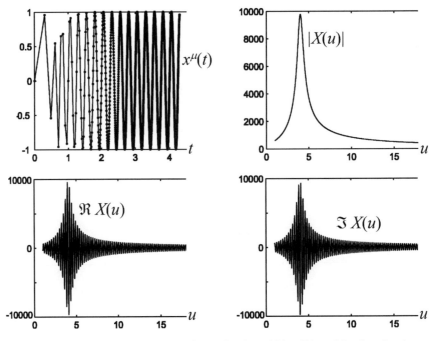

FIGURE 12.18. In $\mathbb{R}$ and $\mathbb{Z}$, the Fourier transform of a sinusoid is a Dirac delta functional centered at the frequency of the sinusoid. If the sinusoid is of finite duration, the transform is a sharp peak centered at the sinusoid frequency. Figures 12.18 through 12.21 illustrate a similar property for the time scale L_1. The upper left plot in this figure is of $x^\mu(t) := x(t)\mu(t) = \cos(8\pi t)\Pi_{20,000}(t)$ on the time scale $\mathbb{L}_1$. The dots denote the values of the function on the time scale and are linearly connected for clarity of presentation. The magnitude of the time scale Fourier transform is shown in the upper right. Since $x^\mu(t)$ is a sinusoid with frequency $f=4$, we expect the transform to peak at $u=f=4$. The real and imaginary parts of the time scale Fourier transform are shown in the bottom left and bottom right of the figure.

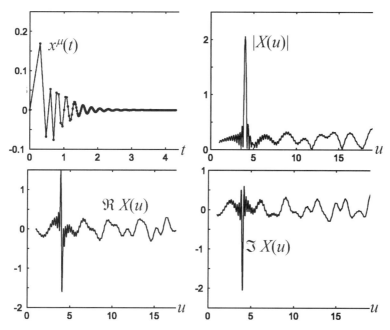

FIGURE 12.19. These plots are the same as in Figure 12.18, except, the transform is of $x(t) = \cos(8\pi t)\, \Pi_{20,000}(t)$. In Figure 12.18, we use x^μ as the sinusoid and here $x(t)$. The magnitude of the time scale Fourier transform, shown in the upper right plot, peaks at the frequency $u = 4$.

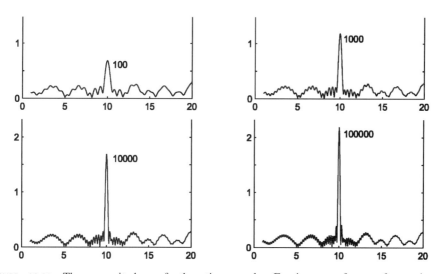

FIGURE 12.20. The magnitudes of the time scale Fourier transform of a sinusoid $x(t) = \cos(2\pi f t)\, \Pi_K(t)$ with frequency $f = 10$ on a time scale $\mathbb{L}_1$ for various durations, K. The longer the duration, the sharper the peak centered around $u = 10$. Fourier transforms on $\mathbb{R}$ and $\mathbb{Z}$ exhibit similar behavior.

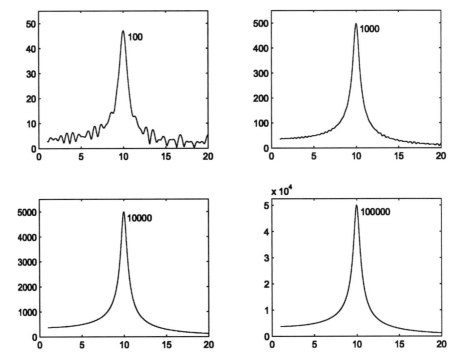

FIGURE 12.21. These plots are the same as in Figure 12.20 except that $x^\mu(t) := x(t)\mu(t)$, rather than $x(t)$, is equal to $\cos(2\pi ft)\,\Pi_K(t)$. As is the case in Figure 12.20, the longer the duration, the sharper the peak centered around $u = 10$.

12.3.3 The Minkowski Sum of Time Scales

A time scale, $\mathbb{T}$, is *closed under Minkowski addition* if, for every $\xi \in \mathbb{T}$ and $\tau \in \mathbb{T}$, the time $t = \xi + \tau \in \mathbb{T}$. The time scales $h\mathbb{Z}$, $h\mathbb{Z}+$, $\mathbb{R}$, $\mathbb{R}_+$, $h\mathbb{Z}_n$, $\mathbb{L}_k$, $\mathbb{Q}_{ab}$, $\mathbb{N}$, and $\mathbb{A}_{\xi\eta}$ are closed under Minkowski addition.

The *additive closure* of a time scale, $\mathbb{T}$, denoted $\mathbb{T}_C$, is the smallest time scale closed under Minkowski addition in which $\mathbb{T}$ is subsumed.

$$\mathbb{T}_C = \{t \mid t = n\xi + m\eta \text{ for all pairs } \xi \in \mathbb{T},\, \eta \in \mathbb{T} \text{ and all } n, m \in \mathbb{N}\}.$$

For example,

- if $\mathbb{T} = \{-1, 0, 1\}$, then $\mathbb{T}_C = \mathbb{Z}$.
- if $\mathbb{T} = \{t \mid 0 \leq t \leq 1\}$, then $\mathbb{T}_C = \mathbb{R}^+$.
- if $\mathbb{T} = \{\frac{4}{5}, 1\}$, then $\mathbb{T}_C = \{\frac{4}{5}, 1, \frac{8}{5}, \frac{9}{5}, 2, \frac{12}{5}, \frac{13}{5}, \frac{14}{5}, 3, \frac{16}{5}, \frac{17}{5}, \frac{18}{5}, \frac{19}{5}, 4\cdots\}$.

The *Minkowski sum*[5] of two time scales $\mathbb{X}$ and $\mathbb{H}$, denoted $\mathbb{X} \oplus \mathbb{H}$, consists of the points $\xi + \tau$ for all $\xi \in \mathbb{H}$ and $\tau \in \mathbb{X}$, i.e., if $\xi \in \mathbb{H}$ and $\tau \in \mathbb{X}$, then $\xi + \tau \in \mathbb{X} \oplus \mathbb{H}$. For example,

- if $\mathbb{X} = \{0, 1\}$ and $\mathbb{H} = \{4, 5, 10\}$, then $\mathbb{X} \oplus \mathbb{H} = \{4, 5, 6, 10, 11\}$.
- if $\mathbb{X} = \{0, 2\}$ and $\mathbb{H} = \{t \mid 0 \leq t \leq 1\}$, then $\mathbb{X} \oplus \mathbb{H} = \{t \mid 0 \leq t \leq 1 \text{ and } 2 \leq t \leq 3\}$.

---

5. This is the same as the dilation operation in Section 12.2.

- if $\mathbb{X} = 2\mathbb{Z}$ and $\mathbb{H} = \mathbb{Z}$, then $\mathbb{X} \oplus \mathbb{H} = \mathbb{Z}$.
- if $\mathbb{X} = \mathbb{L}_-$ and $\mathbb{H} = \mathbb{L}_1$, then $\mathbb{X} \oplus \mathbb{H} = \mathbb{L}_1$.

The time scale containing only the origin, $\mathbb{I} = \{0\}$, is the identity operator for the Minkowski sum since, for any $\mathbb{T}$, we have

$$\mathbb{T} \oplus \mathbb{I} = \mathbb{T}.$$

The $\mathbb{R}$ time scale absorbs any time scale since, for any $\mathbb{T}$,

$$\mathbb{T} \oplus \mathbb{R} = \mathbb{R}.$$

A geometrical interpretation of the Minkowski sum of two time scales is based on the simple geometry in Figure 12.22. A 45° line from the coordinate (ξ, η) intersects the horizontal axis at $\xi + \eta$. Consider, then the time scales are shown in Figure 12.23: $\mathbb{X}$ on the

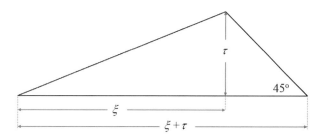

FIGURE 12.22. A right triangle with height τ and base ξ, when adjoined with a 45° right triangle, has a composite base of $\xi + \tau$. This simple geometry allows graphical convolution of the type shown in Figure 12.23.

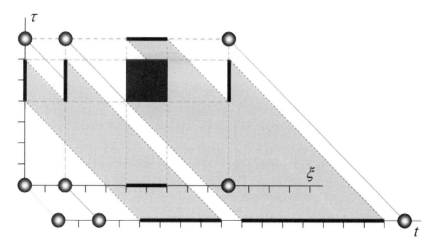

FIGURE 12.23. Geometrical illustration of the Minkowski sum operation. The time scale $\mathbb{X} = \{\tau | \tau = \{0\}, [4, 5], \{6\}\}$ and the time scale $\mathbb{H} = \{\xi \mid \xi = \{0\}, \{2\}, [5, 7], \{10\}\}$. We take the Cartesian product of the time scales on the (ξ, τ) plane as shown. This Cartesian product, when multiplied by $e^{-j2\pi ut}$ and integrated, yields the Fourier transform, $Y(u)$. The family of lines corresponding to $t = \tau + \xi$ are at the 45° angles shown. The shadow of the Cartesian product along these 45° lines contains the set of all $t = \tau + \xi \ \forall \ (\tau, \xi)$. This is the time scale $\mathbb{Y} = \mathbb{X} \oplus \mathbb{H} = \{t \mid t = \{0\}, \{2\}, [4, 8], [9, 16], \{17\}\}$.

τ axis and $\mathbb{H}$ on the ξ axis. The Cartesian product of these time scales shows all allowable combinations of (ξ, τ). All allowable points $t = \xi + \tau$ are shown as the *shadow* of the Cartesian product at 45° as shown. If a point t lies in the shadow of the Cartesian product then there exists a (ξ, τ) such that $t = \xi + \tau$.

12.3.4 Convolution on a Time Scale

Let $x(t)$ on time scale $\mathbb{X}$ have a Fourier transform of $X(u)$ and $h(t)$, on a time scale $\mathbb{H}$, have Fourier transform $H(u)$. Let $Y(u) = X(u)H(u)$ be the time scale Fourier transform of $y(t)$.

$$Y(u) = \int_{t \in \mathbb{Y}} y(t) e^{-j2\pi ut} \Delta t. \tag{12.16}$$

The function $y(t)$ can be deemed to be on time scale $\mathbb{Y} = \mathbb{X} \oplus \mathbb{H}$. To show this, we write

$$Y(u) = X(u)H(u)$$

$$= \int_{\tau \in \mathbb{X}} x(\tau) e^{-j2\pi u\tau} \Delta\tau \int_{\xi \in \mathbb{H}} h(\xi) e^{-j2\pi u\xi} \Delta\xi$$

$$= \int_{\tau \in \mathbb{R}} x_\mathbb{R}(\tau) e^{-j2\pi u\tau} d\tau \int_{\xi \in \mathbb{R}} h_\mathbb{R}(\xi) e^{-j2\pi u\xi} d\xi$$

$$= \int_{\xi \in \mathbb{R}} \int_{\tau \in \mathbb{R}} x_\mathbb{R}(\tau) h_\mathbb{R}(\xi) e^{-j2\pi u(\tau+\xi)} d\tau d\xi$$

$$= \int_{t \in \mathbb{R}} \left[\int_{\tau \in \mathbb{R}} x_\mathbb{R}(\tau) h_\mathbb{R}(t-\tau) d\tau \right] e^{-j2\pi ut} dt$$

$$= \int_{t \in \mathbb{R}} y_\mathbb{R}(t) e^{-j2\pi ut} d\tau dt$$

where

$$y_\mathbb{R}(t) = \int_{\tau \in \mathbb{R}} x_\mathbb{R}(\tau) h_\mathbb{R}(t-\tau) d\tau.$$

and $t = \tau + \xi$. For every $\tau \in \mathbb{X}$ and $\xi \in \mathbb{H}$, there exists a $t = \tau + \xi \in \mathbb{Y} = \mathbb{X} \oplus \mathbb{H}$.

Since $y_\mathbb{R}(t) = 0$ for $t \notin \mathbb{Y}$, a $y(t)$ on a time scale $\mathbb{Y}$ can be constructed to satisfy (12.16). The function $y(t)$, however, can contain Dirac deltas at right dense locations even if neither $x(t)$ or $h(t)$ contains Dirac deltas at right dense locations, e.g. there is a Dirac delta at the right dense location $t = 10$ in Figure 12.23.

12.3.4.1 Convolution on Discrete Time Scales

For discrete time scales $\mathbb{X}$ and $\mathbb{H}$, the discrete time convolution on time scales is

$$y^\mu(t_p) = \sum_{(\tau_n \in \mathbb{X}) \ni (t_p - \tau_n \in \mathbb{H})} x^\mu(\tau_n) h^\mu(t_p - \tau_n). \tag{12.17}$$

Proof.

$$Y(u) = X(u)H(u)$$

$$= \sum_{\tau_n \in \mathbb{X}} x^\mu(\tau_n) e^{-j2\pi u \tau_n} \sum_{\xi_m \in \mathbb{H}} h^\mu(\xi_m) e^{-j2\pi u \xi_m}$$

$$= \sum_{\xi_m \in \mathbb{H}} \sum_{\tau_n \in \mathbb{X}} x^\mu(\tau_n) h^\mu(\xi_m) e^{-j2\pi u (\xi_m + \tau_n)}.$$

Make the substitution $t_p = \tau_n + \xi_m$. Clearly, $t_p \in \mathbb{X} \oplus \mathbb{H} = \mathbb{Y}$. Then

$$Y(u) = \sum_{t_p \in \mathbb{Y}} \left[\sum_{(\tau_n \in \mathbb{X}) \ni (t_p - \tau_n \in \mathbb{H})} x^\mu(\tau_n) h^\mu(t_p - \tau_n) \right] e^{-j2\pi ut}.$$

Since

$$Y(u) = \sum_{t_p \in \mathbb{Y}} y^\mu(t_p) e^{-j2\pi ut},$$

(12.17) follows immediately.

Example 12.3.3. Let $\mathbb{X} = \{\tau_1, \tau_2, \tau_3\} = \{0, 2, 3\}$ and $\mathbb{H} = \{\xi_1, \xi_2\} = \{3, 6\}$. Then

$$\mathbb{Y} = \mathbb{X} \oplus \mathbb{H} = \{t_1, t_2, t_3, t_4, t_5, \} = \{3, 5, 6, 8, 9\}.$$

Using the geometry introduced in Figure 12.23, this dilation is illustrated in Figure 12.24. For $x(\tau) \in \mathbb{X}$ and $h(\xi) \in \mathbb{H}$, we have $y(t) = x(t) * h(t) \in \mathbb{Y}$. Using notation $x^\mu(t_p) = x_p$, we have

$$y_1 = x_1 h_1 \; ; \; y_2 = x_2 h_1; \; y_3 = x_1 h_2 + x_2 h_1$$

$$y_4 = x_2 h_2 \; ; \; y_5 = x_3 h_6$$

The geometry in Figure 12.24 gives this information. For example, the 45° line intersecting at $t_3 = 6$ intersects the Cartesian product at $\{\tau_3, \xi_1\} = \{3, 3\}$ and at $\{\tau_1, \xi_2\} = \{0, 6\}$. Thus, as written above, $y_3 = x_1 h_2 + x_2 h_1$.

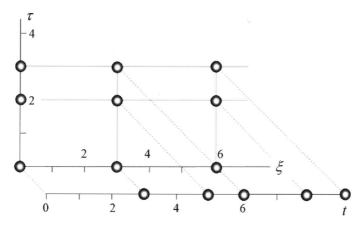

FIGURE 12.24. Geometry for the convolution in Example 12.3.3.

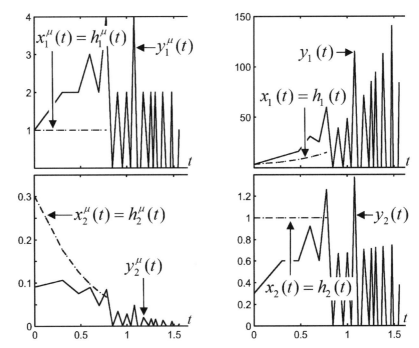

FIGURE 12.25. Example convolutions on an $\mathbb{L}_1$ time scale from Example 12.3.4. The top left plot illustrates the time scale convolution on the time scale $\mathbb{L}_1$ of $x_1^\mu(t) = \Pi_6(t)$ with $h_1^\mu(t) = x_1^\mu(t)$ shown with the broken line. The result of the time scale convolution, $y_1^\mu(t)$, is plotted with a solid line. It is also plotted on a linear scale in Figure 12.28. The top right plots are of $x_1(t) = x_1^\mu(t)/\mu(t)$, $h_1(t) = x_1(t)$ and $y_1(t) = y_1^\mu(t)/\mu(t)$. The two bottom plots are of $x_2(t) = \Pi_6(t)$, $h_2(t) = x_2(t)$ and their time scale convolution, $y_2(t)$. Note that $x_2(t) = x_1(t)/\mu(t)$ and $h_2(t) = h_1(t)/\mu(t)$. The result of the convolution is $y_2(t) \neq y_1(t)/\mu(t)$. The bottom left plot shows the component functions with the subscripts and the right without.

Example convolutions on $\mathbb{L}_1$. For $\mathbb{L}_1$, the time scale convolution in (12.26) is

$$y^\mu(t_p) = \sum_{p/n \in \mathbb{N}} x^\mu(t_n) h^\mu(t_{p/n}) \tag{12.18}$$

or, using the notation in (12.27),

$$y^\mu(t) = x^\mu(t) \stackrel{\mathbb{L}_1}{*} h^\mu(t). \tag{12.19}$$

Example 12.3.4. Examples of the time scale convolution in (12.19) are shown in Figure 12.25. On the top row is convolution corresponding to $x_1^\mu(t) = h_1^\mu(t) = \Pi_6(t)$. The result is $y_1^\mu(t)$. The bottom row shows $y_2(t)$, the convolution when $x_2(t) = h_2(t) = \Pi_6(t)$.

Example 12.3.5. The top row in Figure 12.26 shows convolution corresponding to $x_1^\mu(t) = \cos(8\pi t)\Pi_{100}(t)$ with $h_1^\mu(t) = \Pi_{10}(t)$. The result is $y_1^\mu(t)$. The bottom row shows $y_2(t)$, the convolution when $x_2(t) = \cos(8\pi t)\Pi_{100}(t)$ and $h_2(t) = \Pi_{10}(t)$.

12.3.4.2 Time Scales in Deconvolution

If $x(t) \in \mathbb{X}$ and $h(t) \in \mathbb{H}$, then $y(t) = x(t) * h(t) \in \mathbb{Y}$ where $\mathbb{Y} = \mathbb{X} \oplus \mathbb{H}$. Given time scales $\mathbb{Y}$ and $\mathbb{H}$, is it possible to determine $\mathbb{X}$? Different $\mathbb{X}$'s can give the same $\mathbb{Y}$ for a fixed $\mathbb{H}$, so the answer is no.

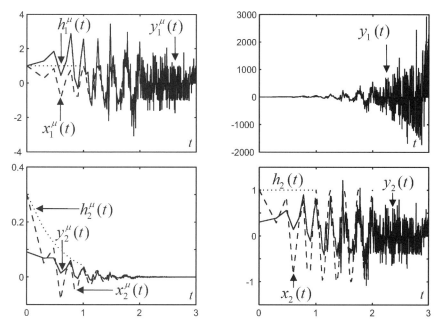

FIGURE 12.26. More example convolutions on an $\mathbb{L}_1$ time scale. See Example 12.3.5.

Example 12.3.6. Let $\mathbb{X}_1 = \{0, 2\}$ and $\mathbb{X}_2 = \{0, 1, 2\}$. If $\mathbb{H} = \{0, 2\}$, then $\mathbb{X}_1 \oplus \mathbb{H} = \mathbb{X}_2 \oplus \mathbb{H} = \{0, 1, 2, 3\}$. Because the inverse is not unique, $\mathbb{X}_1$ can not be determined uniquely from $\mathbb{Y} = \mathbb{X}_1 \oplus \mathbb{H}$ given $\mathbb{Y}$ and $\mathbb{H}$.

From the theorem on distributive properties of mathematical morphological operations in (12.2.6), if $\mathbb{Y} = \mathbb{X}_1 \oplus \mathbb{H} = \mathbb{X}_2 \oplus \mathbb{H}$, then $(\mathbb{X}_1 \cup \mathbb{X}_2) \oplus \mathbb{H} = \mathbb{Y}$. Thus, the union of any solutions is also a solution. This dictates there must be a *biggest* solution equal to the union of all solutions. Denote this set by $\tilde{\mathbb{X}}$. Then $\tilde{\mathbb{X}} \oplus \mathbb{H} = \mathbb{Y}$. We can find this set using the definition of opening and the antiextensive property of opening in (12.11).

$$\mathbb{X} \circ \hat{\mathbb{H}} = (\mathbb{X} \oplus \hat{\mathbb{H}}) \ominus \hat{\mathbb{H}}$$
$$= \mathbb{Y} \ominus \hat{\mathbb{H}}$$
$$\subset \tilde{\mathbb{X}}$$

The conclusion, therefore, is

(a) erosion of the time scale, $\mathbb{Y}$, by $\hat{\mathbb{H}}$, produces a time scale in which the support, $\mathbb{X}$, of the input is subsumed, and
(b) the erosion of $\mathbb{Y}$ by $\hat{\mathbb{H}}$ produces the biggest time scale for which $\mathbb{Y} = \mathbb{X} \oplus \hat{\mathbb{H}}$.

12.3.5 *Additively Idempotent Time Scales*

A time scale, $\mathbb{T}$, is said to be *additively idempotent* if

$$\mathbb{T} \oplus \mathbb{T} = \mathbb{T}.$$

We adopt the acronym AITS for *additively idempotent time scale*. We begin with four theorems.

Theorem 12.3.1. *An AITS time scale must contain the origin. In other words, if $\mathbb{T}$ is an AITS and $\mathbb{T} \in \mathbb{R}^+$ or $\mathbb{T} \in \mathbb{R}^-$, then $\mathbb{I} \in \mathbb{T}$. Proof for $\mathbb{T} \in \mathbb{R}^+$:* Let $\tau_0 = \min\{t_n \in \mathbb{T}\}$. If $t_0 > 0$, then $t_0 \notin \mathbb{T} \oplus \mathbb{T}$, a contradiction.

Theorem 12.3.2. *If $\mathbb{T}$ is closed under Minkowski addition and contains the origin (i.e. $\mathbb{I} \in \mathbb{T}$), then $\mathbb{T}$ is an AITS.*

Proof. If (a) $A \in B$ and (b) $B \in A$, then $A = B$. Since $\mathbb{T}$ is closed under addition, (a) $\mathbb{T} \oplus \mathbb{T} \in \mathbb{T}$. Since $\mathbb{I} \in \mathbb{T}$, we know (b) $\mathbb{I} \oplus \mathbb{T} = \mathbb{T} \in \mathbb{T} \oplus \mathbb{T}$. Therefore $\mathbb{T} = \mathbb{T} \oplus \mathbb{T}$.

Theorem 12.3.3. *If an AITS is scaled, the result is an AITS, i.e., if $\mathbb{T}$ is an AITS, so is $h\mathbb{T}$.*

Theorem 12.3.4. *An AITS other that $\mathbb{I}$ is unbounded, i.e. an AITS containing a $t > 0$ is unbounded from above. An AITS containing a $\tau < 0$ is unbounded from below.*

Proof. For $t > 0$: Let $a \in \mathbb{T}$ and $a > 0$. Since $\mathbb{T}$ is an AITS, $na \in \mathbb{T}$ and $na \to \infty$ as $n \to \infty$.

12.3.5.1 AITS Hulls

The AITS *hull* of a time scale, $\mathbb{T}$, is the smallest AITS, $\mathbb{T}_A$, such that $\mathbb{T} \in \mathbb{T}_A$.

Here are some examples of AITS hulls.

- If $\mathbb{T} = \{0, h\}$ and $h > 0$, then $\mathbb{T}_A = h\mathbb{Z}^+$.
- If $\mathbb{T} = \{-1, 0, 1\}$, then $\mathbb{T}_A = \mathbb{Z}$.
- If $\mathbb{T} = \{ t \mid -1 \leq t \leq 1 \}$, then $\mathbb{T}_A = \mathbb{R}$.

12.3.5.2 AITS Examples

Here are some examples of time scales that are AITS.

1. $\mathbb{I}$.
2. $\mathbb{R}$ and $\mathbb{R}^+$.
3. $h\mathbb{Z}$ and $h\mathbb{Z}^+$.
4. The log time scale, $\mathbb{L}_k$, where $k \in \mathbb{N}$ is arbitrary but fixed. Proof: $[k(n-1)+1] \times [k(m-1)+1] = k(p-1)+1$ where $p = k(n-1)(m-1) + n + m - 1 \in \mathbb{N}$.
5. The time scale

$$\mathbb{L}_- = \{ t_n = \log(n) \mid n = \text{ all positive integers that are not prime}\}.$$

    Proof: Because $\{0\} \in \mathbb{L}_-$ and $\mathbb{I} \oplus \mathbb{L}_-$, we know that $\mathbb{L}_- \oplus \mathbb{L}_-$ contains $\mathbb{L}_-$. No logs of primes are in $\mathbb{L}_- \oplus \mathbb{L}_-$ because $t_n + t_m = \log(nm)$ and nm cannot be prime.

6. Assuming Goldbach's conjecture, the closure of the set of $\{0\}$ and the prime numbers is $\mathbb{Z}_2$. Proof: Goldbach's conjecture is that all even numbers exceeding two can be expressed as the sum of two prime numbers. Thus, all even integers greater than three are in $\mathbb{T}_A$. Since they are prime, the numbers 2 and 3 are also in $\mathbb{T}_A$. All odd numbers, $o > 5$, are in $\mathbb{T}_A$ as the sum of the three and the even number $o - 3$. More generally

$$\mathbb{Z}_N = \{t = n | n = 0, n \geq N\}$$

is an AITS.

7. The AITS hull of $\{0, 7, 11\}$ is

$$\mathbb{A}_{7,11} = \{0,\ 7,\ 11,\ 14,\ 18,\ 21,\ 22,\ 25,\ 28,\ 29,\ 32,\ 33,\ 35,\ \cdots\}.$$

More generally, the AITS hull of $\mathbb{T} = \{0, \xi, \eta\}$ is

$$\mathbb{T}_{\mathcal{A}} = \mathbb{A}_{\xi\eta}.$$

8. An AITS can be mixed. An example is the AITS hull of $\mathbb{T} = \{t \mid t = 0, [a, b]\}$ where $0 < a < b$.

$$\mathbb{T}_{\mathcal{A}} = \mathbb{Q}_{ab}.$$

The Fourier transform of a signal, $x(t)$, on time scale $\mathbb{Q}_{ab}$ is

$$X(u) = \sum_{n=0}^{M-1} x^{\mu}(nb)e^{-j2\pi nbu} + \left[\sum_{n=1}^{M-1} \int_{t=na}^{nb} + \int_{t=Ma}^{\infty}\right] x(t)e^{-j2\pi ut}\, dt$$

where

$$M = \left\lfloor \frac{a}{b-a} \right\rfloor.$$

12.3.5.3 AITS Conservation

Theorem 12.3.5. *If $\mathbb{T}_1$ and $\mathbb{T}_2$ are both AITS', then $\mathbb{T}_3 = \mathbb{T}_1 \oplus \mathbb{T}_2$ is an AITS.*

Proof.
$$\begin{aligned}
\mathbb{T}_3 \oplus \mathbb{T}_3 &= (\mathbb{T}_1 \oplus \mathbb{T}_2) \oplus (\mathbb{T}_1 \oplus \mathbb{T}_2) \\
&= (\mathbb{T}_1 \oplus \mathbb{T}_1) \oplus (\mathbb{T}_2 \oplus \mathbb{T}_2) \\
&= \mathbb{T}_1 \oplus \mathbb{T}_2 \\
&= \mathbb{T}_3
\end{aligned}$$

12.3.5.4 Asymptotic Graininess of $\mathbb{A}_{\xi\eta}$

By asymptotic graininess, we mean, if the limit exists,

$$\mu_{\infty} := \lim_{t \to \infty} \mu(t).$$

In certain instances, the asymptotic graininess is achieved in finite time. We define that time as

$$t_{\mu} = \{\min t\}\ \text{such that}\ \mu(t) = \mu_{\infty}\ \text{for}\ t \geq t_{\mu}.$$

For the time scale $\mathbb{Z}_n$, for example, the asymptotic graininess of $\mu_{\infty} = 1$ begins at $t_{\mu} = n$.

Here we show that the AITS time scale, $\mathbb{A}_{\xi\eta}$, always approaches a constant asymptotic graininess.

Theorem 12.3.6. *Let $L, K \in \mathbb{N}$ be relatively prime[6] and $K < L$. Then $\mathbb{A}_{K,L}$ achieves a graininess of $\mu_{\infty} = 1$ at or before time $t_{\mu} \leq LK$ and maintains that constant graininess thereafter.*

---

6. Two positive integers, K and L, are *relatively prime* if they have no common integer factors other than one. $\{20, 27\}$ are relatively prime. $\{15, 24\}$ are not because they contain the common factor of 3.

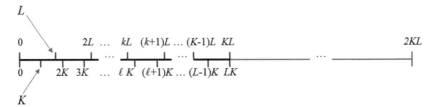

FIGURE 12.27. See Theorem 12.3.6.

Note that when a graininess of one is achieved for K sequential integer times, say $\{p, p+1, \ldots, p+K-1\}$, the graininess will be one thereafter. Since K is in the time scale, we assure unit graininess for the next K values by adding to the sequential integers to form $\{p+K, p+K+1, \ldots, p+2K-1\}$. Continuing this, we are assured that unit graininess occurs at time p and thereafter.

The proof offered is a straightforward result from residue number theory [1296] and is illustrated in Figure 12.27. The numbers K and L are shown on the number line. The time scale $\mathbb{A}_{K,L}$ contains all integer multiples of K and L. The integer multiples, ℓK and kL, are shown in Figure 12.27 up to and including KL. These numbers, $\{\ell K | 1 \leq L\}$ and $\{kL | 1 \leq k \leq K\}$, form an additive basis from which all of the integers on $(KL, 2KL)$ can be formed. We require only K sequential integers to assure henceforth a graininess of one. This is guaranteed to occur, at least, for $\{KL, KL+1, KL+2, \ldots, KL+(K-1)\}$. Thus, the asymptotic graininess of one is achieved at least at time KL.

Elaboration: Define $((n))_m$ as $n \bmod m$. Since L and K are relatively prime, the residues

$$((\ell K))_L; 1 \leq \ell \leq L, \tag{12.20}$$

when arranged by magnitude, are $0, 1, 2, \ldots, L-1$. Similarly, the residues

$$((kL))_K; 1 \leq k \leq K, \tag{12.21}$$

when arranged by magnitude, are $0, 1, 2, \ldots, K-1$. All of the residues correspond to different numbers on $1 \leq n \leq KL$ except the last: $((KL))_K = ((KL))_L = 0$.

Consider the numbers $KL < m < 2KL$. We find $((m))_L = \lambda$ and $((m))_K = \kappa$. There is a a unique ℓ in the interval $1 \leq \ell \mathbb{L}$ such that $((\ell K))_L = \lambda$ and a unique k in the interval $1 \leq k \leq K$ such that $((kL))_K = \kappa$. Residue arithmetic then states

$$m = kL + \ell K.$$

Thus, since $kL \in K\mathbb{T}_{AITS}$ and $\ell K \in K\mathbb{T}_{AITS}$, all m in the interval $KL < m < 2KL$ are in $K\mathbb{T}_{AITS}$. In order to establish arrival at the asymptotic graininess of one, only K of these sequential values is needed. Thus, a graininess of μ_∞ is achieved at $t_\mu \leq NL$ and remains thus thereafter.

To illustrate, let $K = 4$ and $L = 7$. Then (12.20) and (12.21) are

$$\begin{aligned}
&((4))_7 = 4 \quad ((2 \times 4))_7 = 1 \quad ((3 \times 4))_7 = 5 \quad ((4 \times 4))_7 = 2 \\
&((5 \times 4))_7 = 6 \quad ((6 \times 4))_7 = 3 \quad ((7 \times 4))_7 = 0 \\
&((7))_4 = 3 \quad ((2 \times 7))_4 = 2 \quad ((3 \times 7))_4 = 1 \quad ((4 \times 7))_4 = 0
\end{aligned} \tag{12.22}$$

or

$$((4))_7 = 4 \quad ((8))_7 = 1 \quad ((12))_7 = 5 \quad ((16))_7 = 2$$
$$(20))_7 = 6 \quad ((24))_7 = 3 \quad ((28))_7 = 0 \quad (12.23)$$
$$((7))_4 = 3 \quad ((14))_4 = 2 \quad ((21))_4 = 1 \quad ((28))_4 = 0.$$

The mod 7 row contains the residues 0 through 6. The mod 4 row contains the residues 0 through 3. The numbers represented on the top row are $\{4, 8, 12, 16, 20, 24, 28\}$ and, on the bottom, $\{7, 14, 21, 28\}$. Except for $KL = 28$, the numbers are distinct.

Consider $m = 33$. Clearly, $((33))_7 = 5$ and $((33))_4 = 1$. Since, from (12.23), $((12))_7 = 5$ and $((21))_4 = 1$, we obtain $12 + 21 = 33$, the desired answer.

The time scale for this example is

$$\mathbb{A}_{4,7} = \{0, 4, 7, 8, 11, 12, 14, 15, 16, 18, 19, 20, 21, 22, 23, 24, 25, 26, 27, 28, \dots\}.$$

The theorem states asymptotic graininess will begin at least at $t_\mu \leq LK = 28$. It clearly begins earlier at $t_\mu = 18$.

The extension to integers that are not relatively prime also follows from Theorem 12.3.5.3 and leads to the following theorem.

Theorem 12.3.7. *Let $\hat{K}$ and $\hat{L}$ be not relatively prime. Let their common factor be M so that $\hat{K} = MK$ and $\hat{L} = ML$. Then the asymptotic graininess of $\mathbb{A}_{\hat{K},\hat{L}}$ is*

$$\mu_\infty = M$$

and initiates at

$$t_\mu \leq MLK.$$

The time scale of $\mathbb{A}_{\hat{K},\hat{L}}$ is a scaled version of the time scale $\mathbb{A}_{K,L}$. Specifically

$$\mathbb{A}_{\hat{K},\hat{L}} = M \mathbb{A}_{K,L}.$$

From Theorem 12.3.5.3, it follows that an asymptotic graininess of M occurs, at least, at time $t_\mu \leq MLK$.

For example, for $K = 3$, $L = 4$ and $M = 3$, we have

$$\mathbb{A}_{9,12} = \{0, 9, 12, 18, 21, 24, 27, 30, 33, 36, \dots\}.$$

The asymptotic graininess of $\mu_\infty = M = 3$ begins at $t = 18$ before the required bound of $t_\mu \leq MLK = 24$.

Theorem 12.3.8. *The time scale $\mathbb{A}_{1,\frac{L}{K}}$ where $L, K \in \mathbb{N}$ are relatively prime and $K < L$ achieves an asymptotic graininess $\mu_\infty = \frac{1}{K}$. Furthermore, this graininess is achieved at $t_\mu \leq L$.*

The time scale $\mathbb{A}_{1,\frac{L}{K}}$ is a scaled version of $\mathbb{A}_{K,L}$. Specifically

$$\mathbb{A}_{1,\frac{L}{K}} = \frac{1}{K} \mathbb{A}_{K,L}.$$

Since $\mathbb{A}_{K,L}$ achieves and asymptotic graininess of one at time KL, the asymptotic graininess of $\mathbb{A}_{1,\frac{L}{K}}$ is $\frac{1}{K}$ and occurs at $t_\mu = L$.

As an example, for $K = 3$ and $L = 4$,

$$\mathbb{A}_{1,\frac{4}{3}} = \left\{0, 1, \frac{4}{3}, 2, \frac{7}{3}, \frac{8}{3}, 3, \frac{10}{3}, \frac{11}{3}, 4, \ldots\right\}.$$

achieves its asymptotic graininess of $\frac{1}{3}$ is achieved at $t_\mu = 2$. This occurs before the bound of $t_\mu \leq 4$ indicated in Theorem 12.3.8.

Theorems 12.3.5.3 through 12.3.8 are subsumed in the following theorem for all rational time scales.

Theorem 12.3.9. Rational Time Scales. Let $\hat{K}_m$ and $\hat{L}_m \in \mathbb{N}$ for $m = 1, 2$ and let

$$\frac{\hat{L}_1}{\hat{K}_1} < \frac{\hat{L}_2}{\hat{K}_2}.$$

Let M be the common factor of $\hat{K}_1\hat{L}_2$ and $\hat{K}_2\hat{L}_1$ such that

$$ML = \hat{K}_1\hat{L}_2 \text{ and } MK = \hat{K}_2\hat{L}_1.$$

Then the time scale $\mathbb{A}_{\hat{L}_1/\hat{K}_1, \hat{L}_2/\hat{K}_2}$ achieves an asymptotic graininess of

$$\mu_\infty \leq \frac{\hat{L}_1}{K\hat{K}_1} \tag{12.24}$$

at time

$$t_\mu \leq \frac{L\hat{L}_1}{\hat{K}_1}. \tag{12.25}$$

To show this, we scale the time scale

$$\mathbb{A}_{\hat{L}_1/\hat{K}_1, \hat{L}_2/\hat{K}_2} = \frac{\hat{L}_1}{\hat{K}_1}\mathbb{A}_{1, \frac{\hat{K}_1\hat{L}_2}{\hat{K}_2\hat{L}_1}}$$

$$= \frac{\hat{L}_1}{\hat{K}_1}\mathbb{A}_{1, \frac{L}{K}}.$$

From Theorem 12.3.8, the time scale $\mathbb{A}_{1, \frac{L}{K}}$ achieves the asymptotic graininess of $1/K$ at time $t_\mu \leq L$. Thus, $\mathbb{A}_{\hat{L}_1/\hat{K}_1, \hat{L}_2/\hat{K}_2}$ achieves the asymptotic graininess in (12.24) at the time in (12.25).

Example 12.3.7. Consider the time scale $\mathbb{A}_{\frac{3}{4}, \frac{5}{6}}$ so that

$$\hat{L}_1 = 3, \ \hat{L}_2 = 5, \ \hat{K}_1 = 4, \ \hat{K}_2 = 6, \ M = 2, \ L = 10, \ K = 9, \ \mu_\infty = \frac{1}{12}, \ t_\mu \leq \frac{15}{2}.$$

Indeed, keeping the components over their common denominator of 12, we have

$$\mathbb{A}_{\frac{3}{4}, \frac{5}{6}} = \frac{1}{12} \times \{9, 10, 14, 15, 18, 19, 20, 23, 24, 25, 27, 28, 29,$$
$$30, 32, 33, 34, 35, 36, 37, 38, 39, 40, 41, \ldots\}$$

The asymptotic graininess of $\mu_\infty = 1/12$ is achieved at $t_\mu = 33/12 \leq 15/2$.

Theorem 12.3.10. Irrational time scales. *Let x be irrational and $x > 1$. Then the asymptotic graininess of $\mathbb{A}_{1,x}$ is zero and $t_\mu = \infty$.*

Irrational numbers can be expressed more and more accurately as the ratio of large relatively prime numbers. For example, $\pi = 3.14159\ldots \approx \frac{314{,}159}{1{,}000{,}000}$. Since 314,159 is prime, from Theorem 12.3.8, the asymptotic graininess of this approximation is $\mu_\infty =$ one millionth at time $t_\mu \leq 314{,}159$. As the irrational number is approximated more and more closely as the ratio of relatively prime integers, both the numerator and the denominator approach infinity. Thus, $\mu_\infty \to 0$ and the bound for $t_\mu \to \infty$.

From Theorems 12.3.9 and 12.3.10, we have shown that the asymptotic graininess of the time scale $\mathbb{A}_{\xi,\eta}$ is always constant. In general, if either ξ or η are irrational, then, assuming $\eta > \xi$,

$$\mathbb{A}_{\xi,\eta} = \xi \mathbb{A}_{1,\eta/\xi}.$$

The asymptotic graininess is determined by the ratio η/ξ which can be either rational or irrational.

12.3.6 Discrete Convolution of AITS'

Let $x \in \mathbb{D}$ and $h \in \mathbb{D}$ and let $\mathbb{D}$ be a discrete AITS. Then y, the inverse Fourier transform if $Y(u) = X(u)H(u)$, also has time scale $\mathbb{D}$. The time scale convolution in (8.98) becomes

$$y^\mu(t_p) = \sum_{t_p - \tau_n \in \mathbb{D}} x^\mu(\tau_n) h^\mu(t_p - \tau_n); \quad t_p \in \mathbb{D} \tag{12.26}$$

which we will write as

$$y^\mu(t) = x^\mu(t) \overset{\mathbb{D}}{*} h^\mu(t). \tag{12.27}$$

12.3.6.1 Applications

a. **Transformations on a random variable.**
   The time scale $\mathbb{L}_1$ can be used in evaluating the probability mass function of the product of two statistical independent discrete random variables. Let X and H be independent discrete random variables of the lattice type. Let all of the probability mass lie on integers on $[1, \infty)$. Let $Y = XH$ so that $\log Y = \log X + \log H$. If the probability mass function for X is p_n at $n = \{1, 2, 3, \ldots\}$, then the corresponding probability mass function for the random variable X is

$$p_X(t) = \sum_{n=1}^{\infty} p_n \delta[t - n].$$

   The probability mass function for $\log X$ is then

$$x^\mu(t) = \sum_{n=1}^{\infty} x_n^\mu \, \delta[t - \log n].$$

   where $x_n^\mu = p_n$. The probability mass function, $h^\mu(t)$ on $\mathbb{L}_1$, is similarly defined. Since $\log X$ is independent of $\log H$, the probability mass function, $y^\mu(t)$, of their sum, log Y, is given by the time scale convolution on $\mathbb{L}_1$ of $x^\mu(t)$ with $h^\mu(t)$:

$$y^\mu(t) = x^\mu(t) \stackrel{\mathbb{L}_1}{*} h^\mu(t) = \sum_{n=1}^{\infty} y_n^\mu \, \delta[t - \log n], \qquad (12.28)$$

where $y_n^\mu = \sum_{p/n \in \mathbb{N}} x_n^\mu \, h_{p/n}^\mu$. The probability mass function for Y is then

$$p_Y(t) = \sum_{n=1}^{\infty} y_n^\mu \, \delta[t - n]. \qquad (12.29)$$

Example 12.3.8. Two fair dice are rolled. The random variable assigned to a showing of n dots on a die is $\log(n)$. The probability mass function for one die is one sixth for points $1 \leq n \leq 6$ on the AITS corresponding to $\log_{10}(n)$. Thus, $x^\mu(t) = h^\mu(t) = \frac{1}{6}\Pi_6(t)$. The sum of the logs of the two dice, $y^\mu(t)$, is given by the time scale convolution of $x^\mu(t)$ and $h^\mu(t)$ and is shown in the upper left corner in Figure 12.25. The corresponding probability mass function, $p_Y(t)$, is shown in Figure 12.28. The mass at $n = 12$ has four realizations, $\{(3, 4), (4, 3), (2, 6), (6, 2)\}$, and therefore has four times the mass at $n = 25$ which only has one realization: $\{(5, 5)\}$.

b. **Mellin convolution.** A definition of *Mellin convolution* of discrete sequences $x[n]$ and $h[n]$ on $\mathbb{Z}$ is given by

$$y[n] = \sum_{(k \in \mathbb{N}) \ni (\frac{n}{k} \in \mathbb{N})} x[n] h\left[\frac{n}{k}\right].$$

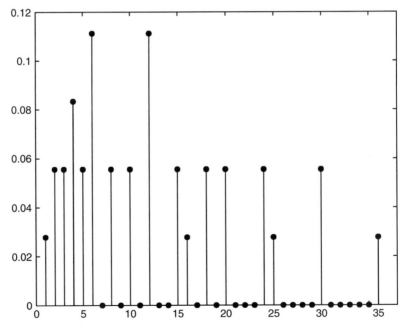

FIGURE 12.28. The probability mass function, $p_Y(t)$, in (12.29) corresponding to the time scale convolution illustrated on the upper left corner of Figure 12.25. The plot is shown as a function of n rather than t. The probability mass is that of the product of the showing on two dice.

For the discrete Mellin transform [891],

$$Y_M(\omega) = \sum_{n=1}^{\infty} y[n] n^{\omega},$$

it is straightforward to show that

$$Y_M(\omega) = X_M(-\omega) H_M(\omega).$$

Mellin convolution is performed when using the AITS $\mathbb{L}_1$. Since $t_p = \log(n)$, the discrete time convolution in (12.17) becomes

$$y^{\mu}(\log(p)) = \sum_{(\log(p)-\log(n)\in\mathbb{L}_1)\ni(\log(n)\in\mathbb{L}_1)} x^{\mu}(\log(n)) h^{\mu}(\log(p) - \log(n))$$

$$= \sum_{(\frac{p}{n}\in\mathbb{N})\cap(n\in\mathbb{N})} x^{\mu}(\log(n)) h^{\mu}\left(\log\left(\frac{p}{n}\right)\right)$$

This is a Mellin convolution for $y[p] = y^{\mu}(t_p)$, $x[n] = x^{\mu}(t_n)$, and $h[m] = h^{\mu}(m)$ where $t_q = \tau_q = \xi_q = \log(q)$.

c. **Time Scale FIR filter.** A 7 tap time scale FIR (finite impulse response filter) for $\mathbb{L}_1 = \{\log(n) | n \in \mathbb{N}\}$ is illustrated in Figure 12.29 for an input of duration 6. An input, x, is on additively idempotent discrete time scale, $\mathbb{D}$. The filter, h, can be implemented using a tap delay line. When h is on $\mathbb{D}$, then the output y will be on time scale $\mathbb{D}$ evaluated in accordance to (12.18). The output for a three input four tap $\log(n)$ FIR filter is

$$\begin{bmatrix} y_1 \\ y_2 \\ y_3 \\ y_4 \\ y_5 \\ y_6 \\ y_7 \\ y_8 \\ y_9 \\ y_{10} \\ y_{11} \\ y_{12} \end{bmatrix} = \begin{bmatrix} h_1 & - & - \\ h_2 & h_1 & - \\ h_3 & - & h_1 \\ h_4 & h_2 & - \\ - & - & - \\ - & h_3 & h_2 \\ - & - & - \\ - & h_4 & - \\ - & - & h_3 \\ - & - & - \\ - & - & - \\ - & - & h_4 \end{bmatrix} \begin{bmatrix} x_1 \\ x_2 \\ x_3 \end{bmatrix}$$

where the matrix entries $\{-\}$ all represent zero. Compare this to the conventional 4 tap 3 input $\mathbb{T} = \mathbb{Z}$ FIR filter.

$$\begin{bmatrix} y_1 \\ y_2 \\ y_3 \\ y_4 \\ y_5 \\ y_6 \end{bmatrix} = \begin{bmatrix} h_1 & - & - \\ h_2 & h_1 & - \\ h_3 & h_2 & h_1 \\ h_4 & h_3 & h_2 \\ - & h_4 & h_3 \\ - & - & h_4 \end{bmatrix} \begin{bmatrix} x_1 \\ x_2 \\ x_3 \end{bmatrix}$$

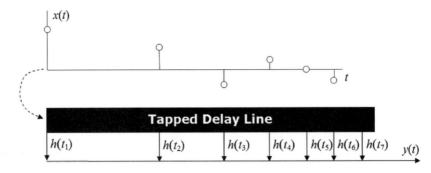

FIGURE 12.29. A time scale FIR filter.

12.3.7 Multidimensional AITS Time Scales

AITS time scales, $\mathbb{T}$, have the property that, if a signal $x \in \mathbb{T}$ and $h \in \mathbb{T}$, then $x * h \in \mathbb{T}$. This can be extended to higher dimensions.

Definition 12.3.2. *An N dimensional time scale*, $\mathbb{T}_N$, *is an arbitrary closed subset of points in* $\mathbb{R}^N$.

As we did in Section 8.2, denote the N-dimensional space with the coordinates $\{t_n \mid 1 \leq n \leq N\}$ which we will represent with a vector, $\vec{t}$. The nth element of the vector is denoted $t_n = (\vec{t}\,)_n$.

Definition 12.3.3. *If a multidimensional time scale* $\mathbb{T}_N$ *is AITS if the sum of all* $\vec{\tau} \in \mathbb{T}_N$ *with all* $\vec{\xi} \in \mathbb{T}_N$ *results in all of the points in* $\mathbb{T}_N$.

Example 12.3.9. The continuous time scale, $\mathbb{T}_N = \mathbb{R}^N$ is AITS.

Example 12.3.10. A periodic replication of points in accordance to a periodicity matrix is an AITS time scale. This is a multidimensional extension of the scalar time scale, $h\mathbb{Z}$. An example is the domain established by the replication of points in Figure 8.18 in Section 8.6.1.

In the figures for the following examples, black indicates inclusion within the time scale where white lies outside of the time scale.

Definition 12.3.4. In 2D, an AITS time scale, $\mathbb{A}_{\vec{p}_1, \vec{p}_2, \vec{p}_3}$ consists of all vectors that are integer multiples of $[0\ 0]^T$, $\vec{p}_1$, $\vec{p}_2$ and $\vec{p}_3$. This is a generalization of the scalar time scale $\mathbb{A}_{\xi\eta}$.

In Figures 12.30 and 12.31, the vectors $\vec{p}_1$, $\vec{p}_2$ and $\vec{p}_3$ are shown circled.[7]

Example 12.3.11. In Figure 12.30, we see $\mathbb{A}_{\vec{p}_1, \vec{p}_2, \vec{p}_3}$ for

$$\vec{p}_1 = [5\ 1]^T,$$
$$\vec{p}_2 = [13\ 13]^T,$$
$$\vec{p}_3 = [2\ 7]^T \qquad (12.30)$$

---

7. For this and the remainder of the examples, elements in the set are black.

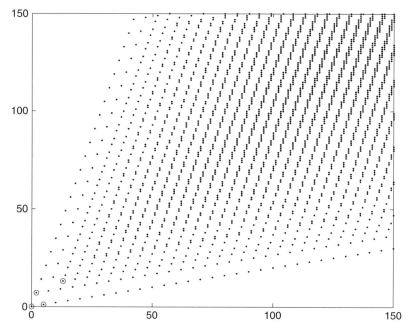

FIGURE 12.30. Figure for Example 12.3.11. The three points in (12.30) and the origin are circled. Using (12.32), the additive closure of these four points is shown here. The result is an AITS time scale. Any two dimensional function defined on this two dimensional time scale when convolved with a second function on this time scale will result in a function on this time scale. The same is true for the Figures 12.31 through 12.35.

Since the vector elements are rational, the equivalent of the two dimensional graininess in the upper right hand corner becomes periodic.

Example 12.3.12. In Figure 12.31, we see $\mathbb{A}_{\vec{p}_1, \vec{p}_2, \vec{p}_3}$ for

$$\vec{p}_1 = [\sqrt{13} \ 0]^T,$$
$$\vec{p}_2 = [\pi \ \pi]^T,$$
$$\vec{p}_3 = [0 \ \sqrt{12}]^T \qquad (12.31)$$

Because the vector components in (12.31) are irrational, the plot becomes black in the upper right hand corner. Contrast this with the rational values in Figure 12.30 where the closure becomes periodic.

Definition 12.3.5. For an arbitrary time scale, $\mathbb{T}_N$, define the *additive closure* of $\mathbb{T}_N$ as

$$\mathbb{T}_N^+ = \oplus_{k=1}^{\infty} \mathbb{T}_N. \qquad (12.32)$$

Theorem 12.3.11. *If $\mathbb{T}_N$ contains the origin, then $\mathbb{T}_N^+$ is AITS.*

Example 12.3.13. Let $\mathbb{T}_N$ be the origin and the rectangle with lower left point (25, 30) and upper right hand point (30, 35). Its additive closure is shown in Figure 12.32.

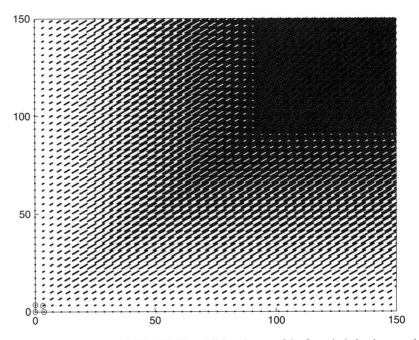

FIGURE 12.31. Figure for Example 12.3.12. The additive closure of the four circled points are shown. The result is AITS.

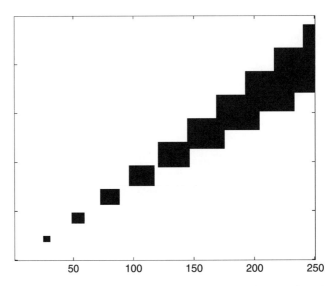

FIGURE 12.32. Figure for Example 12.3.13. The additive closure of the origin and the small black rectangle in the lower left corner is shown. The result is an AITS. The 2-dimensional functions on this support will convolve to a function also on this support.

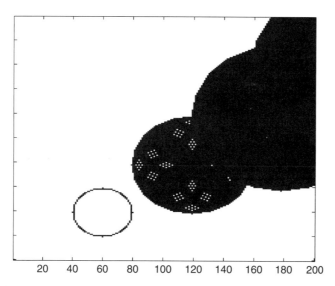

FIGURE 12.33. See Example 12.3.14. Shown is the additive closure of the origin and the annulus in the lower left corner.

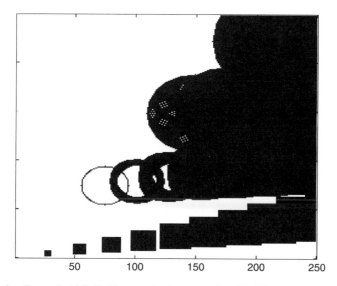

FIGURE 12.34. See Example 12.3.15. The annulus is centered at (75, 75) with a radius of 20. The box has a lower left corner at (3, 25) and an upper right corner at (8, 30). The additive closure is shown.

Example 12.3.14. The additive closure of the origin and the annulus in the lower left corner is shown in Figure 12.33.

Example 12.3.15. The additive closure of a time scale containing a small box, an annulus and the origin is shown in Figure 12.34.

Example 12.3.16. Same as Example 12.34 except that the single point (20, 100) is added.

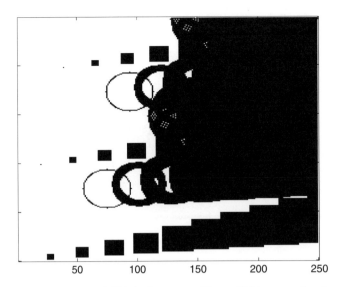

FIGURE 12.35. See Example 12.3.16. This is the same as Figure 12.34, except that the point (20, 100) is added.

12.4 Exercises

12.1. Specify a time scale that is closed under addition, but is not AITS.

12.2. Prove
   (a) the distributive properties in Theorem 12.2.6.
   (b) the extensive/antiextensive components of Theorem 12.2.8.
   (c) the idempotent properties of opening and closing in Theorem 12.2.9.

12.3. Let L_1 and L_2 be lines. Evaluate, in three dimensions, (A) $L_1 \oplus L_2$, (B) $L_1 \ominus L_2$, (C) $L_1 \circ L_2$, and (D) $L_1 \bullet L_2$, when
   (a) the lines are parallel.
   (b) the lines intersect.
   (c) the lines do not intersect and are not parallel.

12.4. (a) Specify an AITS that has both discrete and continuous components.
   (b) Prove that any AITS with right dense components is asymptotically right dense.

12.5. (a) Prove that the dilation of two convex sets is convex.
   (b) Is the erosion of one convex set by another convex?
   (c) The opening?
   (d) The closing?

12.6. The following convex sets of signal are defined in Chapter 11. Evaluate the following mathematical morphological operations. Some results may by empty or $\{0\}$. All results will be convex. Let $A_{\chi(t)} = \{x(t) \mid |x(t)| \leq \chi(t)$ where $\chi(t) \geq 0\}$ and $A_{FS} = \{x(t) | x(t) = 0$ for $|t| > T\}$.
   (a) $A_{CA=1} \oplus A_{CA=3}$.
   (b) $A_{CA=1} \ominus A_{CA=3}$.
   (c) $A_{CA=1} \circ A_{CA=3}$.
   (d) $A_{CA=1} \bullet A_{CA=3}$.
   (e) $A_{\chi(t)} \oplus A_{\chi(t)}$
   (f) $A_{\chi(t)} \ominus A_{\chi(t)}$
   (g) $A_{\chi(t)} \oplus A_{FS}$
   (h) $A_{\chi(t)} \ominus A_{FS}$

12.5 Solutions for Selected Chapter 12 Exercises

12.1.
$$\mathbb{N} \oplus \mathbb{N} = \{2, 3, 4, \cdots\} \subset \mathbb{N}$$
$$\neq \mathbb{N} = \{1, 2, 3, 4, \cdots\}.$$

12.2. Our presentation parallels that of Giardina and Dougherty [506] in which these proofs can be found [578].

13

Applications

> To isolate mathematics from the practical demands of the sciences is to invite the sterility of a cow shut away from the bulls.
> Pafnuty Lvovich Chebyshev [1276]

> Mathematics compares the most diverse phenomena and discovers the secret analogies that unite them.
> Jean Baptiste Joseph Fourier

> I believe there are 15,747,724,136,275,002,577,605,653,961,181,555,468,044,717,914,527,116, 709,366,231,425,076,185,631,031,296 protons in the universe and the same number of electrons.
> Sir Arthur Eddington (1882–1944) [397].

13.1 The Wave Equation, Its Fourier Solution and Harmony in Western Music

Fourier discovered the Fourier series as a solution to a boundary value problem [33, 303, 512, 620] related to the heat wave equation. Fourier's work on heat is still in print [455].

In this section, we derive the wave equation for the vibrating string and show how the Fourier series is used in its solution. The solution, in turn, gives rise to the physics of harmonics used as the foundation of music harmony. We contrast the natural harmony of the overtones to that available from the tempered scale of western music. The tempered scale is able to accurately approximate the beauty of natural harmony using a uniformly calibrated frequency scale.

13.1.1 The Wave Equation

The wave equation is manifest in analysis of physical phenomena that display wave like properties. This includes electromagnetic waves, heat waves, and acoustic waves. We consider the case of the simple vibrating string.

A string under horizontal tension T is subjected to a small vertical displacement, $y = y(x, t)$, that is a function of time, t, and location, x. As illustrated in Figure 13.1, attention is focused on an incremental piece of the string from x to $x + \Delta x$. Under the small

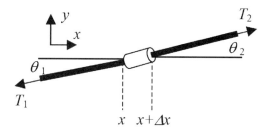

FIGURE 13.1. A section of a vibrating string.

displacement assumption, there is no movement of the string horizontally (i.e., in the x direction), and the horizontal forces must sum to zero.

$$T = T_1 \cos(\theta_1) = T_2 \cos(\theta_2).$$

Let the linear mass density (i.e., mass per unit length) of the string be ρ. The mass of the incremental piece of string is then ρ. The total vertical force acting on the string is $T_2 \cos(\theta_2) - T_1 \cos(\theta_1)$. Using Newton's second law ($F = ma$), we have

$$T_2 \sin(\theta_2) - T_1 \sin(\theta_1) = \rho \Delta x \frac{\partial^2 y}{\partial t^2}.$$

Dividing by the x force equation gives

$$\frac{T_2 \sin(\theta_2)}{T_2 \cos(\theta_2)} - \frac{T_1 \sin(\theta_1)}{T_1 \cos(\theta_1)} = \frac{\rho \Delta x}{T} \frac{\partial^2 y}{\partial t^2}$$

or

$$\tan(\theta_2) - \tan(\theta_1) = \frac{\rho \Delta x}{T} \frac{\partial^2 y}{\partial t^2}.$$

But

$$\tan(\theta_2) = \left.\frac{\partial y}{\partial x}\right|_{x+\Delta x}$$

and

$$\tan(\theta_1) = \left.\frac{\partial y}{\partial x}\right|_{x}.$$

We can therefore write

$$\frac{1}{\Delta x}\left(\left.\frac{\partial y}{\partial x}\right|_{x+\Delta x} - \left.\frac{\partial y}{\partial x}\right|_{x}\right) = \frac{\rho \Delta x}{T} \frac{\partial^2 y}{\partial t^2}.$$

Taking the limit as $\Delta x \to 0$ and applying the definition of the derivative, we arrive at the wave equation.

$$\frac{\partial^2 y}{\partial x^2} = \frac{1}{v^2} \frac{\partial^2 y}{\partial t^2} \qquad (13.1)$$

where

$$v^2 = \frac{T}{\rho} \tag{13.2}$$

is the velocity of the wave.

13.1.2 The Fourier Series Solution

A vibrating string fixed at $x = 0$ and $x = L$ does not move at these points and therefore has boundary conditions

$$y(0, t) = y(L, t) = 0. \tag{13.3}$$

A solution of the wave equation in (13.1) that satisfies these boundary conditions is[1]

$$y(x, t) = \sum_{m=1}^{\infty} A_m \cos\left(\frac{2\pi m v t}{L}\right) \sin\left(\frac{2\pi m x}{L}\right) \; ; \; 0 \leq x \leq L \tag{13.4}$$

where the A_m coefficients are to be determined. This solution can straightforwardly be shown to satisfy the boundary conditions in (13.4) and the wave equation in (13.1).

The coefficients, A_m, are determined by the initial displacement

$$y(x, 0) = f(x)$$

where $f(x)$ is any function satisfying the boundary conditions, i.e., $f(0) = f(L) = 0$. From (13.4),

$$f(x) = y(x, 0) = \sum_{m=1}^{\infty} A_m \sin\left(\frac{2\pi m x}{L}\right) \; ; \; 0 \leq x \leq L.$$

This expression is recognized as a (real) Fourier series. The coefficients are determined by

$$A_m = \frac{2}{L} \int_{x=0}^{L} f(x) \sin\left(\frac{2\pi m x}{L}\right) dx.$$

The lowest, or fundamental, frequency of the vibrating string is

$$u_0 = \frac{v}{L} \tag{13.5}$$

so that (13.4) can also be written as

$$y(x, t) = \sum_{m=1}^{\infty} A_m \cos(2\pi m u_0 t) \sin\left(\frac{2\pi m x}{L}\right) \; ; \; 0 \leq x \leq L \tag{13.6}$$

The frequencies $m u_0$ are harmonics of the fundamental frequency. A collection of simultaneously sounded acoustic harmonics is pleasing to the ear and is the foundation of western music.

---

1. See Exercise 13.1.

13.1.3 The Fourier Series and Western Harmony

The ratio of frequencies in the interval of an octave is 2. The ratio of the frequencies of two adjacent notes on a piano, such as C to C♯, is $2^{1/12}$. There are twelve semitones (or chromatic steps) in an octave producing, as desired, an overall frequency ratio of $\left(2^{1/12}\right)^{12} = 2$. This frequency calibration in western music is the *tempered scale*.

On the other hand true natural harmony, popularized by sixth century BC Greek philosopher Pythagoras required simultaneously sounded frequencies to have ratios equal to the ratios of small whole numbers. The number $2^{1/12}$ and, except for the octave, all western music intervals, cannot be exactly expressed as the ratio of small numbers. Indeed, the frequency ratios cannot be exactly expressed as the ratio of *any* whole numbers. They are irrational. Nevertheless, the tempered scale can be used to generate natural Pythagorean ratios to an accuracy often audibly indistinguishable to a trained listener. The tempered scale, in addition, allows freedom of key changes (modulation) in musical works difficult to instrumentally achieve in the Pythagorean system of harmony. The geometrically spaced frequency intervals of the tempered scale also make perfect use of the logarithmic frequency perception of the human ear. The ability of the tempered scale to generate near perfect Pythagorean harmonies can be demonstrated in a number of ways. A deeper question asks why this remarkable relationship exists. There is, remarkably, no fundamental mathematical or physical truth explaining the observed properties.[2]

Newtonian physics applied to the vibrating strings applies to violins, violas and guitars. Like the vibrating string, the vibrating air columns of bugles, clarinets, trombones and pipe organs obey a wave equation and have nearly identical solutions. Available tones of a string or an air column of fixed length are related by integer multiples (harmonics) of a fundamental frequency. Tonal color is largely crafted by choice of the strength of the components of the harmonics in a tone.

The Mth harmonic of a fundamental frequency, is simply $M + 1$ times the frequency of the reference frequency and can be expressed as the ratio $1 : M + 1$. For a vibrating string or air column, the root can be taken as the lowest frequency allowed by the physics of the imposed boundary constraints. Simultaneous sounding of notes corresponding to the root and its first five harmonics constitute a natural major chord. Notes with frequency ratios of $1 : M + 1$ form pleasing harmony when M is small. More generally, Pythagorean harmony claims two audio tones will harmonize when the ratio of their frequencies are equal to a ratio of small whole numbers (i.e., positive integers). This follows directly from the pleasing harmony of harmonics. For example, the second harmonic (1:3) and the third (1:4) harmonic harmonize. These two notes have a relative ratio of 3:4 and, indeed, form the interval of a natural perfect fourth. This pleasing musical interval is, as required by Pythagoras, the ratio of two small whole numbers.

Western music, on the other hand, is not based on the ratio of small whole numbers. It is, rather, built around twelfth root of two = $2^{1/12} = 1.05946\ldots$ which is the ratio of frequencies between two tones separated by a chromatic step (*a.k.a* a semitone or half step). The frequency, u_n, of a note n chromatic steps from a note with a reference frequency u_0, is

$$u_n = u_0 2^{n/12}. \tag{13.7}$$

---

2. See, however, Exercise 13.2.

The number of chromatic steps, n, between two frequencies, u_α and u_β, follows as

$$n = \log_{2^{1/12}} \left(\frac{u_\beta}{u_\alpha} \right).$$

A measure of frequency deviation is *cents*. There are one hundred cents in a chromatic step. The frequency that lies m cents above a reference frequency u_0, is

$$u_m = u_0 2^{n/1200}. \qquad (13.8)$$

The number of cents, m, between two frequencies, u_α and u_β, follows as

$$m = \log_{2^{1/1200}} \left(\frac{u_\beta}{u_\alpha} \right)$$

Harmonics often lie close to frequencies of the tempered scale. Consider the interval of a perfect fourth. Since a perfect fourth is five semitones, the ratio of frequencies is $\left(2^{1/12}\right)^5$. This seems a far cry from the Pythagorean frequency ratio of 3:4. When, however, the numbers are evaluated, we find, remarkably, that

$$\frac{4}{3} = 1.3333\cdots \approx \left(2^{1/12}\right)^5 = 1.3484\cdots. \qquad (13.9)$$

The difference between these notes is a minuscule audibly indistinguishable 2 cents. The near numerical equivalence in (13.9) illustrates a more general property of the tempered scale of western music. Using the awkward number, $2^{1/12}$, frequencies can be generated whose ratios are nearly identical to ratios of whole numbers.

13.1.4 Pythagorean Harmony

Harmony in western music is based on harmonics that result naturally from solution of the wave equation. According to Pythagoras, tones are harmonious when their frequencies are related by ratios of small whole numbers. The interval of an octave, or diapason, is characterized by frequency ratios of 1:2, 2:1, and 2:4. If any frequency is multiplied by 2^N where N is an integer, the resulting frequency is related to the original frequency by N octaves. For example, A above middle C is currently, by universal agreement, 440 Hz. Then 440/2 Hz = 220 Hz is A below middle C and $440 \times 8 = 3520$ Hz is the frequency of A a total of $N = 3$ octaves above A above middle C. The ratio of 2:3 is the perfect fifth or diapente interval while 3:4 is the perfect fourth or diatesseron interval.

Numbers that can be expressed as ratio of integers are rational numbers. Pythagoras claims harmony occurs between notes when the ratio of their frequencies are small whole numbers. The rules for the first few harmonics follow. The intervals cited are dubbed Pythagorean (or natural) since their relations are determined by ratios of small numbers. After each entry is the note corresponding to a root of middle C, denoted C4. (The *root* is the fundamental frequency.)

0. 1:1 defines the reference note or root. C4.
1. 2:1, with twice the frequency, is an octave above the root. C5.
2. 3:1 is the perfect fifth (2:1) of the first (2:1) harmonic. G5

3. 4:1 is twice the frequency of the first harmonic and is therefore two octaves above the root. C6
4. 5:1 is the major third of the third (4:1) harmonic. E6.
5. 6:1 is twice the frequency of the second (3:1) harmonic and is therefore the perfect fifth of the third (4:1) harmonic. G6.
6. 7:1 is the minor seventh of the third harmonic. B♭ 6.
7. 8:1 is three octaves above the root. C7.
8. 9:1 is the third harmonic of the second harmonic. It is therefore the major second of the seventh harmonic. (D7)
9. 10:1 is an octave above the fourth harmonic, and therefore the major third of the seventh harmonic. (E7).
10. 11:1 (F♯ 7)
11. 12:1 is an octave above the 5th harmonic. (G7) 12. 13:1 (A♭ 7)

There exist variations of harmonic frequencies from the corresponding tempered notes. This variation is shown in Figure 13.2. The fourth column, headed "Ratio", contains the normalized frequencies of the harmonics normalized by the frequency of the root. The next column, labelled "Temper", contains the normalized frequency of the corresponding tempered note. The "Cents" column contains the error between the harmonic and its tempered equivalent. The error for the lower harmonics is small, as is the error between harmonics whose ratios are products of small numbers, e.g., the fifth harmonic where $6 = 3 \times 2$ has an error of only 2 cents and the fifteenth harmonic, with ratio 16:24, has no error whatsoever. Harmonics with larger prime ratios, e.g., the sixth harmonic

Harmonic	Note	ST's	Ratio	Temper	cents
0	C4	0	1	1.000	0.0
1	C5	12	2	2.000	0.0
2	G5	19	3	2.997	-2.0
3	C6	24	4	4.000	0.0
4	E6	28	5	5.040	13.7
5	G6	31	6	5.993	-2.0
6	Bb6	34	7	7.127	31.2
7	C7	36	8	8.000	0.0
8	D7	38	9	8.980	-3.9
9	E7	40	10	10.079	13.7
10	F#7	42	11	11.314	48.7
11	G7	43	12	11.986	-2.0
12	Ab7	44	13	12.699	-40.5
13	Bb7	46	14	14.254	31.2
14	B7	47	15	15.102	11.7
15	C8	48	16	16.000	0.0

FIGURE 13.2. The first fifteen harmonics of the root C4. The tempered note closest to the harmonic is shown. The column ST refers to the number of tempered semitones from the root. The ratio column lists the integer multiple of the root frequency. The tempered column, to be compared to the ratio column, is the frequency ratio when the western tempered scale is used. Cents is the error between the Ratio and Tempered frequencies. One hundred cents is a semitone.

with ratio 7, tend to have larger errors. Indeed, the 10th harmonic corresponding to the prime ratio 11:1, has an error of 48.7 cents, less than 2 cents from being closer to an F than an F♯. The sixteenth harmonic, with prime ratio 1:17, is so far removed from the C4 root it has lost most if not all of its musical harmonic relationship. The seventeenth harmonic, on the other hand, has a ratio of 18 and is exactly an octave above the eighth harmonic.

Note all tempered notes with the same name deviate from the Pythagorean frequency by the same amount. All C's have an error of 0 cents and thus have no deviation. All G's have a deviation of 2 cents, all E's 13.7 cents, etc. The sixth harmonic deviates over 32 cents from the corresponding tempered note. The sixth harmonic deviates over 32 cents from the corresponding tempered note. This difference is enough to be detected by even the untrained musical ear.

The first four harmonics with the root form a natural major chord, e.g., C C G C E. Removing the redundant notes leaves C G and E. Including the next two harmonics yields a seventh chord (C G E B♭) and three more harmonics a ninth chord (C G E B♭ D). We can continue to an eleventh (C G E B♭ D F) and thirteenth (C G E B♭ D F A) chord. All of the numerical chord names are odd numbers simply because even numbers reflect of an octave relationship and add no new notes to the chord. The sixth harmonic, with ratio seven, is, indeed, the new note in the seventh chord (B♭). Likewise, the eighth harmonic with ratio nine, is added to obtain the ninth chord. Continuing further gives deviation. The tenth harmonic (F♯) with ratio eleven is different from new note in an eleventh chord (F). As noted, though, this harmonic is less than two cents from being closer to F than F♯. Similarly, the thirteenth (A), build by adding a major third (four semitones) to the eleventh, deviates from the note closest to eleven times the frequency of the root (A♭) which corresponds, instead, more closely to the addition of a minor third (three semitones).

13.1.4.1 Melodies of Harmonics: Bugle Tunes

Harmonics with ratios of 1:3 through 1:6 are used in the melodies played by bugles, including *Taps* played at military funerals and *Revelry* played to wake soldiers in the morning. The bugle, when unbent, is a simple vibrating air column as illustrated in Figure 13.3. The vibrating lips of the bugle player determine the vibration mode of the air. The vibration modes shown are those used in bugle melodies. The bugle therefore sounds true Pythagorean harmonics and not tempered note intervals.

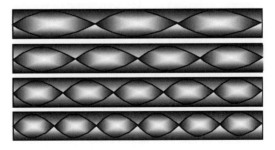

FIGURE 13.3. Vibration modes in an air column for the second (top) through fifth (bottom) harmonics at, respectively, three and six times the frequency of the root. These modes form the four notes for all bugle melodies including *Taps* and *Revelry*.

APPLICATIONS   617

13.1.4.2 Pythagorean and Tempered String Vibrations

A guitar string vibrates similarly to the air column patterns in Figure 13.3. Different notes are sounded only by changing the effective length of the string or, as is the case with generation of harmonics, applying initial conditions that prompt the string not to move at one or more points.

The vibration modes of a string are shown in Figure 13.4. Lightly touching the middle of the string over the twelfth fret bar and plucking results in the string vibrating in two halves. In the wave equation solution in (13.6), lightly touching at the twelfth fret imposes the boundary $A_1 = 0$, so that only higher harmonics can sound. The sounded tone is an octave higher than the string played open. Vibration in two sections continues even after the finger is removed from the string. This is the first harmonic of the note sounded by the open string.

The note sounded in the first harmonic is equivalent to the note sounded by a string of half the length. In other words, the first harmonic in Figure 13.4 can be viewed as two independent strings vibrating, each string being half the length of the open guitar string.

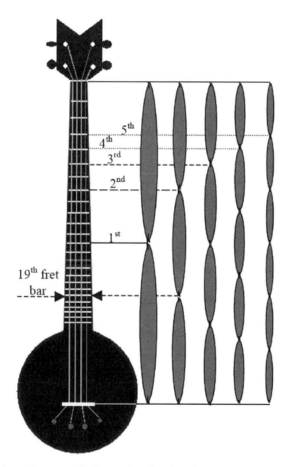

FIGURE 13.4. Vibrating strings can, similar to vibrating air columns, generate harmonics. On a guitar, lightly touching at the twelfth fret and plucking results in the string vibrating in two pieces and sounds the first harmonic of the root tone of the open string. Lightly touching the seventh fret and plucking results in three vibrating string sections and the second harmonic. The process can be repeated, as shown here, for higher order harmonics.

The twelfth fret bar divides the string into two equal pieces. If the string is depressed on the twelfth fret in a conventional matter, the sounded note, after plucking, is the same as the note of the string's first harmonic.

Placing the finger lightly over the seventh fret bar and plucking forces both coefficients A_1 and A_2 to zero in (13.6). The string now vibrates in three equal pieces. (See Figure 13.4.) This is the second harmonic. Besides the bridges that mechanically constrain the string from vibrating, there are two nodes where the string is not moving. One is over the seventh fret bar. While the string is vibrating, this point on the string can be lightly touched and the string continues to vibrate. The same is true if the string is lightly touched over the nineteenth fret bar. This is one of two places, other than the bridges, where the string does not vibrate. Placing the finger on the twelfth fret, on the other hand, interrupts the vibration of the string and the note ceases to sound.

The note sounded in the second harmonic can be viewed as resulting from the vibration of three independent strings each with a length of one third that of the open string. Depressing the nineteenth fret, which leaves one third of the length on the business end of the string, yields this same note, when plucked, as the second harmonic.

Continuing, a finger lightly touching the string over the fifth fret bar will, after plucking, result in the string vibrating in four equal parts. This mode corresponds to the vibration shown second from the top in Figure 13.4 and sounds the third harmonic of the open string's root note. Similarly, lightly touching above the fourth fret gives the fourth harmonic and the third fret gives the fifth.

Note that, by moving the finger between the seventh, fifth, fourth and third fret, the sounded notes are those necessary to play the bugle tunes. Like the vibrating air column, these harmonics can be sounded by applying the proper initial conditions and stimuli. For the bugle, the stimulus is air and the boundary conditions the frequency of the bugler's vibrating lips. For the string, the boundary conditions are imposed through lightly touching the string while the stimulus is a simple pluck. Remarkably, these physically different systems with different physics display similar musical (and physical) properties.

13.1.5 Harmonics Expansions Produce Major Chords and the Major Scale

A structured illustration of the tempered scale's ability to provide Pythagorean harmony comes from derivation of the tones in a tempered scale. We begin with the I, IV and V chords of the major scale. If C is the root, then I is a C major chord, IV is an F chord and V a G. Note G is the closest non-octave harmonic to C. In a dual sense, F has, as its closest non-octave harmonic, C. As before, all frequencies are normalized to the frequency of the root, C. The note C therefore has a normalized frequency of one.

In the construction of the I, IV, V chords, any note's frequency can be multiplied by 2^M without changing the note name as long as M is an integer - negative or positive. For any note, a value of M can always be found to place the notes normalized frequency between 1 and 2. Both 1 and 2 correspond to C's. Placing a note normalized in frequency between 1 and 2 is therefore simply constraining the note to lie in a specified octave.

The C major chord is formed from the first five harmonics of C. Removing the octave harmonics (the first and third), the notes of the C chord are C, G, and E with relative frequencies 1, 3, and 5. The frequency 3 for can be reduced to the desired octave by dividing by 2 to form the normalized frequency 3/2. Similarly, the frequency 5 can be divided by $4 = 2^2$ to form the normalized frequency 5/4. When reduced to the octave between 1 and 2, the notes C, G and E have normalized frequencies of 1, 3/2, and 5/4.

A similar analysis is applicable to the V (G) chord. Beginning with a chord root of 3, the notes in the G chord, corresponding to the second and fourth harmonics, are 9 and 15. Thus, the G, D, and B notes of the G chord have respective frequencies 3, 9 and 15. By choosing an appropriate integer, M, each can be placed in the interval of 1 to 2. The 3, as before, is divided by 2 corresponding to $M = -1$. Using $M = -3$, the frequency of 9 becomes 9/8.

The note B becomes 15/8. The frequencies of the V chord in the frequency interval of 1 to 2 are therefore 3/2, 9/8 and 15/8. Lastly, consider the IV chord. If we construe F as being the note to which C is the second harmonic, then frequency of the F note is 1/3. The notes for the F major chord, F, A, and C, follow as 1/3, 3/3 and 5/3. Each of these can be placed the interval 1 to 2. The result is 4/3, 1, and 5/3.

The results of our analysis are shown in Figure 13.5. The top table has the harmonic construction of the I, IV, and V chords using harmonics followed by reducing the normalized frequencies to the octave between 1 and 2. The middle table contains the corresponding notes when the root note is a C. The last table is a decimal equivalent of the ratio in the top table.

We assign an index of $k = 0$ to C, $k = 1$ to C♯, $k = 2$ to D, etc. Arranging the entries in Figure 13.5 in ascending values of k gives the results shown in Figure 13.6. A logarithmic plot of the ratio values, shown in Figure 13.7, results in a remarkably straight line. Using the entries in Table 7, the standard deviation of the natural 8 note major scale from the tempered scale is a minuscule 8.6 cents.

ratios	IV F	I C	V G
Tonic	4/3	1	3/2
P 5th	1	3/2	9/8
M 3rd	5/3	5/4	15/8

notes	IV F	I C	V G
Tonic	F	C	G
P 5th	C	G	D
M 3rd	A	E	B

decimal	IV F	I C	V G
Tonic	1.3333	1.0000	1.5000
P 5th	1.0000	1.5000	1.1250
M 3rd	1.6667	1.2500	1.8750

FIGURE 13.5. Constructing the Pythagorean 8 note major scale.

note	k	ratio	tempered	cents
C	0	1.0000	1.0000	0
C#	1		1.0595	
D	2	1.1250	1.1225	4
Eb	3		1.1892	
E	4	1.2500	1.2599	-14
F	5	1.3333	1.3348	-2
F#	6		1.4142	
G	7	1.5000	1.4983	2
Ab	8		1.5874	
A	9	1.6667	1.6818	-16
Bb	10		1.7818	
B	11	1.8750	1.8877	-12
C	12	2.0000	2.0000	0

FIGURE 13.6. Comparison of the natural Pythagorean major scale with the tempered major scale.

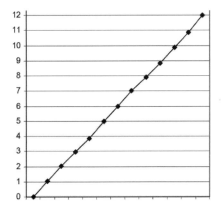

FIGURE 13.7. A logarithmic plot of the Pythagorean ratios in Figure 13.6. If the line were exactly straight, each mark would lie on a horizontal line. (Missing entries in Figure 13.6 were replaced by the geometric mean of adjacent entries. The geometric mean of numbers p and q is $\sqrt{pq}$.)

13.1.5.1 Subharmonics Produce Minor Keys

Consider, then, performing an operation similar to that in Figure 13.5, except subharmonics are used instead of harmonics. The nth subharmonic of a note has a frequency of $1/n$th the root. Rather than being sequential integer multiples of the root as is the case with harmonic, subharmonics are sequential integer divisors of the root. The three notes, F, C, and G, are expanded into their first four subharmonics. For each subharmonic, the note is moved into the octave with normalized frequencies between 1 and 2 by multiplying by 2^M.

Wherein harmonics build major chords, subharmonics build minor chords. With C4 as the root, the note with ratio $\frac{1}{2}$ is clearly an octave lower, or C3. The note with the ratio $\frac{1}{3}$ has C4 as its third harmonic. This note is F2. We quadruple its frequency up two octaves to $\frac{4}{3}$ to place it in the interval between 1 and 2.

APPLICATIONS 621

The next subharmonic, C2, an octave below C3, has a ratio of $\frac{1}{4}$. The ratio $\frac{1}{5}$ is the note having C4 as its fifth harmonic. This is A♭1. To place this in the interval 1 to 2, we multiply by 8 and assign a frequency of $\frac{8}{5}$ to A♭4. The next subharmonic, with a value $\frac{1}{6}$, is an octave below $\frac{1}{3}$ and is therefore F1. The first notes generated by subharmonic expansion of C are C, F, and A♭. These notes constitute an F minor chord. Subharmonics expansions therefore generate minor chords.

Subharmonic expansions are likewise performed for F and G. The result of all three chords is shown in Figure 13.8. The notes generated are those in the natural minor key of F minor. The I, IV and V chords (Fm B♭m and Cm) of the key are generated by the subharmonics.

13.1.5.2 Combining the Harmonic and Subharmonic Expansions Approximates the Tempered Chromatic Scale

Nearly all of the missing notes in the harmonic expansion in Figure 13.5 are present in subharmonic expansion in Figure 13.8. As shown in Figure 13.9, the only missing note in the union of the figures is the tritone, F♯. A logarithmic plot of these entries is shown in Figure 13.10. The standard deviation of the fit is a mere 12.1 cents.

13.1.6 Fret Calibration

From (13.5), the product of the length, ℓ_0, of a string and its fundamental frequency is, from (13.5) and (13.2)

$$u_0 \ell_0 = v$$

$$= \sqrt{\frac{T}{\rho}} \qquad (13.10)$$

ratios	F	C	G
root	4/3	1	3/2
3rd sub	16/9	4/3	1
5th sub	16/15	8/5	6/5

noted	B♭m	Fm	Cm
root	F	C	G
3rd sub	B♭	F	C
5th sub	D♭	A♭	E♭

decimal	B♭m	Fm	Cm
root	1.3333	1.0000	1.5000
3rd sub	1.7778	1.3333	1.0000
5th sub	1.0667	1.6000	1.2000

FIGURE 13.8. Chord expansion using subharmonics. The result is the notes in the key of F minor. The chords are I, IV and V chords of the key of (natural) F minor.

note	k	ratio	tempered	cents
C	0	1.0000	1.0000	0
C#	1	1.0667	1.0595	12
D	2	1.1250	1.1225	4
Eb	3	1.2000	1.1892	16
E	4	1.2500	1.2599	-14
F	5	1.3333	1.3348	-2
F#	6	x	1.4142	x
G	7	1.5000	1.4983	2
Ab	8	1.6000	1.5874	14
A	9	1.6667	1.6818	-16
Bb	10	1.7778	1.7818	-4
B	11	1.8750	1.8877	-12
C	12	2.0000	2.0000	0

FIGURE 13.9. Eleven of the 12 tones of the tempered chromatic scale result naturally from harmonic and subharmonic expansion of the root, perfect fifth and perfect fourth. These entries are a combination of the entries in Figure 13.5 (the harmonic expansion) and Figure 13.8 (the subharmonic expansion). The natural Pythagorean results, as witnessed by the low cents error, are remarkably close to the corresponding tempered frequencies.

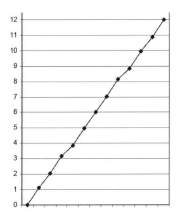

FIGURE 13.10. A logarithmic plot of chromatic intervals in Figure 13.9 as predicted by harmonic and subharmonic expansions within an octave. Ideally, the plot should be a line with every tick exactly intersecting a horizontal line. Although not perfect, the naturally generated notes are nearly identical to their tempered equivalents. (The geometric mean was used to interpolate the tritone.)

Equation 13.10 allows calibration of fret bars. An open string of length ℓ_0 sounds a frequency of u_0, the same string, shortened to length ℓ_n, will produce a frequency u_n. Since the linear mass density and string tension are the same, we conclude

$$u_0 \ell_0 = u_n \ell_n.$$

Substituting the chromatic frequency relationship in (13.7) for u_n followed by simplification leaves the equation for fret spacing.

$$\ell_n = 2^{-n/12} \ell_0.$$

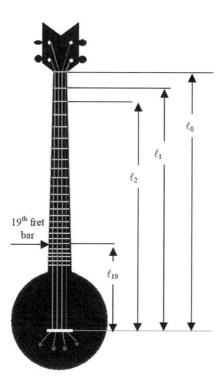

FIGURE 13.11. The length of a string from bridge to bridge is ℓ_0. The length from the bridge to the nth fret bar is $\ell_n = \ell_0 \, 2^{-\frac{n}{12}}$.

Thus, in Figure 13.11, the first fret is placed at a distance $\ell_1 = 2^{-1/12}\ell_0 = 0.9439\ell_0$ from the bottom bridge. The second fret bar is at $\ell_2 = 2^{-1/12}\ell_1 = 2^{-2/12}\ell_0$ and the nineteenth fret bar is at $\ell_{19} = 2^{-19/12}\ell_0$.

The fret calibration aligns closely with the string vibrational harmonics illustrated in Figure 13.4. The twelfth fret in Figure 13.11, corresponding to an octave, divides the string exactly in half. The first overtone in Figure 13.4 is thus achieved. The seventh fret comes within 2 cents of generating the second overtone, etc.

13.1.6.1 Comments

The tempered scale closely mimics the structured naturally occurring whole number Pythagorean frequency ratios while allowing greater flexibility in key changes in compositions and instrument design. There is no foundational mathematical or physical reason the relationship between Pythagorean and tempered western music should exist. It just does. The rich flexibility of the tempered scale and the wonderful and bountiful archives of western music are testimony to this wonderful coincidence provided by nature.

13.2 Fourier Transforms in Optics and Wave Propagation

The Fourier transform is ubiquitous in modelling of physics. In this section, we look as the special case of formation of optical Fourier transforms using monochromatic[3]

---

3. A single wavelength or frequency.

coherent[4] optics. The far field or Fraunhofer diffraction of an aperture is expressed as a two dimensional Fourier transform. The electric field amplitude in the back focal plane of a lens is proportional to the Fourier transform of the electric field amplitude in the front focal plane. Indeed, the field we are introducing in this section is dubbed *Fourier optics* [493, 514, 518, 1344].

Although the analysis presented in this section is specifically for the propagation of optical waves, it is equally effective in the modelling of many diffraction phenomena. The ideas developed in this section, including Fresnel diffraction, Fraunhofer diffraction, and the relation of the far field to Fourier transforms are applicable. Equivalent models are used in acoustics [1214, 1497], sonar [830], microwave engineering and antenna design [290, 1170, 1228, 1525] including beamforming[5] [653, 1546].

13.2.1 Scalar Model for Wave Propagation

There are three commonly used models for describing electromagnetic propagation: (1) ray optics, (2) scalar optics and (3) vector optics. As listed, the mathematical difficulty and result accuracy increase with the number. Ray tracing through elementary imaging systems is typically taught in high school physics. In the fundamental ray optics model, however, the wave nature of light and its ability to constructively and destructively interfere, cannot be straightforwardly described. Scalar, or Fourier, optics does have this ability. The mathematical predictions from scalar optics are much more accurate than that from ray optics. Vector optics, with its roots in Maxwell's equations, must be used, however, to describe concepts like polarization. Scalar optics, as the name suggests, models light as a scalar instead of a vector. Whereas many of the results in ray optics can be derived from scalar optics, the fundamentals of scalar optics can be derived from vector optics. We will derive the fundamental expression for scalar diffraction from Maxwell's equations. The derivation deviates from the more conventional Green's function derivation [493, 514, 514, 1082].

The systems concepts will be applied to the physics of electromagnetic propagation in order to derive the *Rayleigh-Sommerfield scalar diffraction integral*. This integral describes the manner in which monochromatic coherent light propagates.

13.2.1.1 The Wave Equation

Recall Maxwell's equations.

$$\nabla \times \vec{E} = \frac{-\partial \vec{B}}{\partial t}$$

$$\nabla \times \vec{H} = \vec{J} + \frac{\partial \vec{D}}{\partial t} \quad ; \quad \vec{D} = \varepsilon \vec{E}$$

$$\nabla \cdot \vec{D} = \rho \quad ; \quad \vec{B} = \mu \vec{H} \quad (13.11)$$

$$\nabla \cdot \vec{B} = 0.$$

---

4. The light is capable of total constructive and destructive interference.
5. See Section 13.2.5.

The definitions of the various fields, densities and vector operations are those standardly used. In a source free medium, $\vec{J} = \rho = 0$. In a homogeneous[6] isotropic[7] medium such as free space, ε and μ are scalar constants. Thus

$$\nabla \times \vec{E} = -\mu \frac{\partial \vec{H}}{\partial t}$$

$$\nabla \times \vec{H} = \varepsilon \frac{\partial \vec{E}}{\partial t}$$

$$\nabla \cdot \vec{E} = 0$$

$$\nabla \cdot \vec{B} = 0.$$

In general, $\vec{E} = \vec{E}(x, y, z; t)$ and $\vec{H} = \vec{H}(x, y, z; t)$. We will assume that the fields are temporally oscillating at an angular frequency[8] of ω. The resulting time harmonic fields[9] are

$$\vec{E} = \vec{E}_0 e^{-j\omega t} \quad ; \quad \vec{E}_0 = \vec{E}_0(x, y, z)$$
$$\vec{H} = \vec{H}_0 e^{-j\omega t} \quad ; \quad \vec{H}_0 = \vec{H}_0(x, y, z)$$

Optically, this restricts our coherent illumination to a single wavelength, $\lambda = 2\pi c/\omega$ where c is the speed of light in the medium. As a consequence of the time harmonic assumption,

$$\frac{\partial \vec{E}}{\partial t} = -j\omega \vec{E}_0 e^{-j\omega t}$$

and

$$\frac{\partial \vec{H}}{\partial t} = -j\omega \vec{H}_0 e^{-j\omega t}.$$

Thus, Maxwell's equations become

$$\nabla \times \vec{E}_0 = j\omega \mu \vec{H}_0$$
$$\nabla \times \vec{H}_0 = -j\omega \varepsilon \vec{E}_0$$
$$\nabla \cdot \vec{E}_0 = 0$$
$$\nabla \cdot \vec{H}_0 = 0.$$

From the first two equation,

$$\nabla \times \nabla \times \vec{E}_0 = j\omega \mu \nabla \times \vec{H}_0 = \omega^2 \mu \varepsilon \vec{E}_0.$$

Using the vector identity

$$\nabla \times \nabla \times \vec{E}_0 = \nabla(\nabla \cdot \vec{E}_0) - \nabla^2 \vec{E}_0$$

---

6. The same everywhere.
7. Propagation is the same in all directions.
8. Because of exclusive use in this derivation, we use ω in lieu of $2\pi u$ in this section only.
9. An alternate convention uses a positive exponent. This may be done throughout by choosing $\exp(-j\omega t) = \exp(i\omega t)$, i.e., set $j = -i$.

gives, since $\nabla \cdot \vec{E}_0 = 0$,

$$-\nabla^2 \vec{E}_0 = \omega^2 \mu \varepsilon \vec{E}_0$$

or

$$\nabla^2 \vec{E}_0 + \omega^2 \mu \varepsilon \vec{E}_0 = 0 \tag{13.12}$$

where ∇^2 is the Laplacian operator.

$$\nabla^2 = \frac{\partial^2}{\partial x^2} + \frac{\partial^2}{\partial y^2} + \frac{\partial^2}{\partial z^2}.$$

Equation 13.12 is the *electric field wave equation*. Define the *magnitude of the propagation vector* as k where

$$k^2 = \omega^2 \mu \varepsilon = \left(\frac{2\pi}{\lambda}\right)^2 \tag{13.13}$$

where λ is the wavelength. Then the electric field wave equation becomes

$$(\nabla^2 + k^2)\vec{E}_0 = 0. \tag{13.14}$$

13.2.1.2 Solutions to the Wave Equation

We consider two solutions to the wave equation in (13.14). The first treats a propagating wave as a superposition of plane waves and is dubbed the *angular spectrum*. The second treats the wave as a superposition of point sources, also known as *Huygen's spherical wavelets*, and results in Rayleigh-Sommerfield diffraction. In a mathematical sense, the angular spectrum composes a planar cross section coherent wave, $u(x, y)$, as an inverse Fourier transform[10]

$$u(x, y) = \int_{-\infty}^{\infty} \int_{-\infty}^{\infty} U(f_x, f_y) e^{j2\pi(f_x x + f_y y)} df_x df_y. \tag{13.15}$$

whereas the Huygen's wavelets model views the function through the Dirac delta sifting property.

$$u(x, y) = \int_{-\infty}^{\infty} \int_{-\infty}^{\infty} u(\xi, \eta) \delta(x - \xi) \delta(y - \eta) d\xi d\eta. \tag{13.16}$$

This is illustrated in Figure 13.12.

The manner that the angular spectrum plane wave and the Huygen's point sources propagate from plane to plane is governed by the wave equation. The propagation phenomenon is dubbed *diffraction*.

---

10. In this section, we use the more standard optics notation (x, y, z) instead of (t_1, t_2, t_3) used in Chapter 8. Likewise, the frequency variables (f_x, f_y) are used in lieu of (u_1, u_2).

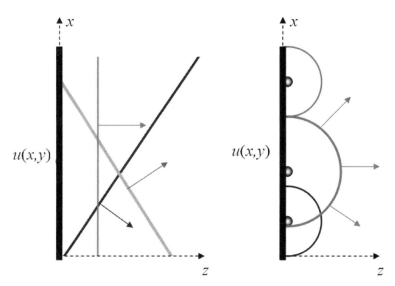

FIGURE 13.12. Scalar diffraction can be viewed from two perspectives. An electric field amplitude, $u(x, y)$, on the $z = 0$ plane can be viewed as emitting a superposition of (LEFT) plane waves travelling in different directions with different weights and phases, or (RIGHT) spherical waves from each point. For the plane wave case, the diffraction is modelled by the two dimensional inverse Fourier transform in (13.15) and results in the angular spectrum model of diffraction. For the spherical waves, the diffraction model uses the Dirac delta sifting property in (13.16) is used to model $u(x, y)$ and results in the Rayleigh-Sommerfeld model of diffraction.

13.2.2 The Angular Spectrum

Assuming $\vec{E}_0$ is linearly polarized in a fixed direction so that $\vec{E}_0 = u_z(x, y)\vec{a}$ where $\vec{a}$ is a unit vector. The wave equation in (13.14) then becomes the scalar equation

$$(\nabla^2 + k^2)u_z(x, y) = 0. \tag{13.17}$$

A plane wave solution to (13.17) is

$$u_z(x, y) = e^{jk(\alpha x + \beta y + \gamma z)} \tag{13.18}$$

where $(\alpha = \cos(\theta_x), \beta = \cos(\theta_y), \gamma = \cos(\theta_z))$ are the *direction cosines* of the propagation. The direction cosines are illustrated in Figure 13.13 and, from the Pythagorean theorem, obey

$$\alpha^2 + \beta^2 + \gamma^2 = 1. \tag{13.19}$$

Since (13.18) is a solution to the wave equation and the wave equation is linear, any superposition of plane waves also solves the wave equation. On the $z = 0$ plane, a solution is[11]

$$u_0(x, y) = \int_\alpha \int_\beta w(\alpha, \beta) e^{jk(\alpha x + \beta y)} d\alpha, d\beta \tag{13.20}$$

---

11. We loose little information by discarding the z component. For a given (α, β), the value of $\pm\gamma$ is specified by (13.19). Thus, except for the ambiguity of a conjugation, the two dimensional slice of a three dimensional electric field defines the field in three dimensions. This property is used in holography where a two dimensional surface is used to reconstruct a three dimensional wavefront.

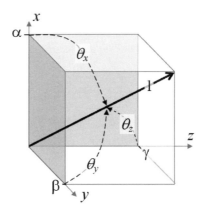

FIGURE 13.13. Direction cosines, (α, β, γ), for a plane wave propagating in the direction of the arrow. The constant phase of the plane is perpendicular to the vector. Since the vector is unit length, the direction cosines (α, β, γ) are the vector's (x, y, z) components.

where the $w(\alpha, \beta)$'s are arbitrary complex weights. Motivated by (13.15), a more meaningful expression of the same result is

$$u_0(x, y) = \int_{\frac{\alpha}{\lambda}} \int_{\frac{\beta}{\lambda}} U_0\left(\frac{\alpha}{\lambda}, \frac{\beta}{\lambda}\right) e^{j2\pi\left(\frac{\alpha}{\lambda}x + \frac{\beta}{\lambda}y\right)} d\left(\frac{\alpha}{\lambda}\right) d\left(\frac{\beta}{\lambda}\right) \quad (13.21)$$

where we have used

$$k = \frac{2\pi}{\lambda}$$

and

$$u_0(x, y) \longleftrightarrow U_0(f_x, f_y).$$

The weight of the plane wave with direction cosines (α, β) contributing to $u_0(x, y)$ is therefore determined by the Fourier transform of $u_0(x, y)$. The function $U_0\left(\frac{\alpha}{\lambda}, \frac{\beta}{\lambda}\right)$ is thus known as the *angular spectrum* of $u(x, y)$.

13.2.2.1 Plane Waves as Frequencies

Equation 13.18 reveals that frequency components in coherent optics can be thought of as plane waves incident on the $z = 0$ plane. This is illustrated in Figure 13.14. To keep the geometry in two dimensions, we set $\theta_y = \frac{\pi}{2}$ so that $\beta = 0$. The propagation direction is at an angle of θ_x as shown. The corresponding direction cosine is α. If the wavelength is λ, then the spacing of the plane wave on the x axis is, as shown, λ/α. Equivalently, this is a spatial frequency of α/λ. This is a visualization of the frequency interpretation of the plane wave used in the inverse Fourier transform in (13.21) where the frequency variable f_x is replaced by α/λ.

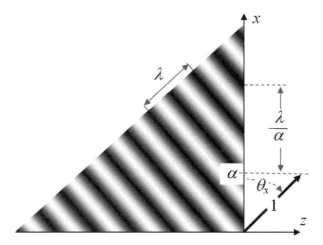

FIGURE 13.14. A plane wave propagating at an angle of θ_x generates a spatial frequency with frequency α/λ on the x axis where $\alpha = \cos(\theta_x)$.

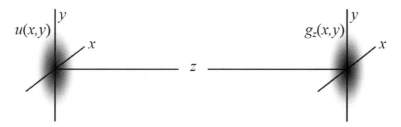

FIGURE 13.15. A coherent monochromatic field amplitude, $u(x, y)$, is on the left hand plane. Diffraction occurs over a distance z to give the field amplitude $g_z(x, y)$.

Here are some observations.

(a) The steeper the plane wave, the higher the frequency.
(b) The smallest period we can achieve is λ. Spatial frequencies higher than $1/\lambda$ cannot be resolved.[12]
(c) Unlike temporal signals, the frequency components from plane waves can be complex. We can have the plane wave corresponding to $e^{jk(\alpha x+\beta y)}$ without its conjugate $e^{-jk(\alpha x+\beta y)}$. Hence, positive frequency components can exist without a conjugately symmetric partner. We also see this must be true since electric field amplitudes like $u_0(x, y)$ in (13.21) are generally complex.

13.2.2.2 Propagation of the Angular Spectrum

Consider the diffraction geometry illustrated in Figure 13.15. A known monochromatic coherent field $u(x, y)$ propagates a distance z where it becomes $g_z(x, y)$. Given the angular spectrum of u, we seek to find the angular spectrum of g_z.

---

12. See Section 13.2.2.3.

A propagating plane wave remains a plane wave, shifting phase by 2π every wavelength. Its direction cosines remain the same. Every plane wave component of the angular spectrum has this property. Using (13.19), it follows that

$$G_z\left(\frac{\alpha}{\lambda}, \frac{\beta}{\lambda}\right) = U\left(\frac{\alpha}{\lambda}, \frac{\beta}{\lambda}\right) e^{jk\sqrt{1-\alpha^2-\beta^2}\,z}. \tag{13.22}$$

The electric field amplitude, $g_z(x, y)$, is determined by an inverse Fourier transform.

13.2.2.3 Evanescent Waves

For a monochromatic wave, there is a fixed wavelength λ. We see in Figure 13.14 that the highest spatial frequency that can be generated by such illumination is

$$f_{\max} = \frac{1}{\lambda}.$$

This is manifest in the angular spectrum expression where the direction cosines must obey

$$-1 \leq \alpha, \beta \leq 1.$$

The integrations in (13.21) are therefore *not* over $(-\infty, \infty)$, but over a circle in the frequency plane with radius $f_{\max}$. Plane waves with spatial frequencies outside of this region are said to be *evanescent* and do not propagate. For evanescent waves, the square root in (13.22) becomes imaginary and, rather than propagating, the wave decays exponentially.

13.2.3 Rayleigh-Sommerfield Diffraction

We now consider an alternative solution to the diffraction problem illustrated in Figure 13.15. In (x, y, z) space, a coherent monochromatic electric field, $u(x, y)$, exists on the $z = 0$ plane. Assuming no other sources, we ask what will the resulting field be at a distance z away? Call this field $g_z(x, y)$. We will assume that only free space is between the two planes. The manner in which the field propagates is determined by the physics of Maxwell's equations. We can directly apply some systems concepts. Note, for example, that the system we seek to analyze is clearly shift invariant or, as called in optics, isoplanatic. As we shift the input about, the output will shift accordingly. System linearity for the electric field, interestingly, manifests itself as a coherence constraint on the monochromatic illumination. In order for the outputs to add, we must allow for constructive and destructive interference. This requires that the light be coherent, such as from a pin hole or laser.

Consider, then, the two plane system illustrated in Figure 13.15. The input is the electric field amplitude on the left plane and the output is the electric field amplitude on the right plane. Under our assumptions, the system is linear since Maxwell's equations are linear. If we double the electric field amplitude on the input plane, we double the electric field amplitude on the output plane (homogeneity). The response of the sum of two input fields is the sum of the responses (additivity). Since the system is both additive and homogeneous, it is linear.[13]

---

13. See Section 3.2.1.3.

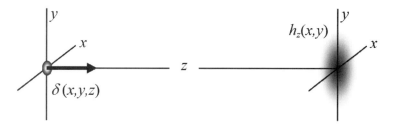

FIGURE 13.16. For coherent monochromatic inputs, diffraction is a linear isoplanatic operation. The system can thus be characterized by the impulse response, $h_z(x, y)$.

Linear shift invariant systems can be characterized by a single impulse response.[14] Given the impulse response, the output corresponding to any input is given by the two dimensional convolution of the input with the impulse response as discussed in Section 8.4.3. Between the planes in Figure 13.15 is free space. On the left we place an impulse of light. The electric field amplitude on the output plane is $h_z(x, y)$. We will use Maxwell's equation to find the impulse response of free space. Then the response to input electric field amplitude can be determined by a convolution. These convolutions are known as *diffraction integrals*.

A point source of light, $\delta(x, y, z)$ is placed at the origin of the input plane. The corresponding electric field falling on the plane a distance z away is then the impulse response of free space, or, as called in optics, the point spread function. It will serve as the kernel in our convolution integral describing diffraction.

Let the impulse response for a fixed z be $\vec{E}_0(x, y, z)$. Clearly, the response of the point source will be spherically symmetric. Thus

$$\vec{E}_0(x, y, z) = E_0(r)\vec{a}_r$$

where

$$r^2 = x^2 + y^2 + z^2$$

and $\vec{a}_r$ is a unit vector. Thus, in imposing the physics in (13.14), we can use the spherically symmetric form of the Laplacian.

$$\nabla^2 = \frac{\partial^2}{\partial x^2} + \frac{\partial^2}{\partial y^2} + \frac{\partial^2}{\partial z^2} = \frac{1}{r^2}\frac{\partial}{\partial r}r^2\frac{\partial}{\partial r}.$$

The electric field wave equation now becomes

$$\left(\frac{1}{r^2}\frac{\partial}{\partial r}r^2\frac{\partial}{\partial r} + k^2\right)E_0(r)\vec{a}_r = 0.$$

Interestingly, it is this point in our derivation where the sense of vector notation leaves. Taking the dot product of both sides with $\vec{a}_r$ gives the scalar equation

$$\left(\frac{1}{r^2}\frac{\partial}{\partial r}r^2\frac{\partial}{\partial r} + k^2\right)E_0(r) = 0. \qquad (13.23)$$

---

14. See Section 3.3.3.

To facilitate a solution to (13.23), define

$$E_0(r) = \frac{\hat{E}_0(r)}{r}.$$

It follows that

$$\frac{1}{r^2}\frac{\partial}{\partial}r^2\frac{\partial}{\partial r^2}\frac{E_0(r)}{r} = \frac{1}{r^2}\frac{\partial}{\partial r}r^2\left(\frac{1}{r}\frac{\partial \hat{E}_0}{\partial r} - \frac{\hat{E}_0}{r^2}\right)$$

$$= \frac{1}{r^2}\left(\frac{\partial \hat{E}_0}{\partial r} + r\frac{\partial^2 \hat{E}_0}{vr^2} - \frac{\partial \hat{E}_0}{\partial r}\right) = \frac{1}{r}\frac{\partial^2 \hat{E}_0}{\partial r^2}.$$

Thus

$$(\nabla^2 + k^2)\frac{\hat{E}_0(r)}{r} = \frac{1}{r}\left(\frac{\partial^2 \hat{E}_0}{\partial r^2} + k^2 \hat{E}_0\right) = 0$$

or, equivalently,

$$\frac{\partial^2 \hat{E}_0}{\partial r^2} + k^2 E_0(r) = 0; \quad r \neq 0.$$

The solution is

$$\hat{E}_0 = C_1 e^{jkr} + C_2 e^{-jkr}$$

where C_1 and C_2 are constants. Equivalently,

$$E_0(r) = C_1 \frac{e^{jkr}}{r} + C_2 \frac{e^{-jkr}}{r} \; ; \; r \neq 0.$$

One of these waves is converging to and the other diverging from the point source. To see which is which, we reimpose the temporal harmonic term.

$$E_0(r)e^{-j\omega t} = C_1 \frac{e^{j(kr-\omega t)}}{r} + C_2 \frac{e^{-j(kr+\omega t)}}{r}.$$

The first term corresponds to a wave travelling in the positive r direction and is therefore the one we desire. The second term corresponds to a converging wave. We discard it by setting $C_2 = 0$. Setting

$$C_1 = \frac{1}{j\lambda}.$$

gives the final desired solution,

$$E_0(x, y, z) = h_z(x, y) = \frac{1}{j\lambda}\frac{\exp(jk\sqrt{x^2 + y^2 + z^2})}{\sqrt{x^2 + y^2 + z^2}} \qquad (13.24)$$

where we have adopted the more conventional point spread function notation, $h_z(x, y)$, for the impulse response of free space.

13.2.3.1 The Diffraction Integral

For a field amplitude of $u(x, y)$ on the $z = 0$ plane in Figure 13.15, we can now compute the resulting field amplitude at a distance z away by convolving $u(x, y)$ with the impulse response of free space. The result is

$$g_z(x, y) = u(x, y) * h_z(x, y)$$

$$= \frac{1}{j\lambda} \int_{-\infty}^{\infty} \int_{-\infty}^{\infty} u(\xi, \eta)$$

$$\times \frac{\exp\left(jk\sqrt{(x-\xi)^2 + (y-\eta)^2 + z^2}\right)}{\sqrt{(x-\xi)^2 + (y-\eta)^2 + z^2}} \, d\xi \, d\eta \quad (13.25)$$

or, in more compact notation,

$$g_z(x, y) = \frac{1}{j\lambda} \int_{-\infty}^{\infty} \int_{-\infty}^{\infty} u(\xi, \eta) \frac{e^{jkR}}{R} d\xi \, d\eta \quad (13.26)$$

where

$$R = \sqrt{(x-\xi)^2 + (y-\eta)^2 + z^2}. \quad (13.27)$$

Equations (13.26) (and (13.25)) is known as the *Rayleigh-Sommerfield diffraction integral*.

The Rayleigh-Sommerfield diffraction integral is analytically intractable for most inputs. In order to gain further insight into the diffraction process, Fresnel and Fraunhofer approximations are made to the Rayleigh-Sommerfield diffraction integral.

13.2.3.2 The Fresnel Diffraction Integral

The Fresnel approximation simplifies the Rayleigh-Sommerfield diffraction integral when z becomes sufficiently large.

The Fresnel Approximation. Consider Figure 13.17 where we have connected the point (ξ, η) on the input plane with the point (x, y) on the output plane. From (13.27), the distance between these two points is R. If the distance z is large with respect to the values of ξ, η, x and y, then $R \approx r$ and we can replace R by z in the denominator of the integrand in (13.26). Since, at optical wavelengths, k is very large, we must be somewhat more careful in the exponential. A small error in R can change kR a significant percentage of 2π and render

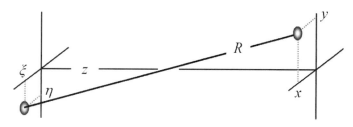

FIGURE 13.17. The variable R is the distance between the coordinate (ξ, η) on the input plane and (x, y) on the output plane.

a poor approximation. Here, we will require a more precise truncated Taylor series expression for R. Since

$$\sqrt{1+\varepsilon^2} \approx 1 + \frac{1}{2}\varepsilon^2; \varepsilon \ll 1,$$

we can make the following approximation

$$R = z\sqrt{1 + \frac{(x-\xi)^2}{z^2} + \frac{(y-\eta)^2}{z^2}}$$

$$\approx z\left[1 + \frac{(x-\xi)^2}{2z^2} + \frac{(y-\eta)^2}{2z^2}\right].$$

With this substitution in the exponent and $R \approx z$ in the denominator, we obtain the *Fresnel approximation* to the Rayleigh-Sommerfield diffraction integral in (13.26).

$$g_z(x,y) = \frac{e^{jkz}}{j\lambda z} \int_{-\infty}^{\infty} \int_{-\infty}^{\infty} u(\xi,\eta) e^{j\frac{k}{2z}[(x-\xi)^2+(y-\eta)^2]} d\xi\, d\eta. \tag{13.28}$$

Impulse Response Interpretation. The Fresnel integral in (13.28) can be written as a convolution.

$$g_z(x,y) = \frac{e^{jkz}}{j\lambda z} u(x,y) * e^{j\frac{k}{2z}(x^2+y^2)}. \tag{13.29}$$

The corresponding impulse response is

$$h_z(x,y) = \frac{e^{jkz}}{j\lambda z} e^{j\frac{k}{2z}(x^2+y^2)}. \tag{13.30}$$

We know that the Rayleigh-Sommerfield impulse response in (13.24) corresponds to a spherical wave since, when we set the phase equal to a constant, we obtain a family of spheres. Specifically, $kr = $ constant $\Rightarrow r = $ constant, which is a family of spheres. Similarly, $\exp(jkz)$ corresponds to a plane wave since $z = $ constant defines a family of planes. For the Fresnel approximation to the impulse response in (13.30), setting the phase equal to a constant results in

$$x^2 + y^2 = \text{constant} \times z - 2z^2.$$

This equation corresponds to a family of paraboloids.[15] Thus, as illustrated in Figure 13.18, we have approximated the spherical impulse response of Rayleigh-Sommerfield diffraction with a parabolic impulse response. This is a good approximation under the Fresnel assumption. In the Fraunhofer approximation, the wavefront is approximated as a plane wave.

13.2.3.3 The Fraunhofer Approximation

The Fraunhofer approximation is a special case of the Fresnel approximation. It requires that we be farther from the diffracting aperture than is required by the Fresnel approximation.

---

15. Set, for example, $y = 0$ to obtain a conventional parabolic conic section.

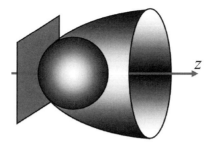

FIGURE 13.18. The impulse response for Rayleigh-Sommerfield diffraction is a spherical wave. Fresnel diffraction approximates this by a parabolic wave and is a good approximation when z, as measured from the center of the sphere, becomes large. Fraunhofer diffraction, which requires a source of finite extent and much larger, z, approximates the sphere with a plane wave.

We can expand the quadratic in the exponential in (13.28) and write

$$g_z(x, y) = \frac{e^{jkz}}{j\lambda z} e^{j\frac{k}{2z}(x^2+y^2)}$$

$$\times \int_{-\infty}^{\infty} \int_{-\infty}^{\infty} u(\xi, \eta) e^{j\frac{k}{2z}(\xi^2+\eta^2)} e^{-j\frac{k}{z}(x\xi+y\eta)} d\xi \, d\eta. \quad (13.31)$$

For the Fraunhofer approximation, set

$$\exp\left(j\frac{k}{2z}(\xi^2 + \eta^2)\right) \approx 1. \quad (13.32)$$

To do so, we will require the *Fraunhofer assumption* that

$$\frac{2z}{k} = \frac{\lambda z}{\pi} \gg (\xi^2, \eta^2) \quad (13.33)$$

where (ξ^2, η^2) denotes the maximum square dimension on the input plane. If, for example, the diffracting aperture is a square with sides of length three, then $(\xi^2, \eta^2) \approx \sqrt{2} \times 3/2$. If (13.33) is true, then so is (13.32) and the Fresnel diffraction integral in (13.31) becomes the *Fraunhofer diffraction integral*.

$$g_z(x, y) = \frac{e^{jkz}}{j\lambda z} e^{j\frac{k}{2z}(x^2+y^2)} \int_{-\infty}^{\infty} \int_{-\infty}^{\infty} u(\xi, \eta) e^{-j\frac{2\pi}{\lambda z}(x\xi+y\eta)} d\xi \, d\eta \quad (13.34)$$

Fields, $g_z(x, y)$, that result from Fraunhofer diffraction are also called *far fields*.

If the Fraunhofer assumption applies, then so does the Fresnel approximation. Both Fraunhofer and Fresnel diffraction are approximations of the Rayleigh-Sommerfield diffraction integral.

Impulse Response Interpretation. The Fraunhofer diffraction integral in (13.34), interestingly, is no longer a convolution integral. It is, rather, of the more general form of a superposition integral. The impulse response corresponding to the Fraunhofer diffraction integral is

$$h(x, y; \xi, \eta) = \frac{e^{jkz}}{j\lambda z} e^{j\frac{k}{2z}(x^2+y^2)} e^{-j\frac{2\pi}{\lambda z}(x\xi+y\eta)}. \quad (13.35)$$

For a point source at the origin, $(\xi, \eta) = (0, 0)$, this is recognized as a plane wave. The plane wave becomes a good approximation for a spherical wave if we are sufficiently far from the origin and are restricted in attention to the region about the leading edge of the spherical wave.

Relation of Fraunhofer Diffraction to Fourier Transformation. The two dimensional Fourier transform of a spatial function $u(\xi, \eta)$, from (8.5), is

$$U(f_x, f_y) = \int_{-\infty}^{\infty} \int_{-\infty}^{\infty} u(\xi, \eta) e^{-j2\pi(f_x \xi + f_y \eta)} d\xi\, d\eta \tag{13.36}$$

where the frequency variables are (f_x, f_y). Using this definition, we can write the Fraunhofer approximation as

$$g_z(x, y) = \frac{e^{jkz}}{j\lambda z} e^{j\frac{k}{2z}(x^2+y^2)} U\left(\frac{x}{\lambda z}, \frac{y}{\lambda z}\right) \tag{13.37}$$

Thus, to within a quadratic phase term, the far field is proportional to the Fourier transform of the diffracting aperture.

Light detectors, such as photo cells, film, and the human eye, respond to the *intensity* of light. The intensity of a field amplitude $u(x, y)$, is given by

$$I(x, y) = |u(x, y)|^2$$

Thus, the intensity of the far field, from (13.37), is

$$I_z(x, y) = |g_z(x, y)|^2$$
$$= \frac{1}{(\lambda z)^2} \left| U\left(\frac{x}{\lambda z}, \frac{y}{\lambda z}\right) \right|^2 \tag{13.38}$$

The relation of the far field to the Fourier transform of the aperture is quite striking. Note that the equivalent of a far field magnification factor here is λz. Thus, when we are in the far field, increasing the distance from the aperture results in the intensity spreading out further.

The Fresnel-Fraunhofer Boundary. Fresnel diffraction obviously does not abruptly change into Fraunhofer diffraction. Identification of the point where Fraunhofer diffraction begins is, at best, a fuzzy measure. Nevertheless, we will attempt to identify that point, being satisfied if we are within an order of magnitude.

The Fraunhofer approximation requires that

$$e^{j\frac{k}{2z}(\xi^2+\eta^2)} \approx 1. \tag{13.39}$$

Since $e^{jz} \approx 1$ for $z < 1$ radian,[16] we wish, from (13.39), to require

$$\frac{k\Xi^2}{2z} = \frac{\pi}{\lambda z}\Xi^2 \approx 1$$

where Ξ is the maximum linear dimension of the aperture. Solving for $z = z_f =$ the *Fresnel-Fraunhofer boundary* gives

$$z_f = \frac{\pi \Xi^2}{\lambda} \tag{13.40}$$

---

16. $e^j = 0.99985 + j.01745 \approx 1$.

For example, consider light from a HeNe laser with a wavelength of $\lambda = 6328$Å. For $\Xi = 1$ mm, $z_f = 5$ meters. For $\Xi = 1$ cm, $z_f = 500$ meters.

The Airy Pattern. For a circular aperture of diameter D, we write $u(x, y) = \Pi\left(\frac{r}{D}\right)$ where $r = \sqrt{x^2 + y^2}$. Using (8.35) in Section 8.4.6, we have the two dimensional Fourier transform pair

$$\Pi\left(\frac{r}{D}\right) \leftrightarrow \frac{1}{2}D^2 \text{jinc}(D\rho/2); \quad \rho = \sqrt{f_x^2 + f_y^2}.$$

Using (13.38), the Fraunhofer diffraction intensity for the circular aperture is

$$I_z(x, y) = \left(\frac{D}{2\lambda z}\right)^2 \text{jinc}^2\left(\frac{Dr}{2\lambda z}\right). \qquad (13.41)$$

This diffraction pattern, shown in Figure 13.19, is known as an *Airy pattern*. The middle circle of the Airy pattern is the *Airy disc*. Its diameter is specified by the first zero crossing of the jinc. Since $J_1(3.83) = 0$, we conclude

$$\text{jinc}^2\left(\frac{Dr}{2\lambda z}\right) = \left[\frac{J_1\left(\frac{\pi D\rho}{\lambda z}\right)}{\frac{Dr}{\lambda z}}\right]^2$$

is zero when

$$\frac{\pi D r}{\lambda z} = 3.83.$$

FIGURE 13.19. The Airy pattern given by the jinc^2 in (13.41). The diameter of the Airy disc is given by (13.43). A mesh plot of jinc(t) is in Figure 8.13.

The diameter of the Airy disc is thus

$$d = 2r$$
$$= \frac{2 \times 3.83 \times \lambda z}{\pi D} \tag{13.42}$$
$$= 2.44 \times \lambda z/D.$$

13.2.4 A One Lens System

Analysis of the one lens system shown in Figure 13.20 is the topic of this section. A monochromatic coherent field, $u(\xi, \eta)$ emanates from the (ξ, η) plane on the left and propagates a distance d_1 to the (α, β) plane on which is placed a thin positive lens with focal length f. The light leaving this plane propagates an additional distance d_2 to the (x, y) plane where we have the output, $g(x, y)$. Two special cases will result from this analysis: a one lens imaging system and a Fourier transformer.

13.2.4.1 Analysis

Our approach is as follows. Fresnel diffraction will be used to find the field amplitude incident on the (α, β) plane. This will be multiplied by the transmittance of the lens. The light leaving the (α, β) plane will then be subjected to Fresnel diffraction over a distance d_2 to produce the resulting output.

From Fresnel diffraction, the light incident on the (α, β) plane is

$$v(\alpha, \beta) = \frac{e^{jkd_1}}{j\lambda d_1} e^{j\frac{k}{2d_1}(\alpha^2+\beta^2)}$$
$$\times \int_\xi \int_\eta u(\xi, \eta) e^{j\frac{k}{2d_1}(\xi^2+\eta^2)} e^{-j\frac{k}{d_1}(\xi\alpha+\eta\beta)} d\xi\, d\eta. \tag{13.43}$$

Let $\hat{v}(\alpha, \beta)$ be the light exiting the (α, β) plane. This is

$$\hat{v}(\alpha, \beta) = t(\alpha, \beta) v(\alpha, \beta)$$

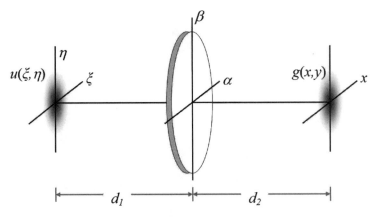

FIGURE 13.20. A one lens system. The output, $g(x, y)$, will be computed by evaluating Fresnel diffraction from the input plane to the lens, multiplying the incident field by the lens transmittance, and computing Fresnel diffraction to the output, (x, y), plane.

where the lens transmittance is[17]

$$t(\alpha, \beta) = e^{jkn\ell_0} e^{-j\frac{k}{2f}(\alpha^2+\beta^2)} \qquad (13.44)$$

where n is refractive index of the lens, f its focal length and ℓ_0 is the maximum lens thickness. We multiply the incident field by the lens transmittance[18] to obtain the field immediately exiting the lens

$$\hat{v}(\alpha, \beta) = \frac{e^{jk(d_1+n\ell_0)}}{j\lambda d_1} e^{j\frac{k}{2}(\frac{1}{d_1}-\frac{1}{f})(\alpha^2+\beta^2)}$$

$$\times \int_\xi \int_\eta u(\xi, \eta) e^{j\frac{k}{2d_1}(\xi^2+\eta^2)} e^{-j\frac{k}{d_1}(\xi\alpha+\eta\beta)} d\xi\, d\eta. \qquad (13.45)$$

Propagating $\hat{v}(\alpha, \beta)$ to the (x, y) plane using Fresnel diffraction gives

$$g(x, y) = \frac{e^{jkd_2}}{j\lambda d_2} e^{j\frac{k}{2d_2}(x^2+y^2)}$$

$$\times \int_\alpha \int_\beta \hat{v}(\alpha, \beta) e^{j\frac{k}{2d_2}(\alpha^2+\beta^2)} e^{-j\frac{k}{d_2}(\alpha x+\beta y)} d\alpha\, d\beta. \qquad (13.46)$$

Substituting (13.45) gives

$$g(x, y) = \frac{e^{jk(d_1+d_2+n\ell_0)}}{-\lambda^2 d_1 d_2} e^{j\frac{k}{2d_2}(x^2+y^2)}$$

$$\times \int_\alpha \int_\beta e^{j\frac{k}{2}(\frac{1}{d_1}+\frac{1}{d_2}-\frac{1}{f})(\alpha^2+\beta^2)} e^{-j\frac{k}{d_2}(\alpha x+\beta y)}$$

$$\times \int_\xi \int_\eta u(\xi, \eta) e^{j\frac{k}{2d_1}(\xi^2+\eta^2)} e^{-j\frac{k}{d_1}(\xi\alpha+\eta\beta)} d\xi\, d\eta\, d\alpha\, d\beta. \qquad (13.47)$$

Interchanging integration order gives

$$g(x, y) = \frac{e^{jk(d_1+d_2+n\ell_0)}}{\lambda^2 d_1 d_2} e^{j\frac{k}{2d_2}(x^2+y^2)} \int_\xi \int_\eta u(\xi, \eta) e^{j\frac{k}{2d_1}(\xi^2+\eta^2)}$$

$$\times \mathcal{I}(x, y; \xi, \eta) d\xi\, d\eta \qquad (13.48)$$

---

17. We have stated the lens transmittance here without prior discussion. The lens transmittance, a pure phase term, describes the phase delay of light going through a slower optical medium. This transmittance is applicable when the lens has a spherical surface - the type that would occur if a thin slice were cut from a large sphere of glass. If the radius of the sphere is large, then the spherical surface can be viewed, as was the case in the Fresnel approximation, as a paraboloid. When applied to lenses, this parabolic approximation is dubbed the *paraxial approximation*. Note, indeed, the phase term in the transmittance in (13.44) is quadratic. For a detailed derivation of the lens transmittance, see any one of the many excellent texts of Fourier optics [493, 514, 518, 1082, 1344].

18. To apply this multiplication, each ray of light entering the lens should exit at the same coordinate it entered. This is approximately achieved when the lens is thin, prompting the term *thin lens approximation*.

where the (α, β) integral is

$$\mathcal{I}(x, y; \xi, \eta) = \int_\alpha \int_\beta e^{j\frac{k}{z}(\frac{1}{d_1}+\frac{1}{d_2}-\frac{1}{f})(\alpha^2+\beta^2)} e^{jk[\alpha(\frac{x}{d_2}+\frac{\xi}{d_1})+\beta(\frac{y}{d_2}+\frac{\eta}{d_1})]} d\alpha\, d\beta. \quad (13.49)$$

We will now analyze two special cases of (13.48).

13.2.4.2 Imaging System

The first special case of the one lens system in Figure 13.20 occurs when the separation between the lens and the planes obeys the familiar *lens law*.

$$\frac{1}{d_i} + \frac{1}{d_o} = \frac{1}{f}$$

The (α, β) integral in (13.49) then becomes

$$\mathcal{I}(x, y; \xi, \eta) = \int_\alpha \int_\beta e^{-j\frac{2\pi}{\lambda}[\alpha(\frac{x}{d_2}+\frac{\xi}{d_1})+\beta(\frac{y}{d_2}+\frac{\eta}{d_1})]} d\alpha\, d\beta$$

$$= \delta\left(\frac{1}{\lambda}\left(\frac{x}{d_2}+\frac{\xi}{d_1}\right)\right) \delta\left(\frac{1}{\lambda}\left(\frac{y}{d_2}+\frac{\eta}{d_1}\right)\right)$$

$$= (\lambda d_1)^2 \delta\left(\xi + \frac{d_1}{d_2}x\right) \delta\left(\eta + \frac{d_1}{d_2}y\right)$$

where we have used the property that $\delta(ax) = \frac{1}{|a|}\delta(x)$. Substituting into (13.48) and using the sifting property of the Dirac delta gives

$$g(x, y) = \frac{e^{jk(d_1+d_2+nl_0)}}{-\lambda^2 d_1 d_2} (\lambda d_1)^2 e^{j\frac{k}{2d_2}(x^2+y^2)}$$

$$\times \int_\xi \int_\eta u(\xi, \eta) e^{j\frac{k}{2d_1}(\xi^2+\eta^2)} \delta\left(\xi + \frac{d_1}{d_2}x\right) \delta\left(\eta + \frac{d_1}{d_2}y\right) d\xi\, d\eta$$

$$= e^{jk(d_1+d_2+nl_0)} \left(\frac{d_1}{d_2}\right) e^{j\frac{k}{2d_2}(x^2+y^2)}$$

$$\times u\left(-\frac{d_1}{d_2}x, -\frac{d_1}{d_2}y\right) e^{j\frac{k}{2d_1}\left[(\frac{d_1}{d_2}x)^2+(\frac{d_1}{d_2}y)^2\right]}$$

$$= e^{jk(d_1+d_2+nl_0)} \left(\frac{d_1}{d_2}\right) e^{j\frac{k}{2d_2}(\frac{d_1}{d_2}+1)(x^2+y^2)} u\left(-\frac{d_1}{d_2}x, -\frac{d_1}{d_2}y\right). \quad (13.50)$$

Define the *magnification* of the system by $M = d_2/d_1$. Then (13.50) becomes

$$g(x, y) = e^{jk(d_1+d_2+nl_0)} e^{j\frac{k}{2d_2}(\frac{1}{m}+1)(x^2+y^2)} \frac{1}{M} u\left(-\frac{x}{M}, -\frac{y}{M}\right). \quad (13.51)$$

This is the output of the imaging system. The relationship becomes more apparent when we write the corresponding intensity.

$$|g(x, y)|^2 = \frac{1}{M^2} \left| u\left(-\frac{x}{M}, -\frac{y}{M}\right) \right|^2. \quad (13.52)$$

This same result can be obtained through geometric ray models of the system and is a familiar result from fundamental optics.

13.2.4.3 Fourier Transformation

The second important case of the one lens imaging system is when $d_2 = f$. Interestingly, the system now performs a Fourier transform in much the same fashion as we saw in Fraunhofer diffraction.

For $d_2 = f$, the integral in (13.49) becomes

$$\mathcal{I}(x, y; \xi, \eta) = \int_\alpha e^{j\frac{k}{2d_1}\alpha^2} e^{jk\alpha(\frac{x}{d_2}+\frac{\xi}{d_1})} d\alpha$$

$$\times \int_\beta e^{j\frac{k}{2d_1}\beta^2} e^{jk\beta(\frac{y}{d_2}+\frac{\eta}{d_1})} d\beta. \tag{13.53}$$

To evaluate this integral, we use a Fourier transform from Table 2.5.

$$\int_{-\infty}^{\infty} e^{jat^2} e^{-j2\pi ut} dt = \sqrt{\frac{j\pi}{a}} e^{-j(\pi u)^2/a}. \tag{13.54}$$

Using this equation[19] to solve (13.53), followed by substitution into (13.48), gives

$$g(x, y) = \frac{\exp(jk(d_1+f+nl_0))}{-\lambda^2 d_1 d_2} (j\lambda d_1) e^{j\frac{k}{2f}(x^2+y^2)}$$

$$\times \int_\xi \int_\eta u(\xi, \eta) e^{j\frac{k}{2d_1}(\xi^2+\eta^2)} \left[e^{-j\frac{k}{2d_1}(\xi^2+\eta^2)} e^{j\frac{d_2 k}{2f^2}(x^2+y^2)} e^{-j\frac{k}{f}(x\xi+y\eta)} \right] d\xi d\eta$$

$$= \frac{e^{jk(d_1+f+nl_0)}}{j\lambda f} e^{j\frac{k}{2f}\left(1-\frac{d_1}{f}\right)(x^2+y^2)} \int_\xi \int_\eta u(\xi, \eta) e^{-j\frac{k}{f}(x\xi+y\eta)} d\xi\, d\eta$$

$$= \frac{\exp(jk(d_1+f+nl_0))}{j\lambda f} e^{j\frac{k}{2f}\left(1-\frac{d_1}{f}\right)(x^2+y^2)} U\left(\frac{x}{\lambda f}, \frac{y}{\lambda f}\right). \tag{13.55}$$

Note that the corresponding intensity in the back focal plane is

$$|g(x, y)|^2 = \frac{1}{(\lambda f)^2} \left| U\left(\frac{x}{\lambda f}, \frac{y}{\lambda f}\right) \right|^2. \tag{13.56}$$

As was the case with Fraunhofer diffraction,[20] we have generated the Fourier transform of the input. Indeed, the single lens system with $d_2 = f$ essentially brings the far field to the back focal plane.

Consider, again, the field amplitude on the back focal plane in (13.55). Setting $d_1 = f$ corresponds to placement of the input on the front focal plane. The field amplitude on the back focal plane now becomes directly proportional to the Fourier transform of the input.

$$g(x, y) = \frac{e^{jk(d_1+f+nl_0)}}{j\lambda f} U\left(\frac{x}{\lambda f}, \frac{y}{\lambda f}\right). \tag{13.57}$$

---

19. Set $a = \frac{k}{2d_1}$ and $2\pi u = k\left(\frac{x}{d_2}+\frac{\xi}{d_1}\right)$ for both integrals.
20. See (13.38).

TABLE 13.1. Summary of the results of the one lens system.

Condition	Result	Equation
$\frac{1}{d_i} + \frac{1}{d_0} = \frac{1}{f}$	imaging system	13.51, 13.52
$d_2 = f$	Fourier transform	13.55, 13.56
$d_2 = f$ and $d_1 = f$	Fourier transform	13.57, 13.56

A summary of the results for the one lens imaging system are summarized in Table 13.1.

13.2.4.4 Implementation of the PGA Using Optics

The Papoulis-Gerchberg algorithm (PGA) is discussed in Section 10.6. Using the Fourier transform property of the lens, Marks *et al.* [895, 896, 907] designed an all optical architecture for performing the PGA. The PGA is illustrated enumerated steps in Figure 10.30. The optical architecture with corresponding enumeration is in Figure 13.21.

Here are the steps as numbered in Figures 10.30 and Figure 13.21.

Step 1. The input is introduced through the aperture (hole) in the (ξ, η) plane. It passes through the lens and the Fourier transform is incident on the (x, y) plane.

Step 2. The Fourier transform is truncated. This is done by the small mirror on the (x, y) plane. Light inside the mirror pupil is reflected back into the system. Light outside of the mirror is lost to the system.

Step 3. The light travels from the mirror on the (x, y) plane back through the lens towards the (ξ, η) plane. This constitutes an inverse Fourier transform.

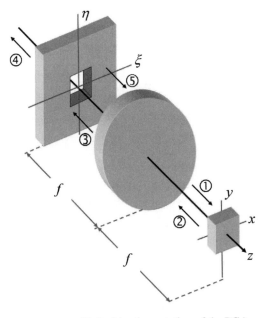

FIGURE 13.21. Optical implementation of the PGA.

Step 4. The middle of the incident light on the (ξ, η) plane goes through the hole and is lost to the system.

Step 5. The hole is in a mirror that reflects light back into the system. The input continues to be input into the system through the hole in the mirror, and the first iteration has been performed.

The PGA iteration continues and converges at the speed of light.

Although an aesthetically pleasing architecture, the optical system suffers from two problems resulting from the highly sensitive nature of PGA restoration.

(1) There is no provision to observe the result of the iteration. A pellicle[21] can be placed in the feedback path to bleed off an image to view. The absorbtion, also present from the lens and imperfect mirrors, will degrade performance.
(2) The optical processor is analog and, like all analog processing, suffers from inexactitude.

13.2.5 Beamforming

The basic goal of beamforming is to synthesize a diffracting aperture to concentrate energy in an steerable beam. The basic beamforming model using a phased array [831, 840] consists of $(2N + 1) \times (2N + 1)$ equally spaced point antennas. If the linear spacing is D units apart on the $z = 0$, then

$$u(x, y) = \sum_{n=-N}^{N} \sum_{m=-N}^{N} \delta(x - nD)\delta(y - mD). \tag{13.58}$$

The *array aperture* is of length $(2N + 1)D$. Using (2.62), the two dimensional Fourier transform is

$$U(f_x, f_y) = (2N + 1)^2 \text{ array}\left((2N + 1)Df_x\right) \text{ array}\left((2N + 1)Df_y\right).$$

From (13.37), the far field is therefore the *array pattern*

$$g_z(x, y) = (2N + 1)^2 \frac{e^{jkz}}{j\lambda z} e^{j\frac{k}{2z}(x^2+y^2)} \text{array}\left(\frac{(2N + 1)Dx}{\lambda z}\right) \text{array}\left(\frac{(2N + 1)Dy}{\lambda z}\right). \tag{13.59}$$

This corresponds to a two dimensional periodic function a square tile.[22] The width of main lobe centered at the origin from zero crossing to zero crossing in x is

$$\Delta = \frac{\lambda z}{(2N + 1)D}. \tag{13.60}$$

This is the beam width. The beam width narrows as the array aperture size increases.

---

21. A thin highly transmissive membrane beam splitter.
22. A tile is illustrated in Figure 8.21. In practice, the far field is not periodic but is best modelled as the array pattern weighted by a *radiation pattern* that attenuates higher order terms.

13.2.5.1 Apodization

Weighting the point antennas can be referred to as *apodization*.[23] In lieu of (13.58), we now have

$$v(x,y) = \sum_{n=-N}^{N} \sum_{m=-N}^{N} w_{nm} \delta(x-nD)\,\delta(y-mD). \qquad (13.61)$$

where w_{nm} is the weight assigned to the nmth antenna. The Fourier transform is a two dimensional periodic trigonometric polynomial.

$$V(f_x, f_y) = \sum_{n=-N}^{N} \sum_{m=-N}^{N} w_{nm} e^{-j2\pi D(nf_x + mf_y)} \qquad (13.62)$$

and the far field is the array pattern

$$\breve{g}_z(x,y) = \frac{e^{jkz}}{j\lambda z} e^{j\frac{k}{2z}(x^2+y^2)} V\left(\frac{x}{\lambda z}, \frac{y}{\lambda z}\right). \qquad (13.63)$$

This far field has a lobe width, Δ, that is determined by the weights, w_{nm}. As is the case with windows, the beam width can be reduced for a fixed array aperture by a wise choice of weights. As is illustrated for windows in Figure 9.4, there is a tradeoff. For a fixed array aperture, as the beam width narrows, the strength of the side lobes, δ, increases.

13.2.5.2 Beam Steering

From the modulation theorem of Fourier transforms, we know that, if $v(x) \longleftrightarrow V(f_x)$, then $v(x)e^{j2\pi f_0 x} \longleftrightarrow V(f_x - f_0)$. This principle can be applied to steer the beam from off center. In lieu of $u(x,y)$ in (13.61), we use the array

$$w(x,y) = v(x,y) e^{j2\pi(\xi x + \eta y)}. \qquad (13.64)$$

Thus

$$W(f_x, f_y) = U(f_x - \xi, f_y - \eta).$$

The far field then becomes

$$\breve{g}_z(x,y) = V\left(\frac{x}{\lambda z} - \xi, \frac{y}{\lambda z} - \eta\right).$$

The far field beam is thus shifted from the origin. It has been steered to be centered at

$$(x,y) = (\lambda z \xi, \lambda z \eta).$$

Since $v(x,y)$ consists of point antennas, the modulation in (13.64) can be performed by using complex weights. We simply make the substitution

$$w_{nm} \longrightarrow w_{nm} e^{j2\pi(\xi n + \eta m)D}.$$

Such antenna's are oft referred to as *phased array antennas* [117, 164, 575, 865]. Receiving antennas can use a similar approach for selecting the direction from which electromagnetic signals are received.

---

23. Apodization in optics is equivalent to windowing in DSP. See Section 9.3.1.

13.3 Heisenberg's Uncertainty Principle

Heisenberg's uncertainty principle states, at the quantum level, the position and momentum (or velocity) of a particle cannot be accurately measured simultaneously. If Δx denotes the uncertainty of position measurement and Δp uncertainty in momentum, then Heisenberg's uncertainty principle states

$$\Delta x \, \Delta p \geq \frac{\hbar}{2} \tag{13.65}$$

where $\hbar = \frac{h}{2\pi}$ and $h = 6.625 \times 10^{-34}$ Joule seconds is Planck's constant. The uncertainties, Δx and Δp, are standard deviations of the probability density functions for the position and momentum.

Heisenberg's uncertainty principle is a manifestation of the uncertainty relationship of Fourier transforms presented in Section 4.3. Schrödinger's equation for a free particle with mass, m, is

$$\frac{j\hbar}{2m} \frac{\partial^2}{\partial x^2} \psi(x, t) = \frac{\partial}{\partial t} \psi(x, t) \tag{13.66}$$

where $\psi(x, t)$ is the wave function.[24] A solution is

$$\psi(x, t) = A e^{-j(kx - 2\pi u_k t)}. \tag{13.67}$$

where A is a constant and $k = 2\pi/\lambda$. Substituting gives the wavelength dependent relationship

$$u_k = \frac{\hbar k^2}{4\pi m}.$$

Since Schrödinger's equation is linear, any superposition of the solutions in (13.67) will also satisfy (13.66). Thus, instead of (13.67), we have the more general wave packet solution

$$\psi_X(x, t) = \int_{k=-\infty}^{\infty} \frac{1}{2\pi} \Psi_V\left(\frac{k}{2\pi}\right) e^{-j(kx - 2\pi u_k t)} dk. \tag{13.68}$$

where $\frac{1}{2\pi} \Psi_V\left(\frac{k}{2\pi}\right)$ is the superposition weighting factor. We choose this notation because, as we will show, it obeys the Fourier transform

$$\psi_X(x, 0) \longleftrightarrow \Psi_V(v). \tag{13.69}$$

We dispose of the arbitrary constant in the solution of Schrödinger's equation by adopting the normalization

$$\int_{-\infty}^{\infty} |\Psi(v)|^2 dv = 1.$$

Parseval's theorem then requires

$$\int_{-\infty}^{\infty} |\psi(x, 0)|^2 dx = 1.$$

Both $\Psi(v)$ and $\psi(x, 0)$ are therefore wave functions and $|\Psi(v)|^2$ and $|\psi(x, 0)|^2$ are probability density functions.

---

24. See Section 4.3.

The de Broglie wavelength of a particle with momentum p is

$$\lambda = \frac{h}{p}$$

or, equivalently,

$$k = \frac{2\pi}{\lambda} = \frac{p}{\hbar}.$$

Changing the integration variable in (13.68) from k to p then gives

$$\psi_X(x, t) = \int_{p=-\infty}^{\infty} \frac{1}{h} \Psi_V\left(\frac{p}{h}\right) e^{-j2\pi(px/h - u_k t)} dp.$$

Setting time $t = 0$ gives

$$\psi_X(x, 0) = \int_{p=-\infty}^{\infty} \frac{1}{h} \Psi_V\left(\frac{p}{h}\right) e^{-j2\pi px/h} dp. \qquad (13.70)$$

We then see the wave functions, $\psi_X(x, 0)$ and $\Psi_V(v)$, as predicted in (13.69), are Fourier transform pairs. The wave function for momentum follows as

$$\psi_P(p) = \frac{1}{h} \Psi_V\left(\frac{p}{h}\right). \qquad (13.71)$$

From the uncertainty relationship in (4.56), we know

$$\sigma_X \sigma_V \geq \frac{1}{4\pi}. \qquad (13.72)$$

From (13.71) we have[25]

$$\sigma_P = h\sigma_V.$$

Substituting into (13.72) gives

$$\sigma_X \sigma_P \geq \frac{\hbar}{2}.$$

For $\sigma_X = \Delta x$ and $\sigma_P = \Delta p$, this is Heisenberg's uncertainty principle between position and momentum in (13.65).

13.4 Elementary Deterministic Finance

The tools of signal analysis are applicable to many areas outside of engineering and science. In this section, we illustrate some applications to problems involving interest in deterministic personal finance. These problems can be solved by

1. writing, by inspection, a describing difference equation, and
2. solving the difference equation using a unilateral z-transform.

---

25. See the derivation between (4.16) and (4.17).

Examples we will present include analysis of

- compound interest on a simple deposit,
- compound interest on periodic deposits,
- payment scheduling of loans, such as mortgages, where premiums are paid periodically, and
- effects of taxes and inflation.

13.4.1 Some Preliminary Math

The definition of the z transform and some of its properties are presented in Section 2.7.4. An important difference equation in deterministic finance is

$$x[n+1] = \xi x[n] + \eta \quad ; \quad x[0] = x_0 \quad ; \quad n \geq 0 \tag{13.73}$$

where both ξ and η are constants. Performing a z transform of both sides using the shift theorem in (2.132) gives

$$zX(z) - zx_0 = \xi X(z) + \frac{\eta}{1 - z^{-1}}$$

where (2.131) has been used with $a = 1$ to compute the transform of $\eta \mu[n]$. Solving for $X(z)$ and expanding using partial fractions give

$$\begin{aligned} X(z) &= z \frac{(z-1)x_0 + \eta}{(z-\xi)(z-1)} \\ &= \frac{1}{\xi - 1} \left[\frac{(\xi - 1)x_0 + \eta}{1 - \xi z^{-1}} - \frac{\eta}{1 - z^{-1}} \right]. \end{aligned} \tag{13.74}$$

From (2.131), the solution to the difference equation in (13.73) is

$$x[n] = \frac{1}{\xi - 1} \left([(\xi - 1)x_0 + \eta]\xi^n - \eta \right) \mu[n]. \tag{13.75}$$

13.4.2 Compound Interest on a One Time Deposit

Interest quotes have two components.

- annual interest and
- the frequency of compounding.

Let r be the annual interest and N the number of times per year compounding occurs. If $N = 12$, as is the case with most savings accounts, compounding is performed monthly.

A one time deposit of d is made in an account that yields an interest of r compounded N times per year. Let $\hat{b}[n]$ be the balance at the end of the nth period. The difference equation describing the accumulating interest is

$$\hat{b}[n+1] = \left(1 + \frac{r}{N}\right) \hat{b}[n] \tag{13.76}$$

with the initial condition $\hat{b}[0] = d$. This is a special case of the difference equation in Equation 13.73 with

$$x[n] \to \hat{b}[n]$$
$$\xi \to 1 + \frac{r}{N}$$
$$\eta \to 0$$
$$x_0 \to d.$$

Making these substitutions in (13.75) gives the balance at the end of the nth compounding period as

$$\hat{b}[n] = d\left(1 + \frac{r}{N}\right)^n.$$

The balance at the end of a year is

$$\hat{b}[N] = d\left(1 + \frac{r}{N}\right)^N \tag{13.77}$$

and at the end of M years is[26]

$$\hat{b}[NM] = d\left(1 + \frac{r}{N}\right)^{NM}. \tag{13.78}$$

13.4.2.1 Yield Increases with Frequency of Compounding

All else constant, the greater the frequency of compounding, the higher the yield. In other words, for a fixed yearly interest, r, and a fixed deposit d, the balance at the end of the year, $\hat{b}[N]$, is an increasing function of N. To show this, we need to show only that the derivative of (13.77), with respect to N is positive. From (13.77),

$$\frac{d\hat{b}[N]}{dN} = d \times \frac{d}{dN} \exp\left(N \ln\left(1 + \frac{r}{N}\right)\right)$$
$$= \hat{b}[N] \times \frac{d}{dN}\left(N \ln\left(1 + \frac{r}{N}\right)\right)$$
$$= \hat{b}[N] \times f(\rho) \tag{13.79}$$

where

$$f(\rho) = \ln(1 + \rho) - \frac{\rho}{1 + \rho}$$

and $\rho = r/N$. Since $\hat{b}[N] \geq 0$ for $N \geq 0$ and $f(\rho) \geq 0$ for $\rho \geq 0$ (and therefore $N \geq 0$), we conclude that

$$\frac{d\hat{b}[N]}{dN} \geq 0.$$

Since equality only occurs for $N = 0$, the function $\hat{b}[N]$ is a strictly increasing function of N.

---

26. To see how long it takes your money to double, see Exercise 13.9.

13.4.2.2 Continuous Compounding

For continuous compounding ($N \to \infty$), the yield at the end of the year for a one time deposit of d is

$$\hat{b}[\infty] = d\, e^r. \tag{13.80}$$

This can be shown by applying (14.3) to (13.77).

13.4.2.3 Different Rates and Compounding Periods - Same Yield

A number of interest rate and compounding period quotes result in the same yearly yield. In other words, for a given r_1, N_1 and N_2, there always exists an r_2 such that the yearly yield is identical. To show this, we use (13.77) and require that $\hat{b}[N_1] = \hat{b}[N_2]$. The result is

$$\left(1 + \frac{r_1}{N_1}\right)^{N_1} = \left(1 + \frac{r_2}{N_2}\right)^{N_2}. \tag{13.81}$$

Solving for r_2 gives

$$r_2 = N_2 \left[\left(1 + \frac{r_1}{N_1}\right)^{\frac{N_1}{N_2}} - 1\right]. \tag{13.82}$$

For example, 12% compounded monthly ($r_1 = 0.12$ and $N_1 = 12$) has the same yield as semi-annual (twice yearly, or $N_2 = 2$) compounding when

$$r_2 = 0.1230$$

or 12.30%.

Fixed rate to instantaneous rate transformation. The equivalent interest rate calculated in (13.82) has interesting asymptotic limits. Using (14.3), we conclude

$$\lim_{N_1 \to \infty} r_2 = N_2 \left(e^{\frac{r_1}{N_1}} - 1\right). \tag{13.83}$$

Similarly,

$$\lim_{N_2 \to \infty} r_2 = \lim_{N_2 \to \infty} \frac{\left(1 + \frac{r}{N}\right)^{\frac{N_1}{N_2}} - 1}{\frac{1}{N_2}}.$$

Applying L'Hopital's rule to this "zero over zero" situation gives

$$\lim_{N_2 \to \infty} r_2 = \lim_{N_2 \to \infty} \frac{\frac{d}{dN_2}\left[\left(1 + \frac{r}{N}\right)^{\frac{N_1}{N_2}} - 1\right]}{\frac{d}{dN_2}\left(\frac{1}{N_2}\right)}$$

$$= N_1 \ln\left(1 + \frac{r_1}{N_1}\right). \tag{13.84}$$

Not surprisingly, solving for r_1 in (13.84) and interchanging subscripts gives (13.83). Equation 13.84 also results from (13.81) when $N_2 \to \infty$. Then

$$\left(1 + \frac{r_1}{N_1}\right)^{N_1} \to e^{r_2}.$$

Solving for r_2 gives (13.84).

13.4.2.4 The Extrema of Yield

The maximum yield for a fixed r and d corresponds to $N = \infty$ and is given by (13.80). The minimum yield, for $N = 1$, follows from (13.76) as

$$\hat{b}[1] = (1 + r)d.$$

Thus

$$1 + r \leq \frac{\hat{b}[N]}{d} \leq e^r.$$

Note that for modest interest rates, the spread is very small since, for $r \ll 1$,

$$e^r \approx 1 + r. \tag{13.85}$$

13.4.2.5 Effect of Annual Taxes

Consider the same problem of evaluating the balance of a one time deposit of d, except that the interest each year is taxed at a rate, t. Let $f[M]$ be the balance after year M before taxation and $c[M]$ be the balance after year M after taxation. The before taxation balance at year $M + 1$ is given by (13.77) with $d \to c[M]$.

$$f[M+1] = c[M]\left(1 + \frac{r}{N}\right)^N.$$

The taxable interest earned in year M is new balance minus the initial balance.

$$i[M] = f[M+1] - c[M]$$

The amount payed in taxes is $t \times i[M]$. The after tax balance is

$$c[M+1] = f[M+1] - t \times c[M].$$

Substituting the previous two equations results in the difference equation gives

$$c[M+1] = \left[(1-t)\left(1 + \frac{r}{N}\right)^N + t\right]c[M].$$

This is a special case of the difference equation in (13.73) with

$$n \to M$$
$$x[n] \to c[M]$$
$$\xi \to (1-t)\left(1 + \frac{r}{N}\right)^N + t$$
$$\eta \to 0$$
$$x_0 \to c[0] = d.$$

Making these substitutions in (13.75) gives the desired result.

$$c[M] = d\left[(1-t)\left(1+\frac{r}{N}\right)^N + t\right]^M. \tag{13.86}$$

Continuous Compounding. Imposing the limit in (14.3) onto (13.86) gives the continuous compounding solution

$$\lim_{N\to\infty} c[M] = d\left[(1-t)e^r + t\right]^M. \tag{13.87}$$

Extrema. As a function of N, (13.86) is minimum for $N=1$ and maximum for $N=\infty$. Thus, from (13.87), the following extrema of yield results.

$$[(1-t)(1+r)+t]^M \le \frac{c[M]}{d} \le \left[(1-t)e^r + t\right]^M.$$

From (13.85), for modest interest rates ($r \ll 1$) and moderate M, these bounds are tight.

Combining the tax and interest rates into an equivalent interest rate. For a given tax rate, t, and compounding frequency, N, an equivalent (smaller) interest rate, r_t, exists. Equating (13.86) and (13.78) gives

$$\left[(1-t)\left(1+\frac{r}{N}\right)\right]^M d = \left(1+\frac{r_t}{N}\right)^{NM} d.$$

Solving for r_t gives

$$r_t = (1-t)\left[\left(1+\frac{r}{N}\right)^N - 1\right]. \tag{13.88}$$

The equivalent instantaneous compounding interest rate from a taxed instantaneous interest rate follows from application of (14.3) to (13.88).

$$\lim_{N\to\infty} r_t = (1-t)(e^r - 1).$$

13.4.2.6 Effect of Inflation

A constant inflation rate can be viewed as a negative interest rate. If u is the rate of inflation, the effect of inflation on d dollars over one year is given by (13.80) making the replacement $r \to -u$.

$$de^{-u}.$$

Over M years, the balance has reduced to

$$\left[de^{-u}\right]^M = de^{-Mu}.$$

For example, if you stuffed $d = \$100$ in your mattress for $M = 3$ years, its purchasing value, at an annual inflation rate of 12%, is diminished to

$$\$100 \times e^{-3\times 0.12} = \$69.77$$

in terms of the purchasing value of money at the time of the initial deposit.

Adjustment for inflation can be assessed after yield is evaluated. Two examples follow.

Compound interest subjected to inflation. If there is an inflation rate of u for M years, its toll on the compounded interest on a single deposit, d, follows from a simple adjustment of (13.78).

$$b_u[NM] = \hat{b}[NM]e^{-Mu}$$
$$= d\left[\left(1 + \frac{r}{N}\right)^N e^{-u}\right]^M. \qquad (13.89)$$

The quantity $b_s[NM]$ is the purchasing value of the balance after M years in terms of the value of the dollar when the original deposit was made. Interpreting this change as a percent is helpful.

$$\frac{b_t[NM]}{d} \times 100\% = \left[\left(1 + \frac{r}{N}\right)^N e^{-u}\right]^M \times 100\%.$$

For 200%, the purchasing value has doubled.

For continuous interest compounding, (13.89) becomes

$$\lim_{N \to \infty} b_s[NM] = de^{r-u}. \qquad (13.90)$$

This can either be derived from (14.3) to (13.89) or by applying the inflation term to (13.80). Equation 13.90 suggests the obvious. If you wish to maintain or increase the value of your deposit, the interest rate must be equal to or exceed the inflation rate.

Inflation *and* tax on a one time deposit. If the effects of inflation are to be applied to taxed interest,[27] the result is simply an inflation adjustment to (13.86).

$$c_u[M] = d\left[\left\{(1-t)\left(1 + \frac{r}{N}\right)^N + t\right\}e^{-u}\right]^M. \qquad (13.91)$$

For continuous interest compounding, the inflation adjusted balance follows from (13.87) as

$$\lim_{N \to \infty} c_u[M] = d\left[\left\{(1-t)e^r + t\right\}e^{-u}\right]^M.$$

13.4.3 Compound Interest With Constant Periodic Deposits

Consider interest at an annual rate of r compounded N times each year when, at each compounding period, a deposit of s is made. If $\hat{b}[N]$ is the balance at the end of n periods,[28] the describing difference equation is

$$\hat{b}[n+1] = \left(1 + \frac{r}{n}\right)\hat{b}[n] + s. \qquad (13.92)$$

---

27. See Exercise 13.9.

28. The notation $\hat{b}$ will be used for the case of constant periodic deposits as opposed to b which denotes the accumulated balance on a single deposit.

APPLICATIONS

Assume the account starts with a balance of $\hat{b}[0] = 0$. Equation 13.92 is then a special case of (13.73) with

$$x[n] \to \hat{b}[n]$$
$$\xi \to 1 + \frac{r}{N}$$
$$\eta \to s$$
$$x_0 \to 0. \qquad (13.93)$$

Substituting these parameters into (13.75) gives

$$\hat{b}[n] = \frac{Ns}{r}\left\{\left(1 + \frac{r}{N}\right)^n - 1\right\}.$$

The balance after one year is thus

$$\hat{b}[N] = \frac{Ns}{r}\left\{\left(1 + \frac{r}{N}\right)^N - 1\right\}. \qquad (13.94)$$

and the balance after M years is[29]

$$\hat{b}[MN] = \frac{Ns}{r}\left\{\left(1 + \frac{r}{N}\right)^{MN} - 1\right\}. \qquad (13.95)$$

13.4.3.1 Continuous Time Solution

For the continuous time solution to this problem, assume y is invested yearly in equal installments. Thus

$$s = \frac{y}{N}.$$

For M years, the balance in (13.95) therefore becomes

$$\hat{b}[MN] = \frac{y}{r}\left\{\left(1 + \frac{r}{N}\right)^{MN} - 1\right\}.$$

Using (14.3) in Appendix 14.3, the balance using continuous time compounding is

$$\lim_{N \to \infty} \hat{b}[MN] = \frac{y}{r}\left(e^{rM} - 1\right).$$

13.4.3.2 Be a Millionaire

From (13.95), the deposit can be evaluated as

$$s = \frac{\frac{r}{N}\hat{b}[MN]}{\left(1 + \frac{r}{N}\right)^{MN} - 1}.$$

---

29. For an example application, see Exercise 13.10.

Thus, if one makes monthly ($N = 12$) deposits at an annual interest rate of 12% (i.e., $r = 0.12$), one becomes a millionare[30] (i.e., $\hat{b}[MN] = \$1,000,000$) for the following year, M, and monthly deposit, s, pairs.

$M =$ years	$s =$ monthly deposit
10	$4347.10
20	$1010.86
30	$286.13
40	$85.00
50	$25.60
100	$0.0652

13.4.3.3 Starting With a Nest Egg

Consider the case where the savings plan is started with an initial balance of b_0. The parameters associated with (13.73) are

$$\hat{b}[n] = x[n]$$

$$\xi = 1 + \frac{r}{N}$$

$$\eta = s$$

$$x_0 = b_0.$$

The variables are the same as in Equation 13.93 except that $x_0 = b_0$. Making the variable substitutions into (13.75) gives

$$\hat{b}[n] = \frac{N}{r}\left\{\left(\frac{r}{N}b_0 + s\right)\left(1 + \frac{r}{N}\right)^n - s\right\}. \tag{13.96}$$

Effect of taxes. Assume that the deposits have already been taxed, but the accumulated interest rate is taxed once yearly at a flat tax rate of t. Let the Mth year start with a balance of $c[M]$. At the end of the year, prior to taxation, the balance, from (13.96) with $b_0 = c[M]$ and $n = N$ is

$$\hat{b}[\text{end of year } M] = \frac{N}{r}\left[\left(\frac{r}{N}c[M] + s\right)\left(1 + \frac{r}{N}\right)^N - s\right].$$

During the year, the total (taxed) deposits were Ns. Thus, the interest earned at the end of year M is

$$i[\text{year } M] = b - Ns.$$

A total of $t \times I$ goes to taxes leaving us with a balance of

$$c[M+1] = b - t \times i$$

$$= \frac{(1-t)N}{r}\left[\left(\frac{r}{N}c[M] + s\right)\left(1 + \frac{r}{N}\right)^N - s\right] + tNs.$$

---

30. Ignoring taxes.

APPLICATIONS 655

This is a special case of the difference equation in (13.73) with the following replacements

$$n \to M$$
$$x[n] \to c[M]$$
$$\xi \to (1-t)\left(1+\frac{r}{N}\right)^N \tag{13.97}$$
$$\eta \to s\varphi \tag{13.98}$$
$$x_0 \to c[0] = 0$$

where we have assumed an initial balance of $c[0] = 0$ and

$$\varphi = N\left\{\frac{1-t}{r}\left[\left(1+\frac{r}{N}\right)^N - 1\right] + t\right\}.$$

Making the appropriate substitutions into (13.75), using the values of ξ and η from (13.97) and (13.98), gives a balance of

$$c[M] = \frac{s\varphi\left(\xi^M - 1\right)}{\xi - 1}.$$

As before, we can solve for the deposit, s.

$$s = \frac{c[m](\xi - 1)}{\varphi\left(\xi^M - 1\right)}.$$

13.4.4 Loan and Mortgage Payments

You borrow β at an annual interest rate of r compounded N times per year. For monthly pay periods, $N = 12$. The balance you owe at month n is equal to the balance of the previous pay period, plus interest, minus the monthly payment, p. The describing difference equation is

$$\hat{b}[n+1] = \left(1 + \frac{r}{N}\right)\hat{b}[n] - p.$$

Note that this is the same difference equation as (13.92) except that s is replaced by $-p$. The solution for an initial condition of $\hat{b}[0] = \beta$ is thus that given in (13.96) with b_0 replaced by β (and s replaced by $-p$).

$$\hat{b}[n] = \frac{N}{r}\left\{\left(\frac{r}{N}\beta - p\right)\left(1 + \frac{r}{N}\right)^n + p\right\}. \tag{13.99}$$

13.4.4.1 Monthly Payment

If the loan is to payed in M years, we require that

$$\hat{b}[MN] = 0.$$

Imposing this condition on (13.99) and solving for the monthly payment, p, gives

$$p = \frac{\frac{r}{N}\beta}{1 - \left(1 + \frac{r}{N}\right)^{-MN}} \approx \frac{\frac{r}{N}\beta}{1 - e^{-rM}}. \tag{13.100}$$

A one million dollar loan at 5% interest ($r = 0.05$) compounded monthly ($N = 12$) has, for 30 years ($M = 30$), a monthly payment of $p = \$5368.22$.

13.4.4.2 Amortization

A portion of the payment in (13.100) goes to interest and a portion to payment of the loan. The balance due at pay period n is $\hat{b}[n]$ given in (13.99). The interest due on this balance is

$$i[n] = \frac{r}{N}\hat{b}[n]$$

$$= \left(\frac{r}{N}\beta - p\right)\left(1 + \frac{r}{N}\right)^N + p.$$

The amount of the payment going to the principle is thus

$$d[n] = p - i[n]$$

$$= \left(\frac{r}{N}\beta - p\right)\left(1 + \frac{r}{N}\right)^N$$

$$= \frac{r}{N}\beta\left(1 - \frac{1}{1 - \left(1 + \frac{r}{N}\right)^{-MN}}\right)\left(1 + \frac{r}{N}\right)^n.$$

13.4.4.3 Monthly Payment Over Long Periods of Time

For loans given over many years, the monthly payment, from (13.100), is

$$p_\infty = \lim_{M \to \infty} \frac{\frac{r}{N}\beta}{1 - \left(1 + \frac{r}{N}\right)^{-MN}}$$

$$= \frac{r}{N}\beta.$$

For modest interest rates, monthly payments for loans taken over a period of 100 years are about the same as those for a thousand years - even though the original loan, β, is the same in both cases. For example, using (13.100), a one million dollar loan at 5% interest compounded monthly has, for 100 years, a monthly payment of $\$4195.23$. For a thousand year mortgage, the monthly payment is $\$4166.67$. Since $p_\infty = \$4166.67$, a ten thousand year mortgage, to the penny, has the same monthly payments due as a thousand year mortgage.

13.5 Exercises

The Wave Equation, Its Fourier Solution and Harmony in Western Music

13.1. Show that the boundary conditions in (13.3) and the wave equation for the vibrating string in (13.1) are satisfied by the Fourier series solution in (13.4).

13.2. Instead of dividing the music octave equally into 12 notes, consider instead 19 notes.

(a) For the Pythagorean ratios for the major scales in Figure 13.5 and the minor scale in Figure 13.8, evaluate how close, in cents, Pythagorean frequencies deviate from the closest of the notes in the 19 notes per octave scale.

(b) What is the maximum deviation, in cents, from an arbitrary frequency to a note in the tempered scale?
(c) Compare your answer to the 12 note tempered scale and comment.
(d) Repeat this for another number of notes per octave. Can you get the deviation from the Pythagorean scale smaller than the 12 or 19 note scale?

Fourier Transforms in Optics and Wave Propagation

13.3. Is the wave equation for the vibrating string in (13.1) the same as the one dimensional wave equation from Maxwell's equations in (13.12) under a time harmonic assumption?

13.4. Show that the plane wave in (13.18) is a solution to the wave equation in (13.17).

13.5. **Phase conjugation**. Consider the angular spectrum plane wave propagation direction illustrated in Figure 13.13.
  (a) Sketch the propagation direction for another plane wave with the same (α, β) direction cosines.
  (b) Sketch the propagation directions for two plane wave with the same $(-\alpha, -\beta)$ direction cosines. Note that these are *conjugate plane waves*.
  (c) Consider, then, the luminous spider in Figure 13.22. Let the coherent wave emanating from the spider be $u(x, y, z)$. Sketch the wavefronts corresponding to the complex conjugate, $u^*(x, y, z)$.

13.6. The Airy pattern in Figure 13.19 is the result of Fraunhofer diffraction of a circle. If we go farther and farther from the source, the Airy pattern simply gets bigger. For small apertures, Fraunhofer diffraction produces the Fourier transform of the aperture. The Fourier transform of a Fourier transform produces the original function with a coordinate reversal. Should not, then, the Fraunhofer diffraction of the Fraunhofer diffraction of the Airy pattern be a circle? Resolve this reasoning paradox.

13.7. Show that the lobe width in (13.60) follows from (13.59).

Heisenberg's Uncertainty Principle

13.8. Can equality be achieved in (13.65)?

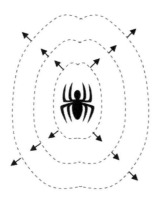

FIGURE 13.22. See Exercise 13.5.

Elementary Deterministic Finance

13.9. Doubling your money on a one time deposit.
  (a) Using (13.78), how many years does it take for your one time deposit to double its value?
  (b) What is this period when the interest rate is $r = 12\%$ and $N = 12$?
  (c) Using (13.91), how many years does it take for your one time deposit to double its value when there are taxes and inflation?
  (d) With an annual taxation of $t = 6\%$ and an inflation rate or $u = 5\%$, how long does it take to double your money? Use the interest rate and compounding frequency in (b).

13.10. You deposit one cent each month at a rate of $r = 12\%$ compounded monthly. How many years does it take to accumulate a million dollars? Assume there is no rounding, no taxation and no inflation.

13.6 Solutions for Selected Chapter 13 Exercises

13.2. Nineteen notes per octave uses an adjacent note frequency ratio of $2^{1/19}$ rather that the $2^{1/12}$ used for the conventional tempered scale.
  (a) In the 19 note system in the right columns of Figure 13.23, values of $2^{k/19}$ are shown for $0 \leq k \leq 19$. The Pythagorean ratios of Num/Den are shown on the left side. The deviation in cents of the Pythagorean ratios to the frequencies in the 19 note system is shown in the right column. The corresponding 12 note tempered scale frequency ratios and their deviation from the Pythagorean scale are also shown.

NOTE NAME	PYTHAGOREAN Num Den	RATIO	12 NOTES k	RATIO	error (cents)	19 NOTES k	RATIO	error (cents)
C	2 1	2.00000	12	2.00000	0	19	2.0000	0
						18	1.92835	
B	15 8	1.87500	11	1.88775	12	17	1.85927	-15
Bb	16 9	1.77778	10	1.78180	4	16	1.79266	14
						15	1.72844	
A	5 3	1.66667	9	1.68179	16	14	1.66652	0
Ab	8 5	1.60000	8	1.58740	-14	13	1.60682	7
						12	1.54926	
G	3 2	1.50000	7	1.49831	-2	11	1.49376	-7
						10	1.44025	
						9	1.38865	
F	4 3	1.33333	5	1.33484	2	8	1.33890	7
						7	1.29094	
E	5 4	1.25000	4	1.25992	14	6	1.24469	-7
Eb	6 5	1.20000	3	1.18921	-16	5	1.20010	0
						4	1.15711	
D	9 8	1.12500	2	1.12246	-4	3	1.11566	-14
Db	16 15	1.06667	1	1.05946	-12	2	1.07569	15
						1	1.03716	
C	1 1	1.00000	0	1.00000	0	0	1.00000	0

FIGURE 13.23. The deviation, in cents, from the Pythagorean musical scale from the evenly spaced 12 notes and 19 notes per octave scales. See Exercise 13.2.

APPLICATIONS 659

(b) 50 cents.
(c) The deviations in the 12 and 19 note systems look similar.

13.5. **Phase conjugation**.
   (c) The wavefronts will be the same, except they will converge towards the spider rather than diverge.

13.6. The Fraunhofer diffraction is not the Fourier transform of the aperture. It is the Fourier transform of the aperture multiplied by a quadratic phase.

13.9. Doubling your money on a one time deposit.
   (a) Set $\hat{b}[NM] = 2d$ in (13.78) and solve for M. The result is

$$M = \frac{\log(2)}{N \log\left(1 + \frac{r}{N}\right)}. \tag{13.101}$$

   (b) Setting $r = 0.12$ and $N = 12$ in (13.101) gives $M = 5.8$ years.
   (c) Solving for M in (13.91) gives

$$M = \frac{\ln(2)}{\ln\left((1-t)\left(1 + \frac{r}{N}\right)^N + t\right) - u}. \tag{13.102}$$

   (d) Substituting the numerical values into (13.102) gives

$$M = \frac{\ln(2)}{\ln\left(0.94\,(1.01)^{12} + 0.06\right) - 0.05} = 11.1 \text{ years}.$$

13.10. Set $\hat{b}[MN] = \$10^6$ in (13.95) and solve for M.

$$M = \frac{\log\left(1 + \frac{r \times 10^6}{Ns}\right)}{N \log(1 + \frac{r}{N})}.$$

Using $s = 0.1$, $r = 0.01$ and $N = 12$ gives $M = 115.7$ years.

14

Appendices

And he made a molten sea, ten cubits from the one brim to the other: it was round all about, and his height was five cubits: and a line of thirty cubits did compass it about. An early estimate of $\pi \approx 3$.

I Kings 7:23.

14.1 Schwarz's Inequality

For two complex functions, $g(t)$ and $h(t)$, Schwarz's inequality[1] is

$$\left| \Re \int_{-\infty}^{\infty} g(t) h^*(t) \, dt \right|^2 \leq \int_{-\infty}^{\infty} |g(t)|^2 \, dt \int_{-\infty}^{\infty} |h(t)|^2 \, dt \qquad (14.1)$$

Equality obviously occurs for $g(t) = \zeta h(t)$ where ζ is real. If this is not true, we have the strict inequality

$$\int_{-\infty}^{\infty} |g(t) + \zeta h(t)|^2 \, dt > 0.$$

Thus, recalling $z + z^* = 2\Re z$, we have

$$\int_{-\infty}^{\infty} |g(t) + \zeta h(t)|^2 \, dt$$

$$= \int_{-\infty}^{\infty} (g(t) + \zeta h(t)) \left(g^*(t) + \zeta h^*(t) \right) dt$$

$$= \zeta^2 \int_{-\infty}^{\infty} |g(t)|^2 \, dt + \zeta \int_{-\infty}^{\infty} \left(g(t) h^*(t) + g^*(t) h(t) \right) dt + \int_{-\infty}^{\infty} |h(t)|^2 \, dt \qquad (14.2)$$

$$= \zeta^2 \int_{-\infty}^{\infty} |g(t)|^2 \, dt + 2\zeta \int_{-\infty}^{\infty} \Re \left(g(t) h^*(t) \right) dt + \int_{-\infty}^{\infty} |h(t)|^2 \, dt$$

$$> 0.$$

---

[1]. Also called the Cauchy Schwarz inequality, the Cauchy inequality, and the Cauchy Bunyakovski Schwarz inequality.

For this inequality to be true, the final quadratic equation in ζ must have no real roots. Using the quadratic equation, the solution of

$$a\zeta^2 + b\zeta + c = 0$$

is

$$\zeta_\pm = \frac{-b \pm \sqrt{b^2 - 4ac}}{2a}.$$

There are no real solutions to the quadratic when the discriminant is negative, i.e. $b^2 - 4ac < 0$. Applying this to (14.2) with

$$a = \int_{-\infty}^{\infty} |g(t)|^2.$$

$$b = 2\int_{-\infty}^{\infty} \Re\left(g(t)h^*(t)\right) dt,$$

and

$$c = \int_{-\infty}^{\infty} |h(t)|^2$$

gives the strict inequality portion of Schwarz's inequality in (14.1).

14.2 Leibniz's Rule

$$\frac{d}{dt}\int_{\xi=\ell(t)}^{u(t)} f(t, \xi)\, d\xi = f(t, u(t)) - f(t, \ell(t)) + \int_{\xi=\ell(t)}^{u(t)} \frac{d}{dt} f(t, \xi)\, d\xi.$$

14.3 A Useful Limit

An important identity is

$$\lim_{a\to\infty} \left(1 + \frac{r}{a}\right)^a = e^r. \qquad (14.3)$$

Proof. We write

$$\lim_{a\to\infty} \ln\left(1 + \frac{r}{a}\right)^a = \lim_{a\to\infty} a\, \ln\left(1 + \frac{r}{a}\right)$$

$$= \lim_{a\to\infty} \frac{\ln\left(1 + \frac{r}{a}\right)}{\frac{1}{a}}.$$

This is a "zero over zero" situation to which we can apply L'Hopital's rule.

$$\lim_{a\to\infty} \ln\left(1 + \frac{r}{a}\right)^a = \lim_{a\to\infty} \frac{\frac{d}{da}\ln\left(1 + \frac{r}{a}\right)}{\frac{d}{da}\left(\frac{1}{a}\right)} = r.$$

This completes the proof.

14.4 Series

14.4.1 Binomial Series

The binomial series is

$$(a+b)^n = \sum_{k=0}^{n} \binom{n}{k} a^k b^{n-k}. \tag{14.4}$$

where the binomial coefficients are

$$\binom{n}{k} := \frac{n!}{k!(n-k)!}.$$

Pascal's triangle is a table of binomial coefficients.

```
n = 1 →                    1    1
    2 →                 1    2    1
    3 →              1    3    3    1
    4 →           1    4    6    4    1
    5 →        1    5   10   10    5    1
    6 →     1    6   15   20   15    6    1
    7 → 1    7   21   35   35   21    7    1
         ↗    ↗    ↗    ↗    ↗    ↗    ↗    ↗
k = 0    1    2    3    4    5    6    7
```

For example, from the table, we read $\binom{7}{3} = 35$.

14.4.2 Geometric Series

The finite geometric series is[2]

$$\sum_{n=0}^{N-1} z^n = \frac{1-z^N}{1-z}. \tag{14.5}$$

If $|z| < 1$, then $\lim_{z \to \infty} z^N = 0$, and

$$\sum_{n=0}^{\infty} z^n = \frac{1}{1-z} \; ; \; |z| < 1 \tag{14.6}$$

---

2. When $z = M$ is a positive integer, the number in (14.5) is dubbed a *repunit* [85]. In base M, these numbers are a string of $N-1$ ones. For $M = 2$, repunits are *Mersenne numbers* [562, 1073]. Mersenne numbers are a popular base for mining large prime numbers of the form $2^M - 1$. The first 39 *Mersenne primes* correspond to

$$M = 2, 3, 5, 7, 13, 17, 19, 31, 61, 89, 107, 127, 521, 607, 1279, 2203, 2281,$$
$$3217, 4253, 4423, 9689, 9941, 11213, 19937, 21701, 23209, 44497,$$
$$86243, 110503, 132049, 216091, 756839, 859433, 1257787,$$
$$1398269, 2976221, 3021377, 6972593, 13466917.$$

The prime 13,466,917, in base 2, is a string of 13,466,916 ones. Primes of the form $2^{2^{M-1}} - 1$ are *double Mersenne primes*. The first few correspond to $M = 2, 3, 5, 7$.

14.4.2.1 Derivation

To show this, let

$$S = \sum_{n=0}^{N-1} z^n.$$

Then

$$S = 1 + z + z^2 + \cdots + z^{N-1}$$
$$zS = z + z^2 + \cdots + z^{N-1} + z^N.$$

Subtracting and solving for S gives (14.5). More generally,

$$\sum_{n=\ell}^{h-1} z^n = \frac{z^\ell - z^h}{1 - z}. \tag{14.7}$$

14.4.2.2 Trigonometric Geometric Series

Using the geometric series, we can write

$$\sum_{n=-N}^{N} z^n = \frac{z^{N+\frac{1}{2}} - z^{-(N+\frac{1}{2})}}{z^{1/2} - z^{-1/2}}.$$

When $z = e^{j2\pi\theta}$,

$$\sum_{n=-N}^{N} e^{j2\pi n\theta} = \frac{\sin(\pi(2N+1)\theta)}{\sin(\pi\theta)}.$$

$$= (2N+1)\,\text{array}_{2N+1}(\theta). \tag{14.8}$$

Similarly

$$\sum_{n=0}^{N-1} e^{j2\pi n\theta} = N\,e^{j\pi(N-1)\theta}\,\text{array}_N(\theta). \tag{14.9}$$

14.5 Ill-Conditioned Matrices

Singular matrices have zero determinants. Matrices close to being singular are *ill-conditioned*. The condition of a matrix is

$$\text{cond}(\mathbf{A}) = \frac{|\lambda_{\max}|}{|\lambda_{\min}|}. \tag{14.10}$$

where the λ's are the eigenvalues of the matrix.

$$\lambda_n \vec{\psi}_n = \mathbf{A}\vec{\psi}_n.$$

The ψ_n's are the corresponding eigenvectors.

Example 14.5.1. Consider the matrix

$$\mathbf{A} = \begin{bmatrix} 0.9562 & 0.6889 & 1.0230 & 0.4380 & 0.8832 \\ 0.6889 & 0.6044 & 0.9112 & 0.2860 & 0.7146 \\ 1.0230 & 0.9112 & 1.3752 & 0.4210 & 1.0710 \\ 0.4380 & 0.2860 & 0.4210 & 0.2087 & 0.3832 \\ 0.8832 & 0.7146 & 1.0710 & 0.3832 & 0.8725 \end{bmatrix}.$$

The condition of this matrix is 1.8874×10^9. When this matrix is multiplied by a vector of ones, the result is

$$\vec{a} = \mathbf{A} \begin{bmatrix} 1 \\ 1 \\ 1 \\ 1 \\ 1 \end{bmatrix} = \begin{bmatrix} 3.9893 \\ 3.2052 \\ 4.8014 \\ 1.7368 \\ 3.9245 \end{bmatrix}. \tag{14.11}$$

The inverse of **A** is

$$\mathbf{A}^{-1} = 10^8 \times \begin{bmatrix} 0.8267 & 1.5639 & -0.5038 & -0.5633 & -1.2519 \\ 0.5816 & 1.1744 & -2.0727 & -2.4462 & 2.0681 \\ 0.0314 & -0.8723 & 0.9632 & 0.5381 & -0.7361 \\ -0.5853 & -2.1954 & 0.3746 & -0.3287 & 2.0751 \\ -1.0947 & -0.5101 & 0.8607 & 2.0575 & -0.4342 \end{bmatrix}. \tag{14.12}$$

As we would expect,

$$\mathbf{A}^{-1} \begin{bmatrix} 3.9893 \\ 3.2052 \\ 4.8014 \\ 1.7368 \\ 3.9245 \end{bmatrix} = \begin{bmatrix} 1.0000 \\ 1.0000 \\ 1.0000 \\ 1.0000 \\ 1.0000 \end{bmatrix}. \tag{14.13}$$

To illustrate the remarkable sensitivity of the ill-conditioned matrix, change the fifth element in the multiplying vector from 3.9245 to 3.9246. Then

$$\mathbf{A}^{-1} \begin{bmatrix} 3.9893 \\ 3.2052 \\ 4.8014 \\ 1.7368 \\ 3.9246 \end{bmatrix} = \begin{bmatrix} -14,318 \\ 23,657 \\ -8419 \\ 23,736 \\ -4966 \end{bmatrix}. \tag{14.14}$$

Compare the dramatic difference of (14.13) and (14.14). By this *very* small change, the results are thrown off by as much as five orders of magnitude.

Insight into the sensitivity can be seen in the large numbers in the matrix's inverse in (14.12). When multiplied by a vector like $\vec{a}$, the resultant vector components are determined by subtracting very large and similar numbers to obtain small numbers. High computational accuracy is required on nearly noiseless data.

14.6 Other Commonly Used Random Variables

There exist numerous other random variable models for use in a plurality of fields. Some are presented in Section 4.2.2. In this section of the Appendix, we list some of the other commonly used. Many of the random variables, such as the chi-squared, F, and Student's t, can be derived from other random variables. For details, consult an introductory text on statistics [613]. Other random variables, such as the Weibull and Pareto, were created due to their descriptive and modelling attributes. Others, such as the Maxwell and Planck's radiation random variables, are descriptive of physical processes.

14.6.1 The Pareto Random Variable

The Pareto probability density function, with parameter a and shift parameter b, is pictured in Figures 14.1 and 14.2. Its probability density function is

$$f_X(x) = \frac{ab^a \mu(x-b)}{x^{a+1}}.$$

The Pareto random variable is used to model many physical phenomena, including the distribution of incomes above a given level and the distribution of the populations of cities above a given population.

In general, for $b > 0$,

$$\int_b^\infty x^\zeta \, dx = \begin{cases} -\dfrac{b^{\zeta+1}}{\zeta+1} & ; \zeta < -1 \\ \infty & ; \zeta \geq -1. \end{cases}$$

The mth moment of the Pareto random variable is therefore

$$\overline{X^m} = ab^a \int_b^\infty x^{m-a-1} \, dx$$

$$= \begin{cases} \dfrac{ab^m}{a-m} & ; a > m \\ \infty & ; a \leq m. \end{cases}$$

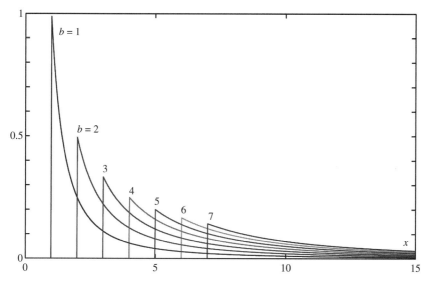

FIGURE 14.1. The Pareto random variable for $a = 1$ and various values of b. $f_X(x) = ab^a \, x^{-a-1} \mu(x-b)$. Each of these random variables has an infinite mean and an infinite variance. Also, see Figure 14.2.

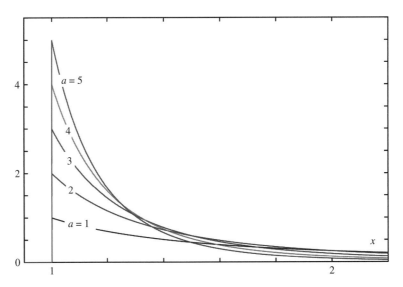

FIGURE 14.2. The Pareto random variable for $b = 1$ and various values of a. Another plot of the Pareto probability density function is in Figure 14.1.

The mean of the Pareto random variable corresponds to $m = 1$.

$$\overline{X} = \begin{cases} \dfrac{ab}{a-1} & ; a > 1 \\ \infty & ; a \leq 1. \end{cases}$$

The second moment, for $m = 2$, is

$$\overline{X^2} = \begin{cases} \dfrac{ab^2}{a-2} & ; a > 1 \\ \infty & ; a \leq 1. \end{cases}$$

The variance follows.

$$\sigma_X^2 = \begin{cases} \dfrac{ab^2}{(a-2)(a-1)^2} & ; a > 2 \\ \infty & ; a \geq 2. \end{cases}$$

14.6.2 The Weibull Random Variable

The Weibull random variable with parameter α scale parameter β, is used largely in reliability theory. The probability density function is

$$f_X(x) = \alpha \beta^{-\alpha} x^{\alpha-1} e^{-\left(\frac{x}{\beta}\right)^\alpha} \mu(x).$$

The mean and variance are

$$\overline{X} = \beta^{\frac{1}{\alpha}} \Gamma\left(1 + \frac{1}{\alpha}\right)$$

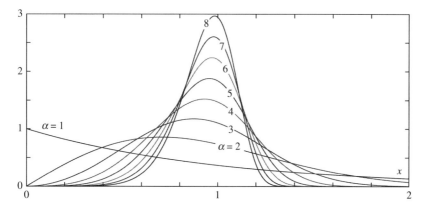

FIGURE 14.3. The probability density function for the Weibull random variable. $\beta = 1$

and

$$\sigma_X^2 = \beta^{\frac{2}{\alpha}} \left[\Gamma\left(1 + \frac{2}{\alpha}\right) - \Gamma\left(1 + \frac{1}{\alpha}\right) \right].$$

Plots are in Figure 14.3 for $\beta = 1$. The exponential random variable is a special case corresponding to $\alpha = 1$.

14.6.3 The Chi Random Variable

The chi random variable with k degrees of freedom is the square root of a chi-squared random variable, equal to the square root of the sum of the squares on k i.i.d. zero mean unit variance Gaussian random variables. Its probability density function is

$$f_X(x) = \frac{2^{1-\frac{k}{2}} x^{k-1} e^{-x^2/2} \mu(x)}{\Gamma(k/2)}.$$

For $k = 1$, the chi random variable is the half normal random variable. For $k = 2$, it is a Rayleigh random variable. The mean and variance are

$$\overline{X} = \frac{\sqrt{2}\,\Gamma\left(\frac{k+1}{2}\right)}{\Gamma(k/2)}$$

and

$$\sigma_X^2 = k - \overline{X}^2.$$

Plots are in Figure 14.4.

14.6.4 The Noncentral Chi-Squared Random Variable

Whereas the chi-squared random variable is the sum of the squares of k zero mean unit variance Gaussian random variables, the noncentral chi-squared random variable, the means are no longer zero. Its probability density function is

$$f_X(x) = \frac{\sqrt{\lambda}\, x^{\frac{k-1}{2}} e^{-\frac{x+\lambda}{2}} I_{\frac{k}{2}-1}\left(\sqrt{\lambda x}\right)}{2\,(\lambda x)^{\frac{r}{4}}}.$$

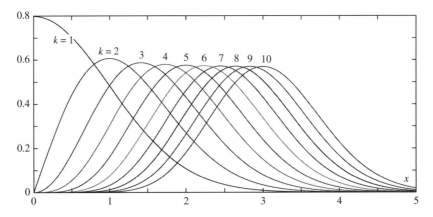

FIGURE 14.4. The probability density function for the chi random variable.

The new parameter is λ. The chi-squared random variable is a special case for $\lambda = 0$.
The mean and variance of the noncentral chi-squared random variable are

$$\overline{X} = \frac{\sqrt{2}\,\Gamma\left(\frac{k+1}{2}\right)}{\Gamma(k/2)}$$

and

$$\sigma_X^2 = k - \overline{X}^2.$$

14.6.5 The Half Normal Random Variable

The half normal (or half Gaussian) random variable, with scale parameter θ, is obtained by taking the magnitude of a zero mean Gaussian random variable. The probability density function for the half normal random variable is

$$f_X(x) = \frac{2\theta}{\pi}\, e^{-\frac{x^2\theta^2}{\pi}}\, \mu(x)$$

and is shown as the $k = 1$ curve in Figure 14.4. The mean and variance are

$$\overline{X} = \frac{1}{\theta}$$

and

$$\sigma_X^2 = \frac{\pi - 2}{2\theta^2}.$$

14.6.6 The Rayleigh Random Variable

The Rayleigh random variable, with scale parameter ς, is a special case of the Weibull random variable. Its probability density function is

$$f_X(x) = \frac{x}{\varsigma^2}\, e^{-\frac{x^2}{2\varsigma^2}}\, \mu(x).$$

For $\varsigma = 1$, the Rayleigh random variable is a chi random variable for $k = 2$. It is obtained by performing the square root of the square of the sum of two independent zero mean Gaussian random variables with identical variances. The mean and variance of the Rayleigh random variable are

$$f_X(x) = \varsigma\sqrt{\frac{\pi}{2}}$$

and

$$\sigma_X^2 = \frac{4-\pi}{2}\varsigma^2.$$

14.6.7 The Maxwell Random Variable

The Maxwell random variable describes the speeds of molecules in thermal equilibrium according to statistical mechanics. The probability density function is

$$f_X(x) = \sqrt{\frac{2}{\pi}}\,\alpha^{3/2}\,x^2 e^{-\frac{\alpha x^2}{2}}\mu(x)$$

where α is a scale parameter. A plot is in Figure 14.5. The mean and variance are

$$\overline{X} = 2\sqrt{\frac{2}{\pi\alpha}},$$

and

$$\sigma_X^2 = \frac{3\pi - 8}{\pi\alpha}.$$

14.6.8 The Log Random Variable

The log normal random variable, as the name implies, is a random variable whose log is a normal, or Gaussian, random variable. With shift parameter θ and scale parameter ς, its probability density function is

$$f_X(x) = \frac{1}{\sqrt{2\pi}\,\varsigma x}\,e^{-\frac{(\log x - \theta)^2}{2\varsigma^2}}\mu(x). \tag{14.15}$$

The mean and variance are

$$\overline{X} = e^{\theta + \frac{\varsigma^2}{2}},$$

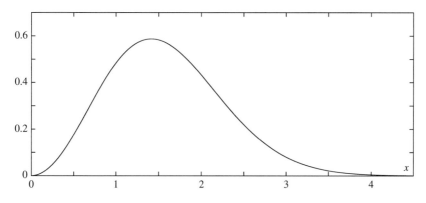

FIGURE 14.5. The probability density function for the Maxwell random variable for $\alpha = 1$.

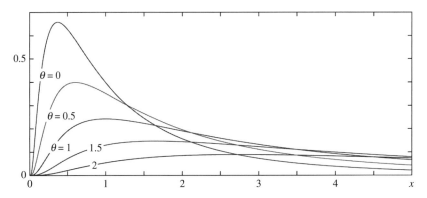

FIGURE 14.6. The probability density function for the log normal random variable in (14.15) for $\varsigma = 1$. Continued in Figure 14.7.

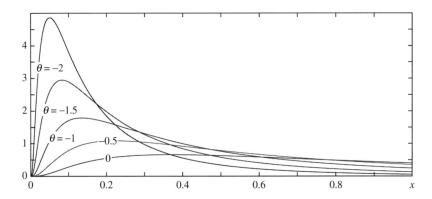

FIGURE 14.7. Continuation of Figure 14.6 for the log normal probability density function.

and

$$\sigma_X^2 = e^{\varsigma^2 + 2\theta}\left(e^{\varsigma^2} - 1\right).$$

Plots are shown in Figures 14.6 and 14.7.

14.6.9 The Von Mises Variable

The Von Mises random variable has a probability density function of

$$f_X(x) = \frac{e^{\beta \cos(x-\alpha)} \Pi\left(\frac{x-\pi}{2\pi}\right)}{2\pi I_0(\beta)}. \tag{14.16}$$

Its characteristic function is the topic of Exercise 4.11. The mean and variance are

$$\overline{X} = \alpha,$$

and

$$\sigma_X^2 = 1 - \frac{I_1(b)}{I_0(b)}.$$

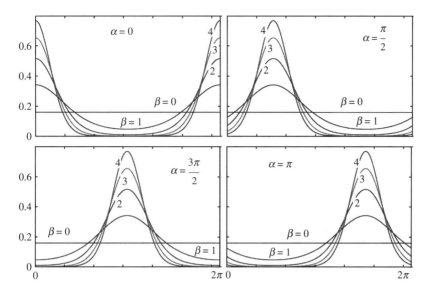

FIGURE 14.8. The probability density function for the Von Mises random variable in (14.16) for various α and β. For $\beta = 0$, a uniform random variable is a special case.

The uniform probability density function on $[0, 2\pi]$ is a special case for $\beta = 0$. Plots are in Figure 14.8 for various α and β.

14.6.10 The Uniform Product Variable

The uniform product random variable is the product of k i.i.d. uniform random variables on $[0, 1]$. Its probability density function is

$$f_X(x) = \frac{(-1)^{k-1} (\log x)^{k-1} \Pi\left(x - \frac{1}{2}\right)}{(k-1)!}. \tag{14.17}$$

As we expect, the uniform random variable is a special case when $k = 1$. Plots are in Figure 14.9. Although for $k > 1$ the density function is ∞ at $x = 0+$, the moments are finite. The mean and variance are

$$\overline{X} = 2^{-k},$$

and

$$\sigma_X^2 = 3^{-k} - 4^{-k}.$$

14.6.11 The Uniform Ratio Variable

The uniform ratio (or uniform quotient) random variable is formed by the ratio of two independent random variables uniform on $[0, 1]$. The probability density function is

$$f_X(x) = \frac{1}{2}\left[\Pi\left(x - \frac{1}{2}\right) + x^{-2}\mu(x - 1)\right].$$

Plots are in Figure 14.10. The tail decays algebraically as x^{-2} so both the first and second moments are infinite.

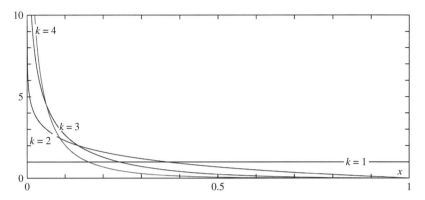

FIGURE 14.9. The probability density function for the uniform product random variable in (14.17) for various k. For $k = 1$, a uniform random variable is a special case.

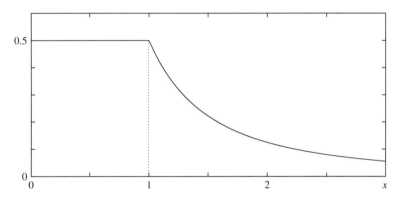

FIGURE 14.10. The probability density function for the uniform ratio random variable. All moments are infinite.

14.6.12 The Logistic Random Variable

The logistic random variable, used in the study of population growth, has a shift parameter η and scale parameter b. Its probability density function

$$f_X(x) = \frac{e^{-\frac{x-\eta}{b}}}{b\left(1 + e^{-\frac{x-\eta}{b}}\right)^2}. \quad (14.18)$$

The mean and variance are

$$\overline{X} = \eta,$$

and

$$\sigma_X^2 = \frac{(\pi b)^2}{3}.$$

Plots of the logistic probability density function are in Figure 14.11.

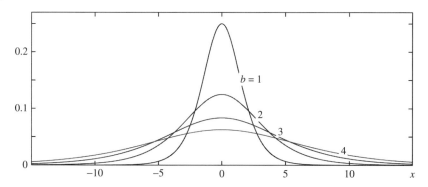

FIGURE 14.11. The probability density function for the logistic random variable in (14.18) for various b and $\eta = 0$.

14.6.13 The Gibrat Random Variable

The Gibrat random variable is a special case of the log normal random variable for $\theta = 0$ and $\varsigma = 1$.

14.6.14 The F Random Variable

The F random variable is obtained by the ratio of a chi-squared random variable with parameter n to an independent chi-squared random variable with parameter m. Its probability density function is

$$f_F(x) = \frac{x^{\frac{n}{2}-1}}{B\left(\frac{n}{2}, \frac{m}{2}\right)} n^{\frac{n}{2}} m^{\frac{m}{2}} (xn+m)^{-\frac{n+m}{2}} \mu(x). \tag{14.19}$$

Plots of the probability density functions are in Figures 14.12 and 14.13.

The mean and the variance of the F random variable are

$$\overline{F} = \frac{m}{m-2}$$

and

$$\sigma_F^2 = \frac{2m^2(m+n-2)}{m(m-2)^2(m-4)}.$$

14.6.15 The Noncentral F Random Variable

The noncentral F random variable is a ratio of two noncentral chi-squared random variables. Its probability density function, in terms of the F probability density function in (14.19), is

$$f_X(x) = e^{\frac{\lambda}{2}} f_F(x) \, {}_1F_1\left(\frac{n+m}{2}, \frac{n}{2}, \frac{\lambda n x}{2(nx+m)}\right).$$

where the confluent hypergeometric function, ${}_1F_1$, in introduced in Exercise 4.9. The F random variable is a special case for $\lambda = 0$.

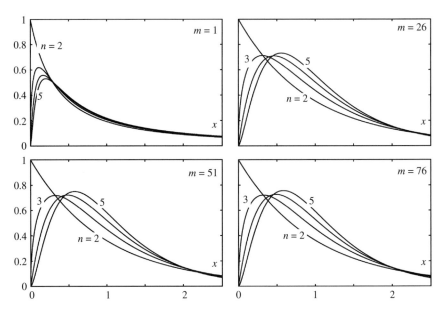

FIGURE 14.12. The probability density function for the F random variable in (14.19) for various values of m and $n = 2, 3, 4, 5$. See also Figure 14.13.

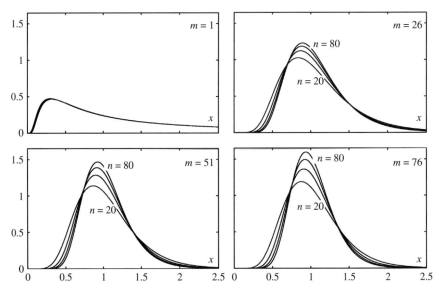

FIGURE 14.13. The probability density function for the F random variable in (14.19) for various values of m and $n = 20, 40, 60, 90$. The plots are graphically indistinguishable for $m = 1$. See also Figure 14.12.

The mean of the noncentral F random variable is

$$\overline{X} = \frac{(\lambda + n)m}{n(m-2)}$$

and the variance is

$$\sigma_X^2 = \frac{2m^2\left[n^2 + (2\lambda + m - 2)n + \lambda(\lambda + 2m - 4)\right]}{n^2(m-4)(m-2)^2}.$$

14.6.16 The Fisher-Tippett Random Variable

The Fisher-Tippett (or extreme value or log-Weibull) random variable.

$$f_X(x) = \frac{1}{\beta} e^{\left(\frac{\alpha - x}{\beta} - e^{\frac{\alpha - x}{\beta}}\right)}. \tag{14.20}$$

Probability density function plots are in Figure 14.14. The random variable's mean is

$$\overline{X} = \alpha + \beta\gamma$$

where the Euler-Mascheroni constant is

$$\gamma = \lim_{n \to \infty} \left(\sum_{k=1}^{n} \frac{1}{k} - \log n\right) = 0.57721566490153286\ldots \tag{14.21}$$

The variance is

$$\sigma_X^2 = \frac{\pi^2 \beta^2}{6}.$$

14.6.17 The Gumbel Random Variable

The Gumbel random variable is a special case of the Fisher-Tippett random variable for $\alpha = 0$ and $\beta = 1$ and therefore corresponds to the $\beta = 1$ curve in Figure 14.14.

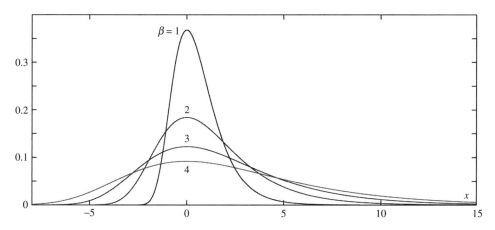

FIGURE 14.14. The probability density function for the Fisher-Tippett random variable in (14.20) for shift parameter $\alpha = 0$ and scale parameter $\beta = 1, 2, 3, 4$.

14.6.18 The Student's t Random Variable

The Student's t random variable is useful for problems associated with inference based on "small" zero mean samples where estimating the mean and variance with sample averages can deviate significantly from the true mean and variance.

$$f_x(X) = \frac{\Gamma\left(\frac{n+1}{2}\right)}{\sqrt{\pi n}\,\Gamma\left(\frac{n}{2}\right)\left(1+\frac{x^2}{n}\right)^{\frac{n+1}{2}}}. \tag{14.22}$$

The plots in Figure 14.15 illustrate that, as $n \to \infty$, the Student's t approaches a zero mean unit variance Gaussian probability density function. The Student's t random variable is zero mean with a variance of

$$\sigma_X^2 = \frac{n}{n-2}.$$

14.6.19 The Noncentral Student's t Random Variable

The noncentral Student's t random variable is a generalization of the Student's t random variable for nonzero mean samples.

$$f_X(x) = \frac{n^{\frac{n}{2}} n!}{2^n e^{\frac{\lambda^2}{2}} (n+x^2)^{\frac{n}{2}} \Gamma\left(\frac{n}{2}\right)}$$

$$\times \left[\frac{\sqrt{2}\lambda x \,_1F_1\left(\frac{n}{2}+1, \frac{3}{2}, \frac{(\lambda x)^2}{2(n+x^2)}\right)}{(n+x^2)\Gamma\left(\frac{n+1}{2}\right)} + \frac{_1F_1\left(\frac{n+1}{2}, \frac{1}{2}, \frac{(\lambda x)^2}{2(n+x^2)}\right)}{\sqrt{n+x^2}\,\Gamma\left(\frac{n}{2}+1\right)} \right].$$

Its mean is

$$\overline{X} = \sqrt{\frac{n}{2}}\,\frac{\lambda \Gamma\left(\frac{n-1}{2}\right)}{\Gamma\left(\frac{n}{2}\right)}$$

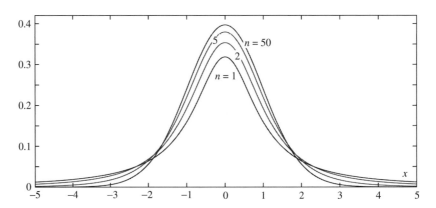

FIGURE 14.15. The probability density function for the Student's t random variable in (14.22) for various n. Due to the central limit theorem, as $n \to \infty$, the Student's t probability density function approaches a zero mean unit variance Gaussian probability density function. In this figure, the $n = 50$ plot is indistinguishable from the Gaussian.

and the variance is

$$\sigma_X^2 = \frac{(\lambda^2 + 1)n}{n-2} - \frac{\lambda^2 n}{2}\left[\frac{\Gamma\left(\frac{n-1}{2}\right)}{\Gamma\left(\frac{n}{2}\right)}\right]^2.$$

The Student's t is a special case for $\lambda = 0$.

14.6.20 The Rice Random Variable

The Rice random variable with parameters $\upsilon \geq 0$ and ς^2, has a probability density function of

$$f_X(x) = \frac{x}{\varsigma^2} e^{-\frac{x^2+\upsilon^2}{2\varsigma^2}} I_0\left(\frac{x\upsilon}{\varsigma^2}\right) \mu(x). \tag{14.23}$$

The Rayleigh random variable is a special case for $\upsilon = 0$. The Rice random variable, also referred to as *Rician*, is used in detection theory [1442]. A plot of its probability density function is in Figure 14.16.

14.6.21 The Planck's Radiation Random Variable

Planck's radiation function, pictured in Figure 14.17, is the probability density function

$$f_X(x) = \frac{15}{\pi^4 x^5}\left(e^{\frac{1}{x}} - 1\right)^{-1} \mu(x). \tag{14.24}$$

14.6.22 The Generalized Gaussian Random Variable

The probability density function of generalized Gaussian random variable the with parameters $k > 0$, variance σ_X^2 and mean $\overline{X}$ is defined by [722]

$$f_X(x) = \frac{k}{2A(k)\Gamma(1/k)} e^{-(|x-\overline{X}|/A(k))^k} \tag{14.25}$$

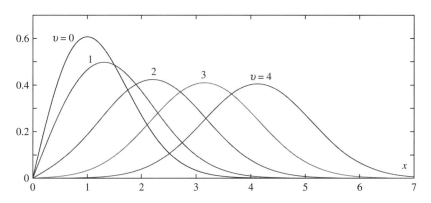

FIGURE 14.16. The probability density function for the Rice random variable in (14.23) for scale parameter $\varsigma = 1$ and parameter $\upsilon = 0, 1, 2, 3, 4$.

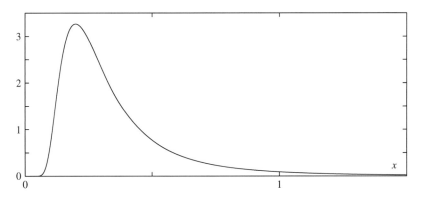

FIGURE 14.17. The probability density function for the Planck's radiation random variable in (14.24).

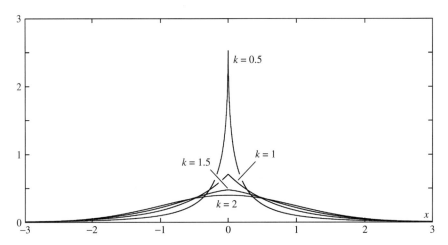

FIGURE 14.18. The generalized Gaussian probability density function in (14.25) for various values of k for $\sigma_X^2 = 1$ and $\overline{X} = 0$. For $k = 1$, we have the Laplace probability density function. For $k = 2$, the Gaussian.

where

$$A(k) = \left(\frac{\sigma_X^2\, \Gamma(1/k)}{\Gamma(3/k)} \right)^{\frac{1}{2}}. \tag{14.26}$$

The Gaussian random variable is a special case for $k = 2$. The Laplace random variable follows from $k = 1$. This is illustrated in Figure 14.18.

The generalized Gaussian random variable is used in detection theory [722].

14.6.23 The Generalized Cauchy Random Variable

The generalized Cauchy random variable. The probability density function for the generalized Cauchy random variable with parameters k and ν is [722]

$$f_X(x) = \frac{\varsigma(k, \nu)}{\left[1 + \frac{1}{\nu} \left(\frac{|x|}{A(k)} \right)^k \right]^{\nu + \frac{1}{k}}} \tag{14.27}$$

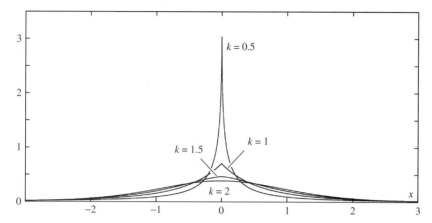

FIGURE 14.19. The generalized Cauchy probability density function in (14.27) for $v = 4$ and $\sigma_X^2 = 1$. (Note: σ_X^2 can not be interpreted as a variance for the generalized Cauchy random variable.)

where $A(k)$ is defined in (14.26) and

$$\varsigma(k, v) = \frac{kv^{-1/k}\,\Gamma\left(v + \frac{1}{k}\right)}{2A(k)\,\Gamma(v)\,\Gamma\left(\frac{1}{k}\right)}.$$

The Cauchy random variable is a special case for $k = 2$ and $v = 1/2$.

The tails of the generalized Cauchy are fat, and decay algebraically in inverse proportion to $|x|^{kv+1}$. The generalized Gaussian random variable has a much faster exponential decay.

The generalized Cauchy random variable, illustrated in Figure 14.19, is used in detection theory [722].

15

References

A

[1] J.B. Abbiss, C. De Mol, and H.S. Dhadwal. Regularized iterative and non-iterative procedures for object restoration from experimental data. *Opt. Acta*, pp. 107–124, 1983.
[2] J.B. Abbiss, B.J. James, J.S. Bayley, and M.A. Fiddy. Super-resolution and neural computing. *Proceedings SPIE, International Society Optical Engineering*, pp. 100–106, 1988.
[3] Edwin A. Abbott. **Flatland: A Romance of Many Dimensions**. Dover, 2007. Originally published in 1884.
[4] Ray Abma and Nurul Kabir. 3D interpolation of irregular data with a POCS algorithm. *Geophysics. Vol. 71(6)*, pp. E91–E97, 2006.
[5] M. Abramowitz and I.A. Stegun. **Handbook of Mathematical Functions**. Dover, New York, 9th ed., 1964.
[6] L.D. Abreu. A q-sampling theorem related to the q-Hankel transform. *Proceedings of the American Mathematical Society, Vol. 133(4)*, pp. 1197–1203, 2005.
[7] Amir D. Aczel. **The Mystery of the Aleph: Mathematics, the Kabbalah, and the Human Mind**. Four Walls Eight Windows, 2000.
[8] H.C. Agarwal. Heat transfer in laminar flow between parallel plates at small Peclet numbers. *Applied Sci. Res., Vol. 9*, pp. 177–189, 1960.
[9] R. Agarwal, M. Bohner, and A. Peterson. Inequalities on time scales: A survey. *Mathematical Inequalities and Applications, Vol. 4(4)*, pp. 535–557, 2001.
[10] R. Agarwal, M. Bohner, D. O'Regan *et al*. Dynamic equations on time scales: a survey. *Journal of Computational and Applied Mathematics, Vol. 141(1–2)*, pp. 1–26, 2002.
[11] T.H. Ahberg, E.N. Nilson, and T.L. Walsh. **The Theory of Splines and Their Applications**. Academic Press, New York, 1964.
[12] C.B. Ahn, J.H. Kim, and Z.H. Cho. High-speed spiral echo planar NMR imaging – I. *IEEE Transactions Medical Imaging, Vol. MI-5*, pp. 2–7, 1986.
[13] S.T. Alexander. Fast Adaptive Filters: A Geometrical Approach. *IEEE ASSP Magazine*, pp. 18–28, 1986.
[14] Z. Alkachouh and M. G. Bellanger. Fast DCT-based spatial domain interpolation of blocks in images. *IEEE Transactions Image Processing., Vol. 9, No. 4*. pp. 729–732, 2000.
[15] J.P. Allebach. Analysis of sampling pattern dependence in time-sequential sampling of spatio-temporal signals. *Journal of the Optical Society America, Vol. 71*, pp. 99–105, 1981.
[16] L.B. Almeida. An introduction to the angular Fourier transform. *Proceedings IEEE International Conf. Acoust., Speech, Signal Process.*, Minneapolis, MN, 1993.
[17] L.B. Almeida. The fractional Fourier transform and time-frequency representations. *IEEE Transactions on Signal Processing. Vol. 42(11)*, pp. 3084–3091, 1994.
[18] L.B. Almeida, Product and Convolution Theorems for the Fractional Fourier Transform. *IEEE Signal Processing Letters, Vol. 4(1)*, pp. 15–17, 1997.

[19] Oktay Altun, Gaurav Sharma, Mehmet U. Celik, and Mark F. Bocko. A set theoretic framework for watermarking and its application to semifragile tamper detection. *IEEE Transactions on Information Forensics and Security, Vol. 1(4)*, pp. 479–492, 2006.

[20] M. Ancis and D.D. Giusto. Reconstruction of missing blocks in JPEG picture transmission. *IEEE Pacific Rim Conference*, pp. 288–291, 1999.

[21] B. Andò, S. Baglio, S. Graziani and N. Pitrone. Threshold Error Reduction in Linear Measurement Devices by Adding Noise Signal. *Proceedings of the IEEE Instrumentation and Measurements Technology Conference*, St. Paul, Minnesota, pp. 18–21, 42–46, 1998.

[22] H.C. Andrews and C.L. Patterson. Singular value decompositions and digital image processing. *IEEE Transactions Acoust., Speech, Signal Processing*, pp. 26–53, 1976.

[23] K. Araki. Sampling and recovery formula in finite Hankel transform based on the minimum norm principle. *Transactions Institute Electron. Commun. Engineering Jpn., Vol. J68A*, pp. 643–649, 1985.

[24] F. Arago. **Joseph Fourier, Biographies of Distinguished Scientific Men**. London, pp. 242–286, 1957.

[25] H. Arsenault. Diffraction theory of Fresnel zone plates. *Journal of the Optical Society America, Vol. 58*, p. 1536, 1968.

[26] H. Arsenault and A. Boivin. An axial form of the sampling theorem and its application to optical diffraction. *Journal Applied Phys., Vol. 38*, pp. 3988–3990, 1967.

[27] H. Arsenault and K. Chalasinska-Macukow. The solution to the phase retrieval problem using the sampling theorem. *Optical Communication, Vol. 47*, pp. 380–386, 1983.

[28] H. Arsenault and B. Genestar. Deconvolution of experimental data. *Can. J. Phys., Vol. 49*, pp. 1865–1868, 1971.

[29] H. Artes, F. Hlawatsch and G. Matz. Efficient POCS algorithms for deterministic blind equalization of time-varying channels. IEEE Global Telecommunications Conference, 2000. GLOBECOM '00. *Vol. 2,27*, pp. 1031–1035, 2000.

[30] S. Chaturvedi, N. Arvind, Mukunda, *et al.* The sampling theorem and coherent state systems in quantum mechanics. *Physica Scripta, Vol. 74(2)*, pp. 168–179, 2006.

[31] R.B. Ash. **Information Theory**. Interscience, New York, 1965.

[32] J.L. Aravena. *Recursive moving window DFT algorithm. IEEE Transactions Computers, Vol. 39*, pp. 145–151, 1990.

[33] Nakhle H. Asmar. **Partial Differential Equations and Boundary Value Problems with Fourier Series (2nd Edition)**. Prentice Hall, 2004.

[34] C. Atzeni and L. Masotti. A new sampling procedure for the synthesis of linear transversal filters. *IEEE Transactions Aerospace & Electronic Systems, Vol. AES-7*, pp. 662–670, 1971.

[35] T. Auba and Y. Funahashi. The structure of all-pass matrices that satisfy directional interpolation requirements. *IEEE Transactions Automatic Control, Vol. 36*, pp. 1485–1489, 1991.

[36] T. Auba and Y. Funahashi. The structure of all-pass matrices that satisfy two-sided interpolation requirements. *IEEE Transactions Automatic Control, Vol. 36*, pp. 1489–1493, 1991.

[37] D.A. August and W.B. Levy. A simple spike train decoder inspired by the sampling theorem. *Neural Computation, Vol. 8(1)*, pp. 67–84, 1996.

[38] Saint Augustine. **DeGenesi ad Letteram, Book II.**

[39] L. Auslander and R. Tolimieri. Characterizing the radar ambiguity functions. *IEEE Transactions Information Theory, Vol. IT-30*, pp. 832–836, 1984.

B

[40] A. Bachl and W. Lukosz. Experiment on super-resolution imaging of a reduced object field. *Journal of the Optical Society America, Vol. 57*, pp. 163–169, 1967.

[41] Sonali Bagchi and Sanjit K. Mitra. **The Nonuniform Discrete Fourier Transform and its Applications in Signal Processing**. Springer, 1998.

[42] A.V. Balakrishnan. A note on the sampling principle for continuous signals. *IRE Transactions Information Theory, Vol. IT-3*, pp. 143–146, 1957.

[43] A.V. Balakrishnan. On the problem of time jitter in sampling. *IRE Transactions Information Theory, Vol. IT-8*, pp. 226–236, 1962.

[44] A.V. Balakrishnan. Essentially bandlimited stochastic processes. *IEEE Transactions Information Theory*, Vol. IT-11, pp. 145–156, 1965.

[45] C. Bardaro, P.L. Butzer, R.L. Stens RL, et al. Approximation error of the Whittaker cardinal series in terms of an averaged modulus of smoothness covering discontinuous signals. *Journal of Mathematical Analysis and Applications*, Vol. 316(1), pp. 269–306, 1 2006.

[46] I. Bar-David. An implicit sampling theorem for bounded bandlimited functions. *Information & Control*, Vol. 24, pp. 36–44, 1974.

[47] I. Bar-David. Sample functions of a Gaussian process cannot be recovered from their zero crossings. *IEEE Transactions Information Theory*, Vol. IT-21, pp. 86–87, 1975.

[48] I. Bar-David. On the degradation of bandlimiting systems by sampling. *IEEE Transactions Communications*, Vol. COM-25, pp. 1050–1052, 1977.

[49] R. Barakat. Application of the sampling theorem to optical diffraction theory. *Journal of the Optical Society America*, Vol. 54, pp. 920–930, 1964.

[50] R. Barakat. Determination of the optical transfer function directly from the edge spread function. *Journal of the Optical Society America*, Vol. 55, pp. 1217–1221, 1965.

[51] R. Barakat. Nonlinear transformation of stochastic processes associated with Fourier transforms of bandlimited positive functions. *International J. Contr.*, Vol. 14, No. 6, pp. 1159–1167, 1971.

[52] R. Barakat. Note on a sampling expansion posed by O'Neill and Walther. *Optical Communication*, Vol. 23, pp. 207–208, 1977.

[53] R. Barakat. Shannon numbers of diffraction images. *Optical Communication*, pp. 391–394, 1982.

[54] R. Barakat and E. Blackman. Application of the Tichonov regularization algorithm to object restoration. *Optical Communication*, pp. 252–256, 1973.

[55] R. Barakat and A. Houston. Line spread and edge spread functions in the presence of off-axis aberrations. *Journal of the Optical Society America*, Vol. 55, pp. 1132–1135, 1965.

[56] R. Barakat and J.E. Cole III. Statistical properties of n random sinusoidal waves in additive Gaussian noise. *Journal Sound and Vibration*, Vol. 63, pp. 365–377, 1979.

[57] V. Bargmann, P. Butera, L. Girardello, and J.R. Klauder. On the completeness of the coherent states, *Rep. Math. Phys.* 2, pp. 221–228, 1971.

[58] H.A. Barker. Synchronous sampling theorem for nonlinear systems. *Electron. Letters*, Vol. 5, p. 657, 1969.

[59] C.W. Barnes. Object restoration in a diffraction-limited imaging system. *Journal of the Optical Society America*, Vol. 56, pp. 575–578, 1966.

[60] H.O. Bartelt, K.H. Brenner, and A.W. Lohmann. The Wigner distribution function and its optical production. *Optical Communication* Vol. 32, pp. 32–38, 1980.

[61] H.O. Bartelt and J. Jahns. Interferometry based on the Lau effect. *Optical Communication*, pp. 268–274, 1979.

[62] M.J. Bastiaans. A frequency-domain treatment of partial coherence. *Opt. Acta*, Vol. 24, pp. 261–274, 1977.

[63] M.J. Bastiaans. A generalized sampling theorem with application to computer-generated transparencies. *Journal of the Optical Society America*, Vol. 68, pp. 1658–1665, 1978.

[64] M.J. Bastiaans. The Wigner distribution function applied to optical signals and systems. *Optical Communication*, Vol. 25, pp. 26–30, 1978.

[65] M.J. Bastiaans. The Wigner distribution function and Hamilton's characteristics of a geometrical-optical system. *Optical Communication*, Vol. 30, pp. 321–326, 1979.

[66] M.J. Bastiaans. Sampling theorem for the complex spectrogram, and Gabor's expansion of a signal in Gaussian elementary signals, in *1980 International Optical Computing Conference*, ed. W.T. Rhodes, *Proceedings SPIE*, Vol. 231, pp. 274–280, 1980 (also published in *Optical Engineering*, Vol. 20, pp. 594–598, 1981).

[67] M.J. Bastiaans. The expansion of an optical signal into a discrete set of Gaussian beams. *Optik*, Vol. 57, pp. 95–102, 1980.

[68] M.J. Bastiaans. A sampling theorem for the complex spectrogram, and Gabor's expansion of a signal in Gaussian elementary signals. *Optical Engineering*, Vol. 20, pp. 594–598, 1981.

[69] M.J. Bastiaans. Gabor's signal expansion and degrees of freedom of a signal. *Opt. Acta, Vol. 29*, pp. 1223–1229, 1982.

[70] M.J. Bastiaans. Optical generation of Gabor's expansion coefficients for rastered signals. *Opt. Acta, Vol. 29*, pp. 1349–1357, 1982.

[71] M.J. Bastiaans. Signal description by means of a local frequency spectrum. *Proceedings SPIE, International Society Optical Engineering, Vol. 373*, pp. 49–62, 1981.

[72] M.J. Bastiaans. Gabor's signal expansion and degrees of freedom of a signal. *Opt. Acta, Vol. 29*, pp. 1223–1229, 1982.

[73] M.J. Bastiaans. On the sliding-window representation in digital signal processing. *IEEE Transactions Acoust., Speech, Signal Processing Vol. ASSP-33*, pp. 868–873, 1985.

[74] M.J. Bastiaans. Error reduction in two-dimensional pulse-area modulation, with application to computer-generated transparencies. *14th Congress of the International Commission for Optics*, ed. H.H. Arsenault, *Proceedings SPIE, Vol. 813*, pp. 341–342, 1987.

[75] M.J. Bastiaans. Error reduction in one-dimensional pulse-area modulation, with application to computer-generated transparencies. *Journal of the Optical Society America A, Vol. 4*, pp. 1879–1886, 1987.

[76] M.J. Bastiaans. Local-frequency descriptions of optical signals and systems. EUT Report 88–E–191 (Faculty of Electrical Engineering, Eindhoven University of Technology, Eindhoven, Netherlands, 1988).

[77] M.J. Bastiaans. Gabor's signal expansion applied to partially coherent light. *Optical Communication, Vol. 86*, pp. 14–18, 1991.

[78] M.J. Bastiaans. On the design of computer-generated transparencies based on area modulation of a regular array of pulses. *Optik, Vol. 88*, pp. 126–132, 1991.

[79] M.J. Bastiaans. Second-order moments of the Wigner distribution function in first-order optical systems. *Optik, Vol. 88*, pp. 163–168, 1991.

[80] M.J. Bastiaans, M.A. Machado, and L.M. Narducci. The Wigner distribution function and its applications in optics. *Optics in Four Dimensions-1980, AIP Conf. Proceedings 65*, eds. (American Institute of Physics, New York, 1980), pp. 292–312, 1980.

[81] Sankar Basu and Bernard C. Levy. **Multidimensional Filter Banks and Wavelets: Basic Theory and Cosine Modulated Filter Banks.** Springer, 1996.

[82] J.S. Bayley and M.A. Fiddy. On the use of the Hopfield model for optical pattern recognition. *Optical Communication*, pp. 105–110, 1987.

[83] M.G. Beaty and M.M. Dodson. The Whittaker-Kotel'nikov-Shannon theorem, spectral translates and Plancherel's formula. *Journal of Fourier Analysis and Applications, Vol. 10(2)*, pp. 179–199, 2004.

[84] Reinhard Beer. **Remote Sensing by Fourier Transform Spectrometry**. John Wiley & Sons, 1992.

[85] A.H. Beiler. Ch. 11 in **Recreations in the Theory of Numbers: The Queen of Mathematics Entertains**. New York: Dover, 1966.

[86] G.A. Bekey and R. Tomovic. Sensitivity of discrete systems to variation of sampling interval. *IEEE Transactions Automatic Control, Vol. AC-11*, pp. 284–287, 1966.

[87] B.W. Bell and C.L. Koliopoulos. Moiré topography, sampling theory, and charged-coupled devices. *Optical Letters, Vol. 9*, pp. 171–173, 1984.

[88] D.A. Bell. The Sampling Theorem. *International Journal of Electrical Engineering, Vol. 18(2)*, pp. 175–175, 1981.

[89] E.T. Bell. **Men of Mathematics**. Touchstone (Reissue edition) 1986.

[90] V.K. Belov and S.N. Grishkin. Digital bandpass filters with nonunform sampling. *Priborostroemie, Vol. 23, No. 3 (in Russian)*, pp. 3–7, 1980.

[91] V.I. Belyayev. Processing and theoretical analysis of oceanographic observations. *Translation from Russian JPRS 6080*, 1973.

[92] John J. Benedetto and Paulo J.S.G. Ferreira. **Modern Sampling Theory.** by (Hardcover - Feb. 16, 2001) Birkhäuser user Boston, 2001.

[93] J.J. Benedetto and Ahmed I. Zayed (Editors). **Sampling, Wavelets, and Tomography**. Pub. Date: December 2003 Publisher: Birkhäuser user Verlag.

[94] M. Bendinelli, A. Consortini, L. Ronchi, and R.B. Frieden. Degrees of freedom and eigenfunctions of the noisy image. *Journal of the Optical Society America, Vol. 64*, pp. 1498–1502, 1974.

[95] M. Bennett, et al. Filtering of ultrasound echo signals with the fractional Fourier transform. *Proceedings of the 25th Annual International Conference of the IEEE Engineering in Medicine and Biology Society, Vol. 1(17–21)*, pp. 882–885, 2003.

[96] N. Benvenuto and G.L. Pierobin. Extension of the sampling theorem to exponentially decaying causal functions. *IEEE Transactions on Signal Processing, Vol. 39(1)*, pp. 189–190, 1991.

[97] N. Benvenuto, L. Franks, F. Hill. Realization of Finite Impulse Response Filters Using Coefficients $+1$, 0, and -1. *IEEE Transactions on Communications*, Oct 1985, Vol.33, Issue 10, pp. 1117–1125.

[98] E.M. Bernat, W.J. Williams, and W.J. Gehring. Decomposing ERP time-frequency energy using PCA. *Clinical Neurophysiology. Vol. 116(6)*, pp. 1314–1334, 2005.

[99] M. Bertero. Linear inverse and ill-posed problems. In P.W. Hawkes, editor, **Advances in Electronics and Electron Physics**. Academic Press, New York, 1989.

[100] M. Bertero, P. Boccacci, and E.R. Pike. Resolution in diffraction-limited imaging, a singular value analysis. II. The case of incoherent illumination. *Opt. Acta*, pp. 1599–1611, 1982.

[101] M. Bertero, C. de Mol, E.R. Pike, and J.G. Walker. Resolution in diffraction-limited imaging, a singular value analysis. IV. The case of uncertain localization or non-uniform illumination of the object. *Opt. Acta*, pp. 923–946, 1984.

[102] M. Bertero, C. De Mol, and E.R. Pike. Analytic inversion formula for confocal scanning microscopy. *Journal of the Optical Society America A*, pp. 1748–1750, 1987.

[103] M. Bertero, C. De Mol, and G.A. Viano. Restoration of optical objects using regularization. *Optics Letters*, pp. 51–53, 1978.

[104] M. Bertero, C. De Mol, and G.A. Viano. On the problems of object restoration and image extrapolation in optics. *Journal Math. Phys.*, pp. 509–521, 1979.

[105] M. Bertero and E.R. Pike. Resolution in diffraction-limited imaging, a singular value analysis. I. The case of coherent illumination. *Opt. Acta*, pp. 727–746, 1982.

[106] M. Bertero and E.R. Pike. Exponential sampling method for Laplace and other dilationally invariant transforms: I. Singular system analysis. *Inverse Problems*, pp. 1–20, 1991.

[107] M. Bertero and E.R. Pike. Exponential sampling method for Laplace and other dilationally invariant transforms: II. Examples in photon correlation spectroscopy and Fraunhofer diffraction. *Inverse Problems*, pp. 21–41, 1991.

[108] A.S. Besicovitch. **Almost Periodic Functions.** Wiley Interscience, New York, 1968.

[109] F.J. Beutler. Sampling theorems and bases in a Hilbert space. *Information & Control, Vol. 4*, pp. 97–117, 1961.

[110] F.J. Beutler. Error-free recovery of signals from irregularly spaced samples. *SIAM Review, Vol. 8, No. 3*, pp. 328–335, 1966.

[111] F.J. Beutler. Alias-free randomly timed sampling of stochastic processes. *IEEE Transactions on Information Theory, Vol. IT-16*, pp. 147–152, 1970.

[112] F.J. Beutler. Recovery of randomly sampled signals by simple interpolators. *Information & Control, Vol. 26, No. 4*, pp. 312–340, 1974.

[113] F.J. Beutler. On the truncation error of the cardinal sampling expansion. *IEEE Transactions Information Theory, Vol. IT-22*, pp. 568–573, 1976.

[114] F.J. Beutler and D.A. Leneman. Random sampling of random processes: stationary point processes. *Information & Control, Vol. 9*, pp. 325–344, 1966.

[115] F.J. Beutler and D.A. Leneman. The theory of stationary point processes. *Acta Math, Vol. 116*, pp. 159–197, September 1966.

[116] V. Bhaskaran and K. Konstantinides. **Image and Video Compression Standards**, Kluwer, 1995.

[117] Arun K. Bhattacharyya. **Phased Array Antennas: Floquet Analysis, Synthesis, BFNs and Active Array Systems.** Wiley-Interscience, 2006.

[118] A.M. Bianchi and L.T. Mainardi and S. Cerutti. Time-frequency analysis of biomedical signals. *Transactions of the Institute of Measurement and Control, Vol. 22(3)*, pp. 215–230, 2000.

[119] H.S. Black. **Modulation Theory**. Van Nostrand, New York, 1953.

[120] R.B. Blackman and J.W. Tukey. The measurement of power spectra from the point of view of communications engineering. *Bell Systems Tech. J., Vol. 37*, p. 217, 1958.

[121] W.E. Blash. Establishing proper system parameters for digitally sampling a continuous scanning spectrometer. *Applied Spectroscopy., Vol. 30*, pp. 287–289, 1976.

[122] V. Blažek. Sampling theorem and the number of degrees of freedom of an image. *Optical Communication, Vol. 11*, pp. 144–147, 1974.

[123] E.M. Bliss. Watch out for hidden pitfalls in signal sampling. *Electron. Engineering, Vol. 45*, pp. 59–61, 1973.

[124] Peter Bloomfield. **Fourier Analysis of Time Series: An Introduction**. Wiley-Interscience, 2000.

[125] R.P. Boas, Jr. **Entire Functions**. Academic Press, New York, 1954.

[126] R.P. Boas, Jr. Summation formulas and bandlimited signals. *Tohoku Math. J., Vol. 24*, pp. 121–125, 1972.

[127] Salomon Bochner and Komaravolu Chandrasekharan. **Fourier Transforms**. Princeton University Press, 1949.

[128] M. Bohner and A. Peterson. **Dynamic Equations on Time Scales: An Introduction with Applications**. Birkhauser, Boston, 2001.

[129] M. Bohner and A. Peterson, **Dynamic Equations on Time Scales**, Birkhäuser, Boston, 2001.

[130] H. Bohr. **Almost Periodic Functions.** Wiley Interscience, New York, 1968.

[131] A. Boivin and C. Deckers. A new sampling theorem for the complex degree of coherence for reconstruction of a luminous isotropic source. *Optical Communication, Vol. 26*, pp. 144–147, 1978.

[132] P. Bonato, S.H. Roy, M. Knaflitz et al. Time-frequency parameters of the surface myoelectric signal for assessing muscle fatigue during cyclic dynamic contractions. *IEEE Transactions of Biomedical Engineering. Vol. 48(7)*, pp. 745–753, 2001.

[133] F.E. Bond and C.R. Cahn. On sampling the zeros of bandwidth limited signals. *IRE Transactions Information Theory, Vol. IT-4*, pp. 110–113, 1958.

[134] F.E. Bond, C.R. Cahn, and J.C. Hancock. A relation between zero crossings and Fourier coefficients for bandlimited functions. *IRE Transactions Information Theory, Vol. IT-6*, pp. 51–55, 1960.

[135] C. de Boor. **A Practical Guide to Splines**. Springer-Verlag, New York, 1978.

[136] E. Borel. Sur l'interpolation. *C.R. Acad. Sci. Paris, Vol. 124*, pp. 673–676, 1877.

[137] E. Borel. Mémoire sur les séries divergentes. *Ann. Ecole Norm. Sup., Vol. 3*, pp. 9–131, 1899.

[138] E. Borel. La divergence de la formule de Lagrange a été établie également. In *Leçons sur les fonctions de variables réelles et les développements en séries de polynômes*. Gauthier-Villars, Paris, 1905, pp. 74–79.

[139] M. Born and E. Wolf. **Principles of Optics**. Pergamon Press, Oxford, 1980.

[140] T. Bortfeld, J. Burkelbach, R. Boesecks, and W. Schlegal. Methods of image reconstruction from projections applied to conformation radiotherapy. *Phys. Med. Biol., Vol. 35*, pp. 1423–34, 1990.

[141] Robert L. Boylestad, and Louis Nashelsky. **Electronic Devices and Circuit Theory** (8th Edition), Prentice Hall, 2001.

[142] A. Boumenir et al. Eigenvalues of periodic Sturm-Liouville problems by the Shannon-Whittaker sampling theorem. *Mathematics of Computation, Vol. 68(227)*, pp. 1057–1066, 1999.

[143] F. Bouzeffour. A q-sampling theorem and product formula for continuous q-Jacobi functions. *Proceedings of the American Mathematical Society, Vol. 135(7)*, pp. 2131–2139, 2007.

[144] G.E.T. Box and G.M. Jenkins. **Time Series Analysis.** *Forecasting and Control* (revised edition). Holden-Day, San Francisco, 1976.

[145] G.D. Boyd and J.P. Gordon. Confocal multimode resonator for millimeter through optical wavelength masers. *Bell Systems Tech. J.*, pp. 489–508, 1961.

[146] G.D. Boyd and H. Kogelnik. Generalized confocal resonator theory. *Bell Systems Tech. J.*, pp. 1347–1369, 1962.

[147] G.R. Boyer. Realization d'un filtrage super-resolvant. *Opt. Acta*, pp. 807–816, 1983.
[148] Ronald Newbold Bracewell. **The Fourier Transform and its Applications**, McGraw-Hill, 1965.
[149] Ronald Newbold Bracewell. **The Fourier Transform and its Applications**, 2nd edition, Revised. McGraw-Hill, New York, 1986.
[150] Ronald Newbold Bracewell. **The Hartley Transform**. Oxford University Press, New York, 1986.
[151] Ronald Newbold Bracewell. **Fourier Analysis and Imaging**. Plenum Publishing Corporation, 2004.
[152] A. Brahme. Optimization of stationary and moving beam radiation therapy techniques. *Radiother. Oncol.*, 12 pp. 129–140, 1988.
[153] Luca Brandolini, Leonardo Colzani, Alex Iosevich, and Giancarlo Travaglini. **Fourier Analysis and Convexity**. Birkhäuser Boston, 2004.
[154] David Brandwood. **Fourier Transforms in Radar and Signal Processing**. Artech House, 2003.
[155] H. Bray, K. McCormick, and R.O. Wells *et al.* Wavelet variations on the Shannon sampling theorem. *Biosystems, Vol. 34(1–3)*, pp. 249–257, 1995.
[156] L.M. Bregman. Finding the common point of convex sets by the method of successive projections. *Dokl. Akad. Nauk. USSR, Vol. 162, No. 3*, pp. 487–490, 1965.
[157] P. Bremer and J.C. Hart. A sampling theorem for MLS surfaces. *Proceedings Eurographics/ IEEE VGTC Symposium on Point-Based Graphics*, pp. 47–54.
[158] Hans Bremermann. **Distributions, Complex Variables, and Fourier transforms**. Addison-Wesley, 1965.
[159] William L. Briggs and Van Emden Henson. **The DFT: An Owners' Manual for the Discrete Fourier Transform**. Society for Industrial Mathematics, 1987.
[160] E.O. Brigham. **The Fast Fourier Transform**. Prentice-Hall, Englewood Cliffs, NJ, 1974.
[161] E.O. Brigham. **Fast Fourier Transform and Its Applications**. Prentice Hall,1988.
[162] L. Brillouin. **Science and Information Theory**. Academic Press, New York, 1962.
[163] B.C. Brookes. Sampling theorem for finite discrete distributions. *Journal of Documentation, Vol. 31(1)*, pp. 26–35, 1975.
[164] Eli Brookner. **Practical Phased Array Antenna Systems.** Artech House Publishers, 1991.
[165] J.L. Brown, Jr. Anharmonic approximation and bandlimited signals. *Information & Control, Vol. 10*, pp. 409–418, 1967.
[166] J.L. Brown, Jr. On the error in reconstructing a non-bandlimited function by means of the bandpass sampling theorem. *Journal Math. Anal. Applied, Vol. 18*, pp. 75–84, 1967; Erratum, *Vol. 21*, p. 699, 1968.
[167] J.L. Brown, Jr. A least upper bound for aliasing error. *IEEE Transactions Automatic Control, Vol. AC-13*, pp. 754–755, 1968.
[168] J.L. Brown, Jr. Sampling theorem for finite energy signals. *IRE Transactions on Information Theory, Vol. IT-14*, pp. 818–819, 1968.
[169] J.L. Brown, Jr. Bounds for truncation error in sampling expansions of bandlimited signals. *IRE Transactions Information Theory, Vol. IT-15*, pp. 669–671, 1969.
[170] J.L. Brown, Jr. Truncation error for bandlimited random processes. *Inform. Sci., Vol. 1*, pp. 261–272, 1969.
[171] J.L. Brown, Jr. Uniform linear prediction of bandlimited processes from past samples. *IEEE Transactions Information Theory, Vol. IT-18*, pp. 662–664, 1972.
[172] J.L. Brown, Jr. On mean-square aliasing error in the cardinal series expansion of random processes. *IEEE Transactions Information Theory, Vol. IT-24*, pp. 254–256, 1978.
[173] J.L. Brown, Jr. Comments on energy processing techniques for stress wave emission signals. *Journal Acoust. Society America, Vol. 67*, p. 717, 1980.
[174] J.L. Brown, Jr. First-order sampling of bandpass signals: a new approach. *IEEE Transactions Information Theory, Vol. IT-26*, pp. 613–615, 1980.
[175] J.L. Brown, Jr. Sampling bandlimited periodic signals – an application of the DFT. *IEEE Transactions Education, Vol. E-23*, pp. 205–206, 1980.
[176] J.L. Brown, Jr. A simplified approach to optimum quadrature sampling. *Journal Acoust. Society America, Vol. 67*, pp. 1659–1662, 1980.

[177] J.L. Brown, Jr. Multi-channel sampling of low-pass signals. *IEEE Transactions Circuits & Systems, Vol. CAS-28*, pp. 101–106, 1981.

[178] J.L. Brown, Jr. Cauchy and polar-sampling theorems. *Journal of the Optical Society America A, Vol. 1*, pp. 1054–1056, 1984.

[179] J.L. Brown, Jr. An RKHS analysis of sampling theorems for harmonic-limited signals. *IEEE Transactions Acoust., Speech, Signal Processing, Vol. ASSP-33*, pp. 437–440, 1985.

[180] J.L. Brown, Jr. Sampling expansions for multiband signals. *IEEE Transactions Acoust., Speech, Signal Processing, Vol. ASSP-33*, pp. 312–315, 1985.

[181] J.L. Brown, Jr. On the prediction of bandlimited signal from past samples. *Proceedings IEEE, Vol. 74*, pp. 1596–1598, 1986.

[182] J.L. Brown, Jr. and O. Morean. Robust prediction of bandlimited signals from past samples. *IEEE Transactions Information Theory, Vol. IT-32*, pp. 410–412, 1986.

[183] J.L. Brown, Jr. Sampling reconstruction of n-dimensional bandlimited images after multilinear filtering. *IEEE Transactions Circuits & Systems, Vol. CAS-36*, pp. 1035–1038, 1989.

[184] J.L. Brown. Summation of certain series using the Shannon sampling theorem. *IEEE Transactions on Education, Vol. 33(4)*, pp. 337–340, 1990.

[185] W.M. Brown. Fourier's integral. *Proceedings Edinburgh Math. Society, Vol. 34*, pp. 3–10, 1915–1916.

[186] W.M. Brown. On a class of factorial series. *Proceedings London Math. Society, Vol. 2*, pp. 149–171, 1924.

[187] W.M. Brown. Optimum prefiltering of sampled data. *IRE Transactions Information Theory, Vol. IT-7*, pp. 269–270, 1961.

[188] W.M. Brown. Sampling with random jitter. *Journal SIAM, Vol. 2*, pp. 460–473, 1961.

[189] N.G. de Bruijn. A theory of generalized functions, with applications to Wigner distribution and Weyl correspondence. *Nieuw Arch. Wiskunde (3), Vol. 21*, 205–280, 1973.

[190] O. Bryngdahl. Image formation using self-imaging techniques. *Journal of the Optical Society America, Vol. 63*, pp. 416–419, 1973.

[191] O. Bryngdahl and F. Wyrowski. Digital holography. Computer generated holograms. In newblock **Progress in Optics**, North-Holland, Amsterdam, pp. 1–86, 1990.

[192] G.J. Buck and J.J. Gustincic. Resolution limitations of a finite aperture. *IEEE Transactions Antennas & Propagation, Vol. AP-15*, pp. 376–381, 1967.

[193] Aram Budak. **Passive and Active Network Analysis and Synthesis**, Waveland Press, 1991.

[194] Y.G. Bulychev and I.V. Burlai. Repeated differentiation of finite functions with the use of the sampling theorem in problems of control, estimation, and identification. *Automation and Remote Control, Vol. 57(4)*, Part 1, pp. 499–509, 1996.

[195] J. Bures, P. Meyer, and G. Fernandez. Reconstitution d'un profil d'eclairement echatillonne par une legne de microphotodiodes. *Optical Communication*, pp. 39–44, 1979.

[196] Tilman Butz. **Fourier Transformation for Pedestrians (Fourier Series)**. Springer, 2005.

[197] P.L. Butzer. A survey of the Whittaker-Shannon sampling theorem and some of its extensions. *Journal Math. Res. Exposition, Vol. 3*, pp. 185–212, 1983.

[198] P.L. Butzer. The Shannon sampling theorem and some of its generalizations. An Overview. in *Constructive Function Theory, Proceedings Conf. Varna, Bulgaria, June 1–5, 1981*; Bl. Sendov et al., eds., Publishing House Bulgarian Academy of Science., Sofia, pp. 258–274, 1983.

[199] P.L. Butzer. Some recent applications of functional analysis to approximation theory. (Proc. Euler Colloquium Berlin, May 1983; Eds: E. Knobloch *et al*). In: Zum Werk Leonhard Eulers. Birkhäuser, Basel, pp. 133–155, 1984.

[200] P.L. Butzer and W. Engels. Dyadic calculus and sampling theorems for functions with multidimensional domain, I. General theory. *Information & Control, Vol. 52*, pp. 333–351, 1982.

[201] P.L. Butzer and W. Engels. Dyadic calculus and sampling theorems for functions with multidimensional domain, II. Applications to dyadic sampling representations. *Information & Control, Vol. 52*, pp. 352–363, 1982.

[202] P.L. Butzer and W. Engels. On the implementation of the Shannon sampling series for bandlimited signals. *IEEE Transactions Information Theory, Vol. IT-29*, pp. 314–318, 1983.

[203] P.L. Butzer, W. Engels, S. Ries, and R.L. Stens. The Shannon sampling series and the reconstruction of signals in terms of linear, quadratic and cubic splines. *SIAM J. Applied Math., Vol. 46*, pp. 299–323, 1986.

[204] P.L. Butzer, W. Engels, and U. Scheben. Magnitude of the truncation error in sampling expansions of bandlimited signals. *IEEE Transactions Acoust., Speech, Signal Processing. ASSP-30*, pp. 906–912, 1982.

[205] P.L. Butzer and G. Hinsen. Reconstruction of bounded signals from pseudo-periodic irregularly spaced samples. *Signal Processing, Vol. 17*, pp. 1–17, 1988.

[206] P.L. Butzer and G. Hinsen. Two-dimensional nonuniform sampling expansions – an iterative approach I, II. *Applied Anal., Vol. 32*, pp. 53–85, 1989.

[207] P.L. Butzer and C. Markett. The Poisson summation formula for orthogonal systems. In *Anniversary Volume on Approximation Theory and Functional Analysis (Proceedings Conf. Math. Res. Institute Oberwolfach, Black Forest;* Eds. Butzer, Stens, Nagy) *Birkhäuser Verlag Basel - Stuttgart - Boston,* = ISNM *Vol. 65*, pp. 595–601, 1984.

[208] P.L. Butzer and R.J. Nessel. Contributions to the theory of saturation for singular integrals in several variables I. General theory. *Nederland Akad. Wetensch. Proceedings Ser. A, Vol. 69* and *Indag. Math., Vol. 28*, pp. 515–531, 1966.

[209] P.L. Butzer and R.J. Nessel. **Fourier Analysis and Approximation Vol. I: One-Dimensional Theory**. Birkhäuser Verlag, Basel, Academic Press, New York, 1971.

[210] P.L. Butzer, S. Ries, and R.L. Stens. The Whittaker-Shannon sampling theorem, related theorems and extensions; a survey. In *Proceedings JIEEEC '83 (Proceedings Jordan - International Electrical and Electronic Engineering Conference, Amman, Jordan)*, pp. 50–56, 1983.

[211] P.L. Butzer, S. Ries, and R.L. Stens. Shannon's sampling theorem, Cauchy's integral formula and related results. In *Anniversary Volume on Approximation Theory and Functional Analysis (Proceedings Conf. Math. Res. Institute Oberwolfach, Black Forest)* Eds. Butzer Stens, and Nagy Basel - Stuttgart - Boston: Birkhäuser Verlag = ISNM *Vol. 65*, pp. 363–377, 1984.

[212] P.L. Butzer, S. Ries, and R.L. Stens. Approximation of continuous and discontinuous functions by generalized sampling series. *Journal Approx. Theory, Vol. 50*, pp. 25–39, 1987.

[213] P.L. Butzer, M. Schmidt, E.L. Stark, and L. Vogt. Central factorial numbers; their main properties and some applications. *Num. Funct. Anal. Optim. Vol. 10*, pp. 419–488, 1989.

[214] P.L. Butzer and M. Schmidt. Central factorial numbers and their role in finite difference calculus and approximation. *Proceedings Conf. on Approximation Theory, Kecskemét, Hungary.* Colloquia Mathematica Societatis János Bolyai, 1990.

[215] P.L. Butzer and W. Splettstößer. *Approximation und Interpolation durch verallgemeinerte Abtastsummen. Forschungsberichte* No. 2515, Landes Nordrhein-Westfalen, Köln-Oplanden, 1977.

[216] P.L. Butzer and W. Splettstößer. A sampling theorem for duration-limited functions with error estimates. *Information & Control, Vol. 34*, pp. 55–65, 1977.

[217] P.L. Butzer and W. Splettstößer. Sampling principle for duration-limited signals and dyadic Walsh analysis. *Inform. Sci., Vol. 14*, pp. 93–106, 1978.

[218] P.L. Butzer and W. Splettstößer. On quantization, truncation and jitter errors in the sampling theorem and its generalizations. **Signal Processing**, *Vol. 2*, pp. 101–102, 1980.

[219] P.L. Butzer, W. Splettstößer, and R.L. Stens. The sampling theorem and linear prediction in signal analysis. *Jahresber. Deutsch Math.–Verein. Vol. 90*, pp. 1–70, 1988.

[220] P.L. Butzer and R.L. Stens. *Index of Papers on Signal Theory: 1972–1989.* Lehrstuhl A für Mathematik, Aachen University of Technology, Aachen, Germany, 1990.

[221] P.L. Butzer and R.L. Stens. The Euler-Maclaurin summation formula, the sampling theorem, and approximate integration over the real axis. **Linear Algebra Applications**, *Vol. 52/53*, pp. 141–155, 1983.

[222] P.L. Butzer and R.L. Stens. The Poisson summation formula, Whittaker's cardinal series and approximate integration. In *Proceedings Second Edmonton Conference on Approximation Theory (Eds., Z. Ditzian, et al.) American Math. Society Providence, Rhode Island:* = *Canadian Math Society Proceedings, Vol. 3*, pp. 19–36, 1983.

[223] P.L. Butzer and R.L. Stens. A modification of the Whittaker-Kotelnikov-Shannon sampling series. *Aequationés Math., Vol. 28*, pp. 305–311, 1985.

[224] P.L. Butzer and R.L. Stens. Prediction of non-bandlimited signals from past samples in terms of splines of low degree. *Math. Nachr., Vol. 132*, pp. 115–130, 1987.

[225] P.L. Butzer and R.L. Stens. Linear prediction in terms of samples from the past; an overview. In **Numerical Methods and Approximation Theory III** (*Proceedings Conf. on Numerical Methods and Approximation Theory*), Nis 18.–21.8.1987; Ed. G.V. Milovanovic, Faculty of Electronic Engineering, University of Nis, Yugoslavia, 1988, pp. 1–22.

[226] P.L. Butzer and R.L. Stens. Sampling theory for not necessarily bandlimited functions; an historical overview. *SIAM Review, Vol. 34*, pp. 40–53, 1992.

[227] P.L. Butzer, R.L. Stens, and M. Wehrens. The continuous Legendre transform, its inverse transform and applications. *International J. Math. Sci., Vol. 3*, pp. 47–67, 1980.

[228] P.L. Butzer and H.J. Wagner. Walsh-Fourier series and the concept of a derivative. **Applicable Analysis**, *Vol. 3*, pp. 29–46, 1973.

[229] P.L. Butzer and H.J. Wagner. On dyadic analysis based on the pointwise dyadic derivative. *Analysis Math., Vol. 1*, pp. 171–196, 1975.

[230] P.L. Butzer, J.R. Higgins, and R.L. Stens. Classical and approximate sampling theorems; studies in the L-P(R) and the uniform norm. *Journal of Approximation Theory, Vol. 137(2)*, pp. 250–263, 2005.

[231] A. Buzo, A.H. Gray Jr., R.M. Gray, and J.D. Markel. Speech coding based upon vector quantization. *IEEE Transactions Acoust., Speech, Signal Processing, Vol. ASSP-28*, pp. 562–574, 1980.

[232] A. Buzo, F. Kuhlmann, and C. Blas. Rate distortion bounds for quotient-based distortions with applications to Itakura-Saito distortion measures. *IEEE Transactions Inform. Theory, Vol. IT-32*, pp. 141–147, 1986.

[233] William E. Byerly. **An Elementary Treatise on Fourier's Series and Spherical, Cylindrical, and Ellipsoidal Harmonics**. Ginn & Co, Boston, 1933.

C

[234] J. Cadzow. An extrapolation procedure for bandlimited signals. *IEEE Transactions Acoust., Speech, Signal Processing, Vol. ASSP-27*, pp. 4–12, 1979.

[235] D. Cahana and H. Stark. Bandlimited image extrapolation with faster convergence. *Applied Optics*, pp. 2780–2786, 1981.

[236] G. Calvagio and D. Munson, Jr. New results on Yen's approach to interpolation from nonuniformly spaced samples. *Proceedings IASSP*, pp. 1533–1538, 1990.

[237] S. Cambanis and M.K. Habib. Finite sampling approximation for non-bandlimited signals. *IEEE Transactions Information Theory, Vol. IT-28*, pp. 67–73, 1982.

[238] S. Cambanis and E. Masry. Truncation error bounds for the cardinal sampling expansion of bandlimited signals. *IEEE Transactions Information Theory, Vol. IT-28*, pp. 605–612, 1982.

[239] L.L. Campbell. A comparison of the sampling theorems of Kramer and Whittaker. *Journal SIAM, Vol. 12*, pp. 117–130, 1964.

[240] L.L. Campbell. Sampling theorem for the Fourier transform of a distribution with bounded support. *SIAM J. Applied Math., Vol. 16*, pp. 626–636, 1968.

[241] L.L. Campbell. Further results on a series representation related to the sampling theorem. *Signal Processing, Vol. 9*, pp. 225–231, 1985.

[242] C. Candan, M.A. Kutay, and H.M. Ozaktas. The discrete fractional Fourier transform. *IEEE Transactions on Signal Processing, Vol. 48(5)*, pp. 1329–1337, 2000.

[243] K.G. Canfield, and D.L. Jones. Implementing time-frequency representations for non-Cohen classes. 1993 Conference Record of The Twenty-Seventh Asilomar Conference on Signals, Systems and Computers, *Vol. 2*, pp. 1464–1468, 1993.

[244] George Cantor and Philip E.B. Jourdain. **Contributions to the Founding of the Theory of Transfinite Numbers**. Cosimo Classics, 2007.

[245] Gianfranco Cariolaro, Tomaso Erseghe, Peter Kraniauskas, and Nicola Laurenti. Multiplicity of Fractional Fourier Transforms and their Relationships. *IEEE Transactions on Signal Processing, Vol. 48(1)*, pp. 227–241, 2000.

[246] A.B. Carlson. *Communications Systems, An Introduction to Signals and Noise in Electrical Communication*, 2nd Ed. McGraw-Hill, New York, 1975.

[247] H.S. Carslaw. **Introduction to the Theory of Fourier's Series and Integrals**. Michigan Historical Reprint Series. Scholarly Publishing Office, University of Michigan, 2005.

[248] W.H. Carter and E. Wolf. Coherence and radiometry with quasihomogeneous sources. *Journal of the Optical Society America Vol. 67*, pp. 785–796, 1977.

[249] D. Casasent. Optical signal processing. In **Optical Data Processing**, D. Casasent, editor, Springer-Verlag, Berlin, pp. 241–282, 1978.

[250] Manuel F. Catedra, Rafael F. Torres, Emilio Gago, and Jose Basterrechea. **CG-FFT Method: Application of Signal Processing Techniques to Electromagnetics**. Artech House, 1994.

[251] W.T. Cathey, B.R. Frieden, W.T. Rhodes, and C.K. Rushforth. Image gathering and processing for enhanced resolution. *Journal of the Optical Society America A*, pp. 241–250, 1984.

[252] A.L. Cauchy. Memoire sur diverses formules d'analyser, *Comptes Rendus, Vol. 12*, pp. 283–298, 1841.

[253] H.J. Caulfield. Wavefront sampling in holography. *Proceedings IEEE, Vol. 57*, pp. 2082–2083, 1969.

[254] H.J. Caulfield. Correction of image distortion arising from nonuniform sampling. *Proceedings IEEE, Vol. 58*, p. 319, 1970.

[255] G. Cesini, P. Di Filippo, G. Lucarini, G. Guattari, and P. Pierpaoli. The use of charge-coupled devices for automatic processing of interferograms. *Journal of the Optical*, pp. 99–103, 1981.

[256] G. Cesini, G. Guattari, G. Lucarini, and C. Palma. An iterative method for restoring noisy images. *Opt. Acta*, pp. 501–508, 1978.

[257] A.E. Cetin, H. Ozaktus, and H.M. Ozaktus. Signal recovery from partial fractional Fourier transform information. First International Symposium on Control, Communications and Signal Processing, pp. 217–220, 2004.

[258] J. Cezanne and A. Papoulis. The use of modulated splines for the reconstruction of bandlimited signals. *IEEE Transactions Acoust., Speech, Signal Processing, Vol. ASSP-36, No. 9*, pp. 1521–1525, 1988.

[259] K. Chalasinkamacukow, H.H. Arsenault. Solution to the phase-retrieval problem using the sampling theorem. *Journal of the Optical Society of America, Vol. 73(12)*, pp. 1875–1875, 1983.

[260] K. Chalasinska-Macukow and H.H. Arsenault. Fast iterative solution to exact equations for the two-dimensional phase-retrieval problem. *Journal of the Optical Society America A, Vol. 2*, pp. 46–50, 1985.

[261] D.C. Champeney. **Fourier Transforms and their Physical Applications**. Academic Press, 1973.

[262] D.C. Champeney. **Fourier Transforms in Physics.** Taylor & Francis, 1985.

[263] C. Chamzas and Wen Xu. An improved version of Papoulis-Gerchberg algorithm on band-limited extrapolation.*IEEE Transactions on Acoustics, Speech, and Signal Processing, Vol. 32(2)*, pp. 437–440, 1984.

[264] S.S. Chang. Optimum transmission of a continuous signal over a sampled data link. *AIEE Transactions, Vol. 79, Pt. II (Applications and Industry)*, pp. 538–542, 1961.

[265] I.J. Chant and L.M. Hastie, A comparison of the stability of stationary and nonstationary magnetotelluric analysis methods. *International Geophysical Journal, Vol. 115(3)*, pp. 1143–1147, 1993.

[266] J.I. Chargin and V.P. Iakovlev. **Finite Functions in Physics and Technology**. Nauka, Moscow, 1971.

[267] D.B. Chase and J.F. Rabolt. **Fourier Transform Raman Spectroscopy: From Concept to Experiment**. Academic 1994.

[268] P. Chavel and S. Lowenthal. Noise and coherence in optical image processing. I. The Callier effect and its influence on image contrast. *Journal of the Optical Society America, Vol. 68*, pp. 559–568, 1978.

[269] P. Chavel and S. Lowenthal. Noise and coherence in optical image processing. II. Noise fluctuations. *Journal of the Optical Society America, Vol. 68*, pp. 721–732, 1978.

[270] Charng-Kann Chen and Ju-Hong Lee. McClellan transform based design techniques for two-dimensional linear-phase FIR filters. *IEEE Transactions on Circuits and Systems I: Fundamental Theory and Applications, Vol. 41(8)*, pp. 505–517, 1994.

[271] D. Chen, L.G. Durand and Z. Guo. Time-frequency analysis of the first heart sound. Part 2: An appropriate time-frequency representation technique. *Medical & Biological Engineering & Computing. Vol. 35(4)*, pp. 311–317, 1997.

[272] D. Chen, L.G. Durand et al. Time-frequency analysis of the first heart sound: Part 3: Application to dogs with varying cardiac contractility and to patients with mitral mechanical prosthetic heart valves. *Medical & Biological Engineering & Computing. Vol. 35(5)*, pp. 455–461, 1997.

[273] D. Chen, L.G. Durand et al. Time-frequency analysis of the muscle sound of the human diaphragm. *Medical & Biological Engineering & Computing. Vol. 35(6)*, pp. 649–652, 1997.

[274] D.S. Chen and J.P. Allebach. Analysis of error in reconstruction of two-dimensional signals from irregularly spaced samples. *IEEE Transactions Acoust., Speech, Signal Processing, Vol. ASSP-35*, pp. 173–179, 1987.

[275] F.R. Chen, H. Ichnose and J.J. Kai et al. Extension of HRTEM resolution by semi-blind deconvolution method and Gerchberg-Saxton algorithm: application to grain boundary and interface. *Journal of Electron Microscopy, Vol. 50(6)*, pp. 529–540, 2001.

[276] G.X. Chen and Z.R. Zhou. Time-frequency analysis of friction-induced vibration under reciprocating sliding conditions. *Digital signal processing. Vol. 17(1)*, pp. 371–393, 2007.

[277] Hong Chen and G.E. Ford. A unified eigenfilter approach to the design of two-dimensional zero-phase FIR filters with the McClellan transform. *IEEE Transactions on Circuits and Systems II: Analog and Digital Signal Processing, Vol. 43(8)*, pp. 622–626, 1996.

[278] K.H. Chen and C.C. Yang. On n-dimensional sampling theorems. *Applied Math. Comput., Vol. 7*, pp. 247–252, 1980.

[279] W. Chen and S. Itoh. A sampling theorem for shift-invariant subspace. *IEEE Transactions on Signal Processing, Vol. 46(10)*, pp. 2822–2824, 1998.

[280] D.K. Cheng and D.L. Johnson. Walsh transform of sampled time functions and the sampling principle. *Proceedings IEEE, Vol. 61*, pp. 674–675, 1973.

[281] Jingdong Chen, Bo Xu, and Taiyi Huang. New features based on the Cohen's class of bilinear time-frequency representations for speech recognition. *Proceedings 1998 Fourth International Conference on Signal Processing, 1998. ICSP '98. Vol. 1*, pp. 674–677, 1998.

[282] Victor C. Chen and Hao Ling. **Time-Frequency Transforms for Radar Imaging and Signal Analysis**. Artech House Publishers, 2002.

[283] K.F. Cheung. The Generalized Sampling Expansion – Its Stability, Posedness and Discrete Equivalent. M.S. thesis, Dept. Electrical Engr., Univ. of Washington, Seattle, Washington, 1983.

[284] K.F. Cheung. Image Sampling Density Reduction Below That of Nyquist. Ph.D. thesis, University of Washington, Seattle, 1988.

[285] K.F. Cheung and R.J. Marks II. Ill-posed sampling theorems. *IEEE Transactions on Circuits and Systems, Vol. CAS-32*, pp. 829–835, 1985.

[286] K.F. Cheung, R.J. Marks II, and L.E. Atlas. Neural net associative memories based on convex set projections, *IEEE Proceedings 1st International Conference on Neural Networks, San Diego*, June 1987.

[287] K.F. Cheung and R.J. Marks II. Papoulis' generalization of the sampling theorem in higher dimensions and its application to sample density reduction. In *Proceedings of the International Conference on Circuits and Systems, Nanjing, China*, 1989.

[288] K.F. Cheung and R.J. Marks II. Image sampling below the Nyquist density without aliasing. *Journal of the Optical Society America A, Vol. 7*, pp. 92–105, 1990.

[289] K.F. Cheung, R.J. Marks II, and L.E. Atlas. Convergence of Howard's minimum-negativity-constraint extrapolation algorithm. *Journal of the Optical Society America A, Vol. 5*, pp. 2008–2009, 1988.

[290] R. Cheung and Y. Rahmat-Samii. Experimental verification of nonuniform sampling technique for antenna far-field construction. *Electromagnetics, Vol. 6, No. 4*, pp. 277–300, 1986.

[291] Francois Le Chevalier. **Principles of Radar and Sonar Signal Processing**. Artech House Publishers, 2002.

[292] T.A.C.M. Claasen and W.F.G. Mecklenbrauker. The Wigner distribution, a tool for time-frequency signal analysis, Part 2: discrete time signals. *Philips J. Res., Vol. 35*, pp. 277–300, 1980.

[293] T.A.C.M. Claasen and W.F.G. Mecklenbrauker. The Wigner distribution, a tool for time-frequency signal analysis, Part 3: relations with other time-frequency signal transformations. *Philips J. Res., Vol. 35*, pp. 373–389, 1980.

[294] D.G. Childers. Study and experimental investigation on sampling rate and aliasing in time-division telemetry systems. *IRE Transactions Space Electronics & and Telemetry*, pp. 267–283, 1962.

[295] H.K. Ching. Truncation Effects in the Estimation of Two–Dimensional Continuous Bandlimited Signals. M.S. thesis, Dept. Electrical Engineering, Univ. of Washington, Seattle, 1985.

[296] P.C. Ching and S.Q. Wu. On approximated sampling theorem and wavelet denoising for arbitrary waveform restoration. *IEEE Transactions on Circuits and Systems II - Analog and Digital Signal Processing, Vol. 45(8)*, pp. 1102–1106, 1998.

[297] P.S. Cho, H.G. Kuterdem, and R.J. Marks II, "A spherical dose model for radiosurgery plan optimization", *Phys. Med. Biol, Vol. 43*, pp. 3145–3148, 1998.

[298] P.S. Cho, Shinhak Lee, R.J. Marks II, Seho Oh, S.G. Sutlief, and M.H. Phillips. Optimization of intensity modulated beams with volume constraints using two methods: cost function minimization and projections onto convex sets. *Medical Physics. Vol. 25(4)*, pp. 435–443, 1998.

[299] H.I. Choi and W.J. Williams. Improved time-frequency representation of multi-component signals using exponential kernels. *IEEE Transactions Acoust., Speech, and Sig. Proceedings, Vol. 37*, pp. 862–871, 1989.

[300] Li Chongrong and Li Yanda. Recovery of acoustic impedance by POCS technique. *1991 International Conference on Circuits and Systems, 1991. Conference Proceedings*, China., Vol. 2, pp. 820–823, 1991.

[301] J.J. Chou and L.A. Piegl. Data reduction using cubic rational B-splines. *IEEE Computer Graphics and Applications Magazine, Vol. 12(3)*, pp. 60–68, 1992.

[302] J. Chung, S.Y. Chung, and D. Kim. Sampling theorem for entire functions of exponential growth. *Journal of Mathematical Analysis and Applications, Vol. 265(1)*, pp. 217–228, 2002.

[303] Ruel V. Churchill. **Fourier Series and Boundary Value Problems**. McGraw-Hill. 2Rev Ed edition, 1963.

[304] Ruel V. Churchill. **Operational Mathematics**. McGraw–Hill, New York, 1972.

[305] T.A.C.M. Claasen and W.F.G. Mecklenbräuker. The Wigner distribution - a tool for time-frequency signal analysis. Part I: Continuous-time signals; Part II: Discrete-time signals; Part III: Relations with other time-frequency signal transformations. *Philips J. Res., Vol. 35*, pp. 217–250, 276–300, 372–389, 1980.

[306] J.J. Clark. Sampling and reconstruction of non-bandlimited signals. In *Proceedings Visual Communication and Image Processing IV, SPIE, Vol. 1199*, pp. 1556–1562, 1989.

[307] J.J. Clark, M.R. Palmer, and P.D. Lawrence. A transformation method for the reconstruction of functions from nonuniformly spaced samples. *IEEE Transactions Acoust., Speech, Signal Processing, Vol. ASSP-33*, 1985.

[308] D. Cochran and J. Clark. On the sampling and reconstruction of time-warped bandlimited signals. *Proceedings IASSP*, pp. 1539–1541, 1990.

[309] L. Cohen. Generalized phase-space distribution functions. *Journal Math. Phys., Vol. 7*, pp. 781–786, 1966.

[310] L. Cohen. Time-frequency distributions - A review. *Proceedings IEEE, Vol. 77*, pp. 941–961, 1989.

[311] D.D. Colclough and E.L. Titlebaum. Delay-Doppler POCS for specular multipath. *Proceedings IEEE International Conference on Acoustics, Speech, and Signal Processing, 2000, Vol. 4*, pp. IV-3940–IV-3943, 2002.

[312] L. Cohen. **Time Frequency Analysis: Theory and Applications**. Prentice-Hall Signal Processing, 1994.

[313] T.W. Cole. Reconstruction of the image of a confined source. *Astronomy and Astrophysics*, pp. 41–45, 1973.

[314] T.W. Cole. Quasi-optical techniques in radio astronomy, In **Progress in Optics**, North-Holland, Amsterdam, 1977, pp. 187–244.

[315] Samuel Taylor Coleridge and Barbara E. Rooke. **The Collected Works of Samuel Taylor Coleridge, Volume 4: "The Friend" by Samuel Taylor Coleridge and Barbara E. Rooke.** Princeton University Press, 1969.

[316] P.M. Colman. Non-Crystallographic symmetry and Sampling Theory. *Zeitschrift fur Kristallographie, Vol. 140(5–6)*, pp. 344–349, 1974.

[317] R.J. Collier, C.B. Burckhardt, and L.H. Lin. **Optical Holography**. Academic Press, Orlando, FL., 1971.

[318] J.M. Combes, A. Grossman, and Ph. Tchamitchian, **Wavelets**, Springer-Verlag, 1989.

[319] P.L. Combettes. The foundation of set theoretic estimation. *Proceedings of the IEEE, Vol. 81*, pp. 182–208, 1993.

[320] P.L. Combettes. Signal recovery by best feasible approximation. *IEEE Transactions on Image Processing, Vol. 2(2)*, pp. 269–271, 1993.

[321] P.L. Combettes. Convex set theoretic image recovery by extrapolated iterations of parallel subgradient projections. *IEEE Transactions on Image Processing, Vol. 6(4)*, pp. 493–506, 1997.

[322] J.W. Cooley and J.W. Tukey. An algorithm for the machine calculation of complex Fourier series. *Mathematics of Computation, Vol. 19(90)*, pp. 297–301, 1965.

[323] J.W. Cooley, P.A.W. Lewis, and P.D. Welch. Historical notes on the fast Fourier transform. *IEEE Transactions on Audio Electroacoustics, Vol. AU-15(2)*, pp. 76–79, 1967.

[324] A. Consortini and F. Crochetti. An attempt to remove noise from histograms of probability density of irradiance in atmospheric scintillation. *Proceedings SPIE*, pp. 524–525, 1990.

[325] J.W. Cooley, P.A.W. Lewis, and P.D. Welch. The finite Fourier transform. *IEEE Transactions Audio Electroacoustics, Vol. AE-17*, pp. 77–85, 1969.

[326] J.W. Cooley and J.W. Tukey. An algorithm for the machine calculations of complex Fourier series. *Math. Comput., Vol. 19*, pp. 297–301, 1965.

[327] W.A. Coppel. et al. Fourier - on the occasion of his two hundredth birthday. *Amer. Math. Monthly, Vol. 76*, pp. 468–483, 1969.

[328] Joshua Corey. **Fourier Series**. Spineless Books, 2005.

[329] I.J. Cox, C.J.R. Sheppard, and T. Wilson. Reappraisal of arrays of concentric annuli as super-resolving filters. *Journal of the Optical Society America, Vol. 72*, pp. 1287–1291, 1982.

[330] Edward J. Craig. **Laplace and Fourier Transforms for Electrical Engineers**. Holt, Rinehart and Winston, 1964.

[331] William Lane Craig. **Time and Eternity: Exploring God's Relationship to Time**. Crossway Books, 2001.

[332] P.W. Cramer and W.A. Imbriale. Speed-up of near-field physical optics scattering calculations by use of the sampling theorem. *IEEE Transactions on Magnetics, Vol. 30(5)*, Part 2, pp. 3156–3159, 1994.

[333] R.E. Crochiere and L.R. Rabiner. Interpolation and decimation of digital signals – a tutorial review. *Proceedings IEEE, Vol. 69*, pp. 300–331, 1981.

[334] R.E. Crochiere and L.R. Rabiner. **Multirate Digital Signal Processing**. Prentice-Hall, Englewood Cliffs, NJ, 1983.

[335] H.Z. Cummins and E.R. Pike, editors. **Photon Correlation Spectroscopy and Velocimetry**. Plenum Press, New York, 1977.

[336] L.J. Cutrona, E.N. Leith, C.J. Palermo, and L.J. Porcello. Optical data processing and filtering systems. *IRE Transactions Information Theory*, pp. 386–400, 1960.

[337] B.S. Cybakov and V.P. Jakovlev. On the accuracy of restoring a function with a finite number of terms of a Kotelnikov series. *Radio Engineering & Electron., Vol. 4*, pp. 274–275, 1959.

[338] R.N. Czerwinski and D.L. Jones. An adaptive time-frequency representation using a cone-shaped kernel. *1993 IEEE International Conference on Acoustics, Speech, and Signal Processing, 1993.* ICASSP-93., *Vol. 4*, pp. 404–407 1993.

[339] R.N. Czerwinski and D.L. Jones. Adaptive cone-kernel time-frequency analysis. *IEEE Transactions on Signal Processing, Vol. 43(7)*, pp. 1715–1719, 1995.

D

[340] J.J. DaCunha. Stability for time varying linear dynamic systems on time scales. *Journal of Computational and Applied Mathematics, Vol. 176(2)*, pp. 381–410, 2005.

[341] J.J. DaCunha. Instability results for slowly time varying linear dynamic systems on time scales. *Journal of Mathematical Analysis and Applications, Vol. 328(2)*, pp. 1278–1289, 2007.

[342] M.I. Dadi and R.J. Marks II. Detector relative efficiencies in the presence of Laplace noise. *IEEE Transactions on Aerospace and Electronic Systesm, Vol. AES-23(4)*, 1987.

[343] W. Dahmen and C.A. Micchelli. Translates of multivariate splines. *Linear Algebra Applications, Vol. 52/53*, pp. 217–234, 1983.

[344] H. Dammann and K. Görtker. High-efficiency in-line multiple imaging by means of multiple and filtering systems. *Optical Communication*, pp. 312–315, 1971.

[345] R. Dandliker, E. Marom, and F.M. Mottier. Wavefront sampling in holographic interferometry. *Optical Communication*, pp. 368–371, 1972.

[346] H. D'Angelo. **Linear Time Varying Systems, Analysis and Synthesis**. Allyn and Bacon, Boston, 1970.

[347] I. Daubechies. The wavelet transform, time-frequency localization and signal analysis. *IEEE Transactions Inform. Theory IT-36*, pp. 961–1005, 1990.

[348] Joseph W. Dauben. Georg Cantor and Pope Leo XIII: Mathematics, Theology, and the Infinite. *Journal of the History of Ideas, Vol. 38(1)*, pp. 85–108, 1977.

[349] Q.I. Daudpota, G. Dowrick, and C.A. Grwated. Estimation of moments of a Poisson sampled random process. *Journal Phys. A: Math. Gen., Vol. 10*, pp. 471–483, 1977.

[350] A.M. Davis. Almost periodic extension of bandlimited functions and its application to nonuniform sampling. *IEEE Transactions Circuits & Systems, Vol. CAS-33(10)*, pp. 933–938, 1986.

[351] J.A. Davis, E.A. Merrill, D.M. Cottrell, and R.M. Bunch. Effects of sampling and binarization in the output of the joint Fourier transform correlator. *Optical Engineering, Vol. 29(9)*, pp. 1094–1100, 1990.

[352] John M. Davis, Ian A. Gravagne, Billy J. Jackson, Robert J. Marks II, and Alice A. Ramos. The Laplace transform on time scales revisited. *Journal of Mathematical Analysis Applications, Vol. 332*, pp. 1291–1307, 2007.

[353] P.J. Davis. **Interpolation and Approximation**. Blaisdell, New York, 1963.

[354] P. Davis and R. Hersh. **The Mathematical Experience**, Boston: Birkhäuser, 1981.

[355] Sumner P. Davis, Mark C. Abrams, and James W. Brault. **Fourier Transform Spectrometry**. Academic Press, 2001.

[356] C. de Boor. On calculating with B–splines. *Journal Approx. Theory, Vol. 6*, pp. 50–62, 1972.

[357] C. de Boor. **A Practical Guide to Splines**. Springer-Verlag, New York, 1978.

[358] L.S. DeBrunner, V. DeBrunner, and Yao Minghua Yao. Edge-retaining asymptotic projections onto convex sets for image interpolation. *Proceedings of the 4th IEEE Southwest Symposium Image Analysis and Interpretation, 2000.* pp. 78–82, 2000.

[359] G. Defrancis. Directivity, super-gain and information. *IRE Transactions Antennas & Propagation, Vol. AP-4*, pp. 473–478, 1956.

[360] P. Delsarte and Y. Gemin. The split Levinson algorithm. *IEEE Transactions Acoust., Speech, Signal Processing, Vol. ASSP-34*, pp. 470–478, 1986.

[361] P. Delsarte and Y. Genin. On the role of orthogonal polynomials on the unit circle in digital signal processing applications. In **Orthogonal Polynomials: Theory and Practice** (ed. by P. Nevai and E. H. Ismail), Kluwer Academic Publishers, Dordrecht, pp. 115–133, 1990.

[362] Juan Arias de Reyna. **Pointwise Convergence of Fourier Series**. Springer, 2002.

[363] Michael B. Marcus and Gilles Pisier. **Random Fourier Series with Applications to Harmonic Analysis**. Annals of Mathematics Studies, Princeton University Press, 1981.

[364] A. Desabata. Extension of a Finite Version of the Sampling Theorem. *IEEE Transactions on Circuits and Systems II- Analog and Digital Signal Processing, Vol. 41(12)*, pp. 821–823, 1994.

[365] Rene Descartes, **Discours de la Methode**. 1637.
[366] P. DeSantis, G. Guattari, F. Gori, and C. Palma. Optical systems with feedback. *Opt. Acta*, pp. 505–518, 1976.
[367] P. DeSantis and F. Gori. On an iterative method for super-resolution. *Opt. Acta*, pp. 691–695, 1975.
[368] P. DeSantis, F. Gori, G. Guattari, and C. Palma. Emulation of super-resolving pupils through image postprocessing. *Optical Communication*, pp. 13–16, 1986.
[369] P. DeSantis, F. Gori, G. Guattari, and C. Palma. Synthesis of partially coherent fields. *Journal of the Optical Society America, Vol. 76*, pp. 1258–1262, 1986.
[370] P. DeSantis and C. Palma. Degrees of freedom of aberrated images. *Opt. Acta*, pp. 743–752, 1976.
[371] A.J. Devaney and R. Chidlaw. On the uniqueness question in the problem of phase retrieval from intensity measurements. *Opt. Acta*, pp. 1352–1354, 1978.
[372] P. DeVincenti, G. Doro, and A. Saitto. Reconstruction of bi-dimensional electromagnetic field by uni-dimensional sampling technique. *Electron. Letters, Vol. 17*, pp. 324–325, 1981.
[373] G.M. Dillard. Method and apparatus for computing the discrete Fourier transform recursively. United States Patent #4,023,028, 1977.
[374] G.M. Dillard. Recursive computation of the DFT with applications to a pulse-doppler radar system. *Comput. and Elec. Engng., Vol. 1(1)*, pp. 143–152, 1973.
[375] G.M. Dillard. Recursive computation of the discrete Fourier transform with applications to an FSK communications receiver. *IEEE National Telecommunication Conference*, San Diego, pp. 263–265, 1974.
[376] Paul Adrien Maurice Dirac, *Scientific American*, 1963.
[377] V.A. Ditkin and A.P. Prudnikiv. **Integral Transforms and Operation Calculus**, Fitmatgiz, Moscow, 1961: English translation, Pergamon Press, New York, 1965.
[378] D.D. Do and R.H. Weiland. Self–poissoning of catalyst pellets. *AICHE Symp. Series, Vol. 77(202)*, pp. 36–45, 1981.
[379] M.M. Dodson and A.M. Silva. Fourier analysis and the sampling theorem. *Proceedings Royal Irish Academy, Vol. 85 A*, pp. 81–108, 1985.
[380] A.H. Dooley. A non-abelian version of the shannon sampling theorem. *SIAM Journal on Mathematical Analysis, Vol. 20(3)*, pp. 624–633, 1989.
[381] R.C. Dorf, M.C. Farren, and C.A. Philips. Adaptive sampling frequency for sampled data control systems. *IRE Transactions Automatic Control, Vol. AC-7*, pp. 38–47, 1962.
[382] C.N. dos Santos, S.L. Netto, L.W.R. Biscainho, and D.B. Graziosi. A modified constant-Q transform for audio signals. *Proceedings IEEE International Conference on Acoustics, Speech, and Signal Processing, 2004. (ICASSP '04). Vol. 2*, pp. 469–472, 2004.
[383] F. Dowla and J.H. McClellan. MEM spectral analysis for nonuniformly sampled signals. In *Proceedings of the 2nd International Conference on Computer Aided Seismic Analysis and Discrimination, North Dartmouth, Mass.*, 1981.
[384] James F. Doyle. **Wave Propagation in Structures: Spectral Analysis Using Fast Discrete Fourier Transforms**. Springer, 2nd edition, 1997.
[385] E. Dubois. The sampling and reconstruction of time-varying imagery with application in video systems. *Proceedings IEEE, Vol. 23*, pp. 502–522, 1985.
[386] Javier Duoandikoetxea and David Cruz-Uribe. **Fourier Analysis**. American Mathematical Society, 2000.
[387] D.E. Dudgeon and R.M. Mersereau. **Multidimensional Digital Signal Processing**. Prentice-Hall, Englewood Cliffs, NJ, 1984.
[388] P.M. Duffieux. **L'integral de Fourier et ses applications a l'optique**. Rennes-Oberthur, 1946.
[389] R.J. Duffin and A.C. Schaeffer. Some properties of functions of exponential type. *Bull. Amer. Math. Society, Vol. 44*, pp. 236–240, 1938.
[390] R.J. Duffin and A.C. Schaeffer. A class of nonharmonic Fourier series. *Transactions America Math. Society*, pp. 341–366, 1952.
[391] J. Dunlop and V.J. Phillips. Signal recovery from repetitive non-uniform sampling patterns. *The Radio and Electronic Engineer, Vol. 44*, pp. 491–503, 1974.

[392] Chen Danmin, L.G. Durand, and H.C. Lee. Application of the cone-kernel distribution to study the effect of myocardial contractility in the genesis of the first heart sound in dog. 17th Annual IEEE Conference on Engineering in Medicine and Biology Society, 1995. IEEE *Vol. 2*, pp. 959–960, 1995.

[393] K. Dutta and J.W. Goodman. Reconstruction of images of partially coherent objects from samples of mutual intensity. *Journal of the Optical Society America, Vol. 67*, pp. 796–803, 1977.

[394] J. Duvernoy. Estimation du nombre d'echantillons et du domaine spectral necessaires pour determiner une forme moyenne par une technique de superposition. *Optical Communication*, pp. 286–288, 1973.

[395] H. Dym and H.P. McKean. **Fourier series and Integrals**. Academic Press, New York, 1972.

[396] H. Dym and H.P. McKean. **Gaussian Processes, Function Theory, and the Inverse Spectral Problem**. Academic Press, New York, 1976.

E

[397] Sir Arthur Eddington. **The Philosophy of Physical Science.** Cambridge, 1939.

[398] R.E. Edwards. **Fourier Series, a Modern Introduction**. Springer; 2d ed edition, 1979.

[399] M. Elad and A. Feuer. Restoration of a single superresolution image from several blurred, noisy, and undersampled measured images. IEEE Transactions on Image Processing, *Vol. 6(12)*, pp. 1646–1658, 1997.

[400] Tryon Edwards, **The New Dictionary of Thoughts-A Cyclopedia of Quotations** (Garden City, NY: Hanover House, 1852; revised and enlarged by C.H. Catrevas, Ralph Emerson Browns, and Jonathan Edwards [descendent, along with Tryon, of Jonathan Edwards (1703–1758), president of Princeton], 1891.

[401] R.E. Edwards. **Fourier Series, A Modern Introduction, Vol. 1**. Holt, Rinehart and Winston, New York, 1967.

[402] Ethel Edwards. **Carver of Tuskegee.** (Cincinnati, Ohio; Ethyl Edwards & James T. Hardwick, 1971) pp. 141–42.

[402a] A. Einstein, J. Mayer, and J. Holmes. *Bite-Size Einstein: Quotations on Just About Everything from the Greatest Mind of the Twentieth Century*, St. Martin's Press, 1996.

[403] P. Elleaume. Device for calculating a discrete moving window transform and application thereof to a radar system. United States Patent #4,723,125, 1988.

[404] P. Elleaume. Device for calculating a discrete Fourier transform and its application to pulse compression in a radar system. United States Patent #4,772,889, 1988.

[405] M.K. Emresoy and P.J. Loughlin. Weighted least-squares implementation of Cohen-Posch time frequency distributions with specified conditional and joint moment constraints. *Proceedings of the IEEE-SP International Symposium on Time-Frequency and Time-Scale Analysis*, pp. 305–308, 1998.

[406] M.K. Emresoy and P.J. Loughlin. Weighted least squares implementation of Cohen-Posch time-frequency distributions. *IEEE Transactions on Signal Processing, Vol. 46(3)*, pp. 753–757, 1998.

[407] M.K. Emresoy and P.J. Loughlin. Weighted least-squares implementation of Cohen-Posch time-frequency distributions with specified conditional and joint moment constraints. *IEEE Transactions on Signal Processing, Vol. 47(3)*, pp. 893–900, 1999.

[408] A.G. Emslie and G.W. King. Spectroscopy from the point of view of the communication theory, II. *Journal of the Optical Society America, Vol. 43*, pp. 658–663, 1953.

[409] W. Engels and W. Splettstößer. A note on Maqusi's proofs and truncation error bounds in the dyadic (Walsh) sampling theorem. *IEEE Transactions Acoust, Speech, Signal Processing, Vol. ASSP-30*, pp. 334–335, 1982.

[410] W. Engels, E.L. Stark, and L. Vogt. Optimal kernels for a sampling theorem. *Journal of Approximate Theory, Vol. 50(1)*, pp. 69–83, 1987.

[411] W. Engels, E.L. Stark, and L. Vogt. On the application of an optimal spline sampling theorem. *Signal Processing, Vol. 14(3)*, pp. 225–236, 1988.

[412] A. Ephremides and L.H. Brandenburg. On the reconstruction error of sampled data estimates. *IEEE Transactions Information Theory, Vol. IT-19*, pp. 365–367, 1973.

[413] A. Erdelyi et al. **Higher Transcendental Functions, Vol. 1**. McGraw-Hill, New York, 1953.
[414] A. Erdelyi et al. **Tables of Integral Transforms II**. McGraw-Hill, New York, 1954.
[415] M.F. Erden, M.A. Kutay, and H.M. Ozaktas. Repeated filtering in consecutive fractional Fourier domains and its application to signal restoration. *IEEE Transactions on Signal Processing, Vol. 47(5)*, pp. 1458–1462, 1999.
[416] T. Ericson. A generalized version of the sampling theorem. *Proceedings IEEE, Vol. 60*, pp. 1554–1555, 1972.
[417] P.E. Eren, M.I. Sezan, and A.M. Tekalp. Robust, object-based high-resolution image reconstruction from low-resolution video. *IEEE Transactions on Image Processing, Vol. 6(10)*, pp. 1446–1451, 1997.
[418] T. Erseghe, P. Kraniauskas, G. Cariolaro. Unified fractional Fourier transform and sampling theorem. *IEEE Transactions on Signal Processing, Vol. 47(12)*, pp. 3419–3423, 1999.
[419] Okan K. Ersoy. **Diffraction, Fourier Optics and Imaging**. Wiley-Interscience, 2006.
[420] A. Erteza and K.S. Lin. On the transform property of a bandlimited function and its samples. *Proceedings IEEE, Vol. 68*, pp. 1149–1150, 1980.
[421] G. Bora Esmer, Vladislav Uzunov, Levent Onural, Haldun M. Ozaktas, and Atanas Gotchev. Diffraction field computation from arbitrarily distributed data points in space. *Signal Processing – Image Communication. Vol. 22(2)*, pp. 178–187, 2007.
[422] H. Eves. **Mathematical Circles Squared**, Boston: Prindle, Weber and Schmidt, 1972.
[423] H. Eves. **Mathematical Circles Adieu**, Boston: Prindle, Weber and Schmidt, 1977.
[424] H. Eves. **Return to Mathematical Circles**, Boston: Prindle, Weber and Schmidt, 1988.

F

[425] G.S. Fang. Whittaker-Kotelnikov-Shannon sampling theorem and aliasing error. *Journal of Approximation Theory, Vol. 85(2)*, pp. 115–131, 1996.
[426] W.H. Fang, H.S. Hung, C.S. Lu, and P.C. Chu. Block adaptive beamforming via parallel projection method. *IEICE Transactions on Communications, Vol. E88B(3)*, pp. 1227–1233, 2005.
[427] N.H. Farhat and G.P. Shah. Implicit sampling in optical data processing. *Proceedings SPIE, Vol. 201*, pp. 100–106, 1979.
[428] A. Faridani. A generalized sampling theorem for locally compact abelian groups. *Mathematics of Computation, Vol. 63(207)*, pp. 307–327, 1994.
[429] Thomas C. Farrar and Edwin D. Becker. **Pulse and Fourier Transform NMR: Introduction to Theory and Methods**. Academic Press, 1971.
[430] H.G. Feichtinger. Wiener amalgam spaces and some of their applications. In *Conf. Proceedings Function spaces, Edwardsville, IL.*, Marcel Dekker, New York, April 1990.
[431] E.B. Felstead and A.U. Tenne-Sens. Optical interpolation with application to array processing. *Applied Optics, Vol. 14*, pp. 363–368, 1975.
[432] M.M. Fenoy and P. Ibarrola. The optional sampling theorem for submartingales in the sequentially planned context. *Statistics and Probability Letters, Vol. 77(8)*, pp. 826–831, 2007.
[433] W.L. Ferrar. On the cardinal function of interpolation–theory. *Proceedings Royal Society Edinburgh, Vol. 45*, pp. 267–282, 1925.
[434] W.L. Ferrar. On the cardinal function of interpolation–theory. *Proceedings Royal Society Edinburgh, Vol. 46*, pp. 323–333, 1926.
[435] W.L. Ferrar. On the consistency of cardinal function interpolation. *Proceedings Royal Society Edinburgh, Vol. 47*, pp. 230–242, 1927.
[436] Thomas Clark Farrar and Edwin D. Becker. **Pulse and Fourier Transform NMR: Introduction to Theory and Methods**. Elsevier Science & Technology, 1971.
[437] John R. Ferraro, Louis J. Basile. **Fourier Transform Infrared Spectroscopy**, Academic Press, 1985.
[438] John R. Ferraro, Louis J. Basile (Editors). **Fourier Transform Infrared Spectra: Applications to Chemical Systems, Vol. 1**. Elsevier Science & Technology Books, 1978.
[439] Paulo J.S.G. Ferreira. Incomplete sampling series and the recovery of missing samples from oversampled band-limited signals. *IEEE Transactions Signal Processing, Vol. 40*, 1992. pp. 225–227.

[440] Paulo J.S.G. Ferreira. Interpolation and the discrete Papoulis-Gerchberg algorithm. *IEEE Transactions on Signal Processing, Vol. 42(10)*, pp. 2596–2606, 1994.

[441] H.A. Ferweda. The phase reconstruction problem for wave amplitudes and coherence functions. In **Inverse Source Problems in Optics**, H.P. Baltes, editor, Springer-Verlag, Berlin, 1978.

[442] H.A. Ferweda and B.J. Hoenders. On the reconstruction of a weak phase-amplitude object IV. *Optik*, pp. 317–326, 1974.

[443] Alfred Fettweis, Josef A. Nossek, and Klaus Meerkötter. Reconstruction of signals after filtering and sampling rate reduction. *IEEE Transactions Acoust., Speech, Signal Processing, Vol. ASSP-33, No. 4*, pp. 893–902, 1985.

[444] M.A. Fiddy and T.J. Hall. Nonuniqueness of super-resolution techniques applied to sampled data. *Journal of the Optical Society America, Vol. 71*, pp. 1406–1407, 1981.

[445] G. Fielding, R. Hsu, P. Jones, *et al.* Aliasing and reconstruction distortion in digital intermediates. *SMPTE Motion Imaging Journal, Vol. 115(4)*, pp. 128–136, 2006.

[446] A.R. Figueiras-Vidal, J.B. Marino-Acebal, and R.G. Gomez. On generalized sampling expansions for deterministic signals. *IEEE Transactions Circuits & Systems, Vol. CAS-28*, pp. 153–154, 1981.

[447] G. Fix and G. Strang. Fourier analysis of the finite element method in Ritz-Galerkin theory. *Studies Applied Math. Vol. 48*, pp. 268–273, 1969.

[448] I.T. Fjallbrandt. Method of data reduction of sampled speech signals by using nonuniform sampling and a time variable digital filter. *Electron. Letters, Vol. 13*, pp. 334–335, 1977.

[449] T.T. Fjallbrandt. Interpolation and extrapolation in nonuniform sampling sequences with average sampling rate below the Nyquist rate. *Electron. Letters, Vol. 11*, pp. 264–266, 1975.

[450] G.C. Fletcher and D.J. Ramsay. Photon correlation spectroscopy of polydisperse samples, I. Histogram method with exponential sampling. *Opt. Acta, Vol. 30*, pp. 1183–1196, 1983.

[451] N.J. Fliege. **Multirate Digital Signal Processing: Multirate Systems - Filter Banks - Wavelets**. Wiley, 2000.

[452] L. Fogel. A note on the sampling theorem. *IRE Transactions, Vol. 1*, pp. 47–48, 1955.

[453] J.T. Foley, R.R. Butts. Sampling Theorem for the Reconstruction of Bandlimited Intensities from Their Spatially Averaged Values. *Journal of the Optical Society of America, Vol. 72(12)*, pp. 1812–1812, 1982.

[454] Gerald B. Folland. **Fourier Analysis and Its Applications**. Brooks Cole, 1992.

[455] Joseph Fourier. **The Analytical Theory of Heat**. Cosimo Classics, 2007.

[456] William T. Fox. **Fourier Analysis of Weather and Wave Data from Lake Michigan.** Williams College, 1970.

[457] J. Frank. Computer processing of electron micrographs. In **Advanced Techniques in Biological Electron Microscopy**, J.K. Koehler, editor, Springer-Verlag, New York, 1973.

[458] G. Franklin. Linear filtering of sampled data. *IRE International Conv. Rec., Vol. 3, Pt. 4*, pp. 119–128, 1955.

[459] L.E. Franks. **Signal Theory**. Prentice-Hall, Englewood Cliffs, NJ, 1969.

[460] D. Fraser. A Conceptual Image Intensity Surface and the Sampling Theorem. *Australian Computer Journal, Vol. 19(3)*, pp. 119–125, 1987.

[461] H. Freeman. **Discrete Time Systems**. John Wiley, New York, 1965.

[462] A.T. Friberg. On the existence of a radiance function for finite planar sources of arbitrary states of coherence. *Journal of the Optical Society America, Vol. 69*, pp. 192–199, 1979.

[463] B.R. Frieden. Image evaluation by the use of sampling theorem. *Journal of the Optical Society America, Vol. 56*, pp. 1355–1362, 1966.

[464] B.R. Frieden. On arbitrarily perfect imagery with a finite aperture. *Opt. Acta*, pp. 795–807, 1969.

[465] B.R. Frieden. Evaluation, design and extrapolation methods for optical signals, based on the use of the prolate functions. In *Progress in Optics*, E. Wolf, editor, North-Holland, Amsterdam, 1971.

[466] B.R. Frieden. Image enhancement and restoration. In **Picture processing and digital filtering**, Springer-Verlag, Berlin, 1975.

[467] B.R. Frieden. Band-unlimited reconstruction of optical objects and spectra. *Journal of the Optical Society America, Vol. 66*, pp. 1013–1019, 1976.

[468] Michael Fripp, Deborah Fripp, and Jon Fripp. **Speaking of Science**. Newnes, 2000.

[469] T. Fujita and M.R. McCartney. Phase recovery for electron holography using Gerchberg-Papoulis iterative algorithm. *Ultramicroscopy, Vol. 102(4)*, pp. 279–286, 2005.

[470] Y.H. Fung, Y.H. Chan. POCS-based algorithm for restoring colour-quantised images. *IEE Proceedings Vision, Image and Signal Processing, Vol. 151(2)*, pp. 119–127, 2004.

[471] Y.H. Fung and Y.H. Chan. A POCS-based restoration algorithm for restoring halftoned color-quantized images. *IEEE Transactions on Image Processing, Vol. 15(7)*, pp. 1985–1992, 2006.

G

[472] N.T. Gaardner. Multidimensional Sampling Theorem. *Proceedings of the IEEE, Vol. 60(2)*, p. 247, 1972.

[473] D. Gabioud. Effect of irregular sampling, especially in the case of justification jitter. *Mitteilungen AGEN, No. 44*, pp. 39–46, 1986.

[474] D. Gabor. Theory of communication. *Journal IEE (London), Vol. 93, Part III*, pp. 429–457, 1946.

[475] D. Gabor. A new microscopic principle. *Nature*, pp. 777–779, 1948.

[476] D. Gabor. Light and information. In **Progress in Optics**, E. Wolf, editor, North-Holland, Amsterdam, 1961.

[477] V. Galdi, V. Pierro and I.M. Pinto. Evaluation of Stochastic Resonance Based Detectors of Weak Harmonic Signals in Additive White Gaussian Noise. *Physical Review E, Vol. 55(6)*, p. 57, 1998.

[478] N.C. Gallagher, Jr., and G.L. Wise. A representation for band-limited functions. *Proceedings IEEE, Vol. 63*, p. 1624, 1975.

[479] H. Gamo. Intensity matrix and degrees of coherence. *Journal of the Optical Society America, Vol. 47*, p. 976, 1957.

[480] H. Gamo. Transformation of intensity matrix by the transmission of a pupil. *Journal of the Optical Society America, Vol. 48*, pp. 136–137, 1958.

[481] H. Gamo. Matrix treatment of partial coherence. In **Progress in Optics**, E. Wolf, editor, North-Holland, Amsterdam, 1964.

[482] George Gamow. **One two three ... infinity: Facts and speculations of science**. New American Library, Mentor book edition (1953). Reprinted by Dover Publications, 1988.

[483] M. Gamshadzahi. Bandwidth reduction in delta modulation systems using an iterative reconstruction scheme. Master's thesis, Department of ECE, Illinois Institute of Technology, Chicago, December, 1989.

[484] X.C. Gan, A.W.C. Liew, and H. Yan. Microarray missing data imputation based on a set theoretic framework and biological knowledge. *Nucleic Acids Research, Vol. 34(5)*, pp. 1608–1619, 2006.

[485] William L. Gans. The measurement and deconvolution of time jitter in equivalent-time waveform samplers. *IEEE Transactions Instrumentation & Measurement, Vol. IM 32(1)*, pp. 126–133, 1983.

[486] William L. Gans. Calibration and error analysis of a picosecond pulse waveform measurement system at NBS. *Proceedings IEEE, Vol. 74, No. 1*, pp. 86–90, 1986.

[487] A.G. Garcia, M.A. Hemandez-Medina. The discrete Kramer sampling theorem and indeterminate moment problems. *Journal of Computational and Applied Mathematics, Vol. 134(1–2)*, pp. 13–22, 2001.

[488] Garcia F.M., I.M.G. Lourtie, and J. Buescu. L_2 nonstationary processes and the sampling theorem. *IEEE Signal Processing Letters, Vol. 8(4)*, pp. 117–119, 2001.

[489] P.H. Gardenier, C.A. Lim, D.G.H. Tan et al. Aperture distribution phase from single radiation-pattern measurement via Gerchberg-Saxton algorithm. *Electronic Letters, Vol. 22(2)*, pp. 113–115, 1986.

[490] N.T. Garder. A note on multidimensional sampling theorem. *Proceedings IEEE, Vol. 60*, pp. 247–248, 1972.

[491] W.A. Gardner. A sampling theorem for nonstationary random processes. *IEEE Transactions Information Theory, Vol. IT-18*, pp. 808–809, 1972.

[492] H. Garudadri and E. Velez. A comparison of smoothing kernels for computing bilinear time-frequency representations. IEEE 1991 International Conference on Acoustics, Speech, and Signal Processing, 1991. ICASSP-91., *Vol. 5.*, pp. 3221–3224, 1991.

[493] J.D. Gaskill. **Linear Systems, Fourier Transforms and Optics**. John Wiley & Sons, Inc., New York, 1978.

[494] M. Gaster and J.B. Roberts. Spectral analysis of randomly sampled signals. *Journal Institute Math. Applic., Vol. 15*, pp. 195–216, 1975.

[495] G.C. Gaunaurd and H.C. Strifors. Signal analysis by means of time-frequency (Wigner-type) distributions-Applications to sonar and radar echoes. *Proceedings of the IEEE. Vol. 84(9)*, pp. 1231–1248, 1996.

[496] W.S. Geisler and D.B. Hamilton. Sampling theory analysis of spatial vision. *Journal of the Optical Society America A, Vol. 3*, pp. 62–70, 1986.

[497] F.J. Geng, D.M. Zhu DM and Q.Y. Lu. A new existence result for impulsive dynamic equations on timescales. *Applied Mathematics Letters, Vol. 20(2)*, pp. 206–212, 2007.

[498] T. Gerber and P.W. Schmidt. The sampling theorem and small-angle scattering. *Journal Applied Crystallography, Vol. 16*, pp. 581–589, 1983.

[499] T. Gerber, G. Walter, and P.W. Schmidt. Use of the Sampling Theorem for Collimation Corrections in Small Angle X-Ray Scatteing. *Journal of Applied Crystallography, Vol. 24*, Part 4, pp. 278–285, 1991.

[500] R.W. Gerchberg. Super-resolution through error energy reduction. *Optica Acta, Vol. 21*, pp. 709–720, 1974.

[501] R.W. Gerchberg and W.O. Saxton. A practical algorithm for the determination of phase from image and diffraction plane pictures. *Optik. 35*, pp. 237–246, 1972.

[502] R.W. Gerchberg. The lock problem in the Gerchberg-Saxton algorithm for phase retrieval. *Optik, Vol. 74(3)*, pp. 91–93, 1986.

[503] Allen Gersho. Asymptotically optimal block quantization. *IEEE Transactions Information Theory, Vol. IT-25*, pp. 373–380, 1979.

[504] A. Gersho and R. Gray. **Vector Quantization and Signal Compression**. Kluwer Academic Publishers, 1992.

[505] G. Gheen and F.T.S. Yu. Broadband image subtraction by spectral dependent sampling. *Optical Communication*, pp. 335–341, 1986.

[506] Charles R. Giardina and Edward R. Dougherty, **Morphological Methods in Image and Signal Processing**, Prentice-Hall, 1988.

[507] R.J. Glauber. Coherent and incoherent states of the radiation field. *Phys. Rev.*, pp. 2766–2788, 1963.

[508] M.J.E. Golay. Point arrays having compact, nonredundant autocorrelations. *Journal of the Optical Society America, Vol. 61*, pp. 272–273, 1971.

[509] M.H. Goldburg and R.J. Marks II. Signal synthesis in the presence of an inconsistent set of constraints. *IEEE Transactions Circuits & Systems, Vol. CAS-32*, pp. 647–663, 1985.

[510] S. Goldman. **Information Theory**. Dover, New York, 1968.

[511] Rafael C. Gonzalez. **Digital Image Processing**. Addison-Wesley Publishing Company, Inc., Reading, MA., 1977.

[512] Enrique A. Gonzalez-Velasco. **Fourier Analysis and Boundary Value Problems**. Academic Press, 1995.

[513] I.J. Good. The loss of information due to clipping a waveform. *Information & Control, Vol. 10*, pp. 220–222, 1967.

[514] J.W. Goodman. **Introduction to Fourier Optics**. McGraw-Hill, New York, 1968. (A revised version is also available [518].)

[515] J.W. Goodman. Imaging with low-redundancy arrays, In **Applications of holography**, Plenum Press, New York, pp. 49–55, 1971.

[516] J.W. Goodman. Linear space-variant optical data processing. In *Optical Information Processing*, S.H. Lee, editor, Springer-Verlag, New York, 1981.

[517] J.W. Goodman. **Statistical Optics**. Wiley, New York, 1985.

[518] J.W. Goodman. **Introduction to Fourier Optics Third Edition**, Roberts & Company Publishers 2004. (This is a revised version of [514].)

[519] G.C. Goodwin, M.B. Zarrop, and R.L. Payne. Coupled design of test signals, sampling intervals, and filters for system identification. *IEEE Transactions Automatic Control, Vol. AC-19*, 1974.

[520] R.A. Gopinath, J.E. Odegard and C.S. Burrus. Optimal Wavelet Representation of Signals and the Wavelet Sampling Theorem. *IEEE Transactions on Circuits and Systems II-Analog and Digital Signal Processing, Vol. 41(4)*, pp. 262–277, 1994.

[521] W.J. Gordon and R.F. Riesenfeld. B-spline curves and surfaces. In **Computer Aided Geometric Design**, R. Barnbill and R.F. Riesenfeld, editors, Academic Press, New York, 1974.

[522] F. Gori. Integral equations for incoherent imagery. *Journal of the Optical Society America, Vol. 64*, pp. 1237–1243, 1974.

[523] F. Gori. On an iterative method for super-resolution. *International Optics Computing Conf., IEEE Catalog no. 75 CH0941-5C*, pp. 137–141, 1975.

[524] F. Gori. Lau effect and coherence theory. *Optical Communication*, pp. 4–8, 1979.

[525] F. Gori. Fresnel transform and sampling theorem. *Optical Communication, Vol. 39*, pp. 293–298, 1981.

[526] F. Gori and R. Grella. The converging prolate spheroidal functions and their use in Fresnel optics. *Optical Communication*, pp. 5–10, 1983.

[527] F. Gori and G. Guattari. Effects of coherence on the degrees of freedom of an image. *Journal of the Optical Society America, Vol. 61*, pp. 36–39, 1971.

[528] F. Gori and G. Guattari. Holographic restoration of nonuniformly sampled bandlimited functions. *Optical Communication, Vol. 3*, pp. 147–149, 1971.

[529] F. Gori and G. Guattari. Imaging systems using linear arrays of non-equally spaced elements. *Phys. Letters, Vol 32A*, pp. 38–39, 1970.

[530] F. Gori and G. Guattari. Nonuniform sampling in optical processing. *Optica Acta, Vol. 18*, pp. 903–911, 1971.

[531] F. Gori and G. Guattari. Optical analog of a non-redundant array. *Phys. Letters Vol. 32A*, pp. 446–447, 1970.

[532] F. Gori and G. Guattari. Use of nonuniform sampling with a single correcting operation. *Optical Communication, Vol. 3*, pp. 404–406, 1971.

[533] F. Gori and G. Guattari. Shannon number and degrees of freedom of an image. *Optical Communication*, pp. 163–165, 1973.

[534] F. Gori and G. Guattari. Degrees of freedom of images from point-like element pupils. *Journal of the Optical Society America, Vol. 64*, pp. 453–458, 1974.

[535] F. Gori and G. Guattari. Eigenfunction technique for point-like element pupils. *Opt. Acta*, pp. 93–101, 1975.

[536] F. Gori and G. Guattari. Signal restoration for linear systems with weighted inputs; singular value analysis for two cases of low-pass filtering. *Inverse Problems*, pp. 67–85, 1985.

[537] F. Gori, G. Guattari, C. Palma, and M. Santarsiero. Superresolving postprocessing for incoherent imagery. *Optical Communication*, pp. 98–102, 1988.

[538] F. Gori and F. Mallamace. Interference between spatially separated holographic recordings. *Optical Communication Vol. 8*, pp. 351–354, 1973.

[539] F. Gori and C. Palma. On the eigenvalues of the sinc kernel. *Journal Phys. A, Math. Gen.*, pp. 1709–1719, 1975.

[540] F. Gori, S. Paolucci, and L. Ronchi. Degrees of freedom of an optical image in coherent illumination, in the presence of aberrations. *Journal of the Optical Society America, Vol. 65*, pp. 495–501, 1975.

[541] F. Gori and L. Ronchi. Degrees of freedom for scatterers with circular cross section. *Journal of the Optical Society America, Vol. 71*, pp. 250–258, 1981.

[542] F. Gori and S. Wabnitz. Modifications of the Gerchberg method applied to electromagnetic imaging, In **Inverse Methods in Electromagnetic Imaging**, D. Reidel Publishing Co., Dordrecht, 1985, pp. 1189–1203.

[543] D. Gottlieb and S.A. Orszag. **Numerical Analysis of Spectral Methods-Theory and Applications.** SIAM Publ., Philadelphia, 1977.

[544] I.S. Gradshteyn, I.M. Ryzhik, and Alan Jeffrey. **Tables of Integrals, Products and Series, 5th Edition**. Academic Press, New York, 1994.

[545] Loukas Grafakos. **Classical and Modern Fourier Analysis**. Prentice Hall, 2003.

[546] I. Grattan-Guinness. Joseph Fourier and the revolution in mathematical physics. *Journal of the Institute of Mathematics and its Applications, Vol. 5*, pp. 230–253, 1969.

[547] I. Grattan-Guinness. **Joseph Fourier, 1768–1830**. London, 1972.

[548] Ian A. Gravagne, John M. Davis, Jeffrey J. DaCunha and R.J. Marks II. Bandwidth Reduction for Controller Area Networks using Adaptive Sampling. *Proc. Int. Conf. Robotics and Automation* (ICRA), New Orleans, LA, pp. 5250–5255, 2004.

[549] R.M. Gray, A. Buzo, A.H. Gray, Jr., and Y. Matsuyama. Distortion measures for speech processing. *IEEE Transactions Acoust., Speech, Signal Processing, Vol. ASSP-28*, pp. 367–376, 1980.

[550] P. Greguss and W. Waigelich. Ultrasonic Holographic Fourier Spectroscopy via Optical Fourier Transforms. *IEEE Transactions on Computers, Vol. C-24(4)*, pp. 412–418, 1975.

[551] U. Grenader and G. Szegö. **Toeplitz Forms and theirApplications**. University of California Press, Berkeley, CA., 1958.

[552] M. Grigoriu. Simulation of Stationary Process via a Sampling Theorem. *Journal of Sound and Vibration, Vol. 166(2)*, pp. 301–313, 1993.

[553] B. Grobler. Practical application of the sampling theorem. *Feingerätetechnik, Vol. 26*, pp. 342–345, 1977.

[554] D. Groutage. The calculation of features for non-stationary signals from moments of the singular value decomposition of Cohen-Posch (positive time-frequency) distributions. *Proceedings of the IEEE-SP International Symposium on Time-Frequency and Time-Scale Analysis*, pp. 605–608, 1998.

[555] D. Groutage and D. Bennink. Feature sets for nonstationary signals derived from moments of the singular value decomposition of Cohen-Posch (positive time-frequency) distributions. *IEEE Transactions on Signal Processing, Vol. 48(5)*, pp. 1498–1503, 2000.

[556] B.Y. Gu, B.Z. Dong, and G.Z. Yang. A New Algorithm for Retrieval Phase Using the Sampling Theorem. *Optik, Vol. 75(3)*, pp. 92–96, 1987.

[557] C.R. Guarino. A new method of spectral analysis and sample rate reduction for bandlimited signals. *Proceedings IEEE, Vol. 69*, pp. 61–63, 1981.

[558] L.G. Gubin, B.T. Polyak, and E.V. Ralik. The method of projections for finding the common point of convex sets. *USSR Comput. Math. Phys., Vol. 7(6)*, pp. 1–24, 1967.

[559] B.K. Gunturk, Y. Altunbasak and R.M. Mersereau. Color plane interpolation using alternating projections. *IEEE Transactions on Image Processing. Vol. 11(9)*, pp. 997–1013, 2002.

[560] B.K. Gunturk, Y. Altunbasak and R.M. Mersereau. Multiframe resolution-enhancement methods for compressed video. *IEEE Signal Processing Letters, Vol. 9(6)*, pp. 170–174, 2002.

[561] S.C. Gupta. Increasing the sampling efficiency for a control system. *IEEE Transactions Automatic Control, Vol. AC-8*, pp. 263–264, 1963.

[562] R.K. Guy. Mersenne Primes, Repunits, Fermat Numbers, Primes of Shape $k \cdot 2^n + 2$ [sic], in **Unsolved Problems in Number Theory, 2nd ed.**, New York: Springer-Verlag, pp. 8–13, 1994.

H

[563] M.K. Habib and S. Cambanis. Dyadic sampling approximations for non-sequency-limited signals. *Information & Control, Vol. 49*, pp. 199–211, 1981.

[564] M.K. Habib and S. Cambanis. Sampling approximation for non-bandlimited harmonizable random signals. *Inform. Sci., Vol. 23*, pp. 143–152, 1981.

[565] A.H. Haddad and J.B. Thomas. Integral representation for non-uniform sampling expansions. In *Proceedings 4th Allerton Conference of Circuit System Theory*, pp. 322–333, 1966.

[566] A.H. Haddad, K. Yao, and J.B. Thomas. General methods for the derivation of sampling theorems. *IEEE Transactions Information Theory, Vol. IT-13*, 1967.

[567] M.O. Hagler, R.J. Marks II, E.L. Kral, J.F. Walkup, and T.F. Krile. Scanning technique for coherent processors. *Applied Optics, Vol. 19*, pp. 1670–1672, 1980.

[568] H. Hahn. Über das Interpolation problem. *Math. Z., Vol. 1*, pp. 115–142, 1918.

[569] E. Hansen. Fast Hankel transform algorithm. *IEEE Transactions on Acoustics, Speech, and Signal Processing. Vol. 33(3)*, pp. 666–671, 1985.

[570] M.W. Hall and R.J. Marks II. Sampling Theorem Characterization of Variation Limited Systems at Reduced Sampling Rates. *Journal of the Optical Society of America, Vol. 68(10)*, pp. 1362–1362, 1978.

[571] J. Hamill, G.E. Caldwell and T.R. Derrick. Reconstructing digital signals using Shannon's Sampling Theorem. *Journal of Applied Biomagnetics, Vol. 13(2)*, pp. 226–238, 1997.

[572] Richard W. Hamming. **Numerical Methods for Scientists and Engineers**. McGraw-Hill, New York, 1962.

[573] Richard W. Hamming. The Unreasonable Effectiveness of Mathematics. *The American Mathematical Monthly, Vol. 87(2)*, 1980.

[574] Richard W. Hamming. **Digital Filters: Third Edition**, Dover Publications, 1998.

[575] R.C. Hansen. **Phased Array Antennas.** Wiley-Interscience, 1998.

[576] R.M. Haralick and L.G. Shapiro. **Computer and Robot Vision**. Addison-Wesley, 1991.

[577] R.M. Haralick, X.H. Zhuang, C. Lin et al. The Digital Morphological Sampling Theorem. *IEEE Transactions on Acoustics, Speech and Signal Processing, Vol. 37(12)*, pp. 2067–2090, 1989.

[578] Robert M. Haralick, **Mathematical Morphology: Theory and Hardware** (Oxford Series in Optical and Imaging Sciences) 2005.

[579] R.D. Harding. **Fourier Series and Transforms**. Taylor & Francis, 1985.

[580] G.H. Hardy and W.W. Rogosinski. **Fourier Series**. Dover Publications, 1999.

[581] J.L. Harris. Diffraction and resolving power. *Journal of the Optical Society America, Vol. 54*, pp. 931–936, 1964.

[582] C.J. Hartley. Resolution of frequency aliases in ultrasonic pulsed doppler velocimeters. *IEEE Transactions Sonics & Ultrasonics, Vol. SU-28*, pp. 69–75, 1981.

[583] R. Hartley and K. Welles II. Recursive computation of the Fourier transform. *Proceedings of the 1990 IEEE International Symposium on Circuits and Systems*, pp. 1792–1795.

[584] Joseph Havlicek. **The Fourier Transform**. Morgan & Claypool Publishers, 2007.

[585] Stephen Hawking. **A Brief History of Time.** Bantam Books, 1996.

[586] S. Haykin. **Introduction to Adaptive Filters**. MacMillan, New York, 1976.

[587] M. Hedley, H. Yan, and D. Rosenfeld. A modified Gerchberg-Saxton algorithm for one-dimensional motion artifact correction in MRI. *IEEE Transactions on Signal Processing, Vol. 39*, pp. 1428–1433, 1991.

[588] Z.S. Hegedus. Annular pupil arrays; application to confocal scanning. *Opt. Acta*, pp. 815–826, 1985.

[589] S. Hein and A. Zakhor. Reconstruction of oversampled band-limited signals from $\sigma \Delta$ encoded binary sequences *IEEE Transactions on Signal Processing, Vol. 42(4)*, pp. 799–811, 1994.

[590] H.D. Helms and J.B. Thomas. Truncation error of sampling theorem expansion. *Proceedings IRE, Vol. 50*, pp. 179–184, 1962.

[591] C.W. Helstrom. An expansion of a signal in Gaussian elementary signals. *IEEE Transactions Information Theory, Vol. IT-12*, pp. 81–82, 1966.

[592] S.S. Hemami et al. Engineering Transform coded image reconstruction exploiting interblock correlation. *IEEE Transactions Image Processing, Vol. 4(7)*, pp. 1023–1027, 1995.

[593] J. Henderson and C.C. Tisdell. Topological transversality and boundary value problems on time scales. *Journal of Mathematical Analysis and Applications, Vol. 289(1)*, pp. 110–125, 2004.

[594] J. Herivel. The influence of Fourier on British mathematics. *Centaurus, Vol. 17(1)*, pp. 40–57, 1972.

[595] John Herivel. **Joseph Fourier: The Man and the Physicist**. Oxford University Press, 1975.

[596] G.T. Herman, ed., **Image Reconstruction from Projections** Springer-Verlag, Berlin, 1979.

[597] G.T. Herman, J. Zheng, and C.A. Bucholtz. Shape-based interpolation. *IEEE Computer Graphics and Applications Magazine, Vol. 12, No. 3*, pp. 69–79, May 1992.

[598] A.O. Hero III and D. Blatt. Sensor network source localization via projection onto convex sets (POCS). *Proceedings IEEE International Conference on Acoustics, Speech, and Signal Processing, Vol. 3*, pp. iii/689–iii/692, 2005.

[599] R.S. Hershel, P.W. Scott, and T.E. Shrode. Image sampling using phase gratings. *Journal of the Optical Society America, Vol. 66*, pp. 861–863, 1976.

[600] J.R. Higgins. An interpolation series associated with the Bessel-Hankel transform. *Journal London Math. Society, Vol. 5*, pp. 707–714, 1972.

[601] J.R. Higgins. A sampling theorem for irregularly spaced sample points. *IEEE Transactions Information Theory, Vol. IT-22*, pp. 621–622, 1976.

[602] J.R. Higgins. **Completeness and Basis Properties of Sets of Special Functions**. Cambridge University Press, Cambridge, 1977.

[603] J.R. Higgins. Five short stories about the cardinal series. *Bull. America Math. Society, Vol. 12*, pp. 45–89, 1985.

[604] J.R. Higgins. Sampling theorems and the contour integral method. *Applied Anal., Vol. 41*, pp. 155–169, 1991.

[605] J.R. Higgins **Sampling Theory in Fourier and Signal Analysis: Foundations.** Oxford University Press, 1996.

[606] J.R. Higgins and Rudolph L. Stens (Editors). **Sampling Theory in Fourier and Signal Analysis: Advanced Topics, Vol. 2**, Oxford University Press, 1999.

[607] S. Hilger. *Ein Maβkettenkalkulül mit Awendung auf Zentrumsmannigfaltigkeitem.* Ph.D. thesus, Univerität Würzburg, 1988.

[608] S. Hilger. Special Functions: Laplace and Fourier Transform on Measure Chains. *Dynamic Systems and Applications, Vol. 8*, pp. 471–488, 1999.

[609] S. Hilger. An Application of Calculus on Measure Chains to Fourier Theory and Heisenberg's Uncertainty Principle. *Journal of Difference Equations and Applications, Vol. 8*, pp. 897–936, 2002.

[610] N.R. Hill and G.E. Ioup. Convergence of the Van Cittert iterative method of deconvolution. *Journal of the Optical Society America, Vol. 66*, pp. 487–489, 1976.

[611] K. Hirano, M. Sakane, and M.Z. Mulk. Design of 3–D Recursive Digital Filters. *IEEE Transactions Circuits and Systems, Vol. 31*, pp. 550–561, 1984.

[612] E.W. Hobson. **The theory of functions of a real variable and the theory of Fourier's series.** Michigan Historical Reprint Series. Scholarly Publishing Office, University of Michigan Library, 2005.

[613] Robert V. Hogg, Allen Craig, and Joseph W. McKean. **Introduction to Mathematical Statistics** (6th Edition). Prentice Hall, 2004.

[614] E. Holzler. Some remarks on special cases of the sampling theorem. *Frequenz, Vol. 31*, pp. 265–269, 1977.

[615] I. Honda. Approximation for a class of signal functions by sampling series representation. *Keio Engineering Rep., Vol. 31*, pp. 21–26, 1978.

[616] I. Honda. An abstraction of Shannon's sampling theorem. *IEICE Transactions on Fundamentals of Electronics Communications and Computer Sciences, Vol. E81A(6)*, pp. 1187–1193, 1998.

[617] W. Hoppe. The finiteness postulate and the sampling theorem of the three dimensional electron microscopical analysis of aperiodic structures. *Optik, Vol. 29*, pp. 619–623, 1969.

[618] H. Horiuchi. Sampling principle for continuous signals with time-varying bands. *Information & Control, Vol. 13*, pp. 53–61, 1968.

[619] Lars Hörmander. **An Introduction to Complex Analysis in Several Variables**. Van Nostrand, Princeton, NJ, 1966.

[620] Lars Hörmander. **The Analysis of Linear Partial Differential Operators I: Distribution Theory and Fourier Analysis**. Springer, 2003.

[621] R.F. Hoskins and J.S. Pinto. Generalized sampling expansion in the sense of Papoulis. *SIAM J. Applied Math., Vol. 44*, pp. 611–617, 1984.

[622] R.F. Hoskins and J.S. Pinto. Sampling expansions and generalized translation invariance. *Journal Franklin Institute, Vol. 317*, pp. 323–332, 1984.

[623] G.H. Hostetter. Generalized interpolation of continuous-time signal samples. *IEEE Transactions Systems, Man, Cybernetics, Vol. SMC-11*, pp. 433–439, 1981.

[624] S.J. Howard. Continuation of discrete Fourier spectra using minimum-negativity constraint *Journal of the Optical Society America, Vol. 71*, pp. 819–824, 1981.

[625] S.J. Howard. Method for continuing Fourier spectra given by the fast Fourier transform. *Journal of the Optical Society America, Vol. 71*, pp. 95–98, 1981.

[626] S.J. Howard. Fast algorithm for implementing the minimum-negativity constraint for Fourier spectrum extrapolation. *Applied Optics, Vol. 25*, pp. 1670–1675, 1986.

[627] Kenneth B. Howell. **Principles of Fourier Analysis**. CRC, 2001.

[628] P. Hoyer. Simplified proof of the Fourier Sampling Theorem. *Information Processing Letters, Vol. 75(4)*, pp. 139–143, 2000.

[629] T.C. Hsiah. Comparisons of adaptive sampling control laws. *IEEE Transactions Automatic Control, Vol. AC-17*, pp. 830–831, 1972.

[630] T.C. Hsiah. Analytic design of adaptive sampling control law in sampled data systems. *IEEE Transactions Automatic Control, Vol. 19*, pp. 39–41, 1974.

[631] Hwei Hsu. **Applied Fourier Analysis**. Harcourt Brace Jovanovich, 1984.

[632] Y. Hu, K.D.K. Luk, W.W. Lu, A. Holmes, and J.C.Y. Leong. Comparison of time-frequency distribution techniques for analysis of spinal somatosensory evoked potential. *Medical & Biological Engineering & Computing, Vol. 39(3)*, pp. 375–380, 2001.

[633] Hua Huang, Xin Fan, Chun Qi, and Shi-Hua Zhu. A learning-based POCS algorithm for face image super-resolution reconstruction. *Proceedings of 2005 International Conference on Machine Learning and Cybernetics, Vol. 8*, pp. 5071–5076, 2005.

[634] E.X. Huang and A.K. Fung, An application of sampling theorem to moment method simulation in surface scattering. *Journal of Electromagnetic Waves and Applications, Vol. 20(4)*, pp. 531–546, 2006.

[635] F.O. Huck, N. Halyo, and S.K. Park. Aliasing and blurring in 2-D sampled imagery. *Applied Optics, Vol. 19*, pp. 2174–2181, 1980.

[636] F.O. Huck and S.K. Park. Optical-mechanical line-scan imaging process: Its information capacity and efficiency. *Applied Optics, Vol. 14*, pp. 2508–2520, 1975.

[637] A. Hughes. Cat retina and the Sampling Theorem-the Relation of Transient and Sustained Brisk Unit Cutoffs Frequency to Alpha Mode and Beta Node Cell Density. *Experimental Brain Research, Vol. 42(2)*, pp. 196–202, 1981.

[638] R. Hulthen. Restoring causal signals by analytic continuation: A generalized sampling theorem for causal signals. *IEEE Transactions Acoust., Speech, Signal Processing, Vol. ASSP-31*, pp. 1294–1298, 1983.

[639] H.E. Hurzcler. The Optional Sampling Theorem for Processes Indexed by a Partially Ordered Set. *Annals Of Probability, Vol. 13(4)*, pp. 1224–1235, 1985.

I

[640] Y. Ichioka and N. Nakajima. Iterative image restoration consideration visibility. *Journal of the Optical Society America, Vol. 71*, pp. 983–988, 1981.

[641] O. Ikeda and T. Sato. Super-resolution imaging system using waves with a limited frequency bandwidth. *Journal Acoust. Society America, Vol. 6*, pp. 75–81, 1979.

[642] Inna V. Ilyina, Alexander S. Sobolev, and Tatyana Yu. Cherezova; Alexis V. Kudryashov. Gerchberg-Saxton Iterative Algorithm for Flexible Mirror Performance. 8-th International Conference on Laser and Fiber-Optical Networks Modeling. pp. 442–445, 2006.

[643] G. Indebetow. Propagation of spatially periodic wavefields. *Opt. Acta.*, pp. 531–539, 1984.

[644] G. Indebetow. Polychromatic self-imaging. *Journal Mod. Opt.*, pp. 243–252, 1988.

[645] Leopold Infeld. **Quest: The Evolution of a Scientist**. Doubleday, Doran (1941). Reprinted Transaction Pub, 1990.

[646] Valéria de Magalhães Iorio. **Fourier Analysis and Partial Differential Equations: An Introduction**. Cambridge University Press, 2001.

[647] H. Ishida, **Fourier Transform–Infrared Characterization of Polymers**. Springer, 1987.

[648] M.E. Ismail and Ahmed I. Zayed. A q-analogue of the Whittaker-Shannon-Kotelnikov sampling theorem. *Proceedings of the American Mathematical Society, Vol. 131(12)*, pp. 3711–3719, 2003.

[649] S.H. Izen. Generalized sampling expansion on lattices. *IEEE Transactions of Signal Processing, Vol. 53(6)*, pp. 1949–1963, 2005.

[650] Leland B. Jackson. **Digital Filters and Signal Processing... with MATLAB Exercises**, Springer; 3 edition 1995.

[651] Dunham Jackson. **Fourier Series and Orthogonal Polynomials**. Dover Publications, 2004.

[652] J.F. James. **A Student's Guide to Fourier Transforms**. Cambridge University Press; 2 edition, 2002.

J

[653] M.A. Jack, P.M. Grant, and J.H. Collins. The theory, design, and applications of surface acoustic wave Fourier-transform processors. *Proceedings of the IEEE, Vol. 68(4)*, pp. 450–468, 1980.

[654] J.S. Jaffe. Limited angle reconstruction using stabilized algorithms. IEEE Transactions on Medical Imaging, *Vol. 9(3)*, pp. 338–344, 1990.

[655] D.L. Jagerman. Bounds for truncation error of the sampling expansion. *SIAM J. Applied Math., Vol. 14*, pp. 714–723, 1966.

[656] D.L. Jagerman. Information theory and approximation of bandlimited functions. *Bell Systems Tech. J., Vol. 49*, pp. 1911–1941, 1970.

[657] D.L. Jagerman and L. Fogel. Some general aspects of the sampling theorem. *IRE Transactions Information Theory, Vol. IT-2*, pp. 139–146, 1956.

[658] J. Jahns and A.W. Lohmann. The Lau effect (a diffraction experiment with incoherent illumination). *Optical Communication*, pp. 263–267, 1979.

[659] A.K. Jain. **Fundamentals of Digital Image Processing**. Prentice-Hall, Englewood Cliffs, NJ, 1989.

[660] A.K. Jain and S. Ranganath. Extrapolation algorithms for discrete signals with application to spectral estimation. *IEEE Transactions Acoust., Speech, Signal Processing, Vol. ASSP-29*, pp. 830–845, 1981.

[661] S. Jang, W. Choi, T.K. Sarkar, M. Salazar-Palma, K. Kyungjung, and C.E. Baum. Exploiting early time response using the fractional Fourier transform for analyzing transient radar returns. *IEEE Transactions on Antennas and Propagation, Vol. 52(11)*, pp. 3109–3121, 2004.

[662] T. Jannson. Shannon number of an image and structural information capacity in volume holography. *Opt. Acta*, pp. 1335–1344, 1980.

[663] A.J.E.M. Janssen. Weighted Wigner distributions vanishing on lattices. *Journal Math. Anal. Applied, Vol. 80*, pp. 156–167, 1981.

[664] A.J.E.M. Janssen. Gabor representation of generalized functions. *Journal Math. Anal. Applied, Vol. 83*, pp. 377–396, 1981.

[665] A.J.E.M. Janssen. Positivity of weighted Wigner distributions. *SIAM J. Math. Anal., Vol. 12*, pp. 752–758, 1981.

[666] A.J.E.M. Janssen. Bargmann transform, Zak transform and coherent states. *Journal Math. Phys., Vol. 23*, pp. 720–731, 1982.

[667] A.J.E.M. Janssen. The Zak transform: a signal transform for sampled time-continuous signals. *Philips J. Res., Vol. 43*, pp. 23–69, 1988.

[668] A.J.E. Janssen, R.N. Veldhuis, and L.B. Vries. Adaptive interpolation of discrete time signals that can be modeled as autoregressive processes. *IEEE Transactions Acoust., Speech, Signal Processing, Vol. ASSP-34*, pp. 317–330, 1986.

[669] P.A. Jansson, editor. **Deconvolution. With applications in Spectroscopy**. Academic Press, Orlando, FL, 1984.

[670] P.A. Jansson, R.H. Hunt, and E.K. Plyer. Resolution enhancement of spectra. *Journal of the Optical Society America, Vol. 60*, pp. 596–599, 1970.

[671] Y.C. Jenq. Digital spectra of nonuniformly sampled signals: Digital look up tunable sinusoidal oscillators. *IEEE Transactions Instrumentation & Measurement, Vol. 37, No. 3*, pp. 358–362, 1988.

[672] Yeonsik Jeong, Inkyeom Kim, and Hyunchul Kang. A practical projection-based postprocessing of block-coded images with fast convergence rate. *IEEE Transactions on Circuits and Systems for Video Technology, Vol. 10(4)*, pp. 617–623, 2000.

[673] A.J. Jerri. On the application of some interpolating functions in physics. *Journal Res. National Bureau of Standards -B. Math. Sciences, Vol. 73B, No. 3*, pp. 241–245, 1969.

[674] A.J. Jerri. On the equivalence of Kramer's and Shannon's generalized sampling theorems. *IEEE Transactions Information Theory, Vol. IT-15*, pp. 469–499, 1969.

[675] A.J. Jerri. Some applications for Kramer's generalized sampling theorem. *Journal Engineering Math., Vol. 3, No. 2*, April 1969.

[676] A.J. Jerri. Application of the sampling theorem to time-varying systems. *Journal Franklin Institute, Vol. 293*, pp. 53–58, 1972.

[677] A.J. Jerri. Sampling for not necessarily finite energy signals. *International J. System Sci., Vol. 4, No. 2*, pp. 255–260, 1973.

[678] A.J. Jerri. Sampling expansion for the L_v^α-Laguerre integral transform. *Journal Res. National Bureau of Standards -B. Math. Sciences, Vol. 80B*, pp. 415–418, 1976.

[679] A.J. Jerri. The Shannon sampling theorem-its various extensions and applications: A tutorial review. *Proceedings IEEE, Vol. 65, No. 11*, pp. 1565–1596, 1977.

[680] A.J. Jerri. Computation of the Hill functions of higher order. *Math. Comp., Vol. 31*, pp. 481–484, 1977.

[681] A.J. Jerri. Towards a discrete Hankel transform and its applications. *Journal of Applied Anal., Vol. 7*, pp. 97–109, 1978.

[682] A.J. Jerri. The application of general discrete transforms to computing orthogonal series and solving boundary value problems. *Bull. Calcutta Math. Society, Vol. 71*, pp. 177–187, 1979.

[683] A.J. Jerri. General transforms hill functions. *Journal Applicable Analysis, Vol. 14*, pp. 11–25, 1982.

[684] A.J. Jerri. A note on sampling expansion for a transform with parabolic cylinder kernel. *Inform. Sci., Vol. 26*, pp. 1–4, 1982.

[685] A.J. Jerri. On a recent application of the Shannon sampling theorem. In *IEEE International Symposium on Information Theory, Quebec, Canada*, 1983.

[686] A.J. Jerri. A definite integral. *SIAM Rev., Vol. 25, No.1*, p. 101, 1983.

[687] A.J. Jerri. Interpolation for the generalized sampling sum of approximation theory. *The AMS Summer Meeting, Albany, NY*, 1983.

[688] A.J. Jerri. Application of the transform–iterative method to nonlinear concentration boundary value problem. *Chem. Engineering Commun., Vol. 23*, pp. 101–113, 1983.

[689] A.J. Jerri. The generalized sampling theorem for transforms of not necessarily square integrable functions. *Journal Math., Math. Sci., Vol. 8*, pp. 355–358, 1985.

[690] A.J. Jerri. **Introduction to Integral Equations with Applications.** Marcel Dekker, New York, 1985.

[691] A.J. Jerri. Part II: The Sampling Expansion – A Detailed Bibliography. 1986.

[692] A.J. Jerri. An extended Poisson type sum formula for general integral transforms and aliasing error bound for the generalized sampling theorem. *Journal Applied Analysis, Vol. 26*, pp. 199–221, 1988.

[693] A.J. Jerri. **The Gibbs Phenomenon in Fourier Analysis, Splines and Wavelet**. Springer, 1998.

[694] A.J. Jerri. A recent modification of iterative methods for solving nonlinear problems. **The Math. Heritage of C.F. Gauss**, 1991, G.M. Rassias, editor, World Scientific Publ. Co., Singapore, pp. 379–404.

[695] A.J. Jerri. **Integral and Discrete Transforms with Applications and Error Analysis.** Marcel Dekker Inc., New York, 1992.

[696] A.J. Jerri and E.J. Davis. Application of the sampling theorem to boundary value problems. *Journal Engineering Math., Vol. 8*, pp. 1–8, 1974.

[697] A.J. Jerri and D.W. Kreisler. Sampling expansions with derivatives for finite Hankel and other transforms. *SIAM J. Math. Anal., Vol. 6*, pp. 262–267, 1975.

[698] A.J. Jerri. Truncation error for the generalized Bessel type sampling series. *Journal Franklin Institute (USA) Vol. 314, No. 5*, pp. 323–328, 1982.

[699] A.J. Jerri, R.L. Herman and R.H. Weiland. A modified iterative method for nonlinear chemical concentration in cylindrical and spherical pellets. *Chem. Engineering Commun., Vol. 52*, pp. 173–193, 1987.

[700] H. Johnen. Inequalities connected with moduli of smoothness. *Mat. Vesnik, Vol. 9(24)*, pp. 289–303, 1972.

[701] G. Johnson and J.H.S. Hutchinson. The limitations of NMR recalled–echo imaging techniques. *Journal Magnetic Resonance, Vol. 63*, pp. 14–30, 1985.

[702] Joint Photographic Experts Group, ISO-IEC/JTC1/SC2/WP8, "JPEG technical specification, revision 8," JPEG-8-R8, 1990.

[703] M.C. Jones. The discrete Gerchberg algorithm. *IEEE Transactions Acoust., Speech, Signal Processing, Vol. ASSP-34*, pp. 624–626, 1986.

[704] D.L. Jones, R.G. Baraniuk. An adaptive optimal-kernel time-frequency representation. *IEEE Transactions on Signal Processing, Vol. 43(10)*, pp. 2361–2371, 1995.

[705] I.I. Jouny. Scattering analysis using fractional Fourier with applications in mine detection. *Proceedings. 2004 IEEE International Geoscience and Remote Sensing Symposium, 2004. IGARSS '04. Vol. 5*, pp. 3460–3463, 2004.

K

[706] M. Kac, W.L. Murdock, and G. Szegö. On the eigenvalues of certain Hermitian forms. *Journal Rat. Mech., Anal.*, pp. 767–800, 1953.

[707] M.I. Kadec. The exact value of the Paley-Wiener constant *Soviet Math. Dokl., Vol. 5*, pp. 559–561, 1964.

[708] R.E. Kahn and B. Liu. Sampling representation and optimum reconstruction of signals. *IEEE Transactions Information Theory, Vol. IT-11*, pp. 339–347, 1965.

[709] T. Kailath. Channel characterization: time-variant dispersive channels. In **Lectures on Communications Systems Theory**, E.J. Baghdady, editor, McGraw-Hill, New York, 1960.

[710] S.C. Kak. Sampling theorem in Walsh-Fourier analysis. *Electron. Letters, Vol. 6*, pp. 447–448, 1970.

[711] Avinash C. Kak, Malcolm Slaney. **Principles of Computerized Tomographic Imaging.** Industrial & Applied Math (2001).

[712] G. Kakoza and D. Munson. A frequency-domain characterization of interpolation from nonuniformly spaced data. In *Proceedings of the International Symposium on Circuits and Systems, Portland, Oregon*, pp. 288–291, 1989.

[713] F. Kamalabadi and B. Sharif. Robust regularized tomographic imaging with convex projections. IEEE International Conference on Image Processing, 2005. ICIP 2005. *Vol. 2*, pp. II - 205–208, 2005.

[714] N.S. Kambo and F.C. Mehta. Truncation error for bandlimited non-stationary processes. *Inform. Sci., Vol. 20*, pp. 35–39, 1980.

[715] N.S. Kambo and F.C. Mehta. An upper bound on the mean square error in the sampling expansion for non-bandlimited processes. *Inform. Sci., Vol. 21*, pp. 69–73, 1980.

[716] E.O. Kamenetskii. Sampling theorem in macroscopic electrodynamics of crystal lattices. *Physical Review E, Vol. 57(3)*, pp. 3556–3562, Part B, 1998.

[717] David W. Kammler. **First Course in Fourier Analysis**. Prentice Hall, 2000.

[718] M. Kato, I. Yamada and K. Sakaniwa. A set-theoretic blind image deconvolution based on hybrid steepest descent method. *IEICE Transactions on Fundamentals of Electronics, Communications and Computer Sciences. Vol. E82A (8)*, pp. 1443–1449, 1999.

[719] K. Kazimierczuk, W. Kozminski, and I. Zhukov. Two-dimensional Fourier transform of arbitrarily sampled NMR data sets. *Journal of Magnetic Resonance, Vol. 179(2)*, pp. 323–328, 2006.

[720] L.M. Kani and J.C. Dainty. Super-resolution using the Gerchberg algorithm. *Optical Communication*, pp. 11–17, 1988.

[721] D. Kaplan and R.J. Marks II. Noise sensitivity of interpolation and extrapolation matrices. *Applied Optics, Vol. 21*, pp. 4489–4492, 1982.

[722] Saleem A. Kassam. **Signal Detection in Non-Gaussian Noise**. Springer-Verlag, 1988.

[723] R. Kasturi, T.T. Krile, and J.F. Walkup. Space-variant 2-D processing using a sampled input/sampled transfer function approach. *SPIE Inter. Optical Comp. Conf., Vol. 232*, pp. 182–190, 1980.

[724] Jyrki Kauppinen and Jari Partanen. **Fourier Transforms in Spectroscopy**. Wiley, 2001.

[725] M. Kawamura and S. Tanaka. Proof of sampling theorem in sequency analysis using extended Walsh functions. *Systems-Comput. Controls, Vol. 9*, pp. 10–15, 1980.

[726] Steven M. Kay. The effect of sampling rate on autocorrelation estimation. *IEEE Transactions Acoust., Speech, Signal Processing, Vol. ASSP-29, No. 4*, p. 859–867, 1981.

[727] R.B. Kerr. Polynomial interpolation errors for bandlimited random signals. *IEEE Transactions Systems, Man, Cybernetics*, pp. 773–774, 1976.

[728] R.B. Kerr. Truncated sampling expansions for bandlimited random signals. *IEEE Transactions Systems, Man, Cybernetics, Vol. SMC-9*, pp. 362–364, 1979.

[729] L.M. Khadra, J.A. Draidi, M.A. Khasawneh and M.M. Ibrahim. Time-frequency distributions based on generalized cone-shaped kernels for the representation of nonstationary signals. *Journal of the Franklin Institute-Engineering and Appled Mathematics. Vol. 335B(5)*, pp. 915–928, 1998.

[730] K. Khare and N. George. Sampling-theory approach to eigenwavefronts of imaging systems.*Journal of the Optical Society of America A-Optics, Image Science and Vision, Vol. 22(3)*, pp. 434–438, 2005.

[731] K. Khare. Bandpass sampling and bandpass analogues of prolate spheroidal functions. *Signal Processing, Vol. 86(7)*, pp. 1550–1558, 2006.

[732] A.A. Kharkevich. Kotelnikov's theorem-a review of some new publications. *Radiotekhnika, Vol. 13*, pp. 3–10, 1958.

[733] Y.I. Khurgin and V.P. Yakovlev. Progress in the Soviet Union on the theory and applications of bandlimited functions. *Proceedings IEEE, Vol. 65*, pp. 1005–1029, 1976.

[734] T. Kida. Generating functions for sampling theorems. *Electronics and Communications in Japan*, (English Translation), *Vol. 65(1)*, pp. 9–18, 1982.

[735] T. Kida. Extended formulas of sampling theorem on bandlimited waves and their application to the design of digital filters for data transmission. *Electron Communication Japan*, Part 1, *Vol. 68(11)*, pp. 20–29, 1985.

[736] Yoon Kim, Chun-Su Park and Sung-Jea Ko. Fast POCS based post-processing technique for HDTV. *IEEE Transactions on Consumer Electronics, Vol. 49(4)*, pp. 1438–1447, 2003.

[737] Yoon Kim, Chun-Su Park, Sung-Jea Ko. Frequency domain post-processing technique based on POCS. *Electronics Letters, Vol. 39(22)*, pp. 1583–1584, 2003.

[738] G.W. King and A.G. Emslie. Spectroscopy from the point of view of the communication theory I. *Journal of the Optical Society America, Vol. 41*, pp. 405–409, 1951.

[739] G.W. King and A.G. Emslie. Spectroscopy from the point of view of the communication theory III. *Journal of the Optical Society America, Vol. 43*, pp. 664–668, 1953.

[740] K. Kinoshita and M. Lindenbaum. Robotic control with partial visual information. *International Journal of Computer Vision, Vol. 37(1)*, pp. 65–78, 2000.

[741] J.G. Kirkwood. Quantum statistics if almost classical ensembles. *Phys. Rev., Vol.44*, pp. 31–37, 1933.

[742] S.J. Kishner and T.W. Barnard. Detection and location of point images in sampled optical systems. *Journal of the Optical Society America, Vol. 62*, pp. 17–20, 1972.

[743] Morris Kline. **Mathematical Thought from Ancient to Modern Times**, Oxford University Press, 1972.

[744] D. Klusch. The Sampling Theorem, Dirichlet Series and Hankel Transforms. *Journal of Computational and Applied Mathematics, Vol. 44(3)*, pp. 261–273, 1992.

[745] I. Kluvanek. Sampling theorem in abstract harmonic analysis. *Mat. Casopis Sloven. Akad. Vied, Vol. 15*, pp. 43–48, 1965.

[746] J.J. Knab. System error bounds for Lagrange estimation of bandlimited functions. *IEEE Transactions Information Theory, Vol. IT-21*, pp. 474–476, 1975.

[747] J.J. Knab. Interpolation of bandlimited functions using the approximate prolate series. *IEEE Transactions Information Theory, Vol. IT-25*, pp. 717–720, 1979.

[748] J.J. Knab. Simulated cardinal series errors versus truncation error bounds. *Proceedings IEEE, Vol. 68*, pp. 1020–1060, 1980.

[749] J.J. Knab. Noncentral interpolation of bandlimited signals. *IEEE Transactions Aerospace & Electronic Systems, Vol. AES-17*, pp. 586–591, 1981.

[750] J.J. Knab and M.I. Schwartz. A system error bound for self truncating reconstruction filter class. *IEEE Transactions Information Theory, Vol. IT-21*, pp. 341–342, 1975.

[751] I.P. Knyshev. The sampling theorem in amplitude, quantization of random signals. *Journal of Communications Technology and Electronics, Vol. 47(12)*, pp. 1361–1363, 2002.

[752] I.P. Knyshev. The sampling theorem in amplitude, quantization of random signals. *Journal of Communications Technology and Electronics, Vol. 47(12)*, pp. 1361–1363, 2002.

[753] Y. Kobayashi and M. Inoue. Combination of two discrete Fourier transforms of data sampled separately at different rates. *Journal Inf. Process, Vol. 3*, pp. 203–207, 1980.

[754] T. Kodama. Sampling theorem application to SAW diffraction simulation. *Electron. Letters, Vol. 16*, pp. 460–462, 1980.

[755] H. Kogelnik and T. Li. Laser beams and resonators. *Proceedings IEEE*, pp. 1312–1329, 1966.

[756] A. Kohlenberg. Exact interpolation of bandlimited functions. *Journal Applied Phys., Vol. 24*, pp. 1432–1436, 1953.

[757] D. Kohler and L. Mandel. Source reconstruction from the modulus of the correlation function: a practical approach to the phase problem of optical coherence theory. *Journal of the Optical Society America, Vol. 63*, pp. 126–134, 1973.

[758] K. Kojima. Sampling Theorem in Diffraction Integral Transform. *Japanese Journal of Applied Physics, Vol. 16(5)*, pp. 817–825, 1977.

[759] Chi-Wah Kok; Man-Wai Kwan. Sampling theorem for ISI-free communication. *Proceedings. IEEE International Symposium on Information Theory*, p. 99, 2003.

[760] Alexander Koldobsky. **Fourier Analysis In Convex Geometry**. American Mathematical Society 2005.

[761] A.N. Kolmogorov. Interpolation und Extrapolation von stationären zufälligen Folgen (Russian). *Bull Acad. Sci. USSR, Ser. Math, Vol. 5*, pp. 3–14, 1941.

[762] B.H. Kolner and D.M. Bloom. Direct electric-optic sampling of transmission-line signals propagating on a GaAs substrate. *Electron. Letters, Vol. 20*, pp. 818–819, 1984.

[763] A. Kolodziejczyk. Lensless multiple image formation by using a sampling filter. *Optical Communication*, pp. 97–102, 1986.

[764] Vilmos Komornik and Paola Loreti. **Fourier Series in Control Theory**. Springer, 2005.

[765] R. Konoun. PPM reconstruction using iterative techniques, project report. Technical report, Dept. of ECE, Illinois Institute of Technology, Chicago, September, 1989.

[766] P. Korn. Some uncertainty principles for time-frequency transforms of the Cohen class. *IEEE Transactions on Signal Processing, Vol. 53(2)*, Part 1, pp. 523–527, 2005.

[767] T.W. Körner. **Fourier Analysis**. Cambridge University Press; Reprint edition, 1989.

[768] A. Korpel. Gabor: frequency, time and memory. *Applied Optics*, pp. 3624–3632, 1982.

[769] Z. Kostic. Low-complexity equalization for $\frac{\pi}{4}$ DQPSK signals based on the method of projection onto convex sets. *IEEE Transactions on Vehicular Technology, Vol. 48(6)*, pp. 1916–1922, 1999.

[770] V.A. Kotelnikov. On the carrying capacity of "ether" and wire in electrocommunications. In *Idz. Red. Upr. Svyazi RKKA(Moscow)*, Material for the first all-union conference on the questions of communications, 1933.

[771] K.T. Kou and T. Qian. Shannon sampling and estimation of band-limited functions in the several complex variables setting. *Acta Mathematica Scientia, Vol. 25(4)*, pp. 741–754, 2005.

[772] H.P. Kramer. A generalized sampling theorem. *Journal Math. Phys., Vol. 38*, pp. 68–72, 1959.

[773] H.P. Kramer. The digital form of operators on bandlimited functions. *Journal Math. Anal. Applied, Vol. 44, No. 2*, pp. 275–287, 1973.

[774] G.M. Kranc. Comparison of an error-sampled system by a multirate controller. *Transactions America Institute Elec. Engineering, Vol. 76*, pp. 149–159, 1957.

[775] E. Krasnopevtsev. Fractional Fourier transform based on geometrical optics *Proceedings KORUS 2003. The 7th Korea-Russia International Symposium on Science and Technology, Vol. 3*, pp. 82–86, 2003.

[776] M.G. Krein. On a fundamental approximation problem in the theory of extrapolation and filtration of stationary random processes (Russian). *Dokl. Akad. Nauk SSSR Vol. 94*, pp. 13–16, 1954. [English translation: *Amer. Math. Society Selected Transl. Math. Statist. Prob. 4*, pp. 127–131], 1964.

[777] R. Kress. On the general Hermite cardinal interpolation. *Math. Comput., Vol. 26*, pp. 925–933, 1972.

[778] H.N. Kritikos and P.T. Farnum. Approximate evaluation of Gabor expansions. *IEEE Transactions Syst., Man, Cybern. SMC-17*, pp. 978–981, 1987.

[779] J.W. Krutch. "The Colloid and the Crystal", in I. Gordon and S. Sorkin (eds.) **The Armchair Science Reader**, New York: Simon and Schuster, 1959.

[780] A. Kumar and O.P. Malik. Discrete analysis of load-frequency control problem. *Proceedings IEEE, Vol. 131*, pp. 144–145, 1984.

[781] H.R. Kunsch, E. Agrell and F.A. Hamprecht. Optimal lattices for sampling. *IEEE Transactions on Information Theory, Vol. 51(2)*, pp. 634–647, 2005.

[782] G.M. Kurajian and T.Y. Na. Elastic beams on nonlinear continuous foundations. *ASME Winter Annual Meeting*, pp. 1–7, 1981.

[783] T.G. Kurtz. The Optional Sampling Theorem for Martengales Indexed by Directed Sets. *Annals of Probability, Vol. 8(4)*, pp. 675–681, 1980.

[784] H.J. Kushner and L. Tobias. On the stability of randomly sampled systems. *IEEE Transactions Automatic Control, Vol. AC-14*, pp. 319–324, 1969.

[785] A. Kutay, H.M. Ozaktas, O. Ankan, and L. Onural. Optimal filtering in fractional Fourier domains. *IEEE Transactions on Signal Processing, Vol. 45(5)*, pp. 1129–1143, 1997.

[786] Goo-Rak Kwon, Hyo-Kak Kim, Yoon Kim and Sung-Jea Ko. An efficient POCS-based post-processing technique using wavelet transform in HDTV. IEEE Transactions on Consumer Electronics, *Vol. 51(4)*, pp. 1283–1290, 2005.

[787] M.S. Kwon, Y.B. Cho and S.Y. Shin. Experimental demonstration of a long-period grating based on the sampling theorem. *Applied Physics Letters, Vol. 88(21)*, Art. No. 211103, 22 2006.

[788] M.S. Kwon and S.Y. Shin. Theoretical investigation of a notch filter using a long-period grating based on the sampling theorem. *Optics Communications, Vol. 263(2)*, pp. 214–218, 2006.

L

[789] B. Lacaze. A generalization of n-th sampling formula. *Ann. Telecomm., Vol. 30*, pp. 208–210, 1975.

[790] P. Lancaster and M. Tiemenesky. **The Theory of Matrices.** Academic Press, Orlando, FL, 1985.

[791] H.J. Landau. On the recovery of a bandlimited signal after instantaneous companding and subsequent band limiting. *Bell Systems Tech. J., Vol. 39*, pp. 351–364, 1960.

[792] H.J. Landau. Necessary density conditions for sampling and interpolation of certain entire functions. *Acta Math.*, pp. 37–52, 1967.

[793] H.J. Landau. Sampling, data transmission and the Nyquist rate. *Proceedings IEEE, Vol. 55*, pp. 1701–1706, 1967.

[794] H.J. Landau. On Szegö's eigenvalue distribution theorem and non-Hermitian kernels. *Journal Analyse Math.*, pp. 335–357, 1975.

[795] H.J. Landau and W.L. Miranker. The recovery of distorted bandlimited signals. *Journal Math. Anal. & Applied, Vol. 2*, pp. 97–104, 1961.

[796] H.J. Landau and H.O. Pollak. Prolate spheroidal wave functions Fourier analysis and uncertainty II. *Bell Systems Tech. J., Vol. 40*, pp. 65–84, 1961.

[797] H.J. Landau and H.O. Pollak. Prolate spheroidal wave functions Fourier analysis and uncertainty III: The dimension of the space of essentially time- and bandlimited signals. *Bell Systems Tech. J., Vol. 41*, pp. 1295–1336, 1962.

[798] H.J. Landau and H. Widom. Eigenvalue distribution of time and frequency limiting. *Journal Math. Anal., Applied, Vol. 77*, pp. 469–481, 1980.

[799] A. Landé. Optik und Thermodynamik, In **Handbuch der Physik**, Springer-Verlag, Berlin, pp. 453–479, 1928.

[800] L. Landweber. An iteration formula for Fredholm integral equations of the first kind. *Amer. J. Math.*, pp. 615–624, 1951.

[801] P.J. La Riviere and X.C. Pan. Interlaced interpolation weighting functions for multislice helical computed tomography. *Optical Engineering, Vol. 42(12)*, pp. 3461–3470, 2003.

[802] Rupert Lasser. **Introduction to Fourier Series**. CRC, 1996.

[803] B.P. Lathi. **Communication Systems**. John Wiley & Sons, Inc., New York, 1965.

[804] K.N. Le, K.P. Dabke and G.K. Egan. Signal detection using time-frequency distributions with nonunity kernels. *Optical Engineering, Vol. 40(12)*, pp. 2866–2877, 2001.

[805] Thuyen Le and M. Glesner. Rotating stall analysis using signal-adapted filter bank and Cohen's time-frequency distributions. *Proceedings of the 2000 IEEE International Symposium on Circuits and Systems*, ISCAS 2000 Geneva. *Vol. 1*, 28–31, pp. 603–606, 2000.

[806] A.J. Lee. On bandlimited stochastic processes. *SIAM J. Applied Math., Vol. 30*, pp. 269–277, 1976.

[807] A.J. Lee. Approximate interpolation and the sampling theorem. *SIAM J. Applied Math., Vol. 32*, pp. 731–744, 1977.

[808] A.J. Lee. Sampling theorems for nonstationary random processes. *Transactions Amer. Math. Society, Vol. 242*, pp. 225–241, 1978.

[809] A.J. Lee. A note on the Campbell sampling theorem. *SIAM J. Applied Math., Vol. 41*, pp. 553–557, 1981.

[810] H. Lee. On orthogonal transformations. *IEEE Transactions Circuits & Systems, Vol. CAS-32*, pp. 1169–1177, 1985.

[811] Y.W. Lee. **Statistical Theory of Communication**. John Wiley & Sons, New York, 1960.

[812] W.H. Lee. Computer-generated holograms: techniques and applications, In **Progress in Optics**, North-Holland, Amsterdam, 1978, pp. 119–232.

[813] A.J. Lee. Approximate Interpolation and Sampling Theorem. *SIAM Journal on Applied Mathematics, Vol 32(4)*, pp. 731–744 1977.

[814] S. Lee, P.S. Cho, R.J. Marks II, and S. Oh. Conformal radiotherapy computation by the method of alternating projection onto convex sets. *Phys. Med. Biol., Vol. 42*, pp. 1065–1086, 1997.

[815] X. Lee, Y.-Q. Zhang, and A. Leon-Garcia. Information loss recovery for block-based Image coding techniques-A fuzzy logic approach. *IEEE Transactions Image Processing, Vol. 4*, pp. 259–273, 1995.

[816] J. Leibrich and H. Puder. A TF distribution for disturbed and undisturbed speech signals and its application to noise reduction. *Signal Processing. Vol. 80(9)*, pp. 1761–1776, 2000.

[817] E.N. Leith and J. Upatnieks. Reconstructed wavefronts and communication theory. *Journal of the Optical Society America*, pp. 1123–1130, 1962.

[818] O.A.Z. Leneman. Random sampling of random processes: impulse processes. *Information & Control, Vol. 9*, pp. 347–363, 1966.

[819] O.A.Z. Leneman. On error bounds for jittered sampling. *IEEE Transactions Automatic Control, Vol. AC-11*, p. 150, 1966.

[820] O.A.Z. Leneman. Random sampling of random processes: optimum linear interpolation. *Journal Franklin Institute, Vol. 281*, pp. 302–314, 1966.

[821] O.A.Z. Leneman and J. Lewis. A note on reconstruction for randomly sampled data. *IEEE Transactions Automatic Control, Vol. AC-10*, p. 626, 1965.

[822] O.A.Z. Leneman and J. Lewis. Random sampling of random processes: Mean square comparison of various interpolators. *IEEE Transactions Automatic Control, Vol. AC-10*, pp. 396–403, 1965.

[823] O.A.Z. Leneman and J. Lewis. On mean-square reconstruction error. *IEEE Transactions Automatic Control, Vol. AC-11*, pp. 324–325, 1966.

[824] A. Lent and H. Tuy. An iterative method for the extrapolation of bandlimited functions. *Journal Math. Anal., Applied*, pp. 554–565, 1981.

[825] J. LeRoux, P. Lise, E. Zerbib, *et al.* A formulation in concordance with the sampling theorem for band-limited images reconstruction from projections. *Multidimensional Systems and Signal Processing, Vol. 7(1)*, pp. 27–52, 1996.

[826] Pete E. Lestrel (Editor). **Fourier Descriptors and their Applications in Biology**. Cambridge University Press, 1997.

[827] Emmanuel Letellier. **Fourier Transforms of Invariant Functions on Finite Reductive Lie Algebras**. Springer, 2005.

[828] L. Levi. Fitting a bandlimited signal to given points. *IEEE Transactions Information Theory, Vol. IT-11*, pp. 372–376, 1965.

[829] N. Levinson. Gap and density theorems. In *Colloq. Pub. 26*. Amer. Math. Society, New York, 1940.

[830] M. Levonen and S. McLaughlin. Fractional Fourier transform techniques applied to active sonar. *Proceedings OCEANS 2003. Vol. 4*, pp. 1894–1899, 2003.

[831] Jian Li and Petre Stoica. **Robust Adaptive Beamforming**. Wiley-Interscience, 2005.

[832] G.B. Lichtenberger. A note on perfect predictability and analytic processes. *IEEE Transactions Information Theory, Vol. IT-20*, pp. 101–102, 1974.

[833] B. Lien and G. Tang. Reversed Chebyshev implementation of McClellan transform and its roundoff error. *IEEE Transactions on Acoustics, Speech, and Signal Processing, Vol. 35(10)*, pp. 1435–1439, 1987.

[834] A. Wee-Chung Liew and Hong Yan. POCS-based blocking artifacts suppression with region smoothness constraints for graphic images. *Proceedings of 2004 International Symposium on Intelligent Multimedia, Video and Speech Processing*, pp. 563–566, 2004.

[835] Y. Linde, A. Buzo and R.M. Gray. An algorithm for vector quantizer design. *IEEE Transactions Commun., Vol. COM-28*, pp. 89–95, 1980.

[836] D.A. Linden. A discussion of sampling theorems. *Proceedings IRE, Vol. 47*, pp. 1219–1226, 1959.

[837] D.A. Linden and N.M. Abramson. A generalization of the sampling theorem. *Information & Control, Vol. 3*, pp. 26–31, 1960.

[838] E.H. Linfoot. Information theory and optical images. *Journal of the Optical Society America*, pp. 808–819, 1955.

[839] E.H. Linfoot. Quality evaluation of optical systems. *Opt. Acta*, pp. 1–14, 1958.

[840] John Litva. **Digital Beamforming in Wireless Communications**. Artech House Publishers, 1996.

[841] C.L. Liu and Jane W.S. Liu. **Linear Systems Analysis**. McGraw-Hill, New York, 1975.

[842] Y.M. Liu. A distributional sampling theorem. *SIAM Journal on Mathematical Analysis, Vol. 27(4)*, pp. 1153–1157, 1996.

[843] S.P. Lloyd. A sampling theorem for stationary (wide sense) stochastic processes. *Transactions America Math. Society, Vol. 92*, pp. 1–12, 1959.

[844] C. Loana, A. Quinquis, and Y. Stephan. Feature extraction from underwater signals using time-frequency warping operators. *IEEE Journal of Oceanic Engineering, Vol. 31(3)*, pp. 628–645, 2006.

[845] Robert P. Loce and Ronald E. Jodoin. Sampling theorem for geometric moment determination and its application to a laser beam position detector. *Applied Optics, Vol. 29, No. 26*, pp. 3835–3843, 1990.

[846] P.P. Loesberg. Low-frequency A-D input systems. *Instrum. Control Syst., Vol. 43*, pp. 124–126, 1970.

[847] A.W. Lohmann and D.P. Paris. Binary Fraunhofer holograms, generated by computer. *Applied Optics*, pp. 1739–1748, 1967.

[848] A.W. Lohmann. The space-bandwidth product applied to spatial filtering and to holography. *IBM Research Paper, RJ-438*, 1967.

[849] A.W. Lohmann and D.P. Paris. Computer generated spatial filters for coherent optical data processing. *Applied Optics*, pp. 651–655, 1968.

[850] A.W. Lohmann. An interferometer based on the Talbot effect. *Optical Communication*, pp. 413–415, 1971.

[851] A.W. Lohmann. Three-dimensional properties of wave-fields. *Optik*, pp. 105–117, 1978.

[852] P.J. Loughlin and G.D. Bernard. Cohen-Posch (positive) time-frequency distributions and their application to machine vibration analysis. *Mechanical Systems and Signal Processing. Vol. 11(4)*, pp. 561–576, 1997.

[853] P.J. Loughlin and K.L. Davidson. Modified Cohen-Lee time-frequency distributions and instantaneous bandwidth of multicomponent signals. *IEEE Transactions on Signal Processing, Vol. 49(6)*, pp. 1153–1165, 2001.

[854] D.G. Luenberger. **Optimization by Vector Space Methods**. Wiley, New York, 1969.

[855] H.D. Luke. Zur Entsehung des Abtasttheorems. *Nachr. Techn. Z., Vol. 31*, pp. 271–274, 1978.

[856] H.D. Luke. The origins of the sampling theorem *IEEE Communications Magazine, Vol. 37(4)*, pp. 106–108, 1999.

[857] W. Lukosz. Optical systems with resolving powers exceeding the classical limit. *Journal of the Optical Society America, Vol. 56*, pp. 1463–1472, 1966.

[858] W. Lukosz. Optical systems with resolving powers exceeding the classical limit, II. *Journal of the Optical Society America, Vol. 57*, pp. 932–941, 1967.

[859] A. Luthra. Extension of Parseval's relation to nonuniform sampling. *IEEE Transactions Acoust., Speech, Signal Processing, Vol. ASSP-36, No. 12*, pp. 1909–1911, 1988.

M

[860] Y.J. Ma and J.T. Sun. Stability criteria of delay impulsive systems on time scales. *Nonlinear Analysis–Theory Mathods & Applications, Vol. 67(4)*, pp. 1181–1189, 2007.

[861] D.M. Mackay. Quantal aspects of scientific information. *Phil. Mag.*, pp. 289–311, 1950.

[862] D.M. Mackay. The structural information-capacity of optical instruments. *Information & Control*, pp. 148–152, 1958.

[863] J. Maeda and K. Murata. Digital restoration of incoherent bandlimited images. *Applied Optics*, pp. 2199–2204, 1982.

[864] H. Maitre. Iterative super-resolution. Some new fast methods. *Opt. Acta*, pp. 973–980, 1981.

[865] Robert J. Mailloux. **Phased Array Antenna Handbook**, Second Edition. Artech House Publishers; 2 edition, 2005.

[866] J. Makhoul. Linear prediction: A tutorial review. *Proceedings IEEE, Vol. 63*, pp. 561–580, 1975.

[867] T. Mallon. **A Book of One's Own**. Ticknor & Fields, New York, 1984.

[868] N.J. Malloy. Nonuniform sampling for high resolution spectrum analysis. In *Proceedings ICASSP'84*, 1984.

[869] L. Mandel and E. Wolf. Coherence properties of optical fields. *Rev. Mod. Phys.*, pp. 231–287, 1965.

[870] L. Mandel and E. Wolf. Spectral coherence and the concept of cross-spectral purity. *Journal of the Optical Society America, Vol. 66*, pp. 529–535, 1976.

[871] Thomas Mann. **The Magic Mountain.** Vintage; 1st Vintage International Edition, 1996.

[872] M. Maqusi. A sampling theorem for dyadic stationary processes. *IEEE Transactions Acoust. Speech, Signal Processing, Vol. ASSP-26*, pp. 265–267, 1978.

[873] M. Maqusi. Truncation error bounds for sampling expansions of sequency band limited signals. *IEEE Transactions Acoust., Speech, Signal Processing, Vol. ASSP-26*, pp. 372–374, 1978.

[874] M. Maqusi. Sampling representation of sequency bandlimited nonstationary random processes. *IEEE Transactions Acoust., Speech, Signal Processing, Vol. ASSP-28*, pp. 249–251, 1980.

[875] M. Maqusi. Correlation and spectral analysis of nonlinear transformation of sequency bandlimited signals. *IEEE Transactions Acoust., Speech, Signal Processing, Vol. ASSP-30*, pp. 513–516, 1982.

[876] I. Maravic and M. Vetterli. Sampling and reconstruction of signals with finite rate of innovation in the presence of noise. *IEEE Transactions on Signal Processing, Vol. 52(8)*, Part 1, pp. 2788–2805, 2005.

[877] E.W. Marchand and E. Wolf. Radiometry with sources of any state of coherence. *Journal of the Optical Society America, Vol. 64*, pp. 1219–1226, 1974.

[878] A. Maréchal and M. Françon. **V Diffraction, structure des images**. Masson, Paris, 1970.

[879] H. Margenau and R.N. Hill. Correlation between measurements in quantum theory. *Progress in Theoretical Physics, Vol. 26*, pp. 722–738, 1961.

[880] C.P. Mariadassou and B. Yegnanarayana. Image reconstruction from noisy digital holograms. *IEE Proceedings F on Radar and Signal Processing, Vol. 137(5)*, pp. 351–356, 1990.

[881] J.W. Mark and T.D. Todd. A nonuniform sampling approach to data compression. *IEEE Transactions Communications, Vol. COM-29*, pp. 24–32, 1981.

[882] W.D. Mark. Spectral analysis of the convolution and filtering of non-stationary stochastic processes. *Journal Sound Vib. Vol. 11*, pp. 19–63, 1970.

[883] J.D. Markel and H.H. Gray, Jr. **Linear Prediction of Speech**. Springer-Verlag, New York, 1982.
[884] R.J. Marks II and T.F. Krile. Holographic representations of space-variant systems: System theory. *Applied Optics, Vol. 15*, pp. 2241–2245, 1976.
[885] R.J. Marks II, J.F. Walkup, and M.O. Hagler. A sampling theorem for space-variant systems. *Journal of the Optical Society America, Vol. 66*, pp. 918–921, 1976.
[886] R.J. Marks II, J.F. Walkup, and T.F. Krile. Ambiguity function display: An improved coherent processor. *Applied Optics Vol. 16*, pp. 746–750, 1977.
[887] R.J. Marks II, J.F. Walkup, and M.O. Hagler. Line spread function notation. *Applied Optics, Vol. 15*, pp. 2289–2290, l976.
[888] R.J. Marks II, J.F. Walkup, and M.O. Hagler. Volume hologram representation of space-variant systems. In **Applications of Holography and Optical Data Processing**, E. Marom, A.A. Friesem, and E. Wiener-Aunear, editors, Oxford: Pergamon Press, pp. 105–113, 1977.
[889] R.J. Marks II, J.F. Walkup, and M.O. Hagler. Sampling theorems for linear shift-variant systems. *IEEE Transactions Circuits & Systems, Vol. CAS-25*, pp. 228–233, 1978.
[890] R.J. Marks II, G.L. Wise, D.G. Haldeman, and J.L. Whited. Detection in Laplace noise. *IEEE Transactions on Aerospace and Electronic Systems, Vol. AES-14*, pp. 866–872, 1978.
[891] R.J. Marks II and J.N. Larson. One-dimensional Mellin transformation using a single optical element. *Applied Optics, Vol. 18*, pp. 754–755, 1979.
[892] R.J. Marks II. Two-dimensional coherent space-variant processing using temporal holography. *Applied Optics, Vol. 18*, pp. 3670–3674, 1979.
[893] R.J. Marks II, J.F. Walkup, and M.O. Hagler. Methods of linear system characterization through response cataloging. *Applied Optics, Vol. 18*, pp. 655–659, 1979.
[894] R.J. Marks II and M.W. Hall. Ambiguity function display using a single 1-D input. in **SPIE Milestone Series: Phase Space Optics**, Markus Testorf, Jorge Ojeda-Castañeda, and Adolf Lohmann, Editors, (The Society of Photo-Optical Instrumentation Engineers, Bellingham, WA, 2006) reprinted from *Applied Optics, Vol. 18(15)*, pp. 2539–2540, 1979.
[895] R.J. Marks II and D.K. Smith. An iterative coherent processor for bandlimited signal extrapolation. *Proceedings of the 1980 International Computing Conference*, Washington D.C., 1980.
[896] R.J. Marks II. Coherent optical extrapolation of two-dimensional signals: processor theory. *Applied Optics, Vol. 19*, pp. 1670–1672, 1980.
[897] R.J. Marks II. Gerchberg's extrapolation algorithm in two dimensions. *Applied Optics, Vol. 20*, pp. 1815–1820, 1981.
[898] R.J. Marks II. Sampling theory for linear integral transforms. *Opt. Letters, Vol. 6*, pp. 7–9, 1981.
[899] R.J. Marks II and M.J. Smith. Closed form object restoration from limited spatial and spectral information. *Opt. Letters, Vol. 6*, pp. 522–524, 1981.
[900] R.J. Marks II and M.W. Hall. Differintegral interpolation from a bandlimited signal's samples. *IEEE Transactions Acoust., Speech, Signal Processing, Vol. ASSP-29*, pp. 872–877, 1981.
[901] R.J. Marks II. Posedness of a bandlimited image extension problem in tomography. *Opt. Letters, Vol. 7*, pp. 376–377, 1982.
[902] R.J. Marks II. Restoration of continuously sampled bandlimited signals from aliased data. *IEEE Transactions Acoust., Speech, Signal Processing, Vol. ASSP-30*, pp. 937–942, 1982.
[903] R.J. Marks II and D. Kaplan. Stability of an algorithm to restore continuously sampled bandlimited images from aliased data. *Journal of the Optical Society America, Vol. 73*, pp. 1518–1522, 1983.
[904] R.J. Marks II. Noise sensitivity of bandlimited signal derivative interpolation. *IEEE Transactions Acoust., Speech, Signal Processing, Vol. ASSP-31*, pp. 1029–1032, 1983.
[905] R.J. Marks II. Restoring lost samples from an oversampled bandlimited signal. *IEEE Transactions Acoust., Speech, Signal Processing, Vol. ASSP-31*, pp. 752–755, 1983.
[906] R.J. Marks II and D. Radbel. Error in linear estimation of lost samples in an oversampled bandlimited signal. *IEEE Transactions Acoust., Speech, Signal Processing, Vol. ASSP-32*, pp. 648–654, 1984.

[907] R.J. Marks II and D.K. Smith. Gerchberg-type linear deconvolution and extrapolation algorithms. In **Transformations in Optical Signal Processing**, W.T. Rhodes, J.R. Fienup, and B.E.A. Saleh, editors, *Vol. 373*, pp. 161–178, 1984.

[908] R.J. Marks II and S.M. Tseng. Effect of sampling on closed form bandlimited signal interval interpolation. *Applied Optics, Vol. 24*, pp. 763–765, Erratum, *Vol. 24*, p. 2490, 1985.

[909] R.J. Marks II. A class of continuous level associative memory neural nets. in **SPIE Milestone Series: Selected Papers in Optical Neural Networks** edited by Suganda Jutamulia (The Society of Photo-Optical Instrumentation Engineers, Bellingham, WA, 1994), pp. 331–336; reprinted from *Applied Optics, Vol. 26*, pp. 2005–2009, 1987.

[910] R.J. Marks II. Linear coherent optical removal of multiplicative periodic degradations: processor theory. *Opt. Engineering, Vol. 23*, pp. 745–747, 1984.

[911] R.J. Marks II. Multidimensional signal sample dependency at Nyquist densities. *Journal of the Optical Society America A, Vol. 3*, pp. 268–273, 1986.

[912] R.J. Marks II. Class of continuous level associative memory neural nets. *Applied Optics*, pp. 2005–2010, 1987.

[913] R.J. Marks II, S. Oh, and L.E. Atlas. Alternating projection neural networks. *IEEE Transactions on Circuits and Systems, Vol. 36*, pp. 846–857, 1989.

[914] R.J. Marks II, L.E. Atlas, and S. Oh. Optical neural net memory. U.S. Patent No. 4,849,940 (assigned to the Washington Technology Center, University of Washington, Seattle), 1989.

[915] R.J. Marks II. **Introduction to Shannon Sampling and Interpolation Theory**. Springer-Verlag, New York, 1991.

[916] R.J. Marks II, Editor, **Advanced Topics in Shannon Sampling and Interpolation Theory**. Springer-Verlag, 1993.

[917] R.J. Marks II, Loren Laybourn, Shinhak Lee, and Seho Oh. Fuzzy and extra-crisp alternating projection onto convex sets (POCS). *Proceedings of the International Conference on Fuzzy Systems* (FUZZ-IEEE), Yokohama, Japan, pp. 427–435, 1995.

[918] R.J. Marks II, Alternating Projections onto Convex Sets, in **Deconvolution of Images and Spectra**, edited by Peter A. Jansson, (Academic Press, San Diego), pp. 476–501, 1997.

[919] R.J. Marks II. *Method and Apparatus for Generating Sliding Tapered Windows and Sliding Window Transforms*. U.S. Patent No. 5,373,460, 1994.

[920] R.J. Marks II, B.B. Thompson, M.A. El-Sharkawi, W.L.J. Fox, and R.T. Miyamoto. Stochastic resonance of a threshold detector: image visualization and explanation. *IEEE Symposium on Circuits and Systems*, ISCAS, pp. IV-521–523, 2002.

[921] R.J. Marks II, Ian A. Gravagne, and John M. Davis, "A Generalized Fourier Transform and Convolution on Time Scales," Journal of Mathematical Analysis and Applications, *Vol. 340(2)*, pp. 901–919, 2008.

[922] R.J. Marks II, Ian Gravagne, John M. Davis, and Jeffrey J. DaCunha. Nonregressivity in switched linear circuits and mechanical systems. *Mathematical and Computer Modelling, Vol. 43*, pp. 1383–1392, 2006.

[923] L. Marple, T. Brotherton, R. Barton, E. Lugo, and D. Jones. Travels through the time-frequency zone: advanced Doppler ultrasound processing techniques. 1993 Conference Record of The Twenty-Seventh Asilomar Conference on Signals, Systems and Computers, 1993. 1–3 *Vol.2*, pp. 1469–1473, 1993.

[924] M. Marques, A. Neves, J.S. Marques *et al.* The Papoulis-Gerchberg algorithm with unknown signal bandwidth. *Lecture Notes in Computer Science*. 4141, pp. 436–445, 2006.

[925] Alan G. Marshall. **Fourier, Hadamard, and Hilbert Transforms in Chemistry**. Perseus Publishing, 1982.

[926] A.G. Marshall and F.R. Verdun. **Fourier transforms in NMR, Optical and Mass Spectrometry**. Elsevier, Amsterdam, 1990.

[927] R.J. Martin. Volterra system identification and Kramer's sampling theorem. *IEEE Transactions on Signal Processing, Vol. 47(11)*, pp. 3152–3155, 1999.

[928] H.G. Martinez and T.W. Parks. A class of infinite-duration impulse response digital filters for sampling rate reduction. *IEEE Transactions Acoust., Speech, Signal Processing, Vol. ASSP-27*, pp. 154–162, 1979.

[929] S.A. Martucci. Symmetric convolution and the discrete sine and cosine transforms. *IEEE Transactions Sig. Processing*, SP-42, 1038–1051 1994.

[930] F. Marvasti. The extension of Poisson sum formula to nonuniform samples. *Proceedings 5th Aachener Kolloquium, Aachen, Germany*, 1984.

[931] F. Marvasti. A note on modulation methods related to sine wave crossings. *IEEE Transactions Communications*, pp. 177–178, 1985.

[932] F. Marvasti. Reconstruction of a signal from the zero-crossings of an FM signal. *Transactions IECE of Japan*, p. 650, 1985.

[933] F. Marvasti. Signal recovery from nonuniform samples and spectral analysis of random samples. In *IEEE Proceedings on ICASSP, Tokyo*, pp. 1649–1652, 1986.

[934] F. Marvasti. Spectral analysis of random sampling and error free recovery by an iterative method. *IECE Transactions of Institute Electron. Commun. Engs., Japan (Section E), Vol. E 69, No. 2*, 1986.

[935] F. Marvasti. A unified approach to zero-crossings, nonuniform and random sampling of signals and systems. In *Proceedings of the International Symposium on Signal Processing and Its Applications, Brisbane, Australia*, pp. 93–97, 1987.

[936] F. Marvasti. Minisymposium on zero-crossings and nonuniform sampling, In *Abstract of 4 presentations in the Proceedings of SIAM Annual Meeting, Minneapolis, Minnesota*, 1988.

[937] F. Marvasti. Analysis and recovery of sample-and-hold signals with irregular samples. In *Annual Proceedings of the Allerton Conference on Communication, Control and Computing*, 1989.

[938] F. Marvasti. Minisymposium on nonuniform sampling for 2-D signals. In *Abstract of 4 presentations in the Proceedings of SIAM Annual Meeting, San Diego, California*, 1989.

[939] F. Marvasti. Rebuttal on the comment on the properties of two-dimensional bandlimited signals. *Journal of the Optical Society America A, Vol. 6, No. 9*, p. 1310, 1989.

[940] F. Marvasti. Reconstruction of 2-D signals from nonuniform samples or partial information. *Journal of the Optical Society America A, Vol. 6*, pp. 52–55, 1989.

[941] F. Marvasti. Relationship between discrete spectrum of frequency modulated (FM) signals and almost periodic modulating signals. *Transactions IECE Japan*, pp. 92–94, 1989.

[942] F. Marvasti. Spectral analysis of nonuniform samples of irregular samples of multidimensional signals. In *6th Workshop on Multidimensional Signal Processing, California*, 1989.

[943] F. Marvasti. Extension of Lagrange interpolation to 2-D signals in polar coordinates. *IEEE Transactions Circuits & Systems*, 1990.

[944] F. Marvasti. The interpolation of 2-D signals from their isolated zeros. *Journal Multidimensional Systems and Signal Processing*, 1990.

[945] F. Marvasti and M. Analoui. Recovery of signals from nonuniform samples using iterative methods. In *IEEE Proceedings of International Conference on Circuits and Systems, Oregon*, 1989.

[946] F. Marvasti, M. Analoui, and M. Gamshadzahi. Recovery of signals from nonuniform samples using iterative methods. *IEEE Transactions Acoust., Speech, Signal Processing*, 1991.

[947] F. Marvasti, Peter Clarkson, Dobcik Miraslov, and Chuande Liu. Speech recovery from missing samples. In *Proceedings IASTED Conference on Control and Modeling, Tehran, Iran*, 1990.

[948] F. Marvasti and Reda Siereg. Digital signal processing using FM zero-crossing. In *Proceedings of IEEE International Conference on Systems Engineering, Wright State University*, 1989.

[949] F.A. Marvasti. Spectrum of nonuniform samples. *Electron. Letters, Vol. 20, No. 2*, 1984.

[950] F.A. Marvasti. Comments on a note on the predictability of bandlimited processes. *Proceedings IEEE, Vol. 74*, p. 1596, 1986.

[951] F.A. Marvasti. *A Unified Approach to Zero-Crossing and Nonuniform Sampling of Single and Multidimensional Signals and Systems*. Nonuniform Publications, Oak Park, IL, 1987.

[952] F.A. Marvasti and A.K. Jain. Zero crossings, bandwidth compression and restoration of nonlinearly distorted bandlimited signals. *Journal of the Optical Society America A, Vol. 3*, pp. 651–654, 1986.

[953] F.A. Marvasti. Extension of Lagrange interpolation to 2-D nonuniform samples in polar coordinates. *IEEE Transactions Circuits & Systems, Vol. 37, No. 4*, pp. 567–568, 1990.

[954] F.A. Marvasti and Liu Chuande. Parseval relationship of nonuniform samples of one and two-dimensional signals. *IEEE Transactions Acoust., Speech, Signal Processing, Vol. 38, No. 6*, pp. 1061–1063, 1990.

[955] F.A. Marvasti. Oversampling as an alternative to error correction codes. *Minisymposium on Sampling Theory and Practice, SIAM Annual Meeting, Chicago, IL*, July 1990. Also in ICC'92, Chicago, IL, 1992.

[956] F.A. Marvasti and T.J. Lee. Analysis and recovery of sample-and -hold and linearly interpolated signals with irregular samples. *IEEE Transactions Signal Processing, Vol. 40, No. 8*, pp. 1884–1891, 1992.

[957] Farokh A. Marvasti. **Nonuniform Sampling: Theory and Practice (Information Technology: Transmission, Processing, and Storage)**. Kluwer Academic/Plenum Publishers, 2001.

[958] E. Masry. Random sampling and reconstruction of spectra. *Inf. Contr., Vol. 19*, pp. 275–288, 1971.

[959] E. Masry. Alias-free sampling: an alternative conceptualization and its applications. *IEEE Transactions Information Theory, Vol. IT–24*, 1978.

[960] E. Masry. Poisson sampling and spectral estimation of continuous time processes. *IEEE Transactions Information Theory, Vol. IT–24*, 1978.

[961] E. Masry. The reconstruction of analog signals from the sign of their noisy samples. *IEEE Transactions Information Theory, Vol. IT–27*, pp. 735–745, 1981.

[962] E. Masry. The approximation of random reference sequences to the reconstruction of clipped differentiable signals. *IEEE Transactions Acoust., Speech, Signal Processing, Vol. ASSP-30*, pp. 953–963, 1982.

[963] E. Masry and S. Cambanis. Bandlimited processes and certain nonlinear transformations. *Journal Math. Anal., Vol. 53*, pp. 59–77, 1976.

[964] E. Masry and S. Cambanis. Consistent estimation of continuous time signals from nonlinear transformations of noisy samples. *IEEE Transactions Information Theory, Vol. IT–27*, pp. 84–96, 1981.

[965] E. Masry, D. Kramer, and C. Mirabile. Spectral estimation of continuous time process: performance comparison between periodic and Poisson sampling schemes. *IEEE Transactions Automatic Control, Vol. AC-23*, pp. 679–685, 1978.

[966] Georges Matheron, **Random Sets and Integral Geometry**, John Wiley & Sons, New York, 1975.

[967] J. Mathews and R.L. Walker. **Mathematical Methods of Physics**, 2nd ed. W.A. Benjamin, Menlo Park, CA, 1970.

[968] D. Maurice. **Convolution and Fourier Transforms for Communications Engineers**. John Wiley & Sons, 1976.

[969] A.C. McBride and F.H. Kerr. On Namias's fractional Fourier transform. IMA *J. Applied Math., Vol. 39*, pp. 159—175, 1987.

[970] James H. McClellan. The Design of Two Dimensional Filters by Transformation. *Proceedings 7th Annual Princeton Conference on Information Sciences and Systems*, pp. 247–251, 1973.

[971] James H. McClellan and David S.K. Chan. A 2-D FIR Filter Structure Derived from the Chebyshev Recursion. *IEEE Transactions on Circuits and Systems, CAS–24, No. 7*, pp. 372–378, 1977.

[972] M.J. McDonnell. A sampling function appropriate for deconnotation. *IEEE Transactions Information Theory, Vol. IT-22*, pp. 617–621, 1976.

[973] K.C. McGill and L.J. Dorfman. High-resolution alignment of sampled waveforms. *IEEE Transactions Biomedical Engineering, Vol. BME–31*, pp. 462–468, 1984.

[974] J. McNamee, F. Stenger, and E.L. Whitney. Whittaker's cardinal function in retrospect. *Math. Comp. Vol. 25*, pp. 141–154, 1971.

[975] A.K. Menon and Z.E. Boutaghou. Time-frequency analysis of tribological systems – part I: implementation and interpretation. *Tribiological International. Vol. 31(9)*, pp. 501–510, 1998.

[976] W.F.G. Mechlenbräuker, Russell M. Mersereau, McClellan Transformation for 2-D Digital Filtering: I-Implementation. *IEEE Transactions on Circuits and Systems, CAS-23, No. 7*, pp. 414–422, 1976.

[977] Alfred Mertins. **Signal Analysis: Wavelets, Filter Banks, Time-Frequency Transforms and Applications.** Wiley, 1999.

[978] F.C. Mehta. A general sampling expansion. *Inform. Sci., Vol. 16*, pp. 41–44, 1978.

[979] P.M. Mejías and R. Martínez Herrero. Diffraction by one-dimensional Ronchi grids: on the validity of the Talbot effect. *Journal of the Optical Society America A, Vol. 8*, pp. 266–269, 1991.

[980] Russell M. Mersereau, W.F.G. Mechlenbräuker, and T.F. Quatieri. McClellan Transformation for 2-D Digital Filtering: I-Design. *IEEE Transactions on Circuits and Systems*, CAS-23, No. 7, pp. 405–414, 1976.

[981] R.M. Mersereau. The processing of hexagonally sampled two dimensional signals. *Proceedings IEEE, Vol. 67*, pp. 930–949, 1979.

[982] R.M. Mersereau and T.C. Speake. The processing of periodically sampled multidimensional signals. *IEEE Transactions Acoust., Speech, Signal Processing, Vol. ASSP-31, No. 31*, pp. 188–194, 1983.

[983] D. Middleton. **An Introduction to Statistical Communication Theory**. McGraw-Hill, New York, 1960.

[984] D. Middleton and D.P. Peterson. A note on optimum presampling filters. *IEEE Transactions Circuit Theory, Vol. CT-10*, pp. 108–109, 1963.

[985] J. Millet-Roig, J.J. Rieta-Ibanez, E. Vilanova, A. Mocholi, and F.J. Chorro. Time-frequency analysis of a single ECG: to discriminate between ventricular tachycardia and ventricular fibrillation. *Computers in Cardiology 1999*, 26–29 pp. 711–714, 1999.

[986] R. Mintzer and B. Liu. Aliasing error in the design of multirate filters. *IEEE Transactions Acoust., Speech, Signal Processing, Vol. ASSP-26*, pp. 76–88, 1978.

[987] R. Mirman. **Point Groups, Space Groups, Crystals, Molecules**. World Scientific, 1999.

[988] S. Mitaim and B. Kosko. Adaptive stochastic resonance. *Proceedings of the IEEE. Vol. 86, No. 11*, 1998.

[989] J.R. Mitchell and W.L. McDaniel Jr. Adaptive sampling technique. *IEEE Transactions Automatic Control, Vol. AC-14*, pp. 200–201, 1969.

[990] J.R. Mitchell and W.L. McDaniel Jr. Calculation of upper bounds for errors of an approximate sampled frequency response. *IEEE Transactions Automatic Control, Vol. AC-19*, pp. 155–156, 1974.

[991] Sanjit K. Mitra. **Digital Signal Processing**, McGraw-Hill Science/Engineering/Math; 3rd edition 2005.

[992] H. Miyakawa. Sampling theorem of stationary stochastic variables in multi-dimensional space (Japanese). *Journal Institute Electron. Commun. Engineering Japan, Vol. 42*, pp. 421–427, 1959.

[993] K. Miyamoto. On Gabor's expansion theorem. *Journal of the Optical Society America, Vol. 50*, pp. 856–858, 1960.

[994] K. Miyamoto. Note on the proof of Gabor's expansion theorem. *Journal of the Optical Society America, Vol. 51*, pp. 910–911, 1961.

[995] W.D. Montgomery. The gradient in the sampling of n-dimensional bandlimited functions. *Journal Electron. Contr., Vol. 17*, pp. 437–447, 1964.

[996] W.D. Montgomery. Algebraic formulation of diffraction applied to self-imaging. *Journal of the Optical Society of America, Vol. 58*, pp. 1112–1124, 1968.

[997] D.R. Mook, G.V. Fisk, and A.V. Oppenheim. A hybrid numerical-analytical technique for the computation of wave fields in stratified media based on the Hankel transform. *Journal Acoustics. Society America, Vol. 76, No. 1*, pp. 222–243, 1984.

[998] N. Moray, G. Synnock, and S. Richards. Tracking a static display. *IEEE Transactions Systems, Man, Cybernetics, Vol. SMC-3*, pp. 518–521, 1973.

[999] Frank Morgan. **Real Analysis and Applications: Including Fourier Series and the Calculus of Variations**. American Mathematical Society, 2005.

[1000] F. Mori and J. De Steffano III. Optimal nonuniform sampling interval and test-input design for identification of physiological systems from very limited data. *IEEE Transactions Automatic Control, Vol. AC-24, No. 6*, pp. 893–900, 1979.

[1001] H. Mori, I. Oppenheim, and J. Ross. *Some Topics in Quantum Statistics: The Wigner Function and Transport Theory*. In **Studies in Statistical Mechanics**, J. de Boer and G.E. Uhlenbeck, eds. North-Holland, Amsterdam, *Vol. 1*, pp. 213–298, 1962.

[1002] Norman Morrison. **Introduction to Fourier Analysis.** Wiley-Interscience, 1995.

[1003] G.V. Moustakides and E.Z. Psarakis. Design of N-dimensional hyperquadrantally symmetric FIR filters using the McClellan transform. *IEEE Transactions on Circuits and Systems II: Analog and Digital Signal Processing, Vol. 42(8)*, pp. 547–550, 1995.

[1004] K. Mullen and P.S. Pregosin. **Fourier Transform Nuclear Magnetic Resonance Techniques**. Academic Press, 1977.

[1005] D.H. Mugler and W. Splettstößer. Difference methods and round-off error bounds for the prediction of bandlimited-functions from past samples. *Frequenz, Vol. 39*, pp. 182–187, 1985.

[1006] D.H. Mugler and W. Splettstößer. Some new difference schemes for the prediction of bandlimited signals from past samples. In *Proceedings of the Conference "Mathematics in Signal Processing"* Bath, 1985.

[1007] D.H. Mugler and W. Splettstößer. Difference methods for the prediction of bandlimited signals. *SIAM Journal Applied Math., Vol. 46*, pp. 930–941, 1986.

[1008] H.D. Mugler and W. Splettstößer. Linear prediction from samples of a function and its derivatives. *IEEE Transactions Information Theory, Vol. IT–33*, pp. 360–366, 1987.

[1009] K. Mullen, P.S. Pregosin. **Fourier Transform Nuclear Magnetic Resonance Techniques: A Practical Approach**. Elsevier Science & Technology Books, 1977.

[1010] N.J. Munch. A Twisted Sampling Theorem. *IMA Journal of Mathematical Control and Information, Vol. 7(1)*, pp. 47–57, 1990.

[1011] D.C. Munson. Minimum sampling rates for linear shift-variant discrete-time systems. *IEEE Transactions Acoust., Speech, Signal Processing, Vol. ASSP–33*, pp. 1556–1561, 1985.

[1012] P.K. Murphy and N.C. Gallagher. A new approach to two-dimensional phase retrieval. *Journal of the Optical Society America, Vol. 72*, pp. 929–937, 1982.

N

[1013] T. Nagai. Dyadic stationary processes and their spectral representation. *Bull. Math. Stat., Vol. 17*, pp. 65–73, 1976/77.

[1014] N. Nakajima and T. Asakura. A new approach to two-dimensional phase retrieval. *Opt. Acta, Vol. 32*, pp. 647–658, 1985.

[1015] V. Namias. The fractional order Fourier transform and its applications to quantum mechanics. *J. Institute Math. Applied, Vol. 25*, pp. 241–265, 1980.

[1016] H. Nassenstein. Super-resolution by diffraction of subwaves. *Optical Communication*, pp. 231–234, 1970.

[1017] A. Nathan. On sampling a function and its derivatives. *Information & Control, Vol. 22*, pp. 172–182, 1973.

[1018] A. Nathan. Plain and covariant multivariate Fourier transforms. *Information & Control, Vol. 39*, pp. 73–81, 1978.

[1019] J.L. Navarro-Mesa, E. Lleida-Solano, and A.A. Moreno-Bilbao. A new method for epoch detection based on the Cohen's class of time frequency representations. *IEEE Signal Processing Letters, Vol. 8(8)*, pp. 225–227, 2001.

[1020] A.W. Naylor and G.R. Sell. **Linear Operator Theory in Engineering and Science**. Springer-Verlag, New York, 1982.

[1021] S.A. Neild, P.D. McFadden, and M.S. Williams. A review of time-frequency methods for structural vibration analysis. *Engineering Structures, Vol. 5(6)*, pp. 713–728, 2003.

[1022] Kazuki Nishi. Generalized comb function: a new self-Fourier function. *IEEE International Conference on Acoustics, Speech, and Signal Processing*, 2004. (ICASSP '04). *Vol. 2*, pp. 573–576.

[1023] J. Ojeda-Castañeda, J. Ibarra, and J.C. Barreiro. Noncoherent Talbot effect: Coherence theory and applications. *Optical Communication, Vol. 71*, pp. 151–155, 1989.

[1024] J. Ojeda-Castañeda and E.E. Sicre. Quasi ray-optical approach to longitudinal periodicities of free and bounded wavefields. *Opt. Acta*, pp. 17–26, 1985.

[1025] J. Von Neumann. **The Geometry of Orthogonal Spaces**. Princeton University Press, Princeton, New Jersey, 1950.

[1026] J. Von Neumann. **Mathematical Foundations of Quantum Mechanics**. Princeton University Press, 1955, Chap. 5, Sect. 4.

[1027] B. Nicoletti and L. Mariani. Optimization of nonuniformly sampled discrete systems. *Automatica, Vol. 7*, pp. 747–753, 1971.

[1028] K. Niederdrenk. **The Finite Fourier and Walsh Transformation with an Introduction to Image Processing**. Vieweg & Sohn, Braunschweig, 1982.

[1029] Y. Noguchi, E. Kashiwagi, E. Kashiwagi *et al.* Ultrasound Doppler sinusoidal shift signal analysis by time-frequency distribution with new kernel. *Japanese Journal of Applied Physics Part 1, Vol. 37(5B)*, pp. 3064–3067, Sp. Iss. SI 1998.

[1030] Y. Noguchi, E. Kashiwagi, K. Watanabe *et al.* Time-frequency analysis of the blood flow Doppler ultrasound signal. *Japanese Journal of Applied Physics Part 1, Vol. 40(5B)*, pp. 3882–3887, 2001.

[1031] Stephen Abbott Northrop. **A Cloud of Witnesses.** (Portland, Oregon: American Heritage Ministries, 1987; Mantle Ministries, 228 Still Ridge, BuIverde, Texas), pp. 147–8.

[1032] K.A. Nugent. Three-dimensional optical microscopy: a sampling theorem. *Optical Communication*, pp. 231–234, 1988.

[1033] H. Nyquist. Certain topics in telegraph transmission theory. *AIEE Transactions, Vol. 47*, pp. 617–644, 1928.

[1034] D. Nguyen and M. Swamy. Formulas for parameter scaling in the McClellan transform. *IEEE Transactions on Circuits and Systems, Vol.33(1)*, pp. 108–109, 1986.

[1035] D. Nguyen and M. Swamy. A class of 2-D separable denominator filters designed via the McClellan transform. *IEEE Transactions on Circuits and Systems, Vol. 33(9)*, pp. 874–881, 1986.

O

[1036] K. Ogura. On a certain transcendental integral function in the theory of interpolation. *Tôhoku Math. J., Vol. 17*, pp. 64–72, 1920.

[1037] S. Oh, R.J. Marks II, L.E. Atlas and J.W. Pitton. Kernel synthesis for generalized time-frequency distributions using the method of projection onto convex sets. SPIE Proceedings 1348, **Advanced Signal Processing Algorithms, Architectures, and Implementation**, F.T. Luk, Editor, San Diego, pp. 197–207, 1990.

[1038] S. Oh and R.J. Marks II. Performance attributes of generalized time-frequency representations with double diamond and cone shaped kernels. *Proceedings of the Twenty Fourth Asimomar Conference on Signals, Systems and Computers*, 5–7 November, 1990, Asilomar Conference Grounds, Monterey, California.

[1039] S. Oh, R.J. Marks II, and Dennis Sarr. Homogeneous alternating projection neural networks. *Neurocomputing, Vol. 3*, pp. 69–95, 1991.

[1040] S. Oh, C.Ramon, M.G. Meyer, A. C. Nelson, and R.J. Marks II. Resolution enhancement of biomagnetic images using the method of alternating projections. *IEEE Transactions on Biomedical Engineering, Vol. 40, No. 4*, pp. 323–328, 1993.

[1041] S. Oh and R.J. Marks II. Alternating projections onto fuzzy convex sets. *Proceedings of the Second IEEE International Conference on Fuzzy Systems*, (FUZZ-IEEE '93), San Francisco, Vol.1, pp. 148–155, 1993.

[1042] S. Oh and R.J. Marks II. Some properties of the generalized time frequency representation with cone shaped kernels. *IEEE Transactions on Signal Processing, Vol.40, No.7*, pp. 1735–1745, 1992.

[1043] S. Oh, R.J. Marks II, and L.E. Atlas. Kernel synthesis for generalized time-frequency distributions using the method of alternating projections onto convex sets. *IEEE Transactions on Signal Processing, Vol. 42, No.7*, pp. 1653–1661, 1994.

[1044] N. Ohyama, M. Yamaguchi, J. Tsujiuchi, T. Honda, and S. Hiratsuka. Suppression of Moiré fringes due to sampling of halftone screened images. *Optical Communication*, pp. 364–368, 1986.

[1045] K.B. Oldham and J. Spanier. **The Fractional Calculus**. Academic Press, New York, 1974.

[1046] A.Y. Olenko and T.K. Pogany. Time shifted aliasing error upper bounds for truncated sampling cardinal series. *Journal of Mathematical Analysis and Applications, Vol. 324(1)*, pp. 262–280, 2006.

[1047] B.L. Ooi, S. Kooi, and Leong M.S. Application of Sonie-Schafheitlin formula and sampling theorem in spectral-domain method. *IEEE Transactions on Microwave Theory and Techniques, 49(1)*, pp. 210–213, 2001.

[1048] E.L. O'Neill. Spatial filtering in optics. *IRE Transactions Information Theory*, pp. 56–62, 1956.

[1049] E.L. O'Neill. **Introduction to Statistical Optics**. Addison-Wesley, Reading, MA, 1963.

[1050] E.L. O'Neill and A. Walther. The question of phase in image formation. *Opt. Acta*, pp. 33–40, 1962.

[1051] L. Onural. Exact analysis of the effects of sampling of the scalar diffraction field. *Journal of the Optical Society of America A – Optics, Image Science and Vision, Vol. 24(2)*, pp. 359–367, 2007.

[1052] Z. Opial. Weak convergence of the sequence of successive approximation for nonexpansive mappings. *Bull. Amer. Math. Society, Vol. 73*, pp. 591–597, 1967.

[1053] A.V. Oppenheim and R.W. Shafer. **Digital Signal Processing**. Prentice-Hall, Englewood Cliffs, NJ, 1975 - updated in [1054].

[1054] A.V. Oppenheim, R.W. Schafer, and J.R. Buck, **Discrete-Time Signal Processing, second edition** (Prentice-Hall, New Jersey, 1999)–an updated version of [1053].

[1055] A.V. Oppenheim. Digital processing of speech. In **Applications of Digital Signal Processing**. Prentice-Hall, Englewood Cliffs, NJ, 1978.

[1056] A.V. Oppenheim, Alan, S. Willsky, and Ian T. Young. **Signals and Systems**. Prentice-Hall, Englewood Cliffs, NJ, 1983.

[1057] J.F. Ormsky. Generalized interpolation sampling expressions and their relationship. *Journal Franklin Institute, Vol. 310*, pp. 247–257, 1980.

[1058] L. Stankovic, T. Alieva, and M.J. Bastiaans. Time-frequency signal analysis based on the windowed fractional Fourier transform. *Signal Processing, Vol. 83(11)*, pp. 2459–2468, 2003.

[1059] Radomir S. Stankovic, Claudio Moraga, and Jaakko Astola. **Fourier Analysis on Finite Groups with Applications in Signal Processing and System Design**. Wiley-IEEE Press, 2005.

[1060] P. Oskoui-Fard, H. Stark. Geometry-free X-ray reconstruction using the theory of convex projections 1988 International Conference on Acoustics, Speech, and Signal Processing, ICASSP–88., pp. 871–874, 1988.

[1061] T.J. Osler. A further extension of the Leibnitz rule for fractional derivatives and its relation to Parseval's formula. *SIAM J. Math. Anal., Vol. 4, No. 4*, 1973.

[1062] L.E. Ostrander. The Fourier transform of spline function approximations to continuous data. *IEEE Transactions Audio Electroacoustics., Vol. AU-19*, pp. 103–104, 1971.

[1063] J.S. Ostrem. A sampling series representation of the gain and refractive index formulas for a combined homogeneously and inhomogeneously broadened laser line. *Proceedings IEEE, Vol. 66*, pp. 583–589, 1978.

[1064] N. Ostrowsky, D. Sornette, P. Parker, and E.R. Pike. Exponential sampling method for light scattering polydispersity analysis. *Opt. Acta, Vol. 28*, pp. 1059–1070, 1981.

[1065] M.K. Ozkan, A.M. Tekalp, and M.I. Sezan. POCS-based restoration of space-varying blurred images. IEEE Transactions on Image Processing, *Vol. 3(4)*, pp. 450–454, 1994.

[1066] H.M. Ozaktas, O. Arikan, M.A. Kutay, and G. Bozdagt. Digital computation of the fractional Fourier transform. *IEEE Transactions on Signal Processing, Vol. 44(9)*, pp. 2141–2150, 1996.

[1067] H.M. Ozaktas, N. Erkaya, and M.A. Kutay. Effect of fractional Fourier transformation on time-frequency distributions belonging to the Cohen class. *IEEE Signal Processing Letters, Vol. 3(2)*, pp. 40–41, 1996.

[1068] Haldun M. Ozaktas, Zeev Zalevsky, and M. Alper. **The Fractional Fourier Transform: with Applications in Optics and Signal Processing**, John Wiley & Sons, 2001.

P

[1069] Hoon Paek, Rin-Chul Kim, and Sang-Uk Lee. On the POCS-based postprocessing technique to reduce the blocking artifacts in transform coded images. *IEEE Transactions on Circuits and Systems for Video Technology, Vol. 8(3)*, pp. 358–367, 1998.

[1070] C.H. Page. Instantaneous Power Spectra. *Jour. Applied Phy., Vol.23*, pp. 103–106, 1952.

[1071] R.E.A. Paley and N. Wiener. **Fourier Transform in Complex Domain**. Colloq. Publications, American Math. Society, New York, *Vol. 19*, 1934.

[1072] F. Palmieri. Sampling theorem for polynomial interpolation. *IEEE Transactions Acoust., Speech, Signal Processing, Vol. ASSP-34*, pp. 846–857, 1986.

[1073] T. Pappas. Mersenne's Number. **The Joy of Mathematics**. San Carlos, CA: Wide World Publ. Tetra, P. 211, 1989.

[1074] A. Papoulis. **The Fourier Integral and Its Applications**. McGraw-Hill, New York, 1962.

[1075] A. Papoulis, **Probability, Random Variables, and Stochastic Processes**, McGraw-Hill, NY, 1965.

[1076] A. Papoulis, **Probability, Random Variables, and Stochastic Processes**, 2nd ed., McGraw-Hill, NY, 1984.

[1077] A. Papoulis. **Probability, Random Variables, and Stochastic Processes**, 3rd ed. McGraw-Hill, New York, 1991.

[1078] Papoulis and S.Unnikrishna Pillai. **Probability, Random Variables and Stochastic Processes**. 4th Revised Edition, McGraw Hill, 2002.

[1079] A. Papoulis. Error analysis in sampling theory. *Proceedings IEEE, Vol. 54*, pp. 947–955, 1966.

[1080] A. Papoulis. Limits on bandlimited signals. *Proceedings IEEE, Vol. 55*, pp. 1677–1681, 1967.

[1081] A. Papoulis. Truncated sampling expansions. *IEEE Transactions Automatic Control*, pp. 604–605, 1967.

[1082] A. Papoulis. **Systems and Transforms with Applications in Optics**. McGraw-Hill, New York, 1968.

[1083] A. Papoulis and M.S. Bertran. Digital filtering and prolate functions. *IEEE Transactions Circuits & Systems, Vol. CT-19*, pp. 674–681, 1972.

[1084] A. Papoulis. A new method of image restoration. *Joint Services Technical Activity Report*, 39, 1973–74.

[1085] A. Papoulis. A new algorithm in spectral analysis and bandlimited signal extrapolation. *IEEE Transactions Circuits & Systems, Vol. CAS-22*, pp. 735–742, 1975.

[1086] A. Papoulis. Generalized sampling expansion. *IEEE Transactions Circuits & Systems, Vol. CAS–24*, pp. 652–654, 1977.

[1087] A. Papoulis. **Signal Analysis**. McGraw-Hill, New York, 1977.

[1088] A. Papoulis. **Circuits and Systems, A Modern Approach**. Holt, Rinehart and Winston, Inc., New York, 1980.

[1089] A. Papoulis. A note on the predictability of bandlimited processes. *Proceedings IEEE, Vol. 73*, pp. 1332–1333, 1985.

[1090] G.Y. Park, C.K. Lee, J.T. Kim, K.C. Kwon, and S.J. Lee. Design of a time-frequency distribution for vibration monitoring under corrosions in the pipe. *Proceedings Key Engineering Materials for Advanced Nondestructive Evaluations.* pp. 321–323, 1257–1261, 2006.

[1091] Jiho Park, D.C. Park, R.J. Marks II, and M.A. El-Sharkawi. Block loss recovery in DCT image encoding using POCS. *IEEE International Symposium on Circuits and Systems*, Scottsdale, Arizona, pp. V245–V248, 2002.

[1092] Jiho Park, R.J. Marks II, D.C. Park, and M.A. El-Sharkawi. Content based adaptive spatio-temporal methods for MPEG repair. *IEEE Transactions on Image Processing, Vol. 13, # 8*, pp. 1066–1077, 2004.

[1093] Jiho Park, Dong Chul Park, R.J. Marks II, and M.A. El-Sharkawi. Recovery of image blocks using the method of alternating projections. *IEEE Transactions on Image Processing*, 2005.

[1094] J.B. Park and D.G. Nishimura, Effects of 3D sampling in (k,t)-space on temporal qualities of dynamic MRI . *Magnetic Resonance Imaging, Vol. 24(8)*, pp. 1009–1014, 2006.

[1095] T.W. Parks and R.G. Meier. Reconstructions of signals of a known class from a given set of linear measurements. *IEEE Transactions Information Theory, Vol. IT–17*, pp. 37–44, 1971.

[1096] E. Parzen. A simple proof and some extensions of sampling theorems. Technical report, Stanford University, Stanford, California, 1956.

[1097] E. Parzen. **Stochastic Processes**. Holden-Day, San Francisco, 1962.

[1098] Blaise Pascal, translated by A. J. Krailsheimer. **Pensées**. Penguin Classics; Reissue edition, 1995.

[1099] C. Pask. Simple optical theory of optical super-resolution. *Journal of the Optical Society America, Vol. 66*, pp. 68–69a, 1976.

[1100] K. Patorski. The self-imaging phenomenon and its applications. In **Progress in Optics**, North-Holland, Amsterdam, 1989.

[1101] Soo-Chang Pei, Min-Hung Yeh, and Tzyy-Liang Luo. Fractional Fourier series expansion for finite signals and dual extension to discrete-time fractional Fourier transform. *IEEE Transactions on Signal Processing, Vol. 47(10)*, pp. 2883–2888, 1999.

[1102] Soo-Chang Pei, Min-Hung Yeh, and Chien-Cheng Tseng. Discrete fractional Fourier transform based on orthogonal projections. *IEEE Transactions on Signal Processing, Vol. 47(5)*, pp. 1335–1348, 1999.

[1103] Haidong Peng, Mohammad Sabati, Louis Lauzon, and Richard Frayne. MR image reconstruction of sparsely sampled 3D k-space data by projection-onto-convex sets. *Magnetic Resonance, Vol. 24(6)*, pp. 761–773, 2006.

[1104] Hui Peng and Henry Stark. Signal recovery with similarity constraints. *Journal of the Optical Society of America A, Vol 6, No.6*, pp. 844–851, 1989.

[1105] L.M. Peng. Sampling theorem and digital electron microscopy. *Progress in Natural Science, Vol. 7(1)*, pp. 110–114, 1997.

[1106] L.E. Pellon. A double Nyquist digital product detector for quadrature sampling. *IEEE Transactions Signal Processing, Vol. 40*, pp. 1670–1681, 1992.

[1107] I. Pesenson. A sampling theorem on homogeneous manifolds. *Transactions of the American Mathematical Society, Vol. 352(9)*, pp. 4257–4269, 2000.

[1108] R.J. Peterka, D.P. Oleary, and A.C. Sanderson. Practical considerations in the implementation of the French-Holden algorithm for sampling of neuronal spike trains. *IEEE Transactions Biomedical Engineering, Vol. BME-24*, pp. 192–195, 1978.

[1109] Terry M. Peters, J. C. Williams, and J. H. Bates, (Editors). **Fourier Transform in Biomedical Engineering** Birkhauser Verlag, 1997.

[1110] D. P. Petersen. Sampling of space-time stochastic processes with application to information and design systems, Thesis, Rensselaer Polytechnic Inst., Troy, N. Y. 1963.

[1111] D.P. Petersen. Discrete and fast Fourier transformations on n-dimensional lattices. *Proceedings IEEE, Vol. 58*, 1970.

[1112] D.P. Petersen and D. Middleton. Sampling and reconstruction of wave number-limited function in n-dimensional Euclidean spaces. *Inform., Contr., Vol. 5*, pp. 279–323, 1962.

[1113] D.P. Petersen and D. Middleton. Reconstruction of multidimensional stochastic fields from discrete measurements of amplitude and gradient. *Information & Control, Vol. 1*, pp. 445–476, 1964.

[1114] D.P. Petersen and D. Middleton. Linear interpolation, extrapolation and prediction of random space-time fields with limited domain of measurement. *IEEE Transactions Information Theory, Vol. IT–11*, 1965.

[1115] J. Peřina. Holographic method of deconvolution and analytic continuation. *Czechoslovak J. Phys., Vol. B21*, pp. 731–748, 1971.

[1116] J. Peřina and J. Kvapil. A note on holographic method of deconvolution. *Optik*, pp. 575–577, 1968.

[1117] J. Peřina, V. Perinova, and Z. Braunerova. Super-resolution in linear systems with noise. *Optica Applicata*, pp. 79–83, 1977.

[1118] E.P. Pfaffelhuber. Sampling series for bandlimited generalized functions. *IEEE Transactions Information Theory, Vol. IT-17*, pp. 650–654, 1971.

[1119] F. Pichler. Synthese linearer periodisch zeitvariabler Filter mit vorgeschriebenem Sequenzverhalten. *Arch. Elek. Übertr. (AEÜ), Vol. 22*, pp. 150–161, 1968.

[1120] W.J. Pielemeier and G.H. Wakefield. A high-resolution time-frequency representation for musical instrument signals. *Journal of the Acoustical Society of America. Vol. 99(4)*, pp. 2382–2396, 1996.

[1121] W.J. Pielemeier, G.H. Wakefield, and M.H. Simoni. Time-frequency analysis of musical signals. *Proceedings of the IEEE. Vol. 84(9)*, pp. 1216–1230, 1996.

[1122] Allan Pinkus and Samy Zafrany. **Fourier Series and Integral Transforms**. Cambridge University Press, 1997.

[1123] H.S. Piper, Jr. Best asymptotic bounds for truncation error in sampling expansions of band limited signals. *IEEE Transactions Information Theory, Vol. IT-21*, pp. 687–690, 1975.

[1124] H.S. Piper, Jr. Bounds for truncation error in sampling expansions of finite energy band limited signals. *IEEE Transactions Information Theory, Vol. IT–21*, pp. 482–485, 1975.

[1125] J.W. Pitton, W.L.J. Fox, L.E. Atlas, J.C. Luby, and P.J. Loughlin. Range-Doppler processing with the cone kernel time-frequency representation communications, computers and signal processing, 1991., *IEEE Pacific Rim Conference on Vol. 2*, pp. 799–802, 1991.

[1126] J.W. Pitton and L.E. Atlas. Discrete-time implementation of the cone-kernel time-frequency representation. *IEEE Transactions on Signal Processing, Vol. 43(8)*, pp. 1996–1998, 1995.

[1127] K. Piwerneta. *A posteriori* compensation for rigid body motion in holographic interferometry by means of a Moiré technique. *Opt. Acta, Vol. 24*, pp. 201–209, 1977.

[1128] E. Plotkin, L. Roytman, and M.N.S. Swamy. Non-uniform sampling of bandlimited modulated signals. *Signal Processing, Vol. 4*, pp. 295–303, 1982.

[1129] E. Plotkin, L. Roytman, and M.N.S. Swamy. Reconstruction of nonuniformly sampled bandlimited signals and jitter error reduction. *Signal Processing, Vol. 7*, pp. 151–160, 1984.

[1130] T. Pogány. On a very tight truncation error bound for stationary stochastic processes. *IEEE Transactions Signal Processing, Vol. 39, No. 8*, pp. 1918–1919, 1991.

[1131] R.J. Polge, R.D. Hays, and L. Callas. A direct and exact method for computing the transfer function of the optimum smoothing filter. *IEEE Transactions Automatic Control, Vol. AC-18*, pp. 555–556, 1973.

[1132] H.O. Pollak. Energy distribution of bandlimited functions whose samples on a half line vanish. *Journal Math. Anal., Applied, Vol. 2*, pp. 299–332, 1961.

[1133] Polybius of Megalopolis, **World History**, translated by H. J. Edwards.

[1134] B. Porat. ARMA spectral estimation of time series with missing observations. *IEEE Transactions Information Theory, Vol. IT-30, No. 6*, pp. 823–831, 1984.

[1135] B. Porter and J.J. D'Azzo. Algorithm for closed-loop eigenstructure assignment by state feedback in multivariable linear systems. *International J. Control, Vol. 27*, pp. 943–947, 1978.

[1136] R.P. Porter and A.J. Devaney. Holography and the inverse source problem. *Journal of the Optical Society America, Vol. 72*, pp. 327–330, 1982.

[1137] M. Pourahmadi. A sampling theorem for multivariate stationary processes. *Journal Multivariate Anal., Vol. 13*, pp. 177–186, 1983.

[1138] F.D. Powell. Periodic sampling of broad-band sparse spectra. *IEEE Transactions Acoustics, Speech, Signal Processing, Vol. ASSP-31*, pp. 1317–1319, 1983.

[1139] M. Pracentini and C. Cafforio. Algorithms for image reconstruction after nonuniform sampling. *IEEE Transactions Acoust., Speech, Signal Processing, Vol. ASSP–35, No. 8*, pp. 1185–1189, 1987.

[1140] S. Prasad. Digital superresolution and the generalized sampling theorem, *Journal of the Optical Society of America A – Optics, Image Science and Vision, Vol. 24(2)*, pp. 311–325, 2007.

[1141] P.M. Prenter. **Splines and Variational Methods**. John Wiley, New York, 1975.

[1142] R.T. Prosser. A multi-dimensional sampling theorem. *Journal Math. Analysis Applied, Vol. 16*, pp. 574–584, 1966.

[1143] R. Prost and R. Goutte. Deconvolution when the convolution kernel has no inverse. *IEEE Transactions Acoust., Speech, Signal Processing*, pp. 542–549, 1977.

[1144] M. Protzmann and H. Boche. Convergence proof for the algorithm by Papoulis and Gerchberg. *Frequenz, Vol. 52(9–10)*, pp. 175–182, 1998.

[1145] E.Z. Psarakis, V.G. Mertzios, and G.P. Alexiou. Design of two-dimensional zero phase FIR fan filters via the McClellan transform. *IEEE Transactions on Circuits and Systems, Vol. 37(1)*, pp. 10–16, 1990.

[1146] E.Z. Psarakis and G.V. Moustakides. Design of two-dimensional zero-phase FIR filters via the generalized McClellan transform. *IEEE Transactions on Circuits and Systems, Vol. 38(11)*, pp. 1355–1363, 1991.

Q

[1147] S. Qazi, A. Georgakis, L.K. Stergioulas, and M. Shikh-Bahaei. Interference suppression in the wigner distribution using fractional Fourier transformation and signal synthesis. *IEEE Transactions on Signal Processing, Vol. 55(6)*, Part 2, pp. 3150–3154, 2007.

R

[1148] S.A. Rabee, B. Sharif, B.S, and S. Sali. Distributed power control algorithm in cellular radio systems using projection onto convex sets 2001 Vehicular Technology Conference. VTC 2001 Fall. IEEE VTS 54th, *Vol. 2*, pp. 757–761, 2001.

[1149] L.R. Rabiner and R.W. Schafer. **Digital Processing of Speech Signals**. Prentice-Hall, Englewood Cliffs, NJ, 1978.

[1150] D. Radbel and R.J. Marks II. An FIR estimation filter based on the sampling theorem. *IEEE Transactions Acoust., Speech, Signal Processing, Vol. ASSP-33*, pp. 455–460, 1985.

[1151] R. Radzyner and P.T. Bason. An error bound for Lagrange interpolation of low pass functions. *IEEE Transactions Information Theory, Vol. IT-18*, pp. 669–671, 1972.

[1152] J.R. Ragazzini and G.F. Franklin. **Sampled Data Control Systems**. McGraw-Hill, New York, 1958.

[1153] Y. Rahmat-Samii and R. Cheung. Nonuniform sampling techniques for antenna applications. *IEEE Transactions Antennas & Propagation, Vol. AP-35, No. 3*, pp. 268–279, 1987.

[1154] P.K. Rajan. A study on the properties of multidimensional Fourier transforms. *IEEE Transactions Circuits & Systems, Vol. CAS-31*, pp. 748–750, 1984.

[1155] A. Ralston and P. Rabinowitz. **A First Course in Numerical Analysis**, 2nd ed. McGraw Hill, New York, 1978.

[1156] Dinakar Ramakrishnan and Robert J. Valenza. **Fourier Analysison Number Fields**. Springer, 1998.

[1157] Jayakumar Ramanathan. **Methods of Applied Fourier Analysis**. Birkhäuser Boston, 1998.

[1158] K.R. Rao and P. Yip. **Discrete Cosine Transform: Algorithms, Advantages, Applications**. Academic Press, Boston, 1990.

[1159] M.D. Rawn. Generalized sampling theorems for Bessel type transforms of bandlimited functions and distributions. *SIAM J. Applied Math., Vol. 49, No. 2*, pp. 638–649, 1989.

[1160] M.D. Rawn. On nonuniform sampling expansions using entire interpolating functions and on the stability of Bessel-type sampling expansions. *IEEE Transactions Information Theory, Vol. 35, No. 3*, 1989.

[1161] M.D. Rawn. A stable nonuniform sampling expansion involving derivatives. *IEEE Transactions Information Theory, Vol. 35, No. 6*, 1989.

[1162] Lord Rayleigh. On the character of the complete radiation at a given temperature. *Phil. Mag.*, series 5, Vol 27, 1889; reprinted in **Scientific Papers**, Cambridge University Press, Cambridge, England, 1902; later published by Dover Publications, New York, *Vol. 3*, p. 273, 1964.

[1163] M.G. Raymer. The Whittaker-Shannon sampling theorem for experimental reconstruction of free-space wave packets. *Journal of Modern Optics, Vol. 44(11–12)*, pp. 2565–2574, 1997.

[1164] M. Reed and B. Simon. **Methods and Modern Mathematical Physics. IV Analysis of Operators**. Academic Press, New York, 1978.

[1165] G.J. Ren and Y.M. Shi. Asymptotic behaviour of dynamic systems on time scales. *Journal of Difference Equations and Applications, Vol. 12(12)*, pp. 1289–1302, 2006.

[1166] F.M. Reza. **An Introduction to Information Theory**. McGraw-Hill, New York, 1961.

[1167] A. Requicha. The zeros of entire functions: Theory and engineering applications. *Proceedings of the IEEE, Vol. 68, No. 3*, pp. 308–328, 1980.

[1168] A. Restrepo, L.F. Zuluaga, and L.E. Pino. Optimal noise levels for stochastic resonance. *Acoustics, Speech, and Signal Processing*, 1997. ICASSP–97, 1997 IEEE International Conference on *Vol. 3*, pp. 2365–2368, 1997.

[1169] D.R. Rhodes. The optimum lines source for the best mean-square approximation to a given radiation pattern. *IEEE Transactions Antennas & Propagation*, pp. 440–446, 1963.

[1170] D.R. Rhodes. A general theory of sampling synthesis. *IEEE Transactions Antennas & Propagation*, pp. 176–181, 1973.

[1171] W.T. Rhodes. Acousto-optical signal processing: Convolution and correlation. *Proceedings IEEE, Vol. 69*, pp. 65–79, 1981.

[1172] W.T. Rhodes. Space-variant optical systems and processing. In **Applications of Optical Fourier Transforms**, H. Stark, editor, Academic Press, New York, 1982.

[1173] S.O. Rice Mathematical analysis of random noise. *Bell System Technical J., Vol. 23*, pp. 282–332, 1944.

[1174] S.O. Rice Mathematical analysis of random noise. *Bell System Technical J., Vol. 24*, pp. 46–156, 1945.

[1175] S. Ries and W. Splettstößer. On the approximation by truncated sampling series expansions. *Signal Processing, Vol. 7*, pp. 191–197, 1984.

[1176] S. Ries and R.L. Stens. A localization principle for the approximation by sampling series. In *Proceedings of the International Conference on the Theory of Approximation of Functions (Kiev, USSR)*, Moscow; Nauka, pp. 507–509, 1983.

[1177] S. Ries and R.L. Stens. Approximation by generalized sampling series. In *Constructive Theory of Functions (Proceedings of Conference at Varna, Bulgaria, 1984)*, Bl. Sendov et al., editor, Sofia, Publishing House Bulgarian Acad. Sci., pp. 746–756, 1984.

[1178] A.W. Rihaczek. Signal energy distribution in time and frequency. *IEEE Transactions Inform. Theory, IT–14*, pp. 369–374, 1968.

[1179] C.L. Rino. Bandlimited image restoration by linear mean-square estimation. *Journal of the Optical Society America, Vol. 59*, pp. 547–553, 1969.

[1180] C.L. Rino. The application of prolate spheroidal wave functions to the detection and estimation of bandlimited signals. *Proceedings IEEE, Vol. 58*, pp. 248–249, 1970.

[1181] J. Riordan.**Combinatorial Identities**. Wiley & Sons, New York, 1968.

[1182] O. Robaux. Application of sampling theorem for restitution of partially known objects. 2. numerical methods. *Optica Acta, Vol. 17(11)*, p. 811, 1970.

[1183] O. Robaux. Use of Sampling theorem for reconstruction of partially known objects. 3. decomposition of spectra. *Optica Acta, Vol. 18(7)*, p. 523, 1971.

[1184] Allan H. Robbins, Wilhelm Miller, **Circuit Analysis: Theory & Practice** (2nd Edition), Delmar Thomson Learning, 1999.

[1185] J.B. Roberts and D.B.S. Ajmani. Spectral analysis of randomly sampled signals using a correlation-based slotting technique. *Proceedings IEEE, Vol. 133*, Pt. F, pp. 153–162, 1986.

[1186] J.B. Roberts and M. Gaster. Rapid estimation of spectra from irregularly sampled records. *Proceedings IEEE, Vol. 125*, pp. 92–96, 1978.

[1187] J.B. Roberts and M. Gaster. On the estimation of spectra from randomly sampled signals – a method of reducing variability. *Proceedings of the Royal Society London, Vol. A371*, pp. 235–258, 1980.

[1188] Gregory S. Rohrer, **Structure and Bonding in Crystalline Materials**. Cambridge University Press, 2001.

[1189] N. Rose. **Mathematical Maxims and Minims**, Rome Press Inc., 1988.

[1190] A. Röseler. Zur Berechnung der optischen abbildung bei teilkohärenter beleuchtung mit hilfe des sampling–theorems. *Opt. Acta, Vol. 16*, pp. 641–651, 1969.

[1191] A. Roshan-Ghias and M.B. Shamsollahi, and M. Mobed et al., Estimation of modal parameters using bilinear joint time-frequency distributions. *Mechanical Systems and Signal Processing. Vol. 21(5)*, pp. 2125–2136, 2007.

[1192] G. Ross. Iterative methods in information processing for object restoration. *Opt. Acta*, pp. 1523–1542, 1982.

[1193] Hugh Ross, **Beyond the Cosmos: The Extra-Dimensionality of God**, Navpress Publishing Group; 2nd Expand edition 1999.

[1194] E. Roy, P. Abraham, S. Montresor, and J.L. Saumett. Comparison of time-frequency estimators for peripheral embolus detection. *Ultrasound in Medicine and Biology, Vol. 26(3)*, pp. 419–423 2000.

[1195] S.H. Roy, P. Bonato, and M. Knaflitz. EMG assessment of back muscle function during cyclical lifting. *Journal of Electromyography and Kinesiology. Vol 8(4)*, pp. 233–245, 1998.

[1196] Y.A. Rozanov. To the extrapolation of generalized random stationary processes. *Theor. Probability Applied, Vol. 4*, p. 426, 1959.

[1197] D.B. Rozhdestvenskii. Sampling and the discretization theorem. *Automation and Remote Control, Vol. 67(12)*, pp. 1991–2001, 2006.

[1198] W. Rozwoda, C.W. Therrien, and J.S. Lim. Novel method for nonuniform frequency-sampling design of 2-D FIR filters. In *Proceedings ICASSP'88*, 1988.

[1199] D.S. Ruchkin. Linear reconstruction of quantized and sampled random signals. *IRE Transactions Communication Systems, Vol. CS–9*, 1961.

[1200] Walter Rudin. **Fourier Analysis on Groups**. Wiley-Interscience, 1990.

[1201] A. Rundquist, A. Efimov, and D.H. Reitze. Pulse shaping with the Gerchberg-Saxton algorithm. *Journal of the Optical Society of America B – Optical Physics, Vol. 19(10)*, pp. 2468–2478, 2002.

[1202] C.K. Rushforth and R.L. Frost. Comparison of some algorithms for reconstructing space-limited images. *Journal of the Optical Society America, Vol. 70*, pp. 1539–1544, 1980.

[1203] C.K. Rushforth and R.W. Harris. Restoration, resolution, and noise. *Journal of the Optical Society America, Vol. 58*, pp. 539–545, 1968.

[1204] F.D. Russell and J.W. Goodman. Nonredundant arrays and postdetection processing for aberration compensation in incoherent imaging. *Journal of the Optical Society America, Vol. 61*, pp. 182–191, 1971.

S

[1205] M.S. Sabri and W. Steenaart. An approach to bandlimited signal extrapolation: the extrapolation matrix. *IEEE Transactions Circuits & Systems, Vol. CAS–25*, pp. 74–78, 1978.

[1206] R. Sahkaya, Y. Gao, and G. Saon. Fractional Fourier transform features for speech recognition *Proceedings. (ICASSP '04). IEEE International Conference on Acoustics, Speech, and Signal Processing, Vol. 1*, pp. I–529–32, 2004.

[1207] R.A.K. Said and D.C. Cooper. Crosspath real-time optical correlator and ambiguity function processor. *Proceedings Institute Elec. Engineering Vol. 120*, pp. 423–428, 1973.

[1208] S.H. Saker. Oscillation of nonlinear dynamic equations on time scales. *Applied Mathematics and Computation, Vol. 148(1)*, pp. 81–91, 2004.

[1209] S.H. Saker. Oscillation of second-order nonlinear neutral delay dynamic equations on time scales. *Journal of Computational and Applied Mathematics, Vol. 187(2)*, pp. 123–141, 2006.

[1210] B.E.A. Saleh. *A priori* information and the degrees of freedom of noisy images. *Journal of the Optical Society America, Vol. 67*, pp. 71–76, 1977.

[1211] B.E.A. Saleh and M.O. Freeman. Optical transformations. In **Optical Signal Processing**. J.L. Horner, editor, Academic Press, New York, 1987.

[1212] F.V. Salina, N.D.A. Mascarenhas, and P.E. Cruvinel. A comparison of POCS algorithms for tomographic reconstruction under noise and limited view. *Proceedings Brazilian Symposium on Computer Graphics and Image Processing, 2002*. pp. 342–346, 2002.

[1213] F.V. Salina and N.D.A. Mascarenhas. A Hybrid Estimation Theoretic-POCS Method for Tomographic Image Reconstruction. 18th Brazilian Symposium on Computer Graphics and Image Processing, 2005. SIBGRAPI 2005. pp. 220–224, 2005.

[1214] M. Salerno, G. Orlandi, G. Martinelli, and P. Burrascano. Synthesis of acoustic well-logging waveforms on an irregular grid. *Geophysical Prospection, Vol. 34, No. 8*, pp. 1145–1153, 1986.

[1215] C. Sanchez-Avila. An adaptive regularized method for deconvolution of signals with edges by convex projections. *IEEE Transactions on Signal Processing, Vol. 42(7)*, pp. 1849–1851, 1994.

[1216] C. Sanchez-Avila and J.A. Garcia-Moreno. An adaptive LSQR algorithm for computing discontinuous solutions in deconvolution problems. *Mathematics and Computers in Simulation, Vol. 50(1–4)*, pp. 323–329, 1999.

[1217] I.W. Sandberg. On the properties of some systems that distort signals – I. *Bell Syst. Tech. J.*, pp. 2033–2046, 1963.

[1218] J. Sanz and T. Huang. On the Gerchberg - Papoulis algorithm. *IEEE Transactions on Circuits and Systems, Vol. 30(12)*, pp. 907–908, 1983.

[1219] J.L.C. Sanz. On the reconstruction of bandlimited multidimensional signals from algebraic sampling contours. *Proceedings IEEE, Vol. 73*, pp. 1334–1336, 1985.

[1220] Vidi Saptari. **Fourier Transform Spectroscopy Instrumentation Engineering**. International Society for Optical Engineering (SPIE), 2003.

[1221] T.K. Sarkar, D.D. Weiner, and V.K. Jain. Some mathematical considerations in dealing with the inverse problem. *IEEE Transactions Antennas & Propagation, Vol. AP-29*, pp. 373–379, 1981.

[1222] K. Sasakawa. *Application of Miyakawa's Multidimensional Sampling Theorem (Japanese)*. Prof. Group on Inform. Theory, Institute Electron. Commun. Engineering Japan, I: no. I, 1960; II: no. 9, 1960; III: no. 2, 1961; IV: no. 6, 1961; V: no. 1, 1962.

[1223] Y. Sasaki and K. Teramoto. Improved satellite ground resolution for sea-ice observation using an inversion method and a priori information. *IEEE Journal of Oceanic Engineering. Vol. 31(1)*, pp. 219–229, 2006.

[1224] P. Sathyanarayana, P.S. Reddy, and M.N.S. Swamy. Interpolation of 2-D signals. *IEEE Transactions Circuits & Systems, Vol. 37, No. 5*, pp. 623–631, 1990.

[1225] K.D. Sauer and J.P. Allebach. Iterative reconstruction of bandlimited images from non-uniformly spaced samples. *IEEE Transactions Circuits & Systems, Vol. 34*, pp. 1497–1505, 1987.

[1226] R.W. Schafer, R.M. Mersereau, and M.A. Richards. Constrained iterative restoration algorithms. *Proceedings IEEE*, pp. 432–450, 1981.

[1227] R.W. Schafer and L.R. Rabiner. A digital signal processing approach to interpolation. *Proceedings IEEE, Vol. 61*, pp. 692–702, 1973.

[1228] S.A. Schelkunoff. Theory of antennas of arbitrary size and shape. *Proceedings IRE*, pp. 493–521, 1941.

[1229] W. Schempp. Radar ambiguity functions, the Heisenberg group, and holomorphic theta series. *Proceedings America Math. Society Vol. 92*, pp. 103–110, 1984.

[1230] W. Schempp. Gruppentheoretische aspekte der signal übertragung und der kardinalen interpolations splines I. *Math. Meth. Applied Sci., Vol. 5*, pp. 195–215, 1983.

[1231] J.L. Schiff and W.J. Walker. A sampling theorem for analytic functions. *Proceedings of the American Mathematical Society, Vol. 99(4)*, pp. 737–740, 1987.

[1232] J.L. Schiff and W.J. Walker. A sampling theorem and wintner results on Fourier coefficients. *Journal of Mathematical Analysis and Applications, Vol. 133(2)*, pp. 466–471, 1988.

[1233] J.L. Schiff and W.J. Walker. A sampling theorem for a class of pseudoanalytic functions. *Proceedings of the American Mathematical Society, Vol. 111(3)*, pp. 695–699, 1991.

[1234] R.J. Schlesinger and V.B. Ballew. An application of the sampling theorem to data collection in a computer integrated manufacturing environment. *International Journal of Computer Integrated Manufacturing, Vol. 2(1)*, pp. 15–22, 1989.

[1235] Dietrich Schlichthärle. **Digital Filters: Basics and Design**. Springer, 2000.

[1236] I.J. Schoenberg. Contributions to the problem of approximation of equidistant data by analytic functions. Part a: On the problem of smoothing or graduation. A first class of analytic approximation formulae. *Quart. Applied Math., Vol. 4*, pp. 45–99, 1946.

[1237] I.J. Schoenberg. Cardinal interpolation and spline functions. *Journal Approx. Theory 2*, pp. 167–206, 1969.

[1238] A. Schönhage. **Approximationstheorie**. de Gruyter, Berlin – New York, 1971.

[1239] Hartmut Schröder and Holger Blume. **One- and Multidimensional Signal Processing: Algorithms and Applications in Image Processing**. John Wiley & Sons, 2001.

[1240] L.L. Schumaker. **Spline Functions: Basic Theory**. Wiley, New York, 1981.

[1241] G. Schwierz, W. Härer, and K. Wiesent. Sampling and discretization problems in X-ray CT. In **Mathematical Aspects of Computerized Tomography**, Springer-Verlag, Berlin, 1981.

[1242] H.J. Scudder. Introduction to computer aided tomography. *Proceedings IEEE, Vol. 66*, pp. 628–637, 1978.

[1243] M. Sedletskii. **Fourier Transforms and Approximations**. CRC, 2000.

[1244] Robert T. Seeley. **An Introduction to Fourier Series and Integrals**. Dover Publications 2006.

[1245] J. Segethova. Numerical construction of the Hill function. *SIAM J. Numer. Anal. Vol. 9*, pp. 199–204, 1972.

[1246] K. Seip. An irregular sampling theorem for functions bandlimited in a generalized sense. *SIAM J. Applied Math., Vol. 47*, pp. 1112–1116, 1987.

[1247] K. Seip. A note on sampling bandlimited stochastic processes. *IEEE Transactions Information Theory, Vol. 36, No. 5*, p. 1186, 1990.

[1248] A. Sekey. A computer simulation study of real-zero interpolation. *IEEE Transactions Audio Electroacoustics, Vol. 18*, p. 43, 1970.

[1249] I.W. Selesnick. Interpolating multiwavelet bases and the sampling theorem. *IEEE Transactions on Signal Processing, Vol. 47(6)*, pp. 1615–1621, 1999.

[1250] M. De La Sen. Nonperiodic sampling and model matching. *Electron. Letters, Vol. 18*, pp. 311–313, 1982.

[1251] M. De La Sen and S. Dormido. Nonperiodic sampling and identifiability. *Electron. Letters, Vol. 17*, pp. 922–925, 1981.

[1252] M. De La Sen and M.B. Pay. On the use of adaptive sampling in hybrid adaptive error models. *Proceedings IEEE, Vol. 72*, pp. 986–989, 1984.

[1253] M.I. Sezan and H. Stark. Image restoration by method of convex set projections: Part II – Applications and Numerical Results. *IEEE Transactions Med. Imaging, Vol MI-1*, pp. 95–101, 1982.

[1254] M. Ibrahim Sezan, Henry Stark and Shu-Jen Yeh. Projection method formulations of Hopfield-type associative memory neural networks. *Applied Optics, Vol. 29, No. 17*, pp. 2616–2622, 1990.

[1255] M. Ibrahim Sezan and Henry Stark. Tomographic image reconstruction from incomplete view data by convex projections and direct Fourier inversion. *IEEE Transactions Medical Imaging, Vol. MI-3, No. 2*, pp. 91–98, 1984.

[1256] C.E. Shannon. A mathematical theory of communication. *Bell Systems Tech. J., Vol. 27*, pp. 379 and 623, 1948.

[1257] C.E. Shannon. Communications in the presence of noise. *Proceedings IRE, Vol. 37*, pp. 10–21, 1949.

[1258] C.E. Shannon and W. Weaver. **The Mathematical Theory of Communication**. University of Illinois Press, Urbana, IL, 1949.

[1259] H.S. Shapiro. **Smoothing and Approximation of Functions**. Van Nostrand Reinhold, New York, 1969.

[1260] H.S. Shapiro. Topics in Approximation Theory. **Lecture Notes in Mathematics**, *Vol. 187*, Springer-Verlag, Berlin - Heidelberg – New York, 1971.

[1261] H.S. Shapiro and R.A. Silverman. Alias free sampling of random noise. *Journal SIAM, Vol. 8*, pp. 225–236, 1960.

[1262] K.K. Sharma and S.D. Joshi. Signal reconstruction from the undersampled signal samples. *Optics Communications, Vol. 268(2)*, pp. 245–252, 2006.

[1263] L. Shaw. Spectral estimates from nonuniform samples. *IEEE Transactions Audio Electroacoustics, Vol. AU–19*, 1970.

[1264] R.G. Shenoy and T.W. Parks. An optimal recovery approach to interpolation. *IEEE Transactions Signal Processing, Vol. 40, No. 6*, pp. 1987–1988, 1992.

[1265] C.E. Shin, S.Y. Chung, and D. Kim. General sampling theorem using contour integral. *Journal of Mathematical Analysis and Application, Vol. 291(1)*, pp. 50–65, 2004.

[1266] J.H. Shin, J.H. Jung, and J.R. Paik. Spatial interpolation of image sequences using truncated projections onto convex sets. *IEICE Transactions on Fundamentals of Electronics, Communications and Computer Sciences, Vol. E82A(6)*, pp. 887–892, 1999.

[1267] Eric Stade. **Fourier Analysis**. Wiley-Interscience, 2005.

[1268] Elias M. Stein and Rami Shakarchi. **Fourier Analysis: An Introduction**. (Princeton Lectures in Analysis. Volume 1). Princeton University Press, 2003.

[1269] Elias M. Stein and Guido Weiss. **Introduction to Fourier Analysis on Euclidean Spaces.** Princeton University Press, 1971.

[1270] A. Stern and B. Javidi. Improved-resolution digital holography using the generalized sampling theorem for locally band-limited fields. *Journal of the Optical Society of America A – Optics, Image Science and Vision, Vol. 23(5)*, pp. 1227–1235, 2006.

[1271] A. Stern. Sampling of linear canonical transformed signals. *Signal Processing, Vol. 86(7)*, pp. 1421–1425, 2006.

[1272] A. Sherstinsky and C.G. Sodini. A programmable demodulator for oversampled analog-to-digital modulators. *IEEE Transactions Circuits & Systems, Vol. CAS-37, No. 9*, pp. 1092–1103, 1990.

[1273] S. Shirani, F. Kossentini, and R. Ward. Reconstruction of baseline JPEG coded images in error prone environments. *IEEE Transactions Image Processing, Vol. 9, No.7*, pp. 1292–1299, 2000.

[1274] F. S. Shoucair. Joseph Fourier's analytical theory of heat: a legacy to science and engineering. *IEEE Transactions on Education, Vol. 32*, pp. 359–366, 1989.

[1275] Emil Y. Sidky, Chien-Min Kao, and Xiaochuan Pan. Accurate image reconstruction from few-views and limited-angle data in divergent-beam CT. *Journal of X-Ray Science and Technology, Vol. 14(2)*, pp. 119–139, 2006.

[1276] G. Simmons, **Calculus Gems**, New York: Mcgraw Hill, Inc., 1992.

[1277] R. Simon. Krishnan and the sampling theorem. *Current Science, Vol. 75(11)*, pp. 1239–1245, 1998.

[1278] G. Sinclair, J. Leach, P. Jordan, *et al.* Interactive application in holographic optical tweezers of a multi-plane Gerchberg-Saxton algorithm for three-dimensional light shaping. *Optics Express, Vol. 12(8)*, pp. 1665–1670, 2004.

[1279] N.K. Sinha and G.J. Lastman. Identification of continuous-time multivariable systems from sampled data. *International J. Control, Vol. 35*, pp. 117–126, 1982.

[1280] D. Slepian. Prolate spheroidal wave functions Fourier analysis and uncertainty IV: Extensions to many dimensions; generalized prolate spheroidal wave functions. *Bell Systems Tech. J., Vol. 43*, pp. 3009–3057, 1964.

[1281] D. Slepian. Analytic solution of two apodization problems. *Journal of the Optical Society America, Vol. 55*, pp. 1110–1115, 1965.

[1282] D. Slepian. Some asymptotic expansions for prolate spheroidal wave functions. *Journal Math., Phys. Vol. 44, No. 2*, pp. 99–140, 1965.

[1283] D. Slepian. On bandwidth. *Proceedings IEEE, Vol. 64*, pp. 292–300, 1976.

[1284] D. Slepian. Prolate spheroidal wave functions, Fourier analysis and uncertainty – V: The discrete case. *Bell Systems Tech. J., Vol. 57*, pp. 1371–1430, 1978.

[1285] D. Slepian and H.O. Pollak. Prolate spheroidal wave functions Fourier analysis and uncertainty I. *Bell Systems Tech. J., Vol. 40*, pp. 43–63, 1961.

[1286] D. Slepian and E. Sonnenblick. Eigenvalues associated with prolate spheroidal wave functions of zero order. *Bell Systems Tech. J., Vol. 44*, pp. 1745–1758, 1965.

[1287] Robert Smedley and Garry Wiseman. **Introducing Pure Mathematics**. Oxford University Press, 2001.

[1288] Alexandre Smirnov. **Processing of Multidimensional Signals**. Springer, 1999.

[1289] D.K. Smith and R.J. Marks II. Closed form bandlimited image extrapolation. *Applied Optics, Vol. 20*, pp. 2476–2483, 1981.

[1290] Julius O. Smith III. **Mathematics of the Discrete Fourier Transform (DFT): with Audio Applications**. Second Edition. BookSurge Publishing, 2007.

[1291] M.J. Smith, Jr. An evaluation of adaptive sampling. *IEEE Transactions Automatic Control, Vol. AC-16*, pp. 282–284, 1971.

[1292] T. Smith, M.R. Smith, and S.T. Nichols. Efficient sinc function interpolation technique for center padded data. *IEEE Transactions Acoust., Speech, Signal Processing, Vol. ASSP-38, No. 9*, pp. 1512–1517, 1990.

[1293] Brian C. Smith. **Fundamentals of Fourier Transform Infrared Spectroscopy**. CRC Press, 1995.

[1294] Ian N. Sneddon. **The Use of Integral Transforms**. McGraw-Hill, New York, 1972.

[1295] Ian N. Sneddon. **Fourier Transforms**. Dover Publications, 1995.

[1296] M.A. Soderstrand. **Residue Number System Arithmetic: Modern Applications in Digital Signal Processing**. IEEE Press, 1986.

[1297] Guy R. Sohie and George N. Maracas. Orthogonality of exponential transients. *Proceedings IEEE, Vol. 76, No. 12*, pp. 1616–1618, 1988.

[1298] E.A. Soldatov and E.V. Chuprunov. The Kotelnikov-Shannon sampling theorem in the structure analysis. *Crystallography Reports, Vol. 44(5)*, pp. 732–733, 1999.

[1299] S.C. Som. Simultaneous multiple reproduction of space-limited functions by sampling of spatial frequencies. *Journal of the Optical Society America, Vol. 60*, pp. 1628–1634, 1970.

[1300] I. Someya. **Waveform Transmission**. Shukyo, Ltd., Tokyo, 1949.

[1301] Young-Seog Song and Yong Hoon Lee. Formulas for McClellan transform parameters in designing 2-D zero-phase FIR fan filters. *IEEE Signal Processing Letters, Vol. 3(11)*, pp. 291–293, 1996.

[1302] M. Soumekh. Bandlimited interpolation from unevenly spaced sampled data. *IEEE Transactions Acoust., Speech, Signal Processing, Vol. ASSP-36, No. 1*, pp. 110–122, 1988.

[1303] M. Soumekh. Reconstruction and sampling constraints for spiral data. *IEEE Transactions Acoust., Speech, Signal Processing, Vol. ASSP-37*, pp. 882–891, 1989.

[1304] Mehrdad Soumekh. **Fourier Array Imaging**. Prentice Hall Professional Technical Reference, 1993.

[1305] J.J. Spilker. Theoretical bounds on the performances of sampled data communications systems. *IRE Transactions Circuit Theory, Vol. CT-7*, pp. 335–341, 1960.

[1306] W. Splettstößer. Über die Approximation stetiger Funktionen durch die klassischen und durch neue Abtastsummen mit Fehlerabschätzungen. Doctoral dissertation, RWTH Aachen, Aachen, Germany, 1977.

[1307] W. Splettstößer. On generalized sampling sums based on convolution integrals. *AEU, Vol. 32, No. 7*, pp. 267–275, 1978.

[1308] W. Splettstößer. Error estimates for sampling approximation of non-bandlimited functions. *Math. Meth. Applied Sci., Vol. 1*, pp. 127–137, 1979.

[1309] W. Splettstößer. Error analysis in the Walsh sampling theorem. In *1980 IEEE International Symposium on Electromagnetic Compatibility, Baltimore, Vol. 409*, pp. 366–370, 1980.

[1310] W. Splettstößer. Bandbegrenzte und effektiv bandbegrenzte funktionen und ihre praediktion aus abtastwerten. Habilitationsschrift, *RWTH Aachen*, 1981.

[1311] W. Splettstößer. On the approximation of random processes by convolution processes. *Z. Angew. Math. Mech., Vol. 61*, pp. 235–241, 1981.

[1312] W. Splettstößer. Sampling series approximation of continuous weak sense stationary processes. *Information & Control, Vol. 50*, pp. 228–241, 1981.

[1313] W. Splettstößer. On the prediction of bandlimited processes from past samples. *Inform. Sci., Vol. 28*, pp. 115–130, 1982.

[1314] W. Splettstößer. Sampling approximation of continuous functions with multi-dimensional domain. *IEEE Transactions Information Theory, Vol. IT-28*, pp. 809–814, 1982.

[1315] W. Splettstößer. 75 years aliasing error in the sampling theorem. In *EUSIPCOM – 83, Signal Processing: Theories and Applications (2. European Signal Processing Conference, Erlangen)*, H.W. Schüssler, editor. North-Holland Publishing Company, Amsterdam - New York - Oxford, 1983, pp. 1–4.

[1316] W. Splettstößer. Lineare praediktion von nicht bandbegrenzten funktionen. *Z. Angew. Math. Mech. Vol. 64*, pp. T393–T395, 1984.

[1317] W. Splettstößer, R.L. Stens, and G. Wilmes. On approximation by the interpolating series of G. Valiron. *Funct. Approx. Comment. Math., Vol. 11*, pp. 39–56, 1981.

[1318] W. Splettstößer and W. Ziegler. The generalized Haar functions, the Haar transform on $\mathbf{r}^+$, and the Haar sampling theorem. In *Proceedings of the "Alfred Haar Memorial Conference,"* Budapest, pp. 873–896, 1985.

[1319] T. Springer. Sliding FFT computes frequency spectra in real time. *EDN Magazine*, September 29, 1988, pp. 161–170.

[1320] Julien Clinton Sprott. **Chaos and Time-Series Analysis**. Oxford University Press, 2003.

[1321] C.J. Standish. Two remarks on the reconstruction of sampled non-bandlimited functions. *IBM J. Res. Develop. Vol. 11*, pp. 648–649, 1967.

[1322] H. Stark. Sampling theorems in polar coordinates. *Journal of the Optical Society America, Vol. 69*, pp. 1519–1525, 1979.

[1323] H. Stark. Polar sampling theorems of use in optics. *Proceedings SPIE, International Society Optical Engineering, Vol. 358*, pp. 24–30, 1982.

[1324] H. Stark, editor. **Image Recovery: Theory and Application**. Academic Press, Orlando, FL, 1987.

[1325] H. Stark, D. Cahana, and H. Webb. Restoration of arbitrary finite-energy optical objects from limited spatial and spectral information. *Journal of the Optical Society America, Vol. 71*, pp. 635–642, 1981.

[1326] H. Stark and C.S. Sarna. Bounds on errors in reconstructing from undersampled images. *Journal of the Optical Society America, Vol. 69*, pp. 1042–1043, 1979.

[1327] H. Stark and C.S. Sarna. Image reconstruction using polar sampling theorems of use in reconstructing images from their samples and in computer aided tomography. *Applied Optics, Vol. 18*, pp. 2086–2088, 1979.

[1328] H. Stark and M. Wengrovitz. Comments and corrections on the use of polar sampling theorems in CT. *IEEE Transactions Acoust., Speech, Signal Processing, Vol. ASSP-31*, pp. 1329–1331, 1983.

[1329] H. Stark and John W. Woods. **Probability, Random Processes, and Estimation Theory for Engineers**. Prentice-Hall, Englewood Cliffs, NJ, 1986.

[1330] H. Stark, J.W. Woods, I. Paul, and R. Hingorani. Direct Fourier reconstruction in computer tomography. *IEEE Transactions Acoust., Speech, Signal Processing, Vol. ASSP-29*, pp. 237–245, 1981.

[1331] H. Stark, J.W. Woods, I. Paul, and R. Hingorani. An investigation of computerized tomography by direct Fourier inversion and optimum interpolation. *IEEE Transactions Biomedical Engineering, Vol. BME-28*, pp. 496–505, 1981.

[1332] H. Stark. Sampling theorems in polar coordinates. *Journal of the Optical Society America, Vol. 69, No. 11*, pp. 1519–1525, 1979.

[1333] H. Stark and Y. Yang, **Vector Space Projections: A Numerical Approach to Signal and Image Processing, Neural Nets, and Optics** Wiley-Interscience, New York, 1998.

[1334] R. Stasinski and J. Konrad. Linear shift-variant filtering for POCS of reconstruction irregularly sampled images. *Proceedings of the 2003 International Conference on Image Processing*, 2003. ICIP 2003. *Vol. 3*, pp. III - 689–692, 2003.

[1335] O.N. Stavroudis. **The Optics of Rays, Wavefronts, and Caustics**. Academic Press, New York, 1972.

[1336] R. Steele and F. Benjamin. Sample reduction and subsequent adaptive interpolation of speech signals. *Bell Systems Tech. J., Vol. 62*, pp. 1365–1398, 1983.

[1337] P. Stelldinger and U. Kothe. Towards a general sampling theory for shape preservation. *Image and Vision Computing, Vol. 23(2)*, pp. 237–248, 1 2005.

[1338] F. Stenger. Approximations via Whittaker's cardinal function. *Journal Approx. Theory, Vol. 17*, pp. 222–240, 1976.

[1339] F. Stenger. Numerical methods based on Whittaker cardinal, or sinc functions. *SIAM Rev., Vol. 23*, pp. 165–224, 1981.

[1340] R.L. Stens. Approximation of duration-limited functions by sampling sums. *Signal Processing, Vol. 2*, pp. 173–176, 1980.

[1341] R.L. Stens. A unified approach to sampling theorems for derivatives and Hilbert transforms. *Signal Processing, Vol. 5*, pp. 139–151, 1983.

[1342] R.L. Stens. Error estimates for sampling sums based on convolution integrals. *Information & Control, Vol. 45*, pp. 37–47, 1980.

[1343] R.L. Stens. Approximation of functions by Whittaker's cardinal series. in General Inequalities 4 (*Proceedings Conf. Math. Res. Institute Oberwolfach*, W. Walter, ed.) Birkhäuser Verlag, Basel, pp. 137–149, 1984.

[1344] E.G. Steward. **Fourier Optics: An Introduction**. Dover Publications, 2004.

[1345] R.M. Stewart. Statistical design and evaluation of filters for the restoration of sampled data. *Proceedings IRE, Vol. 44*, pp. 253–257, 1956.

[1346] D.C. Stickler. An upper bound on aliasing error. *Proceedings IEEE, Vol. 55*, pp. 418–419, 1967.

[1347] G. Strang. **Linear Algebra and Its Applications**. Academic Press, New York, 1980.

[1348] Gilbert Strang and Truong Nguyen. **Wavelets and Filter Banks.** Second Edition. Wellesley College, 1996.

[1349] R.N. Strickland, A.P. Anderson, and J.C. Bennett. Resolution of subwavelength-spaced scatterers by superdirective data processing simulating evanescent wave illumination. *Microwaves, Opt. & Acoust., Vol. 3*, pp. 37–42, 1979.

[1350] G. Stigwall and S. Galt. Non-linear signal retrieval in wide-band photonic time-stretch systems using the gerchberg-Saxton algorithm. International Topical Meeting on Microwave Photonics, 2006. MWP '06. pp. 1–4, 2006

[1351] T. Strohmer and J. Tanner. Fast reconstruction methods for bandlimited functions from periodic nonuniform sampling. *SIAM Journal on Numerical Analysis, Vol. 44(3)*, pp. 1073–1094, 2006.

[1352] J.A. Stuller. Reconstruction of finite duration signals. *IEEE Transactions Information Theory, Vol. IT-18*, pp. 667–669, 1972.

[1353] K.L. Su. **Analog Filters**, Second Edition. Springer, 2002.

[1354] R. Sudhakar, R.C. Agarwal, and S.C. Dutta Roy. Fast computation of Fourier transform at arbitrary frequencies. *IEEE Transactions Circuits & Systems, Vol. CAS-28*, pp. 972–980, 1982.

[1355] R. Sudol and B.J. Thompson. An explanation of the Lau effect based on coherence theory. *Optical Communication*, pp. 105–110, 1979.

[1356] R. Sudol and B.J. Thompson. Lau effect: theory and experiment. *Applied Optics*, pp. 1107–1116, 1981.

[1357] H. Sugiyama. A sampling theorem with equally spaced sampling points of the negative time axis. *Mathematics of Control Signals and Systems, Vol 6(2)*, pp. 125–134, 1993.

[1358] B. Summers, G.D. Cain. An extension to the sampling theorem for lowpass bandlimited periodic time functions using n^{th} order sampling schemes. *IEEE Transactions on Instrumentation and Measurement, Vol. 40(6)*, pp. 990–993, 1991.

[1359] Hong-Bo Sun, Guo-Sui Liu, Hong Gu, and Wei-Min Su. Application of the fractional Fourier transform to moving target detection in airborne SAR. *IEEE Transactions on Aerospace and Electronic Systems, Vol. 38(4)*, pp. 1416–1424, 2002.

[1360] H.R. Sun and W.T. Li. Positive solutions of second-order half-linear dynamic equations on time scales. *Applied Mathematics and Computation, Vol. 158(2)*, pp. 331–344, 2004.

[1361] H. Sun and W. Kwok. Concealment of damaged blocks transform coded images using projections onto convex sets. *IEEE Transactions Image Processing, Vol. 4*, pp. 470–477, 1995.

[1362] W.C. Sun and X.W. Zhou. Sampling theorem for wavelet subspaces: Error estimate and irregular sampling. *IEEE Transactions on Signal Processing, Vol. 48(1)*, pp. 223–226, 2000.

[1363] W.C. Sun and Z.W. Zhou. The aliasing error in recovery of nonbandlimited signals by prefiltering and sampling. *Applied Mathematics Letters, Vol. 16(6)*, pp. 949–954, 2003.

[1364] Y.R. Sun and S. Signell. Effects of noise and jitter in bandpass sampling. *Analog Integrated Circuits and Signal Processing, Vol. 42(1)*, pp. 85–97, 2005.

[1365] Zhaohui Sun, et al. Interactive optimization of 3D shape and 2D correspondence using multiple geometric constraints via POCS. *IEEE Transactions on Pattern Analysis and Machine Intelligence, Vol. 24(4)*, pp. 562–569, 2002.

[1366] H. Sundaram, S.D. Joshi, and R.K.P. Bhatt. Scale periodicity and its sampling theorem. *IEEE Transactions on Signal Processing, Vol. 45(7)*, pp. 1862–1865, 1997.

[1367] D. Sundararajan. **The Discrete Fourier Transform: Theory, Algorithms and Applications**. World Scientific Publishing Company, 2001.

[1368] S.K. Suslov. **An Introduction to Basic Fourier Series**. Springer, 2003.

[1369] K. Swaminathan. Signal restoration from data aliased in time. *IEEE Transactions Acoust., Speech, Signal Processing, Vol. ASSP-33*, pp. 151–159, 1985.

[1370] G.J. Swanson and E.N. Leith. Lau effect and grating imaging. *Journal of the Optical Society America, Vol. 72*, pp. 552–555, 1982.

[1371] S. Szapiel. Sampling theorem for rotationally symmetric systems based on dini expansion. *Opt. Letters, Vol. 6*, pp. 7–9, 1981.

[1372] H.H. Szu and J.A. Blodgett. Wigner distribution and ambiguity function. *Optics in Four Dimensions–1980, AIP Conf. Proceedings* 65, M.A. Machado and L.M. Narducci, eds. American Institute of Physics, New York, pp. 355–381, 1980.

T

[1373] P.D. Tafti, S. Shirani, and X.L. Wu. On interpolation and resampling of discrete data. *IEEE Signal Processing Letters, Vol. 13(12)*, pp. 733–736, 2006.

[1374] S. Takata. J B Fourier in the history of thermal radiation research. *Historia Sci., (2) 2 (3)*, pp. 203–221, 1993.

[1375] M. Takagi and E. Shinbori. High quality image magnification applying the Gerchberg-Papoulis iterative algorithm with DCT. *Systems and Computers in Japan, Vol. 25(6)*, pp. 80–90 1994.

[1376] Y. Tanaka, T. Fujiwara, and H. Koshimizu, et al. OK-Quantization Theory and its relationship to sampling theorem. *Lecture Notes in Computer Science, Vol. 3852*, pp. 479–488 2006.

[1377] B. Tatian. Asymptotic expansions for correcting truncation error in transfer function calculations. *Journal of the Optical Society America, Vol. 61*, pp. 1214–1224, 1971.

[1378] H. Taylor and C.A. Lipson. **Fourier Transforms and X-Ray Diffraction**. G. Bell & Sons Ltd., 1958.

[1379] C. Tellambura. Computing the outage probability in mobile radio networks using the sampling theorem. *IEEE Transactions on Communications, Vol. 47(8)*, pp. 1125–1128, 1999.

[1380] G.C. Temes, V. Barcilon, and F.C. Marshall III. The optimization of bandlimited systems. *Proceedings IEEE, Vol. 61*, pp. 196–234, 1973.

[1381] K. Teramoto. Conjugate gradient projection onto convex sets for sparse array acoustic holography. *IEICE Transactions on Fundamentals of Electronics Communications and Computer Sciences, Vol E80A(5)*, pp. 833–839, 1997.

[1382] Audrey Terras. **Fourier Analysis on Finite Groups and Applications**. Cambridge University Press, 1999.

[1383] Les Thede. **Practical Analog And Digital Filter Design**. Artech House Publishers, 2004.

[1384] M. Theis. Über eine Interpolationsformel von de la Vallée-Poussin. *Math. Z., Vol. 3*, pp. 93–113, 1919.

[1385] C. Thieke, T. Bortfeld, A. Niemierko, and S. Nill. From physical dose constraints to equivalent uniform dose constraints in inverse radiotherapy planning *Medical Physics. Vol. 30(9)*, pp. 2332–2339, 2003.

[1386] G. Thomas. A modified version of Van Cittert's iterative deconvolution procedure. *IEEE Transactions Acoust., Speech, Signal Processing, Vol. ASSP-29*, pp. 938–939, 1981.

[1387] J.B. Thomas. **An Introduction to Statistical Communication Theory**. John Wiley & Sons, Inc., New York, 1969.

[1388] J.B. Thomas and B. Liu. Error problems in sampling representations, Part 5. *IEEE International Conv. Rec., Vol. 12*, pp. 269–277, 1964.

[1389] B.J. Thompson. Image formation with partially coherent light. In **Progress in Optics**, E. Wolf, editor, North Holland, Amsterdam, pp. 169–230, 1969.

[1390] B.J. Thompson. Multiple imaging by diffraction techniques. *Applied Optics*, p. 312, 1976.

[1391] J.E. Thompson and L.H. Luessen, editors. **Fast Electrical and Optical Measurements, Vol. 1**. Martinus Nijhoff Publishers, 1986.

[1392] Sir William Thomson (Lord Kelvin) and Peter Guthrie Tait. **Treatise On Natural Philosophy, By Sir William Thomson And Peter Guthrie Tait**, University of Michigan Library, 2001.
[1393] B. Tian et al. POCS supperresolution image reconstruction using wavelet transform *Proceedings of International Symposium on 2004 Intelligent Signal Processing and Communication Systems*, ISPACS 2004. pp. 67–70, 2004.
[1394] A.N. Tikhonov and V.Y. Arsenine. **Solutions of Ill-Posed Problems**. Wiston/Wiley, Washington, DC, 1977.
[1395] A.F. Timan. **Theory of Approximation of Functions of a Real Variable**. Pergamon Press, Oxford, 1963.
[1396] E.C. Titchmarsh. The zeros of certain integral functions. *Proceedings London Math. Society, Vol. 25*, pp. 283–302, 1926.
[1397] E.C. Titchmarsh. **Introduction to the Fourier Integral**. Oxford University Press, Oxford, 1937.
[1398] E.C. Titchmarsh. *The Theory of Functions,* 2nd ed. University Press, London, 1939.
[1399] E.C. Titchmarsh. **Introduction to the Theory of Fourier Integrals (2nd Edition)**. Clarendon Press, Oxford, 1948.
[1400] D.E. Todd. Sampled data reconstruction of deterministic bandlimited signals. *IEEE Transactions Information Theory, Vol. IT-19*, pp. 809–811, 1973.
[1401] Georgi P. Tolstov. **Fourier Series**. Dover Publications, 1976.
[1402] Lex Tokyo, Yo Sakakibara, Alan Gleason. **Who Is Fourier?: A Mathematical Adventure**, Language Research Foundation, 1995.
[1403] G. Toraldo diFrancia. Super-gain antennas and optical resolving power. *Suppl. Nuovo Cimento*, pp. 426–438, 1952.
[1404] G. Toraldo diFrancia. Resolving power and information. *Journal of the Optical Society America, Vol 45*, pp. 497–501, 1955.
[1405] G. Toraldo diFrancia. Directivity, super-gain and information. *IRE Transactions Antennas & Propagation, Vol. AP-4*, pp. 473–478, 1956.
[1406] G.Toraldo diFrancia. **La diffrazione della luce**. Einaudi Bornghieri, Torino, 1958.
[1407] G. Toraldo diFrancia. Degrees of freedom of an image. *Journal of the Optical Society America, Vol. 59*, pp. 799–804, 1969.
[1408] R.H. Torres. Spaces of Sequences, Sampling Theorem, and Functions of Exponential Type. *Studia Mathematica, Vol. 100(1)*, pp. 51–74, 1991.
[1409] R.T. Torres, P. Pellat-Finet, and Y. Torres. Sampling theorem for fractional bandlimited signals: A self-contained proof. Application to digital holography. *IEEE Signal Processing Letters, Vol. 13(11)*, pp. 676–679, 2006.
[1410] Hans Triebel. **Fractals and Spectra: Related to Fourier Analysis and Function Spaces**. Birkhäuser Basel, 1997.
[1411] Roald M. Trigub and Eduard S. Belinsky. **Fourier Analysis and Approximation of Functions.** Springer, 2004.
[1412] H.J. Trussel, L.L. Arnder, P.R. Morand, and R.C. Williams. Corrections for nonuniform sampling distortions in magnetic resonance imagery. *IEEE Transactions Medical Imaging, Vol. MI-7, No. 1*, pp. 32–44, 1988.
[1413] C.C. Tseng. Design of two-dimensional FIR digital filters by McClellan transform and quadratic programming. *IEE Proceedings on Vision, Image and Signal Processing, Vol. 148(5)*, pp. 325–331, 2001.
[1414] Shiao-Min Tseng. **Noise Level Analysis of Linear Restoration Algorithms for Bandlimited Signals**. Ph.D. thesis, University of Washington, 1984.
[1415] John Hudsion Tiner. **Louis Pasteur – Founder of Modern Medicine**. Milford, MI: Mott Media, Inc., p. 75, 1990.
[1416] S. Tsekeridou and I. Pitas. MPEG-2 error concealment based on block-matching principles. *IEEE Transactions Circuits Syst. Video Technol., Vol. 10*, pp. 646–658, 2000.
[1417] J. Tsujiuchi. Correction of optical images by compensation of aberrations and by spatial frequency filtering. In **Progress in Optics**, E. Wolf, editor. North-Holland, Amsterdam, pp. 131–180, 1963.

[1418] Y. Tsunoda and Y. Takeda. High density image-storage holograms by a random phase sampling method. *Applied Optics, Vol. 13*, pp. 2046–2051, 1974.

[1419] P. Turán. **The Work of Alfréd Rényi**. Matematikai Lapok 21, pp. 199–210, 1970.

[1420] B. Turner. Fourier's seventeen lines problem. *Math. Mag., Vol. 53(4)*, pp. 217–219, 1980.

[1421] J. Turunen, A. Vasara, and A.T. Friberg. Propagation invariance and self-imaging in variable coherence optics. *Journal of the Optical Society America A, Vol. 8*, pp. 282–289, 1991.

[1422] J. Turunen, A. Vasara, J. Werterholm, and A. Salin. Strip-geometry two-dimensional Dammann gratings. *Optical Communication, Vol. 74*, pp. 245–252, 1989.

[1423] Tzafestas **Multidimensional Systems**. CRC, 1986.

U

[1424] Kazunori Uchida. Spectral domain method and generalized sampling theorems. *Transactions of IEICE, Vol. E73, No. 3*, pp. 357–359, 1990.

[1425] T. Umemoto, S. Fujisawa, and T. Yoshida. Constant Q-value filter banks with spectral analysis using LMS algorithm. *Proceedings of the 37th SICE Annual Conference.* (ISICE '98). pp. 1019–1024, 1998.

[1426] M. Unser, A. Aldroubi, and M. Eden Polynomial spline signal approximations: Filter design and asymptotic equivalence with Shannon's sampling theorem. *IEEE Transactions Information Theory, Vol. IT-38, No. 1*, 1992.

[1427] H.P. Urbach. Generalised sampling theorem for band-limited functions. *Mathematical and Computer Modelling, Vol. 38(1–2)*, pp. 133–140, 2003.

[1428] S. Urieli *et al.* Optimal reconstruction of images from localized phase. *IEEE Transactions on Image Processing, Vol. 7(6)*, pp. 838–853, 1998.

V

[1429] P.P. Vaidyanathan and Vincent C. Liu. Classical sampling theorems in the context of multirate and polyphase digital filter bank structures. *IEEE Transactions Acoust., Speech, Signal Processing, Vol. ASSP-36, No. 9*, pp. 1480–1495, 1988.

[1430] P.P. Vaidyanathan. **Multirate Systems And Filter Banks.** Prentice Hall, 1992.

[1431] C. Vaidyanathan and K.M. Buckley. A sampling theorem for EEG electrode configuration. *IEEE Transactions on Biomedical Engineering, Vol. 44(1)*, pp. 94–97, 1997.

[1432] P.P. Vaidyanathan. Generalizations of the sampling theorem: seven decades after Nyquist. *IEEE Transactions on Circuits and Systems I – Fundamental Theory and Applications, Vol. 48(9)*, pp. 1094–1109, 2001.

[1433] Lucia Valbonesi and Ansari, Frame-based approach for peak-to-average power ratio reduction in OFDM, *IEEE Transactions on Communications. Vol. 54(9)*, pp. 1604–1613, 2006.

[1434] P.H. Van Cittert. Zum einfluss der splatbreite auf die intensitat-swerteilung in spektrallinien II. *Z. Phys., Vol. 69*, pp. 298–308, 1931.

[1435] L. Valbonesi and R. Ansari. POCS-based frame-theoretic approach for peak-to-average power ratio reduction in OFDM. *IEEE 60th Vehicular Technology Conference*, 2004. *Vol. 1*, pp. 631–634, 2004.

[1436] E. Van der Ouderaa and J. Renneboog. Some formulas and application of nonuniform sampling of bandwidth limited signal. *IEEE Transactions Instrumentation & Measurement, Vol.IM–37, No. 3*, pp. 353–357, 1988.

[1437] P.J. Van Otterloo and J.J. Gerbrands. A note on sampling theorem for simply connected closed contours. *Information & Control, Vol. 39*, pp. 87–91, 1978.

[1438] G.A. Vanasse and H. Sakai. Fourier spectroscopy. In **Progress in Optics**, E. Wolf, editor. North-Holland, Amsterdam, 1967.

[1439] A. VanderLugt. Optimum sampling of Fresnel transforms. *Applied Optics, Vol. 29, No. 23*, pp. 3352–3361, 1990.

[1440] A.B. VanderLugt. Signal detection by complex spatial filtering. *IEEE Transactions Information Theory, Vol. IT-10*, pp. 2–8, 1964.

[1441] P.J. Vanotterloo and J.J. Gerbrands. Note on a Sampling Theorem for Simply Connected Closed Contours. *Information and Control, Vol. 39(1)*, pp. 87–91, 1978.

[1442] Harry L. Van Trees. **Detection, Estimation, and Modulation Theory, Part I**. Wiley-Interscience; Reprint edition, 2001.

[1443] M.E. Van Valkenburg, **Introduction to Modern Network Synthesis**, John Wiley & Sons, 1960.

[1444] B.J. van Wyk et al. POCS-based graph matching algorithm. *IEEE Transactions on Pattern Analysis and Machine Intelligence, Vol. 26(11)*, pp. 1526–1530, 2004.

[1445] R.G. Vaughan, N.L. Scott, and D.R. White. The Theory of Bandpass Sampling. *IEEE Transactions Signal Processing, Vol. SP–39, No. 9*, pp. 1973–1984, 1991.

[1446] R. Veldhuis. **Restoration of Lost Samples in Digital Signals.** Prentice Hall, New York, 1990.

[1447] R. Venkataramani and Y. Bresler. Sampling theorems for uniform and periodic nonuniform MIMO sampling of multiband signals. *IEEE Transactions on Signal Processing, Vol. 51(12)*, pp. 3152–3163, 2003.

[1448] David Vernon. **Fourier Vision - Segmentation and Velocity Measurement Using the Fourier Transform**. Springer, 2001.

[1449] G.A. Viano. On the extrapolation of optical image data. *Journal Math. Phys., Vol. 17*, pp. 1160–1165, 1976.

[1450] U. Visitkitjakarn, Wai-Yip Chan, and Yongyi Yang. Recovery of speech spectral parameters using convex set projection. *1999 IEEE Workshop on Speech Coding Proceedings*, pp. 34–36, 1999.

[1451] G. Viswanath and T.V. Sreenivas. Cone-kernel representation versus instantaneous power spectrum. *IEEE Transactions on Signal Processing, Vol. 47(1)*, pp. 250–254, 1999.

[1452] H. Voelker. Toward a unified theory of modulation. *Proceedings IEEE, Vol. 54*, p. 340, 1966.

[1453] L. Vogt. Sampling sums with kernels of finite oscillation. *Numer. Funct. Anal., Optimiz., Vol. 9*, pp. 1251–1270, 1987/88.

[1454] M. von Laue. Die freiheitsgrade von strahlenbundeln. *Ann. Phys., Vol. 44*, pp. 1197–1212, 1976.

[1455] J.J. Voss. A sampling theorem with nonuniform complex nodes. *Journal of Approximation Theory, Vol. 90(2)*, pp. 235–254, 1997.

W

[1456] L.A. Wainstein and V.D. Zubakov. **Extraction of Signals from Noise**. Prentice-Hall, Englewood Cliffs, NJ, 1962.

[1457] James S. Walker. **Fourier Analysis**. Oxford University Press, 1988.

[1458] J.F. Walkup. Space-variant coherent optical processing. *Optical Engineering, Vol. 19*, pp. 339–346, 1980.

[1459] G.G. Walter. A sampling theorem for wavelet subspaces. *IEEE Transactions Information Theory, Vol. 38*, pp. 881–884, 1992.

[1460] P.T. Walters. Practical applications of inverting spectral turbidity data to provide aerosol size distributions. *Applied Optics, Vol. 19*, pp. 2353–2365, 1980.

[1461] A. Walther. The question of phase retrieval in optics. *Opt. Acta*, pp. 41–49, 1962.

[1462] A. Walther. Gabor's theorem and energy transfer through lenses. *Journal of the Optical Society America, Vol. 57*, pp. 639–644, 1967.

[1463] A. Walther. Radiometry and coherence. *Journal of the Optical Society America, Vol. 58*, pp. 1256–1259, 1968.

[1464] Danzhi Wang, Li Shujian and Shao Dingrong. The analysis of frequency-hopping signal acquisition based on Cohen-reassignment joint time-frequency distribution. *Proceedings. Asia-Pacific Conference on Environmental Electromagnetics, 2003. CEEM 2003.* pp. 21–24, 2003.

[1465] J.Y. Wang, P.P. Zhu, X. Zheng X, et al. Study on the inside source hologram reconstruction algorithm based on discrete Fourier transform *Acta Physica Sinica, Vol. 54(3)*, pp. 1172–1177, 2005.

[1466] H.S.C. Wang. On Cauchy's interpolation formula and sampling theorems. *Journal Franklin Institute, Vol. 289*, pp. 233–236, 1970.

[1467] Y. Wang and Q.F. Zhu. Error control and concealment for video communication: A review. *Proceedings IEEE Multimedia Signal Processing*, pp. 974–997, 1998.

[1468] Q. Wang and L.N. Wu. Translation invariance and sampling theorem of wavelet. *IEEE Transactions on Signal Processing, Vol. 48(5)*, pp. 1471–1474, 2000.

[1469] Q. Wang and L.N. Wu. A sampling theorem associated with quasi-fourier transform. *IEEE Transactions on Signal Processing, Vol. 48(3)*, pp. 895–895, 2000.

[1470] Yiwei Wang, J.F. Doherty, and R.E. Van Dyck. A POCS-based algorithm to attack image watermarks embedded in DFT amplitudes. *Proceedings of the 43rd IEEE Midwest Symposium on Circuits and Systems*, 2000. *Vol. 3*, pp. 1100–1103, 2000.

[1471] G. Watson **Treatise on the Theory of Bessel Functions**, 2nd ed. Cambridge University Press, Cambridge, 1962.

[1472] H.J. Weaver. **Applications of Discrete and Fourier Analysis.** Wiley, New York, 1983.

[1473] C. Weerasinghe, A. Wee-Chung Liew and Hong Yan. Artifact reduction in compressed images based on region homogeneity constraints using the projection onto convex sets algorithm. *IEEE Transactions on Circuits and Systems for Video Technology, Vol. 12(10)*, pp. 891–897, 2002.

[1474] S.J. Wein. Sampling theorem for the negative exponentially correlated output of lock-in amplifiers. *Applied Optics, Vol. 28(20)*, pp. 4453–4457, 1989.

[1475] A. Weinberg and B. Liu. Discrete time analysis of nonuniform sampling first-and-second order digital phase lock loop. *IEEE Transactions Communications*, pp. 123–137, 1974.

[1476] P. Weiss. Sampling theorems associated with Sturm-Liouville systems. *Bull. Amer. Math. Society, Vol. 63*, p. 242, 1957.

[1477] P. Weiss. An estimate of the error arising from misapplication of the sampling theorem. *Amer. Math. Society Notices 10-351*, (Abstract No. 601–54), 1963.

[1478] F. Weisz. **Summability of Multi-Dimensional Fourier Series and Hardy Spaces**. Springer, 2002.

[1479] A.L. Wertheimer and W.L. Wilcock. Light scattering measurements of particle distributions. *Applied Optics, Vol. 15*, pp. 1616–1620, 1976.

[1480] L.B. White. Signal synthesis from Cohen's class of bilinear time-frequency signal representations using convex projections. *1991 International Conference on Acoustics, Speech, and Signal Processing*, 1991. ICASSP–91. *Vol. 3*, pp. 2053–2056, 1991.

[1481] Robert White. **Chromatography/Fourier Transform Infrared Spectroscopy and its Applications**. CRC, 1989.

[1482] E.T. Whittaker. On the functions which are represented by the expansions of the interpolation theory. *Proceedings Royal Society Edinburgh, Vol. 35*, pp. 181–194, 1915.

[1483] E.T. Whittaker and G.N. Watson. **A Course of Modern Analysis**. Cambridge University Press, Cambridge, 1927, Chaps. 20 and 21.

[1484] J.M. Whittaker. On the cardinal function of interpolation theory. *Proceedings Math. Society Edinburgh, Vol. 2*, pp. 41–46, 1927–1929.

[1485] J.M. Whittaker. The Fourier theory of the cardinal function. *Proceedings Math. Society Edinburgh, Vol. 1*, pp. 169–176, 1929.

[1486] J.M. Whittaker. **Interpolatory Function Theory.** Cambridge Tracts in Mathematical Physics, No. 33, Cambridge University Press, Cambridge, 1935.

[1487] G. Whyte and J. Courtial. Experimental demonstration of holographic three-dimensional light shaping using a Gerchberg-Saxton algorithm. *New Journal of Physics.* 7 No. 117, 2005.

[1488] Roman Wienands and Wolfgang Joppich. **Practical Fourier Analysis for Multigrid Methods**. Chapman & Hall/CRC, 2004.

[1489] N. Wiener. **Extrapolation, Interpolation, and Smoothing of Stationary Time Series**. Wiley and Sons Inc., New York, 1949.

[1490] N. Wiener. **Collected Works, Vol. III** (ed. by P.R. Masani). M.I.T. Press, Cambridge, MA., 1981.

[1491] E. Wigner. On the quantum correction for thermodynamic equilibrium. *Phys. Rev., Vol. 40*, pp. 749–759, 1932.

[1492] Howard J. Wilcox and David L. Myers. **An Introduction to Lebesgue Integration and Fourier Series**. Dover Publications, 1995.

[1493] R.G. Wiley, H. Schwarzlander, and D.D. Weiner. Demodulation procedure for very wideband FM. *IEEE Transactions Commun., Vol. COM–25*, pp. 318–327, 1977.

[1494] R.G. Wiley. On an iterative technique for recovery of bandlimited signals. *Proceedings IEEE*, pp. 522–523, 1978.

[1495] R.G. Wiley. Recovery of bandlimited signals from unequally spaced samples. *IEEE Transactions Communications, Vol. COM–26*, pp. 135–137, 1978.

[1496] Ron R. Williams. **Spectroscopy and the Fourier Transform**. John Wiley & Sons, 1995.

[1497] Earl G. Williams. **Fourier Acoustics: Sound Radiation and Nearfield Acoustical Holography**. Academic Press, 1999.

[1498] W.J. Williams. Reduced interference distributions: Biological applications and interpretations. *Proceedings of the IEEE. Vol. 84(9)*, pp. 1264–1280, 1996.

[1499] N.P. Willis and Y. Bresler. Norm invariance of minimax interpolation *IEEE Transactions Information Theory, Vol. 38, No. 3*, pp. 1177–1181, 1992.

[1500] T. Wilson and C. Sheppard. **Theory and Practice of Scanning Microscopy**. Academic Press, London, 1978.

[1501] Steve Winder. **Analog and Digital Filter Design**, Second Edition, Newnes, 2002.

[1502] D.J. Wingham. The reconstruction of a band-limited function and its Fourier transform from a finite number of samples at arbitrary locations by singular value decomposition. *IEEE Transactions Signal Processing, Vol. SP–40*, pp. 559–570, 1992.

[1503] J.T. Winthrop. Propagation of structural information in optical wave fields. *Journal of the Optical Society America, Vol. 61*, pp. 15–30, 1971.

[1504] E. Wolf. Coherence and radiometry. *Journal of the Optical Society America, Vol. 68*, pp. 6–17, 1982.

[1505] E. Wolf. New theory of partial coherence in the space-frequency domain. Part I: Spectra and cross spectra of steady-state sources. *Journal of the Optical Society America, Vol. 72*, pp. 343–351, 1982.

[1506] E. Wolf. Coherent-mode propagation in spatially bandlimited wave fields. *Journal of the Optical Society America A, Vol. 3*, pp. 1920–1924, 1986.

[1507] H. Wolter. Zum grundtheorem der informationstheorie, insbesondere in der optik. *Physica*, pp. 457–475, 1958.

[1508] H. Wolter. On basic analogies and principal differences between optical and electronic information. In **Progress in Optics**, E. Wolf, editor. North-Holland, Amsterdam, pp. 155–210, 1961.

[1509] John W. Woods. **Multidimensional Signal, Image, and Video Processing and Coding**. Academic Press, 2006.

[1510] P.M. Woodward. **Probability and Information Theory with Applications to Radar**. Pergamon Press Ltd., London, 1953.

[1511] C.R. Worthington. Sampling theorem expressions in membrane diffraction. *Journal of Applied Crystallography, Vol. 21, Part 4*, pp. 322–325, 1988.

[1512] E.G. Woschni. Braking the sampling theorem and its consequences. *Annalen der Physik, Vol. 43(3–5)*, pp. 390–396, 1986.

[1513] J.M. Wozencraft and I.M. Jacobs. **Principles of Communication Engineering**. John Wiley & Sons, Inc., New York, 1965.

[1514] J.S. Wu and J.C.H. Spence. Phase extension in crystallography using the iterative Fienup-Gerchberg-Saxton algorithm and Hilbert transforms. *Acta Crystallographica, Section A. 59*, pp. 577–583, Part 6 2003.

[1515] W.C.S. Wu, Kwan F. Cheung and R.J. Marks II. Multidimensional Projection Window. *IEEE Transactions on Circuits and Systems, Vol. 35, No. 9*, pp. 1168–1172, 1988.

X

[1516] X. Xia and Z. Zhang. A note on difference methods for the prediction of band-limited signals from past samples. *IEEE Transactions Information Theory, Vol. IT–37, No. 6*, pp. 1662–1665, 1991.

[1517] X.G. Xia and Z.Zhang.On sampling theorem, wavelets and wavelet transforms. *IEEE Transactions on Signal Processing, Vol. 41(12)*, pp. 3524–3535, 1993.

[1518] Xg Xia. Extentions of the Papoulis-Gerchberg algorithm for analytic functions. *Journal of Mathematical Analysis and Applications, Vol. 179(1)*, pp. 187–202, 1993.

[1519] Xiang-Gen Xia. On bandlimited signals with fractional Fourier transform. *IEEE Signal Processing Letters, Vol. 3(3)*, pp. 72–74, 1996.

[1520] J. Xian and W. Lin, Sampling and reconstruction in time-warped spaces and their applications. *Applied Mathematics and Computation, Vol. 157(1)*, pp. 153–173, 27, 2004.

[1521] Zelong Xiao, Jianzhong Xu, Shusheng Peng and Shanxiang Mou. Super-Resolution Image Restoration of a PMMW Sensor Based on POCS Algorithm 1st International Symposium on Systems and Control in Aerospace and Astronautics, 2006. ISSCAA 2006. pp. 680–683, 2006.

[1522] Gang Xiong, Zhao Hui-chang, Wang Li-Jun and Liu Ji-Bin. The wide sense time-frequency representation based on the wavelet and its relation with Cohen's class. *ICMMT 4th International Conference on, Proceedings Microwave and Millimeter Wave Technology*, 2004. pp. 677–680, 2004.

[1523] C.Y. Xu, F. Kamalabadi, and S.A. Boppart. Comparative performance analysis of time-frequency distributions for spectroscopic optical coherence tomography. *Applied Optics, Vol. 44(10)*, pp. 1813–1822, 2005.

Y

[1524] A.D. Yaghjian. Simplified approach to probe-corrected spherical near-field scanning. *Electron. Letters, Vol. 20*, pp. 195–196, 1984.

[1525] A.D. Yaghjian. Antenna coupling and near-field sampling in plane-polar cordinates. *IEEE Transactions Antennas & Propagation, Vol. 40*, pp. 304–312, 1992.

[1526] M. Yamaguchi, N. Ohyama, T. Honda, J. Tsujiuchi, and S. Hiratsuka. A color image sampling method with suppression of Moiré fringes. *Optical Communication, Vol. 69*, pp. 349–352, 1989.

[1527] H. Yan and J.T. Mao. The realation of low-frequency restoration methods to the Gerchberg-Papoulis algorithm. *Magnetic Resonance in Medicine, Vol. 16(1)*, pp. 166–172, 1990.

[1528] M. Yanagida and O. Kakusho. Generalized sampling theorem and its application by uniform sampling. *Electronics & Communications in Japan, Vol. 57(11)*, pp. 42–49, 1974.

[1529] G. Yang and B. Shi. A time-frequency distribution based on a moving and combined kernel and its application in the fault diagnosis of rotating machinery. *Key Engineering Materials*. 245-2, pp. 183–190, 2003.

[1530] G.G. Yang and E. Leith. An image deblurring method using diffraction gratings. *Optical Communication, Vol. 36*, pp. 101–106, 1981.

[1531] G.Z. Yang, B.Z. Dong, *et al.* Gerchberg-Saxton and Yang-Gu algorithms for phase retrieval in a nonunitary transform system - a comparison. *Applied Optics, Vol. 33(2)*, pp. 209–218, 1994.

[1532] S.Z. Yang, Z.X. Cheng, and Y.Y. Tang. Approximate sampling theorem for bivariate continuous function. *Applied Mathematics and Mechanics - English Edition, Vol. 24(11)*, pp. 1355–1361, 2003.

[1533] Y. Yang, N.P. Galatsanos, and A. K. Katsaggelos. Projection-Based Spatially-Adaptive Reconstruction of Block Transform Compressed Images. *IEEE Transactions on Image Processing, Vol. 4, No. 7*, pp. 896–908, 1995.

[1534] Y. Yang and N.P. Galatsanos. Removal of compression artifacts using projections onto convex sets and line process modeling. *IEEE Transactions Image Processing, Vol. 10(6)*, pp. 1345–1357, 1997.

[1535] Yongyi Yang, N.P. Galatsanos. Removal of compression artifacts using projections onto convex sets and line process modeling. *IEEE Transactions on Image Processing, Vol. 6(10)*, pp. 1345–1357, 1997.

[1536] K. Yao. **On some representations and sampling expansions for bandlimited signals**. Ph.D. thesis, Dept. of Electrical Engineering, Princeton University, Princeton, NJ, 1965.

[1537] K. Yao. Applications of reproducing kernel Hilbert spaces – Bandlimited signal models. *Information & Control, Vol. 11*, pp. 429–444, 1967.

[1538] K. Yao and J.B. Thomas. On truncation error bounds for sampling representations of bandlimited signals. *IEEE Transactions Aerospace & Electronic Systems, Vol. AES–2*, p. 640, 1966.

[1539] K. Yao and J.B. Thomas. On some stability and interpolating properties of nonuniform sampling expansions. *IEEE Transactions Circuit Theory, Vol. CT–14*, pp. 404–408, 1967.

[1540] K. Yao and J.B. Thomas. On a class of nonuniform sampling representation. *Symp. Signal Transactions Processing* (Columbia Univ.), 1968.

[1541] B. Yegnanarayana, S. Tanveer Fatima, B.T.K.R. Nehru, and B. Venkataramanan. Significance of initial interpolation in bandlimited signal interpolation. *IEEE Transactions Acoust., Speech, Signal Processing, Vol. ASSP-37, No. 1*, pp. 151–152, 1989.

[1542] Shu-jen Yeh and Henry Stark. Iterative and one-step reconstruction from nonuniform samples by convex projections. *Journal of the Optical Society America A, Vol. 7, No. 3*, pp. 491–499, 1990.

[1543] Shu-jeh Yeh and Henry Stark. Learning in neural nets using projection methods. *Optical Computing and Processing, Vol. 1, No. 1*, pp. 47–59, 1991.

[1544] J.L. Yen. On the nonuniform sampling of bandwidth limited signals. *IRE Transactions Circuit Theory, Vol. CT–3*, pp. 251–257, 1956.

[1545] J.L. Yen. On the synthesis of line sources and infinite strip sources. *IRE Transactions Antennas & Propagation*, pp. 40–46, 1957.

[1546] I. Samil Yetik and A. Nehorai. Beamforming using the fractional Fourier transform. *IEEE Transactions on Signal Processing, Vol. 51(6)*, pp. 1663–1668, 2003.

[1547] Changboon Yim and Nam Ik Cho. Blocking artifact reduction method based on non-iterative POCS in the DCT domain. ICIP 2005, *IEEE International Conference on Image Processing, Vol. 2*, pp. 11–14, 2005.

[1548] H. Yoshida and S. Takata. The growth of Fourier's theory of heat conduction and his experimental study. *Historia Sci.,* (2) 1 (1), pp. 1–26, 1991.

[1549] D.C. Youla. Generalized image restoration by method of alternating orthogonal projections. *IEEE Transactions Circuits & Systems, Vol. CAS–25*, pp. 694–702, 1978.

[1550] D.C. Youla and H. Webb. Image restoration by the method of convex projection: Part 1 - theory. *IEEE Transactions Medical Imaging, Vol. MI–1*, pp. 81–94, 1982.

[1551] D.C. Youla. In **Image Recovery: Theory and Applications**, Mathematical theory of image restoration by the method of convex projections, (H. Stark, ed.), Academic Press Inc., Orlando, FL, pp. 29–77, 1987.

[1552] D.C. Youla and V. Velasco. Extensions of a result on the synthesis of signals in the presence of inconsistent constraints. *IEEE Transactions Circuits & Systems, Vol. CAS–33*, pp. 465–468, 1986.

[1553] Robert M. Young. **An Introduction to Nonharmonic Fourier Series**. Academic Press; Revised First Ed., 2001.

[1554] F.T.S. Yu. Image restoration, uncertainty, and information. *Applied Optics, Vol. 8*, pp. 53–58, 1969.

[1555] F.T.S. Yu. Optical resolving power and physical realizability. *Optical Communication, Vol. 1*, pp. 319–322, 1970.

[1556] F.T.S. Yu. Coherent and digital image enhancement, their basic differences and constraints. *Optical Communication, Vol. 3*, pp. 440–442, 1971.

[1557] Gong-San Yu *et al.* POCS-based error concealment for packet video using multiframe overlap information. *IEEE Transactions on Circuits and Systems for Video Technology, Vol. 8(4)*, pp. 422–434, 1998.

[1558] Li Yu, Kuan-Quan Wang, Cheng-Fa Wang and D. Zhang. Iris verification based on fractional Fourier transform. *Proceedings. 2002 International Conference on Machine Learning and Cybernetics*, 2002. *Vol. 3*, pp. 1470–1473, 2002.

[1559] E. Yudilevich and H. Stark. Interpolation from samples on a linear spiral scan. *IEEE Transactions Medical Imaging, Vol. MI–6*, pp. 193–200, 1987.

[1560] E. Yudilevich and H. Stark. Spiral sampling: Theory and application to magnetic resonance imaging, *Journal of the Optical Society America A, Vol. 5*, pp. 542–553, 1988.

Z

[1561] L.A. Zadeh. A general theory of linear signal transmission systems, *Journal Franklin Institute, Vol. 253*, pp. 293–312, 1952.

[1562] L.A. Zadeh. Fuzzy sets. *Information and Control, Vol. 8*, pp. 338–353, 1965; reprinted in J.C. Bezdek, editor, **Fuzzy Models for Pattern Recognition**, IEEE Press, 1992.

[1563] Samy Zafrany. **Fourier Series and Integral Transforms**. Cambridge University Press, 1997.

[1564] J. Zak. Finite translations in solid-state physics. *Phys. Rev. Letters, Vol. 19*, pp. 1385–1387, 1967.

[1565] J. Zak. Dynamics of electrons in solids in external fields. *Phys. Rev., Vol. 168*, pp. 686–695, 1968.

[1566] J. Zak. The kq-representation in the dynamics of electrons in solids. *Solid State Physics, Vol. 27*, pp. 1–62, 1972.

[1567] M. Zakai. Band-limited functions and the sampling theorem. *Information & Control, Vol. 8*, pp. 143–158, 1965.

[1568] A. Zakhor. Sampling schemes for reconstruction of multidimensional signals from multiple level threshold crossing. In *Proceedings ICASSP '88*, 1988.

[1569] A. Zakhor, R. Weisskoff, and R. Rzedzian. Optimal sampling and reconstruction of MRI signals resulting from sinusoidal gradients. *IEEE Transactions Signal Processing, Vol. SP–39, No. 9*, pp. 2056–2065, 1988.

[1570] Z. Zalevsky, D. Mendlovic and R.G. Dorsch. Gerchberg-Saxton algorithm applied in the fractional Fourier or the Fresnel domain. *Optics Letters, Vol. 21(12)*, pp. 842–844, 1996.

[1571] H. Zander. Fundamentals and methods of digital audio technology. II. Sampling process. *Fernseh- und Kino-Tech., Vol. 38*, pp. 333–338, 1984.

[1572] Ahmed I. Zayed. Sampling expansion for the continuous Bessel transform. *Applied Anal., Vol. 27*, pp. 47–65, 1988.

[1573] Ahmed I. Zayed, G. Hinsen, and P. Butzer. On Lagrange interpolation and Kramer-type sampling theorems associated with Sturm-Liouville problems. *SIAM Journal Applied Math, Vol. 50, No. 3*, pp. 893–909, 1990.

[1574] Ahmed I. Zayed, M.A. Elsayed, and M.H. Annaby. On Lagrange interpolations and Kramer sampling theorem associated with self-adjoint boundary-value problems. *Journal of Mathematical Analysis and Applications, Vol. 158(1)*, pp. 269–284, 1991.

[1575] Ahmed I. Zayed. On Kramer sampling theorem associated with general Strum-Liouville problems and Largange interpolation. *SIAM J. on Applied Mathematics, Vol. 51(2)*, pp. 575–604, 1991.

[1576] Ahmed I. Zayed. Kramer sampling theorem for multidimensional signals and its relationship with Largrange - type interpolations. *Multidimensional Systems and Signal Processing, Vol. 3(4)*, pp. 323–340, 1992.

[1577] Ahmed I. Zayed. **Advances in Shannon's Sampling Theory**. CRC, 1993.

[1578] Ahmed I. Zayed. A sampling theorem for signals band-limited to a general domain in several dimensions. *Journal of Mathematical Analysis and Applications, Vol. 187(1)*, pp. 196–211, 1994.

[1579] Ahmed I. Zayed. On the relationship between the Fourier and fractional Fourier transforms. *IEEE Signal Processing Letters, Vol. 3(12)*, pp. 310–311, 1996.

[1580] Ahmed I. Zayed and A.G. Garcia. Sampling theorem associated with a Dirac operator and the Hartley transform. *Journal of Mathematical Analysis and Applications, Vol. 214(2)*, pp. 587–598, 1997.

[1581] Ahmed I. Zayed. Hilbert transform associated with the fractional Fourier transform. *IEEE Signal Processing Letters, Vol. 5(8)*, pp. 206–208, 1998.

[1582] A.H. Zemanian. **Distribution Theory and Transform Analysis**. McGraw-Hill, New York, 1965.

[1583] A.H. Zemanian. **Generalized Integral Transforms**. Interscience, New York, 1968.

[1584] Bilin Zhang, Shunsuke Sato, and P.C. Ching. Speech signal processing using a time-frequency distribution of Cohen's class. *Proceedings of the IEEE-SP International Symposium on Time-Frequency and Time-Scale Analysis*, pp. 624–627, 1994.

[1585] B. Zhang and S. Sato. A time-frequency distribution of Cohen's class with a compound kernel and its application to speech signal processing. *IEEE Transactions on Signal Processing, Vol. 42(1)*, pp. 54–64, 1994.

[1586] C. Zhao and P. Zhao. Sampling theorem and irregular sampling theorem for multiwavelet subspaces. *IEEE Transactions on Signal Processing, Vol. 53(2)*, pp. 705–713, Part 1, 2005.

[1587] Yunxin Zhao, L.E. Atlas and R.J. Marks II. The use of cone-shape kernels for generalized time-frequency representations of nonstationary signals. *IEEE Transactions on Acoustics, Speech and Signal Processing, Vol. 38*, pp. 1084–1091, 1990.

[1588] Di Zhang and Minghui Du. High-resolution image reconstruction using joint constrained edge pattern recognition and POCS formulation. Control, Automation, Robotics and Vision Conference, 2004. *Vol. 2*, 6–9, pp. 832–837, 2004.

[1589] G.T. Zheng, P.D. McFadden. A time-frequency distribution for analysis of signals with transient components and its application to vibration analysis. *Journal of Vibration and Acoustics – Transactions of the ASME. Vol. 121(3)*, pp. 328–333, 1999.

[1590] L. Zhizhiashvili. **Trigonometric Fourier Series and their Conjugates**. Springer, 1996.

[1591] X.W. Zhou and W.C. Sun. On the sampling theorem for wavelet subspaces. *Journal of Fourier Analysis and Applications, Vol. 5(4)*, pp. 347–354, 1999.

[1592] X.W. Zhou and W.C. Sun. An aspect of the sampling theorem. *International Journal of Wavelets, Multiresolution and Information Processing, Vol. 3(2)*, pp. 247–255, 2005.

[1593] Y.M. Zhu. Generalized sampling theorem. *IEEE Transactions on Circuits and Systems - Analog and Digital Signal Processing, Vol. 39(8)*, pp. 587–588, 1992.

[1594] Yang-Ming Zhu, Zhi-Zhong Wang, and Yu Wei. A convex projections approach to spectral analysis. *IEEE Transactions on Signal Processing, Vol. 42(5)*, pp. 1284–1287, May 1994.

[1595] Anders E. Zonst. **Understanding the FFT, Second Edition. Revised.** Citrus Press, 2000.

[1596] Ju Jia Zou and Hong Yan. A deblocking method for BDCT compressed images based on adaptive projections. *IEEE Transactions on Circuits and Systems for Video Technology, Vol. 15(3)*, pp. 430–435, 2005.

[1597] S. Zozor and P.O. Amblard. Stochastic Resonance in Discrete Time Nonlinear AR(1) Models. *IEEE Transactions on Signal Processing, Vol. 47, No. 1*, pp. 108–122, 1999.

[1598] Gary Zukav, **Dancing Wu Li Masters: An Overview of the New Physics**. Harper Perennial Modern Classics, 2001.

Index

A

a.k.a., **ix**, 155, 362, 411, 503, 570
Abbott, Edwin A., 326, 680
Abel transform, 226, **345–348**, 401–402, 406–407
Abel, Niels Henrik, 519
absolute
  convergence, vi, 24
  error, 366
  value, 73
absolutely convergent, **61–62**, 72–73, 266
additive, 9
  basis, 598
  closure, 590, 605–607
  correlation, 462
  noise, 18, **288**, 292–294, 302, 323, 325, 456, 459, 461–463, 518, 682, 699
  sample noise, 265
  systems, 9, **105–106**, 108, 110–113, 135, 142, 630
acoustics, 3–4, 495, 610, 612, 624, 692, 706, 728, 735
acronym table – see tables
additively, 630
additively idempotent time scale, ix, **595**, 603. Also see AITS
Airy
  disc, 637–638
  pattern, 637, 657
  George Biddell, 518
AITS, ix, **595–598**, 601–608. Also see – additively idempotent time scale
  AITS hull, **596–597**
algebraic decay, 591, 596
aliasing, 4, 9, 55, **222**, 236–237, 253–254, 283, 288, 309, 374, 376, 382, 390, 396–397, **451–472**, 483, 489, 684, 686, 691, 692, 697–698, 581, 705, 707, 715, 718–719, 722, 730, 732, 734–735
almost sure convergence, 177
alternating projections onto convex sets, vii, x, 433–435, **495**, 551–553, 716, 721. Also see – POCS

AM, ix, 68, 122, 133. Also see – amplitude modulation
ambient temperature, 106
ambiguity function, 441, 445, 681, 715, 728–729, 735
amortization, 8, **656**
amplitude modulation, ix, 6, 68, **122–123**, 133–134, 204, 276. Also see – AM
analytic signals, xiv, 12, 64, 77, 134
angular
  displacement, 399
  Fourier transform, 126, 680
  frequency, 625
  intervals, 547
  spectrum, **626–630**, 657
  variable, 399
annulus, 342, 572, 574, 607
antenna, 122, 624, 643–644, 684, 686, 691, 703, 714, 726, 729, 736, 741
anti-derivative, 587
antiextensive, **581**, 583, 595, 608
aperiodic, 354, 358, 704
aperture, 4, 415, 532–533, 544, 624, 634–637, 642–644, 657, 659, 687, 698–699
apodization, 415, **644**, 731
Arabshahi, Payman, vii
arbitrary
  but fixed, 586, 596
  closed subset, 604
  complex weights – see complex weight
  constant, 645
  frequency, 424, 657, 734
  function, 248, 429
  integtration, 15
  interval, 564
  number, 385
  period, 14
  point, 501
  phase, 416
  rational time, 424
  signal, 484
  tile shape, 373
  time scale, 605

745

arbitrary (*cont'd*)
  variation, 220
  vector, 506–507, 510, 514, 522
arctan, 22, 91, 228–229, 292, 312–313, 460
arithmetic mean, 14, 439
array
  aperture, 643–644
  beam – see beam array
  detector, 250
  function, xii, 23, **32–35**, 65–66, 69, 90, 99, 100, 103, 154, 169, 266, 267, 272, 276, 454–455, 466, 468, 470, 489, 643, 663
  imaging, 3, 732
  linear – see linear array
  of circles, 378
  of Dirac deltas, 373
  of squares, 392
  pattern, 643–644
  phased – see phased array
  pixel – see pixel array
associative, 115–116, 576
  memory, 7, **516–521**, 525–527, 529, 550, 691, 716, 730
astrologers, 411
asymptotic Gaussian, 189
asymptotic graininess, **597–601**
Atlas, Les E., viii, 436, 551–553, 691, 716, 721, 725, 744
atmospheric gases, 6
audio, 3–4, 68, 122, 234, 439, 613, 695, 743
Augustine, Saint, 411, 681
autoconvolution, 133, 186, 203, 419
autocorrelation, xi, xiii, 15, 194–200, 204, 214, 233, 289–292, 309–312, 316–317, 323, 384, 427, 457, 460–464, 486, 700, 708
average, 124, 160, 176–179, 182, 186, 196, 204, 215–216, 277, 303–304, 306, 439, 517, 519, 533, 676, 682, 698, 737. Also see – sample mean
  image, 179, 517, 519, 533
  power, 124, 196, 204, 215–216, 737
  interpolation noise, 303–394, 306
  sampling density, 277, 698

B

Baby Senator image, 182
back projection, 352
bandlimited, ix, xii, xiv, 8–9, **12**, 39, 49, 61, 68, 71–72, 101–102, 133, 139, 150, 217–222, 224–227, 230–237, 239–244, 250–251, 261–263, 265–266, 271, 273, 275–278, 288, 307, 309, 314, 316, 320, 323–325, 376, 382, 394–395, 397–398, 447–450, 465, 472, 474–475, 477, 479–480, 482–488, 491, 493, 495, 507, 514–517, 535, 537, 550, 682, 685–690, 692, 694, 697–699, 701–703, 706, 708–712, 714–720, 723–737, 740–743
  function, 39, 49, 101, 220, **224–225**, 231, 235, 682, 685, 689, 694, 699, 701, 706, 709–710, 712, 719–720, 725, 726, 732–734, 737, 743. Also see – bandlimited signals
  signal, ix, xii, xiv, 8–9, **12**, 68, 71, 150, 217–219, 226, 231–232, 234–235, 239–240, 242–244, 250–251, 261, 263, 265, 275–277, 288, 307, 316, 320, 323–324, 376, 382, 394, 397, 447–449, 474, 477, 480, 483, 485–487, 495, **507**, 514–517, 535, 537, 550, 685–690, 692, 697, 702–703, 709, 711–712, 714–717, 720, 723, 725, 727–728, 734, 736, 740–742. Also see – bandlimited function
bandpass, ix, **121–123**, 134–135, 267–268, 270–272, 278, 301–302, 369–371, 413, 418, 439, 683, 686, 709, 734, 738
  filter, 121, 123, 134–135, 369, 413, 439, 683
  function, **267–268**, 270, **301–302**
  sampling, **270**, 302, 686, 709, 734, 738
  signal, 268, 270–271, 418, 686
bandstop filter, **121**, 135, 369–370
bandwidth, xii, xiv, 12, 49, 68, 120, 122, 139, 218–219, 221–222, 227, 231–232, 235–236, 239–241, 243, 247–248, 252, 261, 264, 267, 271, 272, 274–276, 276, 280, 289, 293–294, 304, 314, 397–398, 429, 429, 450, 455, 465–467, 470, 473, 475, 477, 482–483, 485–486, 507, 514–515, 535, 537, 685, 699, 702, 705, 713, 716–717, 731, 737, 742
  equivalence, **397**
  interval, 232
Bartlett window, 415, 419
baseband, 68, 122–123, 133–134, 204, 268–269, 277
  signal, 122–123, 133–134, 204, 268, 277
bastinadoed, 413
Bayes, Rev. Thomas, 519
BE, ix. Also see bounded energy signals
beam
  array, 547
  constraint, 555
  cross section, 545–546
  design, 545–546
  dose, 547
  element, 547, 554–558, 560–561
  intensity, 546, 558–559, 562
  splitter, 643
  intensity, 546
  position, 554
  profile, 545–546, 558–559, 561–562
  projection, 546
  steering, **644**
  synthesis, 8
  vector, 547, 556

weights, 445
width, 643–644
beamforming, 8, 126, 415, 495, 624, 643, 697, 713, 742
beat frequency, 429, 435
bel, 413
bell shaped curve, 159
Benedictine abbey of Saint Benoît-sur-Loire, 5
Bernoulli
   random variable, **169**
   trial, **169–170**
   Daniel, 518
   Jacob (Jacques), 518
   Nicolaus (II), 518
Bertrand's paradox, 565
Bessel
   functions, xii, xvi, 9, **37**, 39–41, 43, 64, 69, 89, 92–93, 95–96, 133, 141, 202, 263, 342–345, 401, 406, 415
   Friedrich Wilhelm, 326, 518
beta
   function, xii, 36–37, 164
   probability function, 153, 165, 202
   random variable, 153, 163–165, 202, 209, 210
bias, 122, 124–125, 134, 204, 216, 276, 313–316, 318–319
biased estimate, 313
BIBO, ix, 107, 110–112, 142. Also see – bounded input-bounded output stability
bilateral
   cosine transform, **65**, 83
   Laplace transform, **64**
binary
   image, 182, 570, 575–578, 580, 582, 584
   string, 365
binomial
   coefficients, xiii, 36, **662**
   random variable, **170–172**, 175, 190, 194, 202–203
   series, 170, 203, 481, **662**
bioengineering, 424, 434
bioinformatics, 495
biomedical
   engineering, 3, 724
   signal, 424, 434, 684
bipolar, 517
bit packet, 160
bits, 364
BL, ix. Also see bandlimited signals
Black, H.S., 218, 684
Blackman window, 415, 420, 422, 685
block orthogonal, 414
blurred
   face, 517
   image, **147**, 696, 705, 722, 741
   motion – see motion blur
Bohner, M., 586, 680, 685
Bohr, Niels Henrik David, 151, 685

Boltzmann's constant, xii, 198, 407
Borel, Emile, 218–219, 685
Born-Jordon kernel, 411, 543
Bortfeld, Thomas, 448, 685, 735
Bortfield's geometry, 560–561
boundary. Also see Fresnel – Fraunhofer boundary & inside boundary & outside boundary
   conditions, 127–128, 130, 164, 228, 612, 618, 656
   for the spectrogram, 440
   of a closed convex set, 497
   of a region, 380
   of the Minkowski addition, 571
   wide sense stationary stochastic process, 204
   value problems, 7, 610, 692, 700, 703, 707, 743
bounded. Also see – error bound & lower bound & measureable
bounded signals & unbounded & upper bound
   bandlimited function, 102, 682
   derivative, 512
   energy signals, ix, 305, **511**
   first difference, 512
   input-bounded output stability, ix, **107**, 110, 142. Also see – BIBO
   inverse Fourier transform, 101
   noise level, 482
   signal, ix, **11–12**, 67, 71, 107, 495, **512**, 527, 688
bow tie
   constraint, 428, 539, 543–544
   untruncated, 544
boxcar window, 415, 419–420, 423–424
BPF – see band pass filter
Bracewell, Ronald N., vi, 686
Bradley, Walter L., viii
Brahme's butterfly tumor, 557, 559, 686
Bravais lattice, 327
Bregman, L.M., 495, 686
broadband signal, 122, 700, 725
Brother Ray image, 179, 185
BS, ix. Also see bounded signals
bugle, 613, 616, 618
bunched samples, 252, 257. Also see – interlaced samples
byte, 364, 366, 517, 532

C

CA, ix. Also see constant area signals
Cantor, Georg Ferdinand Ludwig Philipp, 139, 689, 694
capacitors, 106, 116
cardinal series, **49**, 150, 217–220, 222–225, 227, 229, 231–235, 241–243, 245–249, 261, 265–266, 272–274, 277–278, 284–285, 288–290, 294–295, 302, 313–314, 320–321, 323, 449, 461, 484,

cardinal series (*cont'd*) 682, 684, 686, 688–689, 697, 704, 709–710, 718, 722, 729, 733–734, 739
carrier, x, 68, 122–124, 133–134, 276
  frequency, 68, 122, 276
  carrier frequency phase, 124
Cartesian
  coordinates, 15
  Cartesian product, xi, 591–593
CAT, ix, 327
cathode ray tube – see CRT
Cauchy
  autocorrelation, 323
  A.L., 218, 687–688, 690, 738
  - Bunyakovski – Schwarz inequality, **660**
  characteristic function, 168
  inequality, **660**
  principle value, 167
  probability density function, 166–167, 679
  random variable, 153, 166–169, 175–176, 187, 209, 678–679
  - Schwarz inequality, **660**
causal
  comb, 31-**32**, 78
  convolution, 137
  filter, 107, 440, 482
  impulse response, 366
  signal, 8, **12**, 15, 18, 46, 51, 64–65, 107, 111, 137, 140, 420, 482, 486–487, 495, 684, 705
  sine wave, 420
  system, **107–111**, 113–114, 135–136, 142, 144. Also see – nonanticipatory system
  time scale, 586, 599
  window, 420, 423
causality, 6, 107, 111
CD, ix, 4, 107
cellular radio, 495, 726
center of mass, 155, 174
central
  limit theorem, 6, 8, 160, **178–170**, 187–192, 202–203, 211, 676
  slice theorem, **348–350**, 401, 406
  tendency, 155
cents, 610, **614**–616, 621–623, 656–657, 658–659
chain rule, 347, 569
characteristic functions, xiii, xvii, 6, **152–156**, 158–164, 166, 168–171, 174–177, 186–187, 201–203, 205–206, 210–211, 315, 318, 670
Chebyshev's inequality, **157–158**, 178
Chebyshev
  filters, 107
  polynomials, xiv, **42–43**, 50, 69, 89, 96–97, 167, 372
  polynomials of the second kind, 9, **69**
  Pafnuty Lvovich, 519, 610
chemistry, 3, 716
Cheung, Kwan Fai, viii, 516, 691, 726, 740

chi random variable, **667–668**
chi-squared random variable, 153, **162–164**, 175, 190, 201, 206, 664, 667–668, 673
chirp, 432–434, 439, 544, 551–552
  Fourier transform, **439**
Cho, Paul S., viii, 554, 557, 559–562, 692, 711–712, 742
Choi, Jai J., viii
Choi-Williams distribution, 431
chords, 412, 613, 616–621
Christ, v
chromatic
  scale, 8
  step, 411, 613–615, 621–622
chromatography, 4, 739
Churchill, Sir Winston Spencer, 495
circularly symmetric, 341–346, 401, 406
circular
  aperture, 637
  convolution, 51, **53**–54, 118–119, 466
  convolution mechanics, **118–119**
  harmonic, 344–345, 401
  circular pupil, 377, 379, 385
  radius, 395
  spectrum, 387
  support, 377, 379, 391, 394–395
city block metric, 327
clarinets, 613
closed
  convex sets, **496–497**
  under addition, 590, 596, 608
  under independent summation, **174–175**, 189
  under Minkowski addition, 590, 596
closing xi, **575**, 581, 583, 608
  duality, **581**
closure, 497, 590, 596, 605, 606–607
cochlea, 4
Cohen's GTFR, 7, 417, 431, 434, 438, 537, 689, 691, 696, 702, 710, 712, 713, 720, 722, 738–739, 741, 744
Cohen, Leon, 424, 427, 692, 696
coherent
  demodulation, 122–125, 133–134, 268
  demodulator, 122, 124
  field, 629–630, 638, 695, 700
  illumination, 377, 625, 684, 701
  light, 4, 8, 377, 623, 630, 683, 735
  optics, 628, 716
  processor, 445, 702, 713, 715, 738
  sonar, 4
  wave, 626, 657
Colerage, Samuel Taylor, vii
colored noise, 323, 458
column space, 506–507
comb function, xii, 14, 23, 31–32, 34–35, 47, 51, 70–71, 78, 84, 100, 221–222, 235, 243, 253, 270, 273, 356–357, 373, 390, 392, 395–396, 466, 720
  causal – see causal comb

multidimensional – see multidimensional comb lemma
communications engineering, 3, 685
commutative, 115–116, 120, 576
compact
  disk, ix
  support, 150
complement, xii, 369, 506, 574, 578–579
complete, 5, 47–48, 112, 139, 219, 224–225, 234, 238, 245, 267, 273–274, 278, 283, 315, 245, 326, 357, 403, 474, 475, 477, 482, 516, 528, 586, 661, 682, 704, 726
  basis set, **48**, 273–274, 278, 474, 477, 704
complex, 326, 545, 682, 685, 686, 693, 704, 710, 723, 737–738
  arguments, 35–36
  bandlimited signal, 235
  computational – see computational complexity
  conjugate – see conjugate
  exponential, 115–116, 419
  even function, 82
  field amplitude, 629
  Fourier transform, 15, 82
  frequency response, 116
  function, 62, 190, 564, 660
  image, 566
  number, 12
  odd function, 82
  phase – see phase of a complex exponential
  plane wave, 629
  sinusoid, 31, 62, 412
  signal, 239, 495
  target, 558
  Walsh function – see Walsh function
  weights 628, 644
compound interest, 647, 651–652, 655–656
compounding, 647–649, 651–653, 656, 658
computational complexity, 213
computed axial tomography – see CAT
condition, 8, 17, 24, 60, 67, 69, 95–96, 105, 127–128, 130, 136, 153, 164, 179, 186, 195, 224, 228, 248–249, 277, 298–299, 301, 383, 430, 470–472, 485–486, 612, 617–618, 642, 648, 655–656, 663–664, 691, 696, 711
  number, 301, 470–472, 485, **663–664**
cone, 428, 431, 434–438, 441, 444, 504–505, 508, 516, 527, 537, 539, 541, 543–544, 551–552, 693–694, 696, 709, 721, 725, 738, 744
  constraint, 428, 541, 543–544, 551–552
  kernel, 428, 431, **434–438**, 441, 444, 537, 693–694, 696, 709, 721, 725, 738, 744
confluent hypergeometric function, xii, **202**, 209, 673
conformal radiotherapy, 7–8, 495, **545–546**, 554, 557, 449–562, 685, 712

conjugate, 15, 22, 45, 51, 70, 120, 127, 235, 269, 430, 438, 444, 472, 483, 489, 564, 629, 657, 735, 744
  plane waves, **657**
  symmetry, 45, 70, 120, 200, 235, 430, 472, 483, 489, 564, 629. Also see – Hermetian
conservation of contours, 368
constant
  area signal, ix, **508–511**, 533, 545, 549
  graininess, 597
  phase signal, ix, **511–512**, 566, 628
  Q, 415, 439, 695, 737
  addressable memory, 516
continuously sampled signals, 7, 9, **448–450**, 455, 461–463, 477, 480, 483, 715
continuous, 11, 20, 36, 50, 80–81, 115, 120, 203, 224, 398, 449, 468, 564, 511, 586, 546, 608, 685, 689, 711, 716, 722, 732, 742. Also see – continuous time
  circular convolution, 118
  compounding, **649**, **651–653**
  domain, 328
  dose, 549
  functions, 9, 166, 173, 220, 327, 473, 509, 688, 732, 741. Also see continuous signal & continuous time signal & discontinuous function
  image, 328
  interest, 652
  multidimensional, 330
  noise, 324. Also see continuous white noise
  position, 9
  random variable, 153, 173
  sample restoration, 462–464
  sampling, vii, 10, 288, **447–448**, 463, 450, 463, 465, 467, 477, 483
  short time Fourier transform, 441
  signal, 104, 219, 508, 511, 564, 681 682, 690, 692, 704. Also see continuous function & continuous time signal
  space, 326
  stochastic process, xiii, 197. Also see – continuous time stochastic process
  sum, 112
  unit step, xiii
  white noise, 9, **197–198**, 324, 457
continuous time, 120, 196, 200, 425, 501, 718, 731. Also see – continuous
  cone kernel, 441
  convolution, 115, 141
  filter, 135
  Fourier transform, ix, 8, **13–15**, 18, 20, **45**, 50, 54, 81, 583, 587. Also see – CTFT
  GTFR, 417, 427, 434
  impulse response, xii, 121
  periodic function, 67
  rectangle, xiii
  scale, 587, 604

continuous time (*cont'd*)
  signal, ix, xiii–xiv, 6, **11–12**, 14, 50, 71, 193, 417–418, 570, 584, 586, 692, 704, 728–719. Also see continuous function & time signal
  solution, **653**
  stochastic process, 194. Also see – continuous stochastic process
  systems, **112**, 120, 139
  window, 415
contractive
  function, 567, 569
  operators, 208, **558–559**, 562, 569
  iteration, 563, 567, 569
control theory, 3
convergence factor, 26, 32, 66, 83, 87
convex
  cones, 496, **504–505**, 527
  cone hull, **504–505**, 527
  hull, **504–505**, 527
  set, vii, x, 7, 433–435, 438, **495–508**, 510–514, 517, 527–528, 533, 535, 537, 539, 542–543, 545–546, 548, 551–554, 557, 559–564, 567, 608, 686, 691–694, 702–703, 710, 712, 716, 721, 724, 726, 730–731, 735, 738–739
  slab, 508, 567
convolution, vi, xi, xiv, 7, 15, 39, 32, 46, 51, 53–54, 107, **110–112**, 115, 117–119, 126, 130, 132–134, 137, 139, 174, 186–188, 198, 222, 226, 231, 236, 250–251, 263, 331–334, 346–347, 373, 413, 419, 443, 450, 456, 458, 466–467, 469, 570, 572–573, 583, 591, 592–595, 601–603, 631, 634–635, 680, 714, 716–719, 724, 725, 727, 732, 734. Also see – deconvolution
  algebra, **115–116**
  integral, 111, **115**, 117–118, 137, 333, 346, 401, 583, 631, 635, 732
  kernel, 456, 725
  mechanics, **117**, **333–334**
  on a time scale, 592, 594, 601–602
  sum, **115**
  support, **572–573**
  tables – see tables
  theorem, **19**, 32, 333, 347, 413
correlation, 51, 194, 198, 204, 226, 292, 324, 384, 457–458, 684, 693, 698, 703, 710, 714, 727. Also see – autocorrelation
  integral, 15
  length, 324
  theorem, 458
cos, xii, 21–25, 27, 32, 37, 39, 42, 56–58, 65, 67–69, 77–78, 81–84, 87, 90, 92–94, 96–99, 110, 116, 121–124, 132, 198, 204, 214, 228–229, 232, 239, 262–263, 268–271, 276–278, 280, 283–284, 286, 291–292, 302, 306, 310–311, 324, 335, 337, 339, 341–342, 344, 350–352, 362, 367–369, 371, 402, 404, 420–423, 431, 438–439, 442, 445–446, 460, 488, 511, 567, 569, 588–594, 611–612, 627, 629, 670. Also see – cosine
cosech, 23, 25–26, 86, 132, 140. Also see – hyperbolic cosecant
cosh, **24**, 26, 163, 203, 211–212, 291–292, 311, 324. Also see – hyperbolic cosine
cosine ix, xiv, 4, 6, 24, 56, 61–62, 65, 83, 117, 262, 284, 327, 360, 371, 420–424, 446, 627–628, 630, 657, 683, 717, 726–727. Also see – cos
  filter, 421, 423
  series, 422–424
  transform, ix, xiv, 4, 6, **56**, 65, 83, 327, 360, 717, 727
cosinusoid, 123, 268
cotan, 23, 31, 78, 87. Also see – cotangent
cotangent, 23, 32, 78, 87, 304. Also see – cotan
cotanh, 26
countably infinite, 14, 230
CP, ix. Also see constant phase signal
Cramer's rule, 257, 275, 286
cross-correlation, 194, 198, 204, 384
CRT, ix, 395
crystallography, 3, 700, 732, 740
CTFT, ix, **13–16**, 20, 60, 587. Also see – continuous time Fourier transforms
  convergence, 189
  distribution function, xii, **151–152**, 187, 201–202
  spotting, 194
  sum, 51
current, 11, 84, 116, 196
cyclostationary stochastic processes, 323

D

data noise level, 292, 384–386, 402
daVinci, Leonardo, 518
Davis, John M., viii, 583, 694, 702, 716
dB, ix, 12, 68, 118, 366, 386–387, 413, 429, 544. Also see – decibel
DCT, ix, 4, 7, 56–57, 72, 360, 362–366, 527, 680, 723, 735, 742, 744. Also see – discrete cosine transform
  basis functions, **362**
de Broglie wavelength, 646
decade, 53, 414, 737
decibel, ix, 413. Also see – dB
decimation, 387, 392–395, 403, 413, 435–436, 463, 465, 693
deconvolution, 495, 570, 594, 681, 691, 699, 704, 706, 708, 716, 724–725, 729, 735
definite integration, 45
  of derivatives property, 45, 63

degree of aliasing, 253, 462, **467**
degrees Kelvin, 198
Dembski,William A., vii
demagnify, 145
demodulation, 122–125, 134, 204, 268, 740
deMoivre
  Abraham, 518
  -Laplace theorem, 190
derivative. Also see – bounded derivative & fractional derivative Hilger derivative & kernel, xii, **262**, 264–265, 307
  function, 156
  interpolation, **261**, 263, 265, **306–308**, 715
  of a bandlimited signal, 261, 263, 307
  of a bounded function, 101–102
  of a discontinuity, 65
  of a finite energy function, 101
  of the cone kernel, 441, 444
  of the characteristic function, 155, 168, 205
  of the Dirac delta, 232
  of the output, 137
  of the second characteristic function, 156, 162
  of the sinc, xii
  property, 45, 63, 205
  samples, 276, 281, 303–304, 306
  sampling, 302, **304**, 306
  theorem, **18**, 45, 65, 71–72, 86, 102, 141, 154, 192, 232, 262, 347, 401
DeSantis, P., 477, 695
Descartes, Rene, 242, 695
Description de l'Egypte, 5–6
detection theory, 203, 677–679
deterministic
  autocorrelation, xi, 15, 317, 457
  correlation, 15, 457
  random variable, **168**, 177
DFT, ix, 13, 50, **52–57**, 60, 72, 103, 118, 125–126, 359–360, 366, 485, 681, 686, 695, 731, 739. Also see – discrete Fourier transform
  leakage, **55**
diagonal
  elements, 399
  matrix, 337, 340, 359, 377, 549
diamond kernel, **431–432**, 444, 721
diapason, 614
diapente, 614
diatesseron, 614
dice, 151, 602
die, 151, 169, 177, 602
difference equations, 8, 59–60, 523, 646–648, 650, 652, 655, 704, 726
diffraction, 3–4, 7, 495, 623, 626–627, 629–631, 633–639, 641, 657, 659, 681–682, 684, 697, 700, 703, 706, 710, 714, 719, 722, 735, 740–741
  integral, 624, 631, **633–635**, 710

digital integration, 311
diffraction intensity, 63
differential equation – see linear differential equation
digital
  signal processing, ix, 52, 683, 693–694, 729. Also see DSP
  versatile disc, ix. Also see DVD
  video disc, ix. Also see DVD
dilation, xi, **570–573**, 575–576, 578–580, 582, 590, 593, 608
  duality, 576, 578–580
  subset property, 579, 582
dimple, 576
diodes, ix, 106, 687
Dirac delta functions, vi, xii, **20–23**, 31–32, 54–55, 64–66, 83–84, 110, 112–114, 120, 139, 145, 168, 178, 199, 221–222, 232, 283, 356, 373, 429, 443, 456, 458, 542, 588, 592, 626, 627, 640
Dirac delta scaling, **21**
direct interpolation, **246**, 296
direction cosines, **627–628**, 630, 657
directly sampled signals, **271**, 278, 293
Dirichlet
  conditions, **17**, 67, 219
  Johann Peter Gustav Lejeune, 17
discontinuous function, 477, 682, 688, 729. Also see – continuous function
discrete
  cosine transform, ix, 4, 6, **56**, 360, 527, 726–727. Also see – DCT
  Fourier transforms, ix, 10, 13, **52**, 125, 359, 440, 485, 681, 686, 695–696, 710, 731, 735, 738. Also see – DFT10, 11, 46, 599
  periodic nonuniform decimation, **463**, 465
  time convolution, 13, 51, 81, 118, 231, 236, 250–251, 469, 603
  time Fourier transforms, ix, 10, 13, **50–51**, 54, 154, 587. Also see – DTFT
  time periodic signals, 119
  time scale, **586–588**, 592, 603
  time signals, ix, xii–xiv, 7, 11–12, 59, 72, 126, 236, 416, 419, 463, 503, 564, 583–584, 586–587, 692, 706, 722
  time system, 72, **113**, 115, 135–136, 139, 141–142, 698, 720
  time windows, 413
  uniform random variable, **169–170**
discretized
  intensity, 546
  kernel, 547
discriminant, 85, 610–612
dispersion, 155, 157, 174, 177
displaced, 14, 255
displacement, 399, 616
  assumption, 611
  vector, 504, 513
dissipated power, 11

distribution, vi, xii, 20, 151–152, 154, 168, 173–174, 189, 202, 232, 407, 417, 431, 433–435, 537, 551–553, 557, 560–561, 665, 692–683, 686–687, 689, 692, 696, 699–700, 702, 704–706, 709, 711–713, 721–723, 725–727, 735, 738–741, 743–744
distributive, 115–116, 580, 595, 608
dose, 547, 549, 554, 557–558, 560–561, 692
   constraint, 547–549, 554
   contour, 557, 560–561, 735
   distribution, 557, 560
   domain, 547
   matrix, 554, 558
   prescription, 546
   synthesis, 546, 556–557, 560–561
   vector, 547
double
   diamond kernel, 432, 434, 721
   factorial, xiii, **35**, 63, 206
   Mersenne primes, 662
doublet, 232
Dougherty, Edward R., 609, 700
DSP, ix, 644. Also see – digital signal processing
DTFT, ix, **13**, 50–53, 55, 59–60, 82, 142, 359, 368, 466–467, 472, 587. Also see – discrete time Fourier transform
duality theorem, 15, 19, 21, 78, 86, 127, 140, 474
duration limited signals, 495, 507–508, 515, 688, 733
duty cycle, 448, 450, 453, 455, 458–459, 461, 463
DVD, ix, 4
dynamic
   signal, 411
   systems, 106, 694, 704, 526
Dyson, Freeman, 570

E

École Normale, 5
École Polytechnique, 5
Eddington, Sir Arthur, 610, 696
edge detection, 576–577
Edwards, Jonathon, 217, 696
Egyptology, 6
eigenvalues, xiii, 473, 475–476, 481, 492, 523–526, 663, 685, 701, 708, 711, 731
eigenvectors, 523, 663
Einstein, Albert, 17, 104, 151, 288, 696
El-Sharkawi, Mohamed A., viii, 532, 534–536, 538–543, 716, 723
electric
   field amplitude, 624, 627, 629–631
   field wave equation, 626
electromagnetics, 3–4, 112, 690–691
electromagnetic
   signal, 644
   propagation, 624
   wave, 482, 610, 705

engineering, 3, 6, 646, 731
   bioengineering – see bioengineering engineering
   biomedical – see biomedical engineering
   communications –see communications engineering
   math, 20
   microwave, 624
entire functions, 219, 685, 692, 711, 727
envelope, 124–125, 133, 204, 216, 276, 320–321, 326
   demodulation, 124
   detection, 125, 204, 276
   detector, 124, 133
equal to by definition, xi, 14
Erlang random variable, 153, 162–164, 175, 190, 201, 205
erosion, 573–575, 578–581, 584, 595, 608
   duality, **578–579**, 581
   subset property, **579**–580, 584
errata, viii
error
   absolute – see absolute error
   bound, 320, 720, 726, 742
   curve, 525
   free, viii, 684, 717
   interpolation – see interpolation error
   magnitude, 225
   mean square – see mean square error
   normalized – see normalized error
   plot, 518
   quantization – see quantization error
   round off – see round off error
   range, 530
   truncation – see truncation error
Euclidean, 327–328, 404, 724, 731
   norm, 404
Euler's
   formula, **21–22**, 24, 29, 32, 42, 61, 97
   head, 520
Euler, Leonhard, 519, 530, 687
Euler-Mascheroni constant, xii, 675
even, xii, 10, 12–13, 21, 24, 25, 32, 45–46, 61, 63, 65, 81–83, 86, 88–90, 92–94, 96, 99, 139, 162, 168, 207–208, 259, 262–264, 278, 282–284, 286, 292, 304, 310, 345, 386, 420, 432, 435, 441, 444, 457, 472, 563, 567, 596, 616
   functions, xii, 13, 45–46, 65, 81–82, 92–93, 162, 345, 420, 457. Also see – even signal
   function property table – see tables
   moments, 168
   periodic function, 457
   signal, 12–13, 563, 567. Also see – even function
   signal property table – see tables
   window, 432
evenly odd functions, 69, 88, 96, 138, 149

expectation, **152**, 314, 316–317, 384
expected value, xii, **152**, 157, 177–179, 194, 214, 289
exponential, 19, 29, 115–116, 121, 153, 160–162, 164, 171, 175, 203, 208, 419, 633, 635, 667, 679, 684, 692, 695, 698, 722, 732, 736, 739
    decay, 679, 684
    random variable, 153, **160–162**, 164, 175, 667
    Taylor series, 171, 208
extensive, **581**, 583, 608
extrapolation, 8, 328, **447–449**, 462, 473, 475–480, 482, 484–486, 492, 495, 515–516, 684, 689–691, 698, 705–706, 708, 710, 712, 715–716, 723–724, 728, 731, 738–739
    algorithm, **462**, 691, 706, 715–716
    matrix, **484–485**, 708, 728
extrema 17, 65, 67, 91–92, 98–99
    of yield, **650–651**
extreme value random variable, **675**

F

F random variable, **673–675**
fading, **124**, 133
Family, 360–361
far field, 4, 8, 624, 635–636, 643–644, 691
Faraday, Michael, v, 518
fast Fourier transform, ix, 9, 52, 360, 686, 693, 704, 724. Also see – FFT
fault diagnosis, 424, 741
feature extraction, 424, 713
FFT, ix, 9, 52, 126, 360, 373, 436, 438, 690, 733, 744. Also see – fast Fourier transform
filter, ix, 4, 7–9, 59–60, 107, **120–124**, 125–126, 131, 134–135, 142–143, 199–200, 204, 215, 222, 227, 236, 242, 245, 247–250, 252–253, 256–259, 268, 270, 274–275, 278–280, 283, 286, 290–296, 298–301, 306–308, 311, 313–316, 318–319, 322–324, 327, 351–352, 360, 366–371, 373, 385–386, 397, 402, 413–415, 418–424, 436, 438, 439–442, 451, 455, 462, 470, 473–474, 477–478, 489, 507, 516, 537, 603–604, 680–681, 683–684, 687, 691, 693–694, 697–698, 700–701, 703–705, 709–714, 716–723, 725–726, 728–729, 732–738, 740
    bank, 9, **413–415**, 439, 683, 698, 712, 719, 734, 737
filtered
    back projection, **352**
    interpolation, 291–292, 306, 322
    NINV, 298–300
    projection, **351–352**
    signals, **250**, 252, 315
finance, vii, 6–8, 60, **646–647**, 658

finite
    area, xiii, xviii, **11**, 17–18, 64–65, 71, 84, 86, 100–101, 111, 167–168, 224, 231–232, 430, 542
    area constraint, xiii, **430**, 542
    energy signals, xiii, **11**, 20, 39, 48–49, 61, 65, 71–73, 84, 100–101, 217, 224, 231, 235, 239–240, 243, 277–278, 323, 448–449, 474, 493, 502–503, 686, 707, 725, 733
    impulse response, ix, 59, 419, 603. Also see – FIR
    support, 253, 473
FIR, ix, 59, 327, 366, 419, 603–604, 691, 718, 720, 726, 728, 732, 736. Also see – finite impulse response
first. Also see – first order
    derivative, 156, 160–161, 164, 205, 207, 263, 303, 512
    difference, 512
    fret, 623
    harmonics, 613–618
    moment, **155–156**, 167, 671
    orthant, 517
    overtone, 623
    quadrant, 399
    subharmonics, 620
    year graduate student, vi
first order
    aliasing, 451, 456, 459, 461, 483
    decimated sample restoration, **390**, 392, 493
    difference equation, 8
    spherical Bessel function, 263
    spectrum, 483
    statistics, **193–194**
Fisher-Tippett random variable, **675**
fixed point, 498, 516, 559, 562–563
Flatland, 326, 680
flip
    a coin, 151, 169
    and shift, **117–118**, 333, 337
flute, 234
FM, ix, 67, 717, 740. Also see – frequency modulation
for all, **xi**, 11–12, 31, 33, 49, 61, 73, 77, 107, 109, 113, 120, 125, 128, 139, 142, 149, 153, 157, 195, 200, 231, 232, 266, 299, 354, 388–389, 403, 429, 465, 496, 499, 501, 503–504, 507, 517, 546, 558, 563, 566, 569, 579, 581, 586, 590, 600, 616
forward difference, 587
Fourier
    acoustics, 3, 740
    analysis, vi, 3–4, 6–8, 151, 221, 262–263, 416, 458, 570, 583, 695, 698, 708, 711, 731
    array imaging, 3
    - Bessel transform, 343, 345

Fourier (cont'd)
  coefficient, 4, 16, 50, 69–79, 92, 95, 98, 143, 223, 238, 265–266, 276, 283, 323, 357–358, 370, 485, 685, 729. Also see – Fourier series coefficient
  complex – see complex Fourier transform
  descriptors, 3, 712
  Edmie, 4
  integral, 3, 10, 723, 736
  inversion, 3, **19**, 235, 241, 730, 733. Also see – inverse Fourier transform
  Jean Baptiste Joseph, **3–6**, 518, 610
  Joseph (senior), 4
  kernel, 235
  transform matrix, 485
  transform nuclear magnetic resonance, 3, 720
  optics, 3, **624**, 639, 697, 700, 734
  reconstruction, 3, 733
  series, vi, 3, 5, 7, 9, 10, **13–17**, 31, 37, 48–51, 60, 67–70, 92–94, 100, 143, 150, 197, 200, 214, 219, 221–225, 274, 303, 327, 335, 342, 344–345, 352, 356–359, 370–371, 373–374, 402, 408–410, 420, 450, 453, 460, 466, 474, 485, 510, 612–613, 681, 687, 689, 692–696, 703, 706, 710–711, 719, 724–725, 730, 735–736, 739, 743–744
  series coefficient, 14, 31, 60, 67, 93–94, 342, 358, 370, 402. Also see – Fourier coefficient
  spectra, 3–4, 704–705
  spectrometry, 3
  spectroscopy, 3, 702, 737
  theory, 3, 704, 739
  transform, vi–vii, ix, xi–xii, 3–4, 6–10, **13–15**, 18–20, 22, 22–25, 26–32, 34–35, 37, 39, 41, 43–47, 49–52, 54–56, 58, 64–69, 72, 76–78, 81–83, 89, 91, 94, 101, 107, 109–110, 115–116, 120–121, 124–129, 131–133, 136–139, 141, 143–144, 148, 149–154, 160–162, 174, 190, 196, 199–200, 203, 210, 213, 221, 223, 230, 240–241, 249, 251, 253, 269, 273, 287, 291, 315, 327, **330–336**, 340–346, 348–351, 359–361, 373–376, 396–401, 404, 406–407, 411, 413–419, 422–423, 425–426, 428, 435–436, 438–441, 443, 450, 457–458, 460, 474, 477, 482, 485–486, 507–508, 511–512, 516, 566, 568–569, 573, 583, 587–589, 591–592, 597, 601, 623–624, 626–628, 630, 636–638, 641–646, 657, 659, 680–687, 689–690, 693–697, 700, 702–706, 708, 710, 712–713, 716, 718, 720, 722, 724, 726–728, 730–732, 734–735, 738–743
  transform nuclear magnetic resonance, 3, 720
  transformer, 107, **109–110**, 131, 133, 136, 143–144, 638
  transform tables – see tables
  vision, 3, 738

Fox, Warren L.J., viii, 716, 725
fractional
  derivatives, 8, **401**, 722
  Fourier transform, xii, 6, 8, **126–129**, 138–139, 148–150, 680, 684, 689, 690, 697, 706, 708, 710–711, 713, 718, 722, 724, 726, 728, 734, 741–743
Franklin, Benjamin, 519
Fraunhofer
  approximation, **633–636**
  assumption, 635
  diffraction, 4, 624, **635–637**, 641, 657, 659, 684
  Joseph von, 519
frequency, vii, ix–x, xii, xiv, 4, 6–7, 12–14, 19–20, 47, 50–51, 53–55, 60, 64, 67–68, 115–117, 120–124, 126, 133–135, 142, 196–200, 217–218, 220–222, 225, 234–236, 249, 252, 256, 267–268, 275–277, 283, 290, 295, 304, 306, 327, 341–342, 345–346, 348, 351, 360, 365–367, 369–370, 372, 385–386, 389, 402, 411–425, 427–430, 432–437, 439, 441, 447, 450, 454, 467–468, 472, 477, 482–483, 495, 532, 537, 539, 541–542, 551–553, 566, 588–589, 610, 612–616, 618–623, 625–626, 628–630, 636, 647–648, 651, 656–658, 680, 682–685, 689, 691–696, 699–700, 702–703, 705, 708–713, 716–723, 725, 727–728, 732–734, 736, 738–741, 744
  deviation, 68, 614
  domain, 6, 19–20, 51, 53–55, 123, 199, 222, 268, 341, 385–386, 389, 416, 428, 441, 447, 450, 454, 468, 477, 682, 708–709, 740
  marginal, **427–428**, 537, 542
  modulation, ix, **67**. Also see – FM
  resolution, 7, 416–418, 428, 435–436, 438, 537, 539, 541
  response, 60, **115–116**, 120–121, 135, 142, 200, 249, 252, 256, 275, 351, 366–367, 369–370, 372, 402, 719
Fresnel
  approximation, 633–635, 639
  diffraction, 624, **633**, 635–636, 638–639
  integral, 634
  Augustin Jean, 518
  – Fraunhofer boundary, 636
fret, 617–618, 621–623
  bar, 617–618, 622–623
  calibration, **621**, 623
  spacing, 622
full
  modulation, 216
  rank matrix, 506, 517, 547
fundamental frequency, 411, 414, 612–614, 621
fuzzy
  convex sets, 500
  linguistic variable, 500

G

Gabor, Dennis, 424, 699
gamma
 function, xii, **35–36**, 141, 163
 random variable, 153, **161–164**, 175, 190, 201–202, 205–206
 random variable moments, 201, 205
Gauss, Johann Carl Friedrich, 518
Gaussian, xii, 8, 30–31, 43–44, 64, 138, 149, 153, 155, 159–162, 166–167, 175, 178–179, 186–192, 201–203, 207–208, 211, 318–319, 343, 407, 518, 667–669, 676–679, 682, 696, 699, 703, 708
 asymptotic – See asymptotic Gaussian function, **30–31**
 jitter, **318**
 noise, 20, 518, 682, 699, 708
 random variable, **159–162**, 166, 175, 179, 187, 189, 201, 207, 407, 667–669, 677–679
 ratio random variable, **166**
generalized
 Cauchy random variable, **678–679**
 comb function, **70–71**, 720
 Gaussian random variable, 161, **677–679**
 interpolation, **242**, 248, 274, 380, 704, 722
 time-frequency representation, ix, 417, **424**, 436, 721, 744. Also see – GTFR
genius, 3
geometric
 mean, 439, 620, 622
 random variable, 175, 203
 series, 9, 26, 32, 34, 62, 74, 90, 144, 146, 169, 229, 267, 291, 312, 442, 446, 480, 486, 489, **662–623**
geophysics, 424, 434, 495, 680
Gerchberg, R.W., 477, 700
Gerchberg-Papoulis algorithm, see Papoulis-Gerchberg algorithm
Gerchberg-Saxton algorithm, 566, 568, 691, 699–700, 705, 728, 731, 739–741
Giardina, Charles R., 609, 700
Gibb's phenomenon, 9, **17**, **70**, 100, 225, 707
Gibbs, J.Willard, 17
Glad Man image, 568–569
global maximum, 236
God, v, 104, 151, 126, 411, 447, 570, 693, 728
Goertzel's algorithm, 6, 52, **125–126**, 440, 442
Goldbach's conjecture, 596
Goldburg, Marc H., viii, 700
Gori, F., 477, 695, 701
graininess, **586**, 597–601, 605
Gravagne, Ian A., viii, 583, 694, 702, 716
gray scale images, 137, 180, 362, 364–366, 527
greedy limit cycle, 499–500
Green's function, 112, 333, 624
greenhouse effect, 6
group delay, 430

GTFR, ix, 7, **424–432**, 434–439, 441, 443–444, 537, 539, 543. Also see – generalized time-frequency representation
 table – see tables
 kernel, 425–432, 436, 537, 539, 543
 mechanics, **425**
Gubin, L.G., 495, 702
guitars, 613, 617
Gumbel random variable, **675**

H

Hagler, Marion O., viii, 702, 715
Haldeman, Douglas G., viii, 715
half
 Gaussian random variable, **668**
 normal random variable, **668**
 step, 613
 tones, 411
Hall, Michael W., viii, 70, 715
Hamming
 filter, 107
 window, 415, 420, 422
 Richard W., 10, 104, 306, 703
HandbookOfFourierAnalysis.com, vii
Hankel transforms, xii, 134, 226, **343**, 345, 400–401, 406, 680–681, 702, 704, 707, 709, 719
Hann window, 415. Also see – Hanning window
Hanning window, 56, **415**, 420, 422–423, 544
harmonics, 8, 344–345, 401, 610, **612–623**, 625, 632, 657, 686–687, 689, 694, 695, 699, 709, 742
Hartley transform, xiv, **57–58**, 72, 686, 703, 743
Hawking, Stephen, vi, 151, 703
Heisenberg's uncertainty principle, vii, 7–8, 192, 416, **645–646**, 657, 704
Healthy Girl image, 529–530, 534–536, 540, 542–543
HeNe laser, 637
Hermetian, 45, 196, 200, 204, 235, 239, 267, 366. Also see – conjugate symmetry
Hermite
 polynomials, xii, 9, **43**, 68–69, 96, 138, 149
 Charles, 447, 519
 - Gaussian function, xii, **43–44**, 138, 149
Hertz, **12**, 14, 68, 218, 234, 236, 411, 482
Heinrich, 10
heteroassociative memory, **527**
heterodyne, 122, 124, 133, 268–269, 277, 283
heterodyned sampling, **268**
hexagon, 345–355, 376–379, 385–387, 719
Higgins, John Rowland, 218, 689, 704
higher transcendental function, 6. Also see – transcendental function
higher order
 DFT's, 360
 decimation, **393–395**
 kernels, 263
 derivatives, 449

756 INDEX

higher order (*cont'd*)
   derivative sampling, **259**
   harmonics, 617
   spectra, 380
   terms, 315, 643
high pass filters, **121**, 135, 204, 271, 360, 369–370
Hilbert
   David, 120, 519
   spaces, vi, 328, 501–503, 515, 684, 643
   transforms, 8, **46**, 60, 64–65, 77–78, 133, 140, 228, 230–231, 252, 270, 272, 270, 280, 351, 370, 372, 716, 734, 743
Hilger
   derivative, 586–587
   integration, 587
   Stefan, 583, 704
holography, 495, 427, 687, 690, 693, 699–700, 706, 713, 715, 725, 731, 735–736, 740
homogeneity, 112–113, 630, 739
homogeneous systems, 9, **104–105**, 108–111, 135, 142, 625, 630, 690, 721–722, 724
horizontal, 337, 398, 403, 528, 559, 561–562, 565
   axis, 412, 527, 551, 591
   derivative, 146, 147
   flip, 337
   forces, 611
   line, 620, 622, 436
   motion, 138
   movement, 611
   projection, 349
   scaling, 337
   slice, 350, 397
   swath, 418
   tension, 610
Howard's minimum-negativity-constraint, 493, **516**, 691
Howard, S.J., 516, 704–705
Huffman coding, 365
hyperball, 328
hyperbolic
   cosecant, **24**. Also see – cosech
   cosine, **24**. Also see – cosh
   secant, **24–26**, 161–162. Also see – sech
   sechant random variable, **161–162**
   sin, **24**. Also see – sinh
   tangent, **24**
   trig functions, **24–26**
hypercube, 291, 328
hypergeometric function – see confluent hypergeometric function
hyperplane, 446, 503–504
hypersphere, 328, 379, 386, 402, 408, 410

I

ideal gas, 401, 407
idempotent, ix, 497, 501, 506, 583, 595, 603, 608
identical middles, ix, **513**, 515–518, 520, 522, 530
identically distributed, 157, 176, 314
identity
   element, 115
   for Laplace transformation, 229
   matrix, 359, 506
   operator, 591
   vector, 625
iff, **xii**, 57
IIR, ix, **59**, 135, 142, 419–423, 440, 442. Also see – infinite impulse response
ill-conditioned, 298–299, **663–664**
ill-posed, 8, 137, 248, 250, 259, 302–304, 306, 323, 449, 462, 477, 482, 485, 492, 516, 684, 691, 736
ill-posed sampling theorem, 304, 691
IM, ix. Also see – see identical middles
image
   averaged – see average image
   binary – see binary image
   block, 365–366, 532, 534–536, 538–539, 541–543, 723
   coding, 360
   compression, 6–7, 57, 327, 360, **362**, 364, 495, 527
   encoding, 4, 362–363
   JPEG – see JPEG
   library, 517
   optical – see optical image
   processing, vi, 3, 495, 570, 575, 681, 690–691, 700, 706, 721, 730, 733
   restoration, 366, 495, 723, 727, 730, 742
   samples, 385, 691
   synthesis, vii, **495**
   thresholded – see threshold image
   vector, 533
imaginary operator, 229
imaging, 3, 106–107, 112, 327, 377, 533, 544, 624, 638, 640–642, 681–682, 684, 686–687, 691, 694, 700–701, 705–709, 719, 723–724, 728, 732, 735, 737, 743
   aperture, 533, 544
   system, 106, 377, 624, 638, **640–642**, 682, 701, 705, 709
implicit sampling, **277**, 682, 697
impulse
   kernel, 127
   response, ix, xii, 59, 107, **112–118**, 120–122, 125–126, 130–131, 133, 135–137, 139, 142–143, 146, 148, 150, 196, 198–199, 248, 333, 366–368, 418–421, 423, 440, 603, 631–635, 716
incoherent illumination, 377, 706
incomplete
   cosine, 262
   sine, 262
independence, 157

independent, 9–10, 15, **157**, 159, 166, 168, 174–176, 179, 195, 203, 219, 264, 289, 310, 315–316, 342, 401, 447, 477, 499, 558, 601, 617–618, 668, 671, 673
  and identically distributed, 157. Also see – i.i.d.
  random variables, 159, 174, 179, 401, 671
i.i.d., 157, 162, 170, 176, 186–187, 189, 202–203, 211, 667, 671. Also see – independent and identically distributed
index, vi, 41, 48, 53, 95, 310, 474, 501, 517, 619, 639, 688, 705, 711, 722
induction, 145, 205, 481, 492
infinite
  area, 86, 167–168
  mean, 9
  energy, 20, 84, 100, 224, 322
  height, 21, 23, 120
  impulse response, ix, 59, 135. Also see – IIR
  mean, 665
  moment, 168, 202, 209, 671–672
  noise level, 84, 304, 324
  number, 84, 231, 272, 288, 320, 328, 385, 387, 570
  power, 120, 197
  series, 300
  sum, vi, 62, 294, 484, 485
  variance, 665
inflation, 647, 651–652, 658
initial condition, 60, 523, 617–618, 648, 655, 663, 670
inner product, 47, 273, 502–503, 533
inside boundary, 576
instantaneous frequency, 68, 430
integral
  theorem, 87
  transform, 72, 226–227, 273, 695, 697, 707, 710, 715, 725, 743
integrating detector, **250–251**, 276
integration
  by parts, 311
  of derivatives property, 45, 63, 205
  limit, 158, 306, 346, 488
  order, vii, 19, 432
  over a period, 14–15
  over a unit interval, 50–51
  region, **380–381**, 383, 385
integrator, 441
intensity
  modulation, 558
  of light, 636
  of field amplitude, 636
  of the far field, 636
  profile, 558–559, 561–562
interest rates, 649–652, 654–656, 658
interference
  bandwidth, 429
  suppression, 429, 726
interior point, 497, 504

interlaced
  sampling, 252, **257–258**, 303–306
  signal-derivative sampling, **304**, 306
interpolation, vii, 6, 8, 50, 217, 219–220, 222, 225–226, 232, 242–243, 245–249, 252, 256–261, 263, 265–267, 271–278, 281, 284, 287–297, 301–308, 310–316, 322–323, 375, 380–381, 383, 385, 391–392, 397, 402–403, 447–450, 456, 461, 463–465, 467, 470, 473, 476–477, 480, 485–486, 495, 680–681, 685, 688–689, 693–694, 697–698, 702–704, 706–712, 715–717, 721–724, 726, 729–733, 735, 738–740, 742–743
  error, 288, 314, 477
  formula, 219, 222, 245, 247, 261, 306, 315, 738
  function, 8, 225, 242, **247–249**, 252, 256–261, 266–267, 271–278, 284, 287, 294–295, 297, 304–305, 322, 375, 380–381, 385, 392, 397, 403
  kernel, 391
  noise, 243, 261, 289–293, 295–296, 302–304, 306–307, 310, 312–313, 315, 322, 323, 402, 456, 461
  noise level, 289–292, 295, 303, 306, 310, 322, 402, 461
  noise power spectral density, 291
  noise variance, 293–294, 296, 302–304, 306–307, 312–313, 315, 323, 461
  theory, 6, 217, 697, 716, 739
interval interpolation, 447, **463–464**, 473, 476–477, 480, 485–486, 716
invariant – see shift invarian & time invariant
inverse – also see inversion
  Abel transform, 401, 402, 407
  bounded – see bounded inverse Fourier transform
  cosine transform, 65
  DFT, **53–54**, 57, 72, 103, 118, 359, 362, 366, 485
  DTFT, 142
  filter, 27, 279–280, 313, 316, 319
  FFT, 373
  Fourier transform, **14**, 19, 45, 47, 64, 77, 83, 94, 121, 144, 174, 200, 213, 249, 273, 287, 350, 360, 375, 395, 406, 441, 443, 486, 601, 626–628, 630, 642
  Hankel transform, 406
  Hilbert transform, 46, 140
  pseudo – see pseudo inverse
  quantization, 365
  Radon transform, 345, 349, **351–352**, 533
  transposition, 340
inversion, 14, 15, 50–51, 83, 107, 330–331, 400, 480, 684, 729. Also see – inverse
  formula, 223, 238, 250, 406
  of the DFT, 103
  system, see invertible system

inversion (cont'd)
　theorem, **19**
　Fourier – see Fourier inversion
invertible system, **107**, 109–110, 136, 142, 144
irrational
　number, 601, 605, 613
　time scale, **601**
isoplanatic systems, 106, 333, 404, 630–631
iteration relaxation, **558**

J

Jack image, vii, 575
Jacobi, Carl Gustav Jacob, 242, 519–520, 530
Jacobian, 340
Jeremiah image, vii, 180
Jesus Christ, v
jinc function, xii, 23, **37–38**, 63–67, 76–78, 88, 91, 132, 139, 343–344, 358, 385, 406, 637
jitter, xiv, 288, **313–316**, 318–319, 323, 681, 687–688, 699, 712, 725
　density, xiv, 318
　offset, 313
jittered samples, **313–314**, 316, 318
John the Baptist, 5
Johnson-Nyquist noise, 197
Joint Photographic Experts Group, ix, 362, 708. Also see – JPEG
joint probability density function, 157, 316
Josh image, 360–361
Joshua image, viii, 580
JPEG, ix, 4, 57, 362, 527, 681, 708, 731. Also see – JPG
JPG, ix, 7, 327, 362. Also see – JPEG

K

Kaiser window, 415
Kaplan, Dmitry, viii, 708, 715
Kelvin, Lord, 3, 518, 736. Also see – Thomson, William
kernel, x, xii, 72, 128, 226–227, 230, 235, 241, 262–265, 274, 284, 307, 333, 391, 424–438, 441, 444, 456, 537, 539, 543–544, 547, 551–553, 631, 692–694, 696, 699, 701, 707–709, 711–712, 721, 725, 738, 741–742, 744
　constraint, 626, 628
　synthesis, 433–435, 438, 537, **543**, 551–553, 721
Kirkwood-Rihaczek distribution, 431
Kotelnikov, Vladimir A., 218–219
Kral, E. Lee, vii, 702
Kramer's sampling theorem generalization, 6, 50, 218, 239, 242, **273–274**, 278, 699, 706–707, 710, 716, 743
Kramer, H.P., 6, 50, 218, 242, 273, 278, 239, 706, 710
Krile, Thomas F., viii, 445, 702, 708, 715

Kronecker
　delta, xii, **27**, 113, 128
　delta sifting property, **113**, 128
　Leopold, 570
Kuterdem, H.G., viii, 692

L

L'Hopital's rule, 33, 649, 661
L'Hopital, Guillaume François Antoine Marquis de, 518
Lacroix, Sylvestre, 5
Lagrange, Joseph-Louis, 5–6, 218, 518
Lagrangian, 242, 272, 277, 284
　kernel, 284
　interpolation, 242, **272**, 277, 717, 726, 743
Lanczos window, 415
Langer, Rudoph E., 3
Laplace
　autocorrelation, **198**, 291–292, 309, 311–312, 460–464
　jitter, 318
　noise, 161, 203, 694, 715
　Pierre-Simon, 104, 518
　random variable, **161**, 678
　transform, xiv, **18**, 58, 64, 134, 141, **228–230**, 694
Laplacian, **628**, 631
Larson, John N., viii, 715
laser, 630, 637, 705, 710, 713, 722
lattice
　Bravais – see Bravais lattice
　type, **154**, 170–171, 601
law of large numbers, 6, 158, **177–179**
Laybourn, Loren, viii, 716
leakage, **55–57**, 415
leakage-resolution tradeoff, **415–416**
Lebesgue
　Henri Leon, 519
　integral, xiii, 104, 739
　measureable, xiii, 11, 104, 502
LED, ix, 4
Lee, Shinhak, viii, 545, 554, 556–557, 559–562, 692
Legendre
　polynomials, xiii, 9, **40–41**, 50, 68–69, 88–89, 94–96
　Adrien-Marie, 519
Leibniz's rule, 92, 444, 644, **661**, 722
Leibniz, Gottfried Whilhem, 447
Lenore image, 582
lenses, xii, 377, 624, 638–643 710, 738
lens law, **640**
Leonardo da Vinci, 518
library
　matrix, 516–517
　subspace, 520, 522, 530–531
　vector, 516–518, 525–527
lightning, 10, 161

limit
　cycle, 499–500, 543
　point, 498
Linden, D.A., 252, 259, 713
line
　sample, **397**, 403
　spread function, 112, 333, 682, 715
linear
　array, 547, 701
　chirp, 439, 544
　convergence, **523**, 531
　convolution, 118–119
　differential equations with constant coefficients, 135
　integral transform, 72, **226–227**, 715
　manifold, 503, 517
　motion blur, 8, **137–138**, 146
　systems, 6, **105**, 112–114, 129–131, 133, 136, 139, 143, 150, 196, 198, 330, 682, 700–701, 713, 715, 724–725
　time invariant system, x, **106**, 112, **114**, 142, 439. Also see – LTI
　varieties, x, **504**, 507, 509–511, 513–518, 520, 522, 530–531, 533, 563
loan, 647, **655–656**
log normal random variable, **669–701**
log-Weibull random variable, **675**
logistic random variable, **672–673**
long
　column vector, 532
　duration, 418
　history, 424
　period of time, **656**, 417–418
　window, 417. Also see wide window
loose conditions, 179. Also see – mild conditions
Lord Kelvin – see Thomson, William
lost
　block, 7, 530
　image blocks, 527
　information, 532
　pixel, 527
　sample, 7–8, **243–247**, 272, 275, 278, 286, 294–296, 296–302, 323, 327, **382–387**, 402, 486–487, 715, 738
　signal interval, 447
　through dilation, 576
lower
　bound, **308–312**, 512
　frequency, 267, 283
　limit, 439
　sideband, 133
low pass
　filter, ix, 4, 107, **120–121**, 123, 135, 143, 222, 227, 236, 245, 268, 274, 307–308, 369–370, 373, 418, 451, 455, 470, 473–474, 478, 485, 507, 516, 537, 701. Also see – LPF
　filter operator, 478

　sense, 12, 219, 227
　signal, 124, 271, 291, 294, 687
　trigonometric polynomial, 266
low passed kernel, ix, **227**, 230, 235, 241. Also see – LPK
lower sideband, 133
LPF, ix, **123**, 279, 308. Also see – low pass filter
LPK, ix, 227, **229–230**. Also see – low passed kernel
LTI systems, x, **106**, 109–111, 114–118, 120, 125–126, 129–131, 133, 136–137, 139, 142, 198–200, 204, 418–419, 440. Also see – linear time invariant system
Luke, H.D., 218, 713–714
Luther, Martin, 411
LV, x. Also see – linear variety

M

magnification, 108–109, 337–338, 636, 640, 735
magnifier, **107–109**, **130–131**, 136, 144, **404**
magnify, 108, 475
major chord, 613, 616, 618–620
major scale, **618–620**, 636
Mann, Thomas, 217, 714
mantissa, 76
Margenau-Hill distribution, 431, 714
marginal, **427–428**, 441, 537, 541–544, 552–553
　density, 401
marginally stable, 420, 441
Markov, Andrei Andreyevich, 519
Marks
　Connie Lynn J., viii
　Jeremiah J., viii
　Joshua J., viii
　Lenore, viii
　Marilee M., viii
　Ray A., viii
　Robert (Jack), viii
　Robert J., II, iii, 219, 433–436, 445, 495, 532, 534–536, 538–543, 551–554, 557, 559–562, 642, 691–692, 684, 700, 702–704, 708, 712, 715–716, 721, 734, 726, 731, 740, 744
Masquerade image, 529, 538–539
mass density, 611, 622
matrix
　condition number – see condition number
　diagonal – see diagonal matrix
　equation, 276, **505–507**
　extrapolation – see extrapolation matrix
　full rank – see full rank matrix
　idempotent, 506
　identity – see identity matrix
　ill conditioned – see ill conditioned
　inverse, 664
　library – see library matrix
　offset – see offset matrix
　of integers, 402

760 INDEX

matrix (cont'd)
  on ones and zeros, 533
  of zeros, 468
  operation, 452
  nonsingular, 336
  periodicity – see periodicity matrix
  projection – see projection matrix
  quantization – see quantization matrix
  rotation – see rotation matrix
  sampling – see sampling matrix
  scaling – see scaling matrix
  singular – see singular matrix
  square, 405
  sub – see submatrices
  - vector, 53, 226, 230, 468
mathematical morphology, vii, 7, **570**, 703. Also see – morphology
mathematicians, vi, 5, 411, 519, 530–531
Matheron, Georges, 570, 718
maximally packed
  circles, 377–379, 393–394, 403. Also see – maximally packed spheres
  spheres, 380. Also see – maximally packed circles
maximum
  deviation, 528, 657
  dimension, 635–636
  energy, 485, 493
  error, 366
  filter output, 215
  frequency component, 277, 427
  lens thickness, 639
  number of vectors, 526
  pixel value, 569
  power, 204
  value, 12, 33, 178, 236, 558
  width and height, 558
  yield, 650
Maxwell
  - Boltzmann velocity distribution, 407
  random variable, **669**
  James Clerk, 519
Maxwell's equations, **624–625**, 630–631, 657
McClellan
  transforms, 7, 327, **366–373**, 691, 713, 718–721, 726, 732, 736
  James H., 368, 695, 718
mean, 9, 14, 17, 48–49, 65–66, **152–153**, 155–169, 171, 174–178, 186–189, 192, 194–195, 199, 201–206, 211, 223–224, 233, 288–289, 294, 303, 307, 316, 323, 383, 385, 402–403, 439, 456, 474, 499, 502–503, 506, 517–518, 537, 543, 556, 620, 622, 665–673, 675–677, 686, 708, 712, 727
  square, 17, 48–49, 65–66, 223–224, 233, 499, 502–503, 506, 517–518, 537, 543, 556, 686, 708, 712, 727

  square convergence, 49, 65–66, 223–224, 233
  square error, 233, 517, 537, 543, 556, 708
measure theory, vi. Also see – Lesbegue
  measureable, 104, 502
median, 155
Mellin
  convolution, **134**, 602–603
  transfer function, **134**, 141
  transform, xiv, **134**, 141, 226, **603**, 715
memoryless systems, **107–111**, 133, 135, 142
mercury thermometer, 4
Mersenne
  numbers, 9, 662, 723
  primes, 662, 702
Meyer, Michael G., viii, 721
Michelson
  Albert Abraham, 17
  - Morley experiment, 17
middle C, 411, 614, 637
Middleton, D., 218, 373, 719, 724
mild conditions, 186. Also see – loose conditions
million dollars, 8, 656, 658
minimum
  mean square error, 537, 556
  negativity constraint, 493, **516**, 691, 704–705
  norm, 506
  sampling density, 387, **392–395**
  sampling rate, 217, 222, 277, 408, 720. Also see – Nyquist rate & minimum sampling density
Minkowski
  addition, xi, **571–572**, 579, 590, 596
  algebra, **576**
  arithmetic, **570**, 575
  subtraction, xi, **573**
  operatopms, 57
  sum, 571–572, 590–591
minor
  chord, 620–621
  key, **620–621**
  seventh, 615
  scale, 656
  third, 616
mixed random variable, **203**, 211
Miyamoto, Robert T., viii, 716
mode, **155**, 616, 618, 705, 740
modified Bessel functions, xii, 39–40, **64**, 92–93, 202, 415
modulated
  input, **418**
  signal, 124–125, 133, 216
  window, **423**, 441
modulating sinusoid, 8
modulation, ix, 6, 8, 15, 21, 51, 67, 68, 122–125, 133–134, 204, 216, 268, 276, 331, 357, 558, 613, 644, 683–684, 699, 717, 738, 740
  profiles, 558
  theorem, 21, 357, 644, 684, 738

Modulo Man image, 138, 147
moment, xiv, 9, **152**, 154–156, 165, 167–169, 196–197, 200–210, 665–666, 671–672, 683, 694, 696, 699, 705, 713
momentum, 645–646
Monge, Gaspard, 5
Monika image, 137
morphology, vii, 7, **570**, 703
mortgage, 8, 647, **655–656**
motion blur, 8, **137–138**, 146
movie, 4, 32, 326
moving target, 441, 734
multidimensional
  Abel transform, 348
  area, 354
  comb lemma, **356–357**, 373
  convolution, 7, **332–333**
  cosine transform, 327
  differential, xii
  Dirac delta, **356**
  discrete Fourier transform, **359–360**
  Fourier analysis, 4, **330**, 357
  Fourier series, 331, 334, 336, 341, **356**
  Fourier transform, **352**, 399, 425, 726
  function, 327, 341, 360. Also see – multidimensional signal
  parallelogram, 353
  periodic function, 348
  periodicity, 352, **354**
  sampling, 374–375, 402, 716
  sampling theory, vii, 352, **373**, 375, 381, 699, 729
  scaling, 337
  signals, vii, 6–7, 10, 326–328, 359, 368, 695, 712, 717, 719, 729–731, 740, 743. Also see – multidimensional function
  signal analysis, **326**
  symmetry, 6
  system, **404**, 712, 717, 743
  tiles, 380
  time scale, 604
multiplex, 122, 133, 411–412
music, vii, 4, 6–8, 414, 424, 610, 612–614, 616, 618, 656, 658
musical
  notation, xi
  musical score, 4, 7

N

Namias, V., 126, 718, 720
Napier, John, 519
Napoleon, 5–6
Napoleon's Egypt invasion, 6
natural numbers, xiii, 584
narrow
  frequency band, 417
  window, 418. Also see short duration window

negative binomial random variable, 175, 190, **202–203**
Nelson, Alan C., viii, 721
neural networks, 495, 691, 716, 721, 730
neurophysiology, 424, 484
Newton, Isaac, v
Newtonian physics, 613
Newton's second law, 611
NINV, **294–296**, 298–302, 307–311, 318, 322–325, 403, 459, 461–463
noise level, 179, 185, 194, 197–198, 243, 261, 265, 289–292, 295, 301, 303, 306, 308–310, 312, 322, 324, 384–386, 402, 457–458, 461–463, 482, 520, 529–530, 727, 736
NMR, see nuclear magnetic resonance
nonanticipatory system, 107. Also see – causal signal
noncentral
  chi-squared random variable, **667–668**, 673
  F random variable, **673**, 675
  Student's t variable, **676**
nondestructive testing, 424, 434, 723
nonexpansive, 559, 563, 722
  function, **559**, 567
  operator, **558–559**
nonstationary, 436, 690, 699, 708–709, 712, 714, 744
nonuniform sampling, 242, 257, 272, 277, 688, 690–691, 694, 698, 701, 714, 717–719, 725–726, 730, 734, 736–737, 739, 742
norm, xi, 327–328, 404, 486, 506, 510–511, 681, 689, 740
  of a sequence, 501
normal
  random variable, **160**, 667–670, 673
  tissue, 557
normalized, 178, 188, 237, 305, 369, 415, 557, 560–561, 645
  error, 309
  frequencies, 615
  harmonics, 615, 618–620
  interpolation noise, 294–296, 385, 463
  lower bound, 310
  pixels, 517
  probability density function, 163
  random variable, 186
  window, 420
notation table – see tables
notch filter, **121**, 711
nuclear magnetic resonance, 3, 720
null space, xi, **503**, 506, 510–511, 515
Nyquist
  density, 9, 247, 327, **376–382**, 385–387, 390, 392–396, 402, 691, 716. Also see – minimum sampling density
  interval, 252, 257, 259, 303
  Harry, 218, 721

Nyquist (*cont'd*)
  rate, 8, 218–219, **222**, 226, 232, 234–235, 237, 250–252, 258, 260, 269, 275–276, 283, 291, 294, 298–299, 302–303, 306, 323, 376, 381, 396, 402, 461, 475, 698, 711. Also see – minimum sampling rate

O

octave, 411, 414, **613–623**, 656–658
odd
  component, 13, 21, 24, 29, 65
  function, xiii, **13**, 32, 45, 63, 65, 78, 81–83, 86, 88, 96, 166. Also see – odd signal
  integrand, 83, 167, 460
  index, 33
  moments, 207
  signal, **12–13**, 563, 567. Also see – odd function
  summand, 283
  term, 13
offset
  matrix, **389**
  vectors, **388–389**, 510–511
Oh, Seho, viii, 426, 433–435, 500, 537, 551–554, 557, 559–562, 692, 712, 716, 721
ohm, 11, 196–197
opening, xi, 574, 581, 595, 608
  duality, **581**
operator norm, 486
optical image, 385, 393, 690–691, 701, 713, 736, 738
optics, vii, 3–4, 112, 126, 333, 415, 447, 624, 626, 628, 631, 639, 641, 642, 644, 657, 683–684, 693, 697, 700–701, 710–712, 722, 733–734, 737–738
optimal
  detection theory, 203
  periodic replication, 378
optimization, 495, 686, 692, 713, 721, 734–735
orthant, 504, 510, 517
orthogonal, 9, 40–41, 43, 48, 50, 69–70, 89, 95–96, 150, 273, 362–363, 383, 414, 473–475, 484, 502–503, 506, 510, 517, 688, 694, 706–707, 712, 721, 724, 732, 742. Also see – orthonormal
  basis, 9, **48**, 50, 150, 273, 362–363, 383. Also see – orthonormal basis
  complement, 506
  functions, 473, 475
  polynomials, **40**, **50**, 694, 706
orthonormal, **48–50**, 274, 278, 474, 523
  basis, **48–49**, 274, 278. Also see – orthogonal basis
oscillators, 124, 134, 419, 706
oscillator frequency, 134
oscillatory, 41, 233
output noise level, 308–309, 457, 461
outside boundary, 576

oversample, 8, 243–245, 247, 261, 275, 277–278, 285, 286–287, 535, 550, 697, 703, 715, 731
overshoot, 9, 16, 70, 100
oxymoron, 168

P

Page distribution, 431, 723
Papoulis'
  generalization, 6, 242, **252**, 256, 261, 302, 394, 691
  proof, 223
Papoulis
  Athanasios, vi–vii, x, 218, 252, 477, 642, 690, 699, 704, 723, 729, 741
  - Gerchberg algorithm, 477, 514, 516, 690, 697, 716, 726, 735, 741. Also see – PGA
parabolic
  approximation, 639
  conic section, 634
  impulse response, 634
  wave, 635
parallel
  connection, 422
  lines, 608
  planes, 567
  sides, 408
  slices, 397–398
  subspace, 513
parallelogram, 336, 353–354, 376, 393–394, 403, 405, 408
  periodicity, 403
  subtile, 393–394
  tile, 355, 394
parameter estimation, 424
paraxial approximation, 639
Pareto
  random variable, 187, **665–666**
  Vilfredo Federigo Samaso, 518
Paris, 4–6
Park, Dong Chul, viii, 534–543, 723
Park, Jiho, viii, 527–528, 532, 534–543, 723
Parseval's theorem, **16**, **47–49**, 88, 101, 190, 200, 231, 303, 320, 324, **485**, 645, 714, 718, 722
Parzen window, 187, 402, 407, 415, 419–420, 724
Pascal's
  triangle, 662
  wager, v–vi, 151
Pascal
  distribution, 154
  random variable, 202
  Blaise, 151, 288, 519, 570, 724
Pask, C., 449, 724
pattern recognition, 126, 495, 683, 743–744
peak signal-to-noise ratio, x, 366, 532. Also see – PSNR

Pearson III random variable, 153, 175, 190, 202, 210–211
pellicle, 643
pencil beam, 546–547
perfect
 fifth, 614–615, 622
 fourth, 613–614, 622
periodic, 9, 11, 14–17, 21, 31–32, 47, 51, 53–55, 67, 70, 91–93, 118–119, 127, 136, 139, 149–150, 197, 218, 222–223, 227, 238, 243, 266, 303, 327, 342, 344, 352–358, 354–357, 358, 402, 448–450, 455, 457, 453–454, 462–463, 465, 469–470, 472, 484, 448–450, 483, 486, 490, 605, 643, 647, 652, 684–685, 694, 705, 716–718, 730, 734
 bandlimited functions, 218, 266
 continuous sampling, **448–450**, 462–463, 483
 decimation, 465, **387**
 deposits, 647, **652**
 function, xiii–xiv, 9, 14–17, 31, 47, 51, 54–55, 67, 91, 93, 218, 222–223, 342, 352, 356–357, 402, 455, 457, 469–470, 472, 484, 486, 490, 643, 684–685, 734. Also see – periodic signal
 nonuniform decimation, **463**, 465, 470
 replication, 55, 327, 354–356, 358, 378, 604
 sequence, 54, 490
 signal, **11**, 21, 54, 70, 119, 150, 327, 686. Also see – periodic function
 spectrum, 467
 string, 54
 trigonometric polynomial, 644
periodicity, 32, 126, 352–354, 356–359, 373–374, 376–379, 389, 391, 403, 467, 484, 489, 604, 647, 720, 735
 cell, 358, 403
 matrix, **353–354**, 356–357, 359, 373–374, 376–379, 389, 391, 604
 subtile, 389, 392
 vector, **352–355**, 376–377
periodically
 sampled, 448, 719, 725
 spaced, 448
Peterson, A., 587, 680, 685, 719
Peterson, D.P., 195, 330, 335, 625, 719
PGA, **477–482**, 485–487, 515–517, 642–643. Also see – Papoulis-Gerchberg algorithm
phase, vi, 46, 115–117, 120, 123–124, 133, 219, 275, 327, 360–361, 366–367, 416, 495, 511–512, 566, 568, 628, 630, 634, 636, 639, 681, 690–692, 695, 698–700, 702–703, 710, 715, 720, 722, 726, 737–741
 arbitrary – see arbitrary phase
 carrier frequency – see carrier frequency phase
 conjugation, **657**, 659
 constant – see constant phase
 delay, 639

images, 360
 in image characterization, **360**
 of a complex exponential, 419
 of a DFT, 360
 of a Fourier transform, 8, 360–361
 of the frequency response, 120
 plots, 569
 pure – see pure phase
 quadratic – see quadratic phase
 sampling, 219
 shift, 116–117. Also see – shifting phase
 synthecized – see synthesized phase
 term, 639
 zero – see zero phase
phased array, 643–644, 684, 686, 703, 714
Philipp, Harald (Hal), viii
PI, ix. Also see pseudo inverse
PIA, ix, 131. Also see – piecewise invariant approximation
piano, 613
piecewise
 constant integration, 226
 invariant approximation, 130–133, 136, 143, 150
 linear approximation, 226
 linear interpolation, 226, **249**
 linear signal, 277
pin hole, 630
pipe organs, 613
pitch, roll and yaw, **399**
Pitton, James W., viii, 721, 725
pixel, 179, 360, 362, 364, 366, 517–519, 527–530, 532–534, 538–539, 541–542, 544, 546–547, 566, 569, 578–579
 aperture, 533
 array, 547
 block loss, 534, 541
Planck's
 constant, xii, 198, 645
 radiation random variable, 665, **677**-678
Plato, 447
POCS, x, 7, 327, 438, 447, 482, **495–496**, 498–501, 514, 516–518, 520–522, 525–527, 529–533, 535, 537, 539–546, 550–553, 556–566, 568, 680–681, 692, 699, 703, 705, 709, 711, 713, 716, 722–723, 728, 733–734, 736–739, 741–742, 744. Also see – alternating projections onto convex sets
 associate memory, **517–518**, 520–522, 525–526
 convergence, 499–500, 517, 530
 kernel, 553
 restoration, 529, 535, 539–541
 synthesis, 544
point
 masses, 152
 spread function, 112, 330, 333, 404, 631–632
pointwise convergence, 66, 694

Poisson
  approximation, **177**, 193–194
  counting process, 194–195, 199, 204
  point process, **193–195**, 199
  random variable, **171–172**, 176, 190, 201, 204
  Siméon Denis, 518
  sum formula, **47**, 52, 89–90, 234, 236, 248, 253, 267, 303, 308, 317, 688, 717
polar
  coordinates, xi, 15, 46, 341–342, 344–345, 350, 687, 717, 733, 741
  form, 46, 344
  Fourier series, 345
  representaion, **344**
polarization, 624
polarized, 527
Pollak, H.O., 473, 711, 725, 731
Pólya distribution, 154
Polybius of Megalopolis, 413, 725
polymers, 3, 705
popcorn, 194
power, xiii, 11, 47, 50, 76, 84, 89, 120, 124, 142, 187, 192, 196–200, 204, 211, 214–216, 231, 233, 263, 290–291, 308, 324, 427, 460, 490, 685, 703, 723, 726, 736–738, 742
  point presentation, viii
  spectral density, xiii, **196–200**, 204, 214–215, 233, 290–291, 308, 324, 427, 460, 685, 723, 738
  theorem, **47**, 50, 76, 89, 142, 231, 263, 460, 490
prediction, 447–449, 477, 624, 686–689, 714–715, 720, 724, 732, 740
prime
  numbers, 586, 596–599, 601, 662, 702
  ratio, 615–616
probability, vi–vii, xii, xiv, 6, 10, **151–174**, 177–179, 186–193, 201–203, 207–211, 213, 314–316, 319, 401, 510, 564–566, 568, 601–603, 645, 665–679, 693, 723, 733, 735, 740
  density function, xii, xiv, **151–154**, 155, 157–171, 173, 178–179, 187–190, 192, 201–203, 206–207, 209–211, 213, 314–316, 401, 510, 564, 568, 645, 665–679, 693
  mass, **153–154**, 168–174, 177–178, 191, 202–203, 568, 601–602
  mass function, 154, 170–172, 191–192, 203, 601–602
  projection, vii, x, 7–8, 327, 345–359, 351–352, 402, 406, 426, 433–435, 438, 495, 497–502, 504–512, 514, 517, 522, 527–562, 564, 566–567, 585–586, 691–694, 697, 702–703, 706, 708, 710, 712, 716, 721–724, 726, 729–731, 733–735, 738–742, 744. Also see – tomographic projection

matrix, **506**, 517, 522, 548
operator, 497, 504–506, 541, 555–556, 559
prolate spheroidal wave function, x, **48–49**, **473**, 701, 709, 711, 727, 731. Also see – PSWF
propagation, vii, xii, 4–5, 623–629, 657, 687, 694–695, 705–706, 726–727, 729, 736–737, 740–742
  direction, 628, 657
  of the angular spectrum, 629
  vector, xii, **626**
prototype filter, **367**, 370–371, 402
pseudo inverse, x, **506**, 547, 556
PSNR, x, 366, 532. Also see – peak signal to noise ratio
PSWF, x, **473–477**, 480, 484–485, 491. Also see – prolate spheroidal wave function
  coefficient, 475, 484
  expansion, 484
pure
  phase, 639
  tone, 68
Pythagoras, viii, 518, 613–614, 618
Pythagorean
  harmony, **613–614**, 618
  scale, **657–658**
  theorem, **62**, 73–74, 627

Q

Q of a filter, 439
quadratic
  equation, 85, 661
  formula, 76, **85**
  phase, 636, 659
quadrature, 472, 686, 724
quality factor, 439
quantization, 7–8, 364–365, 535, 537, 550, 688–689, 700, 709–710, 735
  error, 7–8, **535**, 537, 550
  level, 535, 537
  matrix, 364
  operator, 535
quantum
  leap, 64
  level, 645
  mechanics, 126, 681, 709, 714, 720–721, 739
quarter note, 411
queuing theory, 160
quotient rule, 156

R

radar, 3, 126, 424, 441, 681, 686, 691–692, 695–696, 700, 706, 714, 729, 740
Radbel, Dmitry, viii, 715, 726
radiotherapy, 7–8, 495, **545–545**, 554, 557, 559–562, 685, 712, 735
Radon transform, 345, **349–352**, 402, 533
Ramon, Ceon, viii, 721
ramp response, **118**, 120

random variable, vi–vii, xii–xiv, 6, 9, **151–179**, 186–187, 189–190, 194–195, 201–207, 209–211, 306, 314, 323, 401, 407, 601–602, 664–679, 723
   Bernoulli – see Bernoulli random variable
   beta – see beta random variable
   binomial – see binomial random variable
   Cauchy – see Cauchy random variable
   chi – see chi random variable
   chi-squared – see chi-squared random variable
   deterministic – see deterministic random variable
   discete uniform – see discete uniform random variable
   Erlang – see Erlang random variable
   exponential – see exponential random variable
   extreme value – see extreme value random variable
   F – see F random variable
   Fisher-Tippett – see Fisher-Tippett random variable
   gamma – see gamma random variable
   Gaussian – see Gaussian random variable
   generalized Cauchy – see generalized Cauchy random variable
   generalized Gaussian – see generalized Gaussian random variable
   geometric – see geometric random variable
   Gumbel – see Gumbel random variable
   half Gaussian – see half Gaussian random variable
   half normal – see half normal random variable
   hyperbolic sechant – see hyperbolic sechant random variable
   independent – see independent random variable
   Laplace – see Laplace random variable
   log normal – see log normal random variable
   log-Weibull – see log-Weibull random variable
   logistic – see logistic random variable
   Maxwell – see Maxwell random variable
   mixed – see mixed random variable
   negative binomial – see negative binomial random variable
   noncentral chi-squared – see noncentral chi-squared random variable
   noncentral F – see noncentral F random variable
   noncentral Student's – see noncentral Student's t trandom variable
   normal – see normal random variable
   normalized – see normalized random variable
   Pareto – see Pareto random variable
   Pascal – see Pascal random variable
   Pearson III – see Pearson III random variable
   Planck's radiation – see Planck's radiation random variable
   Poisson – see Poisson random variable
   Rayleigh – see Rayleigh random variable
   Rice – see Rice random variable
   sechant – see hyperbolic sechant random variable
   triangle – see triangle random variable
   uniform – see uniform random variable
   uniform difference – see uniform difference random variable
   uniform product – see uniform product random variable
   uniform quotient – see uniform quotient random variable
   uniform ratio – see uniform ratio random variable
   Von Mises – see Von Mises random variable
   Weibull – see Weibull random variable
raster, 7, 395–397, 403, 408, 683
   sampling, 7, **395–397**, 403, 408
   scan, 395, 397
rational
   numbers, 17, 424, 601, 605, 692
   time scales, **600**
ray
   of light, 639
   optics, 624
   tracing, 624
Rayleigh
   random variable, **667–669**, 677
   - Sommerfield diffraction, 624, 626–627, 630, 633–635
Rayleigh's theorem, 47
real
   roots, 661
   signals, 15, **45**, 133, 196, 235, 436, **438**, 482, 489, 512, 564
   time filters, 107
   time implementation, 440
rectangle
   function, xiii,14, 21, 23, **28**, 187, 588
   integration, 226, 228
   support, 377
recurrent nonuniform sampling, 242, **257**, 272, 277
regularize, 482, 680, 682, 684, 729
relatively prime, 597–599, 601
relativity, 17
relaxed
   constraint, 429, 539
   interference projection, **541**
   interpolation formula, 222, **247–248**
   nonexpansive operator, 559
relaxation, 558
reliability, 160, **201–202**, 208–209, 666
remote sensing, 3, 495, 683, 708, 728

Rényi, Alfréd, 217, 737
repunit numbers, 9, 662, 702
residue, 598–599
  arithmetic, 598
  number theory, 598, 732
resistor, 11, 106, 116, 196–197
resonant frequency, 215, 439
restoration noise level, 384–386, 457–459, 463, 482
Revelry, 616
reverse integration, 19
Rice random variable, **677**, 727
Riemann
  integral, 104
  Georg Friedrich Bernhard, 519
right dense, 587, 592, 608
ringing Laplace autocorrelation, **198**
robotics, 495, 709
roll, 151, 169, 177, 399, 602
roll-off, 248, 381
Roman empire, 413
Romans, v
root, 45, 235, 239–240, 304, 439, 613–620, 622, 624, 630, 661, 667–668
Ross, Hugh, 326, 628
rotate, 118, 208, 337–341, 349–350, 380, 399–401, 559, 561–562, 565–566, 712, 741. Also see – rotation rotated
  gantry, 559, 561–562
  image, 401
rotating wagon wheel, 4
rotation, 6, 331, **336–341**, 349, 357, 378, 399, 404, 735. Also see – rotate
  and scale theorem, 357
  angle, 340
  in higher dimensions, 339
  in three dimensions, 339
  matrix, 337–339, 399
rounded, 364–365, 535
rounding, 658
round off error, 478, 713, 720

S

Sad Man image, 568–569
Saint Augustine, 411, 681
Saint Benoît-sur-Loire, 5
Saint Joseph, 5
sample
  and hold, **249**, 252, 717–718
  dependency, 243, 327, 716
  mean, **176**
sampled DTFT, 53
sampling
  density, 9, 235, 277, 327, 376, 379, 385, 387–395, 402, 691
  density comparisons, **379**
  matrix, **376–379**, 382, 387–388, 402
  phase, 219
  rate, xiii, 217, 222, 232, 239, 243, 270, 272, 277–278, 289, 291, 298–299, 301, 309–310, 322, 325, 392, 395, 408, 451, 692, 698, 703, 708, 716, 720
  rate parameter, xiii, **243**, 272, 278, 289, 301, 325
  theorem, vi–vii, 4, 6–7, 9, 15, 49–50, 202, 217–223, 227, 231, 234, 236, 242, 252, 256–257, 265, 267, 271, 273, 277–278, 293, 302, 304–306, 310–311, 313, 372–373, 375, 381, 448–449, 462, 474, 495, 680–716, 718–727, 729–741, 743–744
  theorem for periodic bandlimited signals, **218**
  theorem for signal-derivative sampling, **304**
  theorem tables – see tables
  theory for bandpass functions, **267**
Sarr, Dennis, viii, 721
savings accrual, 8, 647
scalar
  diffraction, 627, 722
  optics, **624–625**
  time scale, 604
scaling, 15, 20–21, 117, 156, 161 192, 336–338, 340–341, 399–400, 416, 721
  matrix, 337, 340, 399
  property, 21
  theorem, **20**, 156, 161 192, 406, 416
science, 3, 10, 447, 570, 610, 646, 711, 731
Schrödinger's equation, 192, 645
Schwarz's inequality, 71, 73, 103, 224, 321, 440, **660–661**
sech, 23–27, 66, 86, 132, 140, 153, 161–162
cosech, 23
sechant random variable – see hyperbolic sechant random variable
second
  characteristic function, xiii, **155–156**, 160–162, 164, 166, 171, 174, 177, 205–206, 210
  derivative function, 156
  moment, 165, 168–169, 196–197, **200**, 204, 666, 571
  order aliasing, 254, 451, 479, 471
  order decimation, 403
  order miracle, 10
  order statistics, **193–194**, 29
segmentation, 3, 738
semitones, 411, 613–616
separability, 6, 331, **334–336**, 395, 401. Also see – separable
  theorem, 334–335, 357, 395, 406
separable, 157, 334–335, 401, 721. Also see – separability
  function, **334–335**, 401
set
  subtraction, 576
  translation, 571
Sezan, M. Ibrahim, 495, 697, 722, 730

Shannon
　Claude E., viii, 217–220, 252, 259, 302, 730
　number, 232, 236, 475, 682, 706
　sampling theorem, xi, vii, 6–7, 217–218, 373, 683, 685, 686–689, 695, 697, 701, 703–707, 710, 716, 726, 732, 737, 743
Shannon's proof, 221, 234
Shepp-Logan phantom, 350
shift
　invariant system, **106**, 230, 333, 630–631, 691. Also see – time invariant
　theorem, **15**, 19, 58, 100, 142, 145, 208, 210, 212, **332–333**, 341, 647
shifting phase, 630
short
　duration window, 418. Also see narrow window
　time Fourier transform, 126, **413–419**, 422–423, 435–436, 438–441
shot noise, **199**
sideband, 133
sifting property, **19–20**, 31, 33, 84, 112–113, 128, 443, 626–627, 640
signal
　detection, 424, 708, 712, 737
　integral property, **45**, 77, 87, 200
　level, **194**
　norm, xi, 503
　of samples, **51**, 243, 25
　processing, vii, ix, 3, 10, 52, 424, 434, 496, 722, 729–730
　recovery, vii, 7, 126, **447**, 610, 693, 695, 717, 724, 495
　synthesis, 126, 438, 700, 726, 739
　-to-noise ratio, x, 366, 475, 532
signum function, xiii, **29**. Also see – sgn
sgn, xiii, 14, 17, 22–23, 29–30, 46, 63, 65–66, 77, 83, 101, 140, 166, 230, 239, 275, 278, 280, 351, 370, 441. Also see – signum function
simultaneous
　equations, 7, 253, 256, 258, 330, **514**, 532, 563
　switching tone, 433, 435
　weighted projections, **500–502**, 514
simultaneously
　sample, 9, 242, 260, 303
　sounded harmonics, 612–613
sinc function, 14–15, 17, 21–22, **27–28**, 32–33, 35, 37, 39, 49, 63–67, 70, 76–78, 86–93, 97–98, 101, 107–109, 121–122, 129, 132–133, 137, 139, 141, 150, 153, 159, 219–220, 222–225, 227–234, 236–239, 242–243, 245–247, 249, 251, 253, 258–263, 266–267, 269, 271–274, 277–286, 289–291, 293–298, 301–302, 306–307, 310–311, 314, 319–320, 324, 335–336, 340–341, 357–358, 370–371, 392, 408–410

sin, xii, 17, 21–22, 24, 27, 29, 32–33, 39, 57–58, 63, 66, 69, 77–78, 81–82, 84, 87, 90, 93, 96–100, 124, 218, 220, 228–230, 232, 238, 262–263, 267, 270, 273, 278, 280, 284, 306, 311, 320, 337, 339, 341–342, 350–352, 371, 402, 404, 412, 455, 488–489, 611–612, 663. Also see – sine transform, 65. Also see – sine transform
sine, xiii, xiv, 24, 57, 61, 62, 262, 306, 420
　integral, xiii, **39**, 70, 87, 262, 306
　transform, xiv, 57, 717. Also see – sin transform
　wave, 420, 717
single side band suppressed carrier, 119, 599. Also see – SSSC
sinh, **24–26**, 86, 162, 203, 211–212, 291–292, 311, 324. Also see – hyperbolic sin
sinogram, 402
sinusoid, 3, 6, 31, 116, 125, 276, 412, 417–421, 429, 432, 441, 558–589, 682, 709, 721, 743
slab, 508, 567
Slepian's paradox, 482, 486
Slepian, David, 10, 473, 482, 486–487, 731
sliding window, 413, 416, 683, 716
　Fourier transform, 416
slower roll-off, 381
Smith, David Eugene, 447
Smith, David K., viii, 715–716
Smith, Michael J., viii
sonar, 4, 126, 424, 434, 441, 624, 692, 700, 713
spectra, 3, 52, 71, 222, 224, 232, 237, 240, 247, 253, 258–259, 270, 304, 377, 380–381, 385, 387, 394, 396, 402–403, 451, 453, 467–468, 472, 685, 698, 704, 706, 716, 718, 723, 725, 727, 736, 740. Also see – spectrum
spectral support, 376–377, 379–381, 392–393, 395–398
spectrogram, 4, 7, 415–420, 424, **431–436**, 439–440, 537, 551–553, 683
spectrometry, 3, 683, 694, 716
spectroscopy, 3, 690, 693, 698, 702, 706, 708–709, 729, 732, 737, 739–740
spectrum, 4, 12, 21, 45–46, 51–54, 68, 72, 102, 122–124, 133–134, 197, 219, 222, 232, 234–235, 237, 239, 243, 248, 253–254, 258, 268, 270, 345, 374, 376–377, 380–381, 385–387, 390–391, 396–397, 402, 411, 416, 418–419, 450–452, 456, 467, 472, 477, 483, 489, 516, 626–630, 657, 683, 705, 714, 717, 738. Also see – spectra
speech, 126, 424, 434, 436, 495, 691, 702, 712, 715, 717, 722, 728, 733, 738, 744
spherical
　Bessel functions, xii, 9, **39**, 41, 61, 69, 89, 95–96, 133, 141, 263, 401
　harmonic, 345
spherically symmetric, 341, 343, 401, 631

spinal cord, 545
square
  aperture, 532, 544
  area, 73
  diamond, 388–389, 391–392
  doughnut, **392–393**
  matrices, 405
  root, 235, 239–240, 630, 667–668
  subtiles, 395
  summable, 265
  tile, 395, 643
  wave, 17, 64, 77, 100
SSSC, x, 133. Also see – single side band suppressed carrier
stable, 59, 107, 109–112, 135, 142, 726
  systems, **107**, 420, 441
standard deviation, **155–156**, 177, 204, 318, 619, 621, 645, 518
Stark, Henry, 495, 568, 724, 727, 730, 733, 742–743
stationarity, 289. Also see stationary
stationary, **195–199**, 204, 214, 233, 288, 293, 303, 307, 316, 323, 336, 385, 402, 416, 456, 684, 686, 690, 702, 710, 713–714, 719, 720, 725, 728, 732, 739
  in the wide sense, 195–197, 199, 204, 214, 233, 288–289, 307, 316, 323, 402, 456, 713, 741
  processes, **195–196**, 702, 714, 720, 725, 732
step response, 120
Stirling's formula, 63, 74
stochastic
  bandlimited signal, **316**, 320
  input, **198**
  processes, vi–vii, xiii, 6, 10, **151**, **193–194**, 196–198, 200, 204, 214–215, 233, 289–290, 316, 323, 383, 682, 684, 712–714, 723–725, 730
  representation, 180
  resonance, 6–7, **178–180**, 182, 204, 215, 699, 716, 719, 727, 744
  signal, 316–318
strictly
  convex sets, 497
  positive, 239–240, 277
strong
  convergence, 501
  law of large numbers, 177
structural analysis, 3
structuring element, 575, 577–579, 582, 584
student's t, 664, 676–677
sub-Nyquist sampling, **393**
subharmonics, **620–622**
submatrices, 522
subpixel resolution, 7, **532–533**, 545
subspace, **503–509**, 513–517, 520, 522, 531, 563, 691, 734, 738, 744
subtiles, **388–395**

sufficient condition, 17–18, 105, 136, 383, 430, 486
super resolution, 447, 495, 680–681, 685, 695, 696, 698, 700–701, 705, 708, 714, 720, 724–725, 741
superposition, 106, 113, 115, 127, 129–130, 135, 150, 335, 398, **434**, 635, 696
  integral, **113**, 115, 127, 129–130, 135, 398, 635
  of plane waves, 626–627
  of point sources, 626
  of sinusoids, 6
  of separable functions, 335
  sum, **113**, 330, 398
support, 150, 253, 268, 376–377, 379–381, 385–386, 391–398, 403, 432, 434, 570, 572–573, 575, 689
  domain, 570
symmetric
  Fourier-Bessel transform, 345
  function, **341–346**, 401, 406. Also see – symmetric signal
  kernel, 430, 543, 551–552
  Laplacian, 631
  signal, **12**, 45, 70, 200, 235, 472, 483, 564. Also see – symmetric function
symmetrically spaced, 334
synthesize
  a diffraction grating, 643
  beam, 547, 558–559
  from images, 8
  GTFR kernels, 426, 434
  phase, 568–569
  using POCS, 537, 539, 543, 551–552, 556
system
  operator, xiii, 104, 110, 129, 142
  theory, vi–vii, **104**, 139, 106, 404, 702, 708, 715

T

tables
  acronym, ix–x
  convolution (c)
    continuous time c algebra, 115
    discrete time c algebra, 115
  sampling theorem even and odd function properties, 13
  extrema of sinc, 92
  Fourier transforms (FT)
    characteristic functions, 153–154
    continuous FT theorems, 13
    discrete FT theorems, 13
    FT pairs, 22–23
    FT types, 13
    multidimensional FT properties, 331
  GTFR's
    kernel transforms, 425
    special cases, 431
  notation, xi-xiv
  optical systems summary, 642

probability functions (pf)
  continuous pf's, 1153
  discrete pf's 154
sampling theory
  directly sampled, 272
  history, 218
  noise levels, 462
windows
  commonly used, 415
  cosine transform relationships, 420
tanh, 23–24, 26–27, 66, 86, 140, 162. Also see – hyperbolic tan
taps, 616
target
  border, 558
  detection, 126
  dose, 548–549
  organ, 560–561
  shape, 557–558
  volume, 548
taxes, 647, 651–652, 654, 658
Taylor series, xix, **12**, 22, 24, **61–62**, 73, 155, 171, 186, 202, 208, 210, 259, 261, 265–266, 311, 325, **448–449**, 486–487, 634
  arctan, 312
  cardinal series, 265–266
  characteristic function, 155
  confluent hypergeometric series, 210
  cosine, 24
  exponential, 22
  Gaussian, 208
  log, 61
  sine, 24
  truncated, 634
Taylor, Brook, 519
Tchebycheff – see Chebyshev
television, 4, 395, 495
tempered scale, 7, 610, **613–623**, 657–658
template matching, **517**
TFR, x, **411–413**, 419, 427. Also see – GTFR
thermal
  equalibrium, 669, 739
  noise, **197–198**
  resistors, 106
thermometer, 4
thesaurus, 196
thin lens approximation, **639**
Thompson, Benjamin B., viii, 716
Thomson, William – see Kelvin, Lord
Three Dolls, 576
threshold, 179–180, 475, 681, 743
  detector, 178, 716
  image, 180, 185
tic tac toe, 328–330, 299, 404
tighter integration, 381–381
tile, 354–359, 373–374, 376–381, 383–395, 643. Also see – subtile
time-
  - bandwidth product, 218, **231–232**, 473, 475

  - frequency distribution, 417, 433–435, 537, 551–553, 692, 696, 702, 705, 709, 712–713, 721–723, 727, 738, 741, 744. Also see – time-frequency representation
  - frequency representations, vii, ix–x, 4, 7, 126, 327, **411**, 417, 424, 436, 537, 680, 689, 691–693, 708, 720–721, 725, 741, 744. Also see – TFR & time-frequency distribution
  invariant, x, 6, **106**, 108–112, **114**, 130, 133, 135–137, 142–144 196, 333, 439. Also see – shift invariant
  marginal, **427–428**
  resolution, **416–418**, 428, 436, 439, 537–539, 541–542
  resolution constraint, **428**, 539, 541
  resolution versus frequency resolution trade off, **416–418**, 436, 439
  scale, vii, ix, xi, xiii, 7–8, 326, 328, **570**, 583–592, 594–605, 608, 680, 685, 696, 702–703, 714, 716, 726, 728, 734, 744
  series, 3, 685, 725, 733, 739
  time variant systems, **106**, 112, 130–131, 136, 139, 698, 708
Toeplitz, 452, 702
tomographic
  projection, 7, 327, **345**, 348–349, 406, **533**, 548
  reconstruction, 546, 565
tomography, ix, 3, 7, 327, **345**, 352, 424, 495, 546, 683, 708, 711, 715, 728, 730, 733, 741
tonal color, 613
transcendental
  equation, 563, 567
  function, 6, 697, 721
  relationship, 202
transfer
  function, 60, 134, 248, 308, 421, 423, 440, 442, 682, 708, 725, 735
  kernel, 127, 226–227
transfinite numbers, xii, **139**, 150, 689
transformation function, **367–371**, 373, 402
translation, 218, **404**, 510, 571, 574, 683, 695, 704, 709–710, 739, 743
transpose, 15, 51, 117, 235, 347
  theorem, 347
transposition, 127, 327, 331, 333, **336–338**, 340, 399
trapezoidal integration, **225–228**, 230
triangle
  function, xiii, **32**, 64, 458
  random variable, **159**
trigonometric
  function, **21**, 24, 220
  geometric series, 169, **663**
  polynomials, **266**, 272, 276, 322, 453, 456, 469, 483, 644
tritone, 621–622

trombone, 613
truncated
  cardinal series, 223, 225, 233, 320
  IIR filters, **419–420**, 422–423
  Fourier series, 16–17, 70, 98, 107, 224
  Foutier transform, 642
  PSWF, 474
  signal, 475
  sinusoid response, 420–421
  Taylor series, 634
truncation error, 232, 245, 248, **320–321**, 323, 684, 686, 688–689, 696, 703, 706–709, 714, 725, 735, 742
Tseng, Shiao-Min, viii, 300, 716, 736
tuba, 234
tumor, 545–546, 554, 557, 560–561
twelfth root of two, 613

U

ultrasound, 126, 424, 434, 684, 716, 721
unbiased
  estimate, 314–315, 318–319
  interpolation, **314**, 316
  restoration, **318**
unbounded, 319
  AITS, 596
  degree of aliasing, 462
  Dirac delta, 110
  energy, 304
  from above, 569
  from below, 569
  variance, 306
uncertainty
  principle, vii, 7–8, 192, 416, **645–646**, 657, 704, 710
  relationship, 192, 202, 645–646
unfiltered
  interpolation formula, 291
  NINV, 298–301, 323
  restoration, 295, 298
uniform, 9, 17, 49, 53, 56, 137, 153–154, 159, 164, 169–170, 175, 177–179, 187, 204, 214, 219, 223–225, 231, 232, 304, 319, 376, 385, 461–462, 474, 533, 558, 610, 671–672, 686, 689, 702, 706, 738
  convergence, 49, 223–225
  difference random variable, 159
  jitter, **319**
  motion blur, 137
  probability density function, 671
  product random variable, **671–672**
  ratio, 9, **159**, 164, 169–170, 177–179, 187, 323, 671–672
  quotient random variable, 671
  random variable, 9, **159**, 164, 169–170, 177–179, 187, 323, 671–672
  ratio random variable, 672

unilateral
  cosine transform, **56**, 65
  Laplace transform, 18, 228
unit
  area, 20–21, 23, 156, 188, 354, 376, 564
  circle, 58, 343, 694
  delay, 59
  doublet, 205
  energy, 48, 485, 491
  norm, 510, 511
  periodicity, 368
  gain, 422
  graininess, 598
  interval, 50–51, 188, 205, 403, 405
  length, 628
  multiply, 59
  norm, 510–511
  period, 50, 368
  radius, 328, 343, 408, 410
  slope, 408
  square, 370, 398
  step, xiii, 14, **30**, 58, 65, 111, 120, 133, 140, 228, 346, 419, 430, 586
  variance, 162, 167, 187, 667, 676
  vector, 510–511, 514, 627, 631
unitless, 12, 50
units, 12, 14, 60, 200–201, 204, 312, 328, 354, 398, 423, 439–440, 634
universal
  agreement, 614
  set, 497
upper
  integration limit, 488
  bound, 512, 686, 708, 719, 722, 734
  case letter, 450
  envelope, 124–125
  frequency, 267
  limit, 439, 554
  sideband, 133, 268
upsampling, **236**
unstable, 107, 247, 259, 303, 319, 383, 449

V

Vandermonde determinant, 257, 275, 286
variance, xiii, 153, **155–160**, 162, 164, 166, 167–172, 174–178, 187–189, 192, 194, 201–206, 211, 292–294, 296, 302–304, 306–307, 313, 315–316, 323, 384, 402–403, 463, 665–673, 675–677, 679
Vatican, 139
vectorization, **329**
vector
  arbitrary – see arbitrary vector
  beam – see beam vector
  column, 37, 532. Also – see long column vector
  component, 521, 605, 628, 664
  curl, xi, 332
  displacement – see displacement vector

dose – see dose vector
eigen – see eigenvector
elements, 605
identity, 229, 625
image – see image vector
library – see library vector
matrix – see matrix-vector
maximum – see maximum vector
norm, xi, 327
offset, 517
operator, 229, 625
optics, 624
periodicity – see periodicity vector
propagation – see propagation vector
offset – see offset vector
response – see response vector
stimulus, 527
subtile – see subtile vector
unit – see unit vector
weighted – see weighted vector
velocity
  measurement, 3, 734
  sampling, 9
Venn, John, 518
vertical, 398, 411–412, 528
  bars, 4
  displacement, 611
  force, 610
  line, 348, 515
  scale, 454–455
  slice, 395, 397–398
  swath, 418
vibrating
  air column, 613, 616–617
  lips, 616, 618
  string, 610–613, 617–618, 656–656
vibration
  analysis, 424, 434, 713, 744
  mode, 616–617
vibrational harmonics, 623
video
  disc, 495
  processing, 495
viola, 613
violin, 613
volt, 60, 197, 201, 204
voltage, 11, 84, 116, 151, 196, 439
Von Mises random variable, 202, **670–671**
Von Neumann's alternating projection theorem, **514**
Von Neumann, Johann, 242, 721

W

Walkup, John F., viii, 445 702, 708, 715, 738
Walsh function, 50, 278, 708
Wang, Y., 739
watermarking, 495, 681
WAV file, 107

wave
  equations, vii, **190**, **610–614**, 617, **624**, **626–627**, 656–657
  function, x, 48–49, 190, 192, 473, 645–646, 711, 727, 731
  packet, 645, 726, 412
  propagation, vii, 4, **623–624**, 628–629, 657, 695
waveform, 4, 84, 249, 412, 475, 699–700
wavefront, 627, 634, 657, 659, 733–734
wavelength, xiii, 623, 625–626, 628, 630, 633–634, 637, 645–646, 685, 690
wavelets, xiii, 9, 626, 683, 686, 692–694, 698, 701, 707, 711, 719, 730, 734, 736, 738–739, 741, 744
weak
  convergence, 501, 722
  law of large numbers, 6, 158, 177–178, 501, 722
weather analysis, 3, 698
Webb, H., 495, 733, 742
Weibull random variable, **665–668**, 675
weighted
  area, **510–511**
  Dirac delta, 21
  distance, 501
  fractional Fourier transform, **128–129**, 138–139, 148–150
  function, 511. Also see – weighted signal
  Hermite polynomial, 43
  sum, 468, 501–502, 508
  noise level, 149
  probability, 568
  projection, 501–502
  signal, 510, Also see – weighted function
  vector, 510
weights, 128–129, 138, 149, 451, 453, 501–502, 556, 627–628, 644
Weil, Andre, 411
Welch window, 415, 693
well-posed, 306, 463, 476–477, 484–485, 491, 493
western harmony, 8, **613**
western
  movies, 4
  music, vii, 7–8, **610**, 612–614, 623, 656
  tempered scale, 615
white
  Gaussian noise, 518, 699
  noise, 9, **197–198**, 290, 292–295, 302–303, 309–310, 312, 323–324, 385, 402, 457, 459–460, 463–464, 518, 527, 533, 569, 699
Whited, John L., viii, 715
Whittaker, Edmund Taylor, 218–219, 739
Whittaker, J.M., 218–219, 739
Whittaker-Kotelnikov-Shannon sampling theorem, 217, 683, 689, 697, 705
Whittaker-Kotelnikov-Shannon-Kramer sampling theorem, 217

Whittaker-Shannon sampling theorem, 217, 687–688, 726
wide
  sense cyclostationary, **323**
  sense stationary, 195–197, 199, 204, 214, 233, 288–289, 307, 316, 323, 402, 457, 713
  window, 418. Also see – long windows
Wigner distribution, 431, 434–435, 537, 551–553, 682–683, 687, 692, 700, 706, 720, 726, 735, 739
windows, 56–57, 187, 402, 407, 413–420, 422–424, 431–432, 434–436, 438, 441, 490, 544, 644, 681, 683, 696, 716, 722, 740
  Blackman – see Blackman window
  boxcar – see boxcar window
  Bartlett – see Bartlett window
  causal – see causal window
  commonly used, 56, **415**, 420. Also see – tables
  continuous time – see continuous time windows
  cosine transform relationships – see tables
  discrete time – see discrete time windows
  design, **419**
  even – see even window
  filter, 418–419
  fixed length, 415
  Hamming – see Hamming window
  Hanning – see Hanning window
  Kaiser – see Kaiser window
  Lanczos – see Lanczos window
  long – see long window
  modulated – see modulated window
  narrow – see narrow window
  normalized – see normalized window
  parameter, 439
  Parzen – see Parzen window
  short duration – see short duration window
  sliding – see sliding window
  table – see tables
  typical, 415

Welch – see Welch window
wide – see wide window
Wise, Gary L., viii, 263, 699, 715
Wu, Wen–Chung Stewart, viii, 740
Wunsch, Donald C., viii

X

x axis, 564, 628–629
X-ray, 3, 700, 722, 730–731, 735

Y

Yang, Y., 527, 741
yaw, **399**
Youla, Dante C., 495, 514–515, 742

Z

z transform, xi, xiv, 58–60, 421, 423, 646–647
Zadeh, Lotfi A., 500, 743
ZAM GTFR, 435
zero
  crossing, 33, 38, 66, 84, 637, 643, 682, 685, 717
  input, 113
  initial condition, 60, 523
  locus plot, 445–446
  output, 113
  mean, 162, 166–167, 187, 192, 194, 201, 204, 288–289, 293, 303, 307, 323, 383, 385, 402–403, 456, 518, 667–668, 676
  padding, 236, 732
  phase, 120, 327, 366–367, 691, 726
  sum property, 245
  variance, 177
zeroth order
  sample and hold, 249
  spectrum, 253, 397, 451, 467, 472
  spherical Bessel function, 141
  spherical Hankel transform, 343, 345
  spectrum, 141, 397, 467, 472
Zhao, Yunxin, viii, 436, 744
Zhu, Q. F., 530, 739